MEMBRANE STRUCTURAL BIOLOGY
WITH BIOCHEMICAL AND BIOPHYSICAL FOUNDATIONS

Second Edition

This textbook provides a strong foundation and a clear overview for students of membrane biology and an invaluable synthesis of cutting-edge research for working scientists. The text retains its clear and engaging style, providing a solid background in membrane biochemistry, while also incorporating the approaches of biophysics, genetics, and cell biology to investigations of membrane structure, function, and biogenesis to provide a unique overview of this fast-moving field.

A wealth of new high-resolution structures of membrane proteins are presented, including the Na^+-K^+ pump and a receptor G protein complex, offering exciting insights into how they function. All key tools of current membrane research are described, including detergents and model systems, bioinformatics, protein-folding methodology, crystallography and diffraction, EPR and NMR spectroscopy, and molecular modeling.

This comprehensive and up-to-date text, emphasizing the correlations between membrane research and human health, provides a solid foundation for all those working in this field.

Mary Luckey is Professor Emerita in the Department of Chemistry and Biochemistry at San Francisco State University. She earned her Ph.D. in Biochemistry at the University of California Berkeley with the first identification of an iron transport protein in the bacterial outer membrane. Her postdoctoral work demonstrated the specificity of the *E. coli* maltoporin in proteoliposomes. While continuing research on maltoporin structure and function, she has taught biochemistry for over 25 years, including the graduate-level membrane biochemistry course that provided the impetus for this book.

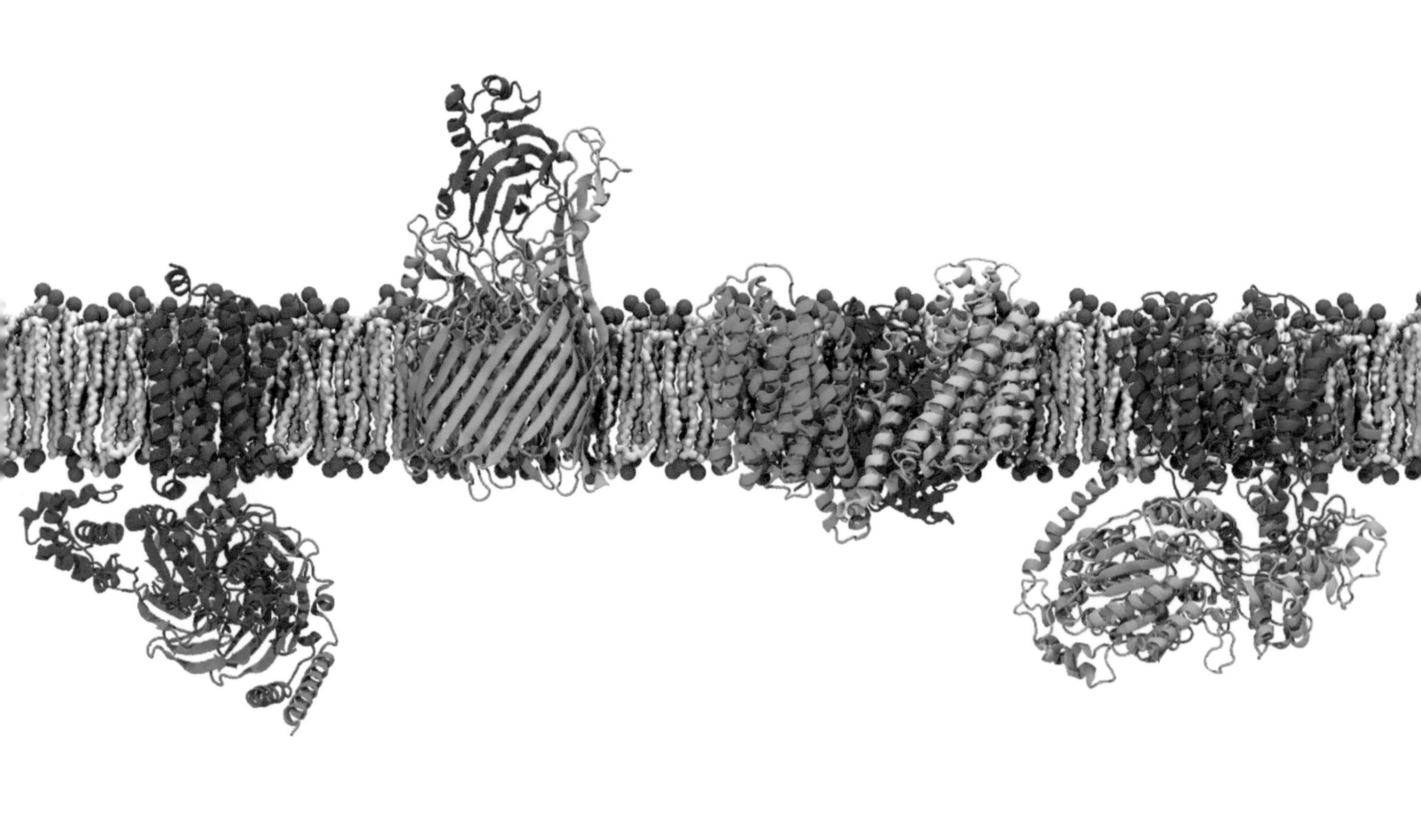

MEMBRANE STRUCTURAL BIOLOGY

WITH BIOCHEMICAL AND BIOPHYSICAL FOUNDATIONS

SECOND EDITION

MARY LUCKEY

San Francisco State University

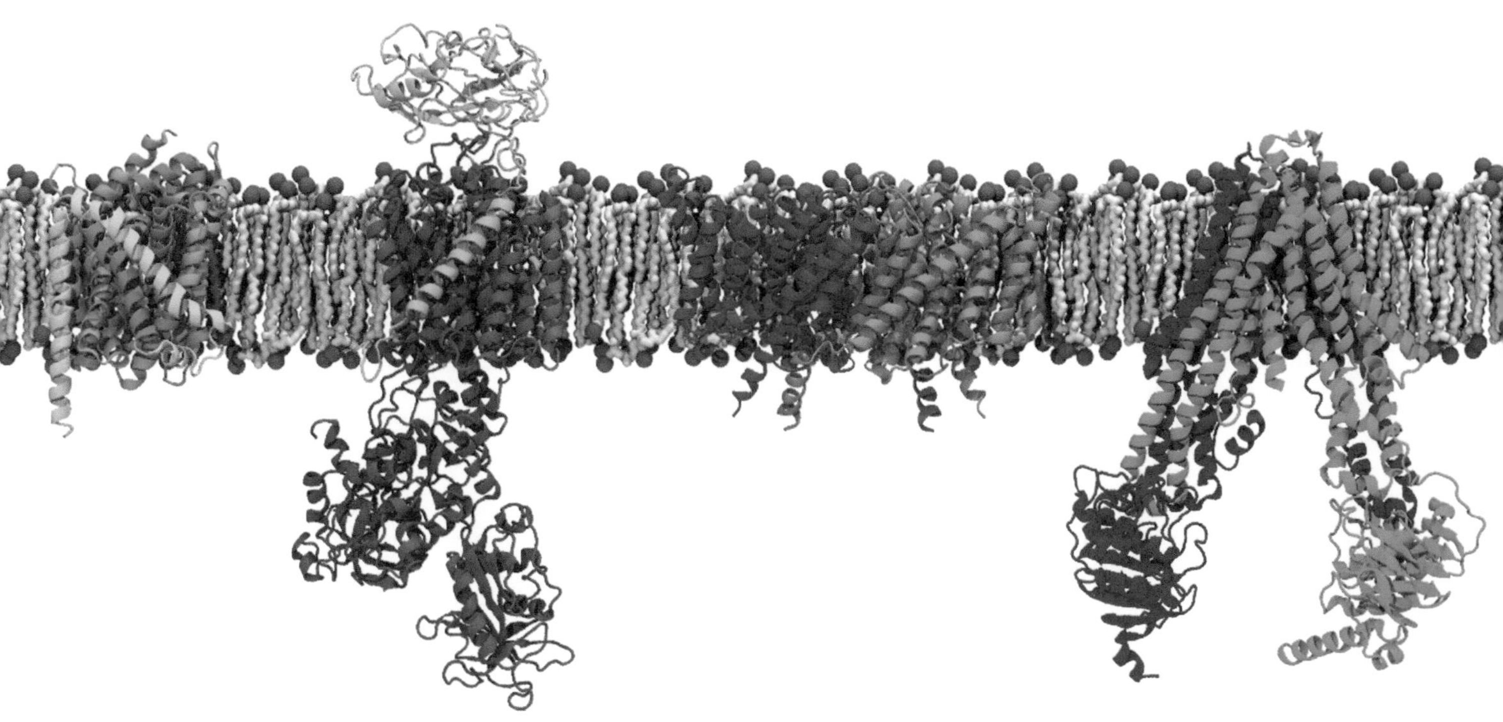

Shaftesbury Road, Cambridge CB2 8EA, United Kingdom

One Liberty Plaza, 20th Floor, New York, NY 10006, USA

477 Williamstown Road, Port Melbourne, VIC 3207, Australia

314–321, 3rd Floor, Plot 3, Splendor Forum, Jasola District Centre, New Delhi – 110025, India

103 Penang Road, #05–06/07, Visioncrest Commercial, Singapore 238467

Cambridge University Press is part of Cambridge University Press & Assessment, a department of the University of Cambridge.

We share the University's mission to contribute to society through the pursuit of education, learning and research at the highest international levels of excellence.

www.cambridge.org
Information on this title: www.cambridge.org/9781107030633

First published 2008
Reprinted 2011 (twice)
Second Edition 2014 (version 4, June 2023)

Printed in the United Kingdom by TJ Books Limited, Padstow Cornwall

Library of Congress Cataloging-in-Publication data
Luckey, Mary, 1948–
Membrane structural biology : with biochemical and biophysical foundations / Mary Luckey. – Second edition.
pages cm
Includes bibliographical references and index.
ISBN 978-1-107-03063-3 (hardback)
1. Membranes (Biology) 2. Membrane lipids. 3. Membrane proteins. I. Title.
QH601.L75 2014
571.6′4–dc23 2013028009

ISBN 978-1-107-03063-3 Hardback

Additional resources for this publication at www.cambridge.org/Luckey2

Cover image:

The title page shows high-resolution structures of membrane proteins incorporated into a simulated lipid bilayer. The proteins are, from left to right, *β*2A in complex with an agonist and a trimeric G-protein, the heme receptor HasR in complex with the heme-binding protein HasA, the trimeric aspartate transporter GltPh that is a homolog for neurotransmitter transporters, the SecYEG translocon in complex with the energizing subunit SecA, the amino acid transporter LeuT that is another homolog for neurotransmitter transporters, the Na^+, K^+ ATPase, the dimeric chloride transporter ClC, and P-glycoprotein, a dimeric transporter that extrudes drugs.

Kindly provided by J. C. Gumbart, Georgia Institute of Technology, and E. Tajkhorshid, University of Illinois.

In memory of Amy L. Davidson, 1958–2013, insightful scientist, meticulous experimentalist, gracious colleague, and good friend.

CONTENTS IN BRIEF

CONTENTS

LIST OF BOXES

PREFACE

The first edition of *Membrane Structural Biology* met a need for a comprehensive presentation of the explosion of information about the structure and organization of biological membranes. It also acknowledged how new techniques and whole new methodologies had changed both how we acquired knowledge of the membrane and how we viewed it. With a foundation derived from basic physical and life sciences, advances in structural biology were depicted through the molecular details of membrane components provided by sophisticated diffraction analysis of fluid lipid bilayers and by high-resolution structures of a variety of membrane proteins. As the book moved from basic membrane biochemistry to detailed examples, it covered a wide range of material at a level appropriate for both students and scientists in the field. I am gratified with the responses from membrane scientists all over the world.

This new edition has been expanded to include over 20 additional membrane proteins visualized in atomic detail. Discovery of superfamilies based on the protein folds shows relationships among membrane proteins, while capture of multiple states begins to disclose mechanisms. Some new topics have been introduced, other topics updated, and yes, sadly, some interesting new findings had to be left out. I hope readers will jump into the literature from the key references provided to learn about the exciting new findings, to study these topics in more detail, and to tackle the larger and more complex systems at the frontiers of membrane research.

While I am responsible for the omissions and any errors, I am indebted to many people who have been generous with their time, reviewing new parts of this edition, as well as those who did so for the first edition. I was assisted when I started writing the book by my former students, Dr. Aram Krauson and Dr. Andréa Dosé. For their comments on specific topics in the first edition I thank Professors Scott Feller, Steve White, Sam Hess, Rosemary Cornell, Ehud Landau, David Hackney, Paula Booth, Bill Plachy, and Hiroshi Nikaido. For comments on new sections of the second edition I thank Poul Nissen, Maike Bublitz, Satinder Singh, Merritt Maduke, Brian Kobilke, Chuck Sanders, Jörg Standfuss, Bill Cramer, Reinhard Krämer, Christine Ziegler, Shelagh Ferguson-Miller, and Eduardo Perozo. I am especially grateful to those who reviewed the entire original manuscript: Professors Lin Randall and Stanley Parsons, and my former students Shyam Bhaskaran, Marla Melnick, and Jared Matt Greenberg. Lin Randall and my former student Chris Chin also read all the new chapters for the second edition.

With much appreciation I thank Professors J. C. Gumbart and Emad Tajkhorshid for the cover figures. In addition I want to thank the many individual scientists who shared their figures of beautiful membrane protein structures, particularly those who prepared figures for this edition: Shelagh Ferguson-Miller, Rosemary Comell, Krzyszt of Palzewski, and Tivadar Orban. For her unflagging enthusiasm and wise editorial help, I thank Dr. Katrina Halliday. Thanks as well to my colleagues and friends who supported my progress writing the book. Finally, I deeply appreciate the affection and encouragement I received from my family, Ariel, SAM, Ryan, donna, Kesa, Amanda, and Dana, with profound gratitude for steadfast love, patience, and support from my husband Paul.

Mary Luckey

INTRODUCTION

1

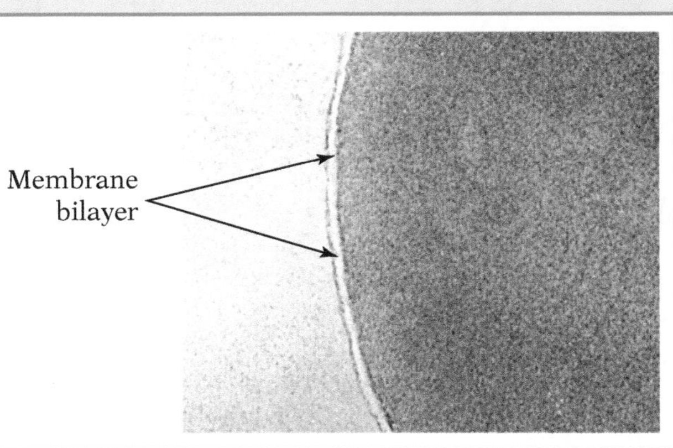

The plasma membrane delineates a cell, defining its borders and determining what can and cannot enter the cytoplasm. Its trilaminar appearance is revealed by electron microscopy after staining with osmium tetroxide. From Nelson, D. L., and M. M. Cox (eds.), *Lehninger Principles of Biochemistry*, 4th ed., W. H. Freeman, 2005, p. 369. © 2005 by W. H. Freeman and Company. Used with kind permission of J. D. Robertson.

Essential for the compartmentalization that defines cells and organisms, biomembranes are fundamental to life. Early membranes played a crucial role in the origin of life as the structures that defined what stayed in and what was kept out of primordial cells. In addition to their compartmentalization function, membranes provide modern cells with energy derived from chemical and charge gradients, organize and regulate enzyme activities, facilitate the transduction of information, and even supply substrates for biosynthesis and for signaling molecules. Some membranes have specialized functions; for example, the brush border membrane lining the intestines absorbs nutrients, the myelin surrounding nerves functions as insulation, and the rod cell membrane of the eye captures light. While prokaryotes either have one cell membrane (Gram-positive) or have inner and outer membranes in the cell envelope (Gram-negative), eukaryotic cells have many membranes (Figure 1.1). In addition to the plasma membrane, eukaryotes have membranes surrounding the nucleus, organelles such as mitochondria, chloroplasts in plants, lysosomes, and, of course, the membrane-based endoplasmic reticulum (ER), Golgi apparatus, and other vesicles involved in intracellular transport. Even some viruses have membrane envelopes. In spite of this variety, much can be generalized about the structure and function of biomembranes.

GENERAL FEATURES OF MEMBRANES

Biological membranes consist of lipids, proteins, and carbohydrates (Figure 1.2). The lipid components include glycerophospholipids (also called phospholipids), sphingolipids, and sterols. The basic unit of the membrane is a bilayer formed by phospholipids and sphingolipids organized in two layers with their polar headgroups along the two surfaces and their acyl chains forming the nonpolar domain in between. Embedded in the lipid bilayer are integral membrane proteins, which cannot be removed without disrupting the membrane. Most of these proteins have one or more transmembrane (TM) segments, and they interact closely with nearby lipids as well as other proteins. In addition, there are peripheral membrane proteins that associate at the surface of the membrane and lipid-anchored proteins that are held into the membrane by covalently attached fatty acids or

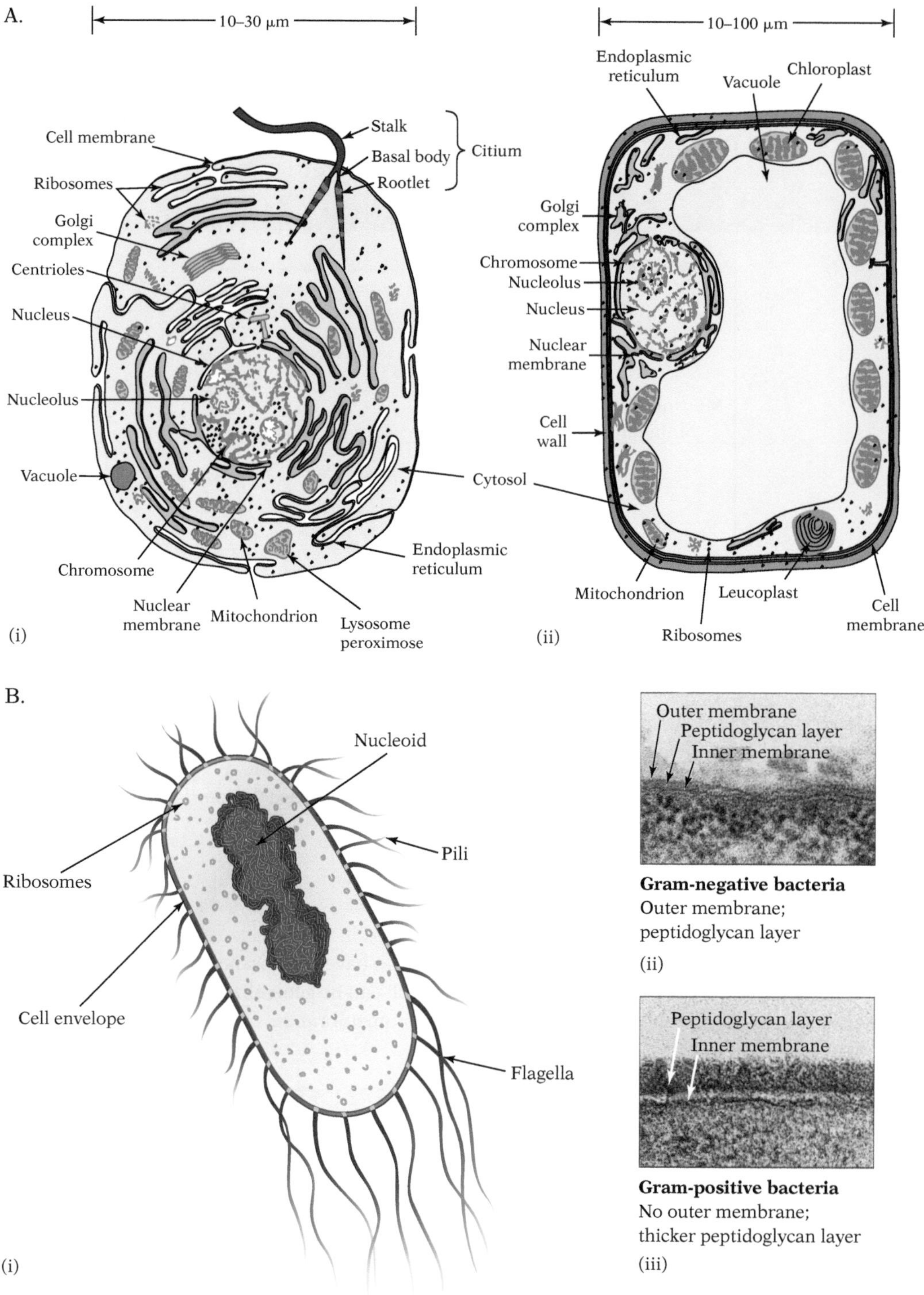

1.1 A variety of types of membranes. (A) Plasma membrane and intracellular membranes in eukaryotic cells are shown in a diagram based on thin-section electron micrographs of generalized animal (i) and plant (ii) cells illustrating the plasma membrane and membrane-bound organelles. Redrawn from Jain, M. K., and R. C. Wagner, *Introduction to Biological Membranes*, 2nd ed., Wiley, 1988, p. 2. © 1998. Reprinted with permission from John Wiley & Sons, Inc. (B) Bacterial membranes are shown in a diagram of a bacterial cell (i) and in thin-section electron micrographs of the cell envelope of Gram-negative (ii) and Gram-positive (iii) bacteria. Redrawn from Nelson, D. L., and M. M. Cox (eds.), *Lehninger Principles of Biochemistry*, 4th ed., W. H. Freeman, 2005, p. 6. © 2005 by W. H. Freeman and Company. Used with permission.

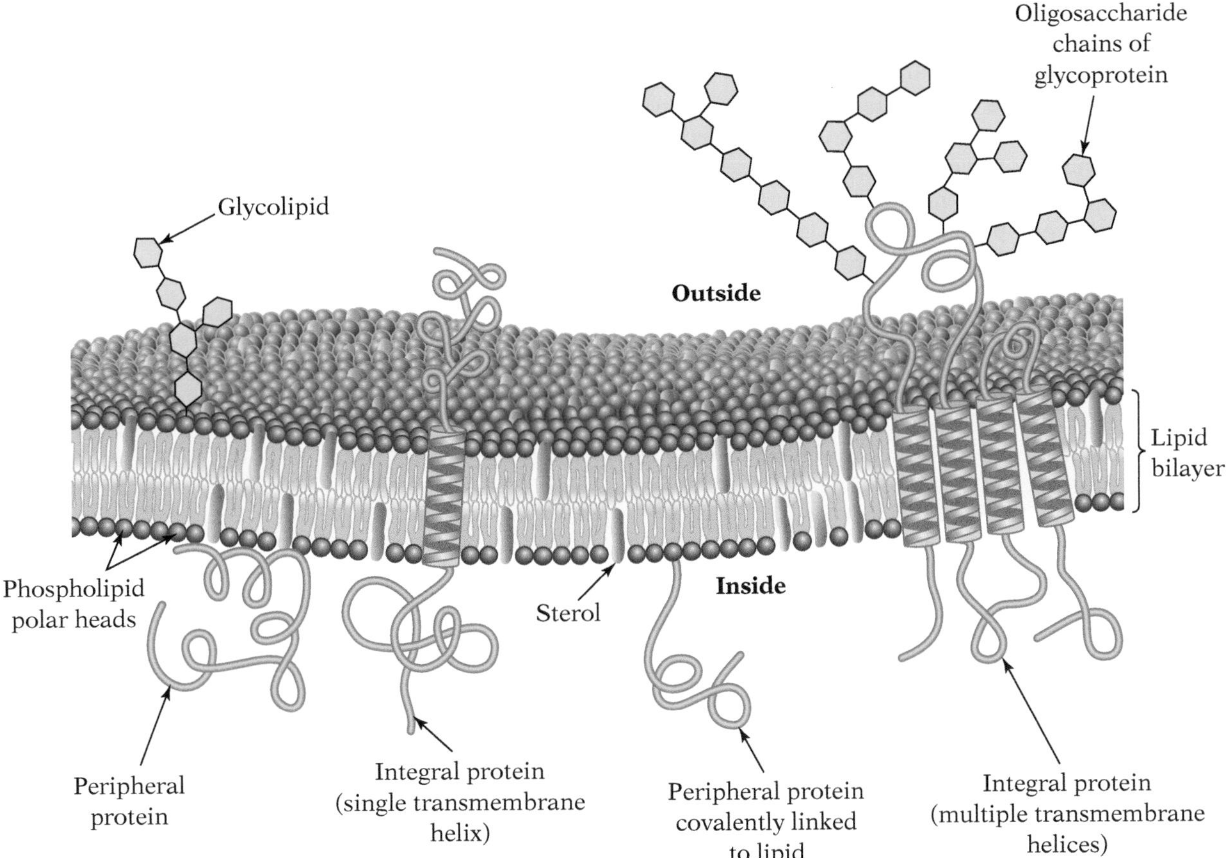

1.2 Membrane components. Membranes contain lipids, proteins, and carbohydrates as glycolipids and glycoproteins. Nelson, D. L., and M. M. Cox (eds.), *Lehninger Principles of Biochemistry*, 4th ed., W. H. Freeman, 2005, p. 372. © 2005 by W. H. Freeman and Company. Used with permission.

lipids. Although carbohydrate membrane constituents serve important functions, these hydrophilic moieties are always on the portions of glycoproteins and glycolipids external to the membrane bilayer. Such glycoconjugates deserve detailed consideration on their own and are not covered in this book.

Membranes are responsible for the selective permeability of cell envelopes that enables cells to take up many nutrients and exclude most harmful agents. The permeability properties are determined by both lipid and protein components of membranes. In general, the lipid bilayer is readily penetrated by nonpolar substances while proteins in the membrane make channels and transporters for ions and hydrophilic substances. This permeability barrier enables the membrane to maintain charge and concentration gradients that are critical to the cell's metabolism. The permeability barrier is maintained during activities such as cell division and exocytosis because the membrane is flexible and self-sealing.

Membranes are also very dynamic structures, with constant activity on their surfaces as well as constant movements in the bilayer, both in the transverse direction across the bilayer and the lateral direction in the plane of this two-dimensional matrix. The latter movements give rise to the fluid nature of the membrane and enable interactions among proteins and between proteins and lipids to provide temporal associations that are important to membrane functions.

Thanks to many, many scientists who have contributed to the enormous progress of the past decades, knowledge of the membrane goes beyond its basic architecture and properties to a multitude of details describing specific elements and functions. While the particular tools and approaches used by biochemists, biophysicists, geneticists, and cell biologists who study the membrane vary greatly, two paradigms[1] provide the framework for understanding their work. The starting point for understanding membrane structure is the hydrophobic effect. A far-reaching paradigm for many areas of chemistry, this principle governs the behavior of membrane components. The specific paradigm for membranes is the Fluid Mosaic Model, a description of membrane properties

[1] Paradigms are scientific models. According to science philosopher Thomas Kuhn, the paradigms of a field of study shape it so thoroughly that they may be unacknowledged and even unobserved by its practitioners. Yet, they determine the assumptions and the tools with which those scientists operate daily.

and organization that has endured for more than three decades.

A description of these paradigms and the classic work on which they are based will lay the groundwork for the rest of this book. Yet today the current paradigm is shifting because of new aspects of membrane organization that have risen to the forefront in the past several years. The importance of transient, specialized regions called membrane rafts affects the contemporary model of cell membranes. The membrane is compartmentalized by protein–lipid and protein–protein interactions. Finally, the organization of many membrane proteins into large assemblies that often involve molecules at the bilayer periphery and beyond indicates that a more complex and comprehensive view is needed for future work.

PARADIGM 1: THE AMPHIPHILIC MOLECULES IN MEMBRANES ASSEMBLE SPONTANEOUSLY DUE TO THE HYDROPHOBIC EFFECT

All biomembranes contain amphiphilic lipid and protein constituents that have both polar and nonpolar parts, and this dual nature of its components is essential to membrane structure. Because proteins are simply polymers of amino acids, their polarity is a function of their amino acid composition; thus they have hydrophobic domains rich in residues with nonpolar side chains and hydrophilic domains generally lacking them. On the other hand, by classification a lipid is quite nonpolar because the definition of lipids is empirical: a lipid is a biological substance that is soluble in organic solvents and has poor solubility in water. Yet all lipids have hydrophilic domains, called their *headgroups*, even when the headgroup is simply a hydroxyl group, as in cholesterol. The structures of lipids vary considerably (as described in Chapter 2) but all provide the amphiphilicity[2] that leads to the formation of distinct phases in aqueous systems, in which the lipids aggregate spontaneously to form polar and nonpolar domains. Mixing a pure lipid with water can result in formation of monolayers, micelles, bilayers, hexagonal arrays, or cubic phases, depending on the nature of the lipid and the method of preparation.

The spontaneous formation of each type of lipidic aggregate depends on the structure and hydrophobicity of the lipid, but it is always driven by the structure of water. In ice each water molecule has four hydrogen bonds worth ~5 kcal/mol each (Figure 1.3). When ice melts ~85% of these hydrogen bonds are preserved, but of course in liquid water they are dynamic, with 10^{11}/sec positional changes. The extensive hydrogen bonding of water accounts for its special properties, such as its high boiling point and high dielectric constant (a measure of the extent to which it shields dissolved ions). It also provides the basis for the hydrophobic effect.

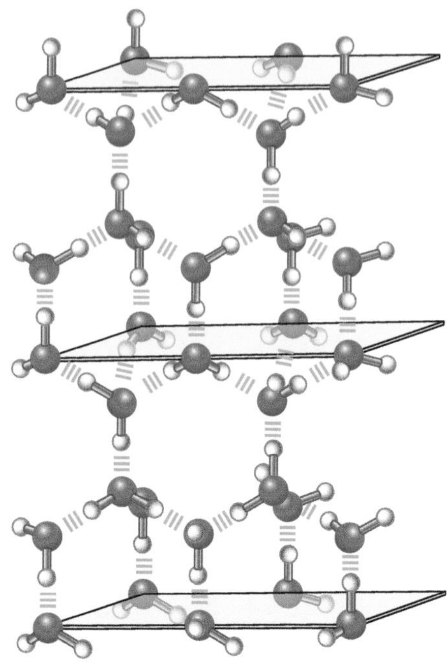

1.3 Importance of hydrogen bonding in the structure of water. In ice, each water molecule forms four hydrogen bonds with its nearest neighbors. In liquid water at room temperature and atmospheric pressure, each water molecule has on average 3.4 hydrogen bonds. Redrawn from Nelson, D. L., and M. M. Cox (eds.), *Lehninger Principles of Biochemistry*, 4th ed., W. H. Freeman, 2005, p. 49.

Insertion of a nonpolar molecule, such as a fatty acid with a long acyl chain, into liquid water reorders the water molecules closest to the hydrocarbon chain to form a hydrogen-bonded cage around the nonpolar moiety. Depending on the size of the nonpolar domain, there may be no net loss of hydrogen bonds so enthalpy does not necessarily have a strong effect. However, as the water molecules rearrange to form the cage around the nonpolar chains, their mobility is drastically reduced, resulting in a large loss of entropy. The best way to lower this entropic cost is to sequester the nonpolar moieties into large aggregates, thus reducing the total surface area of nonpolar material exposed to the aqueous layer and hence decreasing the number of immobilized water molecules. (This is possible because as a sphere increases in size, the volume increases as the cube of the radius while the surface area increases as only the square of the radius, with the result that a larger radius gives a smaller surface area-to-volume ratio.) The end result of this entropic driving force is the separation of the aqueous and lipid molecules into two phases or domains.

[2] *Amphiphilicity* means "having polar and nonpolar domains"; *amphiphilic* and *amphipathic* are used interchangeably.

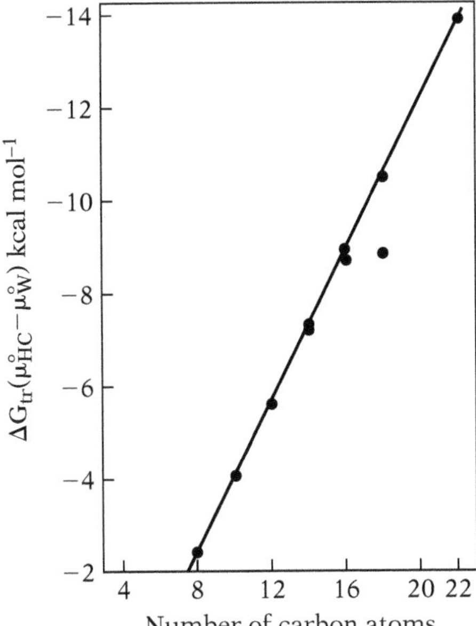

1.4 The free energy of transfer of fatty acids from water to heptane is a function of the chain length. Fatty acids of varying lengths in n-heptane at 23–25°C are equilibrated with dilute aqueous buffer and their activities (μ°) in each phase determined. The *x*-axis gives the number of carbon atoms, and the *y*-axis gives the free energies for transfer. Redrawn from Tanford, C., *The Hydrophobic Effect: Formation of Micelles and Biological Membranes*, 2nd ed., Wiley, 1979, p. 16. © 1979, John Wiley & Sons, Inc.

The nonpolar domain may then be further stabilized by van der Waals forces between the close-packed acyl chains.

The hydrophobicity of a substance is traditionally measured by a partitioning experiment using two solvents, such as heptane and water. From the partition coefficient is calculated the $\Delta G_{transfer}$ for the solute of interest:

$$K_p = K_{eq}\,[solute]_{H_2O}/[solute]_{heptane}$$
$$\Delta G_{tr} = -RT\ln K_{eq},$$

where K_p is the partition coefficient, K_{eq} is the equilibrium constant, and ΔG_{tr} is the free energy change for the transfer from heptane to water.

When the solutes are fatty acids with varying chain lengths, the energy cost is proportional to the chain length: a cost per CH_2 unit of 0.8 kcal mol^{-1} is derived from the plot of ΔG_{tr} versus the chain length (Figure 1.4). Like other structures in biology, the aggregate structures of lipids are stabilized by the cooperative sum of many weak interactions. Thus the thermodynamic stability of the membrane bilayer maximizes water–water interactions outside and acyl chain interactions inside the nonpolar interior while minimizing water–acyl chain interactions that are entropically expensive. The hydrophobic effect explains the energetics of membrane formation but does not address the basic structure of the biological membrane.

PARADIGM 2: THE FLUID MOSAIC MODEL DESCRIBES THE MEMBRANE STRUCTURE

While the Fluid Mosaic Model for the structure of membranes is now familiar to all life science students, the amazing unity it brought to a divided field is not apparent without an appreciation of its historical development. Ben Franklin is credited for early insight into lipidic structures with his calculation of the thickness of an olive oil film on pond water as 25 Å (2.5 nm), the depth of a lipid monolayer on the surface. Then, in 1925, Gorter and Grendel made surface area measurements for a compressed monolayer formed by acetone-extracted lipid from erythrocytes and correctly concluded that the monolayer area covered twice the surface area of the erythrocytes. In 1935 Davson and Danielli used thermodynamic arguments along with measurements of surface tension and permeability to postulate a membrane structure that placed globular proteins on the outer surfaces of a membrane bilayer (Figure 1.5A). This model dominated thinking about membrane structure for the next three decades, with modifications such as changing the protein conformation to extended β-sheets, and led to the concept of a "unit membrane" with a width of 6–8 nm, corresponding to the width of myelin sheath in x-ray diffraction measurements (Figure 1.5B). In 1959 Robertson argued that this unit membrane was common to all biological membranes, citing "railroad track" images from thin section electron microscopy (EM) of tissues stained with osmium tetroxide, which stained the phosphates of phospholipid headgroups and washed proteins out (see Frontispiece). Other staining techniques in use at the time, such as prior crosslinking with glutaraldehyde, produced images in which the full membrane was electron dense.

A challenge to the Davson–Danielli–Robertson model came with the application of freeze-fracture techniques: bumps visible by EM when the membrane was cleaved within the plane of the bilayer were attributed to embedded proteins (Figure 1.6). Support for the interpretation that the bumps were proteins came from their absence in membranes treated with proteases and in samples of myelin sheath, which has very little protein. In studies of the respiratory chain of mitochondria by Benson, and later Green, mitochondrial inner membrane could be separated into lipoprotein subunits and reconstituted to regain activity. These results supported a model in which the lipid is solvent for embedded, globular proteins, consistent with EM images obtained after negative staining with heavy metals that showed subunits (not "railroad tracks") that were unaffected by lipid extraction prior

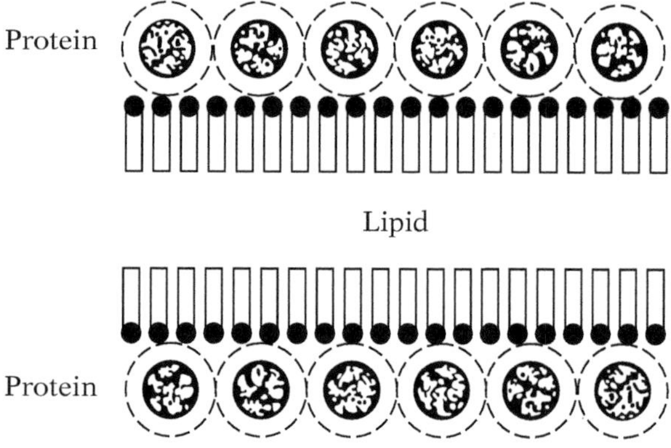

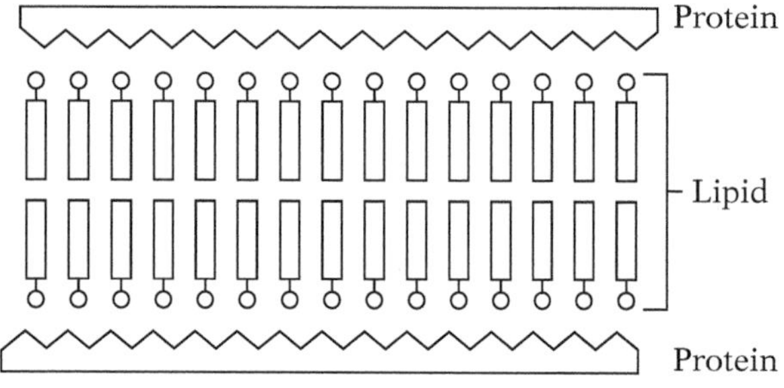

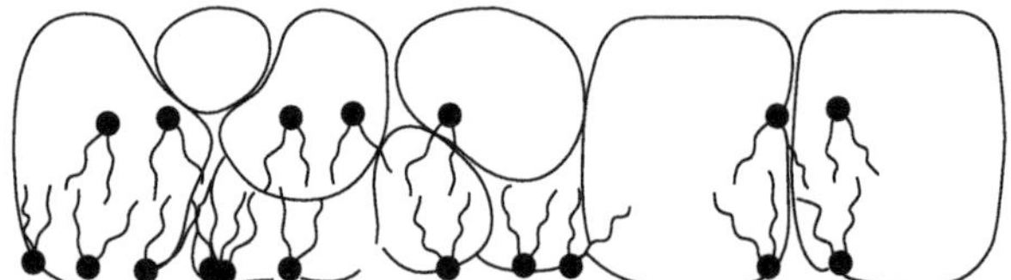

1.5 Early models for the structure of biological membranes. (A) The Davson–Danielli membrane model with layers of globular proteins outside the lipid bilayer. © 1989, Wiley-Liss, Inc., a subsidiary of John Wiley & Sons, Inc. (B) The unit membrane proposed by Robertson had the protein as β-sheets, still outside the lipid bilayer. © 1989, John Wiley-Liss Inc. a subsidiary of John Wiley & Sons Inc. (C) In contrast, the Benson–Green model for the mitochondrial inner membrane showed protein particles that are solvated by lipids and are readily fractionated into complexes. © 1983 by Academic Press. Reprinted with permission from Elsevier. (A) and (B) redrawn from Gennis, R. B., *Biomembranes: Molecular Structure and Function*, Springer-Verlag, 1989, p. 8; (C) redrawn from Aloia, R. C., *Membrane Fluidity in Biology*, vol. I, Academic Press, 1983, p. 119.

to staining. Thus the Benson–Green subunit model was the antithesis of the Davson–Danielli–Robertson model (Figure 1.5C).

Today it is hard to realize the extent of controversy that occurred. As Singer and Nicolson wrote in 1972, "Some investigators who, impressed with the great diversity of membrane compositions and functions, do not think there are any useful generalizations to be made even about the gross structure of cell membranes…." Of course, their now-classic paper on membrane structure did present a general model for the structure of biomembranes – the Fluid Mosaic Model – which is included in every modern biochemistry and biology textbook (Figure 1.7). Their paper should be read in full, for it provides a beautiful example of examining all the biomembrane's properties conducive to testing with available techniques and summarizing the results in a consistent model.

In addition to the thermodynamic principles and EM results discussed above, Singer and Nicolson emphasized the lateral mobility of membrane components. Significant lateral diffusion of membrane proteins had been demonstrated in the elegant Frye–Edidin experiment that followed the mixing of surface antigens in cell fusion experiments (Figure 1.8), and the rates of diffusion of lipids in the plane of the membrane were being measured by fluorescence techniques (discussed in Chapter 2). Singer and Nicolson also described the limited transverse mobility of lipids and the lack of it for proteins; the permeability barrier provided by the membrane; the structure of membrane proteins based on circular dichroism, x-ray diffraction, and labeling experiments (revealing them to be α-helical, globular, and membrane spanning); the assays of certain enzymes that require lipids for activity; and the phase transitions detected with differential calorimetry.

Based on these results, their Fluid Mosaic Model puts forth simple principles: the bulk of the lipid forms the bilayer, which provides the solvent for embedded proteins; most of the proteins are embedded and globular, termed *intrinsic* or *integral* membrane proteins. Some proteins are *extrinsic* (*peripheral*) as they can be removed by washes that change the pH or ionic strength. The bilayer, composed of two lipid layers, or leaflets, is fluid; in fact, it has the viscosity of olive oil, which allows lateral mobility of lipids and some protein components. It is mosaic in that proteins are scattered across it or on its surface. Both lipids and integral membrane proteins are amphipathic, allowing the nonpolar portions of proteins and lipids to interact and the polar portions of proteins and lipids to interact.

This widely accepted model for membrane structure is often abbreviated as a picture of integral proteins floating as icebergs in a sea of lipids, an oversimplification that denigrates the role of the lipids, whose diversity and polymorphic phases provide particular chemical

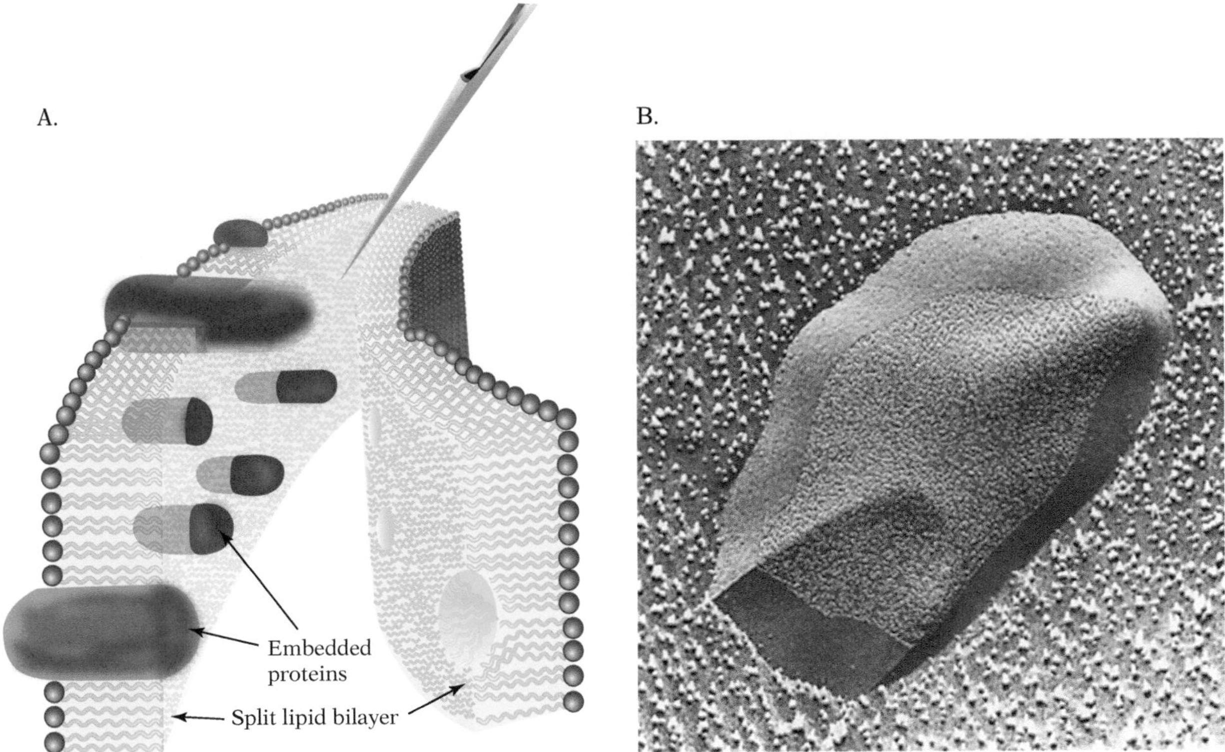

1.6 Visualization of the distribution of proteins in membranes. (A) The freeze-fracture technique reveals the interior of a biological membrane by splitting a frozen membrane sample with a cold microtome knife. Redrawn from Voet, D., and J. Voet, *Biochemistry*, 3rd ed., John Wiley, 2004, p. 405. © 2004. Reprinted with permission from John Wiley & Sons, Inc. (B) Electron microscopy of an erythrocyte plasma membrane split by freeze-fracture shows the inner surface of the membrane is studded with embedded proteins. From Voet, D., J. Voet, and C. W. Pratt, *Fundamentals of Biochemistry*, upgrade ed., John Wiley, p. 247. © 2004. Reprinted with permission of John Wiley & Sons Inc.

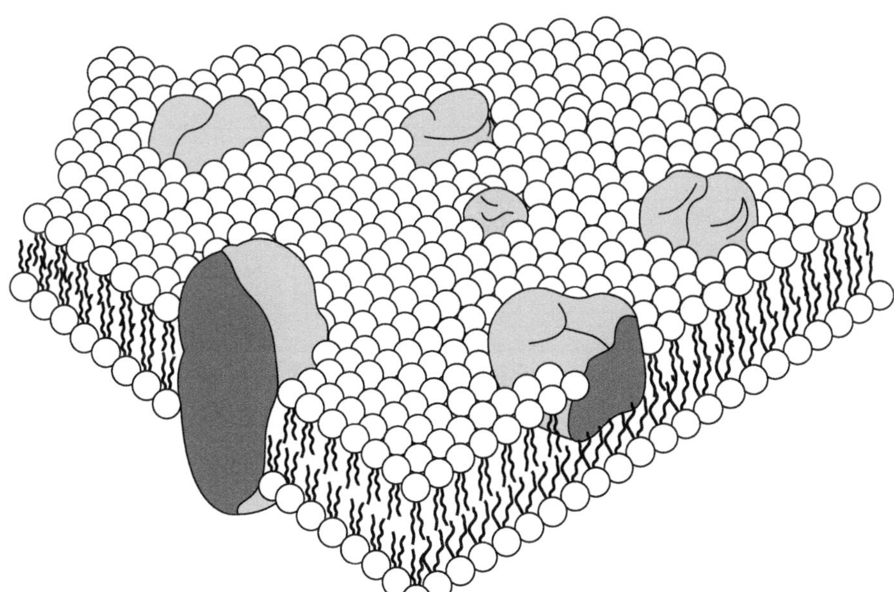

1.7 The Fluid Mosaic Model proposed by Singer and Nicolson. The basic structure of the membrane is a lipid bilayer, with the fatty acyl chains from each leaflet forming a nonpolar interior. Intrinsic proteins are integral to the bilayer, while extrinsic proteins are on its periphery. Redrawn from Singer, S. J., and G. L. Nicolson, *Science*. 1972, 175:720–731.

activities as well as structural domains in that "sea," as the next section asserts. Furthermore, this simple picture obscures the wide variation in membrane composition (not overlooked in the original paper by Singer and Nicolson!). As Table 1.1 shows, the proportion of membrane components varies from ~80% lipid and ~20% protein (myelin) to ~75% protein and ~25% lipid (mitochondrial inner membrane). A rough calculation for the mitochondrial inner membrane suggests that these membranes have on the order of 100 lipid molecules

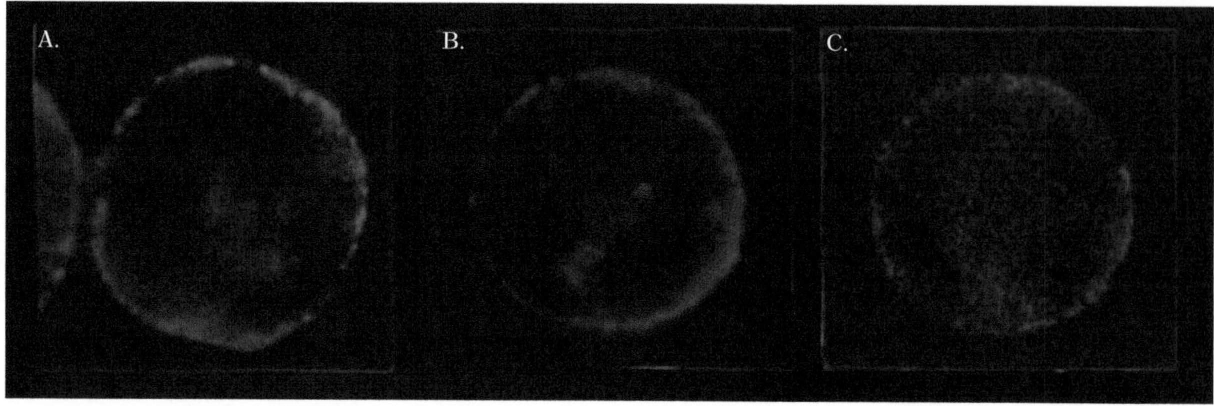

1.8 Diffusion of membrane components after cell fusion. Human and mouse antigens were labeled with red and green fluorescent markers, respectively. Virus-stimulated fusion of the mouse cell and human cell produces a heterokaryon with the two types of antigen on two halves of its surface (A). After 40 minutes the red and green markers have fully diffused so they each cover the surface (B–C). From Frye, L. D., and Edidin, M. J., *Cell Science*. 1970, 7:319–335. © 2004. Reprinted with permission from John Wiley & Sons, Inc., and the Company of Biologists.

per protein. Because it requires at least 40–50 lipid molecules to form a single belt of lipid around a protein, clearly this is not enough lipid to solvate individual proteins and provide a "sea" in which they float. So how does the mitochondrial inner membrane fit the model? First, the total protein given in Table 1.1 includes peripheral proteins. In the mitochondrial inner membrane over half the proteins are peripheral, leaving much less embedded in the lipid bilayer. Second, the protein–protein interactions between integral proteins exclude bulk lipid; thus the lipid solvates the respiratory complexes, not each individual protein. No wonder scientists

TABLE 1.1 COMPOSITION OF MEMBRANE PREPARATIONS BY PERCENT DRY WEIGHT[a]

Source	Lipid	Protein	Cholesterol
Rat liver			
Plasma	30–50	50–70	20
Rough ER	15–30	60–80	6
Smooth ER	60	40	10
Inner mitochondria	20–25	70–80	<3
Outer mitochondria	30–40	60–70	<5
Nuclear	15–40	60–80	10
Golgi	60	40	8
Lysosomes	20–25	70–80	14
Rat brain			
Myelin	60–70	20–30	22
Synaptosome	50	50	20
Rat erythrocyte	40	60	24
Rat rod outer segment	50	40	<3
Escherichia coli	20–30	70	0
Bacillus subtilis	20–30	70	0
Chloroplast	35–50	50–65	0

[a] The percentages by weight of membrane preparations from various eukaryotic and prokaryotic sources are given.
ER, endoplasmic reticulum.
Source: Based on Jain, M. K., and R. C. Wagner, *Introduction to Biological Membranes*, 2nd ed. New York: Wiley, 1988, p. 34. © 1988. Reprinted with permission from John Wiley & Sons, Inc.

who concentrated on this membrane argued strongly for the subunit model!

While much additional work has contributed support for the Fluid Mosaic Membrane, the uniform mixing of bilayer lipids has been challenged by experimental observations of lipid heterogeneity based on the physical measurements of phase separations, as well as the detection of membrane domains with separate functions. Today there is wide acceptance of a shift in the paradigm that allows membranes to have specialized microdomains called *lipid rafts*.

A SHIFT IN THE PARADIGM: BIOMEMBRANES HAVE LATERAL DOMAINS THAT FORM "RAFTS"

In addition to the wide variation in composition shown in Table 1.1, many biomembranes have protein-rich domains and other domains. In fact, some membranes are so rich in a particular protein that they contain quasicrystalline arrays of that protein, such as bacteriorhodopsin in the purple membrane of halobacteria and porins in the outer membrane of Gram-negative bacteria (see Chapter 5). Furthermore, protein-rich domains often need particular lipid species, because some proteins require specific lipids in their boundary layer. The boundary layer of lipids, also called the annulus, is an old concept that is supported by much data from activity assays and electron spin resonance studies, and more recently by x-ray structures (see Chapters 4 and 8). As Singer and Nicolson pointed out, specific lipid–protein interactions play important roles in the annulus. They did not anticipate that such interactions could extend the mosaic nature of the membrane to include functionally important lateral domains selective in terms of *both* protein and lipid components, which was unexpected in view of their emphasis on the fluidity of the bilayer.

Since 1972 a number of new techniques have been developed to measure the fluidity of model membranes. The physical definition of *fluidity* is the inverse of viscosity in an isotropic fluid, a liquid in which movement in all directions is equivalent. This definition does not directly apply to the membrane, which is highly anisotropic with a two-dimensional lipid bilayer as its base. Furthermore, the variation along the membrane normal (perpendicular to the bilayer) means the center is nearly isotropic, but a few angstroms away it is highly ordered, so position-dependent parameters are required. Therefore, measurements of membrane fluidity give results that depend on the method used, the probe for fluidity, and the conditions.

More recently, considerable lateral heterogeneity in lipid bilayers has been detected employing newer techniques such as fluorescence recovery after photobleaching, single-particle tracking, and now mass spectrometry imaging. Characterization of "liquid-ordered" microdomains in biological membranes indicates there are lateral domains with less fluidity, which form transient membrane "rafts" apart from the rest of the fluid bilayer (see "Lateral Domains and Lipid Rafts" in Chapter 2). Rafts are formed in the plasma membrane of many cell types as well as in many intracellular membranes. Although their composition varies, in general they are enriched in cholesterol and sphingolipids, which makes them thicker than the bulk membrane (Figure 1.9). They are also enriched with certain lipid-anchored proteins. Because many raft proteins are involved in signaling and trafficking, their transient associations have profound biological implications.

A VIEW FOR THE FUTURE: DYNAMIC PROTEIN COMPLEXES CROWD THE MEMBRANE INTERIOR AND EXTEND ITS BORDERS

Even with the addition of microdomains of different sizes, lifetimes, and functions, the model of the fluid mosaic membrane is incomplete. While the emphasis on lipid rafts focused attention on the lateral organization of the membrane, a variety of both old and new findings indicate the transverse organization across the plane of the membrane is complex as well. The new view of the membrane acknowledges variation in this transverse direction, encompasses layers outside the bilayer itself, and recognizes the activities going on at its borders.

The important activities occurring at the surfaces, along with striking differences across the bilayer, emphasize the significance of the third dimension of the membrane. Thus the membrane is more than a layer of proteins embedded in a lipid bilayer. Crucial functions are carried out by complexes involving interactions between integral and peripheral proteins at the interfaces. Many of the proteins are oligomers that operate in large assemblies in the membrane. Many large protein complexes operate in very close quarters in normally crowded biomembranes.

To start to describe this complexity, researchers are mapping the microenvironments found along a line extending perpendicular to the plane of the bilayer at different sites along biological membranes. The asymmetry in lipid compositions of the inner and outer leaflets was detected long ago, yet new results show it is associated with complex patterns of lipid trafficking that can turn over components of the plasma membrane each hour.

Below the plasma membrane of eukaryotic cells, the cytoskeleton creates compartments in the membrane by an actin-based meshwork and its associated transmembrane proteins (Figure 1.10). The cytoskeleton has long

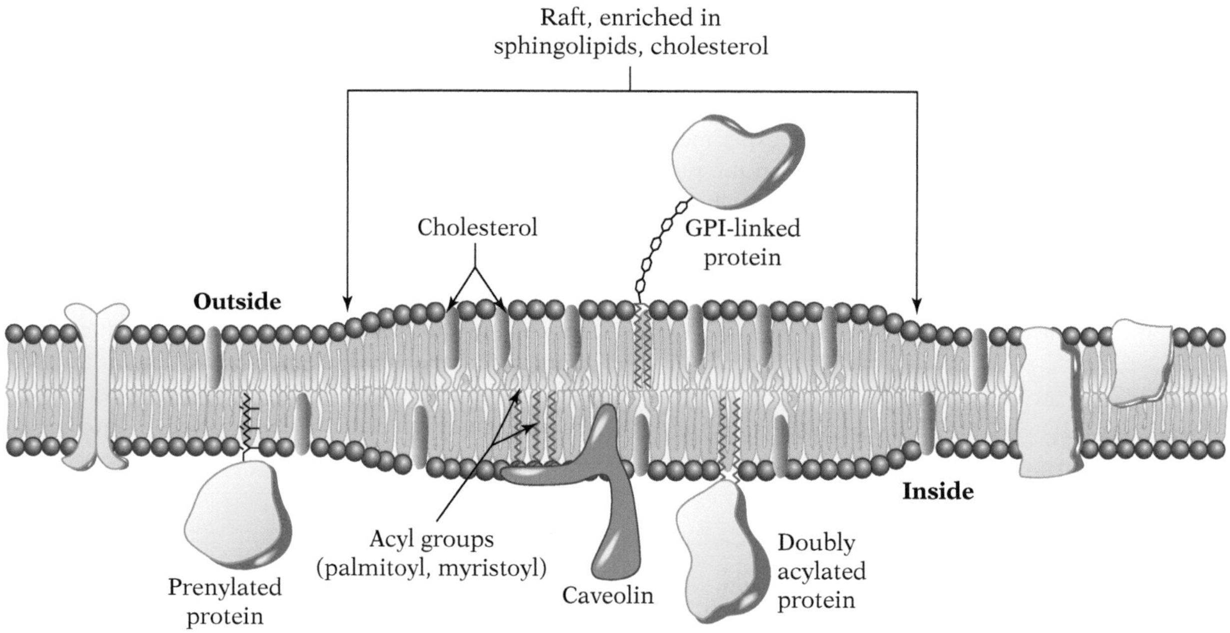

1.9 Lipid rafts. Membranes have stable but transient microdomains that are enriched in cholesterol and sphingolipids, along with glycosylphosphatidylinositol (GPI)-linked proteins and proteins anchored by acyl groups. From Nelson, D. L., and M. M. Cox (eds.), *Lehninger Principles of Biochemistry*, 4th ed., W. H. Freeman, 2005, p. 385. © 2005 by W. H. Freeman and Company. Used with permission.

been known to limit the mobility of some membrane proteins, contributing to differences in protein diffusion rates. Now single-molecule imaging studies reveal the membrane skeleton also affects the diffusion of lipids in the outer leaflet of the membrane. Thus structures beyond the bilayer partition the whole membrane into 30–200 nm compartments (much larger than rafts).

Even the picture of the lipid bilayer itself has been revised from the "lollipop" depiction of lipids in most drawings. Sophisticated analyses of diffraction data and computational modeling (described in Chapter 8) present a new picture of the bilayer in which the nonpolar domain, defined as the center that is free of water, is only about half of its thickness. Each interfacial region, made up of lipid headgroups and amphiphilic domains of proteins and containing some water molecules, contributes another quarter. Furthermore, these regions are the dynamic playgrounds for lipid-metabolizing

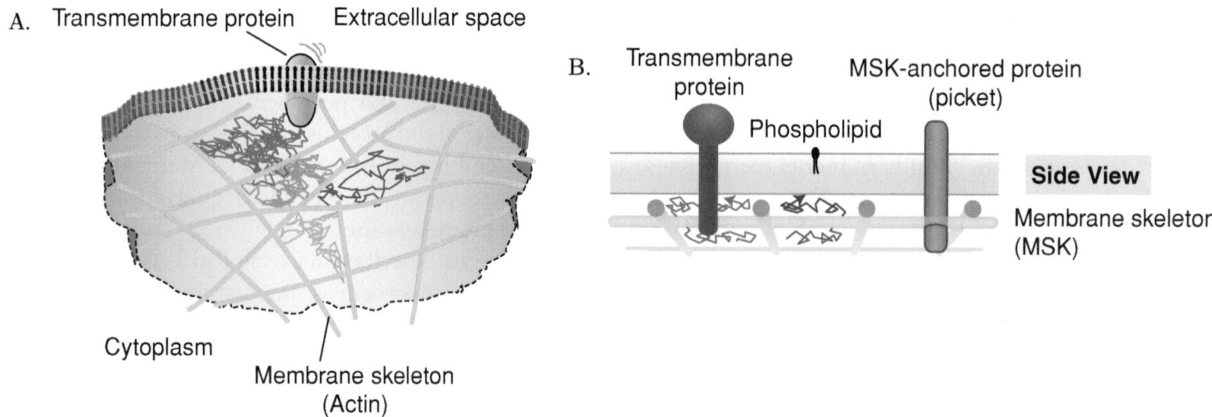

1.10 Plasma membrane domain organization due to the cytoskeleton. Membrane compartments stem from the partitioning of the entire plasma membrane by the actin-based membrane skeleton, as viewed from the cytoplasm (A) and from the plane of the membrane (B). The membrane components are labeled. The diffusion path of a labeled mobile transmembrane protein, CD44 (blue), is colored to show the compartments that confine it for short periods. Similar diffusion paths can be seen for lipids (see Figure 2.13). From Kusumi, A., *et al.*, *TIBS*. 2011, 36:604–615. Reproduced with permission from *Trends Biochem Sci*.

enzymes and other proteins that insert into the bilayer (described in Chapter 4).

The exterior surface of many cell membranes is crowded with peripheral proteins that can form important short-lived complexes with bilayer lipids and membrane proteins. Many of these peripheral proteins have activities that are regulated by binding the membrane in dynamic cycles (see Chapter 4). Their specific binding is mediated by highly conserved motifs and often by divalent cations. Thus, the focus of membrane research has expanded to include the membrane periphery as an additional important layer of the membrane.

Large complexes made up of integral membrane proteins and associated peripheral proteins carry out many functions of the membrane. The typical biomembrane is crowded with protein assemblies, many of which are tightly associated hetero-oligomers (Figure 1.11). Furthermore, some complexes are large enough to span two membranes, either from the same cell or organelle, seen in the double-membrane systems of Gram-negative bacteria and mitochondria; from different cells, such as at gap junctions; or from a cell and a virus, as observed in the fusion events enabling viral penetration. Future challenges in understanding membrane functions include characterizing these larger functional membrane complexes (see Chapter 14).

In short, the biomembrane consists of amphiphilic molecules that respond to the hydrophobic effect by spontaneously assembling into the bilayer. The bilayer is fluid and mosaic with respect to both lipids and proteins and contains dynamic lateral domains, or rafts, which appear to be critical to many biological functions. The central nonpolar region that excludes virtually all water molecules spans about half the bilayer, sandwiched between the two interfacial layers, where considerable activity takes place. The surface of the bilayer is often crowded with peripheral proteins, some coming and going and others providing mechanical support. Many functions of the membrane are carried out by molecular assemblies that are large multicomponent complexes, some of which even span two membranes.

To study this marvelous, multifaceted biological structure in some detail and to appreciate the stunning molecular structures of membrane constituents now emerging requires an initial understanding of the physical and chemical properties of the membrane bilayer. Thus, this book starts with the fundamental properties of membrane components and ends with examples of membrane proteins whose high-resolution structures give insights into their functions. Chapter 2 takes a close look at the structure and function of membrane lipids, moving from the diversity and properties of membrane lipids to the properties of domains and the formation of lipid rafts. It concludes with the examination of the role of nonbilayer-forming lipids in biological membranes. Chapter 3 describes the tools needed for *in vitro* characterization of membrane constituents. It opens with the properties of the detergents that are so important in purification of membrane components and then surveys the old and new model systems available for studying lipid aggregates and for reconstituting membrane proteins. Chapter 4 portrays the different types of proteins at or in the bilayer. It describes amphitropic proteins, including peripheral proteins and lipid-anchored proteins, and the dynamics of their binding to the interface. It considers peptides and proteins that can insert into bilayers, a group that includes some ionophores and toxins. Finally, it examines the constraints placed by the

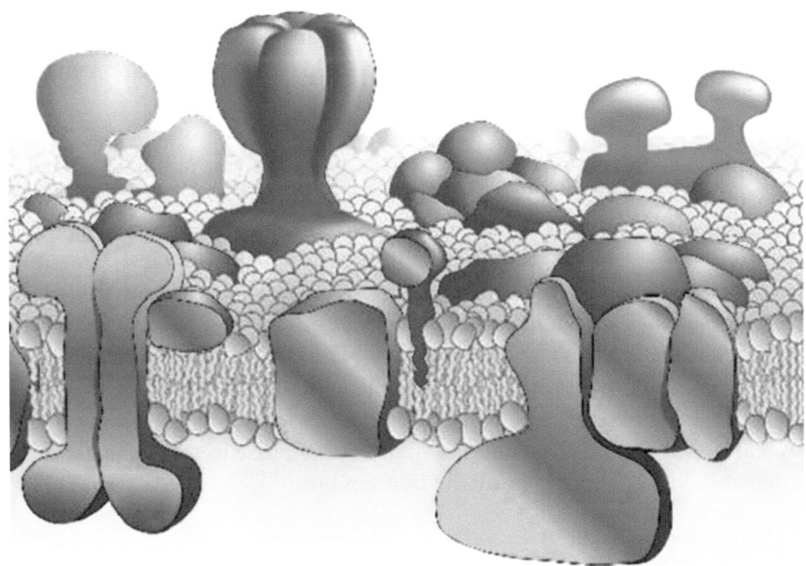

1.11 Peripheral proteins and complexes. Membranes encompass not only the bilayer but also peripheral proteins and, in the case of plasma membranes, the cytoskeleton. Typically crowded with proteins, membranes contain many complexes that extend beyond the bilayer. From Engelman, D., *Nature*. 2005, 438:578–580. © 2005. Adapted by permission of Macmillan Publishers Ltd.

bilayer on the general features of integral membrane proteins. Then Chapter 5 gives an in-depth view of the two major classes of integral protein structures – bundled α-helices and β-barrels – and presents the proteins that are paradigmatic examples of each class. It ends with a reminder that the known atomic structures represented in these classes at present account for only a small proportion of the membrane proteins in the genomes.

With this foundation of the structural principles that determine the characteristics of membrane constituents, the book turns to biochemical, biophysical, and proteomic studies of membrane proteins. Chapter 6 describes the three major (and overlapping) classes of membrane protein functions – enzymes, transport proteins, and receptors – with examples of each. Then it covers the bioinformatics tools that are used to predict structures and families of membrane proteins for genomic analysis and describes proteomics techniques designed to reveal inter-subunit and other protein–protein interactions. Chapter 7 focuses on folding and biogenesis of membrane proteins, starting with studies on the folding of purified membrane proteins and moving to the complex systems involved in their biogenesis in cells. It concludes with a look at human diseases that result from misfolding membrane proteins.

Chapter 8 presents diffraction and simulation techniques that give analyses and represen tations of the fluid membrane. It moves from structures of lipids in crystalline arrays to liquid crystallography and other techniques that describe the fluid bilayer based on diffraction data. It shows how computational modeling using molecular dynamics gives simulations of increasingly complex lipid bilayers. Then it describes lipids observed in high-resolution x-ray structures of membrane proteins and ends with a brief discussion of techniques used to obtain suitable crystals of membrane proteins.

The last part of the book tours the growing field of membrane structural biology with descriptions of representative membrane proteins of known structures. Each of the last chapters describes a class of membrane proteins grouped by function. Because of overlapping functions, such as receptors that make ion channels, these classes are not distinct: there are ion channels in Chapter 10 as well as Chapter 12 and enzymes in Chapter 13 as well as Chapter 9. Blurring the distinctions further, it turns out that one group of related "chloride channels" described in Chapter 12 includes both channels and transporters!

Chapter 9 depicts membrane enzymes of widely varying structures and functions. Although they carry out their catalytic functions within the membrane milieu, they employ chemical mechanisms similar to those of soluble enzymes due to convergent evolution. Chapter 10 emphasizes developments in two exciting areas of receptor research, signaling and neurochemistry, with structures of G protein-coupled receptors and neurotransmitter receptors.

Chapter 11 presents structures of a variety of transporters. In this area with the largest number of structures, it is interesting that many appear to share a basic mechanism – alternating access – to accomplish translocation of substrates. Chapter 12 describes channels small and large, from aquaporins to gap junctions. And Chapter 13 examines different mechanisms for energy transduction, with a focus on the complexes of the respiratory chain and the ATP synthase.

To conclude, Chapter 14 looks at the increasing complexity of problems for membrane structural biology in the future as it tackles more complex proteins – with increasing emphasis on human proteins, more multiprotein assemblies, and more dynamic processes. Such membrane protein complexes involved in vital functions in the body include those in the nuclear pore and those presenting antigens during the immune response. The tools developed today will be applied to address these and many other challenging questions.

Molecular characterization of the membrane lagged behind other fields of biochemistry, in part because of the difficulties in obtaining high-resolution structural information on its components. Today, structural insights into the membrane are taking off, fueled by the recent growth in the number of membrane proteins whose structures have been solved. As always in biochemistry, the first structures provide models that suggest possibilities for structures and mechanisms still unknown, those involved in similar as well as more complicated membrane functions. From the fundamental principles covered in the first chapters to the descriptions of the high-resolution structures in the last part, this book gives a solid grasp of the field needed for an appreciation of the progress to come.

Membrane research is leading to significant understanding of basic processes involved in differentiation and development, immunology, neurochemistry, and nutrition. It is also essential to medical research, covering topics such as membrane fusion in viral infection, roles of various oncogenes in cancer development, efflux mechanisms causing multidrug resistance, and drug design for a large variety of targets. Membrane components are the targets of the majority of newly developed pharmaceuticals, and defects in membrane components are the causes of many human diseases. It is an exciting time to study the biomembrane.

FOR FURTHER READING

Bagatolli, L. A., J. H. Ipsen, A. C. Simonsen, and O. G. Mouritsen, An outlook on organizations of lipids in membranes: searching for a realistic connection with the organization of biological membranes. *Prog Lipid Res*. 2010, 49:378–389.

Edidin, M., Lipids on the frontier: a century of cell-membrane bilayers. *Nat Rev Mol Cell Biol*. 2003, 4:414–418.

Engelman, D. M., Membranes are more mosaic than fluid. *Nature*. 2005, 438:578–580.

Kuhn, T. S., *The Structure of Scientific Revolutions*, 3rd ed. Chicago, IL: University of Chicago Press, 1996.

Kusumi, A., K. G. Suzuki, R. S. Kasai, K. Ritchie, and T. K. Fujiwara, Hierarchical mesoscale domain organization of the plasma membrane. *TIBS*. 2011, 36:604–615.

Lindner, R., and H. Y. Naim, Domains in biological membranes. *Exp Cell Res*. 2009, 315:2871–2878.

Simons, K., and E. Ikonen, Functional rafts in cell membranes. *Nature*. 1997, 387:569–572.

Singer, S. J., and G. L. Nicolson, The Fluid Mosaic Model of the structure of cell membranes. *Science*. 1972, 175:720–731.

Tanford, C., *The Hydrophobic Effect: Formation of Micelles and Biological Membranes*, 2nd ed. New York: Wiley, 1980.

THE DIVERSITY OF MEMBRANE LIPIDS

2

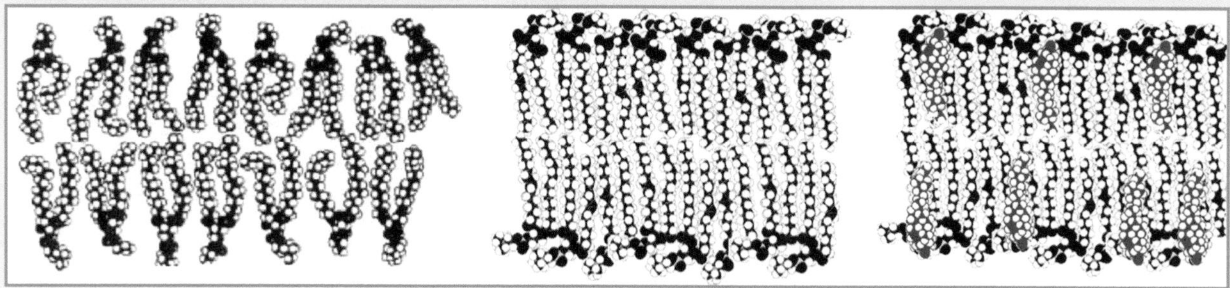

The versatility of lipids as a function of their composition and temperature results in different phases in a biomembrane. Glycerophospholipids with unsaturated hydrocarbon chains (left) tend to be in fluid phases called liquid-crystalline (L_α) or liquid disordered (L_d). Sphingomyelin with long, saturated hydrocarbon chains (middle) tends to form more solid phases (L_β or s_o). Adding a sterol such as cholesterol (right) produces a special phase called liquid ordered (L_o) that is found in lipid rafts. From van Meer, G., *et al.*, *Nat Reviews Mol Cell Biol*. 2008, 9:112–124. Copyright © 2008, Nature Publishing Group. Reprinted by permission of Macmillan Publishers Ltd.

To understand biological membranes requires a detailed knowledge of their components. It is appropriate to start with the lipids that make up the bilayer, because they are not just a solvent providing the "sea" in which membrane proteins float. If this were the case, a few lipidic species would suffice to provide the amphiphilic base of the bilayer and some variation of shapes for packing it. Instead, the diversity of membrane lipids is amazing. A typical biomembrane contains more than 100 species of lipids, which vary in general structure and in the length and degree of saturation of their fatty acyl chains. This chapter begins with the properties and modulation of the acyl chains and then reviews the structural features of the major complex lipids. It describes the properties of lipid aggregates, including their polymorphism and phase separations, which are the basis of their ability to form lateral microdomains. It addresses the characteristics of lipid rafts in membranes. The chapter ends by considering lipid diversity first in *Escherichia coli*, with studies that suggest non-bilayer-forming lipids are needed to maintain the elasticity of the membrane, and finally noting variations due to synthesis and trafficking in eukaryotic cells.

THE ACYL CHAINS

When lipids are extracted from a cell with organic solvent, such as a 2:1 mixture of chloroform and methanol, free fatty acids make up only about 1% of the total; most fatty acids are bound covalently in complex lipids. At present more than 500 different species of fatty acids have been identified.

Table 2.1 shows the fatty acids normally occurring in the lipids of biological membranes, grouped as saturated or unsaturated (those with at least one carbon–carbon double bond). The acyl chains are typically 10–24 carbons long, with an even number of carbons resulting from their synthesis from the precursor acetyl-coenzyme A.

The systematic names of fatty acids are rarely used. Instead they may be referred to by common names or by symbols indicating their chain length and degree of saturation. For example, stearic acid is C18:0, a saturated 18-carbon chain, drawn in Figure 2.1A with the acyl chain fully extended. If it is part of a complex lipid, the 18-carbon saturated acyl chain is stearoyl. If the chain is monounsaturated (with one double bond), it is oleoyl, because oleic acid is C18:1 (Δ9), an 18-carbon chain with a double bond between carbons 9 and 10 (Figure 2.1B). A single *cis* double bond makes a 30° bend, or kink, in the acyl chain, as shown in the figure. Naturally occurring unsaturated fatty acids have *cis* double bonds. Partial dehydrogenation of dietary fats produces *trans* fatty acids such as elaidic acid, C18:1 (Δ9) *trans* (Figure 2.1C), which are found in processed foods and enter the human body (to the apparent detriment of health).

TABLE 2.1 SOME NATURALLY OCCURRING FATTY ACIDS: STRUCTURE, PROPERTIES, AND NOMENCLATURE[a]

Carbon skeleton	Structure[b]	Systematic name[c]	Common name (derivation)	Melting point (°C)	Solubility at 30°C (mg/g solvent)	
					Water	Benzene
12:0	$CH_3(CH_2)_{10}COOH$	*n*-Dodecanoic acid	Lauric acid (Latin *laurus*, "laurel plant")	44.2	0.063	2600
14:0	$CH_3(CH_2)_{12}COOH$	*n*-Tetradecanoic acid	Myristic acid (Latin *myristica*, nutmeg genus)	53.9	0.024	874
16:0	$CH_3(CH_2)_{14}COOH$	*n*-Hexadecanoic acid	Palmitic acid (Latin *palma*, "palm tree")	63.1	0.0083	348
18:0	$CH_3(CH_2)_{16}COOH$	*n*-Octadecanoic acid	Stearic acid (Greek *stear*, "hard fat")	69.6	0.0034	124
20:0	$CH_3(CH_2)_{18}COOH$	*n*-Eicosanoic acid	Arachidic acid (Latin *Arachis*, legume genus)	76.5		
24:0	$CH_3(CH_2)_{22}COOH$	*n*-Tetracosanoic acid	Lignoceric acid (Latin *lignum*, "wood" + *cera*, "wax")	86.0		
16:1 (Δ9)	$CH_3(CH_2)_5CH{=}CH(CH_2)_7COOH$	*cis*-9-Hexadecenoic acid	Palmitoleic acid	0.5		
18:1 (Δ9)	$CH_3(CH_2)_7CH{=}CH(CH_2)_7COOH$	*cis*-9-Octadecenoic acid	Oleic acid (Latin *oleum*, "oil")	13.4		
18:2 (Δ9, 12)	$CH_3(CH_2)_4CH{=}CHCH_2CH{=}CH(CH_2)_7COOH$	*cis*-,*cis*-9,12-Octadecadienoic acid	Linoleic acid (Greek *linon*, "flax")	–5		
18:3 (Δ9, 12, 15)	$CH_3CH_2CH{=}CHCH_2CH{=}CHCH_2CH{=}CH(CH_2)_7COOH$	*cis*-,*cis*-,*cis*-9,12,15-Octadecatrienoic acid	α-Linolenic acid	–11		
20:4 (Δ5, 8, 11, 14)	$CH_3(CH_2)_4CH{=}CHCH_2CH{=}CHCH_2CH{=}CHCH_2CH{=}CH(CH_2)_3COOH$	*cis*-,*cis*-,*cis*-,*cis*-5,8,11,14-Eicosatetraenoic acid	Arachidonic acid	–49.5		

[a] The symbol for fatty acids gives the number of carbon atoms, followed by the number of carbon–carbon double bonds. For unsaturated fatty acids, the notations in parentheses denote the positions of their double bonds. For example, Δ9 denotes a double bond between C9 and C10. All the double bonds in these fatty acids have *cis* configuration.

[b] All acids are shown in their nonionized form. At pH 7, all free fatty acids have an ionized carboxylate. Note that numbering of carbon atoms begins at the carboxyl carbon.

[c] The prefix *n* indicates the normal unbranched structure. For instance, *dodecanoic* simply indicates 12 carbon atoms, which could be arranged in a variety of branched forms; *n-dodecanoic* specifies the linear, unbranched form.

Source: Data from Nelson, D. L., and M. M. Cox, *Lehninger Principles of Biochemistry*, 4th ed. New York: W. H. Freeman, 2005. © 2005 by W. H. Freeman. Used with permission.

The fully extended chain in the illustration of stearic acid (Figure 2.1A) results when all the dihedral angles along the hydrocarbon chain are 180°, which is the favored angle of rotation around the single C–C bonds in fatty acids. This angle, called the torsion angle, is evident when bonds between four neighboring C atoms are considered (Figure 2.2). The most favored value of the torsion angle of a hydrocarbon chain is 180° and is called *trans* or *anti*; a second favored value is around 60° and is called *gauche*. In the liquid state, hydrocarbon chains sample different torsion angles with rotation around their C–C bonds; rotation is more restricted around C=C double bonds.

Phospholipids in biological membranes often have a saturated acyl chain on C1 of the glycerol moiety and an unsaturated chain on C2 (see Figure 2.5). Polyunsaturated fatty acids, including omega-3 fatty acids such as α-linolenic acid (C18:3, Δ9, 12, 15), whose last C=C is three carbons from the end of the chain, are important nutritional precursors to molecules such as prostaglandins and isoprenes. As minor constituents of membrane lipids, polyunsaturated acyl chains are

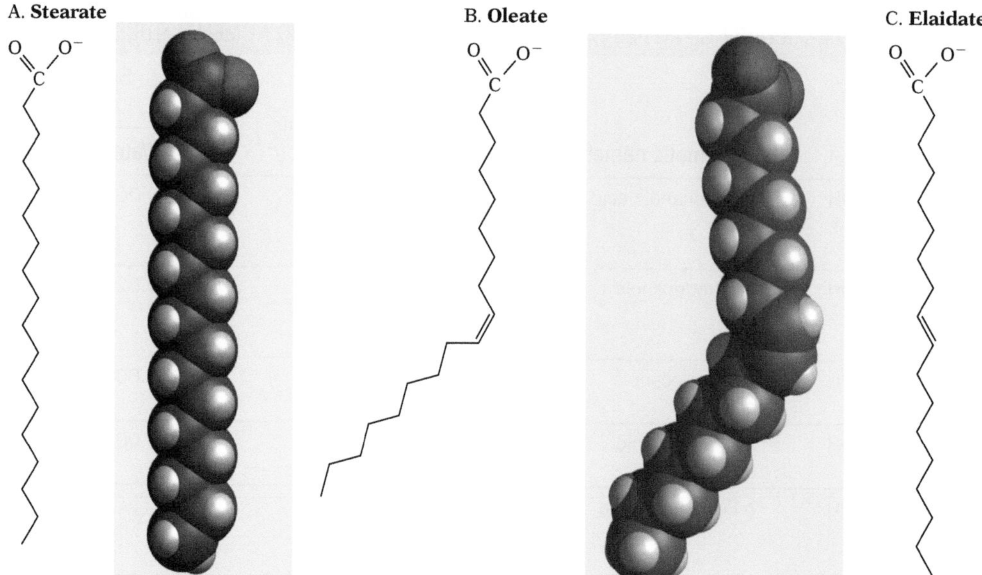

2.1 Saturated and unsaturated fatty acids with 18-carbon chains. With the carboxyl group deprotonated at neutral pH, stearic acid becomes stearate (C18; (A)), oleic acid becomes oleate (C18:1, Δ9 *cis*; (B)), and elaidic acid becomes elaidate (C18:1 Δ9 *trans*; (C)). While the *trans* double bond does not affect the conformation of the acyl chain, the *cis* double bond introduces a kink in the chain, as illustrated by the space-filling model for oleic acid. Redrawn from Nelson, D. L., and M. M. Cox (eds.), *Lehninger Principles of Biochemistry*, 4th ed., W. H. Freeman, 2005, p. 345. © 2005 by W. H. Freeman and Company. Used with permission.

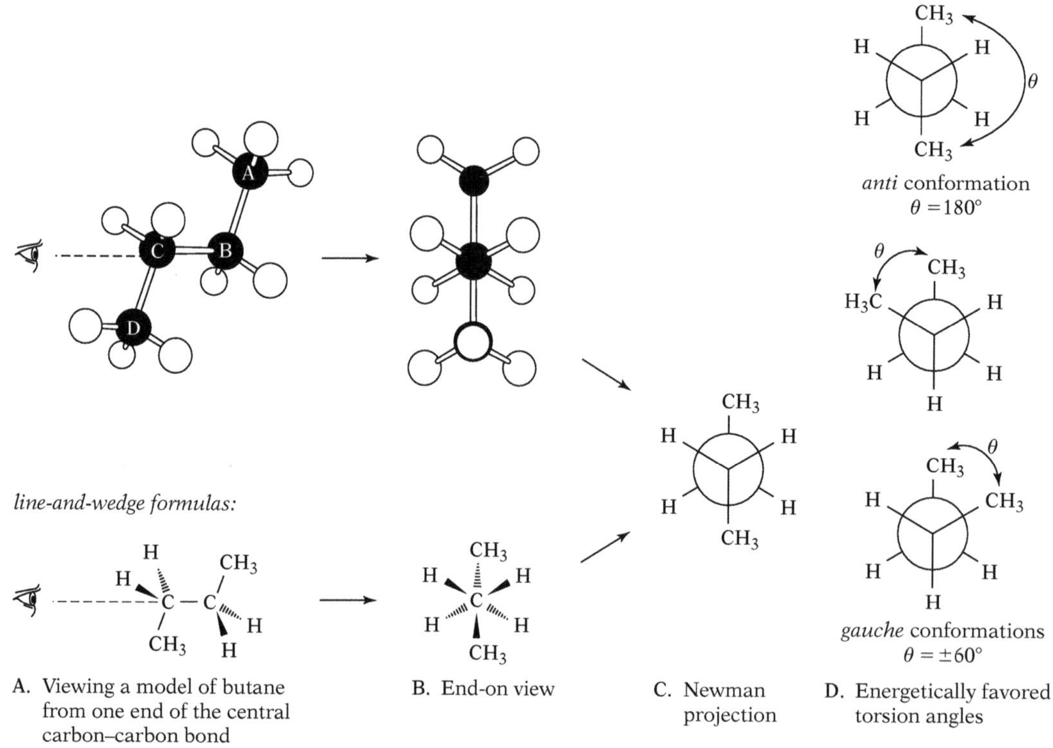

2.2 Bond torsion angles in hydrocarbon chains. Four atoms of a hydrocarbon chain, labeled ABCD, may be represented by butane, shown in (A). The torsion angle is the dihedral angle defined by the planes between atoms ABC and atoms BCD. This angle may be visualized from the end-on view (B) and is diagrammed in the Newman projection (C). The most stable torsion angles for hydrocarbon chains are called *anti* and *gauche*, shown for butane in (D). A long hydrocarbon has considerable flexibility with varying torsion angles along the chain. Redrawn from Loudon, G. M., *Organic Chemistry*, 4th ed., Oxford University Press, 2002. (A) © 2002 by Oxford University Press, Inc. Reproduced by permission of Oxford University Press, (C–D) © 2002 by Oxford University Press, Inc. Reproduced by permission of Oxford University Press, Inc. USA.

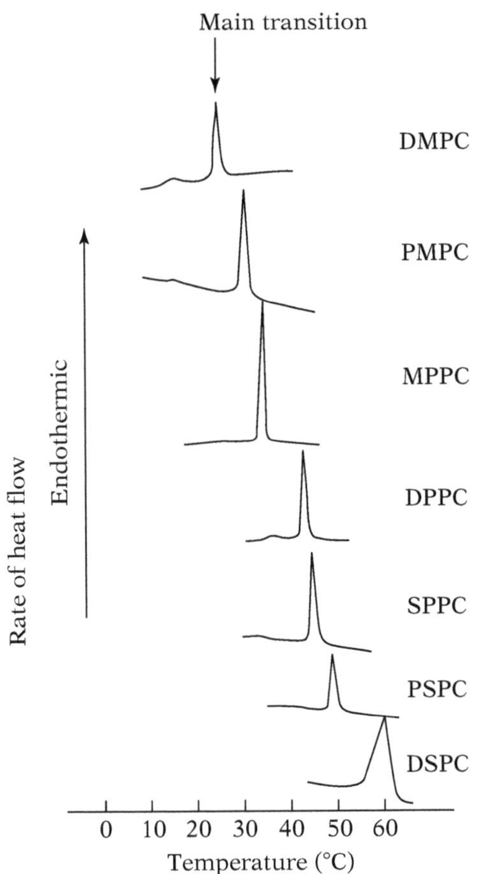

2.3 Detection of the gel-to-liquid crystal transition in different phosphatidylcholine (PC) molecules with acyl chains varying from 14 carbons (dimyristoyl PC) to 18 carbons (distearoyl PC). Differential scanning calorimetry is used to measure the heat consumption as the temperature is increased. The peaks correspond to the enthalpic "melting" events, which occur at higher temperatures as the chain lengths increase. When the PC contains two acyl chains of different lengths, its melting temperature is midway between that of the two PC molecules having identical chains of the two types. Redrawn from Keough, K. M., and P. J. Davis, *Biochemistry*. 1979, 18:1453. © 1979 by American Chemical Society.

bulky yet highly flexible and affect membrane elasticity (see below); they are usually absent in bacteria. Unusual fatty acids are found in the membranes of some organisms. Some bacteria have branched, hydroxylated, or iso–acyl chain fatty acids. *E. coli* membranes can have ~25% cyclopropane-containing fatty acids. In some marine organisms, an odd number of carbons is common.

Pure fatty acids undergo a sharp phase transition when they are melted. This transition is detected by the increased heat uptake measured by differential scanning calorimetry (DSC), as illustrated in Figure 2.3. For the acyl chains of a lipid, such as phosphatidylcholine, this transition reflects the change from the gel state ("solid") along the chains to the liquid crystalline state ("fluid") as the temperature is increased. In the gel state, the acyl chains are fairly ordered, with a high *trans/gauche* ratio. In the liquid crystalline state, the rotational freedom decreases the *trans/gauche* ratio and allows many more configurations of the hydrocarbon chains (see chapter Frontispiece). From the trend in Figure 2.3, as well as the melting temperatures (T_m) of the fatty acids listed in Table 2.1, it is evident that T_m increases with increasing chain length and decreases with increasing unsaturation. In the gel phase, the extended saturated acyl chains pack closely, stabilized by van der Waals forces. Fatty acids with longer chains have higher T_m, because with more extensive contact areas, more heat is required to disrupt this structure. The kinks introduced by double bonds disrupt this order, destabilizing the gel phase so less heat is required to melt unsaturated fatty acids.

Many examples of the adaptation of unicellular organisms to their environments illustrate the functional importance of these phase transitions. In *E. coli* growing at 37°C, the ratio of saturated to unsaturated fatty acids is about 1:1; when growing at 17°C, it changes to about 1:2 (see Table 2.2). When growing at high temperatures, bacterial membranes are enriched in saturated and longer acyl chains, as is especially evident in thermophilic bacteria such as those found in the hot springs at Yellowstone National Park, with ambient temperatures around 85°C. Marine organisms – both bacteria and deep-sea fish – have adapted to high pressures by increasing the proportion of unsaturated fatty acids to enable them to maintain the fluidity of their membranes. In general, organisms vary the composition of acyl chains in their membranes to achieve a state that is fluid but not too fluid, indicating that the acyl chains may be important determinants of phase polymorphism (see below), even in a complex mixture that would not exhibit a simple "melting" point.

COMPLEX LIPIDS

Lipids found in biomembranes fall into three main classes (Figure 2.4):

- glycerophospholipids (often called phospholipids)
- sphingolipids (including sphingophospholipids)
- sterols and linear isoprenoids.

In addition, glycolipids may be considered a separate class, although they consist of either glycerophospholipids or sphingolipids with oligosaccharide headgroups. Their importance in human health and disease is evident: the blood groups A, B, and O are determined by the glycosphingolipids on cell surfaces, and several hereditary diseases, such as Tay-Sachs disease, result from abnormal accumulation of glycosphingolipids.

TABLE 2.2 FATTY ACID COMPOSITION OF *E. COLI* CELLS CULTURED AT DIFFERENT TEMPERATURES[a]

	Percentage of total fatty acids[b]			
	10°C	20°C	30°C	40°C
Myristic acid (14:0)	4	4	4	8
Palmitic acid (16:0)	18	25	29	48
Palmitoleic acid (16:1)	26	24	23	9
Oleic acid (18:1)	38	34	30	12
Hydroxymyristic acid	13	10	10	8
Ratio of unsaturated to saturated[c]	2.9	2.0	1.6	0.38

[a] The values are given in weight percent of total lipid.

[b] The exact fatty acid composition depends not only on growth temperature but on growth stage and growth medium composition.

[c] Ratios calculated as the total percentage of 16:1 plus 18:1 divided by the total percentage of 14:0 plus 16:0. Hydroxymyristic acid was omitted from this calculation.

Source: Data from Marr, A. G., and J. L. Ingraham, Effect of temperature on the composition of fatty acids in *Escherichia coli*. *J Bacteriol*. 1962, 84:1260–1267. Reprinted with permission from Nelson, D. L., and M. M. Cox, *Lehninger Principles of Biochemistry*, 4th ed. New York: W. H. Freeman, 2005. © 2005 by W. H. Freeman and Company. Used with permission.

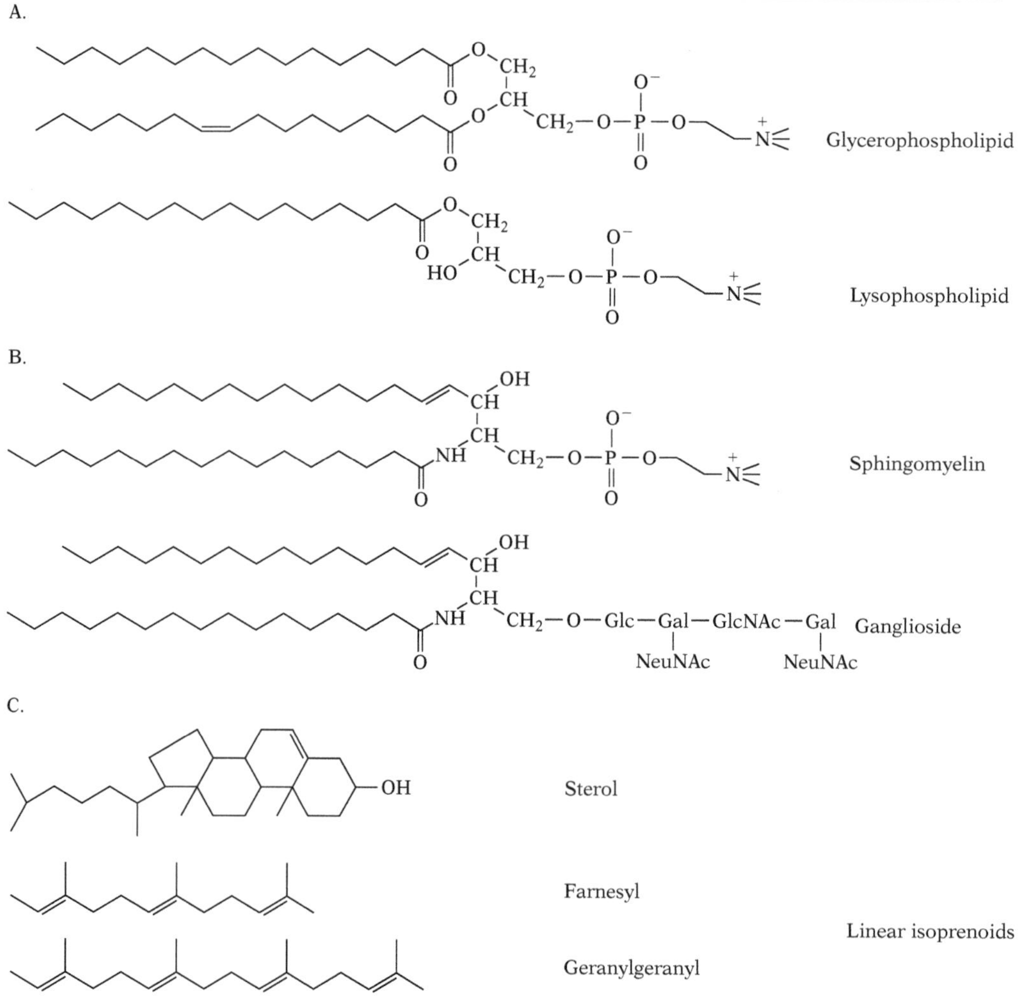

2.4 Three major classes of membrane lipids. The structures of representative glycerophospholipids (A), sphingolipids (B), and sterols and linear isoprenoids (C) are shown. With a sugar headgroup on a sphingomyelin base, a ganglioside is an example of a glycolipid. (A) and (B) redrawn from Gennis, R. B., *Biomembranes*, Springer-Verlag, 1989, p. 24; (C) redrawn from Nelson, D. L., and M. M. Cox (eds.), *Lehninger Principles of Biochemistry*, 4th ed., W. H. Freeman, 2005, p. 355.

The ganglioside GM_2 that accumulates in patients with Tay-Sachs disease is a sphingolipid with galactose (Gal), glucose (Glc), N-acetylneuraminic acid (NeuNAc), and N-acetylglucosamine (GlcNAc) moieties.

Phospholipids

A glycerophospholipid, commonly known simply as a phospholipid (PL), is built on a glycerol molecule, which becomes chiral when derivatized to glycerol-3-phosphate. The backbone of membrane PLs is the l isomer, called *sn*-glycerol 3-phosphate (*sn*, for stereochemical numbering, is used instead of d and l or *R* and *S*).

With fatty acyl chains in ester linkage on carbons 1 and 2 it becomes phosphatidic acid (PA). Esterification of PA with another alcohol creates the following PLs: phosphatidylcholine (PC), phosphatidylethanolamine (PE), phosphatidylserine (PS), phosphatidylglycerol (PG), and phosphatidylinositol (PI) (Figure 2.5). In addition, PG can link through its glycerol headgroup to PA to form diphosphatidylglycerol (CL for its common name, cardiolipin). These abbreviations are coupled with the abbreviated common names for the acyl chains: DOPC is thus dioleoylphosphatidylcholine and MPoPS is 1-myristyl 2-palmitoleoylphosphatidylserine. (See Appendix I for a list of abbreviations.) The phosphate groups and headgroups are the polar portions, and the acyl chains are the nonpolar parts of these amphiphilic molecules. In many PLs of biological membranes, the acyl chain on C1 is saturated and 16 or 18 carbons long, while that on C2 is frequently unsaturated and often longer.

Although phospholipids do act as solvent for membrane proteins and define the polar and nonpolar domains of the bilayer, they also have important chemical, biological, and physical properties. The anionic PLs (PS, PI, PG, and CL) have a net negative charge at physiological pH, while the zwitterionic PLs (PE and PC) are neutral. PE and PS contain reactive amines that can participate in hydrogen bonding. PI, PC, and cardiolipin (CL) are relatively bulky, which affects their packing in bilayers. When a phospholipid loses one acyl chain through the action of a phospholipase, it becomes a

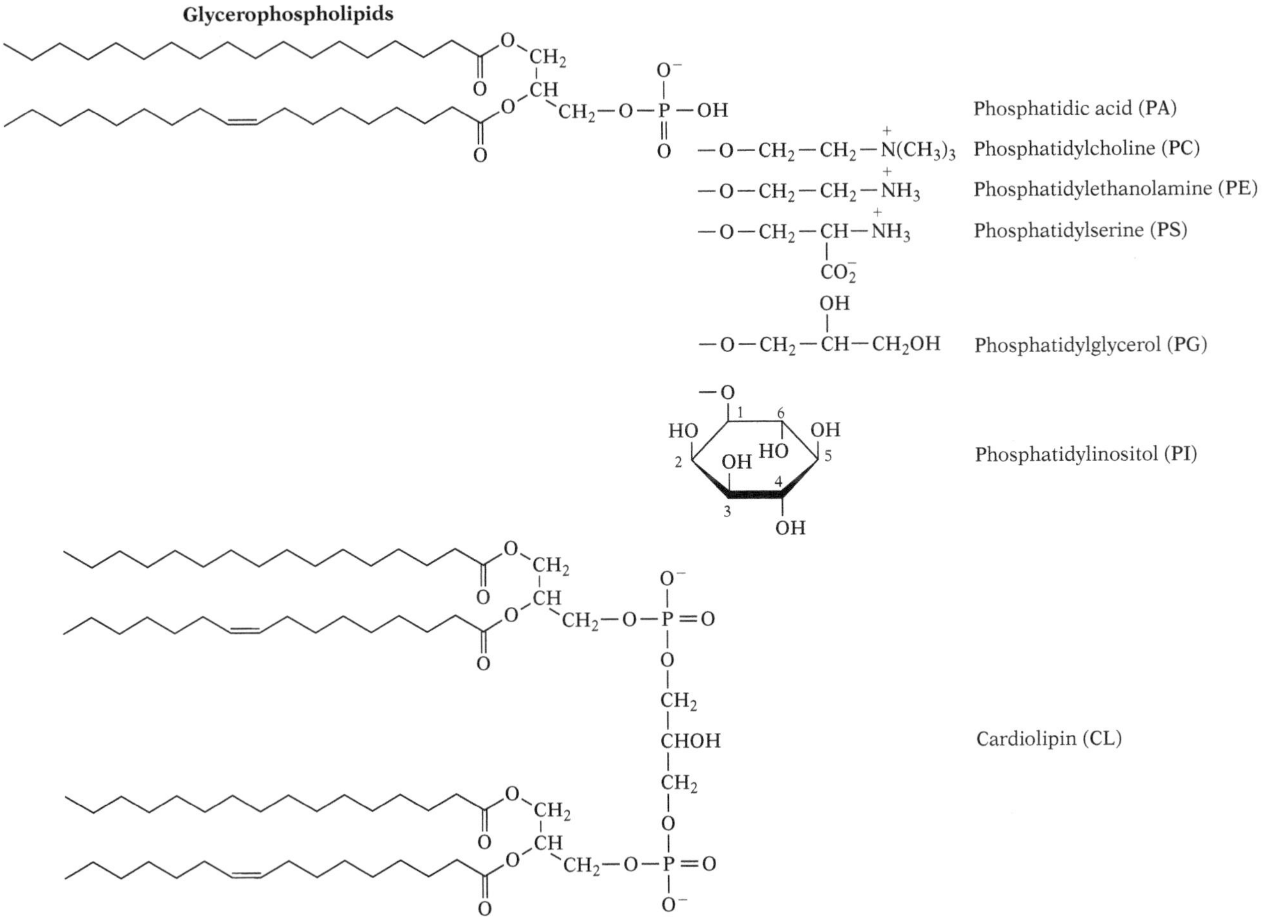

Glycerophospholipids

Phosphatidic acid (PA)
Phosphatidylcholine (PC)
Phosphatidylethanolamine (PE)
Phosphatidylserine (PS)
Phosphatidylglycerol (PG)
Phosphatidylinositol (PI)
Cardiolipin (CL)

2.5 Structures of glycerophospholipids. The common glycerophospholipids in biological membranes contain one of the polar headgroups shown. In addition, they vary greatly in the length and saturation of their acyl chains, although in general the acyl chain on C1 is saturated and the acyl chain on C2 is unsaturated. Redrawn from Gennis, R. B., *Biomembranes*, Springer-Verlag, 1989, p. 25.

lysophospholipid with increased water solubility that gives it surfactant (detergent) activity. Phospholipids provide sources of second messengers for signaling across the membrane and enhance the activity of membrane enzymes and transport proteins (see Chapter 4). Their degree of unsaturation contributes to the elasticity of the membrane, which influences insertion and sequestering of proteins (see Figure 2.18 and the discussions of folding studies in Chapter 7).

Archaea have a unique set of phospholipids that have ether linkages to their phytanyl chains (instead of ester bonds to acyl chains) as they are derived from archaeol. Archaeol is 2,3-di-O-phytanyl-*sn*-glycerol, and its phytanyl groups are 20-carbon branched-chain isoprenoids (3,7,11,15-tetramethyl hexadecanoic acid) (Figure 2.6). They also have unusual headgroups and differ in stereochemistry, esterified to the phosphate headgroup or sulfated glycolipid on the *sn*1 instead of the *sn*3 carbon. Some archeal lipids even have the two archaeol groups fused with headgroups on both ends (Figure 2.6B).

Sphingolipids

Sphingolipids are built not on a glycerol backbone but on sphingosine, a long-chain amino alcohol, to which a fatty

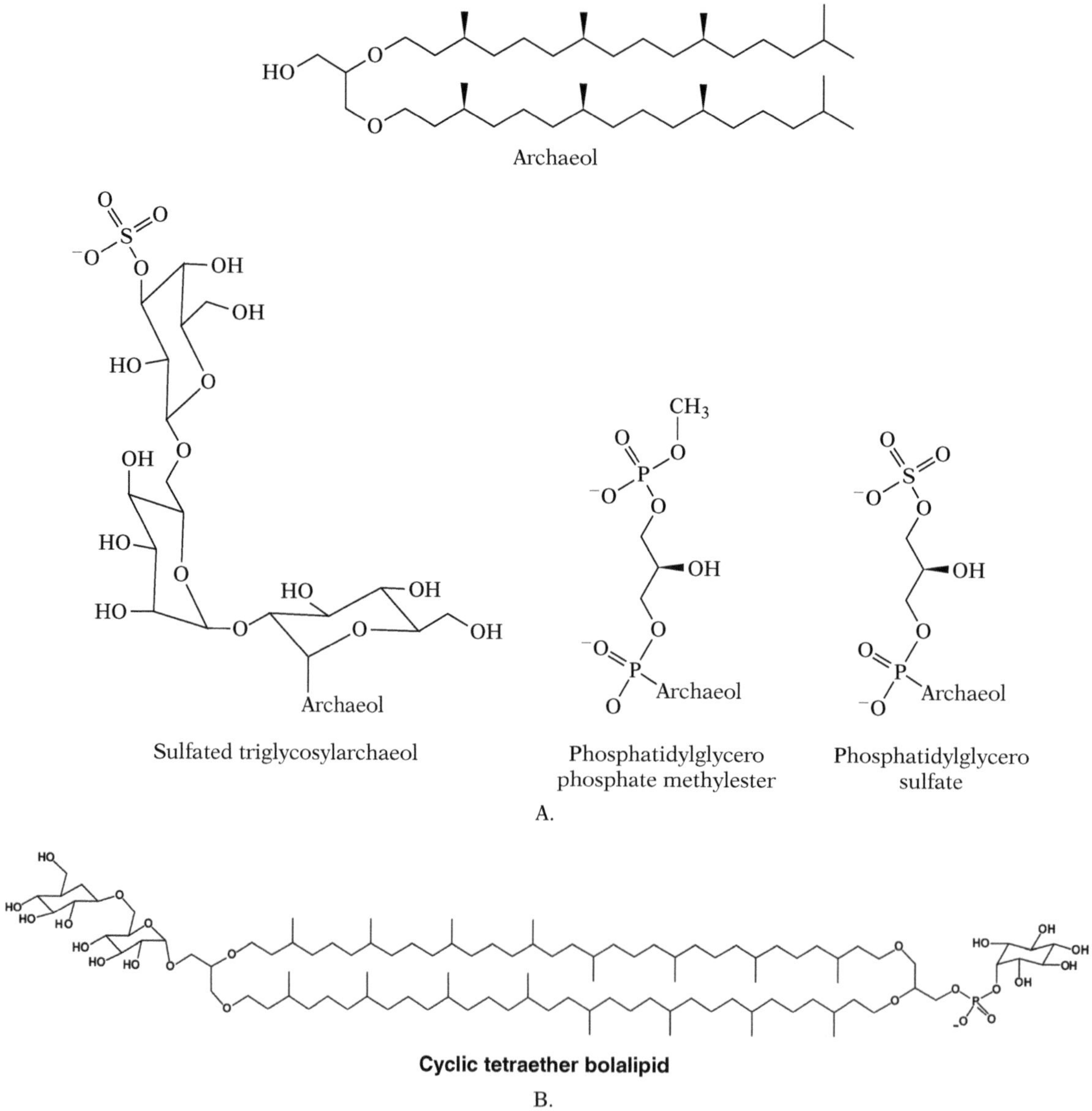

Archaeol

Sulfated triglycosylarchaeol

Phosphatidylglycero phosphate methylester

Phosphatidylglycero sulfate

A.

Cyclic tetraether bolalipid

B.

2.6 Archael lipids. Thermophilic archaebacteria typically have ether-linked isoprenoid lipids. **(A)** The three major polar lipids found in the purple membrane of *Halobacterium salinarium* are derived from archaeol (2,3-di-O-phytanyl-*sn*-glycerol, shown above). **(B)** Some of the lipids span both leaflets of the bilayer and contain two polar headgroups, such as this cyclic tetraether bolalipid from *Archaeoglobus fulgidis*, an extreme thermophile. **(A)** Redrawn from Lee, A. G., *Biochim Biophys Acta*. 2003, 1612:1–40; **(B)** from Sanders, C. R., and K. F. Mittendorf, *Biochemistry*. 2011, 50:7858–7867.

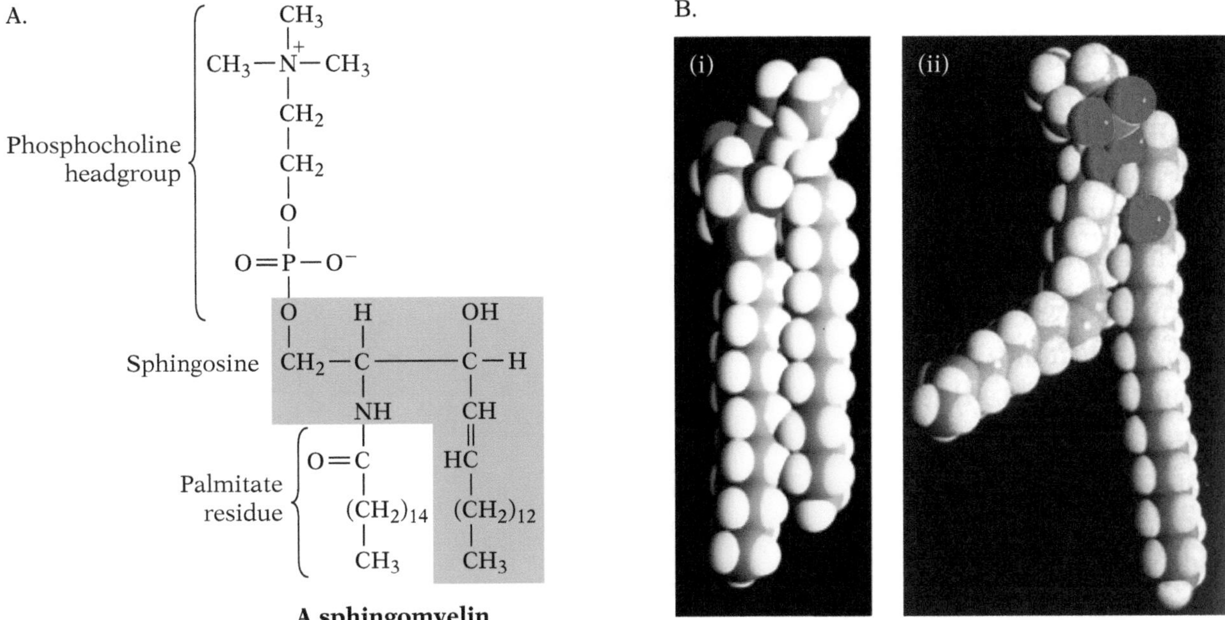

2.7 Structure of a sphingomyelin. (A) Sphingosine is shown with a palmitoyl chain in amide linkage and a phosphocholine headgroup. (B) Comparison of space-filling models of this sphingolipid (i) with SOPC, the glycerophospholipid 1-stearoyl-2-oleoyl phosphatidyl choline (ii), reveals how very similar they are in spite of the lack of an unsaturated chain in the sphingolipid. Redrawn from Voet, D., and J. Voet, *Biochemistry*, 3rd ed., John Wiley, 2004, pp. 386–387. © 2004. Reprinted with permission from John Wiley & Sons, Inc.

acyl chain is attached in amide linkage (Figure 2.7A). The most common sphingolipids are sphingomyelins, which are sphingophospholipids with either phosphocholine or phosphoethanolamine headgroups, giving them overall shapes much like those of PC and PE (Figure 2.7B). Other sphingolipids have headgroups made up of sugars or oligosaccharides, providing the great diversity of structure of cerebrosides (with monosaccharide headgroups) and gangliosides (with oligosaccharides). The ability of the amide bond and the hydroxyl group of sphingolipids to hydrogen bond at the membrane–water interface allows specific interactions with PL headgroups, the hydroxyl group of cholesterol, or other polar groups. In mammalian cell membranes, the fatty acid is generally saturated, with 16 to 24 carbons, making both "chains" of the sphingolipids fully saturated. A small fraction of the 24-carbon chains have a single C=C far along the chain (C24:1 Δ15), so they still pack tightly together. Sphingolipids are important components of nerve membranes, and their carbohydrate moieties are vital in cell recognition and differentiation.

Sterols and Linear Isoprenoids

A third major class of biological lipids includes compounds derived from five-carbon units called isoprene (2-methyl-1,3-butadiene). The linear isoprenyl groups farnesyl (C15) and geranylgeranyl (C20) are used to anchor certain proteins to the bilayer (see Chapter 4). Dolichols are long (C90) polyisoprenoid lipids needed to attach sugars to membrane proteins in the ER of animal cells. Moreover, all steroids are derived from cyclized polyisoprene precursors that have 30 carbon atoms.

The dominant sterol in animal membranes is cholesterol (Figure 2.8). Other eukaryotes have different sterols in their membranes (ergosterol in yeast and fungi, sitosterol and stigmasterol in plants), while prokaryotes have essentially none. The cholesterol content of various cell membranes varies from 0% to ~25% (see Table 1.1). Eukaryotic cells have as much as 90% of their cholesterol in the plasma membrane, maintained by a dynamic cholesterol supply route from the ER. Both the plasma membrane and intracellular membranes have cholesterol-rich domains (see below). Experimental results following the effects of depletion of cellular cholesterol by treatment with cyclodextrin suggests that many different functions in eukaryotic cells involve cholesterol.

Pure cholesterol cannot form a bilayer, and excess cholesterol (beyond 50–60 mol %[1]) precipitates out of PL bilayers. X-ray diffraction detection of the crystalline precipitate establishes precise cholesterol solubility limits of 66 mol % in PC and 51 mol % in PE bilayers. In calorimetric studies of mixtures of cholesterol and pure phospholipid, the melting transition of the lipid

[1] *Surface concentration terms* are either mole fraction ([specific lipid]/[total lipid]) or mol % (mole fraction × 100). Thus 60 mol % cholesterol would be 60 moles of cholesterol per 100 total moles; typically this would be in a mixture with 40 moles of other lipids.

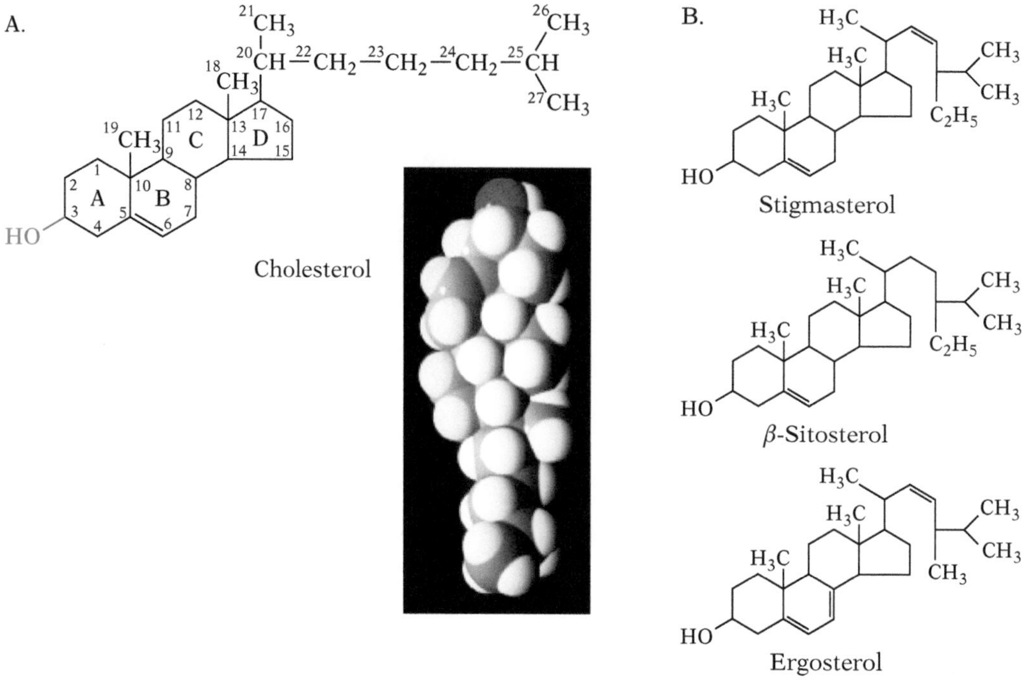

2.8 Sterols found in biological membranes. (A) The structure and space-filling model of cholesterol, a major component of animal membranes. Redrawn from Voet, D., and J. Voet, *Biochemistry*, 3rd ed., John Wiley, 2004, p. 389. © 2004. Reprinted with permission from John Wiley & Sons, Inc. (B) Different sterols occur in membranes of other organisms: plants have stigmasterol and β-sitosterol, whereas yeast and fungi have ergosterol.

broadens and is eventually eliminated as the percentage of cholesterol is increased. This phenomenon has been attributed to the endothermic dissolution of condensed cholesterol–phospholipid complexes of defined stoichiometries. These cholesterol–PL complexes form cooperatively and are described by $[C_qP_p]_n$, in which q molecules of cholesterol complex with p molecules of PL, with n denoting the cooperativity.

The interactions between cholesterol and other lipids have been characterized by nuclear magnetic resonance (NMR) and simulated using molecular dynamics (see Chapter 8). The rigid portion of the sterol imposes conformational order on neighboring lipids, while the larger headgroups of the phospholipids form "umbrellas" over their cholesterol neighbors (Figure 2.9). Because its rigid tetracycle can align better next to saturated acyl chains, cholesterol exhibits a strong preference for saturated sphingomyelin over unsaturated PLs. The presence of cholesterol increases bilayer thickness, tight packing of acyl chains, and compressibility, while it decreases the translational diffusion rates of PLs.

THE LIPID BILAYER MATRIX

These diverse lipid molecules form bilayers with properties that reflect their individual structures and collective interactions. A basic description of the bilayer as a matrix starts with the well-characterized structure of phospholipid molecules in a bilayer. It then considers the diffusion of phospholipids within the two dimensions of the bilayer: the rapid lateral movements that generate the fluidity of the bilayer and the relative lack of rapid transverse movements that allows it to be asymmetric.

Structure of Bilayer Lipids

Details of the structure of PL bilayers have been obtained with x-ray crystallography, small angle neutron and x-ray scattering, NMR, and molecular modeling. X-ray crystallography of pure PLs crystallized from aqueous acid shows structures in which the headgroups are bent in a position almost parallel to the plane of the bilayer, with the acyl chains aligned with each other as a result of a bend in the acyl chain at C2 (Figure 2.10). The position of the headgroup and the angle of chain tilt vary for different PL species. Of course, these structures do not represent the structure of lipids in the fluid membrane. Precise measurements of their dimensions in the fluid phase are obtained with the combined analysis of neutron and x-ray scattering, since in deuterated water neutron scattering gives the bilayer thickness while x-ray scattering resolves the headgroup–headgroup distance. Variation of acyl chain length and saturation and the nature of the headgroup affect lipid dimensions. The data also show the effect of temperature: the lipid area

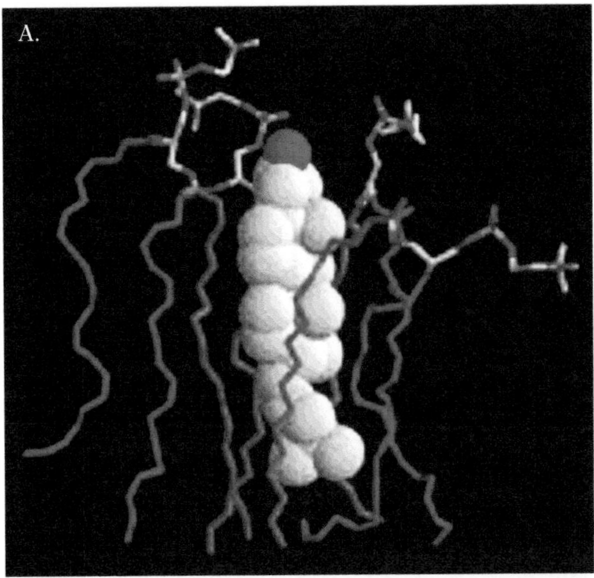

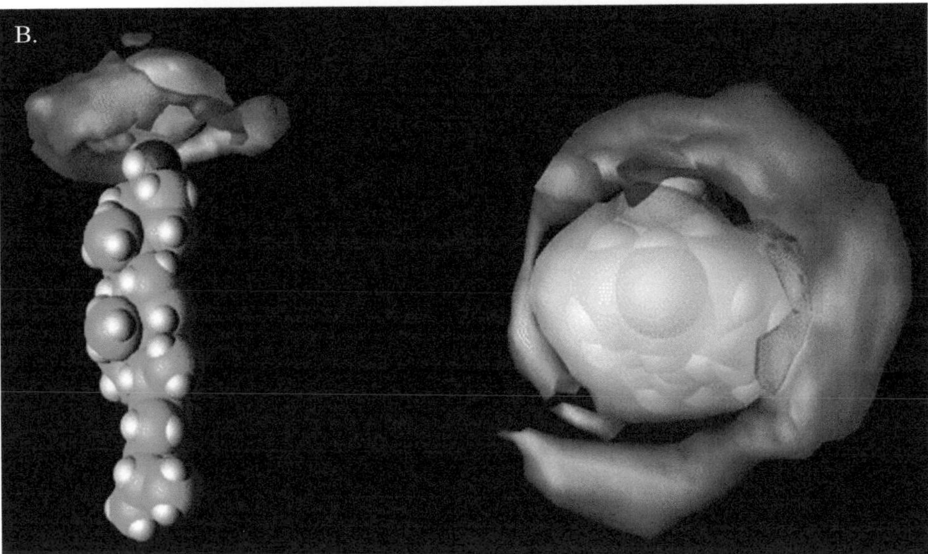

2.9 Close packing of cholesterol with phospholipids. Molecular dynamics portrays the interactions between cholesterol and phospholipids and reveals a close association between saturated acyl chains and the rigid tetracycle of the sterol, whereas the bulky PL headgroup can cover the small (OH) headgroup of cholesterol in the manner of an umbrella. (A) Side view of a simulated cholesterol–PL complex. The PL in the model is 1-stearoyl-2-docosahexaenoyl-PC, which has a polyunsaturated omega-3 fatty acid on C2. The saturated acyl chains of the PL extend past the sterol. Saturated acyl chains (blue) pack along the smooth faces of the cholesterol, whereas polyunsaturated acyl chains (red) have a much lower affinity for it. Even with oleoyl on C2 (not shown), the kink from the double bond at C9–10 interferes with this close packing. On the other hand, sphingomyelin usually has saturated chains of 16–24 carbons, and when unsaturated the double bond occurs at C15, well below the rigid portion of cholesterol, which explains the preference of cholesterol for sphingomyelin. (B) Side and top views showing probability functions for the headgroup of PC in dynamic complex with cholesterol, illustrating the umbrella effect. The probability density for phosphate is gray, and that for choline is orange. As the PC orientation varies, the headgroup completely covers the sterol. From Pittman, M. C., *et al.*, *Biochemistry*. 2004, 43:15318–15328. © 2004 American Chemical Society.

increases and the bilayer thickness decreases as temperature is increased.

NMR experiments show that in the fluid membrane the acyl chains are not rigid. Motion along the acyl chain can be probed by NMR using ^{2}H labels at different positions along the acyl chains. The deuterium NMR spectra of different DMPC (dimyristoyl phosphatidylcholine) preparations labeled in different positions clearly show a gradation of mobility along the chain (Figure 2.11). The spectrum for DMPC with ^{2}H at the terminal

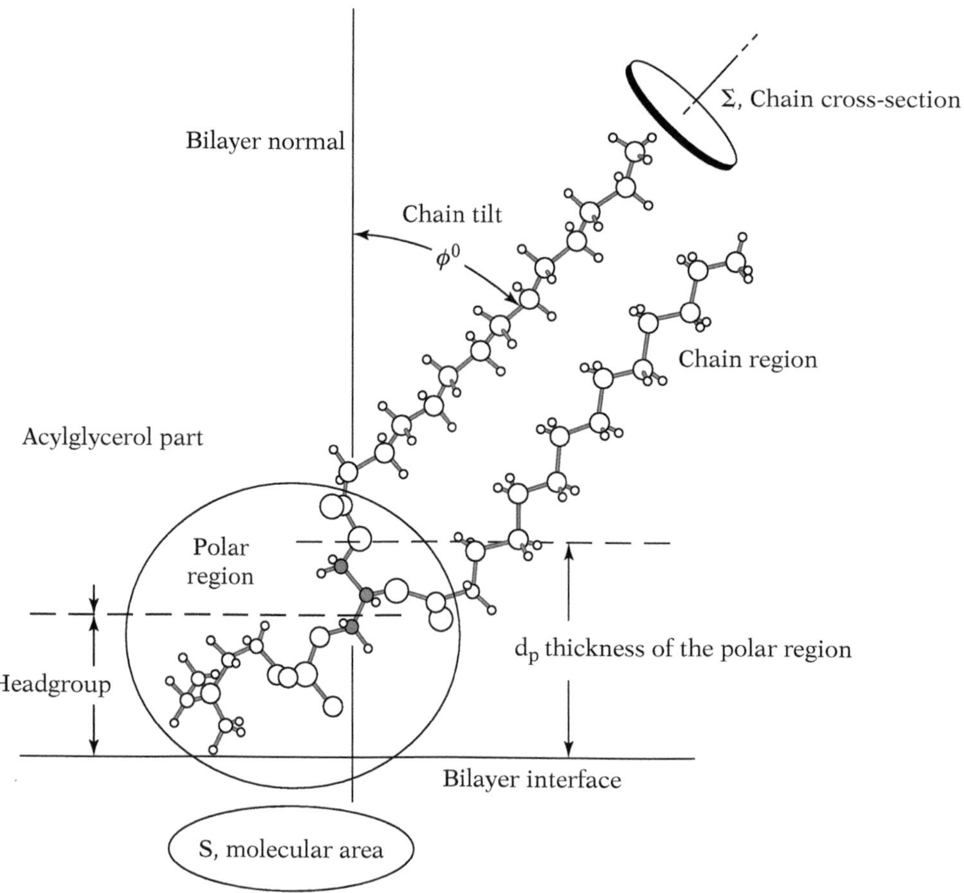

2.10 The structure of a PC molecule determined with x-ray crystallography. The acyl chains are fully extended (all *anti* dihedral angles) and are tilted from the bilayer normal. The polar region (circled) includes the headgroup and the glycerol moiety. Other PL molecules give a different chain tilt and thickness of the polar region, depending on the way their polar headgroups pack in the crystal. Redrawn from Gennis, R. B., *Biomembranes*, Springer-Verlag, 1989, p. 38. © 1981 by Elsevier. Reprinted with permission from Elsevier.

methyl residue is narrow, reflecting the disorder near the center of the bilayer, in contrast to the broadened peaks observed when the ^{2}H is closer to the interface and less mobile. Similar results are obtained using probes suitable for fluorescence depolarization and recovery (see Box 2.1) and electron paramagnetic resonance (see Box 4.2).

Molecular modeling of aqueous lipid systems illustrates the mobility along the acyl chains in bilayers. In these simulations, the freedom of *trans/gauche* rotations of the carbon–carbon bonds of long chains makes the chains highly flexible, with the result that they are clearly not aligned in a snapshot of a simulated bilayer (Figure 2.12). The mobility of the acyl chains is revealed by molecular dynamics (see Chapter 8). The angular rotations along hydrocarbon chains of individual molecules contribute to the fluidity of the membrane that is so essential to life it is maintained by organisms living under extreme conditions (as discussed above). Another important source of fluidity is the movement of whole molecules within the two-dimensional matrix of the bilayer.

Diffusion of Bilayer Lipids

Lipid molecules move within the bilayer in three different modes: rotational, lateral, and transverse. Rotational diffusion, the spinning of a single molecule around its axis, affects a lipid's interactions with its nearest neighbors but does not alter its position. Lateral diffusion occurs when neighboring molecules exchange places via Brownian motion; it enables lipids to travel within a monolayer. Transverse diffusion is the exchange of lipid molecules between leaflets and is commonly called "flip-flop."

The fluid aspect of the Fluid Mosaic Model is primarily due to lateral diffusion of membrane components. One technique used to measure the rates of lateral diffusion of lipids is called FRAP – fluorescence recovery after photobleaching (see Box 2.1). The diffusion rates obtained for freely diffusing lipid molecules are very fast, typically around 10^{-7} to 10^{-8} cm^2/sec. This is fast enough for a single lipid molecule in the erythrocyte plasma membrane to go around the whole cell in

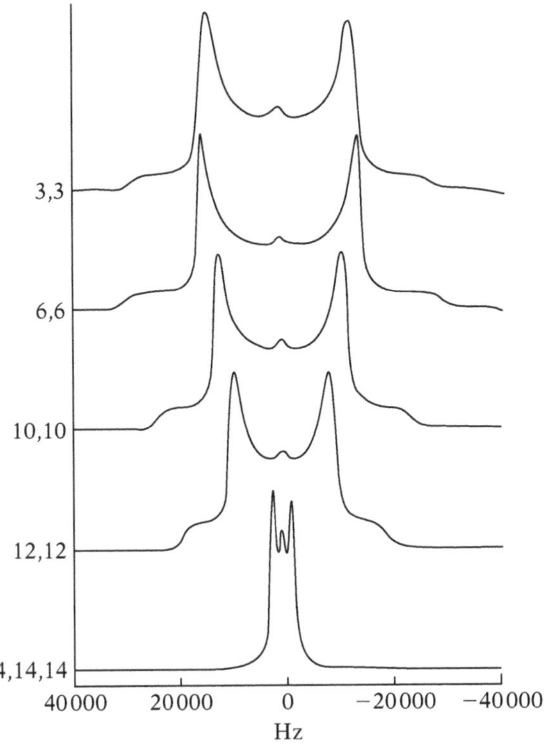

2.11 Deuterium NMR of DMPC with ^{2}H at different positions on the acyl chains as indicated on the left. The deuterium located closer to the center of the bilayer experiences greater disorder, giving a sharper peak, than deuterium located near the interface. From Gennis, R. B., *Biomembranes*, Springer-Verlag, 1989, p. 53. © 1989. Reprinted with permission from E. Oldfield.

seconds, which would randomize the positions of lipids in the bilayer. However, measured rates of lateral diffusion of PLs in the plasma membrane are significantly retarded (by factors up to 100) compared with rates in model membranes. Recently this discrepancy has been explained by powerful methods that track the movements of single lipid molecules.

Observation of lateral diffusion by single-particle tracking uses very fast video fluorescence microscopy to follow the motion of a single fluorescent lipid molecule on the surface of a cell. This method reveals an irregular path for lipid motion in the plane of the membrane, called "hop diffusion" because the tagged molecule stays in a confined region for milliseconds before hopping to a new region (Figure 2.13). Such compartments that restrict lateral diffusion are observed in plasma membranes from many cell types and appear to be formed by contacts with the cytoskeleton (see Figure 1.10). Within the compartments, the diffusion rate is as fast as the rate observed *in vitro*, while the overall rate on the cell surface is slowed by the movement between compartments, attributed to higher viscosity around the proteins that make the barriers.

While lateral diffusion in the membrane is fast, transverse diffusion is very slow. Lipid molecules do not readily flip from one leaflet to the other because it is energetically unfavorable for their polar headgroups to pass through the nonpolar center of the bilayer. When the rate of transverse exchange (flip-flop) of lipid molecules between the two leaflets of the bilayer is measured in model membranes called liposomes (see Chapter 3),

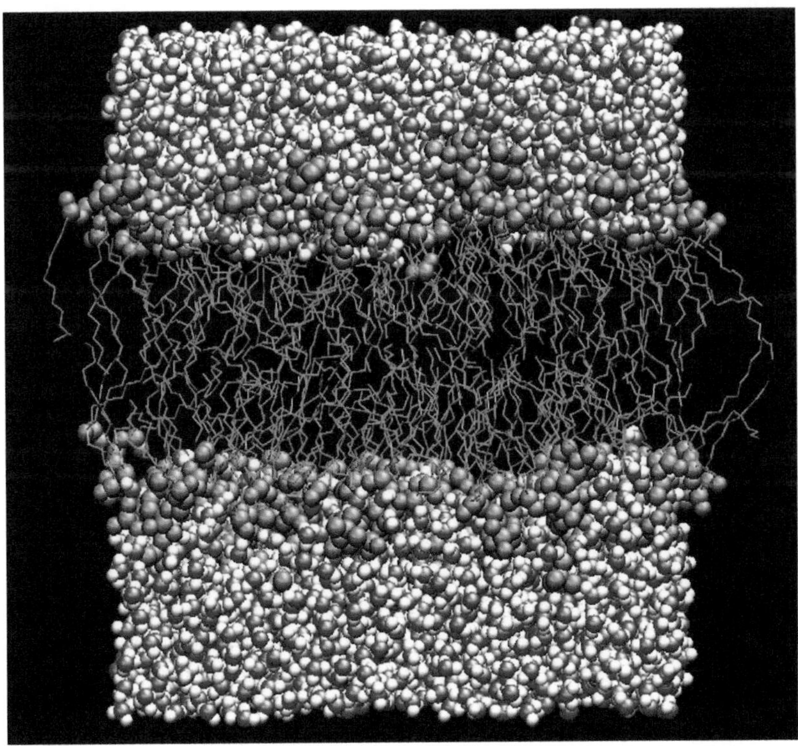

2.12 A snapshot from a simulated model of a fully hydrated DMPC bilayer. This molecular dynamics simulation shows clearly the disorder among the acyl chains (green). The starting point for the simulation was the lipid configuration from the x-ray crystal structure, which was then "heated" to a constant temperature and pressure. Note the penetration of water molecules (blue and white) into the extensive interfacial regions, while water is absent from the nonpolar center. Phospholipid headgroups are orange, water hydrogens are white, water oxygens are blue, and phospholipid hydrocarbon chains are green. From Chiu, S. W., *et al.*, *Biophys J.* 1995, 69:1230–1245. © 1995, Elsevier. Reprinted with permission from Elsevier.

BOX 2.1 FLUORESCENCE TECHNIQUES

A number of spectroscopic methods make use of probes or derivatized biomolecules that fluoresce; that is, they absorb energy to reach an excited electronic state and then emit radiation (photons) when they return to the ground state. The excitation wavelength, which is the wavelength of incident light required to excite the fluorescent molecule, depends on the nature of the fluorophore, the energy-absorbing group. The emission spectrum is the variation of fluorescence intensity with the wavelength of the emitted light, and is always at a lower frequency (higher wavelength) than the excitation spectrum. This difference between the excitation wavelength and the emission wavelength increases the sensitivity of fluorescence 100-fold over absorption spectroscopy simply because the detector on the instrument is set for the emission wavelength and does not pick up background from the light source. Because the excited fluorophore can interact with surrounding solvent molecules before emission, both the intensity of the emission and its maximum wavelength (λ_{max}) are sensitive to the environment of the molecule. In general, movement of the fluorescent group from a nonpolar environment to an aqueous milieu will decrease the intensity and shift the λ_{max} to a higher wavelength.

A variety of techniques employ fluorescence in membrane research. Fluorescence depolarization measures rotational diffusion and thus quantitates viscosity (the inverse of fluidity). It requires excitation with plane-polarized light and observation through analyzing polarizers to resolve the fluorescence intensity into parallel and perpendicular components. Fluorescence recovery after photobleaching (FRAP) uses fluorescence microscopy to follow fluorescence intensity over time after a laser beam destroys the fluorophores in a small observation area. The rate of recovery of fluorescence is a measure of the rate of diffusion of unbleached molecules into the bleached area. FRAP was used to investigate the lateral diffusion of lipids in cell membranes using fluorescent probes attached to PL headgroups. Within milliseconds, the bleached patch of membrane recovered its fluorescence as unbleached lipid molecules diffused into it and bleached lipid molecules diffused away (see Figure 2.1.1).

Many experimental designs make use of specific quenching processes to provide additional information on the location of the fluorophore. Effective quenching decreases the fluorescence intensity. Quenchers such as oxygen; heavy ions, including iodide and bromide; or other paramagnetic molecules remove the excited state energy upon collision with the fluorescent molecule. An example is the use of phospholipids with bromide bound to carbon atoms at different positions of the acyl chains to probe the depth of the bilayer occupied by a fluorophore (as described in Chapter 7). Fluorescence (Förster) resonance energy transfer (FRET) involves a second type of quenching process that does not require collisions. In this process, the fluorophore interacts through a short distance (~10–80 Å [1–8 nm]) with an acceptor molecule with similar electronic properties. As the excited donor molecule passes its energy to the acceptor, it does not emit a photon. Because the acceptor is now excited, it can decay to the ground state, emitting a photon with a longer λ_{max}. FRET experiments can detect the decrease of the donor fluorescence emission or the increase of the acceptor fluorescence emission. If the fluorescent probe is not close enough to an acceptor, neither emission will change.

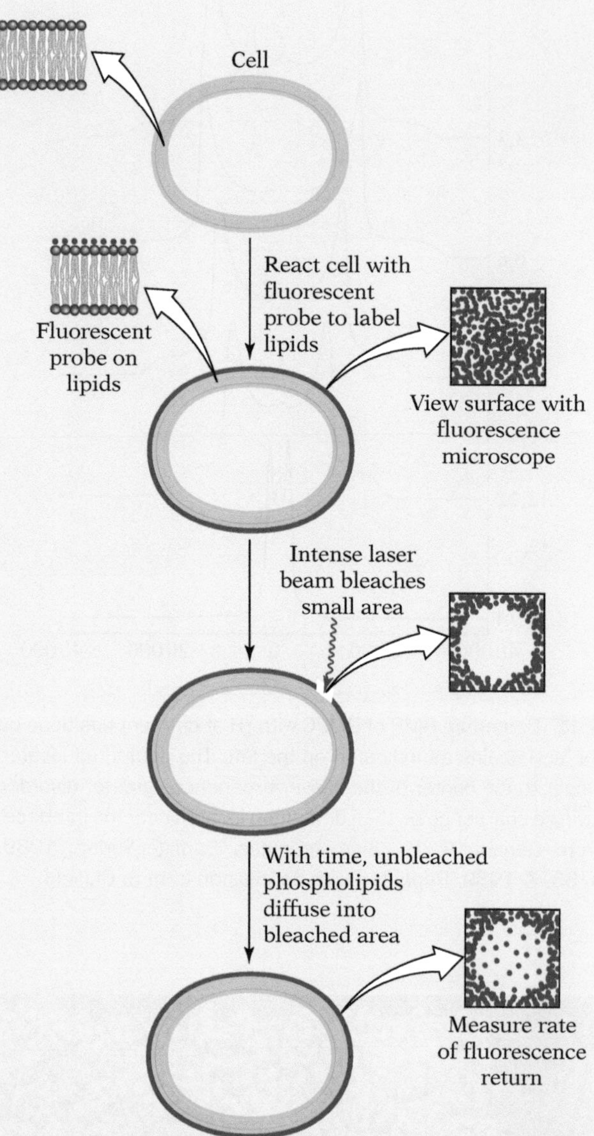

2.1.1 Measurement of lateral diffusion of lipids by FRAP. From Nelson, D. L., and M. M. Cox., *Lehninger Principles of Biochemistry*, 4th ed., W. H. Freeman, 2005. © 2005 by W. H. Freeman and Company. Used with permission.

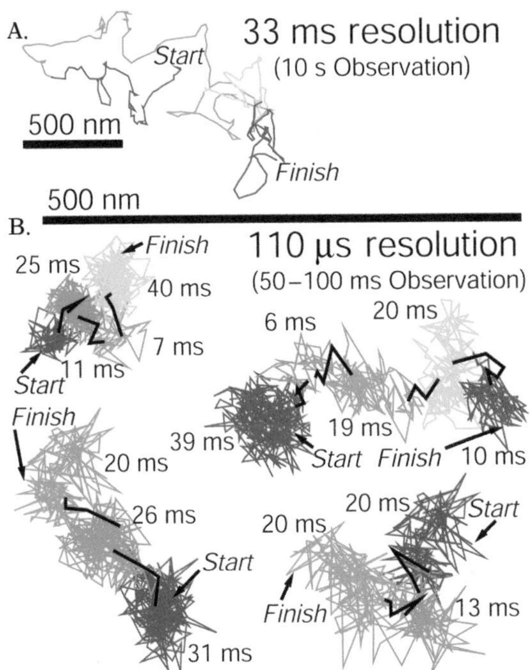

2.13 Hop diffusion of individual lipid molecules. Computerized time-resolved single-particle tracking was used to follow the diffusion path of a single gold-labeled DOPE molecule on the surface of the cell. (A) At a resolution of 33 msec, the path appears to be simple Brownian diffusion. Each color represents 60 step periods or two seconds. (B) At a resolution of 110 μs, the pattern of movement reveals the phenomenon of hop diffusion as the lipid hopped from one region to the next. Each color indicates confinement within a compartment, with black for intercompartmental hops. The residency time for each compartment is indicated. From Murase, K., *et al.*, *Biophys J.* 2004, 86:4075–4093. © 2004, Elsevier. Reprinted with permission from Elsevier.

the half-times are several hours or more in the absence of proteins or other defects. A pH gradient can stimulate some lipids, such as PG and PA, to cross the bilayer. Some biological processes, such as incorporation of newly synthesized lipids into membranes, require a much faster rate of transbilayer movement. For this purpose, lipid transfer is carried out by flippases, enzymes that catalyze flip-flop at the cost of adenosine triphosphate (ATP) hydrolysis.

Lipid Asymmetry and Membrane Thickness

Because the transverse exchange of lipids between monolayers is slow, the lipids do not readily equilibrate between them. In addition, many cells use ATP-driven transport of lipids catalyzed by flippases to maintain different lipid compositions in the inner and outer leaflets of their membranes. Analysis of this asymmetry has used phospholipases that cannot permeate the membrane and therefore only hydrolyze lipid substrates from the outer leaflet, as well as chemical labeling with nonpenetrating agents such as trinitrobenzenesulfonic acid (TNBS). A good example of lipid asymmetry (and the first observed) is the erythrocyte membrane, whose outer leaflet is enriched in sphingomyelin and PC, while the inner leaflet contains most of the PE and nearly all of the PS found in the membrane (Figure 2.14). Furthermore, the same phospholipid species may have acyl chains in the outer leaflet different from those in the inner leaflet. The concentration of sphingolipids is typically six-fold higher in the outer leaflet of membranes than in the inner leaflet; in contrast, cholesterol is commonly distributed in both leaflets of eukaryotic membranes. The consequences of lipid asymmetry are being explored *in vitro* with techniques that produce supported bilayers with asymmetric lipid composition (see Chapter 3).

Numerous *in vitro* studies have shown that the two leaflets of a lipid bilayer can be coupled together by interdigitation produced when some of the acyl chains

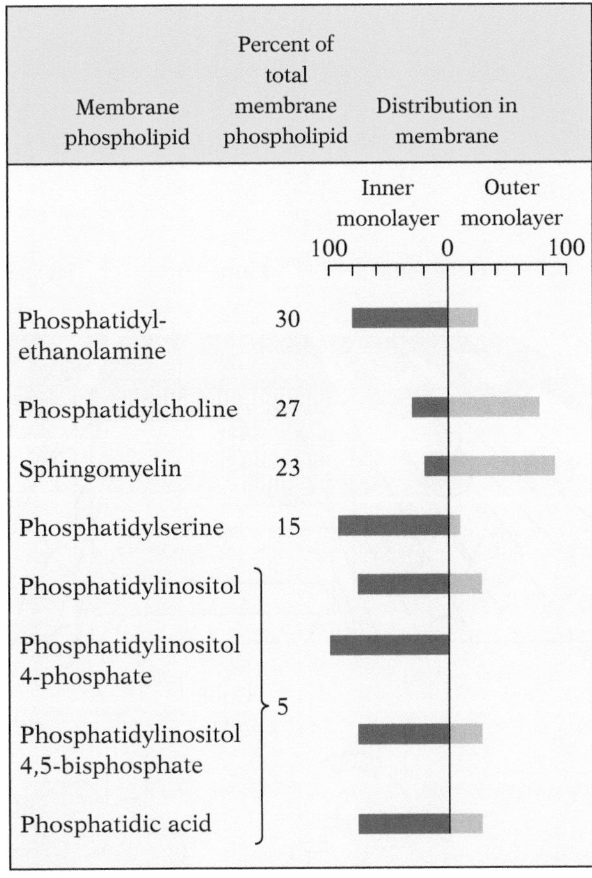

2.14 The asymmetric distribution of lipids in erythrocyte membranes. The graph shows the content of each lipid type expressed as mol % in the inner and outer leaflets. Redrawn from Nelson, D. L., and M. M. Cox (eds.), *Lehninger Principles of Biochemistry*, 4th ed., W. H. Freeman, 2005, p. 373. © 2005 by W. H. Freeman and Company. Used with permission.

extend past the bilayer midplane, pushing their terminal methyl groups into the opposing leaflet. This may result from chain length asymmetry within individual lipid molecules (when a lipid bears one acyl chain that is much longer than the other), as frequently occurs in sphingolipids. Because the two monolayers become physically coupled, interdigitation may be observed as a distinct phase in calorimetric studies and as line broadening in the ^{31}P-NMR spectrum (see below). An important consequence of interdigitation is a decrease in the bilayer thickness, because it allows the two monolayers to approach each other more closely.

The thickness of the lipid bilayer is strongly influenced by the length and degree of saturation of acyl chains. In addition, much experimental evidence indicates that cholesterol increases the thickness of a lipid bilayer. Thickening by cholesterol is attributed to stabilizing the neighboring acyl chains in their most extended conformations (with all *anti* dihedral angles), thus increasing their effective length (see Figure 2.9A). However, a challenge to this view is presented by experiments that measure the thickness of various cell membranes by x-ray scattering and find thickness is not correlated with their cholesterol content but rather seems to be strongly influenced by the protein content (see "Hydrophobic Mismatch" in Chapter 4).

LIPID POLYMORPHISM

The bilayer is only one of the possible lipid aggregates that form spontaneously when amphiphilic lipids are mixed with water. Different compositions and/or changes in conditions bring about polymorphic (from *poly+morph*, "many shapes") phase changes. The three general categories of lipid phases are lamellar, hexagonal, and cubic (Figure 2.15). The familiar bilayer of the membrane is lamellar, yet the presence of lipid components that prefer hexagonal phase has turned out to be crucial for many of its functions (see below). The cubic phase has received much interest as a likely intermediate in membrane fusion and a medium for crystallization of membrane proteins.

Lamellar Phase

The commonly observed lamellar states are called L_α and L_β:

L_α = lamellar liquid crystalline, also called L_d (or *ld*, liquid-disordered)
L_β = lamellar gel, also called S_o (ordered solid)

The highly ordered, dehydrated PL arrays characterized by x-ray crystallography (Figure 2.10) are in a third lamellar phase: L_c = lamellar crystalline.

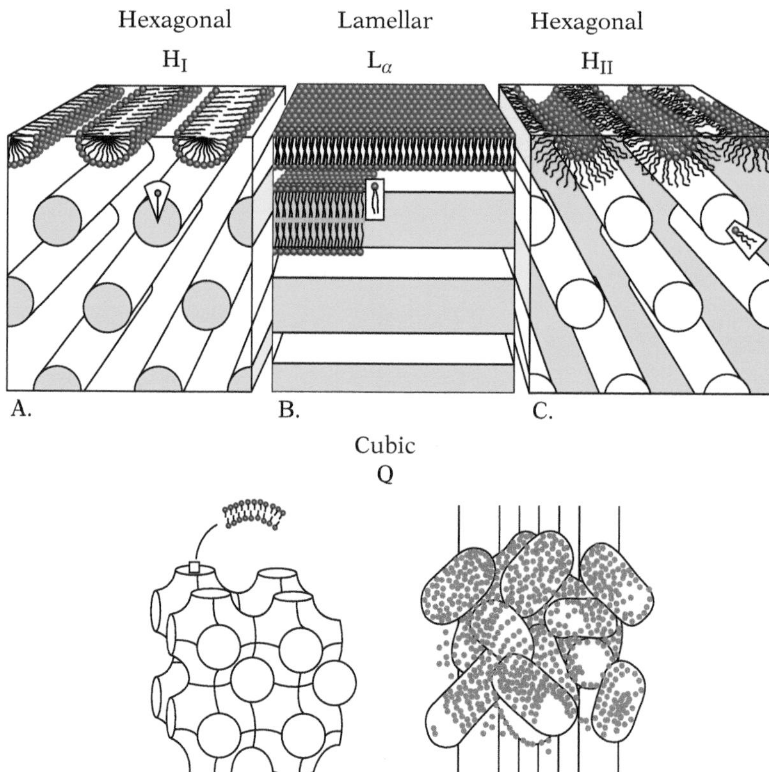

2.15 Structures of lamellar, hexagonal, and cubic phases, the most common polymorphic states observed with membrane phospholipids. Lamellar phase (B) is L_α. Hexagonal phase is either normal – H_I (A), with nonpolar regions inside the tubes – or inverted – H_{II} (C), with polar groups and water inside. Cubic phases are three-dimensional systems of lipid channels or networks interpenetrated by water channels, represented by the bicontinuous type (D) and the micellar type (E) that occur in excess water. Redrawn from Lindblom, G., and Rilfors, L., *Biochem Biophys Acta*. 1989, 988:221–256. © 1989 by Elsevier. Reprinted with permission from Elsevier.

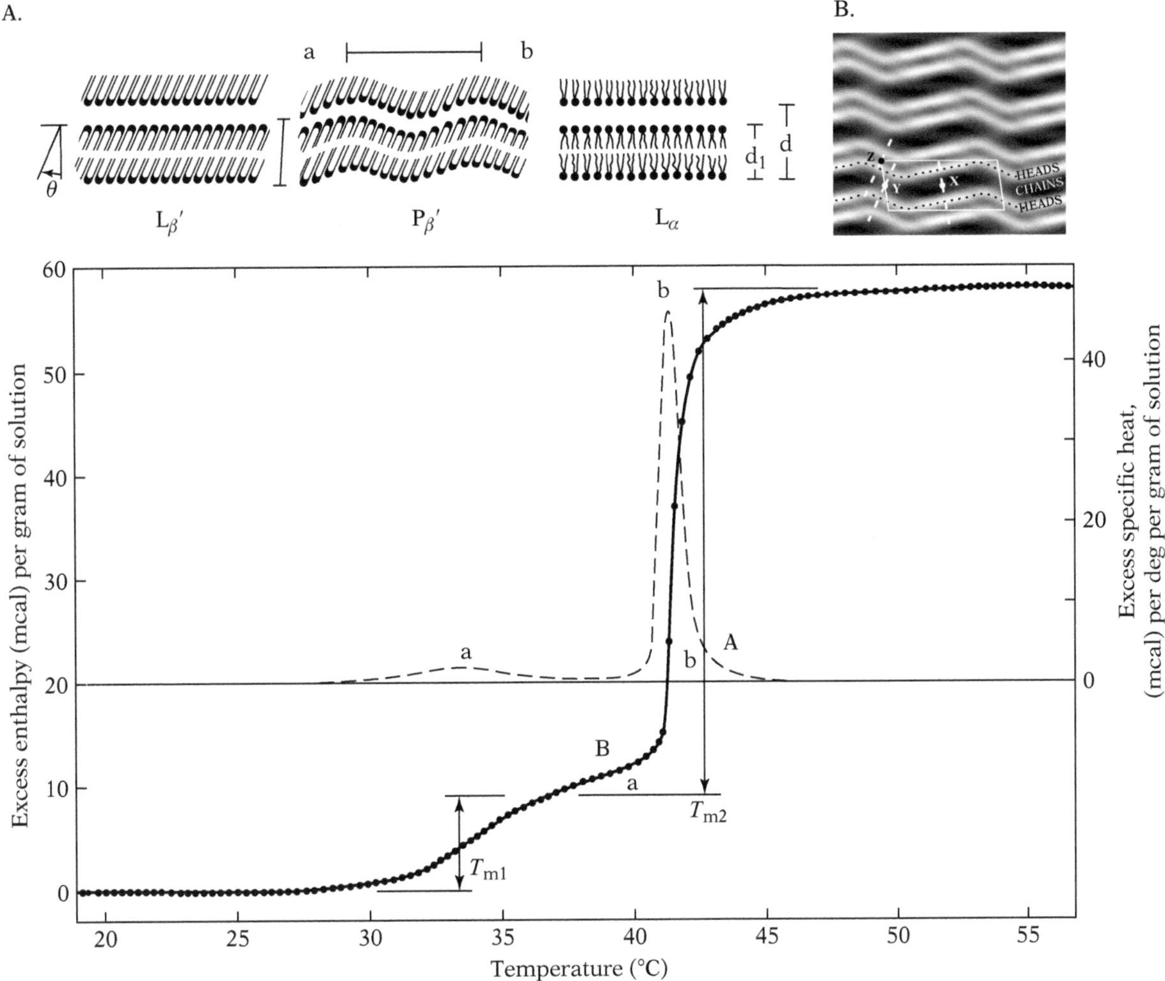

2.16 Differential scanning calorimetry of an aqueous dispersion of DPPC. (A) The excess enthalpy (heat taken up) of the sample compared with a reference is measured as the temperature is raised. DPPC exhibits two phase transitions. The first is a pretransition called T_{m1}, which produces the ripple phase $P_{\beta'}$, followed by the transition T_{m2} to L_{α}, as diagrammed schematically above the graph. © 1976, American Chemical Society. Main image. © 1972, by American Society for Biochemistry & Molecular Biology. (B) Electron density map for the ripple phase of DMPC with 25% water at 18°C at high resolution achieved by x-ray diffraction. From the dimensions of the unit cell (drawn in white), the rippling repeat period is 142 Å and the lamellar repeat is 58 Å. However, the thickness of the hydrocarbon varies, being different at X and Y; the thickness of the water layer between bilayers is shown at Z. Redrawn from Nagle, J. F., and S. Tristram-Nagle, *Curr Opin Struct Bio*. 2000, 10:474–480. © 2000 by Elsevier. Reprinted with permission from Elsevier.

The melting temperatures that were described earlier for transitions from "solid" (gel) to "fluid" (liquid crystalline) states characterize the L_{β} to L_{α} transition (Figure 2.16). Additionally, some pure PLs exhibit minor transitions between L_{β} and L_{α}, with phases called $L_{\beta'}$, in which the chains are tilted, and P_{β} and $P_{\beta'}$, which are ripple phases seen with pure PC. Figure 2.16B shows why the term *ripple* is appropriate.

Hexagonal Phase and the Amphiphile Shape Hypothesis

Hexagonal phases consist of hexagonally packed arrays of lipids in long cylindrical tubes. They have two topologies (see Figure 2.15A and C):

H_{I} Cylinders with nonpolar centers and polar groups and water outside.

H_{II} Cylinders with polar groups and water inside, nonpolar groups outside.

Because its orientation is opposite the usual bilayer orientation, H_{II} phase is called *inverted*. Lipids that prefer hexagonal phases are important constituents of biological membranes, as described below.

What determines whether lipids form lamellar or hexagonal phases under certain conditions? The amphiphile shape hypothesis suggests the lipid aggregates formed by aqueous dispersions of pure PLs and other amphiphilic compounds reflect the general shape of the individual molecules (Figure 2.17).

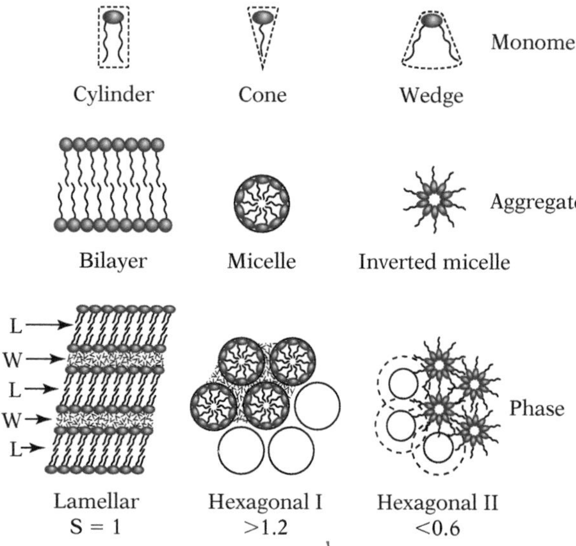

2.17 The amphiphile shape hypothesis: the general relationship between lipid shape and aggregates, which influences lipid polymorphism. The shape parameter, S, is the ratio of the volume to the area of the polar headgroup multiplied by the length. When S = 1, the lipid is roughly cylindrical and in aqueous dispersion can form stable bilayers (lamellar phase). When S > 1, the lipid tends to organize into micelles or H_I; when S < 1, the lipid forms inverted micelles or H_{II}. L, lipid; W, water. Redrawn from Jain, M. K., and R. C. Wagner, *Introduction to Biological Membranes*, 2nd ed., Wiley, 1988, p. 53.

Bilayer-forming lipids such as PCs that favor lamellar aggregates are roughly cylindrical in shape, with similar cross-sectional areas for the headgroup and the acyl chains. This can be described by a shape parameter, S, calculated as

$$\frac{\text{cross-sectional area of polar headgroup} \times \text{lipid length}}{\text{lipid volume}}$$

For a fairly cylindrical lipid, S equals ~1. A lipid that is conical (S > 1) or wedge shaped (S < 1) introduces curvature, leading to the formation of nonlamellar phases. For example, DOPE, with its small headgroup and unsaturated acyl chains, is conical and forms the H_{II} phase at temperatures at which the more cylindrical DOPC is lamellar. The physical situation in a bilayer is more dynamic than that implied by the amphiphile shape hypothesis because of the mobility of acyl chains above the temperature of their L_β to L_α transition. Bilayer-forming lipid molecules in the fluid L_α phase are not confined to cylindrical spaces, as illustrated by the bilayer snapshot in Figure 2.12.

Thermodynamics provides a more complete account of the shape concept. The free energy per lipid molecule differs in lamellar and hexagonal phases, in which the molecule occupies different volumes. This shape-dependent free energy has four components: hydrocarbon-packing energies, the elastic bending of the lipid monolayers, hydration, and electrostatic potentials.[2] The hydrocarbon-packing energies, due to the hydrophobic effect, depend on the extent of hydration. In the H_{II} phase with water molecules sequestered inside the cylindrical tubes, increasing the water-to-lipid ratio increases the hydrocarbon packing free energy. The elastic bending of lipid molecules describes their tendencies to form curved monolayers. A lipid monolayer in lamellar phase is essentially flat, whereas in hexagonal phase it is tightly rolled into cylinders. When the decrease in elastic energy that results from curling the layers competes favorably with the increase due to packing the hydrocarbon chains, the system undergoes the L_α to H_{II} transition to lower the total free energy.

The elastic bending of a monolayer is described in terms of R, the radius of curvature of the lipid/water interface (Figure 2.18A, B). R_o is the intrinsic value of R for each lipid species – that is, the radius of curvature it would reach at equilibrium in the absence of other forces. When other forces drive the lipid molecules into a phase with a different R value, R – R_o is an indication of how far they lie from their intrinsic curvature. Factors that widen the splay of the lipid tails (e.g., temperature or unsaturation) make R_o more negative, while factors that increase the effective headgroup area (e.g., size and charge of the headgroup) make R_o less negative. In general, PC species have larger R_o values and remain in the L_α phase at higher temperatures while PE species have lower R_o values and go into H_{II} phase at those temperatures. A mixture of the two has an intermediate R_o value. However, an aqueous mixture of DOPE and DOPC will adopt a lamellar phase with just over 20 mol % of the bilayer-forming lipid. When a bilayer restricts the tendency of the lipids of each monolayer to curve, the forces exerted on the lipid create a state of "curvature frustration." The frustration is the result of the lateral pressures pushing apart the acyl chains and/or the polar headgroups within the bilayer, countered by the hydrophobic effect stabilizing it (Figure 2.18C).

Their intrinsic curvature explains a role for nonbilayer-forming lipids in the membrane. The average R_o of a biological membrane made up of a variety of lipids should not be too small – to avoid destabilizing its lamellar structure – but also should not be too large – to allow the transient local perturbations of lamellar structure needed for fusion, endocytosis, or a similar event. Thus membranes of many organisms contain a significant fraction of nonbilayer lipids, which the organism varies homeostatically to control the curvature of its membranes (see below).

Cubic Phase

In addition to lamellar and hexagonal phases, lipids can form cubic phases (see Figure 2.15D, E). Like hexagonal

[2] See Gruner (1985) for a quantitative treatment.

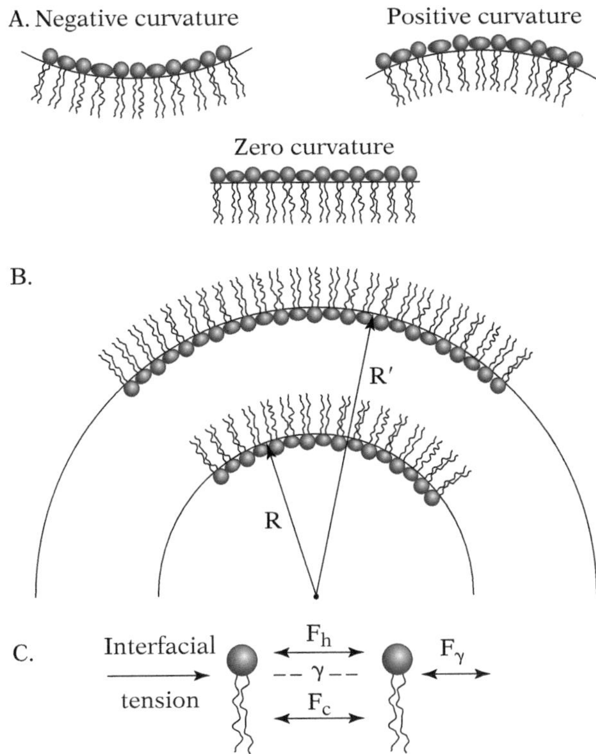

2.18 Curvature of a lipid monolayer. (A) R is the radius of curvature of the lipid–water interface, which is defined as positive for H_I phase (right) and negative for H_{II} phase (left). (B) A larger value of R produces less curvature than a smaller value of R. (C) The zero curvature of one leaflet of a lamellar phase is the result of a balance of forces: F_c is the lateral pressure pushing the chains apart due to motions of bond rotation, and F_h is the lateral pressure in the headgroup region that consists of steric, hydrational, and electrostatic effects, in addition to some positive contributions from hydrogen bonds. F_γ is the result of the hydrophobic effect at the interface, where the interfacial tension minimizes the hydrocarbon–water contacts. Redrawn from Lee, A. G., *Biochim Biophys Acta*. 2004, 1666:62–87. © 2004 by Elsevier. Reprinted with permission from Elsevier.

phases, cubic phases are type I (positive curvature, acyl chains inside) and type II (negative curvature, acyl chains outside). Cubic phases have a much greater variety of three-dimensional structures as they are formed from cubic packing of rod-like elements, resulting in discontinuous phases. Various geometries of periodic minimal surfaces are formed by different lipids in aqueous solvents. Only two of the cubic phases – Q^{224} and Q^{227} – can exist in excess water. Q^{224} is bicontinuous, containing two networks of rods, each with tetrahedral joints providing connections (Figure 2.15D). The walls of the rods are curved bilayers, with water on either side. Q^{227} has quasi-spherical micelles packed into cubes (Figure 2.15E). Cubic phases can have less curvature frustration than lamellar phases and less packing constraints than hexagonal phases. Therefore cubic phases can form in the transition between L_α and H_{II} phases and are likely to be intermediates in membrane fusion processes.

Cubic phases have received much attention because they provide fertile environments for crystallizing membrane proteins (see Chapter 8). The bicontinuous system of the Q^{224} type cubic phase with a continuous lipid bilayer separating a pair of interpenetrating aqueous channels allows diffusion of proteins within the bilayer to promote crystal growth. The lipidic cubic phase crystallization system typically consists of monoolein (a monoacylglycerol) in water (40% wt/wt) at 20°C (Figure 2.19).

MISCIBILITY OF BILAYER LIPIDS

The polymorphism of mixtures of pure lipids is summarized in diagrams that show the phases of a two-component system as a function of temperature and the mole fraction of one component (see Box 2.2). Such diagrams are constructed using data obtained with a variety of probes that can detect the immiscibility of two phases, such as electron paramagnetic resonance (EPR, see Box 4.2), which detects the mobility of spin probes, compounds carrying unpaired electrons. Early examples, constructed using EPR data on the mobility of TEMPO, a small lipid-soluble spin probe whose nitroxide group has an unpaired electron, are shown for different aqueous binary dispersions of two PLs (Figure 2.20). The phase diagrams for mixtures of DEPC and DPPC (Figure 2.20A) and for DEPC and DSPC (Figure 2.20B) indicate they are completely miscible, with ideal mixing in both fluid and gel states and with a solid state coexisting with a fluid state at intermediate temperatures and compositions. This ideal mixing suggests that the two lipid molecules are completely interchangeable and is limited to PLs with acyl chains that differ by less than four methylene residues. The phase diagrams for DOPC plus DPPE (Figure 2.20C) and for DEPC plus DPPE (Figure 2.20D) show immiscibility in fluid plus gel states and even immiscible fluid states (in Figure 2.20D).

An intermediate situation occurs when the two lipids are miscible in the fluid state and immiscible in the gel state, as seen in mixtures of DMPC and DEPC probed with FRAP. The fluorescence probe NBD-DLPE, a PE labeled with the nitrobenzoxadiazolyl group, partitions almost exclusively in the liquid crystalline phase. When added to bilayers consisting of varying proportions of DMPC and DSPC, the extent of recovery after photobleaching has revealed the discontinuity between the gel and liquid crystalline phases. Similar discontinuities have been observed with the techniques of single-particle tracking and optical tweezers.

Ternary phase diagrams for aqueous mixtures of cholesterol and two other lipids have been constructed

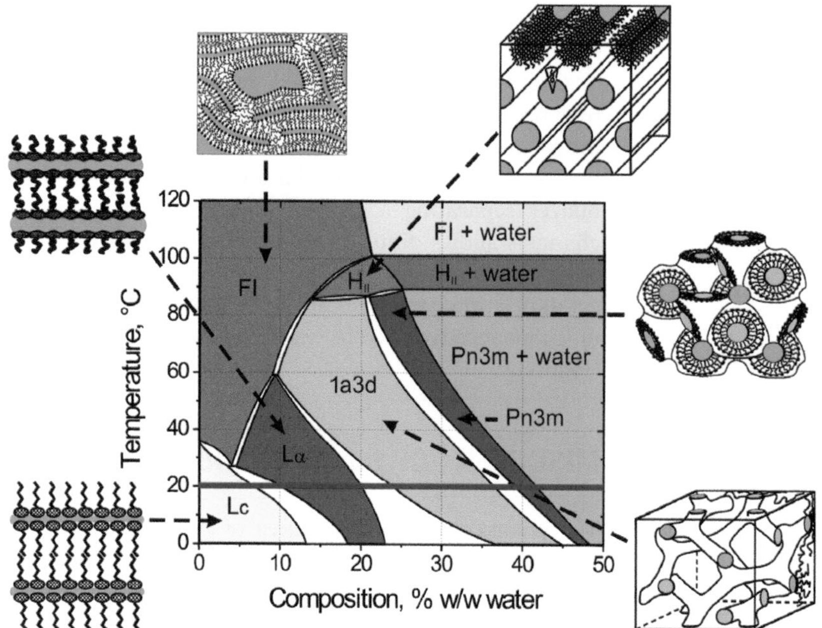

2.19 Phase diagram of the monoolein–water system used for cubic phase crystallization. The phase diagram shows the lipid phases that occur at different temperatures as a function of the water content (shown as % wt/wt water) (see Box 2.2). Two types of lipidic cubic phases, designated space group Pn3m (purple region) and 1a3d (light green region), occur around 30–40% water. The 20°C isotherm (blue line) starts in lamellar phases, Lc (yellow) at low water content and L_α (red) at around 20% water. At 50% water two phases coexist, the Pn3m cubic phase and an aqueous phase (blue). In the representations of the various phase states, the colored areas represent water. From Caffrey, M, and Cherezov, V., *Nature Prot*. 2009, 4:706–731. © 2009, Nature Publishing Group. Reprinted by permission of Macmillan Publishers Ltd.

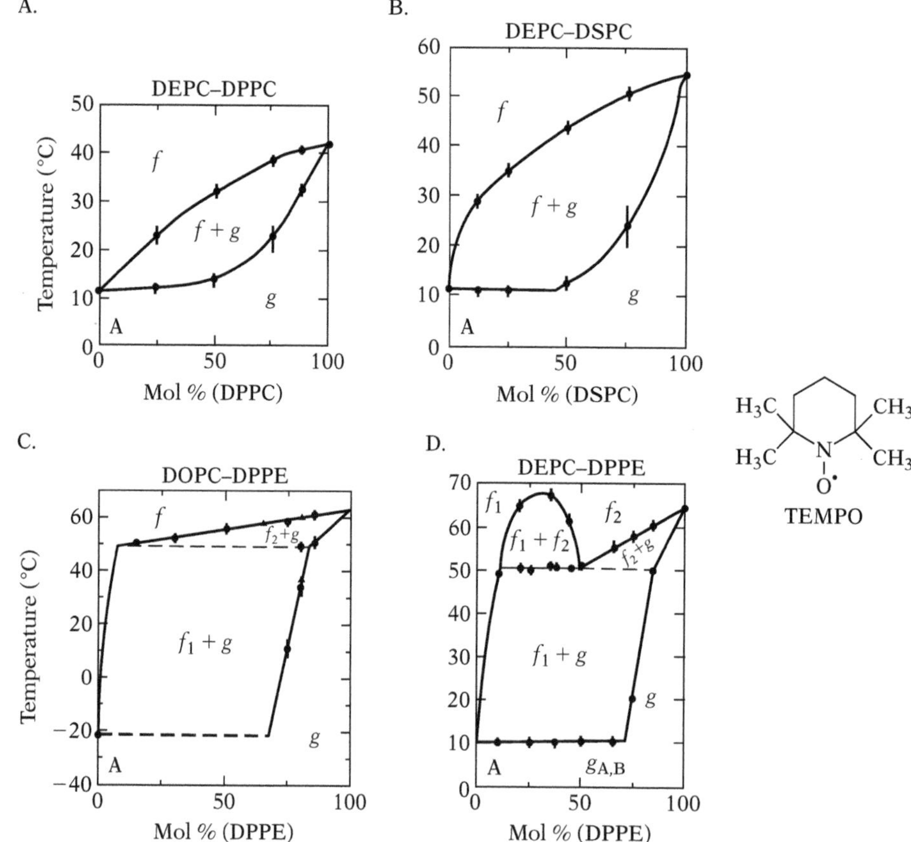

2.20 Phase diagrams of aqueous binary PC and PE mixtures that were determined from the mobility of the lipid-soluble nitroxide spin probe TEMPO (inset). (A) and (B) The DEPC–DPPC and DEPC–DSPC mixtures show regions of fluid (*f*), gel (*g*), and liquid–solid (*f* + *g*) miscible phases as described in Box 2.2. (C) The DOPC–DPPE mixture shows two different liquid phases coexisting with a solid ($f_1 + g$ and $f_2 + g$). (D) The DEPC–DPPE mixture shows these regions and an additional one consisting of immiscible liquid phases ($f_1 + f_2$). Redrawn from Wu, S. H., and H. M. McConnell, *Biochemistry*. 1975, 14:847–854. © 1975 by American Chemical Society. Reprinted with permission from American Chemical Society.

BOX 2.2 PHASE DIAGRAMS

Simple phase diagrams show the phases of a pure substance as a function of temperature and pressure. To learn more about phases in biological systems, which are more complex but are normally at a constant pressure, phase diagrams can be applied to simple mixtures, showing their phases with temperature on one axis and the composition of the mixture – in terms of the mole fraction *X* of one component – on the other. When the two substances in the mixture are miscible in both solid and fluid phases, as are many combinations of two similar phospholipids, the diagram looks like the example in Figure 2.2.1.

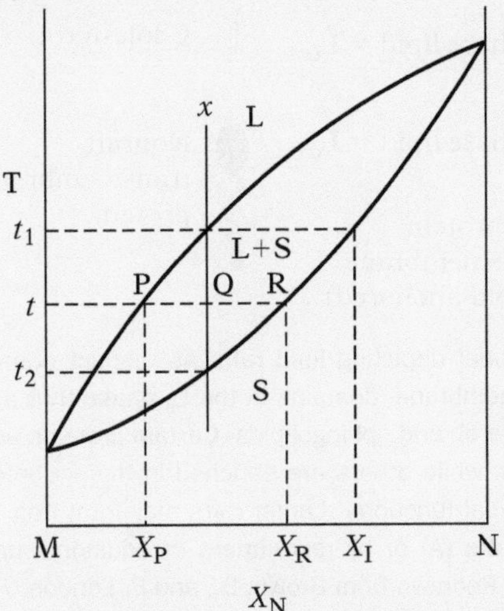

2.2.1 Phase diagram for a mixture of two substances, designated M and N, that are miscible in both liquid (L) and solid (S) phases. From Lee, A. G., *Biochim Biophys Acta.* 1977, 472:237–281. © 1977 by Elsevier. Reprinted with permission from Elsevier.

The *x*-axis gives the mole fraction of the component N, which has the higher melting point. At higher temperatures, the mixture is liquid at all mole fractions of N (L region, above both curves); at lower temperatures, it is solid (S region, below both curves). Between the curves is a region where both solid and liquid coexist (L + S). As the temperature is lowered for a particular composition – for example, starting at position *x* – the liquid pool is depleted of the higher melting point component. At point Q between t_1 and t_2, both phases are present, with the overall mole fraction of *x*. However, the tie line from P to R gives the mole fraction in each phase: X_P is the mole fraction of N in the liquid, and X_R is the mole fraction of N in the solid. This type of phase diagram is observed in parts (A) and (B) of Figure 2.20. If the two phospholipids are sufficiently different, they may be immiscible in the liquid phase or in both liquid and solid phases, as observed in parts (C) and (D) of Figure 2.20.

in many labs using confocal fluorescence microscopy, fluorescence resonance energy transfer (FRET, see Box 2.1), NMR, and EPR. In addition to the phases L_α and L_β (lamellar liquid crystalline and lamellar gel) and crystalline lipid, these techniques detect a "liquid-ordered" state, called L_o, which occurs as a result of the close interactions between cholesterol and PLs or sphingolipids (see Figure 2.9). In L_o the acyl chains of the lipid are extended and tightly packed as in the L_β state, but they exhibit rates of lateral diffusion close to that of lipids in L_α. To contrast with the L_o state, L_α is also called the liquid-disordered state (L_d). Cholesterol confers rigidity to the bilayer by forming condensed complexes with phospholipids (see above).

The ternary phase diagram for the mixture of DOPC, DPPC, and cholesterol shows seven regions (Figure 2.21). At low concentrations of cholesterol, the transition from L_d (L_α), which is rich in DOPC, to s_o (L_β), rich in DPPC, goes through an intermediate area where both L_d and s_o are present. At higher concentrations of cholesterol the L_o phase occurs. Even higher concentrations of cholesterol produce crystalline cholesterol in a cholesterol-saturated lipid lamellar phase or a nonlamellar phase.

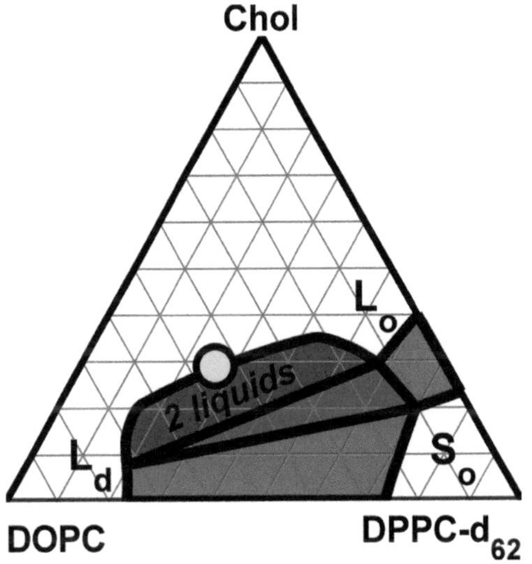

2.21 Ternary phase diagram for mixtures of DOPC, DPPC, and cholesterol. Using perdeuterated DPPC (DPPC-d_{62}) phase boundaries were mapped by ^{2}H-NMR at temperatures from 10°C (shown) to 60°C. The coexistence of two fluid phases, L_d and L_o (blue region), includes a critical point (yellow circle) at the highest miscibility transition temperature, where fluctuations lead to peak broadening. At low cholesterol three phases (L_d, L_o and S_o, green triangle) coexist, connected to two separate regions of solid–liquid coexistence (L_o-S_o, right, and L_d-S_o, bottom, red). Each axis is divided into 10 mol % increments. Redrawn from Veatch, S. L., *et al.*, Critical fluctuations in domain-forming lipid mixtures, *PNAS* 2007, 104:17650–17655. © 2007 National Academy of Sciences U.S.A.

At intermediate mole fractions of cholesterol are two areas where more than one phase coexists. The upper area has coexistence of two fluid phases, both ordered and disordered phases (L_d and L_o), while at lower cholesterol, all three phases (L_d, L_o, and s_o) coexist. Coexistence of L_d and L_o phases can be observed directly by fluorescence microscopy in various membrane systems (see Chapter 3).

Physical studies of lipid mixtures clearly indicate that nonideal mixing occurs as a function of composition and temperature and demonstrate the possibility that different lipid domains coexist in bilayers composed of much more complex lipid mixtures. Recently a great deal of attention has focused on the properties and roles of segregated lipid domains in biomembranes, generally called lipid rafts.

LATERAL DOMAINS AND LIPID RAFTS

While the Fluid Mosaic Model for membrane structure emphasizes the fluidity of the bulk lipid phase of the membrane, allowing random diffusion of its components not bound by the cytoskeleton, Singer and Nicolson did acknowledge the possibility of small membrane domains. Particular cases of lateral organization in the membrane have long been recognized. For example, budding of membrane-enclosed viruses occurs in select regions of the host membranes, and epithelial cells have different lipid (and protein) compositions in their apical and basolateral domains, that is, in the portions of their cytoplasmic membranes that provide their external (exposed) and internal surfaces on opposite ends of the cells. The question of how different lipids are delivered to the two distinct surfaces of epithelial cells led to the concept of lipid rafts, lateral regions distinguished from the bulk lipid of the bilayer that are involved in lipid trafficking as well as protein targeting and other important biological functions (Figure 2.22).

The surface of a plasma membrane on a eukaryotic cell is studded by close to a million lipid rafts, and evidence is emerging for rafts in intracellular membranes as well. Rafts may only reside in the outer leaflet of the membrane, as mixtures of lipids imitating the composition of the inner leaflet do not form rafts *in vitro*. Although there is no direct evidence for co-localized raft formation in the inner and outer leaflets of the bilayer, it appears that the L_o phase in one leaflet induces L_o in the other, possibly because the two leaflets are coupled by interdigitation of the longer acyl chains. Certain proteins concentrate in rafts, whereas others are excluded from them (see below); in fact, the clustering of some proteins seems to induce raft formation or at least increase the size of rafts.

The lipids in rafts have physical properties different from those of the bulk lipids. Domains enriched in glycosphingolipids and cholesterol are several angstroms thicker than the rest of the bilayer. Indeed they look like rafts on the surface of the fluid bilayer when their increased thickness is detected by atomic force microscopy, producing images that validate the name *lipid rafts* (Figure 2.23). The raft lipids, with a preponderance of saturated acyl chains, are in the L_o state, and are thus more ordered than the bulk lipids. Coexistence of L_o and L_d phases has now been demonstrated in raft-imitating model membranes of varying compositions, such as cholesterol, palmitoyl-sphingomyelin, and POPC in 1:1:1 molar ratios. Segregated lipid domains are also observed in monolayers and bilayers made with the lipids extracted from epithelial brush border membranes from the apical microvilli of the absorptive cells lining the intestine. Treatment of these membranes with methyl-β-cyclodextrin, which removes cholesterol, abolishes the domains. β-cyclodextrin appears to disrupt rafts on the surface of whole cells as well, but these experiments must be interpreted with caution as cholesterol depletion has many other effects on cells

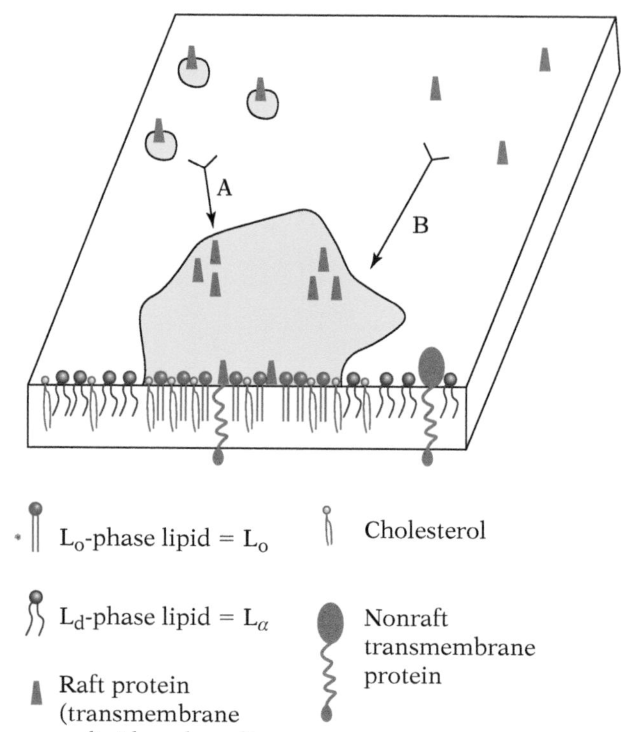

2.22 Model depicting lipid rafts as distinct domains of the plasma membrane, domains in the L_o phase that are enriched in cholesterol, and sphingolipids. Certain proteins are excluded from rafts, while others are enriched in them, conferring upon them special functions. Larger rafts may form from coalescing smaller rafts (A) or by recruitment of additional proteins and lipids (B). Redrawn from Brown, D., and E. London, *J Biol Chem*. 2000, 275:17221–17224. © 2000, by American Society for Biochemistry & Molecular Biology.

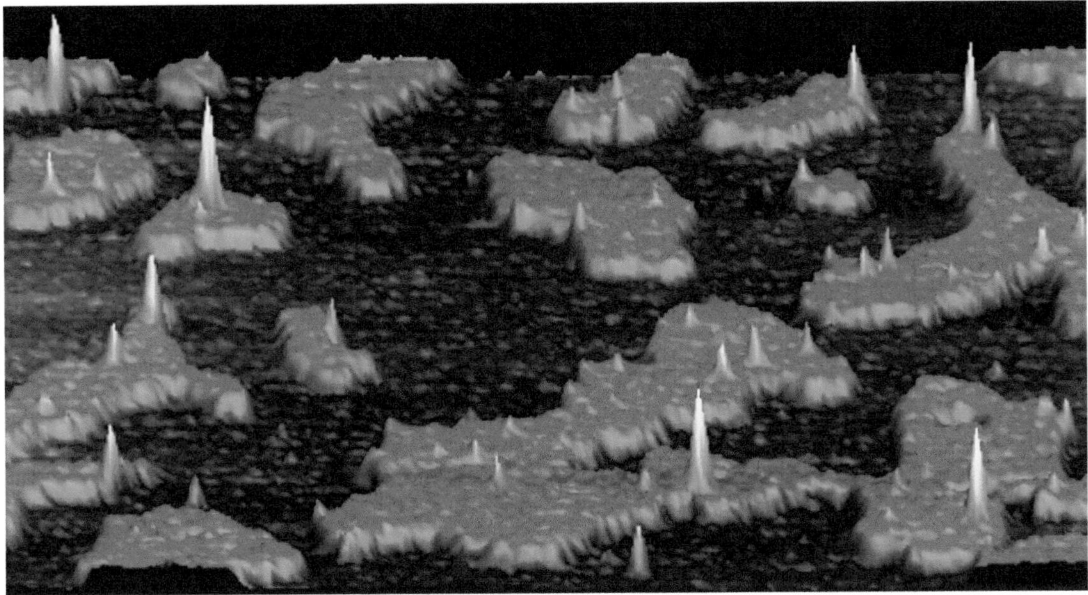

2.23 Observation of rafts by atomic force microscopy, which detects the increased thickness of the L_o domains. Kindly provided by Michael Edwardson, University of Cambridge. Reprinted by permission of the author.

and β-cyclodextrin has been shown to retard diffusion of membrane proteins independent of its effect on cholesterol.

Lipid rafts are diverse in terms of their composition (and therefore functions), their lifetimes, and their sizes. Because membrane microdomains of different sizes form dynamically, methods to detect them employ different time scales and different length scales. The results must be integrated into a consistent model, combining observations from model membranes and from cell membranes. Depending on the method of observation used, raft sizes vary from 10 nm to 200 nm. When immunolabeling techniques are used, antibody interactions can stimulate fusion of microdomains to create the larger domains. Very small microdomains (containing as few as 25–50 lipid molecules) have been detected in model membranes using fluorescence-quenching techniques, in which incorporation of lipid-linked bromines identifies quencher-rich and quencher-poor domains that form in response to addition of cholesterol or sphingomyelin. Rafts can now be detected with a very new application of mass spectrometry that provides quantitative images of lipid bilayer components with spatial resolution of less than 100 nm.

The size of small raft domains is limited by bilayer curvature and by the domain boundary at the interface between the domain and the surrounding bulk lipid. The energy per unit length of this interface is referred to as the *line tension*, and it physically determines the sizes and shapes of the domains. If a large portion of the membrane is occupied by small rafts, then raft boundaries must be extensive. In this case, the line tension is small, entropy dominates, and the domains are small. If the line tension is large, it favors fusion into larger domains because when many small rafts merge into a large one, the total length of the raft boundary is reduced, thus reducing the boundary's energy.

Large rafts appear to encompass smaller heterogeneous domains within them and likely form by coalescence of preexisting structures (Figure 2.24). In cells, clustering of smaller rafts into larger domains is probably stimulated by particular proteins (see Chapter 4) and may have important functional or regulatory consequences. These clusters are probably transient, resulting from very dynamic formation and growth. High-speed video microscopy has recorded the formation and dissolution of small rafts (with diameters ~50 nm, containing ~3000 lipid molecules and probably only 10–20 protein molecules) with lifetimes of less than 1 msec. It is thought that certain raft proteins stimulate raft formation, as suggested by models that posit that "lipid shells" around such proteins come together to make rafts. In this case, raft heterogeneity is a consequence of the particular lipid–protein and lipid–lipid interactions that trigger their formation.

Detergent-Resistant Membranes

The proposition that rafts exist in cell membranes was given a huge boost by methods for extraction of an L_o fraction of biological membranes. Treatment of mammalian cells with the nonionic detergent Triton X-100 (see Chapter 3) produces a Triton-insoluble low-density membrane fraction. These detergent-resistant

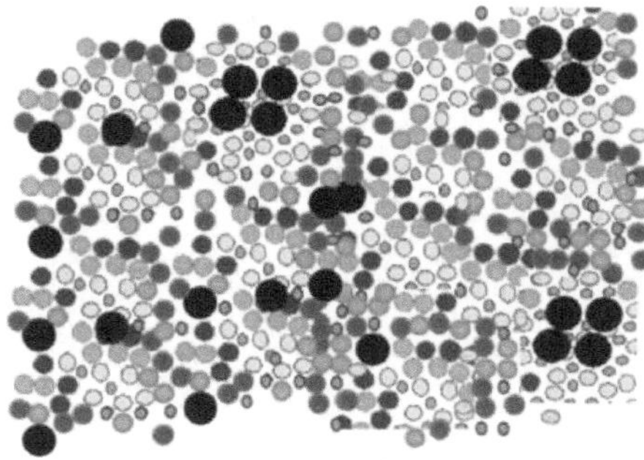

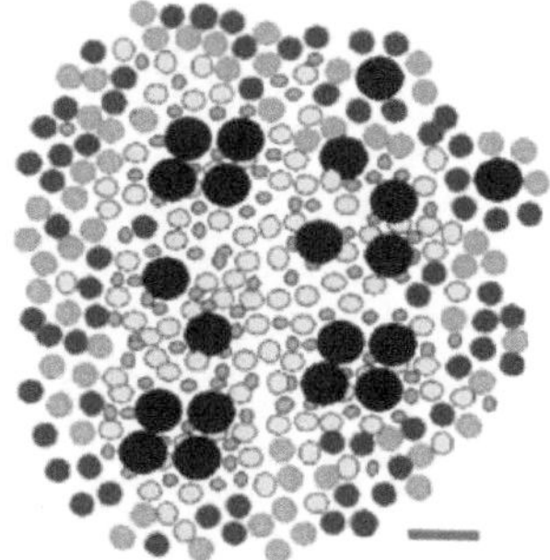

2.24 A model for the clustering of small, preexisting membrane domains into larger rafts. The clustering of small domains into relatively large rafts is actively organized in both space and time. The affinities between some lipids and proteins form preexisting structures (left) that can coalesce into rafts (right). The red and pink circles represent different nonraft lipid species, the yellow circles represent raft lipids, the green circles represent cholesterol, and the larger black circles represent GPI-linked raft proteins. The scale bar is ~5 nm. Redrawn from Mayor, S., *Traffic*. 2004, 5:231–240. © 2004, John Wiley and Sons.

membranes (DRMs) are rich in cholesterol and sphingolipids, observed to be in the L_o state, and enriched in fatty acid- or GPI-linked proteins, as are rafts. According to x-ray diffraction measurements, they are 9 Å thicker than nonraft lipid bilayers. Because the DRMs are isolated at low temperatures, it is difficult to assess how much of this membrane fraction would have been in the L_o phase at growth temperatures, and this adds uncertainty regarding the specificity of the procedure for raft extraction. Raft proteins are used as markers, most successfully in the case of caveolae (see below). Other questions focus on the type and amount of detergent used. Titrated addition of Triton X-100 can disrupt preexisting L_o domains, as well as induce formation of L_o domains. Several different detergents produce variations in the insoluble fraction, compared with the "raft selectivity" of Triton X-100; the heterogeneity of the resulting DRMs may not reveal anything more than the selectivity of the detergents for subsets of lipids and proteins. Non-detergent methods to isolate rafts employ sonication, which may cause undesirable mixing of membrane components and, in fact, give a DRM fraction enriched in arachidonic acid-containing plasmalogens, which are polyunsaturated lipids that would be excluded from typical rafts.

DRMs contain specific types of proteins, most of which are also detected in lipid rafts by other techniques. These raft proteins include GPI-linked proteins; doubly acylated proteins, such as tyrosine kinases of the Src family and Gα subunits of heteromeric G proteins; and certain TM proteins (see Chapter 4). The proteins enriched in lipid rafts suggest they have special biological functions, such as signal transduction and protein trafficking. Logically, the segregation of signaling proteins in rafts could speed the rates of their interactions with other raft proteins and slow their interactions with nonraft proteins. Furthermore, treatment with β-cyclodextrin to disrupt rafts by removing cholesterol causes defects in some signaling pathways. Among the tyrosine kinases associated with rafts are the receptors for epidermal growth factor (EGF), and treatment of fibroblasts in culture with EGF causes its receptor to leave the raft. The dynamic process of signal transduction could thus take advantage of constant change in raft constituents as well as size.

Protein trafficking through the Golgi apparatus appears to use rafts to sort membrane proteins, because domains rich in cholesterol and sphingomyelin form in the Golgi and become vesicles targeted for the plasma membrane. One of the factors in protein localization could be the thickness of the TM domains (see Chapter 4), because the increased thickness of these domains could help "raft proteins" partition into them. Thus proteins with shorter TM domains could be excluded from rafts and retained in the Golgi, while those with longer TM domains partition into rafts that move to the plasma membrane.

A minor fraction of the DRMs contain caveolae, small invaginations in the membrane associated with the protein caveolin in addition to other raft proteins. Caveolae may function in both protein trafficking and signal transduction. Although caveolae are considered lipid rafts, they are a special case, because caveolin inserts from the cytoplasmic side, which is not enriched in raft lipids, and then oligomerizes to force an inward curvature of both leaflets of the membrane (Figure 2.25). A similar process may occur when membrane domains segregate for viral envelope formation. The envelope of influenza virus is enriched in cholesterol and sphingomyelin compared with the host membrane from which it is formed, and its formation involves the clustering of specific glycoproteins before budding. It is believed that a matrix protein, M1, docks at the inner leaflet, selectively binds the glycoproteins, and polymerizes to induce the curvature that triggers the budding process.

Whether lipid rafts are defined as domains of L_o phase floating in the L_d membrane, Triton X-100 insoluble

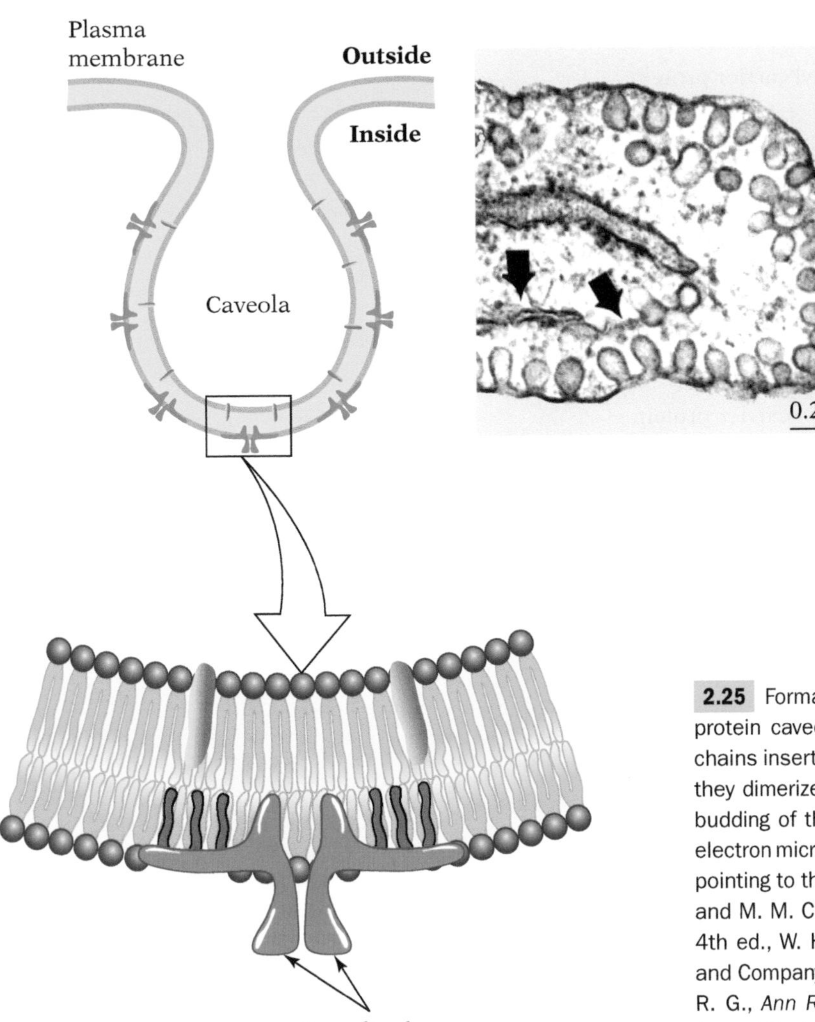

2.25 Formation of caveolae by insertion of the specific protein caveolin. Caveolin monomers linked to three acyl chains insert into the membrane from the cytoplasm. When they dimerize, they force an inward curvature that leads to budding of the membrane. The inset shows a thin-section electron micrograph of caveolae from fibroblasts (with arrows pointing to the ER). Main figure: Redrawn from Nelson, D. L, and M. M. Cox (eds.), *Lehninger Principles of Biochemistry*, 4th ed., W. H. Freeman, 2005. © 2005 by W. H. Freeman and Company. Used with permission. Inset: From Anderson, R. G., *Ann Rev Biochem*. 1998, 67:199–225. © 1998 by Annual Reviews. Reprinted with permission from the Annual Review of Biochemistry, www.annualreviews.org.

membrane fractions of low density, or heterogeneous microdomains of membrane containing proteins involved in signaling and trafficking, they clearly have profound effects on the nature of the membrane. The presence of rafts may also contribute to the need for lipid diversity, to support the dynamic formation of such stable but fluid lipid domains segregated from the bulk bilayer lipids and to maintain permeability barriers at their boundaries.

DIVERSITY OF LIPIDS

The typical biological membrane contains hundreds of lipid species when their particular acyl chains are considered, and the types of complex lipids predominant in membranes from different sources vary significantly. An obvious example is the absence of sterols and sphingolipids from prokaryotic membranes. Among phospholipid species, PG and CL are found in bacterial membranes but not in eukaryotic membranes other than mitochondria. *E. coli* normally lacks PC but grows well when carrying the *Rhodobacter sphaeroides* gene for the PE methylase, producing PC to levels as high as 20% of its total PL. In fact, the lipid composition of the *E. coli* cytoplasmic membrane varies drastically in different mutants and/or different conditions.

Genetic studies of the effects of different lipid compositions make use of mutations affecting the PL biosynthetic pathway of *E. coli*. In wild type *E. coli*, the normal PL composition is 70% to 80% PE, 20% to 25% PG, and ≤5% CL, all of which are produced from phosphatidic acid in a few enzymatic steps (Figure 2.26). The anionic lipids are essential, as *psgA*⁻ null mutants that cannot make PG and CL are not viable. The negatively charged membrane is required for initiation of DNA synthesis, attributable to the binding of PG by the DnaA protein. Other examples of proteins that have affinity for anionic PLs include type I topoisomerases, some DNA polymerases, SecA protein (which is involved in protein translocation, see Chapters 4 and 7), and some signaling proteins.

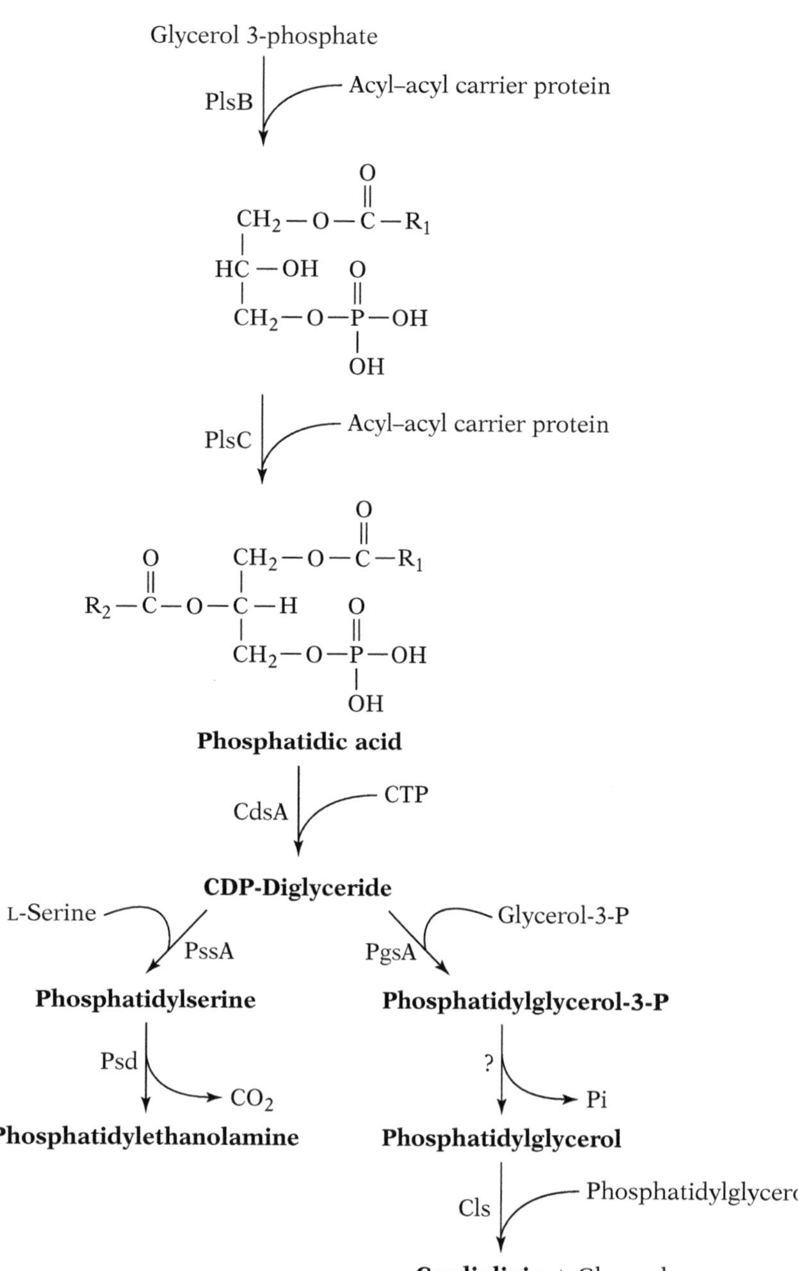

2.26 The pathway for phospholipid biosynthesis in *E. coli*. All PLs are formed from the activated precursor cytidine 5′-diphosphate (CDP)-diacylglycerol, from which PE is synthesized via PS, and PG and CL are synthesized on a second branch of the pathway. Note that the *pssA*$^-$ mutant cannot make PS or PE. CTP, cytidine 5′-triphosphate. Redrawn from Cronan, J. E., *Annu Rev Microbiol*. 2003, 57:203–224. © 2003 by Annual Reviews. Reprinted with permission from the Annual Review of Microbiology, www.annualreviews.org.

On the other hand, PE does not seem to be essential in spite of its normal abundance: *pssA*$^-$ null mutants, with <0.1% PE in their membranes, are able to grow in rich medium supplemented with divalent cations, under which conditions the membrane is 90% PG and CL. However, these mutants grow poorly on defined minimal media and show defects (transport and motility problems, filament formation, and early entry into stationary phase) that indicate zwitterionic lipids are needed for normal membrane functions. Finally, mutants deficient in PE regulate their CL content, allowing them to maintain the proportion of nonlamellar lipids. Indeed, mutants lacking both PE and CL are not viable, indicating that the membrane requires nonbilayer-forming lipid. Lipid diversity appears to be essential to reach the desired R_o value and maintain the general state of the bilayer in a narrow range between the lamellar gel state and the inverted phase ($L_\beta < L_\alpha < H_{II}$).

The role of PE in maintaining the polymorphism of *E. coli* membrane lipids is illustrated by NMR studies of ^{31}P-labeled phospholipids carried out in whole cells, with purified membrane vesicles or with extracted lipids, because the shift in the ^{31}P-NMR powder spectra can distinguish lamellar and H_{II} phases. When the total *E. coli* lipids are first extracted, they are lamellar (Figure 2.27). After incubation at 42°C, they shift to hexagonal phase, giving an NMR spectrum that resembles that of isolated PE from the same strain. If 5% lysophospholipid, whose

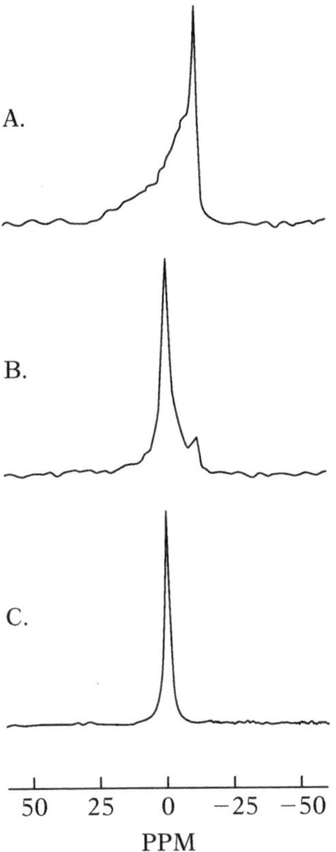

2.27 Detection of lamellar and hexagonal phases in *E. coli* lipids by ^{31}P-NMR. The lipids were extracted from a fatty acid auxotroph grown at 37°C on oleic acid, in which over half the acyl chains are C18:1. The initial NMR signal indicates the lipid is lamellar (A). After incubation at 42°C for 15 minutes (B) and 90 minutes (C), the signal shifts to that of an isotropic lipid and now resembles the NMR spectrum for H_{II}-phase PE. Redrawn from de Kruijff, B., *et al.*, in R. M. Epand (ed.), *Lipid Polymorphism and Membrane Properties*, vol. 44 of *Current Topics in Membranes*, Academic Press, 1997, pp. 477–515. © 1997 by Academic Press. Reprinted with permission from Elsevier.

inverted cone shape complements the shape of PE, is added to the total lipid extract, the shift to H_{II} at higher temperature is abolished. In other words, *E. coli* lipids dominated by PE are lamellar at growth temperatures and convert to H_{II} at higher temperatures. Because PE isolated from other sources prefers H_{II} even at the growth temperature, *E. coli* must control the L_α–H_{II} transition temperature (T_{LH}) by the degree of saturation of its acyl chains. In this way *E. coli* membrane lipids, which are up to 80% PE, stay between T_m (L_β–L_α) and T_{LH} (L_α–H_{II}) to achieve the desired fluidity.

The zone from T_m to T_{LH} transitions is important for many organisms that keep their membranes close to T_{LH}. Clearly if the whole membrane passes beyond T_{LH}, the loss of lamellar lipids would be deleterious because of the loss of the permeability barrier. However, this does occur in certain sites in mammals, such as tight junctions in polarized cells and networks of myelin in lung tissue. The possible roles for nonlamellar membrane structures in processes such as membrane fusion, cell division, and gene transfer keep interest in these transitions high. As discussed above, nonbilayer lipids are essential for the curvature of the membrane; the resistance to curvature creates pressure gradients that may be triggers for mechanosensitive channels and membrane-inserting peptides such as gramicidin and melittin (see Chapters 4 and 12).

In eukaryotic cells the lipid composition of membranes varies throughout the cell (Figure 2.28). A few differences are due to the sites of lipid biosynthesis, for example cardiolipin is found only in the mitochondria, where it is synthesized. The major phospholipids are synthesized in the endoplasmic reticulum (ER) and then trafficked to other organelles and the plasma membrane. Some membrane lipids are also produced in the Golgi and mitochondria. Among the phospholipids there are quite striking differences in the proportion of PI and PS in different organelles. In addition there is a ten-fold variation in the ratio of sterol to phospholipid. (Most specialized lipids with roles in signaling, such as the phosphorylated phosphatidyl inositols, are synthesized from components of the plasma membrane and released into the cell.) Lipid transport in the cell occurs through a process of budding and fusing of membrane vesicles as well as nonvesicular transport through extensions of the ER, and also may involve lipid transfer proteins. Current research is focused on how lipid trafficking is regulated to produce such nonrandom sorting of phospholipids.

CONCLUSION

This chapter describes the diversity of lipid structures and the polymorphic phase behavior it enables. The properties of these remarkable amphiphiles allow them to assemble spontaneously and to form the fluid bilayer that characterizes the structure of all biomembranes. From the need for nonlamellar lipids in effecting membrane elasticity to the specialized functions of lipid rafts involved in cellular communication and intracellular trafficking, membrane lipids do much more than provide the structure of the bilayer and the solvent for membrane proteins. Applications of sophisticated biophysical and biochemical techniques will undoubtedly continue to reveal their complex and crucial roles.

The portrayal of the nature and diversity of lipids is essential for understanding membrane structure and function. Of course, the other major contributors to membrane properties are the proteins of the membrane, which have to be isolated from the membrane to characterize them. Chapter 3 describes the detergents

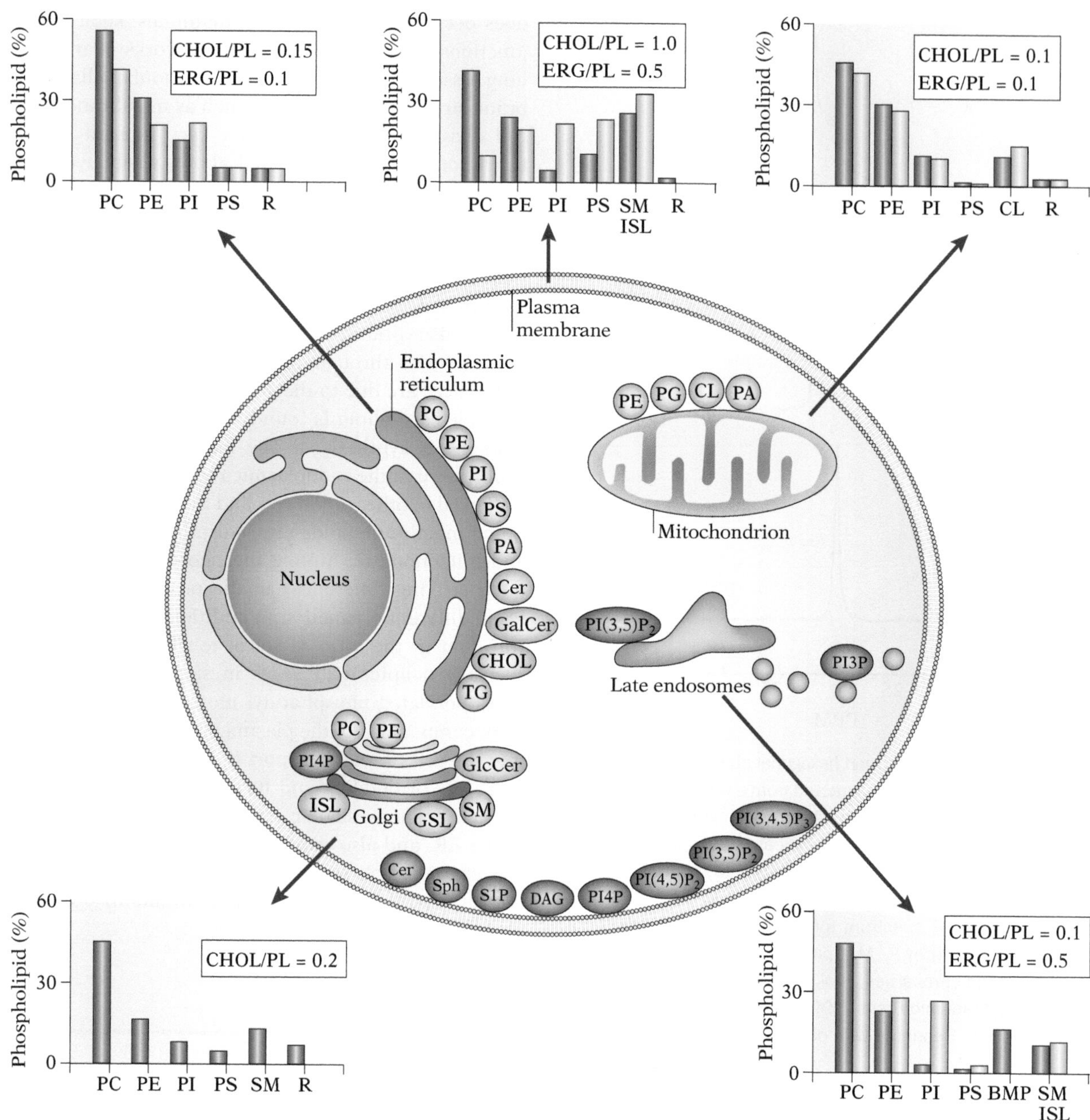

2.28 Lipid composition of membranes and sites of lipid synthesis in eukaryotic cells. A eukaryotic cell (center diagram) contains the plasma membrane and membranes surrounding the different organelles, mitochondria, endoplasmic reticulum, golgi, and endosomes. The graphs give the percentage of the total phospholipid in mammals (blue) and yeast (light blue), along with the molar ratio of cholesterol (CHOL, in mammals) and ergosterol (ERG, in yeast) in each. The types of lipids synthesized in the different organelles are abbreviated as follows on dots (blue for the major phospholipids and red for lipids involved in signaling): phosphatidylcholine (PC), phosphatidylethanolamine (PE), phosphatidylinositol (PI), phosphatidylserine (PS), phosphatidic acid (PA), cardiolipin (CL), ceramide (Cer), galactosylceramide (GalCer), sphingomyelin (SM), triacylglycerol (TG), glycosphingolipids (GSLs), inositol sphingolipid (ISL, in yeast), diacylglycerol (DAG), PG, phosphatidylglycerol; PI(3,5)P_2, phosphatidylinositol-(3,5)-bisphosphate; PI(4,5)P_2, phosphatidylinositol-(4,5)-bisphosphate; PI(3,4,5)P_3, phosphatidylinositol-(3,4,5)-trisphosphate; PI4P, phosphatidylinositol-4-phosphate; R, remaining lipids; S1P, sphingosine-1-phosphate; Sph, sphingosine, and BMP (bis(monoacylglycero)phosphate). From van Meer, G., *et al.*, *Nat Reviews Mol Cell Biol*. 2008, 9:112–124. © 2008, Nature Publishing Group. Reprinted by permission of Macmillan Publishers Ltd.

used for this process and the model systems used to reconstitute membrane mimetics. The importance of lipids will reemerge when their interactions with membrane proteins are described in Chapter 4 and the detailed structures of lipid bilayers are examined in Chapter 8.

FOR FURTHER READING

Reviews

Daleke, D. L., Regulation of transbilayer plasma membrane phospholipid asymmetry. *J Lipid Res*. 2003, 44:233–242.

Dowhan, W., Molecular basis for membrane phospholipid diversity: why are there so many lipids? *Annu Rev Biochem*. 1997, 66:199–232.

Edidin, M., The state of lipid rafts: from model membranes to cells. *Annu Rev Biophys Biomol Struct*. 2003, 32:257–283.

Jacobson, K., O. G. Mouritsen, and R. G. Anderson, Lipid rafts: at a crossroad between cell biology and physics. *Nat Cell Biol*. 2007, 9:7–14.

Lingwood, D. and K. Simons, Lipid rafts as a membrane-organizing principle. *Science*. 2010, 327:46–50.

McConnell, H. M., and A. Radhakrishnan, Condensed complexes of cholesterol and phospholipids. *Biochim Biophys Acta*. 2003, 1610:159–173.

McConnell, H. M., and M. Vrljic, Liquid–liquid immiscibility in membranes. *Annu Rev Biophys Biomol Struct*. 2003, 32:469–492.

Ohvo-Rekila, H., B. Ramstedt, P. Leppimaki, and J. P. Slotte, Cholesterol interactions with phospholipids in membranes. *Prog Lipid Res*. 2002, 41:66–97.

Simons, K., and W. L. C. Vaz, Model systems, lipid rafts and cell membranes. *Annu Rev Biophys Biomol Struct*. 2004, 33:269–295.

van Meer, G., D. R. Voelker, and G. W. Feigenson, Membrane lipids: where they are and how they behave. *Nat Reviews Mol Cell Biol*. 2008, 9:112–124.

Searchable database for fatty acids: http://sofa.bfel.de

Database of biologically relevant lipids sponsored by the *National Institute of General Medical Sciences*: www.lipidmaps.org/data/structure

Seminal Papers

Anderson, D. M., S. M. Gruner, and S. Leibler, Geometrical aspects of the frustration in the cubic phases of lyotropic liquid crystals. *Proc Natl Acad Sci USA*. 1988, 85:5364–5368.

Baumgart, T., S. T. Hess, and W. W. Webb, Imaging coexisting fluid domains in biomembrane models coupling curvature and line tension. *Nature*. 2003, 425:821–824.

Chapman, D., Phase transitions and fluidity characteristics of lipids and cell membranes. *Q Rev Biophys*. 1975, 8:185–235.

Feigenson, G. W., and J. T. Buboltz, Ternary phase diagram of dipalmitoyl-PC/dilauroyl-PC/cholesterol: nanoscopic domain formation driven by cholesterol. *Biophys J*. 2001, 80:2775–2788.

Gruner, S. M., Intrinsic curvature hypothesis for biomembrane lipid composition: a role for nonbilayer lipids. *Proc Natl Acad Sci USA*. 1985, 82:3665–3669.

Jain, M. K., and H. B. White, III, Long range order in biomembranes. *Adv Lipid Res*. 1977, 15:1–60.

Pan, J., F. A. Herbele, S. Tristram-Nagle, *et al.*, Molecular structures of fluid phase phosphatidylglycerol bilayers as determined by small angle neutron and x-ray scattering. *Biochim Biophys Acta*. 2012, 1818:2135–2148.

TOOLS FOR STUDYING MEMBRANE COMPONENTS DETERGENTS AND MODEL SYSTEMS

3

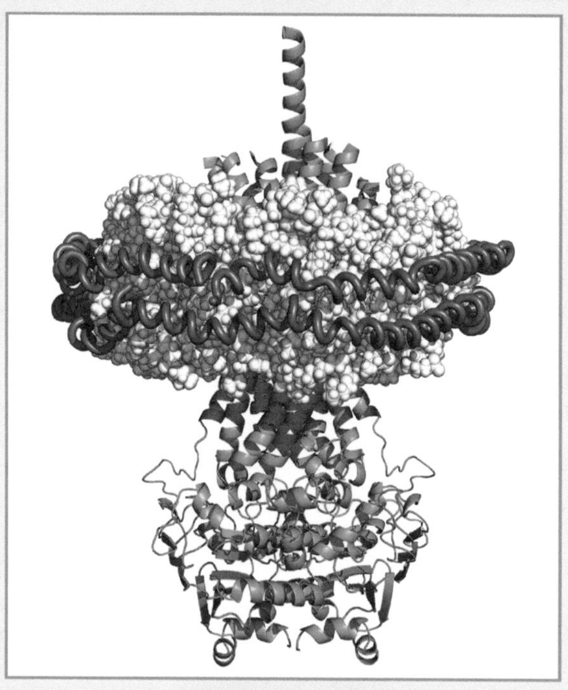

Nanodiscs for membrane protein reconstitution. A model shows a cytochrome P450 (green) incorporated in a nanodisc consisting of a lipid bilayer (gray) surrounded by an amphipathic protein (blue) derived from apolipoprotein. Many different proteins have now been reconstituted in nanodiscs to study their functions such as enzyme activities, ligand binding, lipid dependence, interactions with other proteins, as well as their structure (by NMR). Kindly provided by Prof. William Atkins of the University of Washington. Reprinted by permission of the author.

While progress in biochemistry, biophysics, and structural biology relies on studies of purified biological components, the purification of membrane components is complicated by their amphipathic nature. First, their removal from the membrane usually requires disruption of the lipid bilayer. Once removed, they tend to aggregate in aqueous buffers due to their low solubility in water. And finally, study of the functions of many membrane components requires their insertion back into a reconstituted membrane. The critical tools that allow *in vitro* characterization of membrane components are detergents and model membranes. Detergents are used to solubilize membrane components, removing them from the lipid bilayer and preventing their aggregation. This chapter begins with an overview of detergents, emphasizing their mechanisms of action in solubilizing membrane components.

The goal of reconstitution is to insert the purified membrane component into a good mimic of the biological membrane, usually a lipid bilayer. Because many aqueous lipid mixtures spontaneously assemble in lamellar phase (see Chapter 2), model bilayers tend to form spherical vesicles called liposomes. Liposomes are just one type of model system used for *in vitro* characterization of membrane components. Even for that one type, the nature of the lipid vesicles depends on how they are made and determines their suitability for different experimental techniques. By necessity, these models are simpler than biological membranes, which contain hundreds of lipid species; depending on the objective, a single lipid species can often suffice. Model membranes used to study lipid properties offer control over the stoichiometry in a mixture of two or three types of lipids.

The availability of widely varying model membrane systems is critical in the characterization of purified membrane components because they allow the use of different experimental tools. Some membrane mimetics allow the application of the powerful techniques that have provided a wealth of information about soluble proteins, while others enable use of methods based on select properties of the membrane, such as electrical conductance or pressure effects. The latter part of this chapter surveys the characteristics of the different model membrane systems from classic systems such as black films to new technologies such as nanodiscs. Examples are provided to illustrate the uses of these systems in membrane research.

DETERGENTS

Detergents are defined as water-soluble surfactants, which makes them amphiphiles that are effective in the solubilization of membrane components. Their solubility in water is much greater than that of most lipids; for example, sodium dodecylsulfate (SDS) has a monomer solubility of 10^{-2} M, compared with the solubilities of DPPC and cholesterol of 10^{-10} M and 10^{-8} M, respectively. Occasionally, detergents are considered synonymous with surfactants, because they reduce the surface tension of a liquid when dissolved in it (see Box 3.1); however, the water solubility of detergents is essential for their role in disrupting membranes.

Purification of a membrane protein typically begins with solubilization of the membrane with a detergent, after which one or more types of chromatography in detergent separate the desired protein from the others. The purpose of the detergent is to prevent aggregation of the membrane protein as it is removed from its lipid environment. During these procedures, one detergent can replace another, often to maintain mild (nondenaturing) conditions. Thus it is essential to be familiar with the different detergents, their properties, and their modes of action.

> **BOX 3.1 SURFACTANTS AND SURFACE TENSION**
>
> By definition, surfactants are substances that reduce surface tension. Surface tension results from the cohesive forces between liquid molecules that are unopposed by other molecules at an air–liquid interface. These unbalanced forces produce the tendency for a liquid to minimize its surface area, which is why a drop of liquid is spherical.
>
> Surface tension is measured as the work required to break a liquid film, and has the units dynes per centimeter (1 dyne = 10^{-5} newtons). With its extensive hydrogen bonding, water at 20°C has a relatively high surface tension, 72.8 dynes/cm compared with 22.3 dynes/cm for ethyl alcohol.

Types of Detergents

A variety of detergents are in common use for membrane biochemistry. While most detergents are synthetic, there are natural compounds with detergent activity, such as the bile salts that solubilize dietary fats in the intestine, and saponins, a varying group that includes the heart stimulant digitonin. Dozens of detergents are commercially available (examples are shown in Figure 3.1). They may be ionic (e.g., SDS, cetyltrimethylammonium bromide [CTAB]), zwitterionic (e.g., lauryldimethylamine oxide [LDAO], which varies from SDS in its headgroup), or nonionic (e.g., Triton X-100, octylphenol linked to a hydrophilic polyoxyethylene headgroup with an average of 9–10 repeats). Most synthetic detergents have a polar or ionic headgroup and a nonpolar hydrocarbon tail, while some, like sodium cholate, sodium deoxycholate, and 3-[3-(cholamidopropyl) dimethyl-ammonio]-1-propanesulfonate (CHAPS) are derivatives of the bile salts, with a tetracyclic structure similar to that of cholesterol. Some commonly used nonionic detergents, such as octyl β-D-glucoside (OG) and dodecyl β-D-maltoside (DDM), have sugar headgroups.

The wide number of detergents available allows researchers to choose from many options, testing several for best results. While nonionic detergents like Triton X-100 are relatively mild and can stabilize membrane proteins, some ionic detergents, notably SDS, bind strongly to proteins and usually denature them. The bile salt derivatives are much less denaturing than ionic detergents having linear nonpolar chains with the same headgroups. Short-chain nonionic detergents, including OG, denature some proteins, in which case DDM, with its longer chains, is a good alternative. DDM has become the most widely used detergent in membrane protein research. Triton X-100 has also been widely used because it allows retention of biological activities and is generally very effective at solubilization; however, its aromatic ring interferes with spectroscopic studies involving ultraviolet (UV) absorbance, fluorescence, and circular dichroism. To address this limitation, a reduced form of Triton X-100 is available.

Several of the nonionic detergents, including Triton X-100, are synthesized by condensation of ethylene oxide with the parent alcohols, resulting in polydisperse mixtures containing variable chain lengths. Although the manufacturers give the average chain lengths, they do not indicate the variability, which fits a nearly

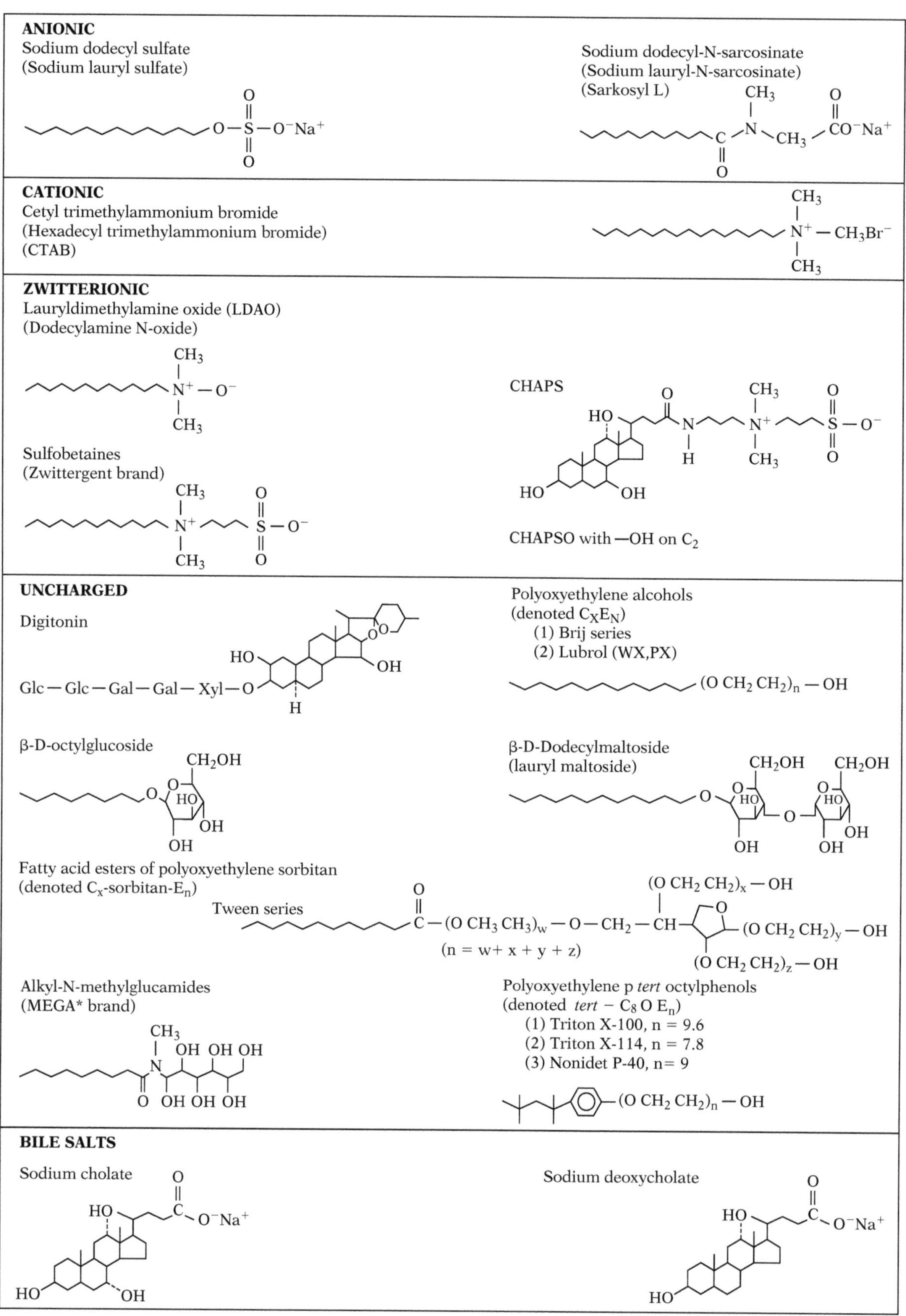

3.1 Structures of some detergents used to solubilize membrane components. From Gennis, R. B., *Biomembranes*, Springer-Verlag, 1989, pp. 90–91.

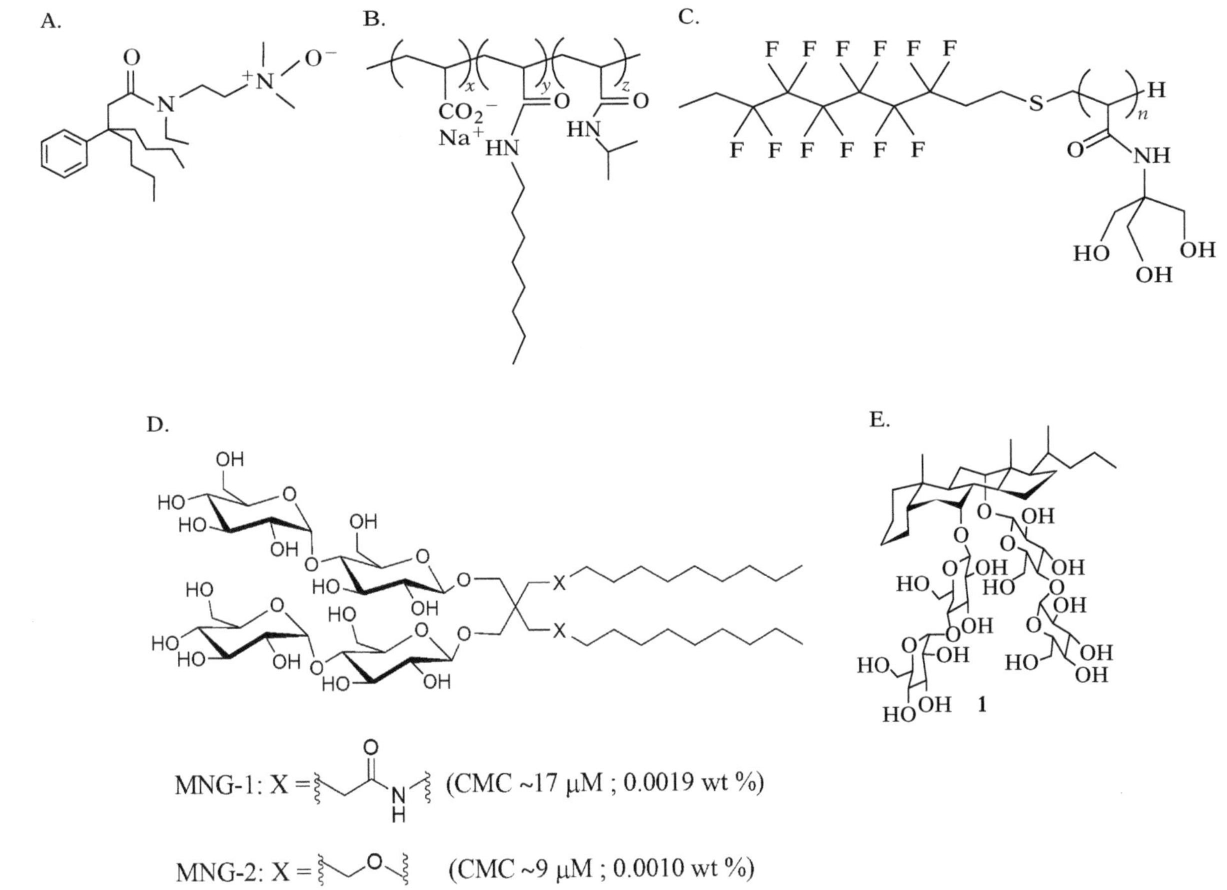

MNG-1: X = (CMC ~17 μM ; 0.0019 wt %)

MNG-2: X = (CMC ~9 μM ; 0.0010 wt %)

3.2 Some new alternatives to detergents. (A) The tripod amphiphile that extracts bacteriorhodopsin. Redrawn with permission from Yu, S. M., *et al.*, *Prot Sci*. 2000, 9:2518–2527. (B) An example of an amphipol called A8-35 with variation in the polymeric backbone giving $x = 35\%$, $y = 25\%$, and $z = 40\%$. Redrawn with permission from Gohon, Y., *et al.*, *Anal Biochem*. 2004, 334:318–334. (C) A hemifluorinated surfactant called HF-TAC [$C_2H_5C_6H_{12}C_2H_4$-S-poly-Tris-(hydroxymethyl)aminomethane] with a single hemifluorinated hydrocarbon chain. Redrawn with permission from Breyton, C., *et al.*, *FEBS Lett*. 2004, 564:312–318. (D) The flexible MNG amphiphiles have links to acyl chains at the central tetrahedral carbon with amide, ester or carbon–carbon (not shown) bonds. From Chae, P. S., *et al.*, *Nature Meth*. 2010, 7:1003–1008. (E) Di-β-D-maltoside cholane is an example of a facial amphiphile that places the polar groups on one side and also lacks the carboxylic group of cholic acid. From Zhang, Q., *et al.*, *Ange Chem Intl Ed Eng*. 2007, 46:7023–7025.

Poisson distribution of oxyethylene chain lengths from one to >20, and the resulting heterogeneity can be quite deleterious, especially in crystal formation. For homogeneous alternatives, a series of alkyl polyoxyethylenes of defined chain length (C_XE_N, where X is the number of C atoms in the alkyl group and N is the number of oxyethylene monomers in the headgroup) is used.

Commercially available detergents may have problems of impurities; for example, SDS often contains *n*-dodecanol and polyoxyethylene-based detergents may contain peroxide and aldehyde, which necessitates additional purification steps or the purchase of "protein-grade" or "especially purified" quality. A few detergents, including sodium cholate and β-octylglucoside, can be purified by crystallization prior to use.

The common detergents in Figure 3.1 do not always suffice to allow purification while preventing aggregation or denaturation of the protein of interest, so alternatives to the traditional detergents are constantly being designed and tested (Figure 3.2). One of the tripod amphiphiles (Figure 3.2A), which have three hydrophobic tails and a hydrophilic headgroup, has been used to solubilize bacteriorhodopsin. The amphipols (Figure 3.2B) are small amphiphilic polymers of various structures that form water-soluble complexes with membrane proteins, presumably by wrapping around their nonpolar domains. A8-35, a polyacrylate chain with octyl-amine and isopropylamine chains, is an amphipol that has been shown to be effective with numerous membrane proteins. Hemifluorinated surfactants (Figure 3.2C) are unlike detergents because the perfluorinated regions of their chains are hydrophobic but still lipophobic. The nonfluorinated tails enable them to interact with membrane proteins while presumably allowing the interactions between the proteins and lipids to remain intact in aqueous solutions.

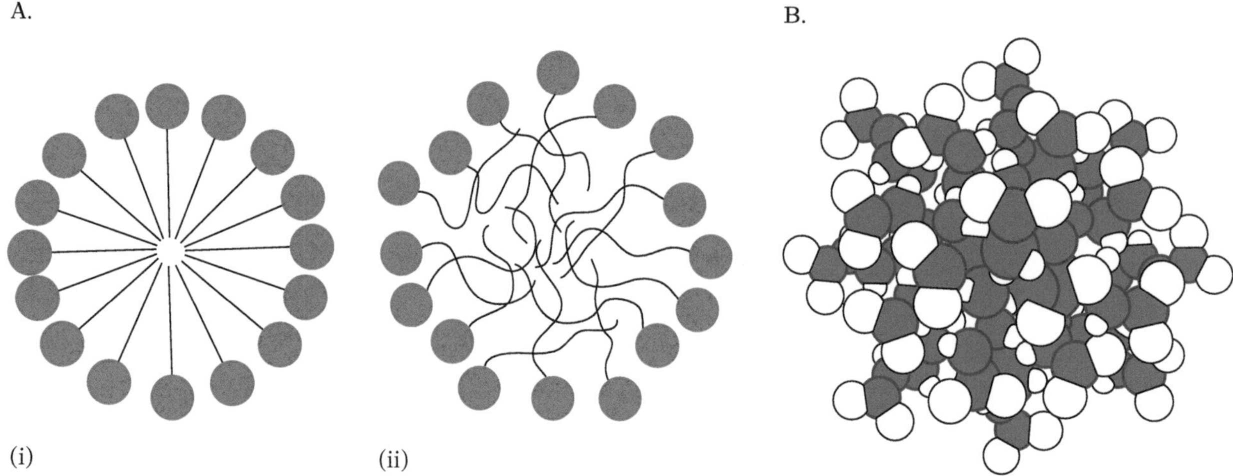

3.3 (A) Cross-sectional views of detergent micelles. The old view (i) incorrectly portrays the chains as ordered like spokes, whereas they are actually disordered and fluid (ii), resulting in an uneven surface. Redrawn with permission from Menger, F. M., *et al.*, *J Chem Educ*. 1998, 75:93, 115. (B) Model of the surface of a micelle, showing the uneven surface at the detergent–water interface. Redrawn from Lindman, B., *et al.*, in J.-J. Delpuech (ed.), *Dynamics of Solutions and Fluid Mixtures by NMR*, Wiley & Sons, 1995, p. 249. © 1995, John Wiley & Sons Ltd.

Perhaps because of the extensive use of DDM, some of the newest types of detergent incorporate maltose derivatives. CHOBIMALT is simply a cholesterol-maltoside (not shown) that enhances thermal stability. Maltose neopentyl glycol (MNG) amphiphiles are based on a central quaternary carbon derived from neopentyl glycol that allows two hydrophilic and two lipophilic units to be linked, providing much conformational flexibility (Figure 3.2D). An advantage of the MNG amphiphiles is their very low critical micellar concentrations (see below), compared to conventional detergents with comparable chain lengths. Their chemistry can be varied to have an amide or ether linkage, as well as different length chains. An alternative approach produces facial amphiphiles by putting the maltosyl groups on one side of cholate (Figure 3.2E). Like the other bile salt derivatives, facial amphiphiles use a different mechanism of detergent solubilization (see Figure 3.8), although their critical micellar concentration (CMC) is similar to that of DDM and much lower than that of cholate, CHAPS, and CHAPSO. Given their advantages, use of the new detergents will facilitate research on membrane proteins.

Mechanism of Detergent Action

The action of most detergents involves micelle formation. Micelles are roughly spherical assemblies of surfactant molecules, in which most of the nonpolar tails are sequestered from the aqueous environment in a disorganized (liquid-like) hydrophobic interior. Thus the chains are not fully extended like the spokes of a wheel, and the radius of the micelle is 10%–30% smaller than the fully extended monomer (Figure 3.3A). Furthermore, the surface is rough and heterogeneous rather than smoothly covered by polar headgroups: NMR studies of SDS micelles revealed that only one-third of the surface was covered by hydrophilic headgroups (Figure 3.3B). At high concentrations of detergent, micelles change shape to become elliptical or rod-like; this occurs at lower concentrations for surfactants with weakly polar headgroups. Micelles of small detergents exhibit even more fluctuations in shape as they can deform, split, and fuse over time.

Micelle formation is a direct consequence of the degree of amphiphilicity of surfactants. The surfactant molecules that form micelles are more water soluble than most lipids but still contain nonpolar groups with a propensity to form hydrophobic domains. They also tend to have conical shapes with bulky headgroups relative to their nonpolar groups (see Figure 2.17). In addition to detergents, lysophospholipids (phospholipids lacking one acyl chain) form micelles, as do PLs with very short acyl chains (e.g., PC with 4–9 carbon chains) under certain conditions.

Self-association of detergents into micelles is strongly cooperative and occurs at a defined concentration called the CMC (Table 3.1). Below the CMC, the amphipath dissolves as monomers; as its concentration increases beyond the CMC, ideally the monomer concentration is unchanged while the concentration of micelles increases (Figure 3.4). The CMC can be detected by measuring surface tension or other aqueous properties, such as conductivity or turbidity (Figure 3.5). Micelle formation is dynamic, allowing constant interchange between constituents of aggregates and soluble monomers. For ionic surfactants, it is strongly affected by ionic strength (see Table 3.2).

Micelle formation is also a function of temperature. The critical micellar temperature (CMT) is defined as the temperature above which micelles form (Figure 3.6). The

TABLE 3.1 PROPERTIES OF MICELLES DERIVED FROM SOME COMMONLY USED DETERGENTS

Detergent	Monomeric MW	Critical micelle concentration (M)	Aggregation number
Octyl-*β*-D-glucoside	292	2.5×10^{-2}	27
Dodecyl-maltoside	528	1.7×10^{-4}	140
Lauryldimethylamine oxide (LDAO)	229	2.2×10^{-3}	75
Lauramido-*N*,*N*-dimethyl-3-*n*-propylamineoxide (LAPAO)	302	3.3×10^{-3}	–
Dodecyl-*N*-betaine (zwittergent 3-12)	336	8×10^{-2}	87
Tetradecyl-*N*-betaine (zwittergent 3-14)	350	6×10^{-3}	130
Myristoylphosphoglycerocholine	486	9×10^{-5}	–
Palmitoylphosphoglycerocholine	500	1×10^{-5}	–
3-[(3-cholamidopropyl)-dimethylammonio]-1-propanesulfonate (CHAPS)	615	5×10^{-3}	–
Deoxycholic acid	393	3×10^{-3}	22
Cholic acid	409	1×10^{-2}	4
Taurodeoxycholic acid	500	1.3×10^{-3}	20
Glycocholic acid	466	–	6
Sodium dodecylsulfate (SDS) in 50 mM NaCl		8×10^{-3}	62
Dodecylammonium Cl^{-2}		15×10^{-3}	55
Ganglioside GM_1		10^{-9}	150
PEG-dodecano		1×10^{-4}	130
Polyoxyethylene glycol detergents			
C_8E_6	394	1×10^{-2}	32
$C_{10}E_6$	422	9×10^{-4}	73
$C_{12}E_6$	450	8.2×10^{-5}	105
$C_{12}E_8$	538	8.7×10^{-5}	120
$C_{12\,\&\,14}\,E_{9.5}$ (Lubrol PX)	620	–	100
$C_{12}E_{12}$	710	9×10^{-5}	80
$C_{12}E_{23}$ (Brij 35)	1200	9×10^{-5}	40
$C_{16\&18}E_{17}$ (Lubrol WX)	1000	–	90
tert-p-$C_8ØE_{9.5}$ (Triton X-100)	1625	3×10^{-4}	140
tert-p-$C_8ØR_{7.8}$ (Triton X-114)	540	2×10^{-4}	–
C_{12} sorbitan E_{20} (Tween 20)	1240	6×10^{-5}	–
$C_{18:1}$ sorbitan E_{20} (Tween 80)	1320	1.2×10^{-5}	60
Cetyltrimethylammonium bromide (CTAB)	364	9.2×10^{-4}	169

MW, molecular weight.
Source: Jain, M. K., and R. C. Wagner, *Introduction to Biological Membranes*, 2nd ed. New York: Wiley, 1988, p. 71.

Krafft point, also called the cloud point, is the temperature at which a turbid solution of surfactant becomes clear due to the formation of micelles.

Thus the Krafft point falls at the intersection of the lines for the CMT and the CMC, and at the Krafft point the temperature dependence of solubility rises steeply as the result of micelle formation. At the Krafft point, insoluble crystalline detergent is in equilibrium with monomers and micelles, so if the temperature is lowered, the detergent crystallizes out of solution. A familiar illustration is the precipitation of SDS in aqueous solutions below 4°C (its Krafft point). The CMT for nonionic surfactants and the common bile salts is below 0°C.

TABLE 3.2 EFFECT OF IONIC STRENGTH ON MICELLE FORMATION BY IONIC SURFACTANTS IN AQUEOUS SOLUTIONS AT 25°C

Surfactant	Medium	CMC (mM)	N	p/N
Anionic				
Sodium *n*-octylsulfate	H_2O	130		
Sodium *n*-decylsulfate	H_2O	33		
Sodium *n*-dodecylsulfate (SDS)	H_2O	8.1	58	0.18
Sodium *n*-dodecylsulfate	0.1 M NaCl	1.4	91	0.12
Sodium *n*-dodecylsulfate	0.2 M NaCl	0.83	105	0.14
Sodium *n*-dodecylsulfate	0.4 M NaCl	0.52	129	0.13
Cationic				
n-Dodecyltrimethylammonium bromide	H_2O	14.8	43	0.17
n-Dodecyltrimethylammonium bromide	0.0175 M NaBr	10.4	71	0.17
n-Dodecyltrimethylammonium bromide	0.05 M NaBr	7.0	76	0.16
n-Dodecyltrimethylammonium bromide	0.10 M NaBr	4.65	78	0.16

N, aggregation number; p, micellar charge.
Source: Jones, M. N., and D. Chapman, *Micelles, Monolayers, and Biomembranes*. New York: Wiley-Liss, 1995, p. 68.

The size of detergent micelles is usually described by the aggregation number (N), the average number of surfactant molecules per micelle, although for some situations the molecular weight or hydrodynamic radius is reported (Table 3.1). The aggregation numbers given in the literature are averages, and the size distribution may be quite large. Micelle size can be determined by light scattering, ultracentrifugation, viscometry, and gel filtration. It varies widely, reflecting the size of the nonpolar domain: N increases with increasing

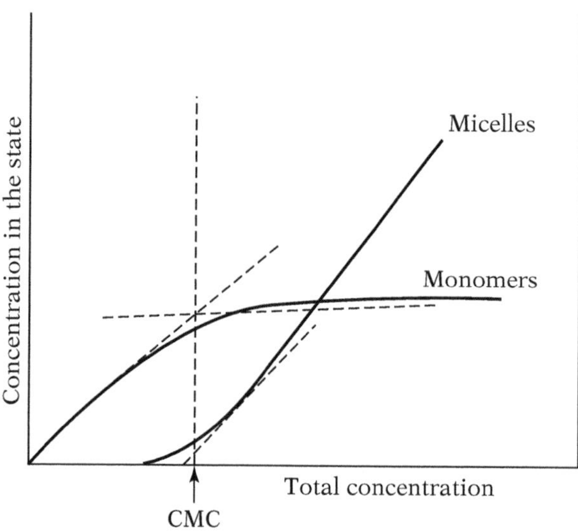

3.4 The critical micellar concentration. As detergent (or surfactant) is added to an aqueous solvent, the concentration of dissolved monomers increases until the critical micellar concentration (CMC) is reached. At that concentration, micelles form. Further addition of detergent increases the concentration of micelles without appreciably affecting the concentration of monomers. Redrawn with permission from Helenius, A., and K. Simons, *Biochim Biophys Acta*. 1975, 415:38.

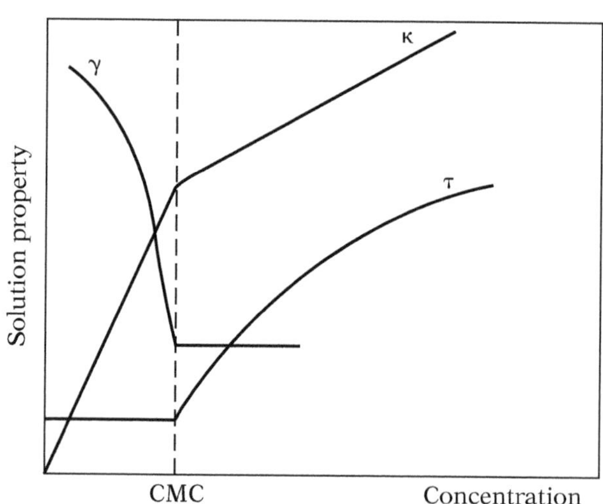

3.5 Variation in surface tension (γ), specific conductivity (κ), and turbidity (τ) as a function of detergent concentration. The schematic plots show the dependence on concentration of detergent (surfactant) in solution of properties commonly used to find the CMC. (Note that conductivity only applies to ionic surfactants.) At the CMC, denoted by the dashed line, there is a break in the line for each property. Redrawn from Jones, M. N., and D. Chapman, *Micelles, Monolayers and Biomembranes*, Wiley-Liss, 1995, p. 65. © 1995. Reprinted with permission from John Wiley & Sons, Inc.

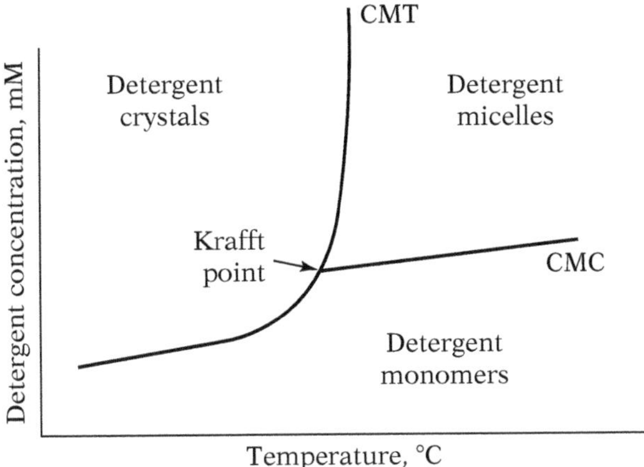

3.6 Detergent phase diagram. At temperatures below the Krafft point, the detergent exists as monomers at very low concentrations and insoluble crystals at higher concentrations. Raising the temperature increases the monomer concentration until the critical micellar temperature (CMT) is reached, when micelles form. At (and above) that temperature, the solution clears because the only two phases present are micelles and monomers. The Krafft point falls at the intersection of the lines for the CMT and the CMC, where the temperature dependence of solubility rises steeply due to micelle formation. Redrawn from Helenius, A., and K. Simons, *Biochim Biophys Acta*. 1975, 415:37. © 1975 by Elsevier. Reprinted with permission from Elsevier.

tail length for a series of surfactants in which only the hydrocarbon chain length is varied. For ionic surfactants, N is strongly dependent on the ionic strength of the aqueous medium (see Table 3.2), as well as the kind of counterions available to shield the charged headgroups.

Membrane Solubilization

Detergents are used to extract membrane lipids and proteins into an aqueous suspension. When a low concentration of detergent is added to a membrane, the detergent molecules intercalate into the bilayer. When a higher concentration is added, the detergent disrupts the bilayer and forms mixed micelles containing lipid, protein, and detergent (Figure 3.7). Mixed micelles vary considerably in structure and size. The detergent concentration must be kept above its CMC to maintain the mixed micelles. Sometimes adding still higher concentrations displaces the lipid completely, producing detergent–protein complexes free of lipid. Thus both the detergent concentration and the detergent-to-protein ratio are important variables that influence how a particular membrane protein will be extracted from the membrane. The behavior of the membrane protein in further purification and characterization steps will depend on detergent–protein and detergent–detergent interactions,

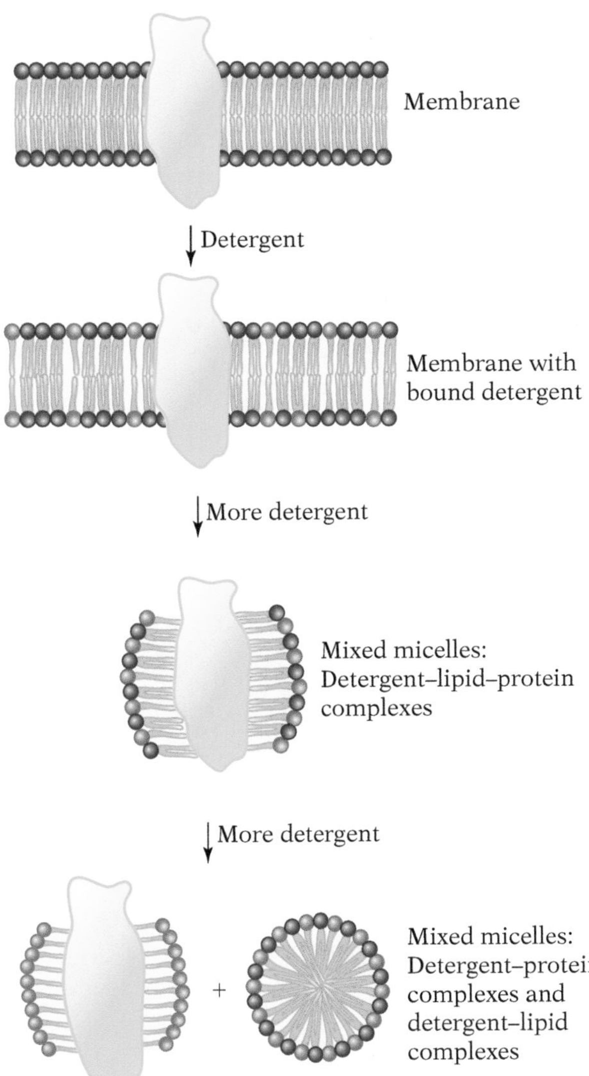

3.7 The stages in membrane solubilization. This schematic illustration follows the addition of increasing amounts of detergent to a membrane. Initially, integral membrane proteins are embedded in the lipid bilayer. At low concentrations of detergent, some detergent molecules penetrate the bilayer but do not disrupt it. As more detergent is added, disruption of the bilayer results in mixed micelles containing detergent, lipid and protein. At even higher detergent concentrations, most of the lipid is removed from the protein, producing detergent–protein complexes, along with detergent–lipid complexes.

along with detergent–lipid and lipid–protein interactions if lipid remains.

The amount of a particular detergent that solubilizes the membrane is roughly proportional to its CMC. Bile-type detergents solubilize segments of the membrane as detergent/bilayer sandwiches (Figure 3.8). The success of an extraction procedure is determined by checking the amount and composition of the desired component (usually protein) in the supernatant following sedimentation

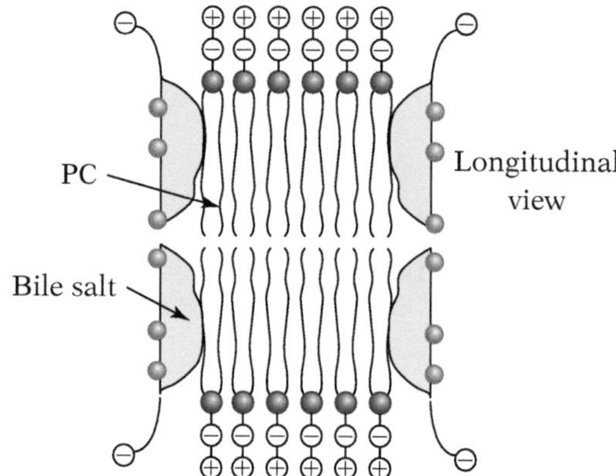

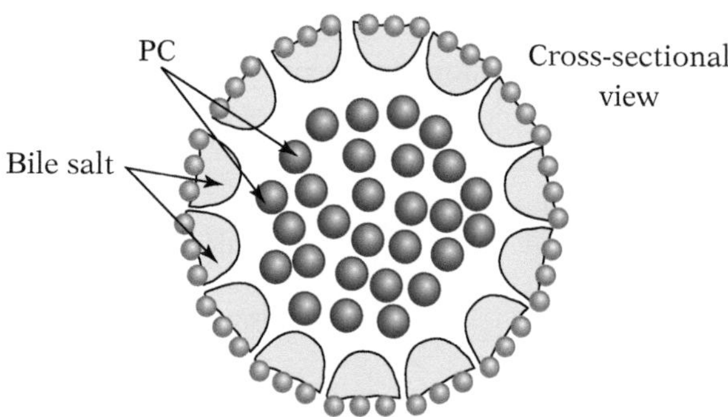

3.8 Schematic illustration of the structure of mixed micelles of bile salts and phospholipid sandwiches of bile salt detergent with lipids. Redrawn from Jones, M. N., and D. Chapman, *Micelles, Monolayers and Biomembranes*, Wiley-Liss, 1995, p. 97. © 1980, American Chemical Society.

of the membrane. The ratio of phospholipid to protein solubilized can indicate whether proteins leak from the membrane before it is completely disrupted, revealing whether a detergent concentration is sufficient to disrupt the membrane or only to solubilize segments of it (Figure 3.9).

For reconstitution experiments, it is often desirable to replace the solubilizing detergent with lipids to better imitate biological conditions. Methods for detergent removal include dialysis, gel filtration, adsorption to polystyrene beads, and pH changes. Detergents that have a low CMC, like Triton X-100 (CMC = 0.24 mM), are much more difficult to remove by dialysis than detergents like OG (CMC = 25 mM) because so little of the detergent is present as monomers. Gel filtration is most effective when there is a large difference in the sizes of the detergent micelle and the detergent–protein mixed micelle, and thus works best with detergents with small values of N (and high CMCs). Adsorption to polystyrene beads (e.g., SM_2 Bio-Beads) is effective for most detergents, including Triton X-100, OG, DDM, cholate, CHAPS, and $C_{12}E_8$. Of course, it is a problem if the protein of interest also adsorbs to the beads, in which case the beads can be placed outside a dialysis chamber. Some ionic detergents, such as cholate and deoxycholate, precipitate at about one pH unit below their pK_as (5.2 and 6.2, respectively), which greatly simplifies their removal provided the mildly acidic conditions do not harm the protein being studied. SDS can be precipitated after exchange of sodium for potassium, as potassium dodecylsulfate is insoluble at room temperature. For some hardy proteins (such as bacterial porins), complete removal of detergent is effected by precipitating the protein with organic solvents.

Although detergents have been widely used in purifying proteins for crystallization studies, their disorder prevents their resolution in the resulting high-resolution structures obtained by x-ray diffraction. To visualize the detergent in protein–detergent complexes, low-resolution images of the structure of detergent domains in crystals can be obtained by neutron diffraction with H_2O/D_2O contrast variation (see Box 8.1 on neutron diffraction). In this procedure, several crystals are prepared that vary in their H_2O/D_2O content, giving relative contrasts to the protein and detergent with respect to the solvent to allow visualization of individual components. Discrete belts of detergent around the nonpolar portion of the protein are clearly detected by neutron diffraction studies of membrane proteins such as the β-barrel OmpF porin (Figure 3.10). In the case of OmpF

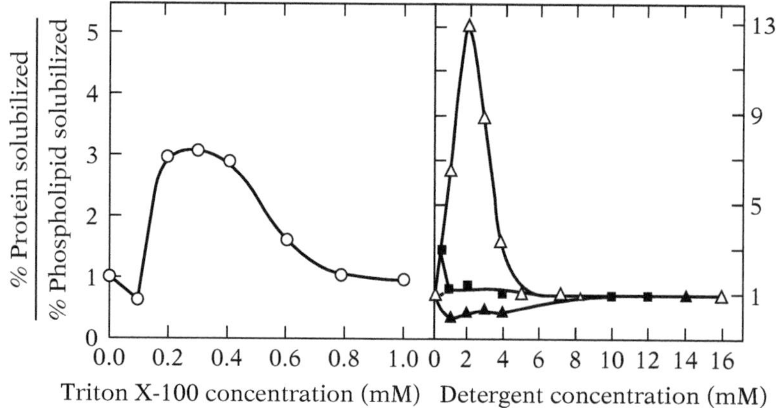

3.9 Ratio of protein to phospholipid solubilized from epithelial cells by four different detergents. A spike at the low detergent concentrations indicates protein leaking out before the lipid is solubilized. ○, Triton X-100; ■, sodium dodecylsulfate; Δ, dodecyltrimethylammonium bromide; ▲, sodium cholate. Redrawn from Jones, M. N., and D. Chapman, *Micelles, Monolayers and Biomembranes*, Wiley-Liss, 1995, p. 148. © 1991 by Elsevier. Reprinted with permission from Elsevier.

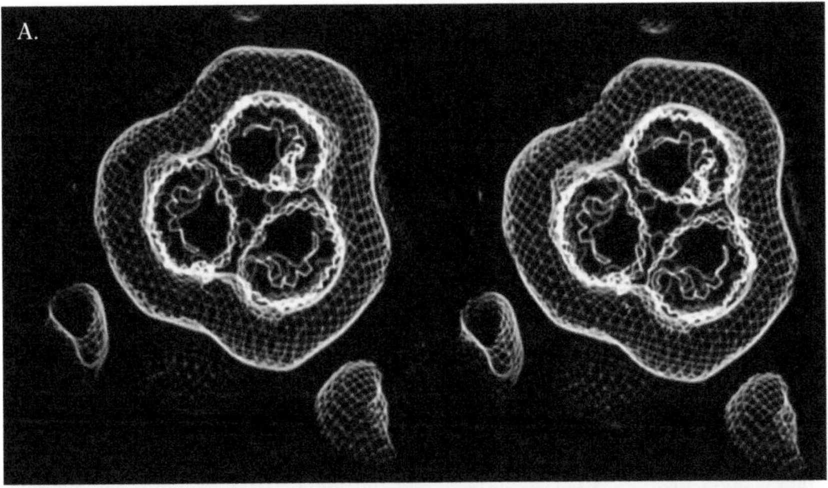

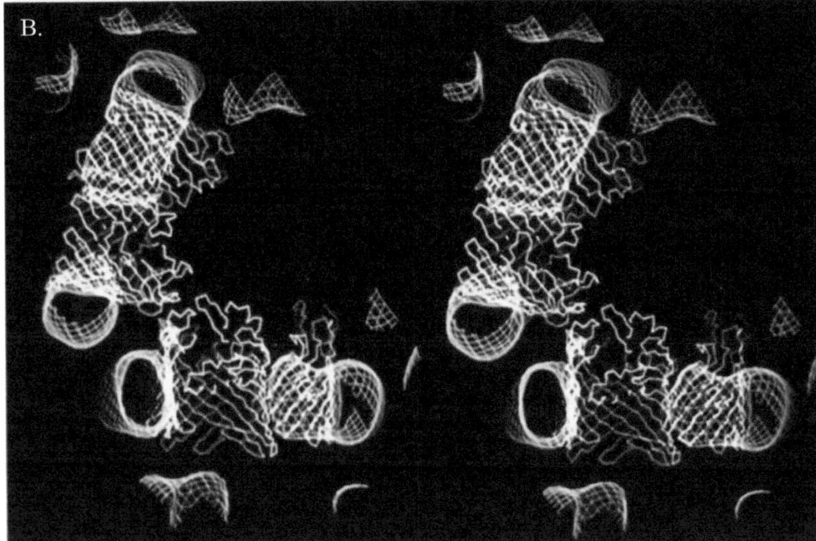

3.10 Belts of detergents around purified membrane proteins. The positions of detergent molecules in solutions of detergent-solubilized proteins are revealed in neutron diffraction density maps obtained at different H_2O/D_2O ratios to provide contrast variation. This image obtained with OmpF porin in C_{10}DAO also shows Cα traces for protein obtained from the x-ray structure (protein is pink and detergent is green). From Pebay-Peyroula, E., *et al.*, *Structure*. 1995, 3:1053. © 1995 by Elsevier. Reprinted with permission from Elsevier.

protein, the "hardness" of the detergent torus affected the observed shape: With softer detergents, such as OG, the belts of detergents fuse with those of their nearest neighbors. Clearly the size and shape of the detergent molecules – along with smaller additives, such as heptane – are crucial to the success of the crystallization process (see Chapter 8), which has led to much interest in new detergents.

Lipid Removal

Although it is rare for detergent extraction to completely remove bound lipid from membrane proteins, lipid removal may lead to loss of biological activity. There are dozens of examples of proteins (including succinate dehydrogenase and other components of the electron transport chain, several ATPases, numerous transferases, and receptors) that are inactivated when stripped of lipid by detergent or organic solvent and are reactivated by addition of lipid (see Table 4.2 for more examples). The lipid requirement may be quite specific, such as the absolute requirement of β-hydroxybutyrate dehydrogenase extracted from mitochondrial inner membrane for PC. Even for the cases of a general lipid requirement, it is clear that a portion of the lipids in a biomembrane associates dynamically with membrane proteins. This boundary lipid, which differs in mobility from the bulk lipid of the bilayer, may be functionally important and can be studied in model membranes.

MODEL MEMBRANES

The functions of the membrane and of many of its components are lost upon its disruption, necessitating reconstitution of the membrane in an *in vitro* model system for studies to elucidate mechanisms of transport and energy transduction, to measure enzyme kinetics and ion flows, and to explore phase changes and microdomains.

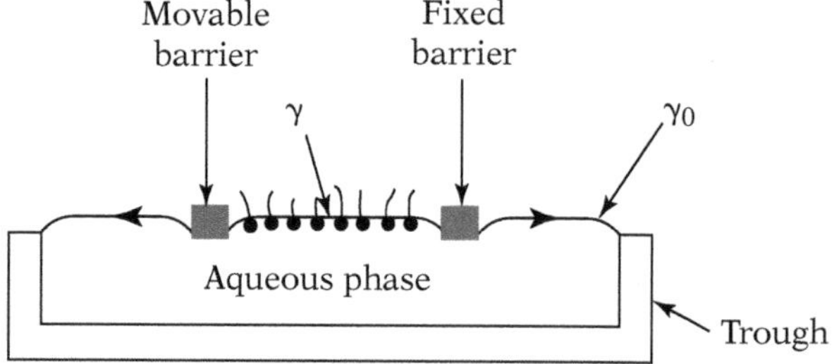

3.11 Formation of a monolayer in the Langmuir trough. The movable barrier allows the area covered by the monolayer to vary. The surface tension, γ, between the two barriers relative to the aqueous surface tension, γ_0, gives rise to surface pressure on the barriers. Redrawn from Jones, M. N., and D. Chapman, *Micelles, Monolayers and Biomembranes*, Wiley-Liss, 1995 p. 26. © 1995. Reprinted with permission from John Wiley & Sons, Inc.

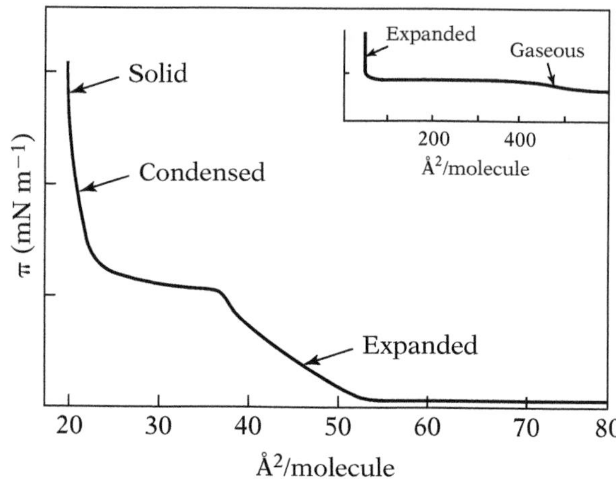

3.12 A surface pressure-versus-area isotherm for a monolayer of a fatty acid at the air–water interface. A diagram of the surface pressure (π) versus the area per molecule shows the relationship between π and the area of the molecules forming the monolayer. The isotherm reveals phase changes: The monolayer approaches a solid state at high pressure, changes to liquid states at lower pressures, and changes to gaseous states at very low pressures, as shown in the inset. Redrawn from Jones, M. N., and D. Chapman, *Micelles, Monolayers and Biomembranes*, Wiley-Liss, 1995, p. 27. © 1995. Reprinted with permission from John Wiley & Sons, Inc.

The availability of a wide variety of model membrane systems is fortunate as no single system is suitable for all the membrane components being characterized or for all the techniques used to study them. Hundreds of papers describe the applications of each classical system, while the promise of the newest membrane mimetics is just begining to be realized.

Monolayers

Amphipathic lipid molecules with sufficiently large hydrophobic portions will line up at an air–water interface with their hydrophobic tails in the air. Such monolayers are commonly formed in a Langmuir trough, a container with a movable barrier on one side that allows control of the area and measurement of the pressure of the monolayer (Figure 3.11). The surface pressure (π) is created by the difference between the surface tension of the monolayer (γ) and that of the air–water interface (γ_o): $\pi = \gamma_o - \gamma$. The high surface tension of water means that it takes work to cover the area of the air–water interface. To decrease that work, the monolayer spreads over the surface, putting pressure on the movable barrier. Inward movement of the barrier to decrease the area increases the surface pressure of the monolayer.

In forming monolayers, the composition is controlled and the amount of lipid is known. A surface pressure-versus-area isotherm, showing the relationship between π and the area of the molecules forming the monolayer, reflects phase changes (Figure 3.12). Highly compressed molecules are so condensed they approach a solid state, whereas at very low pressures the molecules are so spread out they do not interact and are considered to be in a two-dimensional gaseous state. Between these extremes, the monolayer is in a fluid phase described as liquid. The effect of chain length on the phase of the monolayer is revealed in the pressure–area isotherms for a series of monolayers composed of PC with varying acyl chains (Figure 3.13A). Two fluid phases, called L_E (liquid expanded) and L_C (liquid condensed), are evident in the curve for DPPC, but only when the temperature is held at values between 15°C and 30°C (Figure 3.13B) indicating the L_E/L_C transition is a function of the state of the acyl chains.

Monolayers provide a means to study the effects on lipids of factors such as pH, ionic strength, and addition of multivalent versus monovalent ions. Monolayers have also been used to examine the intercalation of membrane-active peptides, such as gramicidins (see "Proteins and Peptides That Insert into the Membrane" in Chapter 4) and signal peptides. In spite of their obvious limitations as membrane mimics, monolayers can give useful physiological information, such as an understanding of respiratory distress syndrome in newborns. Pulmonary surfactant is a complex of proteins and phospholipids that coats the surface of alveoli and small bronchioles to reduce surface tension and confer compressibility to lungs (Figure 3.14). The lipid is predominantly DPPC with a significant fraction of PG. Monolayer studies indicate that compression squeezes out the PG and the remaining PC-rich surfactant in the L_C state is less resilient in responding to pressure. Thus the lack of PG in the surfactant of premature babies can trigger alveoli collapse, resulting in respiratory failure.

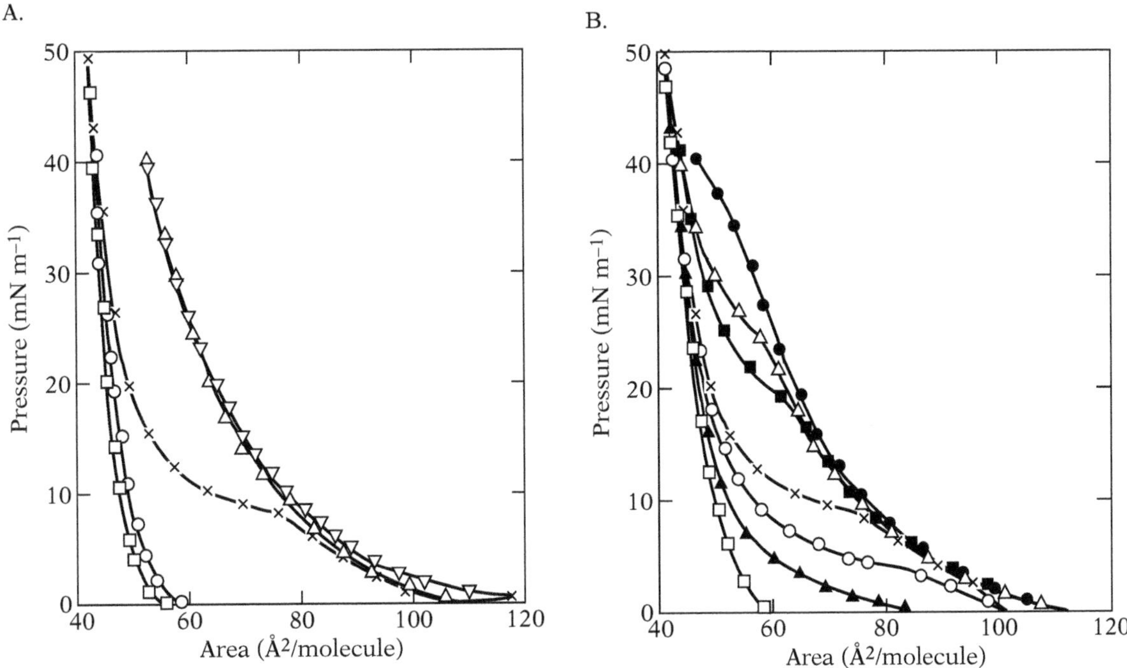

3.13 Phospholipid phase changes in monolayers. Surface pressure-versus-area isotherms reveal the effects of chain length and temperature on lipid monolayers. (A) When monolayers are made of phosphatidylcholine with saturated acyl chains of varying lengths (□, dibehenoyl [C22]; o, distearoyl [C18]; x, dipalmitoyl [C16]; Δ, dimyristoyl [C14]; and ∇, dicapryloyl [C10]) phospholipid behavior in the monolayers correlates with their melting temperatures. The PLs with longer chains and higher melting temperatures form liquid condensed (L_C) monolayers at 22°C, while the PLs with shorter chains and lower melting temperatures form liquid extended (L_E) monolayers at 22°C. (B) For a particular PL, increased temperature changes the monolayer state from L_C to L_E, shown in surface pressure-versus-area isotherms for DPPC in 0.1 M NaCl at varying temperatures. (•, 34.6°C; Δ, 29.5°C; ■, 26.0°C; x, 21.1°C; o, 16.8°C; ▲, 12.4°C; □, 6.2°C.) Redrawn from Phillips, M. C., and D. Chapman, *Biochim Biophys Acta*. 1968, 163:301. © 1968 by Elsevier. Reprinted with permission from Elsevier.

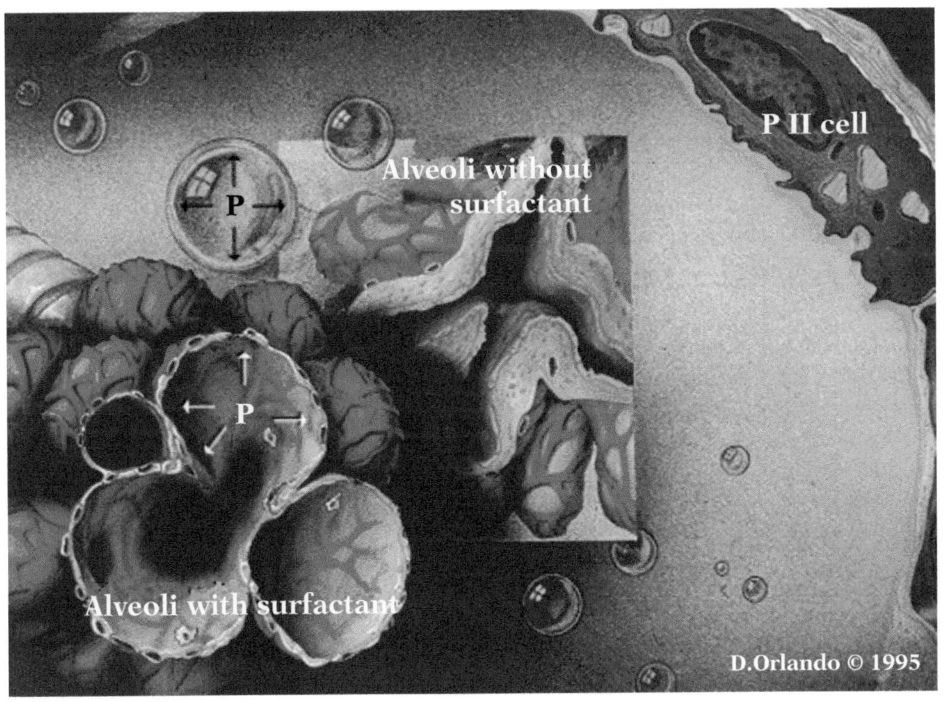

3.14 Alveoli with and without pulmonary surfactant. The presence of natural surfactants enables the alveoli to withstand changes in pressure (P) in the lungs. Without surfactant the alveoli cannot expand due to high surface tension. From Johns Hopkins School of Medicine Interactive Respiratory Physiology, http://oac.med.jhmi.edu/res_phys/Encyclopedia/Surfactant/Surfactant.html © 1995 by Daphne Orlando. Reprinted by permission of Daphne Orlando.

Example (Briggs, M. S., and L. M. Gierasch, Exploring the conformational roles of signal sequences: synthesis and conformational analysis of lambda receptor protein wildtype and mutant signal peptides. Biochemistry. 1984, **23**:3111–3114; Briggs, M. S., *et al.*, Conformations of signal peptides induced by lipids suggest initial steps in protein export. *Science*. 1986, 233:206–20.)

Monolayers were used to determine the conformation of signal peptides as they interacted with membrane lipids. Signal peptides are the N-terminal extensions of newly synthesized proteins targeted for secretion from the cytoplasm (see Chapter 7). Their internal sequence of about 10–15 nonpolar amino acids suggested they could insert into the membrane. Synthetic signal peptides were found to lay on the surface of a monolayer with high surface pressure and to insert into the monolayer when the pressure was lowered. Circular dichroism revealed the inserted peptides have an α-helical structure (Figure 3.15). This early evidence for insertion of α-helices into the membrane was given functional significance by the much lower tendency of mutant signal peptides to form α-helices in monolayers.

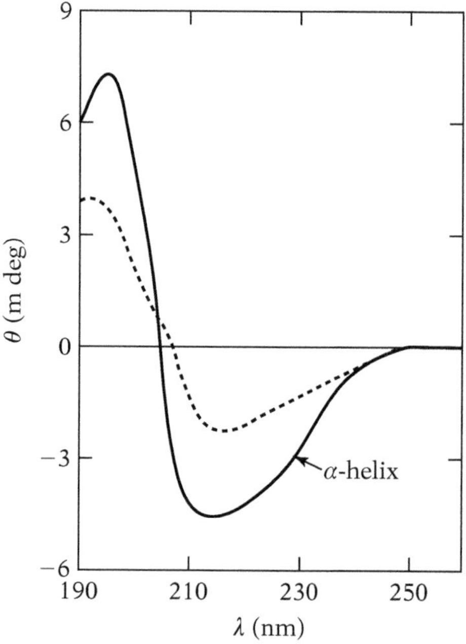

3.15 Circular dichroism spectra for synthetic signal peptide interacting with a monolayer of POPE:POPG in a 2:1 ratio at high pressure (dashed line) or low pressure (solid line). θ, molar ellipticity; λ, wavelength. Redrawn from Briggs, M. S., *et al.*, *Science*. 1986, 233:206–208. © 1986, American Association for Advancement of Science. Reprinted with permission from AAAS.

Planar Bilayers

While a monolayer reveals pressure effects of membrane constituents, a planar bilayer separating two aqueous compartments is used to study electrical properties because it offers access for electrodes on both sides of the membrane. Pure lipid bilayers are not permeable to ions (which is why the myelin membrane is a good insulator), so the introduction of molecules that form ion channels can be closely monitored. These systems mimic the electrophysiological aspects of the cell membrane with its electrochemical potential and numerous ion channels, and they have been the technique of choice for biophysical studies of ion channels (see Box 3.2).

Historically, planar bilayers were called "black films" because they appear black when made on a Teflon sheet. The lipid is dissolved in organic solvent (such as hexane, decane, or hexadecane) and is painted on a tiny orifice (about 1 mm in diameter) in the Teflon sheet, which is then inserted between two aqueous compartments. As the solvent dissipates and the lipid gradually drains or "thins," a region of single bilayer stretches across part of the orifice and can last for a few hours (Figure 3.16). Peptides, small proteins, and other lipids will diffuse into the bilayer when added to one of the aqueous compartments, although the incorporation of larger proteins may be problematic. By inserting electrodes in the two compartments, a voltage may be applied across the bilayer and

BOX 3.2 ELECTROPHYSIOLOGY

Current (I) measures the flow of charge from one place to another in the units amperes (amps). The flow of charge responds to a potential difference (V, in volts), requires a conducting path, and obeys Ohm's law: $V = IR$, where R is the resistance of the path, with the units ohms. The conductance is the reciprocal of the resistance, giving the ease of the flow of current, and is measured in siemens.

An electrochemical gradient exists across any membrane that separates two compartments containing different amounts and kinds of ions. The potential difference between the compartments drives the movement of ions, provided there is a pathway such as a pore or ion channel. Many excitable cells (i.e., nerve, sensory, and muscle cells) have a potential across the plasma membrane in the resting state of around –60 mV, which creates an enormous electric field across the membrane (calculated to be 200,000 V cm^{-1} for a membrane that is 3 nm thick). In *E. coli*, the electric potential measured across the cytoplasmic membrane is around –95 mV. Many *in vitro* measurements utilize application of a "voltage clamp" in which the experimenter applies a voltage to control the potential across the membrane and then measures the current.

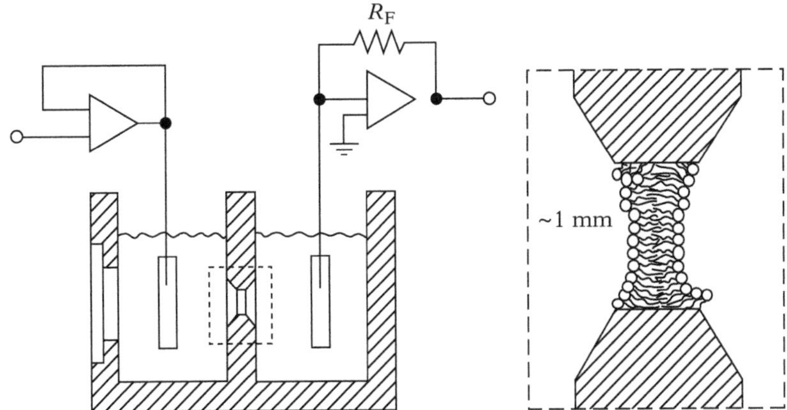

3.16 Schematic diagram of a black film separating two aqueous compartments with electrodes. Inset shows the planar bilayer region of the film. Redrawn from Jain, M. K., *Introduction to Biological Membranes*, John Wiley, 1988, p. 95. © 1998. Reprinted with permission from John Wiley & Sons, Inc.

current may be measured. This system permits precise measurements of ion flows, such as those observed when OmpF porin inserts into the black film, and even detects the closing of single channels (Figure 3.17). However, the electrical capacity of a black film is considerably lower than that of cell membranes, implying there are structural differences, which have been attributed to the presence of excess solvent.

Montal and Mueller introduced a significant new method for the formation of a planar lipid bilayer by apposing two monolayers derived from air–water interfaces (Figure 3.18). The organic solvent introduced with the lipid is evaporated from each monolayer prior to forming the bilayer, and the electrical capacity of the bilayer thus formed matches that of biomembranes, supporting the claim that the bilayer is

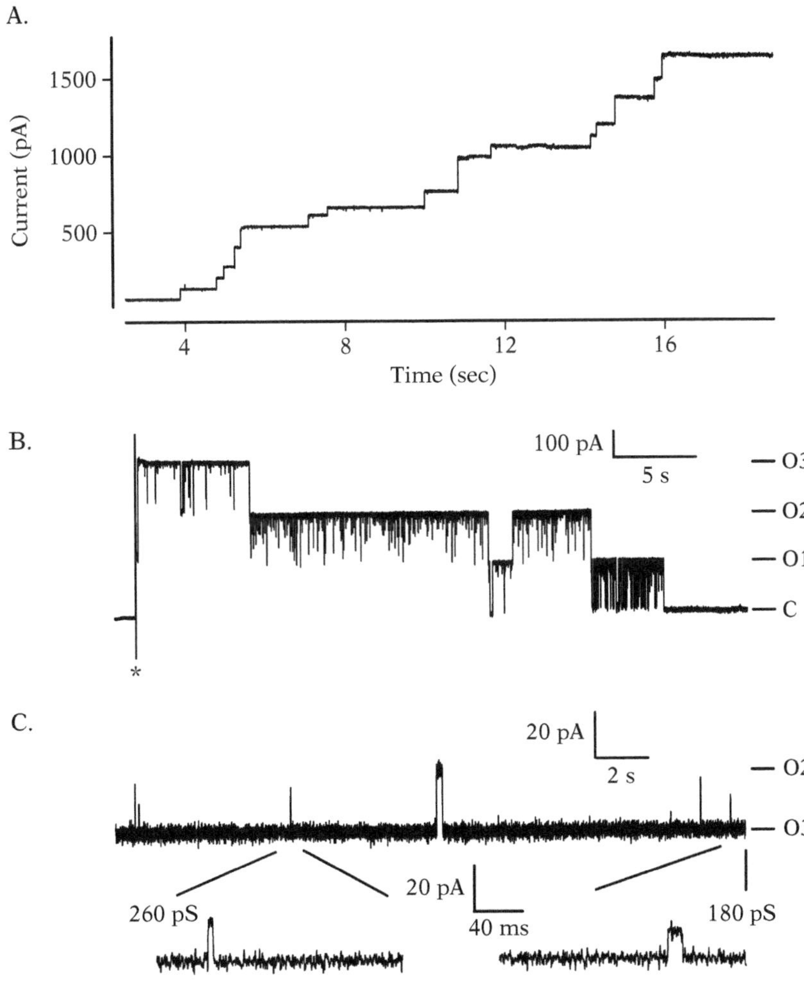

3.17 Conductance data obtained with pure OmpF porin added to the *cis* side of black films. OmpF is a channel-forming protein described in detail in Chapter 5. **(A)** When OmpF inserts into a planar lipid bilayer at a membrane potential of +50 mV, a stepwise current increase results from sequential insertion events of open porin trimers (pA, picoamperes). **(B)** After insertion of a single trimer is complete, channel closures occur in response to a membrane potential of +140 mV. These closures provide evidence of the voltage gating of OmpF channels and typically occur in three steps believed to correspond to the three channels in OmpF trimers. **(C)** Small fluctuations occur at −70 mV, indicating fast flickering events involving subconductance states (monomeric conductance of 380 picosiemens [pS] marked by tick marks). From Basle, A., *et al.*, *Biochim Biophys Acta*. 2004, 1664: 100–107. © 2004 by Elsevier. Reprinted with permission from Elsevier.

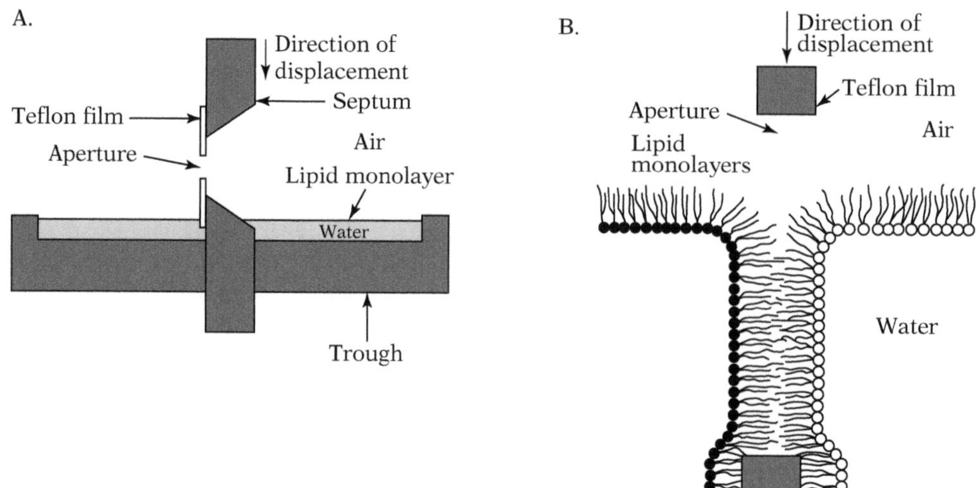

3.18 Technique introduced by Montal and Mueller for making a planar lipid bilayer by apposing two monolayers. (A) The apparatus has two compartments containing solutions into which electrodes can be inserted. The compartments are separated by a movable partition called a septum, which has a Teflon film with a tiny aperture. After monolayers are formed at the air–water interfaces of the two compartments, the septum is moved down, allowing the two monolayers to come together to form a bilayer across the aperture. M. Montal and P. Mueller, Formation of Biomolecular Membranes from Lipid Monolayers and a Study of their Electrical Properties, *Proc Natl Acad Sci USA* 1972; 69:3561–3566. © 1972 National Academy of Sciences, U.S.A. (B) A close-up view of the bilayer formed across the aperture in the Montal–Mueller apparatus, with polar headgroups from one monolayer shaded to indicate the asymmetry achieved in the bilayer. Redrawn with permission from Jain, M. K., *Introduction to Biological Membranes*, John Wiley, 1988, p. 95. © 1998. Reprinted with permission from John Wiley & Sons, Inc.

Example (Armah, C. N., *et al.*, The membrane-permeabilizing effect of avenacin A-1 involves the reorganization of bilayer cholesterol. *Biophys J.* 1999, **76**:281–290).

To investigate hemolytic pore formation by saponins such as digitonin and the avenacins (a group of fungicidal steroid glycoside plant natural products), Montal–Mueller planar lipid bilayers were formed in an optical chamber that allowed simultaneous measurements of both conductance and fluorescence. The data show that an increase in conductivity of bilayers coincides with a decrease in the lateral mobility of NBD-cholesterol determined by FRAP (Figure 3.19). Monolayer studies showed that although insertion of the saponin into the bilayer does not require cholesterol (not shown), pore formation occurs only in the presence of cholesterol. The results indicate that saponin–cholesterol complexes form pores in the bilayer, consistent with their biological activity.

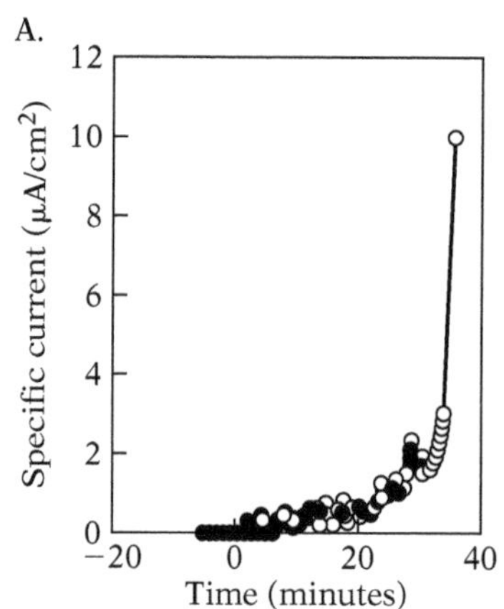

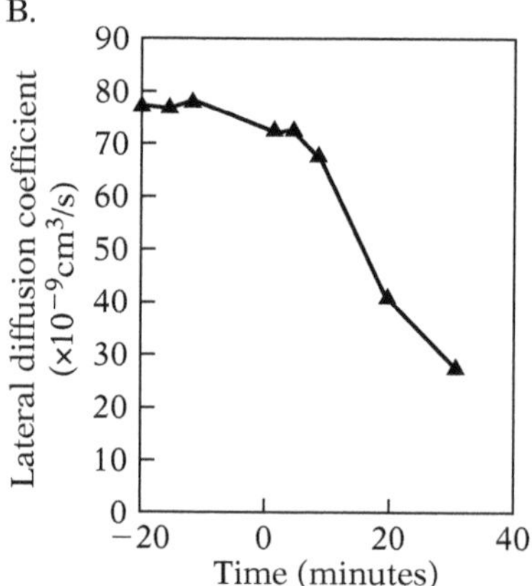

3.19 Effect of saponin treatment on conductivity and on lateral diffusion in Montal–Mueller bilayers consisting of POPC:DOPE:cholesterol (7:3:10) containing 1 mol % NBD-cholesterol. Avenacin A-1 (1.0 μM) was added at time zero. (A) Conductivity increases after ~20 minutes, which is not observed without addition of a saponin (data not shown). (B) The decrease of the rate of lateral diffusion of NBD-cholesterol observed by FRAP corresponds with the increase of conductivity. Redrawn from Armah, C. N., *et al.*, *Biophys J.* 1999, 76:281–290. © 1999, Elsevier.

solvent-free. One clear advantage of the Montal–Mueller system is that it allows independent manipulation of the lipid contents of the two leaflets. Schindler introduced a further modification by spreading the monolayers from preformed lipid vesicles, making it easier to incorporate proteins and to direct their introduction from one leaflet specifically. Recent advances in the design of the compartment to allow both optical and electrical measurements have enabled the determination of electrical properties to be correlated with other physical characteristics of the bilayer, such as fluidity.

Patch Clamps

Although technological improvements have enhanced their sensitivity, traditional planar bilayers have exhibited a high level of noise, making it difficult to detect currents flowing through single channels whose amplitude is very small (typically 1 or 2 picoamperes [pA]). A crucial advance that allowed detection of single ion channels in cell membranes was recognized by the award of the Nobel Prize in Physiology or Medicine to Erwin Neher and Bert Sakmann in 1991. Their patch clamp technique accesses a tiny portion of a membrane in a way that allows electrical measurements to be made and has now been used with a variety of membrane systems. To clamp a patch of membrane bilayer, a portion is sucked into a polished tip of a glass pipette, sealing off an area of 1–6 μm^2. The pipette can remain attached or, if the seal is sufficiently stable, the pipette may be withdrawn to excise the patch of membrane (Figure 3.20). With a gigaohm seal (having an electrical resistance higher than 10^9 ohms), the patch clamp allows detection of current down to the level of picoamperes (10^{-12} amperes). Because of its sensitivity and small size, the conductance behavior of a small number of channels may be monitored through the patch, revealing single-channel opening and closing (Figure 3.21).

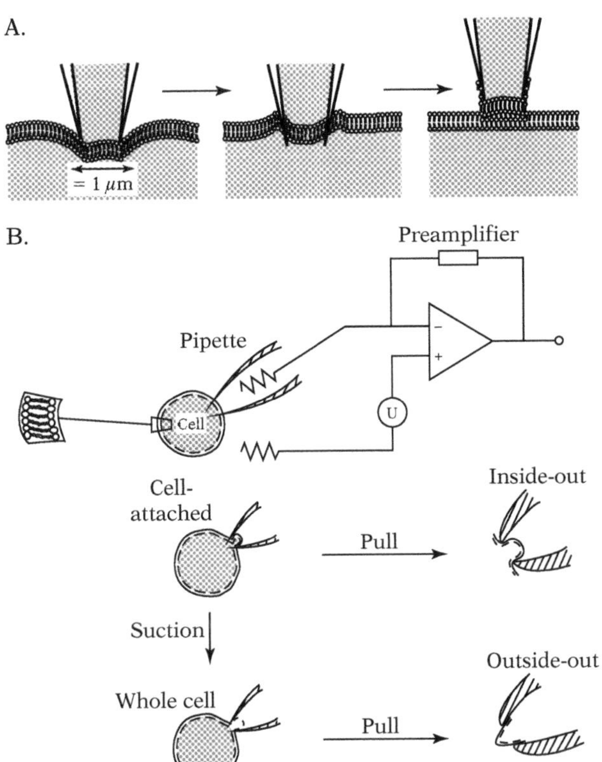

3.20 The patch clamp method. (A) Excision of a portion of bilayer by a patch clamp. A clean fire-polished pipette is pressed against the cell membrane to form a gigaohm seal. As it is withdrawn, the membrane reseals. Conductance can be measured across the patch on the cell surface or across the excised patch. (B) Methods for patch recording with cell-attached patch or excised patches in the inside-out and outside-out configurations. The cell membrane is represented by a solid line with a dotted line, and the dotted line indicates the inner surface of the membrane in the two patch configurations. Redrawn with permission from Jain, M. K., *Introduction to Biological Membranes*, John Wiley, 1988. © 1998. Reprinted with permission from John Wiley & Sons, Inc.

Supported Bilayers

Planar lipid bilayers that sit on glass, quartz (mica), or gold supports allow direct observation of their surface using atomic force microscopy (AFM), immunolabeling, fluorescence, and other spectroscopic techniques. They can be made by fusion of lipid vesicles of the desired composition on the surface of the support in an aqueous environment or by sequential deposition of monolayers,

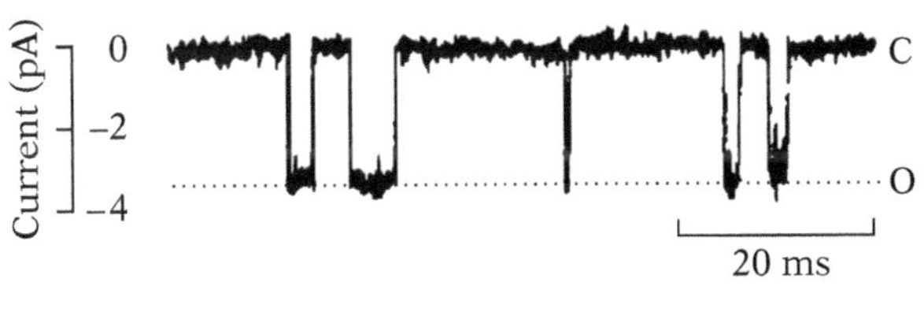

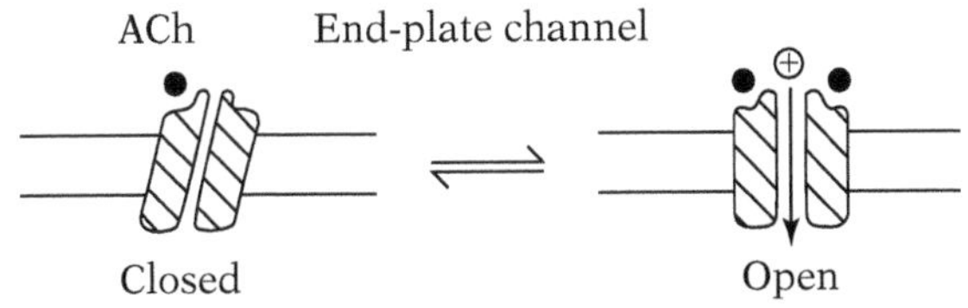

3.21 Early single-channel recording by Neher and Sakmann showing the current through rat muscle activated by acetylcholine (ACh). The diagram under the tracing illustrates how the channel opens reversibly in response to ACh binding. Redrawn from the Nobel lecture by Bert Sakmann, December 9, 1992. © The Nobel Foundation 1991.

Example (Neher, E., and B. Sakmann, Single-channel currents recorded from membrane of denervated frog muscle fibres. *Nature*. 1976, **260**:799–802, and Nobel lectures.)

The first patch clamp studies were performed on denervated frog and rat muscle and measured ion flow through individual acetylcholine receptors (AChRs) of the neuromuscular junction (see Chapter 6). The recorded currents indicated that the acetylcholine-activated channels exist in only two conductance states, open and closed, and showed that bursts of current through open channels result when acetylcholine binds. The observation of two classes of currents, which differ in amplitude and duration, enabled detection of two isoforms of the receptor, now known to be fetal and adult AChRs (Figure 3.22).

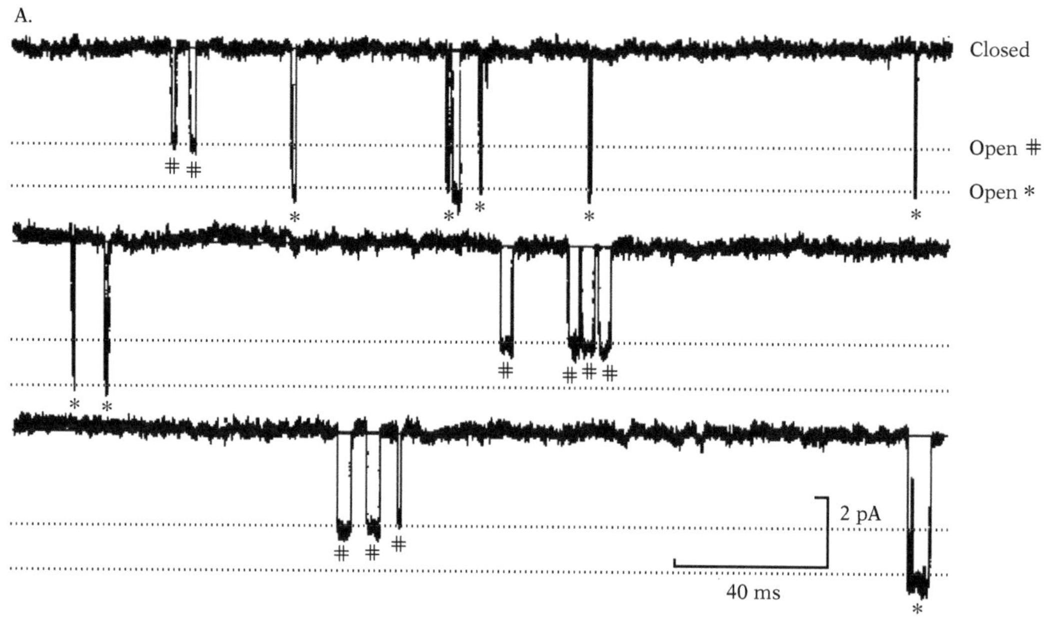

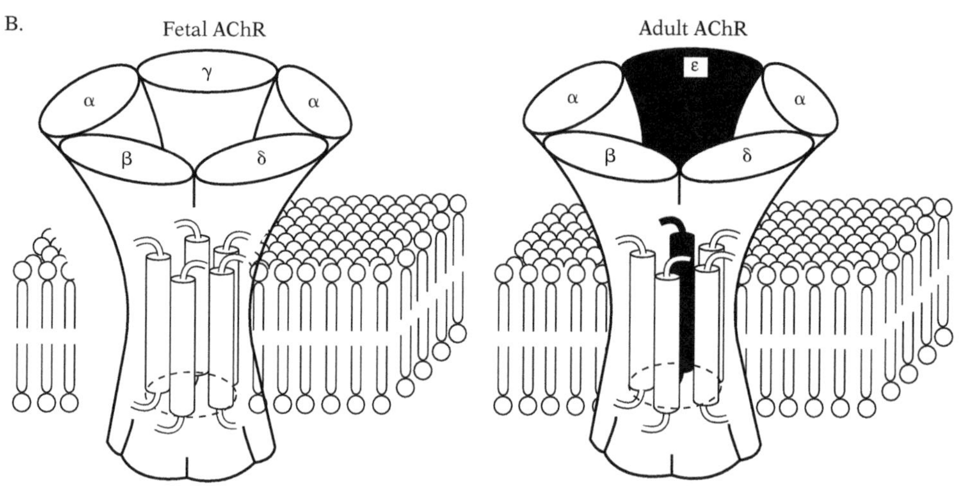

3.22 Characterization of two forms of the acetylcholine receptor. (A) Patch clamps observed in postnatal rat muscle fiber treated with 0.5 μM ACh revealed two classes of currents through the acetylcholine receptor (AChR; marked with * and #) that differ in amplitude and duration. (B) The two types of AChR responsible for different classes of currents have been shown to be fetal and adult isoforms, which differ in subunit composition as shown. Redrawn from the Nobel lecture by Bert Sakmann, December 9, 1992. © The Nobel Foundation, 1991.

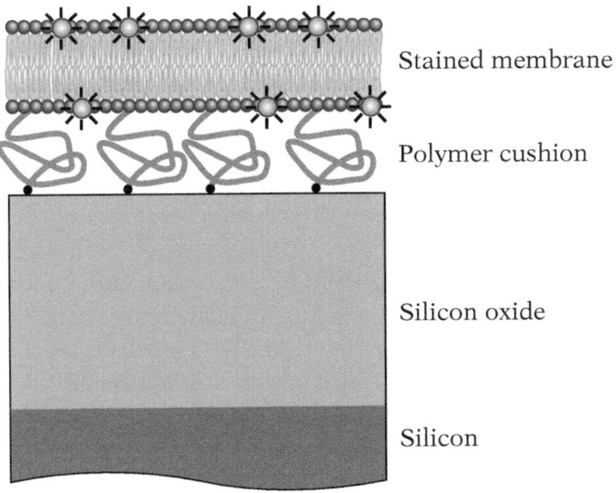

3.23 Diagram of a tethered polymer-supported planar bilayer, representing a 3400-Da polyethylene glycol covalently attached to a silicon oxide surface and at the other end to a phospholipid, which doubles the distance from the bilayer to the support. From Kiessling, V., and L. K. Tamm, *Biophys J.* 2003, 84:408–418. © 2003, Elsevier.

which allows asymmetric bilayers to be formed. Incorporation of specific cell surface receptors enables the supported bilayers to be used as biosensors for cell populations or ligands. However, artifacts due to the presence of the solid support arise when incorporating components with soluble portions that can adhere to the support. Also, the distance between the support surface and the supported lipid bilayer is typically about 2 nm (and water filled). Thus membrane-spanning proteins incorporated in supported bilayers can contact the support, losing their lateral mobility and possibly affecting their functional properties as well. A variety of new strategies address this problem by employing polymerized lipids or tethered polymers (even stretches of DNA) to alter the distance to the support. A cushion provided by polyethylene glycol linked to a lipid on one end and a reactive silane to attach to the glass support on the other end increased the distance to around 4 nm (Figure 3.23), which allowed successful reconstitution of green fluorescent protein-labeled membrane proteins.

Example (Crane, J. M., and L. K. Tamm, Role of cholesterol in the formation and nature of lipid rafts in planar and spherical model membranes. *Biophys J.* 2004, 86:2965–2979.)

Addition of cholesterol to binary lipid mixtures (PC + sphingomyelin) stimulates formation of an L_o phase, which has been of great interest as the basis of domain formation in lipid rafts (see Chapter 2). The use of various lipid-linked dyes that show different preferences for L_o and L_d phases (and are excluded from gel phase) allowed observation of domain formation in response to cholesterol in planar lipid bilayers supported on quartz slides. In the absence of cholesterol, the PC/sphingomyelin bilayers exhibit coexistence of gel and fluid phases; with addition of cholesterol, the gel phase disappears and the rounded domains of L_o phase appear (Figure 3.24). The extent of L_o phase correlated with the mobility of the dyes revealed by FRAP measurements.

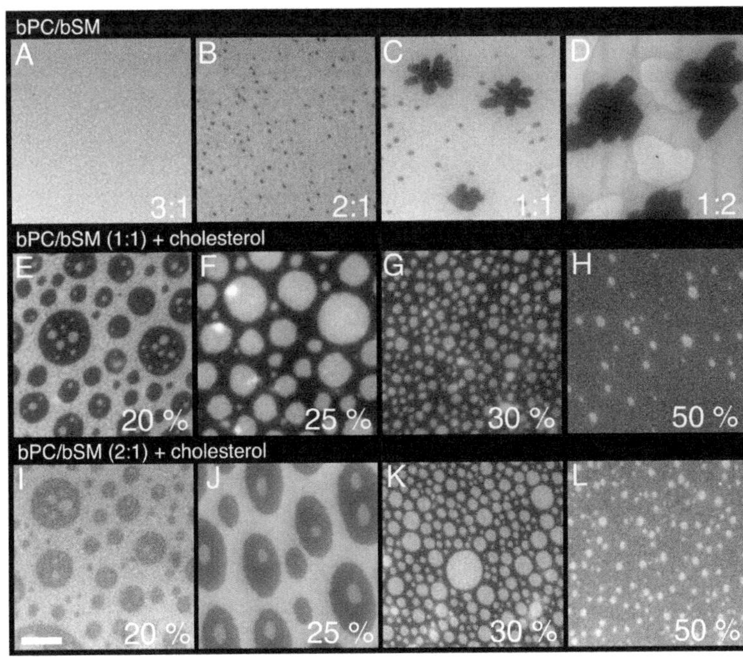

3.24 Fluorescence micrographs of PC/sphingomyelin (SM)-supported planar bilayers stained with rhodamine-DOPE at different ratios of PC/SM and different concentrations of cholesterol. The top row has different ratios of PC/SM, as labeled, in the absence of cholesterol. The middle row has PC/SM of 1:1 with different cholesterol concentrations as labeled, and the bottom row is the same, with PC/SM of 2:1. The irregularly shaped domains in the absence of cholesterol correspond to regions of gel phase (top row: C, D). As the cholesterol concentration increases, the lighter regions corresponding to L_o domains increase. At 20% all three domains (gel, fluid, and L_o) coexist. The white bar represents 10 μm. From Crane, J. M., and L. K. Tamm, *Biophys J.* 2004, 86:2965–2979. © 2004, Elsevier. Reprinted with permission from Elsevier.

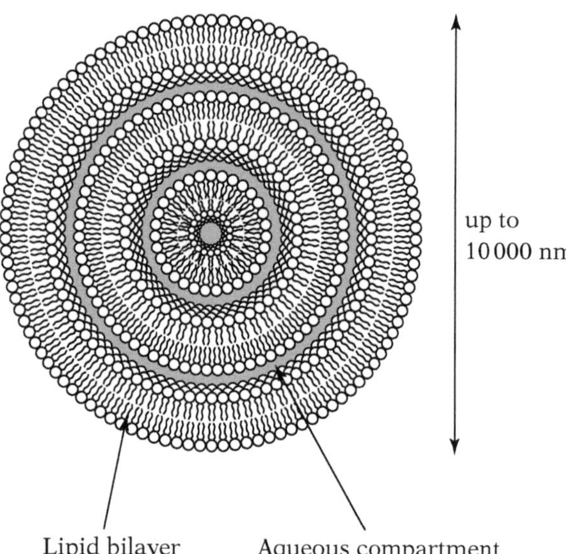

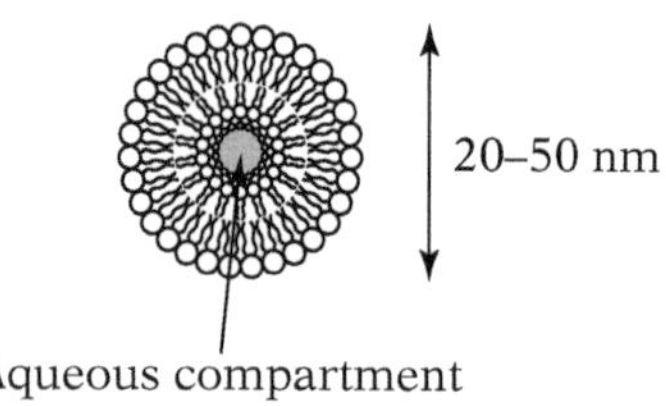

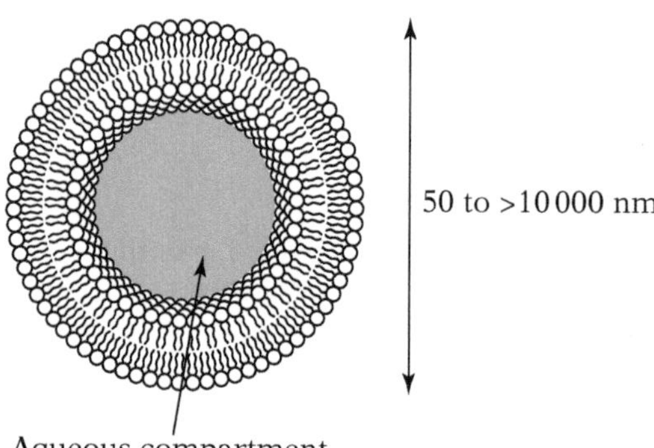

3.25 The structures and dimensions of three types of liposomes. Multilamellar liposomes (MLVs) have many more layers than indicated. Comparison of small unilamellar liposomes (SUVs) and large unilamellar liposomes (LUVs) reveals the difference in curvature that results in more loosely packed acyl chains in SUVs. Redrawn with permission from Jones, M. N., and D. Chapman, *Micelles, Monolayers and Biomembranes*, Wiley-Liss, 1995, p. 119. © 1995. Reprinted with permission from John Wiley & Sons, Inc.

Liposomes from SUVs to GUVs

In contrast to planar lipid bilayers, liposomes are closed bilayer vesicles that do not allow access to the inner compartment, although they can be preloaded with diverse compounds from the dispersion medium in which they are formed. Liposomes result when bilayer-forming lipids are mechanically dispersed in aqueous suspensions due to the tendency for bilayer edges to seal so the acyl chains are not exposed to water. Depending on the method used, they may be unilamellar (encapsulated by a single bilayer) or multilamellar. They are also classified by size, and the large variation in size effects large differences in curvature (Figure 3.25). The lipids that form vesicles are roughly cylindrical in shape and generally have more than 11 carbons in their acyl chains.

Liposomes are used to study effects of particular lipids, such as cholesterol, in mixtures of lipid components. In addition, liposomes are used in many studies of membrane protein function, folding, and assembly. When proteins are reconstituted into them, the vesicles are called *proteoliposomes*. Proteoliposomes are judged by numerous criteria, such as homogeneity regarding size and number of lamellae when visualized with thin section EM. Proteins should be distributed fairly evenly and oriented nonrandomly to mimic their incorporation in biomembranes. Like other liposomes, proteoliposomes should have low permeability to ions, giving them the ability to maintain a charge gradient. Such characteristics may depend on variation of the lipid-to-protein ratio.

Multilamellar Vesicles

Multilamellar vesicles (MLVs) contain concentric spheres of lamellae and may be made by simply shaking a thoroughly dried lipid film into an aqueous solution. They are usually polydisperse, with diameters from 0.2 to 10 μm, and have as many as 20 concentric layers of membranes. Their internal volume is unknown but quite small. To increase the internal volume, they can be converted to vesicles with one to four lamellae by extrusion through polycarbonate filters. They have been used in studies of lipid phase transitions by DSC and in many studies of enzyme and peptide binding. Their osmotic sensitivity allows quantitative measurements of solute uptake rates from turbidity changes.

Example (Luckey, M., and H. Nikaido, Specificity of diffusion channels produced by lambda phage receptor protein of *Escherichia coli*. *Proc Natl Acad Sci USA*. 1980, 77:167–171.)
The liposome swelling assay follows absorbance changes as the MLVs respond to osmotic pressure; when the liposomes are suspended in hypotonic solutions, swelling results from water entry that pushes the concentric bilayers further apart, causing

a decrease in light scattering. In isotonic solution, swelling does not occur without solute uptake, so MLVs can be used to measure the rate of uptake, which is more sensitive than methods that measure the extent of uptake after reaching equilibrium. This liposome swelling assay was crucial to discovering the specificity of maltoporin (LamB protein) because it could distinguish among uptake rates of disaccharides (Figure 3.26).

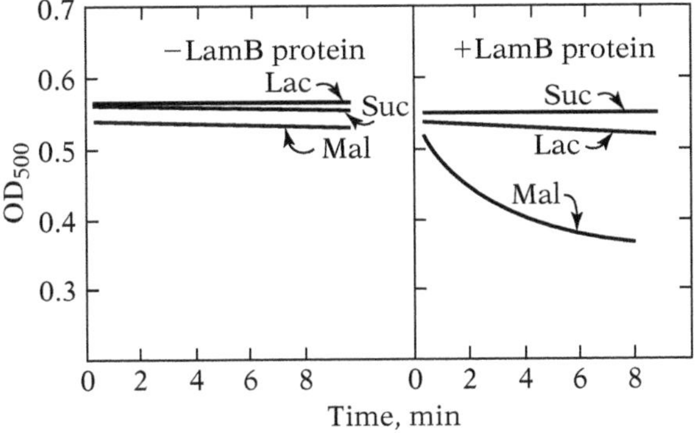

3.26 The liposome swelling assay follows a decrease in optical density (OD) at 500 nm upon mixing MLVs with permeant solutes. The optical density tracings in the control panel show very little change observed in the absence of maltoporin (LamB protein). When MLVs have incorporated purified maltoporin (3 μg/mg PL), the rate of uptake of maltose is significantly higher than that of lactose and sucrose. Redrawn with permission from Luckey, M., and H. Nikaido, *Proc Natl Acad Sci USA*. 1980, 77:167–171.

Small Unilamellar Vesicles

Small unilamellar vesicles (SUVs) with diameters of 20–50 nm result from extensive sonication of MLVs. They are also made by extrusion through polycarbonate filters of defined pore size and can be further sized by gel filtration. Another method to make SUVs is by injection of lipids in organic solvent into aqueous media, followed by removal of the organic solvent. SUVs are very asymmetric due to their extreme curvature. For example, SUVs of PC are 22 nm in diameter and have 1900 and 1100 molecules in the outer and inner leaflet, respectively. Although the acyl chains are less tightly packed than in larger liposomes, the extreme curvature of SUVs makes it difficult to incorporate proteins.

Example (Beschiaschvili, G., and J. Seelig, Peptide binding to lipid bilayers: nonclassical hydrophobic effect and membrane-induced pK shifts. *Biochemistry*. 1992, 31:10044–10053.)

This study employed SUVs made using POPC with and without POPG, which had diameters of around 30 nm, to compare the binding of a small amphiphilic peptide to its binding in larger vesicles. Since the acyl chains are less tightly packed in the small vesicles than in the larger vesicles, they have less lateral tension, which is related to membrane elasticity.

High-sensitivity titration calorimetry was used to measure the heats of reaction for binding the cyclic peptide, an analog of somatostatin called SMS that is an amphiphilic peptide with a positive charge. Monolayer studies had already indicated that SMS intercalated into lipid with little change in its conformation according to circular dichroism, and ^{2}H NMR studies had shown it could diffuse rapidly on the surface when bound.

The binding enthalpy for SMS was −7.3 kcal mol^{-1} for SUVs, independent of pH or lipid composition, in contrast to the enthalpy of binding to larger vesicles of −1.4 kcal mol^{-1}. The difference in ΔG for binding to the two classes of vesicles was less than 1 kcal mol^{-1}, indicating there is a large enthalpy–entropy compensation, which means entropy is not important in SUV binding but is the driving force for binding the larger vesicles. The entropy effect is explained in terms of the increased area of the bilayer that accompanies peptide binding, which lessens the internal bilayer tension of the more tightly packed acyl chains in larger vesicles.

Large Unilamellar Vesicles

Large unilamellar vesicles (LUVs) with diameters from 50 nm to 10 μm can be made by freeze–thaw methods that induce fusion of SUVs. Since the introduction of commercial extruders that force the liposomes under nitrogen pressure through polycarbonate filters of defined pore size, more uniform LUV sizes have been obtained, especially with repeated extrusions. Other procedures make proteoliposomes in this size range by mixing the protein in detergent with an excess of lipid and then gently removing most of the detergent by dialysis or dilution. Dialysis is slow (often requiring days) and is successful for detergents with high CMCs and small aggregation numbers, such as OG, sodium cholate, and CHAPS. Dilution to well below the detergent CMC is rapid; micelles break up and proteins (along with detergent monomers) incorporate into the lipid vesicles, which are collected by centrifugation. The rate of detergent removal affects how well the proteins are distributed. Gel filtration is used to size the vesicles, as well as to separate proteoliposomes from excess detergent. LUVs have the advantage of large encapsulated volumes, up to 50 L mol^{-1} of lipid, but their disadvantages include heterogeneous size distributions and fragility of larger vesicles.

Example (Costello, M. J., *et al.*, Morphology of proteoliposomes reconstituted with purified lac carrier protein from *Escherichia coli*. *J Biol Chem*. 1984, 259:15579–15586; Costello, M. J., *et al.*, Purified lac permease and cytochrome-o oxidase are functional as monomers. *J Biol Chem*. 1987, 262:17072–17082.) Lactose permease (also called the LacY protein; see Chapter 10) was originally reconstituted by OG dilution followed by freeze–thaw/sonication to make proteoliposomes that were examined by freeze–fracture EM. At a molar protein-to-lipid ratio of 1:2500, the majority of the proteoliposomes had diameters of 30–150 nm and exhibited fairly even distributions of protein particles. Quantitation of the size and distribution of the lactose permease with variation of protein/lipid ratios led to the conclusion that the protein was incorporated as a monomer. To answer the question of the quaternary state of lactose permease during active transport, proteoliposomes were reconstituted with both lactose permease and cytochrome o (a terminal oxidase of the *E. coli* respiratory chain) and energized by providing ubiquinol, generating an electrical gradient of about −130 mV. Alternatively, proteoliposomes containing the lactose permease were suspended in buffers containing high levels of K^+ and later treated with valinomycin, which carries K^+ across the membranes to dissipate the K^+ gradient. Changes in energized states did not lead to dimerization of the lactose permease, proving the reconstituted lactose permease was capable of both passive and active lactose transport as a monomer. For many subsequent experiments, LUVs were used to compare the activities of lactose permease variants made when every residue of LacY protein was altered by mutation.

Short-Chain/Long-Chain Unilamellar Vesicles

Short-chain/long-chain unilamellar vesicles (SLUVs) form spontaneously from aqueous suspensions of long-chain phospholipid (saturated PC, PE, and sphingomyelin with acyl chain lengths of at least 14 carbon atoms) mixed with small amounts of short-chain lecithin (acyl chain lengths of 6–8 carbons). They range in diameter from >10 nm to 100 nm, depending on the ratio of short-chain to long-chain components (increasing short-chain PL produces smaller vesicles). Inclusion of cholesterol can increase the size of the SLUV. While they have not been of general importance, SLUVs have been employed in functional studies of lipolytic enzymes because they are superior as substrates for the water-soluble phospholipases (phospholipase C and phospholipase A_2; see Chapter 4).

Giant Unilamellar Vesicles

Giant unilamellar vesicles (GUVs) are 5–300 μm in diameter. These giant liposomes are cell-size vesicles that are large enough to insert a microelectrode or to visualize surface sections by optical microscopy. They can be manipulated by micropipettes to test their elastic compressibility by their adherence to other vesicles (Figure 3.27). While they are generally viewed as excellent membrane mimetics, their large internal volume may be a disadvantage. Also, proteoliposomes of this size are very fragile.

GUVs can be made by slowly hydrating lipid at low ionic strength and high lipid concentration, followed by sedimentation through sucrose to eliminate MLVs and amorphous material. Alternatively, a homogeneous population of <100 μm diameter can be prepared by electroswelling, applying a voltage to the solution of lipids in 100 mM sucrose at 60°C. Preparation of GUVs at high ionic strengths (comparable to physiological salt concentrations) requires 10% to 20% of a charged PL and millimolar concentrations of Mg^{2+} or Ca^{2+}. GUVs may also be made from native membrane with addition of lipid: for *E. coli* membrane, the optimal lipid concentration is 90 mg ml^{-1}. To better incorporate membrane proteins into GUVs, a fusion technique has been devised.

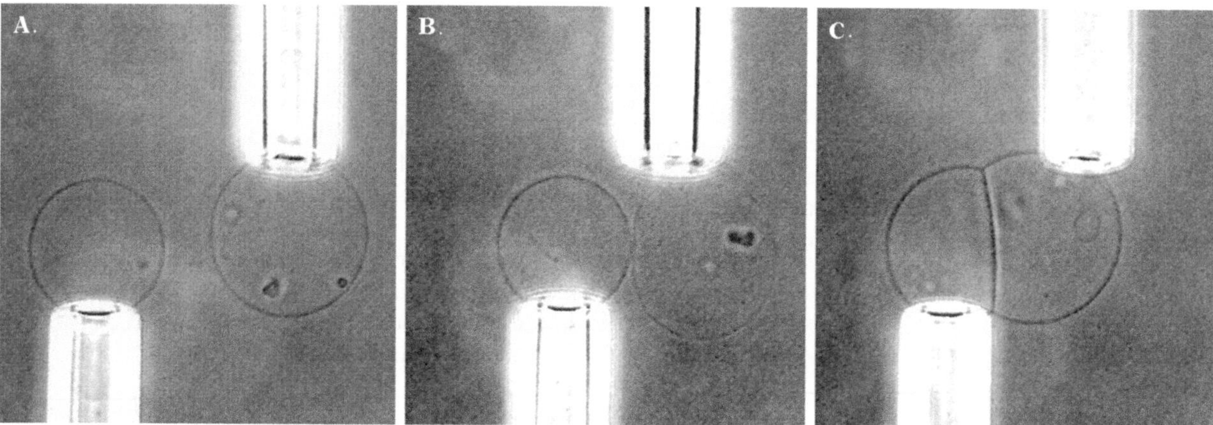

3.27 A series of pictures showing the effect of Ca^{2+} on the elastic compressibility of GUVs manipulated with micropipettes. The suction pressure is sufficient to hold the vesicles firmly on the pipette tips (A). When the suction pressure is reduced as they are brought together, the vesicles in 3 mM Ca^{2+} (B) do not adhere to each other, while in 0.12 M sucrose (C), they do. From Akashi, K., et al., *Biophys J*. 1998, 74:2973. © 1998, Elsevier. Reprinted with permission from Elsevier.

First LUVs are prepared with the proteins and then coupled to fusion-inducing peptides. One such fusogenic peptide is a short α-helix called WAE; since it is negatively charged, a positively charged target is incorporated in the GUV to facilitate docking of the LUVs. Thousands of LUVs dock onto the surface of a GUV, and after a few minutes they fuse, as demonstrated by free diffusion of the lipids between them.

Example (Korlach, J., *et al.*, Characterization of lipid bilayer phases by confocal microscopy and fluorescence correlation spectroscopy. *Proc Natl Acad Sci USA*. 1999, 96:8461–8466.)
Two fluorescent probes were incorporated into GUVs of DLPC/DPPC/cholesterol. The probe DiI-C_{20} (1,1′-dieicosanoyl-3,3,3′,3′-tetramethylindolcarbocyanine perchlorate) partitions preferentially (3:1) in the L_o phase and the probe Bodipy-PC (2-(4,4-difluoro-5,7-dimethyl-4-bora-3a,4a-diaza-s-indacene-3-pentanoyl)-1-hexadecanoyl-sn-glycero-3-phosphocholine) partitions preferentially (4:1) in the L_d phase. Their appearance in complementary regions of the images obtained by confocal microscopy facilitated phase assignments for a set of GUVs of varying compositions and gave conclusive evidence for the coexistence of separate lipid phases (see Chapter 2).

Mixed Micelles and Bicelles

Because micelles have little resemblance to bilayers, they are not generally considered to be good membrane mimetics. Even so, mixed micelles of phospholipid and detergent have been used in a multitude of studies of membrane proteins, especially when either detergent removal led to denaturation or detergent inclusion gave better activity. Micelles have been used to determine quaternary structure of membrane proteins by gel filtration and electrophoresis. Mixed Triton X-100 micelles are especially useful for kinetic analysis of enzymes with phospholipid substrates, because variations in PL concentration (up to 15 mol %) have little effect on the micelle structure. Finally, due to their small size, micelles form isotropic solutions that are advantageous for NMR studies of membrane-associated peptides and small proteins.

Bilayered micelles, which are called bicelles, provide a lipid environment that more closely imitates the membrane and avoids the severe curvature of micelles. Bicelles are discoidal lipid aggregates composed of long-chain phospholipid and either detergent or short-chain phospholipid. The center of the disc is a lipid bilayer with its edges stabilized by the detergent or short-chain lipid (Figure 3.28). Bicelles made with detergent have a much lower detergent content than mixed micelles. Varying the long-chain lipid alters features of the bilayer, changing both headgroups (typically DMPC doped with

A.

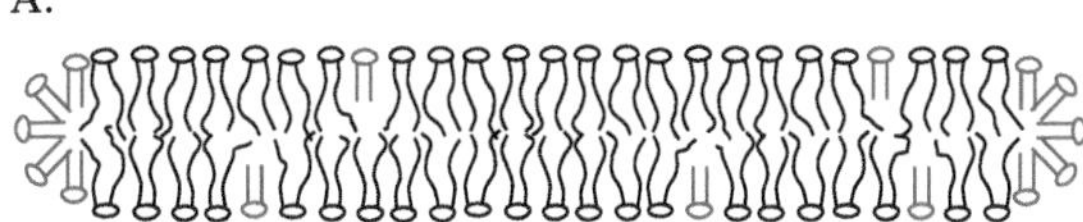

B.

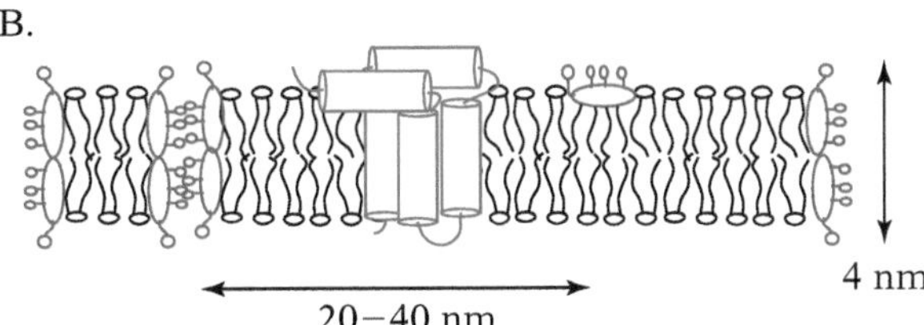

3.28 Schematic cross-sections of bicelles. Bicelles contain a mixture of long-chain phospholipids, such as DMPC, and either short-chain PLs, such as DHPC (A), or bile salt detergents, such as CHAPSO (B). The planar region is composed mainly of long-chain PLs, while the rim is formed by a monolayer of the short-chain PL (in (A)) or the bile salt detergent (in (B)). A polytopic membrane protein is shown incorporated in the bicelles in (B). Redrawn from Sanders, C. R., and R. S. Prosser, *Structure*. 1998, 6:1227–1234. © 1998 by Elsevier. Reprinted with permission from Elsevier.

DMPS or DMPG) and length of acyl chains (DMPC, DPPC, and DLPC). The size of the bicelles is dependent on both the ratio of long-chain to short-chain PL (q) and the total concentration of PL (c_L). When q > 3 and c_L is 15–20%, the bicelle diameter is 500 Å and the bicelles orient in strong magnetic fields, allowing them to be used for solid-state NMR. When q < 1 and c_L is 5–15%, the bicelle diameter is only 80 Å and these form isotropic suspensions suitable for solution NMR. Bicelles have been used to incorporate proteins for crystallization and to mimic lipid rafts by incorporating cholesterol or sphingomyelin.

Example (Czerski, L., and C. R. Sanders, Functionality of a membrane protein in bicelles. *Anal Biochem*. 2000, 284:327–333.)
Kinetic analysis of the integral membrane protein diacylglycerol kinase (DAGK, which functions to phosphorylate the lipid diacylglycerol using Mg-ATP; see "Membrane Enzymes" in Chapter 6) shows its activity is optimal in mixed micelles containing decyl maltoside and cardiolipin and decreases in bicelles. The DAGK activity in bicelles is dependent on lipid composition, demonstrating a preference for DMPC or DPPC with 3-([3-cholamidopropyl]dimethylammonio)-2-hydroxy-1-propanesulfonate (CHAPSO). The kinetic data show that decreased activity of DAGK in bicelles results from a reduced V_{max} rather than changes in K_m, suggesting little perturbation at the substrate-binding site. The enzyme activity exhibited by DAGK in bicelles validates the use of this system for NMR studies.

Blebs and Blisters

The goal of most reconstitution systems is to reproduce the membrane environment in a model system. A different approach is to use protrusions from the membrane, which are called blebs or blisters. Because they allow experimentalists to examine portions of membranes still attached to living cells, blebs are model membranes with a different set of advantages. Blebs are composed of the physiological mixture of lipids, providing a native environment for other constituents free of detergents. They preserve the asymmetric orientation of membrane proteins and may also preserve the distribution of lipids in inner and outer leaflets. Because they are still connected to the cell, they may be reached by diffusible intracellular compounds that modulate important membrane properties. Finally, they allow comparisons of different cells to test the effect of specific mutations or of up-regulation or down-regulation of membrane components.

The formation of such protrusions of the plasma membrane is one of the many changes induced by eukaryotic cell injury. When a bleb bursts, the loss of the permeability barrier triggers the onset of cell death, but the events leading up to rupture are reversible. Cell surface blebbing has been observed with numerous cell types and may be caused by mechanical or chemical (e.g., depletion of ATP with potassium cyanide, injection of polar organic solvents, addition of iodoacetic acid) treatments, as well as bacterial infection of macrophages. The absence of actin and tubulin from blebs formed on oocytes of *Xenopus laevus* clearly indicates the bleb membrane is detached from the cell cytoskeleton (Figure 3.29). For this reason, the lateral mobility of certain integral membrane proteins measured by FRAP reveals faster diffusion rates in membrane blebs than in intact cells. Mass spectrometry shows a large number of (if not most) cell membrane lipids are found in the lipid composition of vesicles derived from blebs. Blebs have also been studied by confocal fluorescence microscopy, immunocytochemistry, EM, and patch clamp techniques.

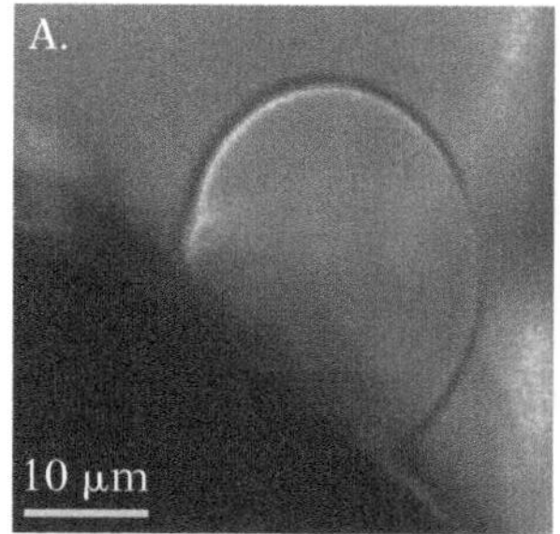

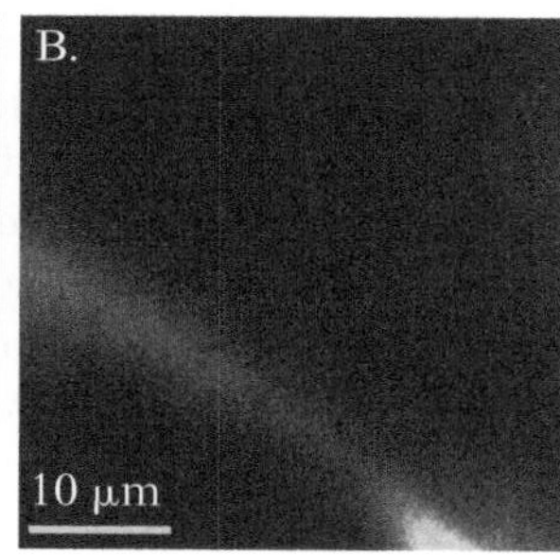

3.29 Bleb on *Xenopus laevus* oocyte surface viewed by confocal microscope. After using hypertonic stress to induce blebbing, the sample was treated with NBD-phallacidin, a fluorescent dye that stains actin. The bleb is readily visualized in the light image (A) and is absent in the fluorescence image (B), where the plasma membrane from which it derived is clearly stained. Similar results are obtained with a stain for tubulin, indicating the bleb lacks an associated cytoskeleton. Redrawn with permission from Zhang, Y, *et al.*, *J Physiol.* 2000, 523:117–130. Copyright © 2000, John Wiley and Sons.

Example (Baumgart, T., *et al.*, Large-scale fluid/fluid phase separation of proteins and lipids in giant plasma membrane vesicles. *Proc Natl Acad Sci USA*. 2007, 104:3165–3170.)
Membrane microdomains were observed in blebs induced by treating cultured fibroblast cells and leukemia cells with polar organic solvents. Imaging of two fluorescent dyes, naphthopyrene that partitions preferentially in L_o phase and rhodamine-B-DOPE that partitions preferentially into L_d phase, shows that L_o–L_d fluid–fluid phase coexistence occurs in blebs at room temperature as well as at 4°C (Figure 3.30). The variation in shape of the observed regions provides evidence for phase boundary line tension. Selective partitioning into L_o domains of the blebs by lipid-anchored proteins associated with rafts, obtained using antibodies against specific membrane proteins and green fluorescent protein/membrane protein chimeras, supports the hypothesis that the ordered regions of the blebs are rafts, which makes it the first physical demonstration of rafts in biological membranes on the micrometer scale.

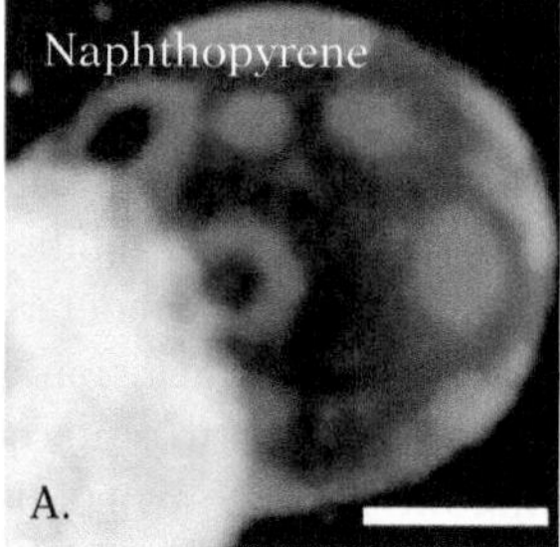

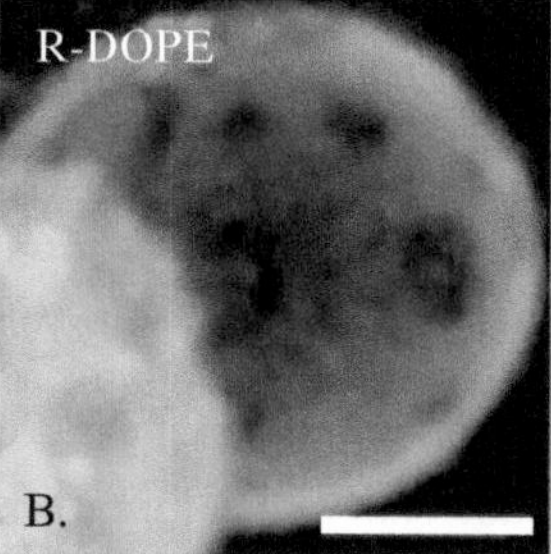

3.30 Evidence for coexisting L_d and L_o domains in blebs derived from mammalian plasma membranes. Direct visualization of large lateral domains in membrane blebs of cultured mammalian cells at 25°C is achieved by labeling with specific dyes, naphthopyrene in (A) and rhodamine-DOPE in (B). Previous experiments have shown that naphthopyrene preferentially labels the L_o phase and rhodamine-DOPE preferentially labels the L_d (L_α) phase. Blebs are induced with injection of 4% (V/V) ethanol. Scale bars = 5 μm. From Baumgart, T., *et al.*, *Proc Natl Acad Sci USA*. 2007, 104:3165–3170. © 2007 National Academy of Sciences, U.S.A.

In pathogenic bacteria, such as *Neisseria gonorrhoeae*, *Pseudomonas aeruginosa*, and *Borrelia burgdorferi* (Lyme disease), membrane blebs occur as part of the growth cycle and lead to the shedding of membrane vesicles that probably have a role in the spread of infection. These blebs have been visualized by various microscopic techniques but have not been extensively used as model membranes. Bleb-like structures called blisters have been made on giant liposomes containing reconstituted *E. coli* membrane fractions for patch clamp studies. Treatment with 20 mM Mg^{2+} induces collapse of the liposomes, followed by the emergence of blisters of 50–100 μm diameters that are stable for several hours.

Example (Iyer, R., and A. H. Delcour, Complex inhibition of OmpF and OmpC bacterial porins by polyamines. *J Biol Chem.* 1997, 272:18595–18601; Samartzidou, H., and A. H. Delcour, Distinct sensitivities of OmpF and PhoE porins to charged modulators. *FEBS Lett.* 1999, 444:65–70.)

Application of the patch clamp technique to *E. coli* outer membrane blisters gives much higher resolution than electrophysiological studies of purified outer membrane proteins in planar lipid bilayers and reveals very fast and cooperative gating between open and closed states that is modulated by physiological compounds. The opening of the cation-selective channels through OmpF porin can be inhibited by polyamines, while the anion-selective PhoE channel can be blocked by ATP (Figure 3.31A). Because the outer membrane fractions seem to incorporate a preferred orientation in the blisters, they reveal the opposite voltage dependence of OmpF and PhoE channels (Figure 3.31B).

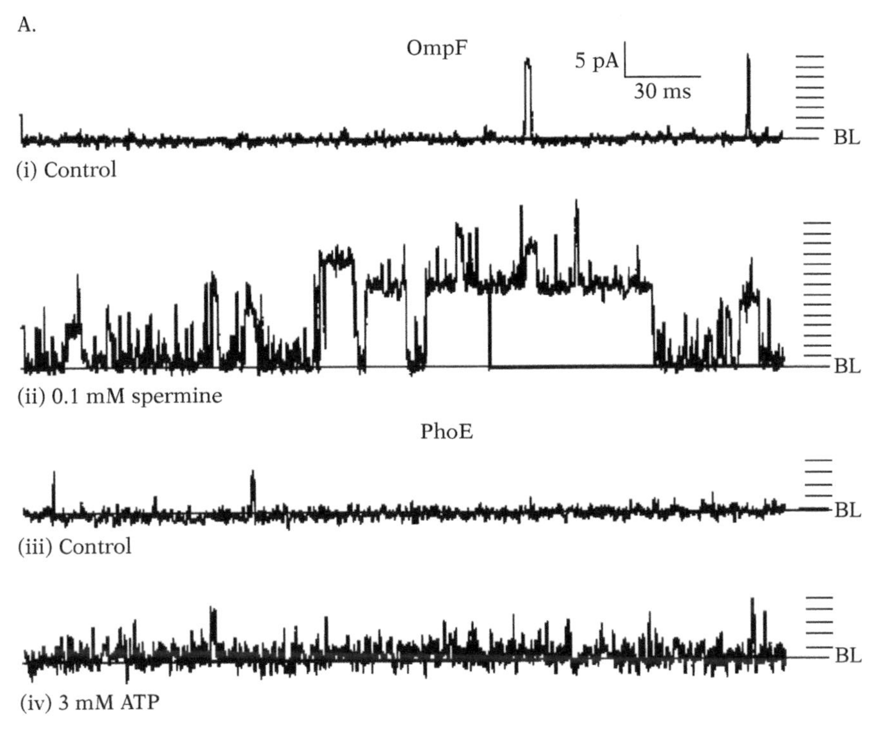

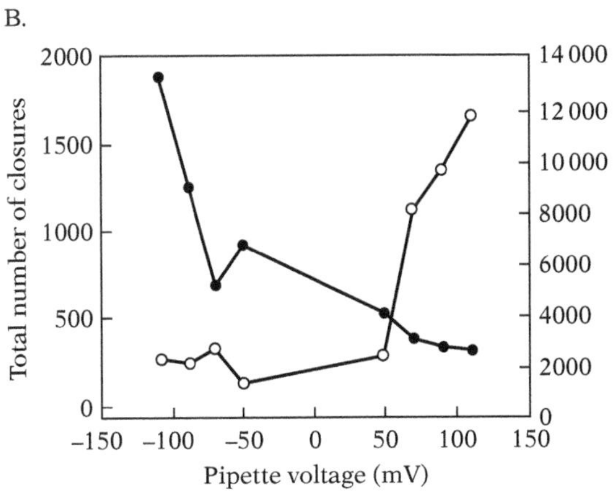

3.31 Electrophysiological tracings obtained by patch clamps on blisters on the surface of *E. coli* show the effects of physiological modulators of porin activities. **(A)** OmpF porin with and without 0.1 mM spermine (i, ii) and PhoE protein with and without 3 mM ATP (iii, iv). The current level corresponding to a large number of open pores is indicated by the baseline (BL) on the right. Redrawn from Samartzidou, H., and A. H. Delcour, *FEBS Lett.* 1999, 444:65–70. © 1999 by the Federation of European Biochemical Societies. Reprinted with permission from Elsevier. **(B)** Opposite voltage dependence was observed for OmpF (•) and PhoE (○). Redrawn from Samartzidou, H., and A. H. Delcour, *EMBO J.* 1998, 17:93–100. © 1999 by the Federation of European Biochemical Societies. Reprinted with permission from Elsevier.

Nanodiscs

An effective new membrane mimetic is the nanodisc, a lipid bilayer delineated by a protein scaffold that is small enough to solubilize a single protein molecule: one nanodisc contains approximately 160 PL molecules in a circular bilayer approximately 10 nm in diameter (Figure 3.32). The membrane scaffold protein (MSP) is derived from apoliprotein A-I, an amphipathic protein that forms a helical belt wrapping around lipid–cholesterol complexes in high-density serum lipoproteins. MSPs are engineered with insertions or deletions to vary their length and therefore determine the size of the nanodiscs. Lipid composition can be varied with different headgroups and/or acyl chains, and the lipids in nanodiscs have been shown to undergo normal phase transitions as a function of temperature. The structural organization of nanodiscs has been probed by atomic force microscopy, NMR, and small angle x-ray scattering. Nanodiscs behave much like a soluble protein and can even be frozen or lyophilized.

The lipid:detergent ratio is critical for formation of nanodiscs, and cholate is the most successful detergent, although it can be used in combination with another detergent. With the right size MSP, nanodiscs containing membrane proteins assemble spontaneously upon detergent removal. When they solubilize membrane proteins in the circle of bilayer lipids, nanodiscs allow access from both sides of the membrane. This has been important in studies of the interactions between integral and peripheral membrane proteins, such as G protein-coupled receptors and their G proteins (see Chapter 10), and between the SecYEG translocon and its soluble partners (see Chapter 7). Nanodiscs can isolate membrane complexes to show cooperativity among homo-oligomers or interaction between partners, such as a cytochrome P450 and its reductase (see Chapter 6).

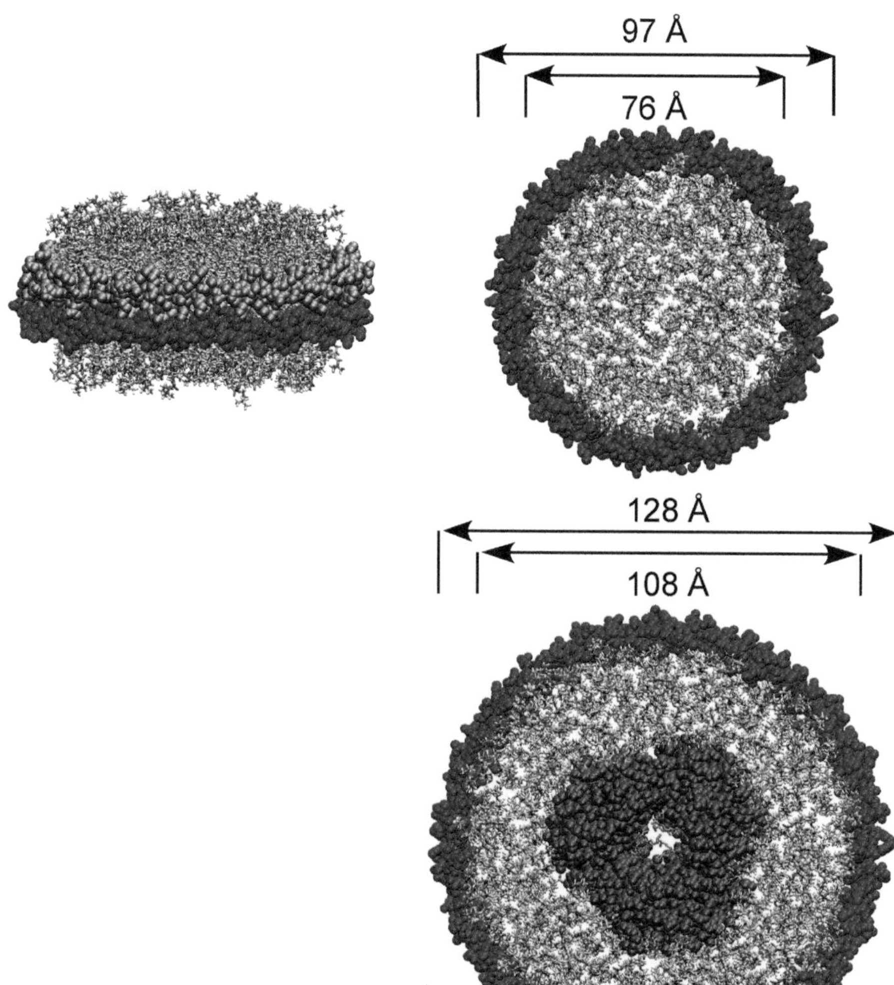

3.32 Illustration of the structure of nanodiscs. Nanodiscs composed of MSP1 and phospholipid are visualized from the side and from the bottom (top panel) with two MSPs (orange and blue) encircling phospholipids (cyan and red). A longer variant of MSP1 is used for the nanodisc containing a bacteriorhodopsin trimer (pink), viewed from the bottom (lower panel). From Bayburt, T. H., and S. G. Sligar, *FEBS Letts*, 2010, 584:1721–1727. © 2010, Elsevier.

Example (Denisov, I. G. and S. G. Sligar, Cytochromes P450 in nanodiscs. *BBA*. 2011, 1814:223–229.)
CYP3A4, a cytochrome P450 that is involved in steroid biosynthesis in humans, cycles through seven intermediate states during catalysis. Previous attempts to measure the effect of substrate binding on the redox properties of membrane-bound P450s were hampered by aggregation. Incorporated into nanodiscs, CYP3A4 monomers cooperatively bind three molecules of testosterone, exciting a shift in the ferric spin state equilibrium (Figure 3.33A). To characterize the activities of the substrate-bound states, nanodiscs were prepared with CYP3A4 and cytochrome P450 reductase (CPR) in a 1:1 ratio (Figure 3.33B). In these nanodiscs the mechanism of testosterone metabolism was studied by spectral titration of the intermediates, kinetics of NADPH consumption, and kinetics of product formation.

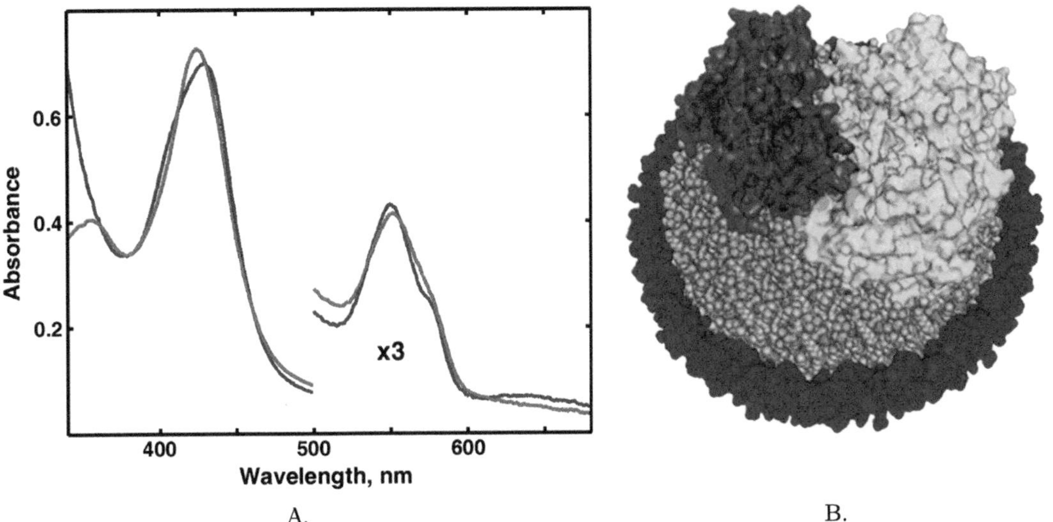

3.33 Characterization of a membrane-bound cytochrome P450 in nanodiscs. (A) The UV–VIS absorption spectra of the oxy-ferrous complex of CYP3A4 in nanodiscs show a slight shift in the position of the Soret band in the presence of testosterone (red) and bromocriptine (blue). (B) A molecular model of a 1:1 functional complex of CYP3A4 (red) and the cytochrome P450 reductase (yellow) is based on the x-ray structures of the substrate proteins and molecular dynamics simulations of the POPC bilayer, with an ideal α-helix for the MSP. From Denisov, I. G., and S. G. Sligar, *BBA*. 2011, 1814:223–229. Copyright © 2011, Elsevier.

CONCLUSION

The challenges of studying membranes in the laboratory can be met with a thorough grasp of detergent action and appreciation of both old and new kinds of detergents. In addition, an array of membrane mimetics enables researchers to apply a wide variety of techniques to studies of membrane components. Most of the information on lipid–lipid interactions described in the last chapter was obtained using such systems. Furthermore, reconsituted systems are indispensable for the exploration of protein–lipid interactions, as covered in the next chapter.

FOR FURTHER READING

Jones, M. N., and D. Chapman, *Micelles, Monolayers and Biomembranes*. New York: Wiley-Liss, Inc., 1995.

Detergents

Chae, P. S., S. G. Rasmussen, R. R. Rana, *et al.*, Maltose neopentyl glycol (MNG) amphiphiles for solubilization, stabilization and crystallization of membrane proteins. *Nature Meth*. 2010, 7:1003–1008.

Garavito, R. M., and S. Ferguson-Miller, Detergents as tools in membrane biochemistry. *J Biol Chem*. 2001, 276:32403–32406.

Helenius, A., and K. Simons, Solubilization of membranes by detergents. *Biochim Biophys Acta*. 1975, 415:29–79.

Linke, D., Detergents: an overview. *Methods in Enzymol*. 2009, 463:603–617.

Popot, J.-L., Amphipols, nanodiscs, and fluorinated surfactants: three non-conventional approaches to studying membrane proteins in aqueous solutions. *Annu Rev Biochem*. 2010, 79:737–775.

Membrane Mimetics

Bayburt, T. H., and S. G. Sligar, Membrane protein assembly into nanodiscs. *FEBS Letts*. 2010, 584:1721–1727.

Denisov, I. G., Y. V. Grinkova, A. A. Lazarides, and S. G. Sligar, Directed self-assembly of monodisperse phospholipid bilayer nanodiscs with controlled size. *J Am Chem Soc*. 2004, 26:3477–3487.

Montal, M., and P. Mueller, Formation of bimolecular membranes from lipid monolayers and study of their electrical properties. *Proc Natl Acad Sci USA*. 1972, 69:3561–3566.

Neher, E., and B. Sakmann, Single-channel currents recorded from membrane of denervated frog muscle fibres. *Nature*. 1976, 260:779–802.

Sanders, C. R., and G. C. Landis, Reconstitution of membrane proteins into lipid-rich bilayered mixed micelles for NMR studies. *Biochemistry*. 1995, 34:4030–4040.

Sanders, C. R., and R. S. Prosser, Bicelles: a model membrane system for all seasons? *Structure*. 1998, 6:1227–1234.

Other references given in examples and figures in text.

PROTEINS IN OR AT THE BILAYER

4

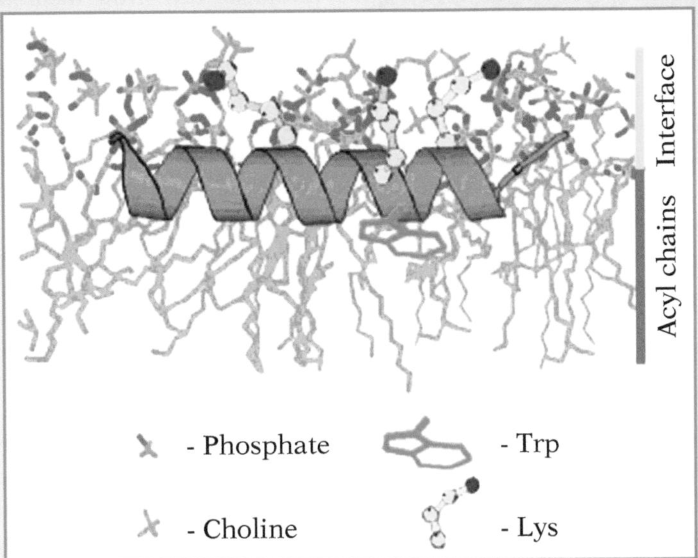

Protein insertion into lipid bilayers involves polar and nonpolar interactions, exemplified by the interaction of an 18-residue α-helix from colicin A with the membrane interfacial region and nonpolar interior. Three lysine residues (yellow) extend toward the aqueous milieu, while a tryptophan residue (green) reaches toward the center of the bilayer. From Zakharov, S. D., and W. A. Cramer, *Biochimie*. 2002, 84:472. © 2002 by Société Française de Biochimie et Biologie Moléculaire. Reprinted with permission from Elsevier.

The chemical and physical properties of the lipids described in Chapter 2 make it clear that the lipid bilayer provides a special milieu for proteins. Its constraints affect the structure, function, and regulation of proteins that dock on it or assemble in it to perform jobs such as energy transduction, nutrient transport, and signaling. The variety among proteins that interact in some way with the membrane is quite astonishing and provides fascinating examples of how these proteins are structured to work in their environment.

This chapter first defines the classes of proteins that are found in or at the bilayer. It describes many examples of peripheral proteins, as well as some lipid-anchored proteins, before looking at what modulates the interactions of these proteins with membrane lipids. Next it focuses on molecules that insert into the membrane, including examples of toxins, colicins, and ion-carrying peptides. Then a look at the special qualities of the nonpolar milieu of the membrane explains many characteristics of integral membrane proteins. The chapter ends with studies of protein–lipid interactions in membranes.

CLASSES OF PROTEINS THAT INTERACT WITH THE MEMBRANE

A typical biomembrane contains many species of proteins, some embedded in the lipid bilayer and others on its surface. The Fluid Mosaic Model described in Chapter 1 distinguished between extrinsic and intrinsic membrane proteins by how easily they could be isolated from the membrane: Extrinsic (peripheral) proteins can be removed by washes of the membrane, while the extraction of intrinsic (integral) proteins requires disruption of the membrane. Most integral membrane

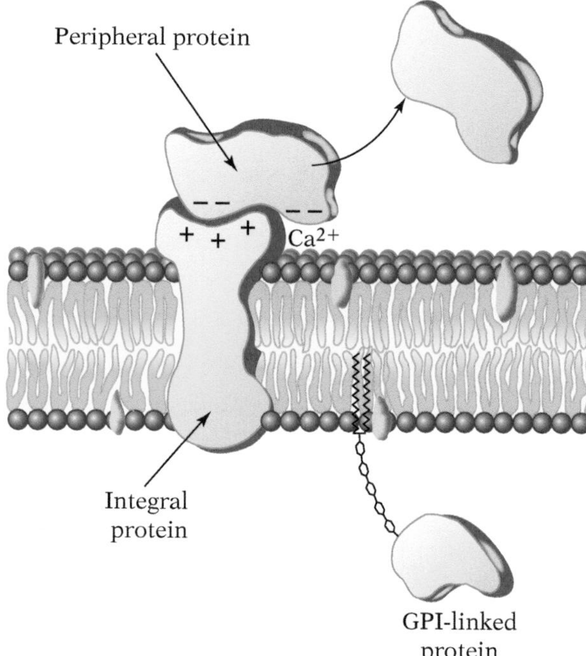

4.1 Schematic diagram showing classes of membrane and membrane-associated proteins: peripheral proteins, TM proteins and lipid-anchored proteins. The peripheral protein has electrostatic interactions with the membrane protein, as well as with membrane lipids (mediated by calcium ions). Redrawn from Nelson, D. L., and M. M. Cox, *Lehninger Principles of Biochemistry*, 4th ed., W. H. Freeman, 2005, p. 374. Adaptation © 2005 by W. H. Freeman and Company. Used with permission. Original figure © 1976 by Annual Reviews. Reprinted with permission from the Annual Review of Biochemistry, www.annualreviews.org.

proteins have at least one TM peptide segment: *Bitopic* proteins have only one TM span, while *polytopic* proteins have more than one. In addition, there are *monotopic* proteins that insert but do not span the membrane and lipid-anchored proteins held in the membrane by covalently linked lipid groups, whether or not the polypeptide is membrane spanning (Figure 4.1). Those peripheral proteins that bind the membrane weakly and reversibly and are regulated by that binding are called *amphitropic proteins* and are involved in many important biological functions. A few soluble proteins and many peptides insert into membrane bilayers to accomplish their functions, often coming from the outside and causing deleterious consequences to the cell or organism. Finally, some proteins blur the boundaries between classes since they are found in the cytoplasm, on the membrane periphery, and even inserted into membrane structures and can be fully characterized only when the dynamics of their localization are understood. To discuss the characteristics of proteins capable of existing in or interacting with the membrane, each of these classes will be described, giving attention to how the proteins and lipids interact.

PROTEINS AT THE BILAYER SURFACE

Extrinsic/Peripheral Membrane Proteins

The surface of some biomembranes is almost crowded with peripheral proteins in dynamic interplay with the lipid bilayer, with integral membrane proteins, and with each other. Typically, extrinsic membrane proteins bind to the surface of the membrane via electrostatic interactions, interacting with anionic lipids or with charged groups on other proteins, or both. Thus they sediment with the membrane and then can be isolated from it by treatment with buffers that vary the pH or increase the ionic strength; after sedimentation of the membrane under the new conditions, the extrinsic proteins remain in the supernatant and can be purified as soluble proteins. In addition to electrostatic interactions, hydrophobic interactions with the acyl chains often contribute to the association of peripheral proteins with the membrane, as described below.

The classic example of a peripheral protein is cytochrome *c*, whose cluster of lysine residues enables it to bind to anionic lipids of the bilayer, as well as to acidic residues on the surfaces of cytochrome bc_1 and cytochrome *c* oxidase. The first use of ion exchange chromatography was in the purification of this highly basic protein: 20 amino acids of the 104-residue human cytochrome *c* are lysine and arginine residues. Nine highly conserved lysine residues form a ring around the only exposed edge of its heme group (Figure 4.2).

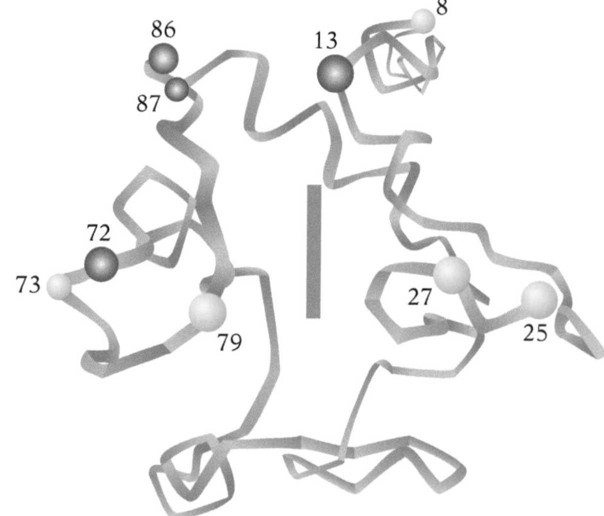

4.2 Cluster of basic residues on surface of cytochrome *c*. Some lysine residues (dark blue balls) are strongly protected against acetylation when complexed with cytochrome *c* oxidase or reductase Polypeptide chain (green ribbon), heme cofactor (red bar), lysine residues (blue balls). Redrawn from Voet, D., and J. Voet, *Biochemistry*, 2nd ed., John Wiley, 1995, p. 579. © 1995. Reprinted with permission from John Wiley & Sons, Inc.

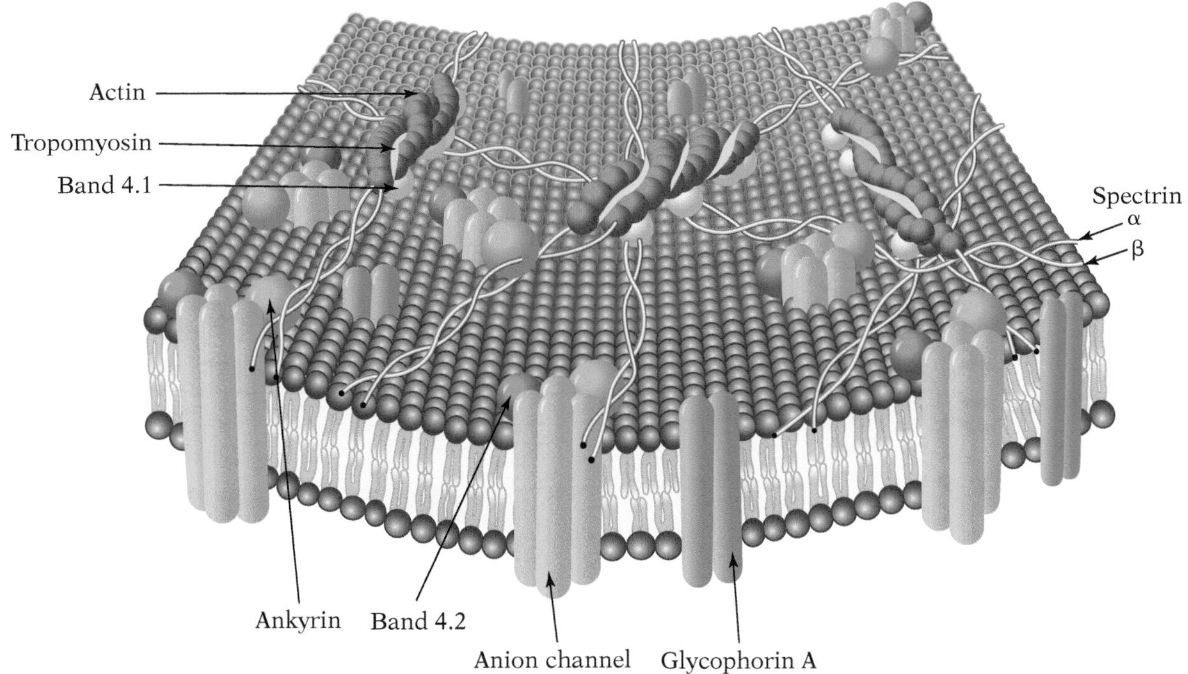

4.3 Role of ankyrin in attaching various components of the cytoskeleton. Ankyrin (green spheres) binds spectrin (pink, helical filaments) and numerous TM proteins (orange). Redrawn from Voet, D., and J. Voet, *Biochemistry*, 2nd ed., John Wiley, 1995, p. 302. © 1995. Reprinted with permission from John Wiley & Sons, Inc.

Differential labeling experiments have shown that the complex between yeast cytochrome *c* and cytochrome *c* oxidase shields these lysine residues of cytochrome *c* and also shields three acidic residues of cytochrome *c* oxidase. Further evidence that this is the interaction site between cytochrome *c* and the other cytochromes with which it reacts derives from the x-ray structure of a complex of cytochrome *c* and cytochrome *c* peroxidase (which together reduce organic hydroperoxides) that reveals specific ion pairs between Lys73 and Lys87 of cytochrome *c* and Glu290 and Asp34 of the peroxidase. In the absence of other proteins, cytochrome *c* binds to bilayers formed with anionic lipid and has been used in many studies of the interactions between peripheral proteins and bilayer lipids (see below).

Electrostatic interactions are also paramount in the interactions of myelin basic protein with the myelin sheath. Myelin basic protein is a 21-kDa peripheral membrane protein that exists in isomers of different net charge. It is involved in the electrical activity of the myelin sheath and is implicated in the pathology of the autoimmune disease multiple sclerosis. Like cytochrome *c*, myelin basic protein requires anionic lipids for membrane binding and has been used to explore the energetics of the association between peripheral proteins and the bilayer (described below). Binding of one myelin basic protein immobilizes 18 lipid molecules of the bilayer, as probed with EPR (see Box 4.3).

Another familiar example of a peripheral protein is ankyrin, a 200-kDa globular protein that mediates the linkage of the cytoskeleton to the plasma membrane by binding to spectrin and numerous integral membrane proteins, including ion channels and cell adhesion molecules (Figure 4.3). The membrane-binding domain of ankyrin contains 24 ANK repeats, organized into four six-repeat folding domains. Each domain consists of a bundle of stacked antiparallel α-helices connected by β-turns, whose tips make contacts with other proteins (Figure 4.4). While ANK repeats have been found with assorted other domains in over 1500 proteins, ankyrin is composed almost entirely of them. Different isoforms of ankyrin have different combinations of the ANK repeat domains to make up distinct, high-affinity sites for protein binding. A hereditary ankyrin deficiency severely weakens erythrocytes, leading to anemia, jaundice, and eventually hemolysis.

An important family of peripheral proteins is the diverse group of phospholipases, water-soluble enzymes that cleave the phospholipids of the bilayer. Some phospholipases are key regulatory enzymes, producing second messengers. For example, phospholipase C cleaves the glycerophosphate ester bond to liberate a phosphorylated alcohol and diacylglycerol, which activates protein kinase C (PKC). Phospholipase A_2 (PLA_2) releases the *sn*-2 fatty acid, which can be converted to arachidonic acid and then to the eicosanoids – prostaglandins, thromboxanes, and leukotrienes – which trigger inflammation and other physiological reactions. For these peripheral enzymes, binding the membrane bilayer is mediated in part by binding their

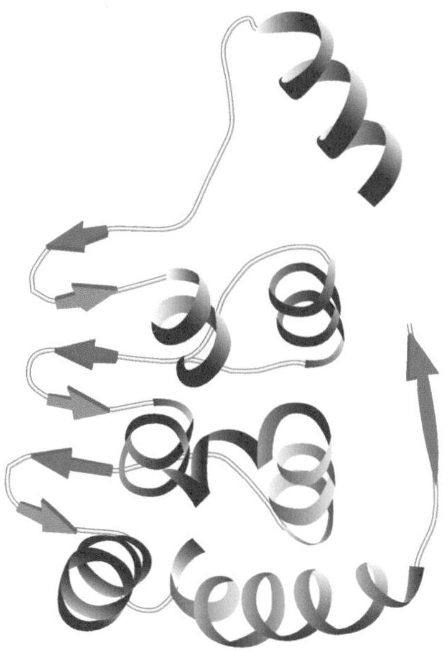

4.4 Structure of ANK repeats, each composed of 33 amino acids that make up a pair of α-helices (purple) connected by a tight β-turn (green arrows). ANK repeats bind to target proteins with the aligned β-turns and the surface of the helical bundle. Three repeats from the yeast transcription factor Swi6 are shown. From *Cell Signaling Technology Catalog*, 2000, with permission from Pawson lab (http://pawsonlab.mshri.on.ca/index.php?option=com_content&task=view&id=142&Itemid=64). © 2000. Reprinted with permission from Cell Signalling Technology.

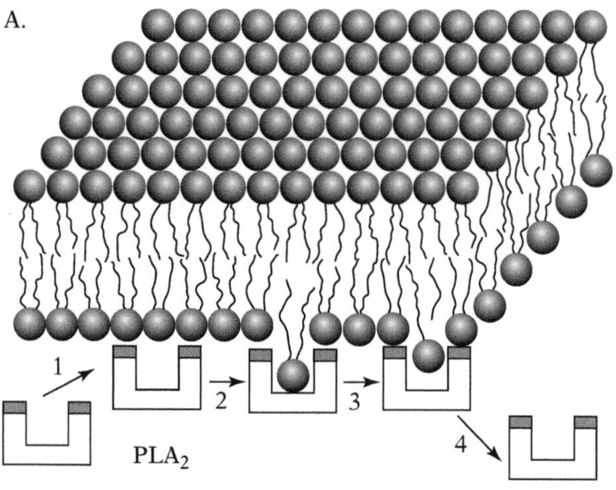

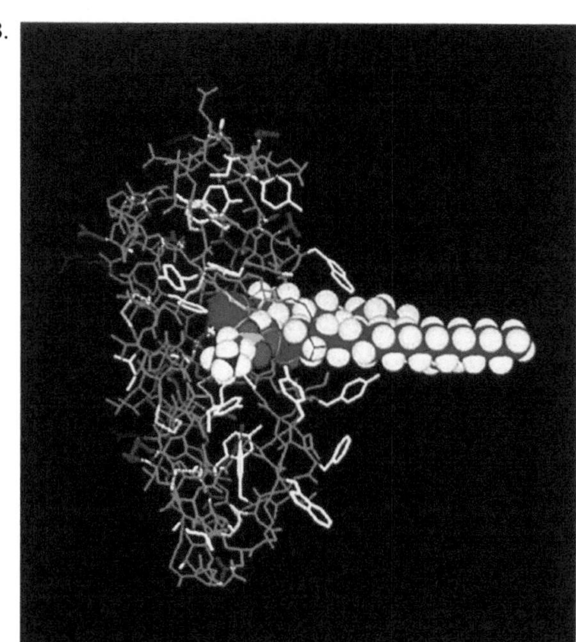

4.5 Membrane binding mediated by substrates. (A) Model for PLA_2 binding to bilayer surface via substrate binding. PLA_2 binds to the membrane periphery (1) and then binds a phospholipid molecule in its active site (2). It cleaves the acyl chain (3) and then diffuses from the bilayer (4). Redrawn from Seaton, B. A., and M. F. Roberts, in K. Merz and B. Roux, *Biological Membranes*, Birkhauser, 1996, p. 363. © 1996 by Springer-Verlag. With kind permission of Springer Science + Business Media B. V. (B) X-ray crystal structure of PLA_2 from cobra venom binding to DMPE. The bound Ca^{2+} is magenta and the substrate, DMPE, is represented with a space-filling model in the active site. The PL substrates in a biological membrane would typically be 4–6 carbons longer and would be integrated into a leaflet as diagrammed in (A). From Dennis, E. A., *J Biol Chem*. 1994, 269:13057–13064. © 1994, American Society for Biochemistry & Molecular Biology.

substrates (Figure 4.5A). The x-ray crystal structure of PLA_2 binding to DMPE illustrates its "lipid clamp," a highly specific binding pocket for lipid (Figure 4.5B). Crystal structures of PLA_2 with and without a phosphonate transition state analog have revealed the movement of a "flap" that appears to uncover the substrate binding site as well as a hydrophobic patch that seals the enzyme to the bilayer surface once the PL substrate is inside (Figure 4.6). A similar flap mechanism is observed in many other lipases and in some annexins.

Annexins are a large family of proteins that require Ca^{2+} to bind to membranes and carry out a variety of functions. Their affinity for Ca^{2+}, which enables them to mediate Ca^{2+} signaling, is low in the absence of lipids and high (μM Kd) in the presence of anionic lipids. Members of the annexin family are structurally related proteins with four or eight repeats of a right-handed superhelical binding site for Ca^{2+}, which together form a curved disc that binds the surface of the membrane (Figure 4.7). The dissimilar N-terminal domains of different annexins form an outward-facing concave side of the disc. They are "interaction domains" that enable annexins to form complexes with other proteins. Some annexins form complexes in a way that allows them to bring two membranes into close contact (although not to

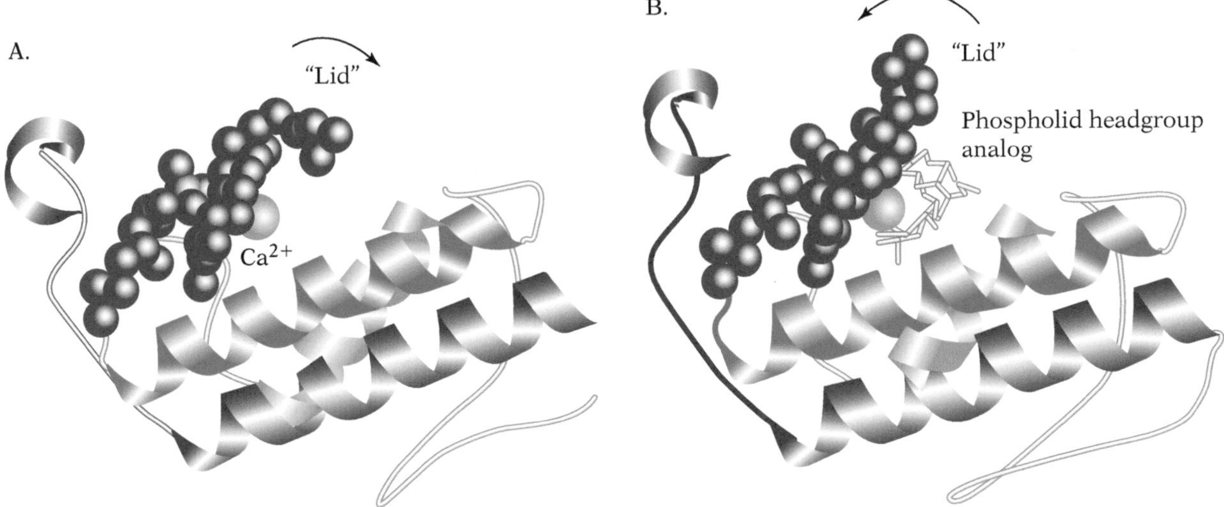

4.6 Model for flap movement in PLA_2. (A) In the unliganded state, the left side of the substrate binding site forms the flap, or lid region, which can close over the bound Ca^{2+}. (B) With the phosphonate transition state analog bound, the lid is held in place over it, which would seal the enzyme to the membrane if the substrate were a lipid from the bilayer. From Seaton, B. A., and M. F. Roberts, in K. Merz and B. Roux, *Biological Membranes*, Birkhauser, 1996, pp. 361–362. © 1996 by Springer-Verlag. With kind permission of Springer Science + Business Media B. V.

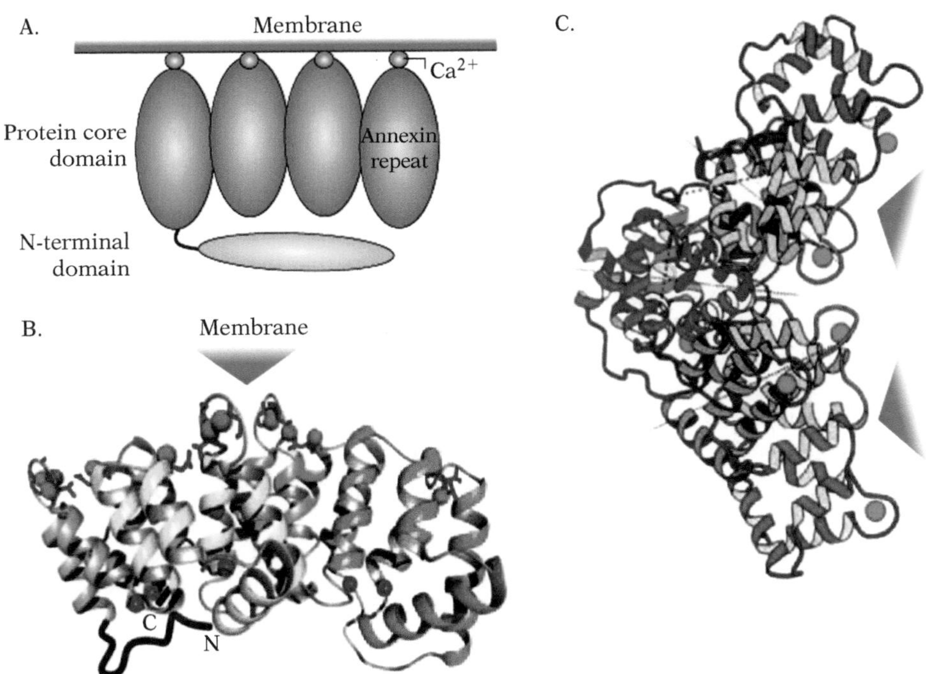

4.7 General structure of annexins. (A) Schematic drawing of an annexin attached to a membrane surface through bound Ca^{2+} ions (blue). Four ANK domains are shown, with the N-terminal domain on their surface away from the membrane. (B) Structural model of annexin core based on alignment of over 200 annexin sequences. Each ANK repeat is colored differently, with N and C termini black. Oxygens involved in binding Ca^{2+} are red, and nitrogen atoms of highly conserved basic residues are blue. (C) Annexin A6, with eight ANK repeats in two halves connected by a flexible linker (green helix), is capable of binding two membranes (at arrows). From Gerke, V., *et al.*, *Nat Rev Mol Cell Biol*. 2005, 6:449–461. © 2005. Reprinted by permission of Macmillan Publishers Ltd.

fuse without the help of fusogenic proteins; Figure 4.7C). Some participate in actin cytoskeleton attachment and others may play a role in cholesterol-dependent raft formation, while others have been observed to form two-dimensional ordered arrays on model membranes. Clearly their dynamic roles in lipid organization are important to many of their intracellular functions, such as endocytosis and lipid trafficking. Some annexins

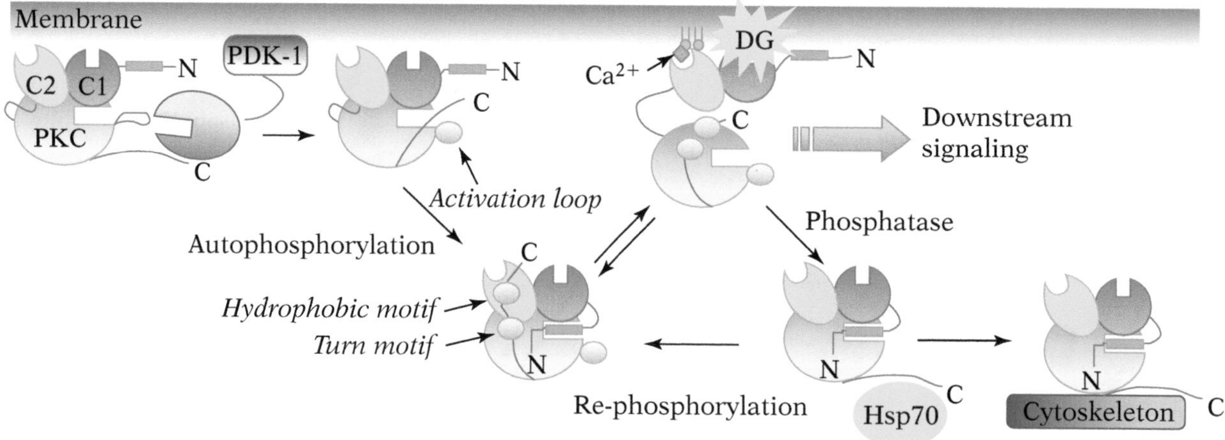

4.8 Model showing interaction of PKC with the membrane in response to elevated Ca^{2+} levels. Ca^{2+} binds to the C2 domain to tether PKC to the membrane with a low affinity, allowing the C1 domain to find and bind DAG, giving PKC a high affinity for the membrane and expelling its pseudosubstrate (N-terminal portion of the peptide chain), at the same time exposing the C terminus for autophosphorylation. Signaling activity is further controlled by phosphorylation and dephosphorylation, as well as association with Hsp70 and the cytoskeleton. For structural details of the C1 and C2 domains, see Figure 4.16. Here the C2 domain is yellow and the C1 domain is orange on the blue PKC; the pseudosubstrate is green. Redrawn from Newton, A. C., *et al.*, *Biochem J.* 2003, 370:361–371. Copyright © 2003 by the Biochemical Society. Reprinted with permission from Portland Press Ltd.

also have extracellular roles, such as suppression of inflammation and coagulation. Interest in annexins is high because of their roles in immune functions, their use as early markers for apoptosis, and their increased expression in certain tumors.

Amphitropic Proteins

Many of these peripheral proteins are considered amphitropic proteins because their activity is regulated by the change from a water-soluble form to a membrane-bound form. This regulation may affect their catalytic function and/or their access to substrates, as well as their assembly into complexes (sometimes linking them to the cytoskeleton). While most amphitropic proteins, such as phospholipase C and PKC, come from inside the cell and are vital in signal transduction, some are extracellular proteins such as blood clotting factors and apolipoproteins involved in transport of lipids through the blood.

Activation of amphitropic proteins upon binding the membrane can be due to the proximity of effector molecules or of substrate, as in the case of the phospholipases. When restricted to the two-dimensional plane of the membrane, the effective concentration of the substrate (or effector) goes up approximately 1000-fold, calculated by comparing the volume of a sphere to the volume of surface phase: $(4/3)\pi r^3/4\pi r^2 d$, assuming a spherical cell radius r of 10 μm and a surface thickness d of 1–10 nm. This concentration effect is also very important in facilitating interactions between membrane-bound proteins.

Alternatively, activation can be due to structural changes induced in the proteins that relieve autoinhibition, as observed in PKC. PKC has two well-conserved membrane-binding domains – C1, which binds diacylglycerol (DAG) or phorbol esters, and C2, which binds Ca^{2+} – each triggering conformational changes. It appears that the initial interaction between Ca^{2+}-bound C2 and anionic lipids brings PKC close to the membrane to allow C1 to penetrate and bind DAG (Figure 4.8; see also Figure 4.16A, B). Once both regulatory domains have bound, PKC undergoes a structural rearrangement that releases a pseudosubstrate N-terminal group from its active site and undergoes phosphorylation to produce the catalytically active enzyme. The C1 and C2 domains are two of a small set of membrane-binding domains that are used for signaling and sub-cellular targeting by many other proteins (described below).

Lipid-Anchored Proteins

Lipid modification gives an opportunity to target proteins to the membrane (and even to specific intracellular membranes in eukaryotes) as well as to stabilize their interactions at the bilayer. The lipids can be fatty acids, terpenes, or glycosylphosphatidylinositol (GPI; Figure 4.9). The acyl chains are typically myristoyl groups in amide linkage to N-terminal glycine residues and palmitoyl groups covalently linked to serine or cysteine residues. Covalent modification with myristate occurs on nearly 100 different proteins, not all of which bind membranes. Peripheral membrane proteins with mutations at their myristoylation site no longer bind membranes,

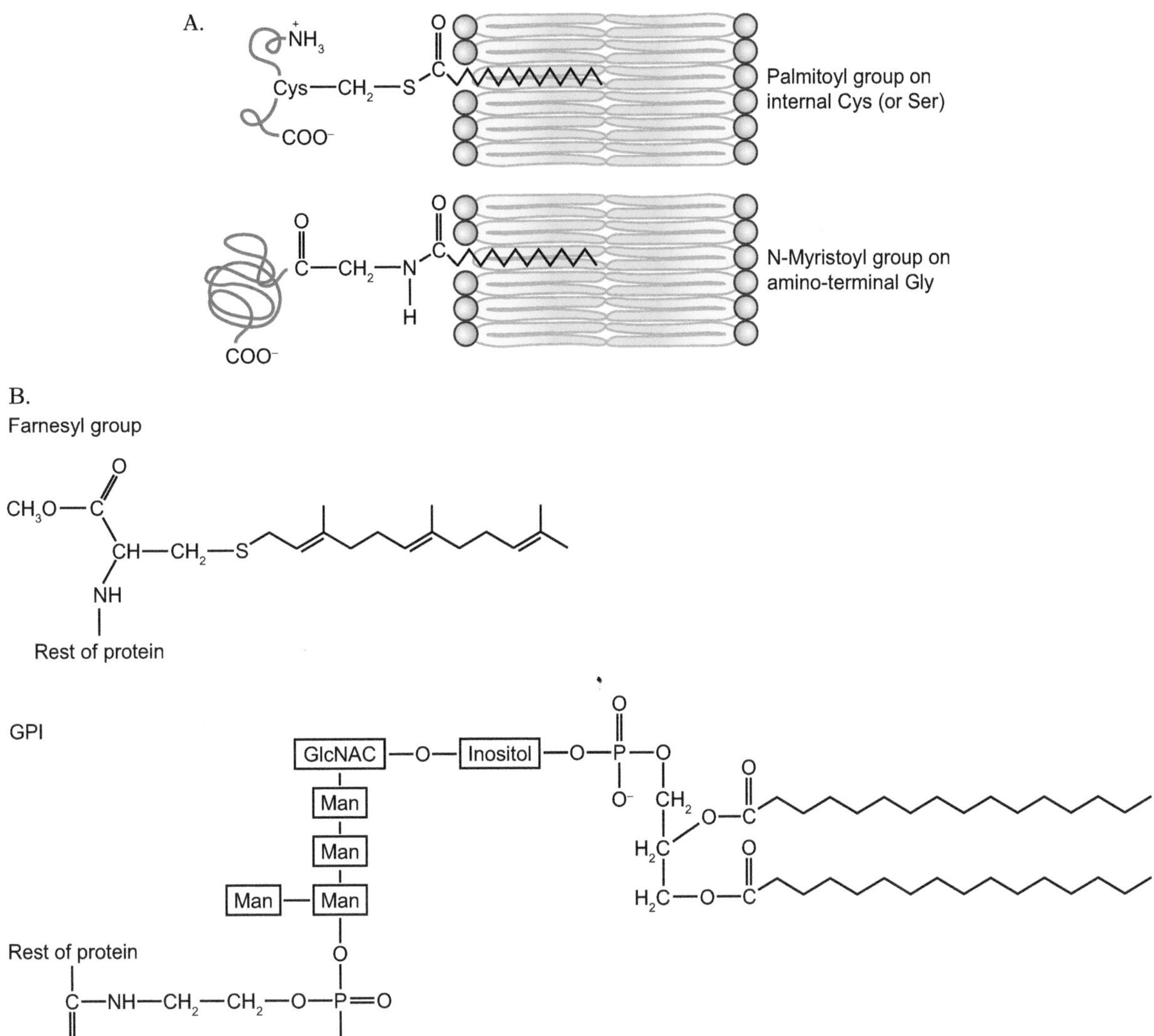

4.9 Lipid-anchored proteins. **(A)** Proteins may be anchored to the membrane by covalent modification with acyl chains that insert into the bilayer, shown here with palmitate and myristate. Myristoyl groups are typically linked by amide bonds with terminal amino groups, while palmitoyl groups are more often linked by ester or thioester bonds to internal Ser or Cys residues. **(B)** Other lipid anchors include terpenes such as farnesyl or geranylgeranyl (not shown) groups and glycosylphosphatidylinositol (GPI). In the plasma membrane the proteins covalently attached to acyl chains and terpenes are found on the cytoplasmic surface, while proteins attached to GPI are on the extracellular surface.

indicating the myristoyl group is required for membrane binding. Often these proteins have more than one lipid molecule bound because binding a single lipid anchor is not sufficient to maintain a stable attachment to the bilayer. The second acyl chain is frequently palmitate. Detergent-resistant membranes (thought to correspond to lipid rafts, see Chapter 2) are enriched with proteins modified with two or more acyl chains, and loss of one of these abolishes raft targeting. Terpenes, derived from isoprene, are either farnesyl or geranylgeranyl groups in thioether linkages to cysteine residues and are generally observed on proteins at intracellular membranes. Many members of the Ras–guanine triphosphatase (GTPase) superfamily have farnesyl adducts and are localized to the plasma membrane. Modification with isoprenyl groups tends to exclude a protein from lipid rafts, presumably since these groups would disrupt the packing of acyl chains in the L_O phase.

Examples of lipid-anchored proteins are found in bacteria, such as lipoproteins in *E. coli* and penicillinase in *Bacillus licheniformis*, as well as in eukaryotes, such as the catalytic subunit of cyclic adenosine monophosphate (cAMP) protein kinase, the G protein complex (Gα is acylated and Gγ is isoprenylated), the Src family of tyrosine kinases, and the Ras superfamily of small GTPases. While some lipid-modified proteins such as rhodopsin and the transferrin receptor are membrane spanning (and will be covered in other sections of this book), many are soluble except for the lipid anchor and thus are held on the membrane surface.

Proteins linked by GPI are common in animal cell membranes, and in fact account for ~0.5% of all eukaryotic proteins. Normally components of plasma membranes, they can also be found in different internal membranes following endocytosis. They play many important roles, including essential functions in embryonic development in animals, and they are also essential for viability in lower eukaryotes such as fungi. Proteins destined for attachment to GPI are synthesized as precursors with GPI-anchor sites, called ω-sites, near their carboxyl termini. The GPI moiety is synthesized by a number of glycosyl transferases. On the luminal side of the ER membrane, a GPI transamidase cleaves the peptide bond in the precursor and forms an amide bond to the ethanolamine of the GPI moiety. When these proteins do not get their GPI anchor, they remain soluble.

While GPI anchors have a common core structure, they differ in their sugar and fatty acid composition, including both saturated and unsaturated acyl chains (Figure 4.10). Furthermore, some lipid moieties are modified after attachment to the protein; for example, in many yeast GPI-anchored proteins DAG is replaced with ceramide. The GPI anchor of prion is modified with sialic acid, which could increase its lateral mobility in the membrane. Most likely these differences provide additional information for targeting GPI-linked proteins.

The role of the GPI anchor as a sorting tag was suggested by early studies of the membranes of epithelial cells whose apical and basolateral surfaces differ in composition. GPI-anchored proteins were found concentrated at apical surfaces. Furthermore, if typically basolateral membrane proteins were modified with a GPI tag, they also localized in the apical surface. GPI-anchored proteins are found in DRMs (see Chapter 2), along with sphingolipids and cholesterol. The presence of GPI-anchored proteins in DRMs depends on the presence of cholesterol, since they are solubilized by detergent treatment if the cells are first treated with saponin, which complexes cholesterol. These results have led to the concept that lipid rafts are enriched in GPI-anchored proteins, where they interact with tyrosine kinases and other signaling complexes.

Several techniques have been used to probe the association of GPI-anchored proteins with rafts. Results from FRET measurements carried out with varying protein concentrations indicate that some portion of GPI-anchored proteins are not randomly distributed, although the proportion depends on the study. Dynamics were revealed by single-particle tracking of a gold-linked GPI-anchored protein that recorded "transient confinement zones" of 200–300 nm diameter in which the particles were trapped for 10% to 15% of the time. It is likely that such large zones consist of smaller domains, each of which contains a few GPI-anchored proteins. Whether and how the proteins are targeted to these domains is still unclear, but it is possible that the nature of the GPI anchor is one determinant of raft localization since GPI

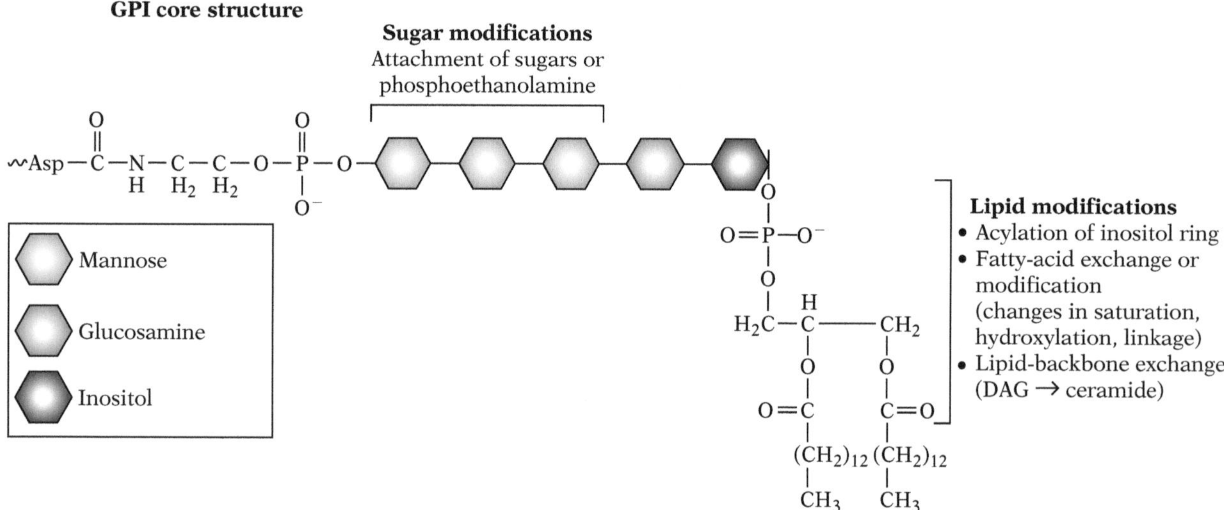

4.10 Core structure of GPI anchors, showing locations of sugar and lipid modifications. This structure has three mannose residues and a glucosamine on the phosphatidylinositol, but it can be varied by addition of extra sugars or ethanolamine phosphates to the mannose residues, acylation of the inositol, changes in the fatty acids (length, saturation, hydroxylation) or their linkages to the glycerol backbone, or remodeling of the entire DAG to ceramide. Note also the amide linkage between the ethanolamine moiety of GPI and the carboxyl group created by cleavage at the ω-site of the precursor protein. Redrawn from Mayor, S., and H. Riezman, *Nat Rev Mol Cell Biol*. 2004, 5:110–119. © 2004. Reprinted by permission of Macmillan Publishers Ltd.

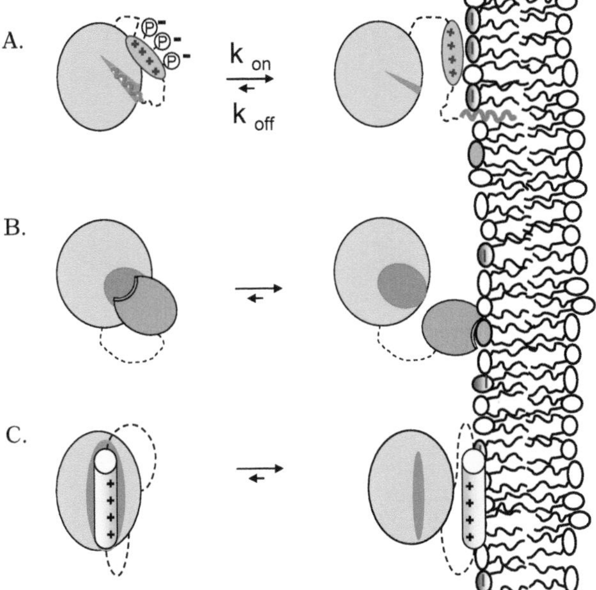

4.11 Different modes of binding amphitropic proteins to membranes to regulate their activities. (A) Electrostatic interactions between a polybasic motif (blue) on the protein and anionic lipid coupled with insertion of a lipid covalent anchor (orange) that is sequestered in the protein interior in the protein's soluble form. Polyphosphorylation antagonizes membrane binding by charge neutralization. (B) Binding via a lipid clamp (violet), a binding pocket with an affinity for a specific lipid headgroup. In many cases the lipid clamp functions as an inhibitory domain in the protein's soluble form. (C) Insertion into the bilayer of an amphipathic α-helix, which can also be auto-inhibitory in the protein's soluble form. Kindly provided by Professor Rosemary Cornell, modified from Johnson, J. E., and R. B. Cornell, *Mol Membr Bio*. 1999, 16:217–235.

anchors containing unsaturated acyl chains are expected to be much less compatible in L_o phase than those containing saturated chains (see "Lateral Domains and Lipid Rafts" in Chapter 2).

Reversible Interactions of Peripheral Proteins with the Lipid Bilayer

The reversible associations between peripheral proteins and membranes are accomplished by a number of mechanisms, singly or in combination (Figure 4.11).

An important mechanism is electrostatic attraction between charged groups on the peripheral protein and charged groups on either lipid or protein components of the membrane, exemplified by cytochrome *c* and myelin basic protein, described above. As will be seen below, changes that affect the net charge on these peripheral proteins alter their affinity for the membrane, setting up an "electrostatic switch" that can control their binding. A second mechanism is the insertion of a lipid anchor, which occurs when proteins with covalently bound acyl chains can markedly shift the position of the chains in the aqueous and membrane-bound forms. In the water-soluble form of such a protein, the acyl chain is sequestered into the protein interior; then, at the membrane, it projects from the protein to embed into the closer leaflet of the membrane. In the third mechanism, the protein has a binding site for a particular lipid that is often specific for the headgroup. In this case, association of the protein with the membrane depends on its affinity for the lipid as well as the concentration of the lipid. To modulate the binding to a high-affinity site, the protein may have a "flap" that covers the binding site in the aqueous form, as described above for PLA_2. The fourth mechanism is insertion of an amphipathic helix in the interfacial region of the bilayer, requiring a large conformational change in the protein. While the peptide usually occupies the polar headgroup region of the membrane, in some cases it may penetrate the non-polar region, in which case it generally does not extend more than halfway into the leaflet to which it binds (see Frontispiece). Before describing these interactions in more detail, the effects on membrane lipids of binding peripheral protein will be considered.

Effects of Peripheral Protein Binding on Membrane Lipids

Thermodynamic studies reveal the impact of binding a peripheral protein on the general properties of lipids in the bilayer, exemplified by DSC measurements of lipid phase transitions in the presence of increasing amounts of protein. These calorimetry experiments measure the heat capacity of the system, which is the heat input divided by the temperature change, as described in Chapter 2. Both cytochrome *c* binding to DMPG bilayers and myelin basic protein binding to DMPS bilayers produce shifts in T_m, although the shifts are in the opposite directions (Figure 4.12). Binding increasing amounts of cytochrome *c* results in increased broadening of the phase transition, as well as a shift of T_m up to +5°C, chiefly due to the effect of protein binding on the surface electrostatics of the bilayer. When the chains melt, the charge density is less since the area per lipid is larger, which means there is a lower electrostatic binding energy between cytochrome *c* and the membrane. (In addition, when cytochrome *c* binds, the strong ionic attractions to anionic headgroups can cause those phospholipids to cluster, as described in the next paragraph.) Titration calorimetry indicates the binding of cytochrome *c* is endothermic; therefore, it is driven by entropy. The opposite shift is observed with myelin basic protein, where the binding is driven more by enthalpy, most likely because in addition to the electrostatic interaction with anionic lipid headgroups, myelin basic protein has

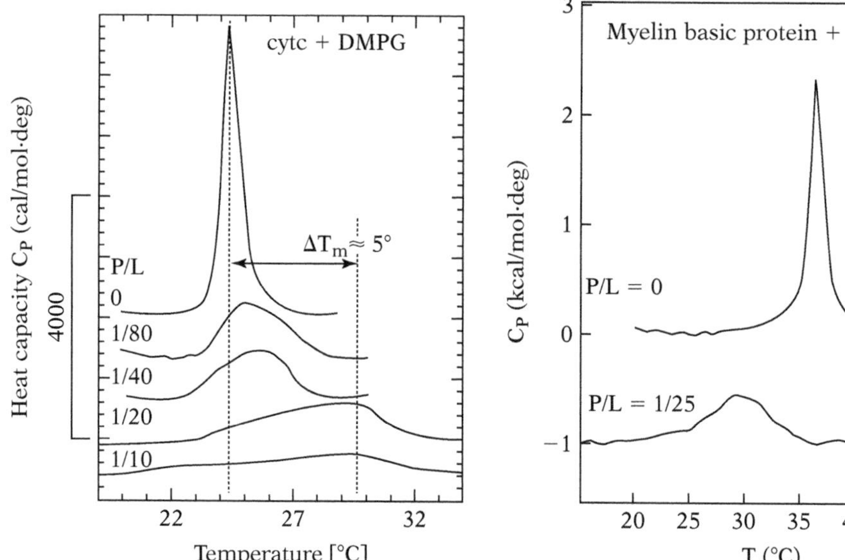

4.12 Effect of peripheral proteins on heat capacity of lipid bilayers. Changes in heat capacity detected by DSC illustrate the influence of binding peripheral proteins on the chain melting of the lipid. (A) Binding cytochrome c to DMPG bilayers was measured at five different degrees of surface saturation indicated by the protein:lipid ratio (P/L). Increasing surface saturation broadens the transition and shifts the heat capacity maximum (T_m), until at ~100% saturation it shifts by 5°, corresponding to a change of –4.3 kcal mol^{-1}. From Ramsay, G., *et al.*, *Biochemistry*. 1986, 25:2265–2270. © 1996, Elsevier. (B) Binding myelin basic protein to DMPS bilayers broadens and lowers (by 1.6 kcal mol^{-1}) the heat capacity of the phase transition. Redrawn from Heimburg, T., and R. L. Biltonen, *Biophys J.* 1996, 70:84–96. © 1986, American Chemical Society.

hydrophobic segments that intercalate into the nonpolar domain of the bilayer

In addition to the effect on heat capacity, protein binding to the bilayer can affect the organization and state of the lipids. How much the lipids rearrange when the protein binds depends on the relative affinities of the different lipid species for the protein, as well as the mixing properties of the lipids, that is, whether the interactions between lipids are favorable or unfavorable (Figure 4.13). This subject is of great relevance to the mechanism of formation of lipid rafts. Measurements of cytochrome *c* binding to the surface of DOPG bilayers at different ionic strengths fit quite well to predicted isotherms for nonlocalized binding with little interaction between proteins, with the effective charge determined to be +3.8 for cytochrome *c* (Figure 4.14; see Box 4.1). The calculations give a lipid stoichiometry of 12 lipids per protein, like the stoichiometry determined by crystallography.

Interactions between Peripheral Proteins and Lipids

Binding of peripheral proteins typically involves electrostatic interactions. If the binding shows an absolute requirement for anionic lipid and is abolished in high ionic strength, it can be attributed solely to electrostatic interactions. Studies of peptides whose binding to the membrane is strictly electrostatic indicate that these peptides actually remain about 3 Å from its surface. When basic peptides bind to acidic lipid vesicles, the contribution from each positive charge on the peptide is approximately –1.4 kcal. The strength of the association is not dependent on the chemical nature of the anionic lipids (e.g., PG, PI, or PS) or of the amino acids (Lys or Arg), but obviously it is diminished by deletion of one or more of the basic amino acids. Reversible reactions that decrease the net charge provide opportunities for modifying the force of the electrostatic interaction (see below).

Quite often, binding of peripheral proteins to the membrane surface involves hydrophobic interactions in addition to electrostatic interactions. When a peripheral protein inserts one or more acyl chains to anchor into the bilayer, the hydrophobic interaction makes a significant contribution to the binding. Studies with model acylated peptides show the binding energy is proportional to the chain length, with –0.8 kcal mol^{-1} per CH_2 group, mirroring Tanford's partitioning data for fatty acids (see Figure 1.4). Myristoylated peptides bind electrically neutral liposomes with ΔG = –8 kcal mol^{-1}, which indicates that 10 methylenes of the 14 in the chain are embedded in the nonpolar domain. The corresponding binding constant (K_a) of 10^4 M^{-1} is insufficient to maintain stable binding of the protein to the membrane. Additional binding strength is provided by a second acyl chain or prenyl group, or by another mechanism, such as a group of basic residues that participate in electrostatic interactions. The latter mechanism

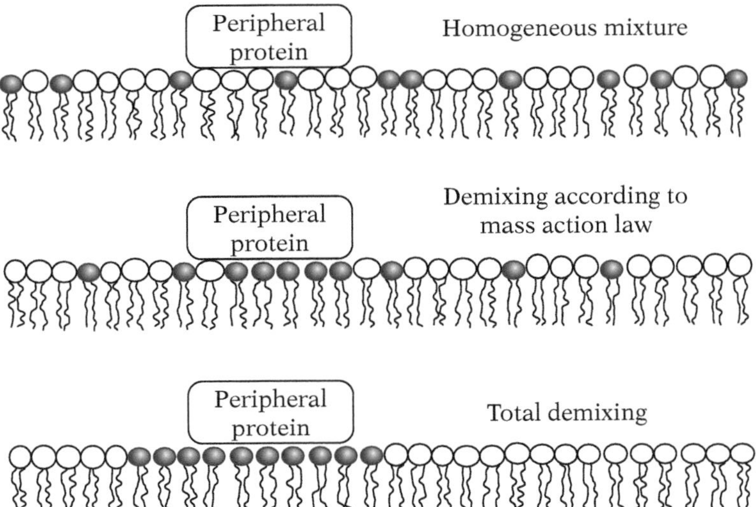

4.13 Schematic representation of the different effects of peripheral proteins on lipid organization. Two different species of PL are indicated by the light and dark headgroups. Binding of a peripheral protein can cause (1) no lipid rearrangement (top); (2) redistribution of lipids with no preferential interaction between them, according to different affinities for protein (middle); and (3) concentration of certain lipids in a domain that binds protein, resulting in total demixing (bottom). Redrawn from Heimburg, T., and D. Marsh, in K. Merz and B. Roux (eds.), *Biological Membranes*, Birkhauser, 1996, p. 410. © 1996 by Springer-Verlag. With kind permission of Springer Science + Business Media B. V.

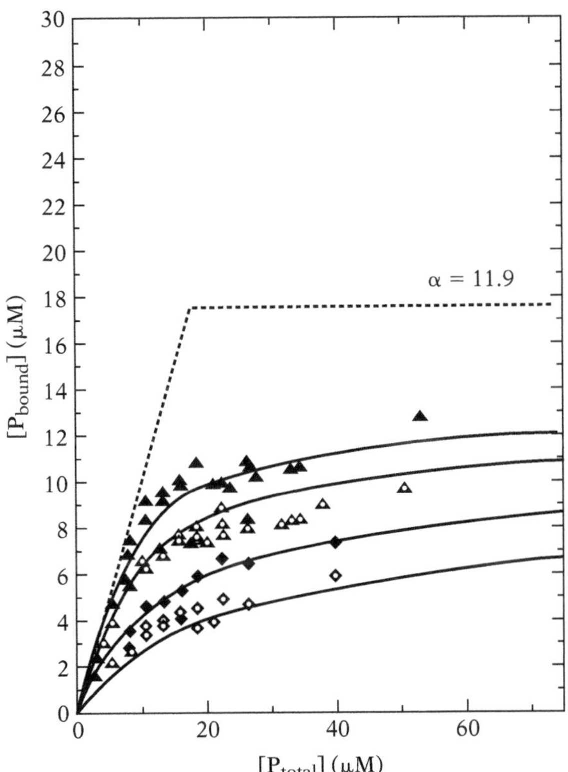

4.14 Binding isotherms for cytochrome c binding to DOPG at neutral pH, 20°C, which plot the concentration of protein bound (*y*-axis) as a function of the total protein (*x*-axis) at ionic strengths of 41.9, 54.4, 79.4, and 104.4 mM (from top to bottom). Redrawn from Heimburg, T., and D. Marsh, in K. Merz and B. Roux (eds.), *Biological Membranes*, Birkhauser, 1996, p. 415. Copyright © 1995, Elsevier. Reprinted with permission from Elsevier.

is used by the signaling kinase Src, which is myristoylated on its N-terminal glycine and has three lysine and three arginine residues in its membrane-binding N-terminal domain (Figure 4.15). Measurements of the binding of the 15-residue peptide corresponding to this domain (myristate-GSSKSKPKDPSQRRR[1]) to vesicles containing 2:1 PC:PS show the myristoylated peptide binds with $K_a = 10^7\ M^{-1}$, while the nonmyristoylated Src peptide binds with $K_a = 10^3\ M^{-1}$. Recalling that the K_a for insertion of the myristate is $10^4\ M^{-1}$, it is clear that the hydrophobic and electrostatic energies add (i.e., the binding constants multiply).

In addition to its contribution to the binding energy, the insertion of a lipid chain can help localize the basic residues of the peripheral protein to the appropriate sites on the membrane. In the case of the Src peptide, there is only 1–3 nm between the lipid-anchored N terminus and the cluster of basic residues. In other peripheral proteins the distance is greater, yet the anchoring greatly increases the probability that the oppositely charged groups will come in close proximity.

A good example of combining both hydrophobic and electrostatic interactions to bind the membrane is provided by the C1 and C2 domains of PKC, described above. The C1 domain has a well-characterized binding site for DAG and phorbol esters (analogs of DAG that promote tumor formation), and the C2 domain has a Ca^{2+}-binding site that interacts with anionic lipids. The C1 domains do not interact with membranes lacking DAG and show

1 Single-letter amino acid codes, often used for motif names and for peptide sequences, are given in Appendix II.

BOX 4.1 BINDING OF LIGANDS TO SURFACES

The Langmuir isotherm gives the simplest mathematical model to describe the binding of a ligand to defined binding sites on a surface and can approximate the binding of peripheral proteins to the surface of a membrane. It characterizes the binding of the surface, S, by the ligand, L, as S + L → SL by describing the binding constant, K, in terms of bound and free sites:

$$K = [S_b]/([S_f]\,[L]),$$

where S_f are free binding sites and S_b are sites with ligand bound, and [L] is the concentration of free ligand (protein). Solving this equation for [L] gives

$$[L] = [S_b]/(K\,[S_f]).$$

Then, if θ is the fraction of occupied sites, $\theta=[S_b]/([S_b]+[S_f])$. Solving this equation for $[S_b]$ / $[S_f]$ and substituting that into the equation derived for [L] gives the Langmuir isotherm,

$$[L] = \frac{1}{K} \cdot \frac{\theta}{1-\theta}.$$

In the Langmuir-type isotherm the proteins bind to a surface with fixed, independent binding sites (Figure 4.1.1, top). However, the Langmuir isotherm is a fair approximation for protein binding to a bilayer only in the condition of very little protein binding ($\theta << 1$) as it does not account for steric interactions between proteins that become likely at higher amounts of binding. Furthermore, the binding sites for peripheral proteins on a membrane are dynamic and delocalized (Figure 4.1.1, bottom). When they bind to a continuous surface, the proteins can rearrange on the surface and interact with neighboring proteins. The mathematics to describe such delocalized binding utilizes the Gibbs absorption isotherm, along with Gouy–Chapman and Debye–Hückel theories to model the electrostatic energy of binding.

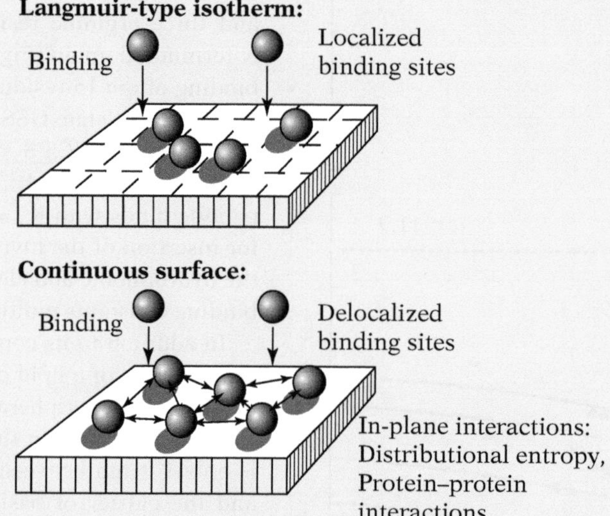

4.1.1 Different models for binding of peripheral proteins to a surface. Redrawn from Heimburg, T., and D. Marsh, in K. Merz and B. Roux (eds.), *Biological Membranes*, Birkhauser, 1996, p. 408. © 1996 by Springer-Verlag. With kind permission of Springer Science and Business Media.

The Gibbs absorption isotherm treats the lateral interactions between proteins on the surface as a two-dimensional pressure, $\Pi(i)$, where i is the number of proteins bound. Then

$$d\,\Pi(i)/d\ln[L] = ikT/n\,\Delta A,$$

where n is the number of proteins that saturate the surface and ΔA is the excluded area per protein. [L] is still the free protein concentration. In the simplest case,

$$\Pi(i) = ikT/(n-i)\,\Delta A,$$

(the Volmer equation of state).

BOX 4.1 *(continued)*

After substituting and integrating, the isotherm becomes

$$[L] = \frac{1}{K} \cdot \frac{\theta}{1-\theta} \exp \frac{\theta}{1-\theta},$$

with $\theta = i/n$. This equation fits the binding to continuous surfaces with delocalized sites much better than the Langmuir isotherm and allows for a higher binding constant.

To calculate the free energy of binding of a charged ligand to a membrane with consideration of surface electrostatics, both the Gouy–Chapman and Debye–Hückel theories are used. The Gouy–Chapman (and later Gouy–Chapman–Stern) theory describes quantitatively the electrical potential energy as a function of the distance from the bilayer surface, assuming the charges on the surface are spread out rather than localized. An important result is the existence of an electrostatic double layer created by the balance between the entropic drive for ions to randomize and their electrical attraction to the surface. The Debye–Hückel theory gives the distribution of electric potential around an ion in solution, describing the thermal and electrical forces around the ion. Using Gouy–Chapman to calculate the electrostatic free energy of the bilayer surface and Debye–Hückel for the electrostatic energy of the free peripheral protein (the ion) gives a complex mathematical expression for the change in total electrostatic free energy when the charged protein binds to the surface. From the fit of empirical data to the theoretical curves, it appears that most peripheral proteins bind to membranes with some demixing (nonrandom rearrangement) according to the law of mass action, as shown in the intermediate diagram of Figure 4.13.

strict specificity for the physiological stereoisomer. The binding has been characterized by fluorescence (using phorbol ester analogs), NMR, and x-ray crystallography. Lipid binding displaces water from the pocket and increases the hydrophobicity of the protein surface, thus increasing its affinity for the membrane. The C2 domain binds two to three calcium ions coordinated by aspartate residues and carbonyl groups, which stabilizes a loop region of its structure and increases the affinity for the membrane.

Domains Involved in Binding the Membrane

C1 and C2 domains of PKC have now been observed in many proteins involved in cell signaling. In addition, there are two other dominant types of domains utilized

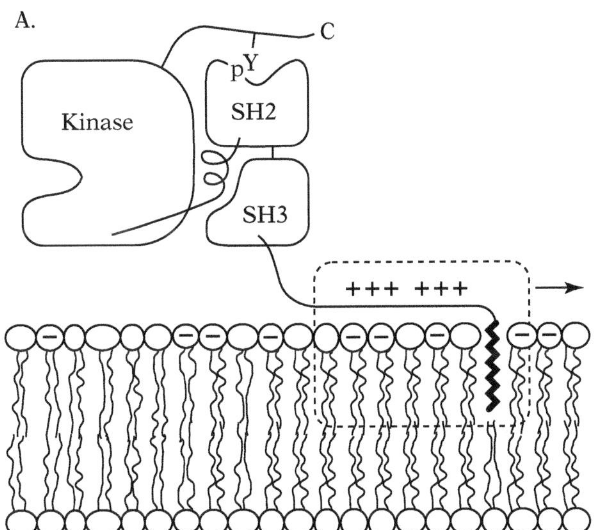

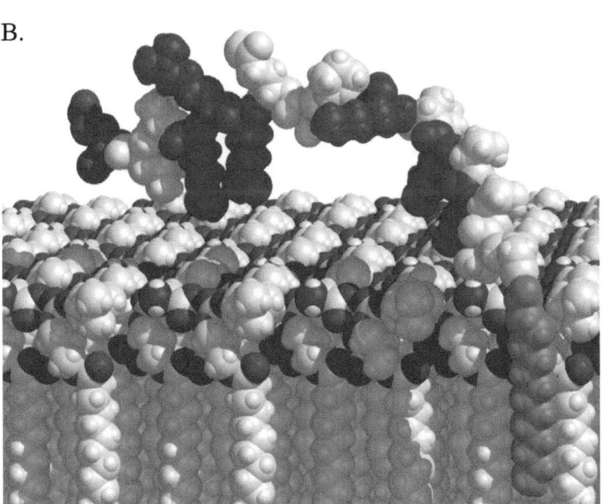

4.15 Models of the protein–lipid interactions involved in binding the N-terminal Src peptide to lipid vesicles. **(A)** An illustration of the domain structure of c-Src shows the myristate chain on the N terminus, near a cluster of basic residues that interact with anionic lipids in the bilayer. **(B)** A space-filling representation of the portion surrounded by a dashed line in **(A)** illustrates the docking of the myristoylated 18-residue peptide in a PC:PS (2:1) bilayer. The model shows the insertion of the myristoyl group (green) and the interactions of the six basic residues (blue) of the N-terminal peptide with phosphatidyl serine headgroups (acidic residues red and the nitrogen atoms blue). Insertion of the acyl chain confines the peptide, increasing the chance of forming electrostatic interactions between the basic residues with anionic lipids. From Murray, D., *et al.*, *Structure*. 1997, 5:985–989. © 1997, Elsevier. Reprinted with permission from Elsevier.

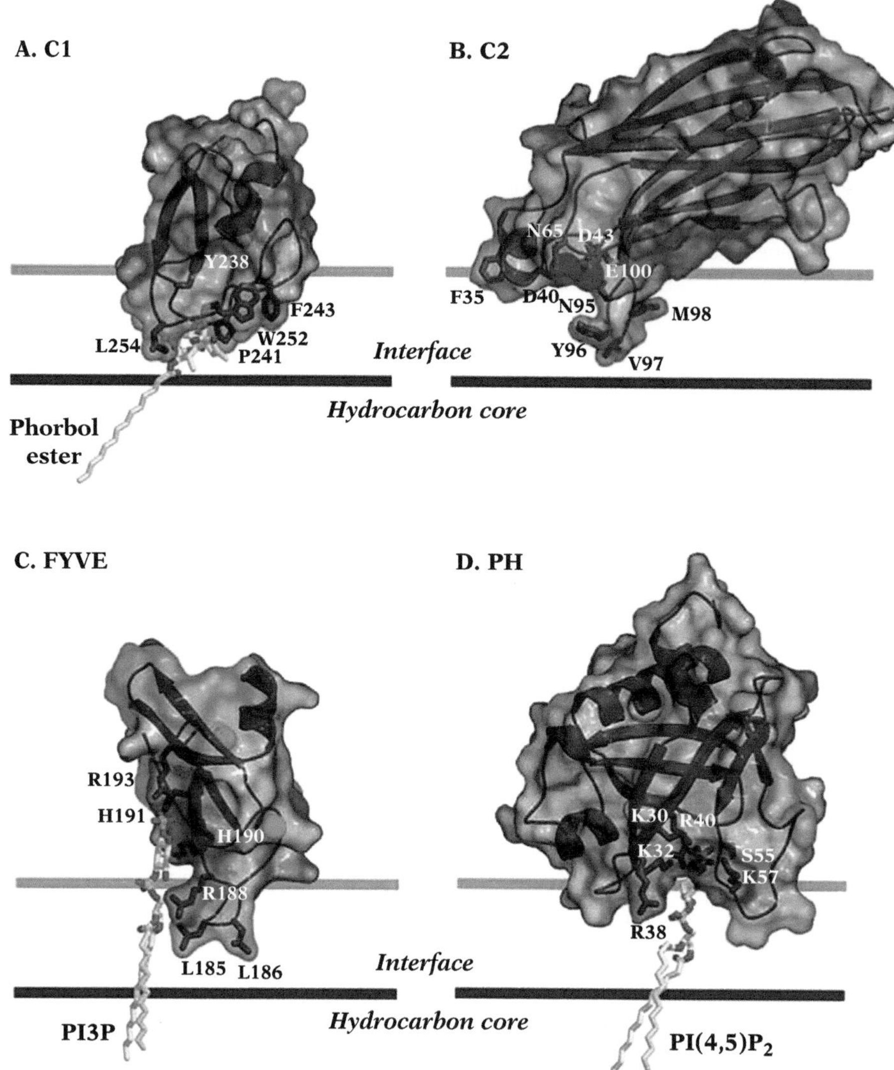

4.16 Four types of membrane-binding domains found in hundreds of peripheral proteins involved in signal transduction. The x-ray structures reveal major features of four membrane-binding domains: (A) C1 domain from protein kinase C; (B) C2 domain from PLA_2; (C) FYVE domain from Vps27p (a yeast protein for endosomal maturation); and (D) PH domain of phospholipase C. Selective residues that make contact with the surface are labeled, and specifically recognized lipids are modeled. PI3P is phosphatidylinositol 3-phosphate and $PI(4,5)P_2$ is phosphatidylinositol 4,5-bisphosphate. The membrane leaflet is divided into an interfacial zone and the hydrocarbon core. Hydrophobic residues are colored green and basic residues are blue. From Hurley, J. H., and S. Misra, *Annu Rev Biophys Biomol Struct*. 2000, 29:49–79. © 2000 by Annual Reviews. Reprinted with permission from the Annual Review of Biophysics and Biomolecular Structure, www.annualreviews.org.

for membrane binding by peripheral proteins: FYVE zinc-binding domains, named for their amino acid motif, bind to polyphosphorylated inositol, and pleckstrin homology (PH) domains bind phosphoinositides and are essential in PI3-kinase signaling pathways. These C1, C2, FYVE, and PH domains have now been identified in hundreds of proteins involved in signal transduction and membrane trafficking, enabling them to bind the membrane and thereby regulating their localization and activity.

Each domain type consists of a specific binding site (usually for a lipid ligand) that is often flanked by additional, less specific membrane-binding sites. The nonspecific binding is important in increasing the overall affinity for the membrane and/or making an additional point of contact to define the stereoselectivity of the specific binding site. The weak interactions can also be enhanced by the combination of two (or more) domains in the protein or by the oligomerization of protein subunits, which allows the temporal variation needed in regulatory systems.

Examples of each of the four types of domains have been characterized by x-ray crystallography (Figure 4.16), NMR, fluorescence, mutagenesis, and localization of green fluorescent protein fusions. The C1 domain, now identified in >200 different proteins, has ~50 amino acids making two small β-sheets with a short α-helix built around two 3-Cys-1-His clusters that bind Zn^{2+} very tightly. As described for PKC (above), this domain binds DAG and phorbol esters. The binding occurs at the tip of the domain, unzipping the two β-strands to expose the binding site. This groove is located in a hydrophobic end of the domain that is adjacent to a ring of basic residues positioned so that membrane penetration by the hydrophobic tip allows the basic ring to contact the membrane surface. Most C1 domains are not targeted to the membrane if they lack specific binding sites for DAG, although there are a few atypical C1 domains that do not require DAG. Most PKCs and DAG kinases have pairs of C1 domains, and in some, DAG is an allosteric activator.

C2 domains, identified by a conserved sequence motif that binds Ca^{2+} reversibly, have been found in >400 proteins, including many involved in signal transduction, inflammation, synaptic vesicle trafficking, and membrane fusion. The C2 domain is a β-sandwich like the immunoglobulin fold, with the Ca^{2+}-binding sites formed by three loops at a tip analogous to the antigen-binding site. It is not easy to generalize about this domain. There are five different C2 domain structures that fall into two permutations of this fold. Some are hydrophobic enough to penetrate the membrane, while others are not; most require acidic PLs, while the C2 domain of the c isoform of PLA_2 prefers neutral lipids, especially PC. There are even C2 domains that do not bind calcium!

The FYVE domain, identified in ~60 proteins, is noted for its specificity in binding phosphatidylinositol 3-phosphate (PIP_3), which enables it to target proteins to endosomal membranes that are enriched in PIP_3. This domain consists of 70–80 residues, forming two small double-stranded β-sheets and an α-helix, with a conserved RKHHCR motif that binds PIP_3. It also contains eight Cys or seven Cys with one His that coordinate two Zn^{2+} ions, two Leu residues at one end that protrude into the membrane, and a few less-conserved Lys residues that probably contact the membrane surface.

The PH domains bind different phosphoinositides with varying degrees of specificity and thus respond to signaling that interconverts phosphoinositides having different phosphorylation patterns. Found in 500 proteins, this domain consists of two curved β-sheets of three or four strands capped by an α-helix. It has different subsites that bind phosphate groups to make up the substrate-binding site of varying affinity and specificity. It has a positively charged face that interacts with acidic lipids in the membrane; mutations that strengthen this interaction can result in constitutive activation, while mutations that decrease it can lead to loss of function. Some PH domains also participate in protein–protein interactions and some provide allosteric regulation. Like the other domains, adjacent portions of the PH domain contribute nonspecific interactions to give variable interplay with the membrane.

A less common mechanism for binding to the membrane is the insertion of an amphipathic α-helix parallel to the plane of the bilayer, observed in a miscellaneous group of proteins including DnaA, adenosine diphosphate (ADP)-ribosylation factor, vinculin, epsin, several regulators of G protein signaling, and cytidyltransferase (CT). The hydrophobic interactions between the hydrophobic face of the helix and the nonpolar core of the bilayer provide the driving force for insertion, which is opposed by the resulting perturbation of lipid packing. Formation of the α-helix concomitant with insertion can provide a significant additional driving force (see below).

CT is a well-characterized representative of this group of amphitropic proteins. It carries out the transfer of a cytidyl group from cytidine triphosphate (CTP) to phosphocholine in the key regulatory step for the synthesis of PC. Studies employing circular dichroism indicate that its ~60-residue membrane-binding domain changes from a mix of β-strand, β-turn, and disordered conformations to α-helix upon binding the membrane. The amphipathic helix has basic residues on one face and a mixture of acidic and basic residues on another, with a center strip dominated by acidic residues. A nonpolar face contains a total of 18 hydrophobic residues giving a hydrophobic surface area of ~2500 Å^2. Interestingly, three glutamate residues positioned at the interfacial region contribute to the selectivity of CT for anionic lipids because they become protonated in the low pH milieu at the surface of anionic, but not zwitterionic, membranes (Figure 4.17). Studies of CT binding to multilamellar vesicles (MLVs) suggest that the first step of membrane association is electrostatic: CT binds when the MLVs are in gel phase, but can only insert its amphipathic helix upon raising the temperature above the transition to liquid crystalline phase.

Curvature

Insertion of an amphipathic helix is one way to generate membrane curvature, which is essential for many cellular processes, including division, vesicle trafficking, and extension of neurons. Another domain found in hundreds of proteins that influences curvature is called the BAR domain, named for some of the first proteins in which it was identified, Bin/amphiphysin/Rvs. BAR domains dimerize to form concave modules with intrinsic curvature (Figure 4.18). Patches of basic residues on BAR domains allow them to form electrostatic interactions with negatively charged phospholipids. Variations of BAR domains include N-BAR with an additional amphipathic helix, F-BAR with a nearby coiled coil, and I-BAR, which are nearly flat with an inverse curvature. BAR domains are also found in proteins containing some of the other

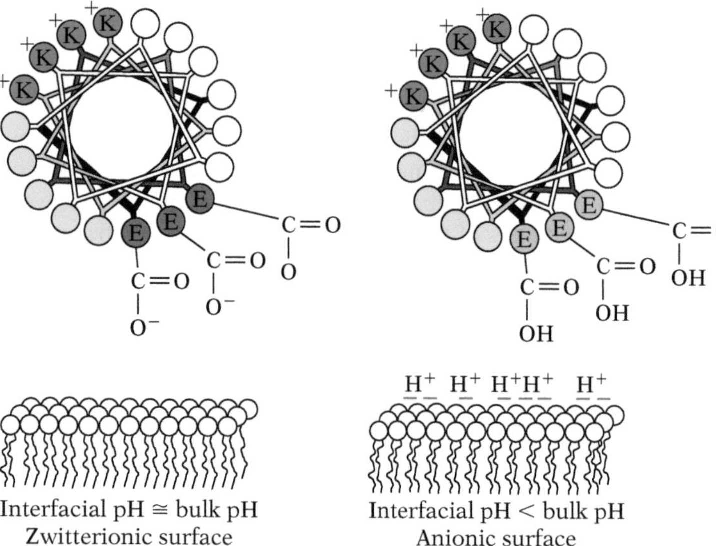

4.17 Helical wheel of the amphipathic helix of cytidyltransferase. Since the pH at the surface of anionic (but not zwitterionic) membranes is lower than the bulk pH due to the attraction of protons to the negative surface, the probability of protonation of three Glu residues increases, which effectively increases the hydrophobicity of the surface of the peptide. Redrawn from Johnson, J. E., *et al.*, *J Biol Chem*. 2003, 278:514–522. © 2003, by the American Society for Biochemistry & Molecular Biology.

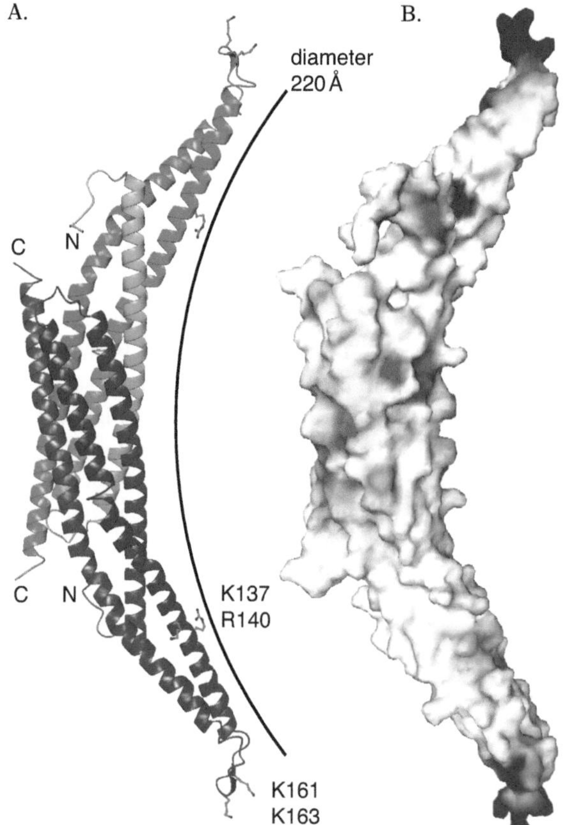

4.18 The structure of the BAR domain from amphiphysin. The BAR domain from the *Drosophila* amphiphysin is a crescent-shaped antiparallel homodimer shown in ribbon representation ((A) with purple and green monomers) and as a surface colored by electrostatic potential ((B) red, negative; blue, positive). The residues that make positive patches on the convex surface (marked on the lower monomer in (A)) form electrostatic interactions with anionic lipids in a bilayer with curvature shown by the line. From Peter, B. J., *et al.*, *Science* 2004, 303:495–499. © 2004, American Association for the Advancement of Science. Reprinted with permission from AAAS.

membrane-binding domains such as PH domains that confer lipid specificity.

BAR domains both induce and stabilize bilayer curvature but do not have strong effects by themselves. BAR domain proteins work with other proteins in membrane remodeling. A well-characterized example is the formation of clathrin coats for endocytosis, which utilizes the N-BAR proteins endophilin and amphiphysin, along with other BAR variants and other endocytotic and cytoskeletal proteins to accomplish the deep invagination required for vesicle fission. As might be expected, these processes are tightly regulated and targeted in response to signaling pathways.

Modulation of Binding

Since membrane binding regulates the activity of amphitropic proteins and thus controls many key processes of the cell, modulation of reversible binding to the membrane is crucial. It is accomplished by one or more of several mechanisms that respond to signaling kinases, altered levels of ions or effector molecules, or changes in local compositions of the membrane that are sometimes linked to trafficking, which is the targeted movement of specific molecules to particular regions or organelles. A simple mechanism is the disruption of the electrostatic interactions between basic groups of the protein and anionic lipids by the addition of negative charges when the protein is phosphorylated on serine or tyrosine residues in the membrane-binding region. In the example of the N-terminal region of Src, the phosphorylation site is a serine located between the clusters of lysine and arginine residues. Because the phosphate can be cleaved by phosphatases, the reversible change in charge provides an "electrostatic switch" that affects membrane binding.

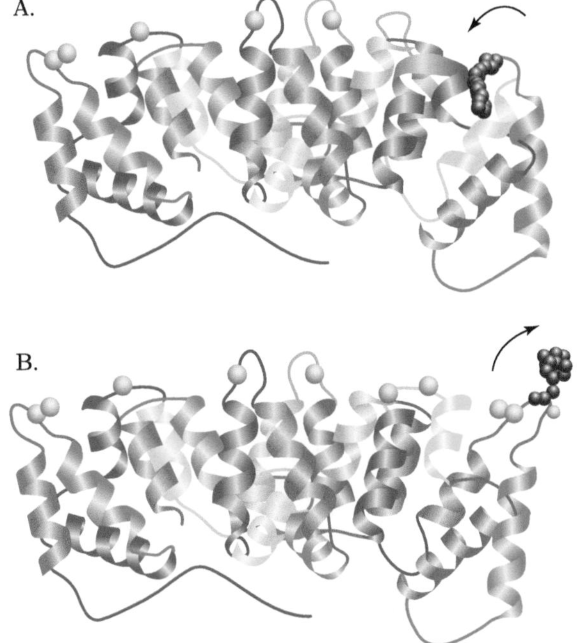

4.19 Model for the Ca^{2+}-triggered extrusion of a Trp residue in annexin V. The ribbon diagrams show annexin V viewed from the side with the membrane-binding surface face up and the mobile Trp residue at the right. Ca^{2+} ions are blue spheres. (A) If Ca^{2+} is absent from domain 3, the Trp side chain is buried. (B) When Ca^{2+} binds domain 3, the Trp side chain emerges from its buried position to interact with the lipid bilayer. From Seaton, B. A., and M. F. Roberts, in K. Merz and B. Roux (eds.), *Biological Membranes*, Birkhauser, 1996, p. 385. © 1996 by Springer-Verlag. With kind permission of Springer Science + Business Media B. V.

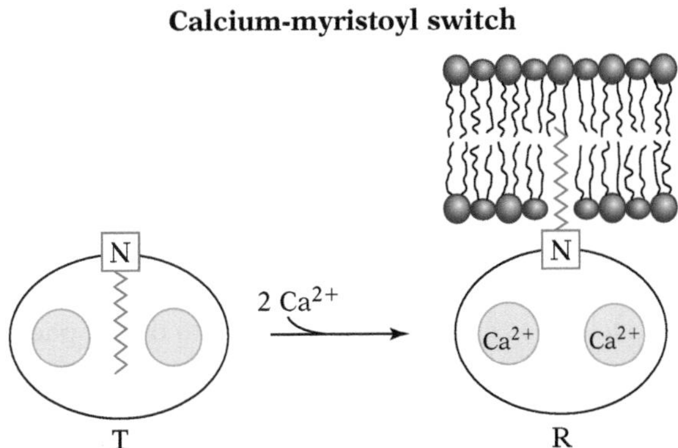

4.20 Diagram of the calcium-myristoyl switch. A conformational change in recoverin enables it to extrude its bound myristoyl group upon binding calcium. The protruding acyl chain interacts with the lipid bilayer and activates the protein to prolong the photoresponse of rod and cone cells. From Seaton, B. A., and M. F. Roberts, in K. Merz and B. Roux, *Biological Membranes*, Birkhauser, 1996, p. 393. © 1996 by Springer-Verlag. With kind permission of Springer Science + Business Media B. V.

Another mode of regulation is by the enzymatic addition and removal of an acyl chain. In these cases, the first lipid anchor is permanently added in a posttranslational modification and barely provides enough energy for the protein to bind to the bilayer. Modification with a second acyl chain (or prenyl group) doubles the hydrophobic interactions with the bilayer, converting a weak association into a strong one; thus its addition and removal regulate localization to the membrane. For example, after the α subunit of the G_s protein complex is myristoylated in a posttranslational modification that remains for the lifetime of the protein, specific addition and removal of a palmitate chain by acyltransferase and thioesterase activities control its affinity for the membrane.

A third major type of regulation involves binding ligands such as nucleotides or calcium. The binding of guanosine triphosphate (GTP) to the G protein ADP-ribosylation factor causes a conformational change that exposes its amphipathic helix, which then inserts into the bilayer. Calcium binding can trigger structural changes in the protein that affect its membrane-binding surface, as seen in annexin V. The crystal structure of annexin V shows a flattened molecule with opposing convex and concave faces. EM studies suggest the convex side flattens on the surface of the membrane, allowing multiple Ca^{2+}-binding loops to contact the surface. Ca^{2+} binding triggers the rotation of a single tryptophan side chain, moving it from a buried position to a protrusion that intercalates into the bilayer (Figure 4.19). Binding calcium has a very different effect on the retinal protein recoverin. When recoverin binds Ca^{2+}, its conformational change induces the extrusion of a bound myristoyl group that becomes a membrane anchor (Figure 4.20). This "myristoyl switch" has been observed in other peripheral membrane proteins that are involved in signal transduction.

Finally, those amphitropic proteins that bind a specific lipid component of the membrane, such as DAG or polyphosphorylated inositols, are subject to temporal changes in the concentrations of their ligands in the bilayer, and these concentrations are controlled by the phospholipases in response to signaling. Furthermore, the different concentrations of the specific lipid ligands in various membranes of eukaryotic cells can contribute to membrane-selective targeting, as seen in the enrichment of endosomal membranes for PIP_3.

PROTEINS AND PEPTIDES THAT INSERT INTO THE MEMBRANE

While a few amphitropic proteins have small segments that insert into the nonpolar domain of the bilayer, there are soluble proteins and peptides that insert more extensive portions to cross the bilayer, often via major conformational changes. Insertion of these TM domains is generally governed by the same forces that drive the insertion of TM segments of integral membrane proteins, described in detail below. Besides the hydrophobic and electrostatic forces, insertion of a peptide across the bilayer involves the perturbation of the acyl chains in the membrane, immobilization of the peptide, and possibly its unfolding or refolding. Clearly the state and lateral tension of the lipid are important: the enthalpy of peptide insertion into small unilamellar vesicles was greater for insertion into tightly packed vesicles than loosely packed ones. How soluble proteins and peptides such as toxins, colicins, and ionophores move into the membrane milieu to accomplish their functions is fascinating.

Toxins

Some protein toxins that come from outside the cell and affect cytosolic targets provide their own mechanism of translocation across the membrane. Decades of research on diphtheria toxin (DT) established the AB model, in which part A of the toxin carries out its catalytic attack on an intracellular target and part B carries out its translocation. DT has three domains, a catalytic C domain at the N terminus, a receptor-binding R domain at the C terminus, and between them a T domain involved in translocation (Figure 4.21). After binding to a receptor on the cell surface, DT is cleaved into the A fragment consisting of the C domain and the B fragment containing both T and R domains; A and B are linked by two disulfide bonds. Because DT is internalized by endocytosis, it enters the cell in acidic membrane-bound compartments called endosomes. The low pH of the endosomes triggers acid-induced conformational changes that enable the T domain to insert into the endosomal membrane and translocate the A domain into the cytosol, where it carries out ADP-ribosylation of elongation factor 2, inhibiting protein synthesis and leading to cell death. The T domain of DT is a bundle of ten α-helices, two of which are hydrophobic and insert as a helical hairpin that spans the membrane. In planar lipid bilayers at low pH (≤6), DT as well as its B fragment and isolated T domain have all been observed to form cation-selective channels, but their structures in the membrane and the mechanism for translocation of the A domain are not known.

The family of AB toxins has grown to include cholera toxin, pertussis toxin, tetanus and botulinum neurotoxins, and anthrax toxin, although it is structurally more complicated than the others. Anthrax toxin is made up of three different proteins, PA (protective antigen), EF (edema factor), and LF (lethal factor). PA, the antigen for the anthrax vaccine, binds to a receptor on the cell surface and is then cleaved, releasing a 20-kDa fragment. The remaining 63-kDa PA fragment oligomerizes to form a ring-shaped heptamer called a prepore, which binds up to three molecules of EF and/or LF (Figure 4.22). Following endocytosis, low pH triggers a conformational

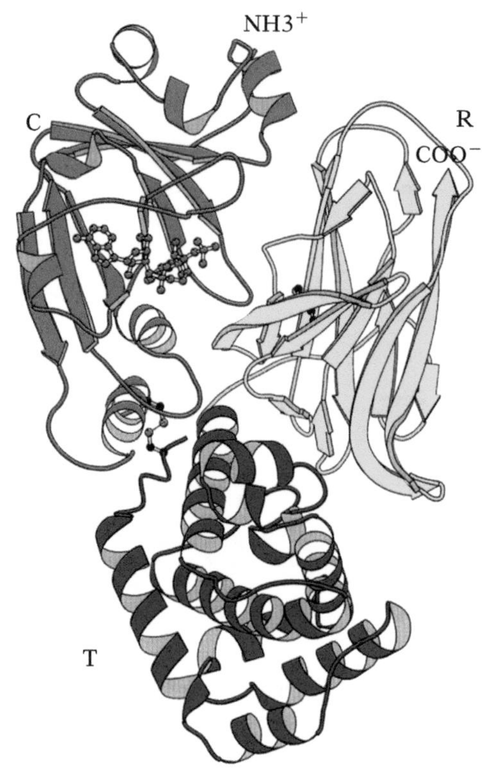

4.21 High-resolution structure of diphtheria toxin. Diphtheria toxin has three domains: the C (catalytic), R (receptor-binding), and T (translocation) domains. The active site cleft of the C domain contains the endogenous dinucleotide ApUp. Two helices of the T domain insert into the membrane as a helical hairpin. From Collier, R. J., *Toxicon*. 2001, 39:1793–1801. © 2001 by The International Society on Toxinology. Reprinted with permission from Elsevier.

change in PA, which forms a pore that translocates EF and LF into the cytosol, where they exert their lethal inhibitory actions.

While the x-ray structure of the PA prepore was determined over 15 years ago, structural analysis of the PA pore was prevented by aggregation until it was stabilized by the chaperone GroEL or by insertion into vesicles or nanodiscs. The low-resolution EM structure shows a mushroom-shape with a 100 Å-long stem and 15 Å-wide lumen; it fits a model of a 14-stranded β-barrel, formed when each subunit inserts a β-hairpin into the membrane (Figure 4.23, see also Figure 4.24C). When reconstituted in planar bilayers the PA pore has cation channel activity that is blocked by adding excess LF or EF, and driven by a pH gradient it translocates LF and EF. Ingenious biotin/streptavidin experiments measured the length of the PA channel by the minimum number (33) of residues of the LF N terminus needed to emerge from the pore.

The anthrax toxin PA pore is very similar to the pore formed by α-hemolysin (αHL), a hemolytic toxin secreted by *Staphylococcus aureus* in a soluble form. A major

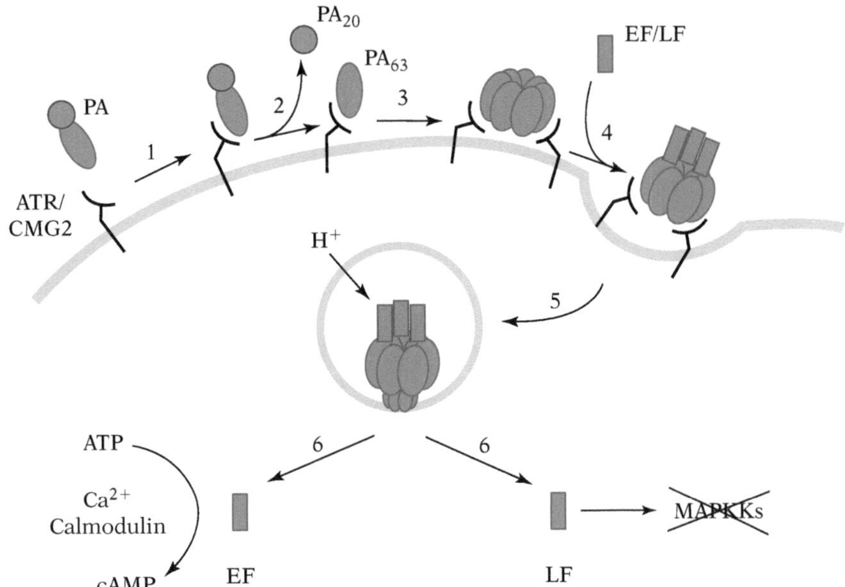

4.22 Steps in the internalization of anthrax toxin. Anthrax toxin is made up of three proteins, PA (protective antigen), EF (edema factor), and LF (lethal factor). (1) PA binds to a receptor, either ATR or CMG2. (2) Cleavage by a furin protease removes PA_{20}. (3) PA_{63} self-associates to form the heptameric prepore. (4) Up to three molecules of EF and/or LF bind to the prepore. (5) Endocytosis brings the complex to an acidic intracellular compartment. (6) The low pH triggers conversion of the prepore to a pore, and EF and LF are translocated to the cytosol, where EF catalyzes the formation of cAMP and LF proteolytically inactivates MAPKKs. Redrawn from Collier, R. J., and J. A. T. Young, *Annu Rev Cell Dev Biol*. 2003, 19:45–70. © 2003 by Annual Reviews. Reprinted with permission from the Annual Review of Cell and Developmental Biology, www.annualreviews.org.

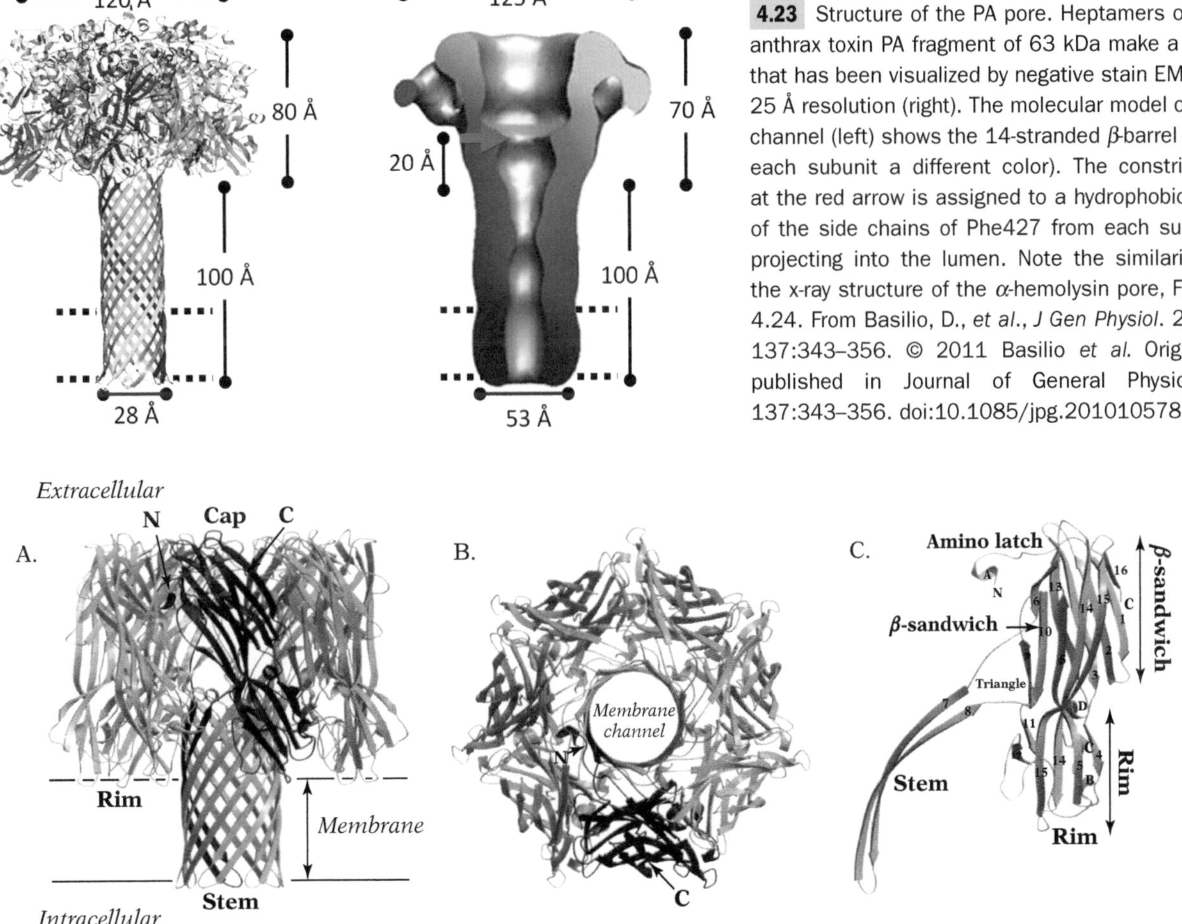

4.23 Structure of the PA pore. Heptamers of the anthrax toxin PA fragment of 63 kDa make a pore that has been visualized by negative stain EM with 25 Å resolution (right). The molecular model of the channel (left) shows the 14-stranded β-barrel (with each subunit a different color). The constriction at the red arrow is assigned to a hydrophobic ring of the side chains of Phe427 from each subunit projecting into the lumen. Note the similarity to the x-ray structure of the α-hemolysin pore, Figure 4.24. From Basilio, D., *et al.*, *J Gen Physiol*. 2011, 137:343–356. © 2011 Basilio *et al.* Originally published in Journal of General Physiology. 137:343–356. doi:10.1085/jpg.201010578

4.24 The x-ray structure of the α-hemolysin heptameric pore, viewed from the side (A) and the top (B). In (C), one protomer is shown in the open-pore configuration; before insertion, the extended β-strands of the stem are folded into the rest of the β-sandwich. From Gouaux, E., *J Struct Biol*. 1998, 121:110–122. © 1998 by Elsevier. Reprinted with permission from Elsevier.

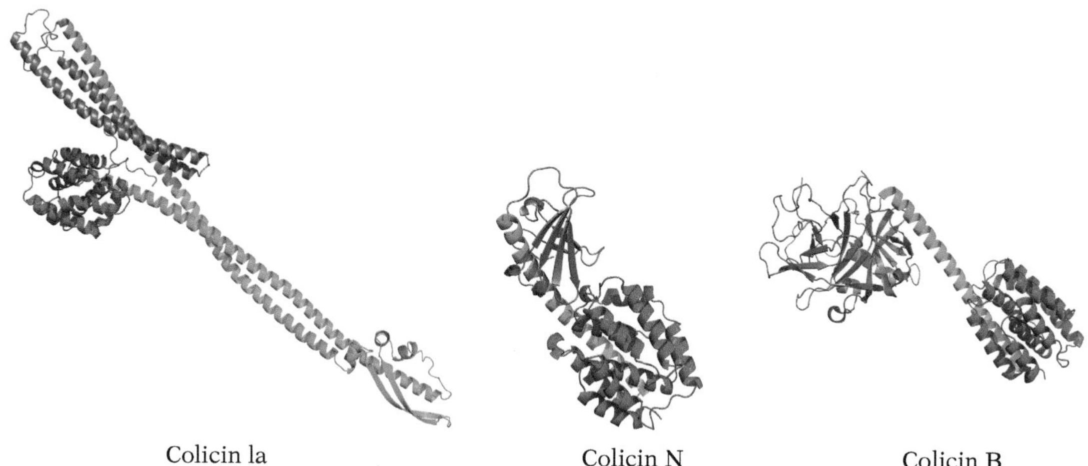

4.25 X-ray structures of three pore-forming colicins. The ribbon diagrams of three colicins that form pores, colicins Ia, N, and B, show the typical domain structure with the T domain (blue), R domain (green), and C domain (red). The R domain forms a central helix or helical pair that elongates the molecule: the length of the coiled coil of colicin Ia is 160 Å! From Jakes, K. S. and W. A. Cramer, *Ann Rev Genetics*. 2012, 46:209–231. Copyright © 2012 by Annual Reviews. Reprinted with permission from the Annual Review of Genetics, www.annualreviews.org.

difference is that αHL requires no catalytic domain or other inhibitors; it simply lyses human erythrocytes and other cells by pore formation, causing leakage. Each 33.4-kDa polypeptide inserts into the membrane before forming the heptameric prepore at the surface of the bilayer. Each monomer in the prepore extends a β-hairpin, forming a β-barrel pore that is 52 Å long and 26 Å in diameter, with an inner diameter that is only ~15 Å in the narrowest part (Figure 4.24). The rest of the β-sheet structure forms a much wider cap, with hydrophobic residues at the rim that contacts the membrane bilayer. This mushroom-like structure was first observed for aerolysin, another β-barrel channel-forming toxin, which is a virulence factor of *Aeromonas* bacteria that cause gastrointestinal disease and wound infections.

Colicins

Colicins are bacteriocins, a class of toxins synthesized and released by bacteria to kill competing microorganisms. More than one-third of *E. coli* strains produce colicins. These bacteria harbor plasmids that encode specific colicins, along with specific immunity proteins that ensure their own protection from the lethal action of the colicins. Colicins enter their target cells utilizing outer membrane receptors and either the Tol or Ton intermembrane translocation systems (see Chapter 11). Those in the channel-forming subfamily then insert into the cytoplasmic membrane to form voltage-gated channels that leak ions at a very great rate (>10^6 ions channel^{-1} sec^{-1}), depolarizing the membrane and eventually killing the cell.

Colicins are proteins of around 60 kDa organized into three functional domains: the N-terminal T domain, which mediates translocation across the cell envelope; the R domain for receptor binding; and the C-terminal domain, called C for cytotoxicity. The R domain makes a central helix or pair of helices connecting the other two structural domains, which can produce a very elongated shape, as seen in three of the channel-forming colicins (Figure 4.25). Although the average length of the C domain α-helices is ~13 amino acids – clearly not enough to span the bilayer – isolated C domains from many different channel-forming colicins have been shown to form voltage-gated pores in planar bilayers. Experiments with biotin-labeled single cysteine mutants of the C domains of both colicin Ia and colicin A indicate that a large portion of the peptide chain (115 and 70 amino acids, respectively) moves to the opposite side of the membrane during insertion and voltage-gated channel opening. Again, the mechanistic details of pore formation are not known.

Peptides

Peptides that insert into membranes include many antimicrobial peptides and peptide toxins – some well characterized, like melittin of bee venom, and others newly identified, such as the peptide toxins of spiders and sea anemones. Numerous antimicrobial peptides are under study not only for their role in the natural immune defense of mammals, but also their potential as valuable new therapeutics either by themselves or as transporters, as in the case of the family called "Trojan peptides," which can deliver other agents into cells. Other peptides capable of membrane insertion are the ionophores, named for their affinity for ions, which can be highly specific. While some ionophores simply chelate the ion, surrounding it with a lipid-soluble coat, others insert into the bilayer to make ion channels. Alamethicin and gramicidin are two widely studied examples of

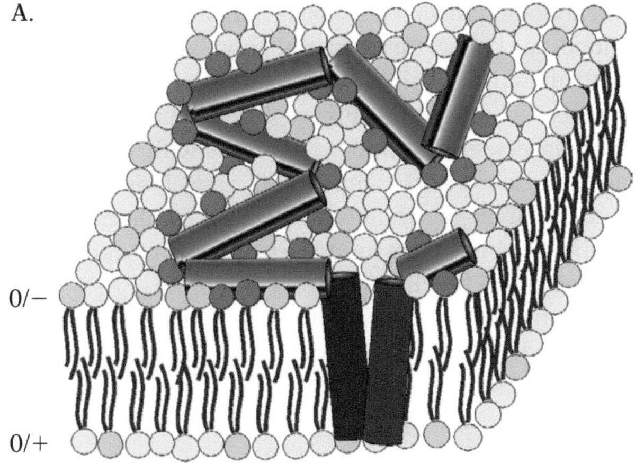

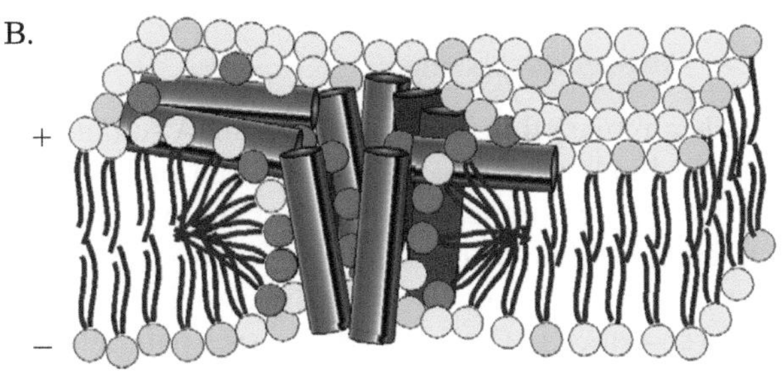

4.26 Two mechanisms for pore formation by inserted peptides: the barrel-stave model (A) and the toroidal model (B). Note how peptide insertion affects the lipid bilayer differently in the two Mechanisms. Redrawn from Yang, L., *et al.*, *Biophys J.* 2001, 81:1475–1485. © 2001, Elsevier.

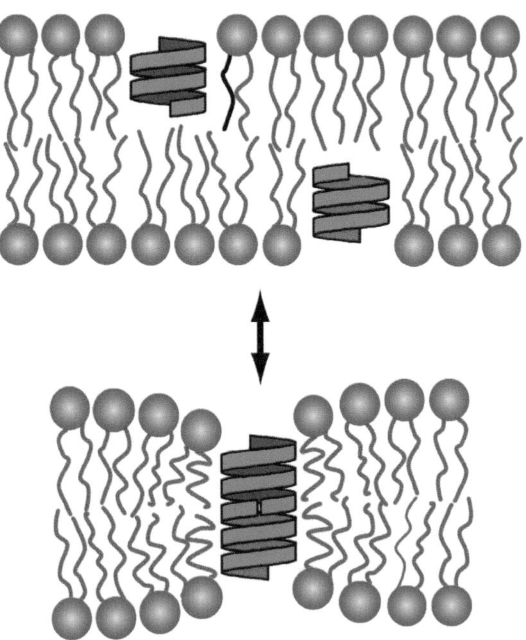

4.27 Dimer formation of gramicidin A. The dominant form of dimers is an end-to-end β-helix. Such a dimer makes a conducting channel in a bilayer. Many simulations have studied the effect of the dimer on the width of the lipid bilayer: shown here is a local deformation of the bilayer due to hydrophobic mismatch (see below). From Khandelia, H., *et al.*, *BBA.* 2008, 1778:1528–1536. © 2008, Elsevier. Reprinted with permission from Elsevier.

channel-forming ionophores that have provided many insights for the understanding of protein ion channels (see Chapter 10). In addition, many synthetic peptides have been designed to insert into membrane bilayers.

The wealth of data on peptide insertion into model membranes from studies using conductance, Fourier transform infrared (FTIR) spectroscopy and oriented circular dichroism, solid-state NMR, neutron diffraction, and other techniques emphasizes the importance of the lipid composition, temperature, extent of hydration, and the peptide-to-lipid ratio. In general, the peptides can insert in one of two orientations, parallel to the plane of membrane or perpendicular to it, and then permeabilize the membrane by one of four possible mechanisms. In the Carpet mechanism, peptides bind as α-helices to the membrane surface and embed in the headgroup region oriented parallel to the bilayer plane. Although they remain in this orientation, at high concentrations they disrupt the membrane integrity (without forming pores). In two of the insertion mechanisms, the peptides bind in parallel orientation but oligomerize when a critical concentration is reached, changing their orientation to approximately perpendicular to the bilayer and resulting in pore formation. There are two mechanisms of oligomeric pore formation (Figure 4.26).

In the barrel-stave model, the peptides span the two leaflets of the bilayer to line the pore like the staves of a barrel. In the other mechanism, the peptides form a toroidal pore when the insertion of helices stimulates the lipid monolayer to bend back on itself like the inside of a torus (a mathematical term for a surface containing a hole). In contrast to the barrel-stave model, the toroidal pore has a continuous bending of a PL monolayer, stabilized by the peptides. In a fourth mechanism, the peptides insert to span the two leaflets of the membrane and then form an open pore when they align as dimers (see Figure 4.27).

Melittin is an amphipathic α-helix of 26 amino acids, with five basic residues along its polar side and a bend of ~120° at its internal proline. Because of its large pore size (4.2 nm inner diameter and 7.7 nm outer diameter), the hydration and temperature dependence of pore formation, and the lack of discrete conductance steps, melittin is considered to form a toroidal hole. In contrast, a solid-state NMR study of the antimicrobial peptide ovispirin supports the Carpet mechanism for its membrane disruption by showing it remains parallel to the plane of the membrane, even at high peptide-to-lipid ratios.

Alamethicin is a 20-residue peptide with eight helix-stabilizing amino isobutyric acid residues, only one charged amino acid (Glu18), an internal proline (Pro14), an acetylated N terminus, and an alcohol (phenylalaninol) at the C terminus. An amphipathic α-helix that can bend at Pro14, it forms voltage-gated ion channels of the barrel-stave configuration, as indicated by much experimental evidence. Its single-channel conductance is characterized by multiple discrete states, suggesting the channel is oligomeric and changes its conductance state when a single alamethicin molecule joins or leaves the aggregate. The pore dimensions determined by neutron scattering give it a thickness that matches the helix diameter. Solid-state NMR results indicate that in the nonconductive state, the helices are tilted by 10° to 20° from the bilayer normal, suggesting a possible "pre-aggregate" state that leads to oligomerization and an open channel.

Gramicidin A, which forms channels that are specific for monovalent cations, is composed of 15 nonpolar amino acids of alternating L and D configurations. Gramicidin A can form β-helices, which are helical structures made up of β-sheets twisted into cylinders, with hydrogen bonding of the backbone N–H and carbonyl groups roughly parallel to the axis of the helix. The β-helices have hydrophobic exteriors since with alternating L- and D-amino acids all the side chains are on the outside. The N and C termini are blocked, so with the lack of polar side chains, the most polar part of the molecule is the peptide backbone; indeed the ion path in gramicidin A involves the carbonyl oxygens. Different gramicidin structures are observed depending on the solvent (lipid, organic solvent, or detergent) and ions present. Detailed structures obtained by x-ray crystallography and NMR have been classified as either a double β-helical pore that can span the bilayer or a β-helical dimer, both of which can have open and closed states. Since conditions under which the double helix forms a conducting pore are very limited, the helical dimer is probably the major conducting form and exemplifies the fourth mechanism for peptide channel formation (Figure 4.27). There are actually several species of natural gramicidins with slight differences in amino acid composition, as well as numerous synthetic analogs. Because of its self-associating dimer, gramicidin A has been found to be a suitable nanodevice for membrane biochips.

The need for novel antimicrobial medicines has stimulated design of new peptide antibiotics utilizing targeted high throughput screening of peptide libraries. The power of combinatorial chemistry now allows peptide design to target specific membrane proteins, leading to several candidates for new drugs.

SecA: Protein Acrobatics

SecA is a remarkable peripheral protein. A soluble protein that binds to anionic lipids and to a protein complex in the membrane, it carries out the power stroke of a nanomachine. As the motor for the translocon that exports proteins across the bacterial cytoplasmic membrane (described in Chapter 7), it inserts a pair of helices into the complex to move the protein substrates along, fueled by its hydrolysis of ATP. It also binds a cytoplasmic chaperone called SecB and exhibits some chaperone activity on its own, which may explain its high concentration (8 µM inside *E. coli*). SecA is composed of four domains: a nucleotide-binding domain; a regulator of ATPase domain; a substrate-binding domain; and a domain that binds lipids and SecB (Figure 4.28). Different structures of SecA reveal different relationships

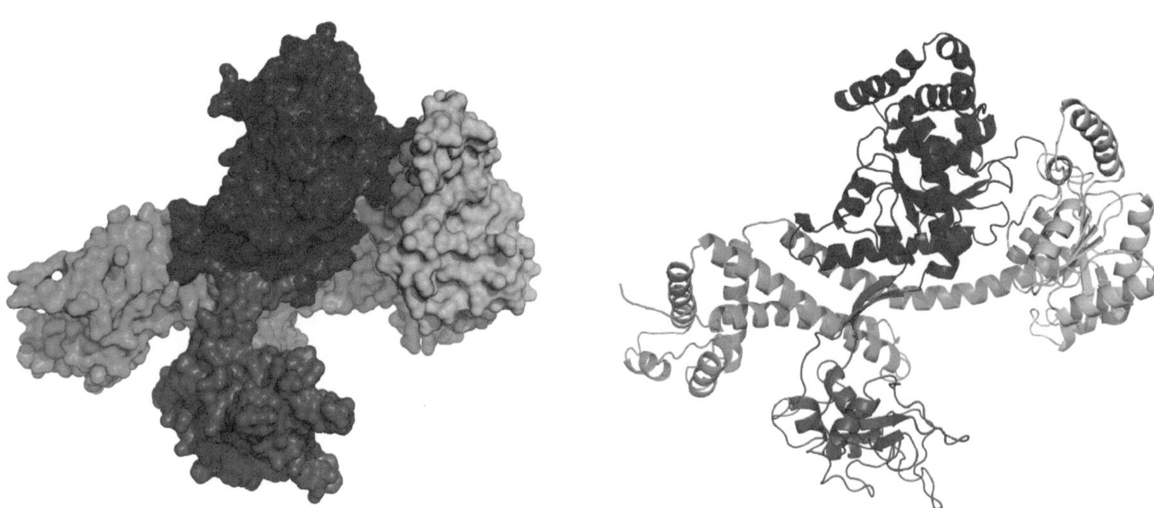

4.28 The domain structure of SecA. The x-ray structure of SecA from *E. coli* is shown in space-filling (left) and ribbon (right) representations. The domains of SecA are the nucleotide-binding domains (dark blue), the regulatory domain of the ATPase (cyan), the substrate-binding domain (magenta), and the C-domain that binds lipids and SecB (green). While SecA is shown here as a monomer, different x-ray structures show SecA dimers of quite different organization. From Sardis, M. F., and A. Economou, *Mol Microbiol*. 2010, 76:1070–1081. © 2010 Blackwell Publishing Ltd.

between the domains due mainly to a swivel of the protein-binding domain. The differences can be interpreted to imply open and closed states of SecA that allow the protein-binding domain to move toward the nucleotide-binding domain during translocation. In addition, under different conditions SecA dimers form with different dimer interfaces, which may relate to its different functions. Even with a high-resolution structure of SecA with the translocon (see Chapter 7), intriguing questions remain about the mechanism of SecA action.

PROTEINS EMBEDDED IN THE MEMBRANE

The operational definition of integral (or intrinsic) membrane proteins implies that they are embedded in the membrane, since disruption of the membrane is required to solubilize them. The exceptions are those peripheral proteins that are held in the membrane by two or more lipid anchors, binding them to the membrane with enough strength to require its disruption for their release (see above). Embedded membrane proteins include monotopic proteins, which insert into the membrane but do not span it, and proteins with one or more TM segments.

Monotopic Proteins

There are only a few well-characterized examples of monotopic proteins. Some enzymes involved in lipid metabolism access their substrates by integrating into one leaflet of the membrane. Structures have been solved for three of these that are important targets of pharmaceutical drugs: prostaglandin H_2 synthase (described in Chapter 9), squalene-hopene cyclase, and fatty acid amide hydrolase. A high-resolution structure is available for another monotopic protein with clinical importance: monoamine oxidase, which binds to the outer membrane of mitochondria. It is important for inactivation of several neurotransmitters, such as serotonin and dopamine, as well as catabolism of monoamines ingested in foods.

Another example of monotopic proteins is the caveolins, which associate with rafts to form caveolae (see Chapter 2). Caveolin is inserted into the plasma membrane from vesicles derived from the Golgi and remains in the inner leaflet, strongly immobilized by associations with the cytoskeleton. It forms dimers that bind cholesterol and form a striated coat as the membrane invaginates for endocytosis (see Figure 2.25).

Integral Membrane Proteins

Most integral membrane proteins have one or more TM segments. Bitopic proteins span the bilayer one time and are classified by their topology as type I, with their N terminus outside, or type II, with their C terminus outside. A bitopic protein may also be called single-spanning (not to be confused with monotopic). Type III proteins, with multiple membrane spans connected by loops, are called polytopic or multi-span membrane proteins. When several bitopic integral membrane proteins oligomerize with interacting TM segments, they are called type IV (Figure 4.29).

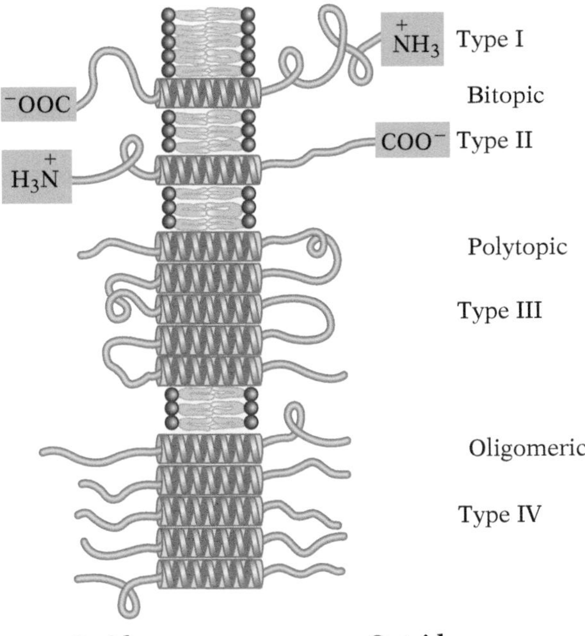

4.29 Classification of integral membrane proteins by topology. Both type I and type II are bitopic proteins with only one TM helix. Type I has the N terminus outside while type II has it inside. Type III proteins have multiple TM segments in a single polypeptide. In contrast, type IV proteins are oligomers assembled from several polypeptides, each having one TM helix. From Nelson, D. L., and M. M. Cox, *Lehninger Principles of Biochemistry*, 4th ed., W. H. Freeman, 2005, p. 374. © 2005 by W. H. Freeman and Company. Used with permission.

The chapters at the end of this book present many examples of detailed structures of integral membrane proteins. A database of high-resolution structures of membrane proteins is maintained by the Stephen White laboratory. The possible folds of integral membrane proteins are dictated by the process of export to and assembly in the bilayer, as well as by the stability of the embedded protein. These constraints on their folds may explain why nearly all of the type III integral membrane proteins whose detailed structures have been solved are of two structural types, bundles of α-helices and β-barrels, described in detail in the next chapter. Since most of these structures were solved by x-ray crystallography, it is also possible that crystallization methods favor these two structural types. With known atomic structures for fewer than 10% of all predicted integral membrane proteins, a more varied menu of structures, such as the β-helix, will likely be revealed in future work. As many of the following chapters will illustrate, the TM α-helix is currently the focal point for most researchers dedicated to understanding the nature of these intriguing proteins.

The structures of membrane-spanning proteins must cope with many chemical and physical differences from soluble proteins, as indicated in a list of important environmental factors (Table 4.1). Notable among these are the lack of homogeneity and isotropy; the presence in most membranes of gradients of pH, electric field, pressure, dielectric constant, and redox potential; the paucity of solvent; and the very low dielectric constant. Some properties vary as a function of the depth in the membrane (see Chapter 8). Together these factors significantly alter the ΔG° of functions associated with folding processes. Overall, it is much harder to break a main-chain hydrogen bond, ionize a side chain, or break a salt bridge of a protein domain in the membrane interior than in the cytosol; of course, it is easier to expose a hydrophobic group, as well as to bring subunits in close proximity.

The membrane milieu strongly favors the formation of secondary structure in TM segments. The low dielectric constant of the nonpolar domain of the membrane and the scarcity of water molecules favor formation of hydrogen-bonded secondary structure. This is readily shown by computing the change in free energy of transfer (ΔG_{tr}) for peptide bonds with and without H-bonds:

	ΔG_{tr} from water to alkane
Non-H-bonded –NH–C=O	+6.4 kcal mol^{-1}
H-bonded –NH–C=O	+2.1 kcal mol^{-1}

Therefore the per-residue cost of disrupting H-bonds in the membrane is ~4 kcal mol^{-1}. For a TM segment of 20 amino acids, this is 80 kcal mol^{-1} driving the formation of an α-helix!

Analysis of the solved integral membrane protein structures and mutagenesis of particular residues in them have led to some generalizations about the locations of amino acids in the nonpolar membrane domain:

(1) Nonpolar amino acids are typically found in membrane-spanning α-helices with their side chains pointing into the hydrophobic interior of the bilayer. This is expected from thermodynamic arguments and was tested when the small bitopic protein phospholamban from sarcoplasmic reticulum was engineered to replace nonpolar residues in its TM helix with polar ones, resulting in a water-soluble analog.

(2) Acidic and basic amino acids either (i) remain uncharged due to the effect of the low dielectric environment on their pK_as, (ii) form ion pairs that neutralize their charges, or (iii) play a special role, for example, in transport of protons or electrons or

TABLE 4.1 PROPERTIES OF THE CYTOSOLIC AND MEMBRANE ENVIRONMENTS THAT AFFECT PROTEINS

Property	Cytosol	Plasma membrane[a]
Solvent chemical homogeneity	Yes	No
Chemical groups available	HOH, ions, –SH	$-CH_3$, $-CH_2-$, = CH–
Isotropy	~Yes	No
pH gradient	No	Yes[b]
Electric field (V m^{-1})	~0	~2 × 10^{6}[b]
Pressure gradient	No	Yes
Dielectric constant gradient	No	Yes
Redox potential gradient	No	Yes[b]
Volume or surface occupancy [protein/solvent (%)][c]	~17	~35
Separation between two proteins:		
Distance (Å)	~50	~30–35
Intervening solvent molecules	~15–20	~4
Exchange time between solvent molecules (s)[d]	~10^{-11}	~10^{-7}
Viscosity at 20°C (η; N s m^{-2})	0.001	0.1
Dimensions	3	~2
Translational diffusion:[e]		
D_{lat} (m^2 s^{-1})	~10^{-10}	~10^{-11}
Average range explored in 1 μs (x; Å)	~250	~50
Dielectric constant (ε)	80	~2
ΔG° (kcal mole^{-1}) for:		
Breaking a main-chain H-bond	~0	+4–6
Deprotonating a Glu side chain (pH 7)	–4	>30
Opening a salt bridge	<1	60
Exposing one Å^2 of hydrophobic surface	+0.025	~0
Exposing a Leu side chain to the solvent	+2.8	~0
Associating two 50-kDa proteins (T Δ S at 20°C)	85	5

[a] For properties that vary as a function of the depth in the membrane, the data correspond to those at the membrane center.
[b] In most but not all membranes.
[c] Estimated from data for the cytosol and plasma membrane of an *E. coli* cell. Calculations for the cytosol assume a 1:2.5 w/w ratio of RNA to protein; calculations for the plasma membrane assume the average integral protein (often an oligomer) to comprise ~12 TM helices and to have about half of its volume buried into the membrane. Estimates published in the literature vary from 17% to 50%.
[d] In pure solvent.
[e] For a middle sized protein (~50 kDa) in either pure water or pure lipids; in the cytosol and in real membrances, diffusion coefficients vary with the distance range considered.

Source: Popot, J. L., and D. M. Engelman, Helical membrane protein folding, stability, and evolution. *Annu Rev Biochem*. 2000, 69:881–922.

in binding a cofactor such as heme or retinal. Polar residues are not completely excluded from the nonpolar region, since the partitioning of many hydrophobic side chains into the interior is so favorable it can overcome the cost of including a few less-favorable groups. In addition, the charged residues found in TM segments can move their polar groups toward the interface by snorkeling, adopting configurations that orient their polar atoms toward the interface to escape from the hydrophobic membrane core. Snorkeling can be quantified by measuring the displacement of the polar atom(s) from the β-carbon of the amino acid, which shows that the largest snorkeling distances are achieved by lysine residues. However, side chains of Arg, Tyr, Asp, Glu, Asn, and Gln also snorkel, in addition to Trp residues at the interfaces.

(3) Hydrogen-bond formers often use H-bonds to link their side chains to backbone carbonyl groups. These can provide "caps" for the ends of helices, as well as stabilizing the interactions between helices in a helical bundle or oligomer. Even the hydrogens on the main-chain α-carbons can form hydrogen bonds with backbone or side chain oxygen atoms. Several of these bonds, each with about half the strength of a conventional hydrogen bond, can stabilize a helix in the nonpolar bilayer interior and are often found in glycine-, alanine-, serine-, and threonine-rich packing interfaces.

(4) Of the amino acids considered to be helix breakers, glycine and proline are found frequently in TM helices, often at conserved positions. Glycine residues are important for allowing the close packing that minimizes interhelical distances in bundles of α-helices (see below). While proline residues are rare in α-helices of soluble proteins, they are common in TM α-helices and are often located near the center of the bilayer. Inclusion of proline in an α-helix leaves one carbonyl of the helix without an intrachain H-bond. The restricted backbone angles of the peptide chain at proline make a kink in the helix, a bend in the chain of ~120° in the direction away from the missing backbone H-bond. Interestingly, mutagenesis of bacteriorhodopsin showed that substitution of alanine for proline does not remove the kink, indicating that the tertiary structure of the integral membrane proteins maintains the helix distortion. Similarly, engineering a single proline into a TM helix of a polytopic protein is not enough to bend the helix and disrupt the tertiary packing. Conservation of prolines at kinks in many integral membrane proteins led to the suggestion that the TM segments with kinks that do not have prolines evolved from TM segments with proline, since homologous proteins do have proline in the position of the kinks.

(5) Aromatic amino acids, especially Trp and Tyr, play a special role at the interface of the hydrophilic and nonpolar domains in both α-helical and β-barrel integral membrane proteins. NMR studies with model indole compounds reveal that this is not due to their dipole moment or H-bonding ability, but rather to their flat rigid ring and aromaticity that lead to complex electrostatic interactions with the hydrocarbon core.

The well-characterized integral membrane proteins that are bundles of α-helices typically have an even number of helices, with the notable exception of the family of seven-helix bundles that are involved in signal transduction, such as bacteriorhodopsin (see Chapter 5) and the G protein-coupled receptors (see Chapter 10). Analysis of predicted membrane proteins from 26 genomes (see Chapter 6) shows the number of predicted TM helices is distributed over all integers from 2 to 13, with the occurrence decreasing as the number increases, except for spikes at 4, 7, and 12. When all the inner membrane proteins of *E. coli* are similarly analyzed, by far the highest incidence is for bundles of 12 predicted helices, with the next-largest groups having two and six predicted TM helices and significant numbers with 4, 5, and 10 predicted TM helices.

Distortions from the classical α-helix are not uncommon in the TM segments seen in the x-ray structures (Figure 4.30). Many of the distortions probably have a functional role; alternatively, they may serve to facilitate folding by preventing off-pathway intermediates (see Chapter 7). The kinks described above are one class of distortions seen often in TM helices. Another type is a π-bulge, where one backbone carbonyl is not H-bonded, such as a site involved in retinal binding in bacteriorhodopsin (see Chapter 5). Helix unwinding is a third kind of distortion, seen in the Ca^{2+} pump from sarcoplasmic reticula (see Chapter 9), where the unwinding frees up carbonyls to coordinate Ca^{2+}. Half-helices, where two short helices stack end-to-end to span the bilayer, have been observed, for example, in the glycerol facilitator and the aquaporins (see Chapter 12). Finally, stretches of 3_{10} helices, which differ from α-helices in the hydrogen bonding pattern (with bonds between i, $i + 3$ residues instead of i, $i + 4$ residues), are seen in portions of TM segments of some potassium channels.

A number of approaches have been taken to study helix–helix interactions. In model peptides that form TM helices, inserting glutamine in the middle of the TM sequence drives formation of oligomers. Indeed, any amino acid capable of acting simultaneously as both donor and acceptor of hydrogen bonds (Asp, Glu, Asn, Gln, and His) promotes oligomerization, while serine, threonine, or tyrosine does not. In polytopic integral

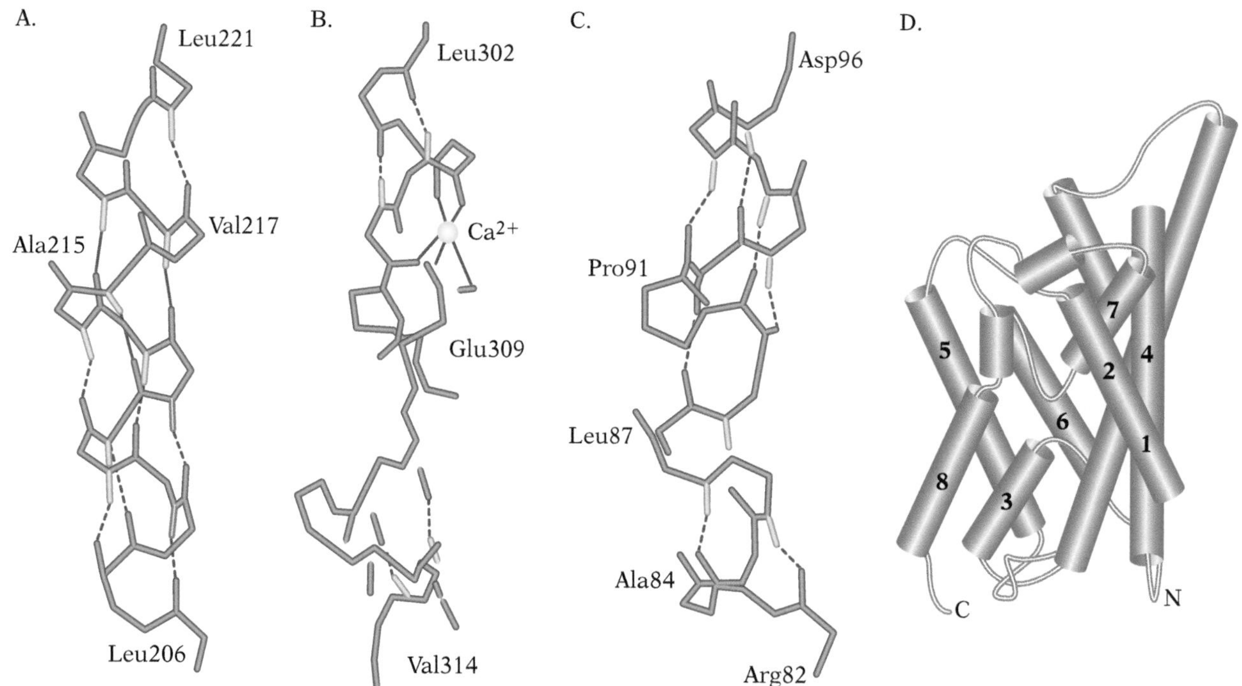

4.30 Distortions of α-helices in TM segments of integral membrane proteins. (A) π-bulge at Ala215 in helix G of bacteriorhodopsin, which causes the peptide plane to tilt away from the helix axis locally. (B) Unwinding of helix M4 in the calcium ATPase of the sarcoplasmic reticulum exposes backbone carbonyl groups that participate in coordinating Ca^{2+}. (C) Proline kink in helix C of bacteriorhodopsin, resulting in a lack of hydrogen bond to the carbonyl of Leu87. (D) Half-helices in the glycerol facilitator, numbered 3 and 7. From Ubarretxena-Belandia, I., and D. M. Engelman, *Curr Opin Struct Biol.* 2001, 11:370–376. © 2001 by Elsevier. Reprinted with permission from Elsevier.

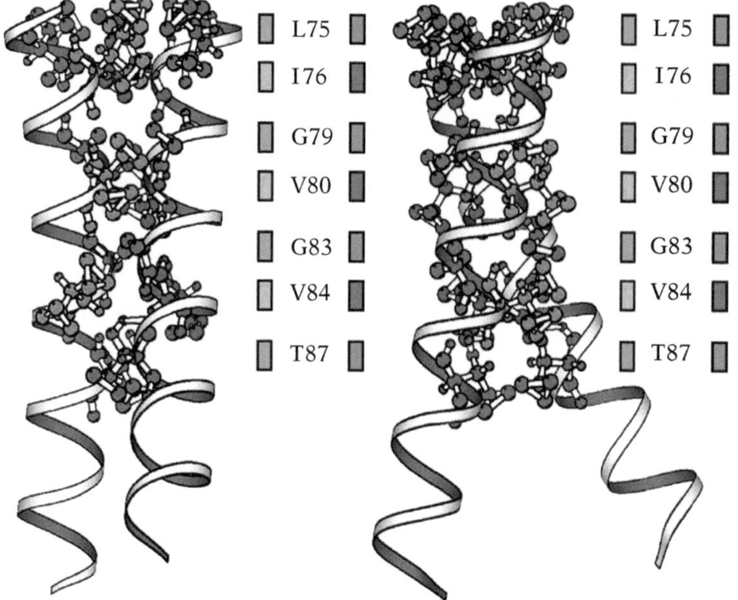

4.31 The TM helices in a dimer of glycophorin. Two views of the dimer show the intermolecular contacts with residues colored as shown in the legend on the right. The two views of the TM helices differ by 90°. From Arkin, I. T., *Biochim Biophys Acta.* 2002, 1565:347–363. © 2002 by Elsevier. Reprinted with permission from Elsevier.

membrane proteins, interstrand H-bonds play an important role in the tertiary structure. For example, there is at least one hydrogen bond between each pair of helices in bacteriorhodopsin (see Chapter 5).

Glycophorin A, the primary sialoglycoprotein of human erythrocyte membranes, has a single TM helix with a critical GxxxG amino acid sequence that is needed for the helix–helix interaction of dimer formation (Figure 4.31). Finding this sequence in numerous other TM peptides has defined a TM-oligomerization motif of GxxxG, along with the less common GxxxA, which clearly reflects the importance of small amino acids at positions buried between the helices. In glycophorin A the critical sequence is LIxxGVxxGVxxT.

Two helices cross at an angle of 40°, making a right-handed coiled coil in which the helices mesh closely by "knob-into-hole" interactions. The knobs are formed by isoleucine and valine residues and the holes by glycine residues. These interactions bring the helices close enough for important van der Waals interactions along the coiled coil (see Figures 4.31 and 7.5). The structure of the dimeric transmembrane domain of glycophorin A was solved by solution NMR (see Box 4.2). In addition, the detailed analysis of glycophorin A by saturation mutagenesis and computational approaches makes it a good model for type IV proteins as well as for type III integral membrane proteins with multiple TM helices.

BOX 4.2 NMR DETERMINATION OF MEMBRANE PROTEIN STRUCTURES

Solution NMR is a method of spectroscopy that detects nuclear-spin reorientation in an applied magnetic field for molecules or complexes (such as micellar membrane proteins) that are tumbling rapidly in solution. It works because those nuclei with spin (such as ^{1}H, ^{2}H, ^{13}C, ^{15}N) have magnetic moments that become oriented in the applied magnetic field, interact with nearby electrons and nuclei, and precess at different frequencies. The resulting signals are characterized by their intensity, chemical shift (position on the frequency scale), splitting due to coupling with other nuclei, and relaxation rates. In addition, nuclei that interact through space also produce the nuclear Overhauser effect (NOE). NOEs give distance information: for example, hydrogen atoms within 2.5 Å produce a strong NOE, while if separated by 5 Å they give a weak NOE. In one-dimensional NMR spectra, Fourier transformation of the intensity versus time signal following pulsed excitation gives plots of intensity vs. frequency that can be readily interpreted for small compounds. However, two- and even three- or four-dimensional spectra are needed for macromolecules. A two-dimensional heteronuclear correlation spectrum with a ^{1}H signal as one axis and a ^{15}N or ^{13}C signal as the other is produced in a heteronuclear single quantum coherence (HSQC) experiment and shows a field of contours corresponding to peaks for each heteroatom-attached proton in the molecule (Figure 4.2.1). An important variation of this technique, called transverse relaxation optimized spectroscopy (TROSY), for which Kurt Wuthrich was awarded the Nobel Prize in Chemistry in 2002, is frequently used in NMR studies of membrane proteins. Structural

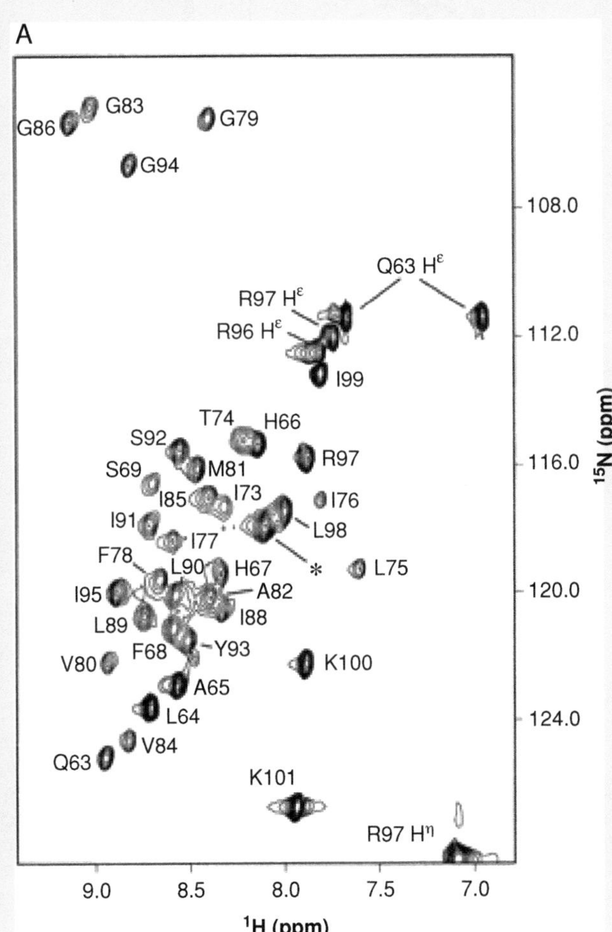

4.2.1 ^{1}H-^{15}N heteronuclear single quantum coherence (HSQC) spectrum of the TM segment of glycophorin A in 5% dodecylphosphocholine (DPC). The spectrum recorded at 500 MHz shows a single correlation of each backbone amide proton (identified by residue number) except those of Glu70 and Glu72, although the signals for Thr87 and Arg96 overlap(*). Each peak arises from the interaction of the nitrogen and amide proton along the peptide backbone or in any NH-containing side chains. It is typical for the side chain of Trp to produce a peak near the bottom left corner (shifted downfield) and for the amide peaks of glycine to appear near the top, as seen here. From Mackenzie, K. R., *et al.*, *Science* 1997, 276:131–133. © 1997, American Association for Advancement of Science. Reprinted with permission from AAAS.

(continued)

BOX 4.2 *(continued)*

determination requires that a protein's NMR spectrum be assigned (matching peaks to specific nuclei in the protein) and that structural restraints in the forms of measured internuclear distances, torsion angles, etc., be collected that are then used in structural calculations.

Solution NMR requires that the molecules tumble rapidly. For a membrane protein the solution particle size reflects not only the peptide or protein molecular weight, but whether oligomers form and the size of the micelle, mixed micelle, or bicelle used as a membrane mimetic. However, with TROSY-based methods, various isotope labeling protocols, and the high field strength (>700 Mhz) of newer instruments it is now possible to acquire data for membrane protein complexes up to 100 kDa or even higher, allowing impressive advances to be made in NMR determination of membrane protein structures.

Numerous factors are important in the acquisition of good NMR data. The system must be stable for several hours to allow repeated recording of the spectra to enhance the signal-to-noise ratio. Because the signals are weak, NMR requires high concentrations of sample (0.1 mM or higher), and obtaining large quantities of uniformly isotopically labeled and highly pure membrane protein can be a significant obstacle. The membrane protein is solubilized in detergent, and its spectra may then be recorded in detergent micelles. The choice of detergent is critical (Figure 4.2.2). The detergents most commonly used for NMR include SDS, LDAO, DDM, DPC (dodecylphosphocholine), and other lyso-PLs such as LMPC (lyso-myristoyl phosphatidyl choline). While bicelles

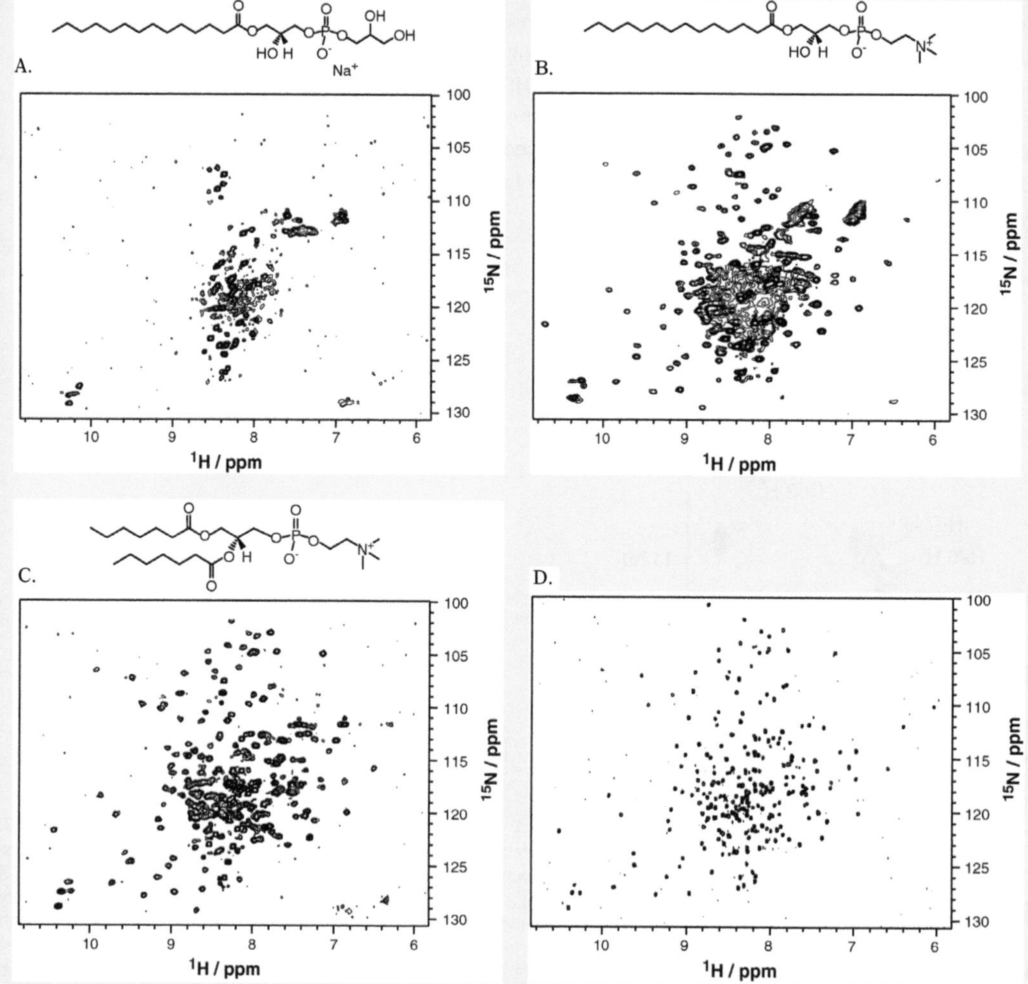

4.2.2. Effect of detergent choice on NMR spectra. TROSY spectra of sensory rhodopsin II solubilized in different detergents were recorded at 800 MHz for three hours at either 40°C or 50°C. The structures of detergents are shown. **(A)** LMPG at 40°C, **(B)** LMPC at 50°C, **(C)** DHPC at 40°C, and **(D)** DHPC at 50°C. From Nietlispach, D., and A., Gautier, *Curr Opin Struct Biol.* 2011, 21:497–508. © 2011, Elsevier. Reprinted with permission from Elsevier.

BOX 4.2 *(continued)*

provide regions of phospholipid bilayer that can preserve features of the protein structure, their use in solution NMR is limited to compositions with a high detergent:lipid ratio to enable high tumbling rates. Recent success has been achieved with bicelles made of DHPC and DMPC (or their ether-linked analogs for improved stability at higher temperatures).

The first multi-span integral membrane proteins to have their structures determined in the 1990s by NMR were subunit c of the F_1,Fo-ATPase (in an organic solvent mixture) and the glycophorin A homodimer. This was followed by determination of a number of outer membrane *β*-barrel porins (see Box 5.1). Striking progress now allows structure determination of helical membrane proteins up to 350 amino acids in length, such as archael rhodopsins (Figure 4.2.3). Finally, it should be noted that NMR spectroscopy is useful to measure dynamics within the protein structure, as well as to give information on ligand binding (affinity and binding site location) and about protein–lipid interactions.

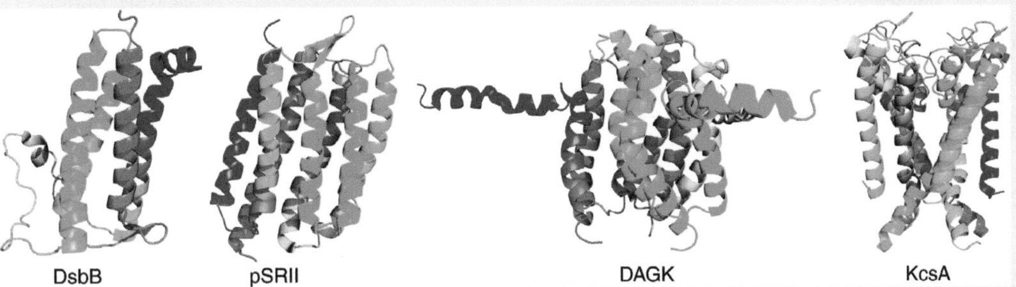

4.2.3 Examples of multispanning *α*-helical membrane protein structures determined by solution NMR. From the left, DsbB is the disulfide bond formation protein; pSRII is the senory rhodopsin, a GPCR (see Chapter 10); DAGK, diacylglycerol kinase is an enzyme that forms a domain-swapped trimer (see Chapter 6); and KcsA is a potassium channel that forms a tetramer (see Chapter 12) and has been solved as a water-soluble analog. From Nietlispach, D., and A. Gautier, *Curr Opin Struct Biol.* 2011, 21:497–508. Copyright © 2011, Elsevier. Reprinted with permission from Elsevier.

Solid-state NMR methods have the advantage that they can be applied to membrane proteins in bilayers. While these methods are not yet as well-advanced as solution NMR methods, they nevertheless have been used to determine the structure of some small membrane proteins such as the M2 viral ion channel and gramicidin, with recent progress indicating that soon much larger membrane proteins can be characterized by these emerging methods.

For further study

Freeman, R., *Magnetic Resonance in Chemistry and Medicine*, Oxford University Press, 2003.

Kim, H. J., S. C. Howell, W. D. Van Horn, Y. H. Jeon, and C. R. Sanders, Recent advances in the application of solution NMR spectroscopy to multi-span integral membrane proteins. *Prog Nuc Magn Res Spec.* 2009, 55: 335–360.

Nietlispach, D., and A. Gautier, Solution NMR studies of polytopic *α*-helical membrane proteins. *Curr Opin Struct Biol.* 2011, 21:497–508.

Qureshi, T., and N. K. Goto, Contemporary methods in structural determination of membrane proteins by solution NMR. *Top Curr Chem.* 2012, 326:123–186.

PROTEIN–LIPID INTERACTIONS

The properties of membrane proteins are best understood in the context of their lipid surroundings. Indeed, the contacts between integral membrane proteins and lipids must be very tight to maintain the seal of the membrane as a permeability barrier. The presence of a protein has essentially no effect on distant lipids, but has a large effect on the shell or ring (annulus) of lipids that surround it, forming the interface between it and the rest of the membrane. These lipids are called annular or boundary lipids and can be distinguished experimentally from the bulk lipids of the bilayer. In addition to bulk lipids and annular lipids, there is a third class of lipids comprising those which are tightly bound in crevices or between subunits of the proteins. These lipids are called nonannular lipids (not to be confused with bulk lipids) or lipid cofactors, as they are frequently required for activity. Experiments with purified proteins have shown that many membrane proteins require specific lipids or classes of lipids to stably bind or insert into bilayers, while numerous enzymes require specific lipids for

TABLE 4.2 SPECIFIC LIPID REQUIREMENTS OF MEMBRANE PROTEINS AND ENZYMES ASSESSED BY VARIOUS TECHNIQUES

A Lipid specificity for reactivation of delipidated enzymes

Enzyme	Source	Delipidation by	Reactivation by	Reference
Cytochrome *c* oxidase	Bovine heart mitochondria	PLA_2	Cardiolipin, not PE, not PC	Sedlak and Robinson, *Biochemistry* 1999, 38:14966
β-hydroxybutyrate dehydrogenase	Bovine heart mitochondria	Phospholipase A	Only PCs	Sandermann *et al.*, *J Biol Chem* 1986, 261:6201
Sarcoplasmic reticulum Ca^{2+}-ATPase	Rabbit skeletal muscle	Cholate extraction	Phosphatidylinositol-4-phosphate	Starling *et al.*, *J Biol Chem* 1995, 270:14467
Monoamine oxidase	Rat brain mitochondria	PLA_2	PI, negatively charged PLs	Huang and Faulkner, *J Biol Chem* 1981, 256:9211

B Lipid specificity in reconstitution of membrane proteins

Protein	Source	Reconstituted in	Specific requirements	Reference
Acetylcholine receptor	*Torpedo californica*	DOPC vesicles	Cholesterol, PA	Fong and McNamee, *Biochemistry* 1986, 25:830
Rhodopsin	Bovine retinal rod	Egg PC or DOPC/DOPE supported bilayers	PE (favors activated M2 state)	Alves *et al.*, *Biophys J* 2005, 88:198

C Lipid requirement of amphitropic proteins for binding and activation

Protein	Source	Binding to	Specific requirements	Reference
Protein kinase C	Rat brain	PC/PS LUVs	DAG and PS, anionic lipids	Slater *et al.*, *J Biol Chem* 1994, 269:4866
MARCKS	Mouse	PC/PS monolayer	PIP2	Wang *et al.*, *J Biol Chem* 2001, 276:5013
Phosphocholine-cytidylyltransferase	Rat	PC LUVs or SUVs	Anionic lipids, unsaturated PE, DAG	Arnold and Cornell, *Biochemistry* 1996, 35:9917; Davies *et al.*, *Biochemistry* 2001, 40:10522

activity (Table 4.2). The activity of Ca^{2+}-ATPase, for example, increases as lipid is added up to 30 moles of lipid per mole of ATPase. Both the nature of interactions with annular lipids and the influences of the physical state of the lipid bilayer on the functions of membrane proteins have been extensively studied using EPR, fluorescence quenching and energy transfer, and molecular dynamics simulations. Additional information can be gleaned from those lipids detected in high-resolution structures obtained by x-ray crystallography described in Chapter 8.

EPR is especially suited for studying boundary lipids because it can readily detect two populations of membrane lipids (Figure 4.32; Box 4.3). Since the mobility of the acyl chains is greatest near the center of the membrane (see Figure 2.11), incorporation of a nitroxyl spin label close to the terminal methyl of the chain gives an EPR spectrum with quite narrow line widths in a pure lipid bilayer. The rotation around the C–C bonds (*trans–gauche* isomerism, see Chapter 2) is fast (~10^{-10} seconds) and averages out, so the spectrum results from the axial rotation of the lipid molecule as a whole (10^{-8}–10^{-9} seconds). With proteins present, the axial rotation of a spin-labeled lipid molecule in the annular layer is hindered. Because it is relatively immobilized, it produces broader line widths, resulting in a second component most easily seen in the "outer wings" of the EPR spectrum. The selectivity of a protein for annular lipids can be determined from the relative intensity of the peaks in the outer wings, as observed for a series of spin-labeled lipids reconstituted with myelin proteolipid protein (Figure 4.33).

The selectivity for different lipids is a reflection of different exchange rates, since there is constant exchange of lipids between the bulk and the annular layer. For the exchange equilibrium, $L_NP + L^* \leftrightarrow L_{N-1}L^*P + L$, the lipid association constant, K_r, is ([L*P] [L]) / ([LP] [L*]). (Typically the concentration of the spin-labeled lipid, [L*], is less than 1 mole % of [L].) The reference lipid in these studies is PC, the most abundant PL in most animal cell

BOX 4.3 ELECTRON PARAMAGNETIC RESONANCE

Electron paramagnetic resonance (EPR), also called electron spin resonance (ESR), detects the orientation of unpaired electrons on paramagnetic molecules placed in a strong magnetic field. EPR can be used for biochemical substances containing paramagnetic transition metals or free radicals. To broaden its applicability, researchers employ free radical probes that are spin-labeled analogs of the molecules of interest, such as DPPC with a nitroxyl group. Such a compound, phosphatidylcholine carrying a spin probe on carbon-14 of the acyl chain (14PCSL), is shown in Figure 4.3.1A.

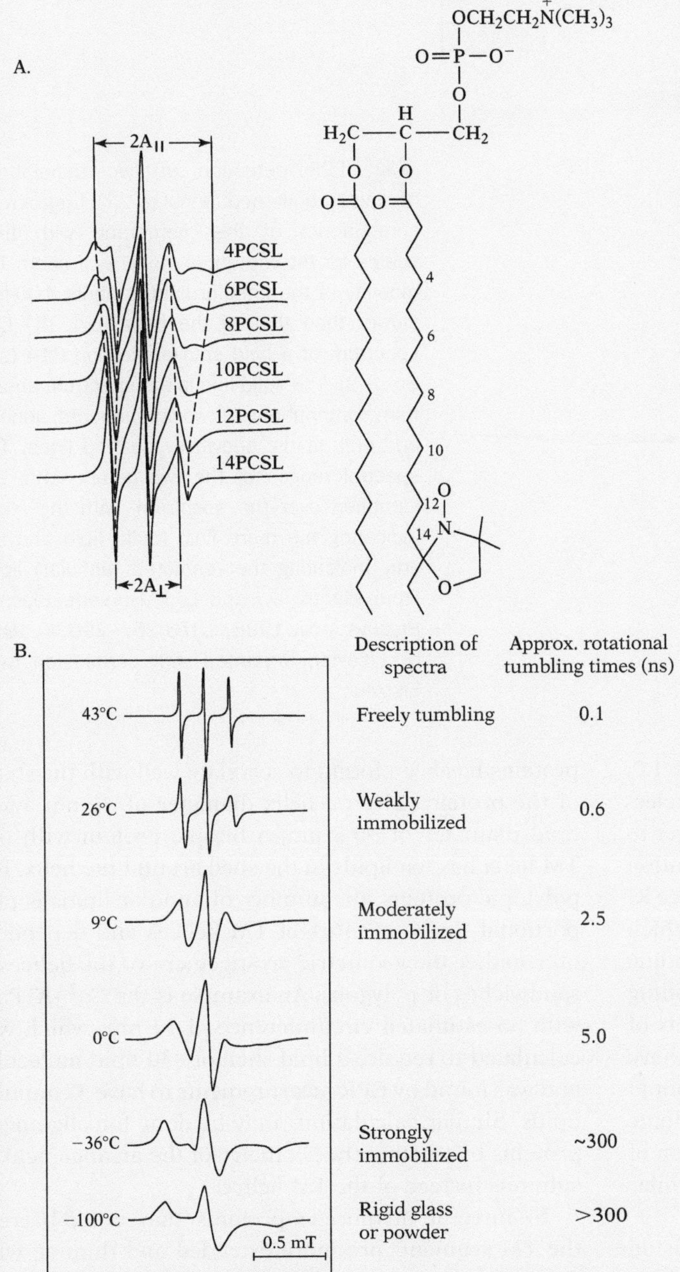

4.3.1 (A) Dependence of the EPR spectra of nitroxide-labeled DPPC on the location of the spin probe. (B) Effect of temperature on the mobility of a spin-labeled PC. Both redrawn from Campbell, I. D., and R. A. Dwek, *Biological Spectroscopy*, Benjamin Cummings, 1984, pp. 197, 192. © 1984 by Iain D. Campbell and Raymond A. Dwek. Reprinted with permission from the authors.

The absorption spectrum created by irradiation of the sample in the magnetic field is displayed as a first derivative, characterized by its intensity, line width, γ-value (for positions), and multiplet structure. The nitroxide group gives three lines, whose line widths (related to the spin relaxation times, T1 and T2) indicate the mobility of the molecule carrying the unpaired electron, which can vary from freely tumbling to strongly immobilized. In a series of PCSL probes, the position of the probe on the acyl chain determines its mobility. Placing the spin probe near the terminal methyl group gives relatively narrow lines due to the angular fluctuations from rotations around C–C bonds all along the chain, as shown in Figure 4.3.1A. The sharp lines that result are typical of isotropic mobility. Of course, in a lipid bilayer, the spin probes do not rotate isotropically, but their large fluctuations average the orientational anisotropy when they are labeled near the end of the acyl chains. Cooling the lipid sample will slow the movement of the spin probe, broadening the lines, until it eventually produces a rigid glass or powder, as shown in Figure 4.3.1B. When the spin label is not free to tumble in all directions, it has anisotropic motion. The frequent *trans–gauche* isomerizations along the acyl chains give motional averaging, while the axial rotation of the lipid is slower and is characterized by the rotational correlation times $\tau_{R\parallel}$ and $\tau_{R\perp}$, as shown in Figure 4.3.1A.

In addition to its application in studies of lipid–protein interactions described in this chapter, EPR is also used to probe protein conformations with a procedure called site-directed spin labeling (SDSL). The first step in these studies is the use of site-directed mutagenesis to create a single reactive site, typically by replacing an individual residue in a protein with cysteine. Then reaction with a sulfhydryl-reactive spin label positions a spin-labeled side chain at that site to provide information about structure, orientation, and conformational changes in membrane proteins.

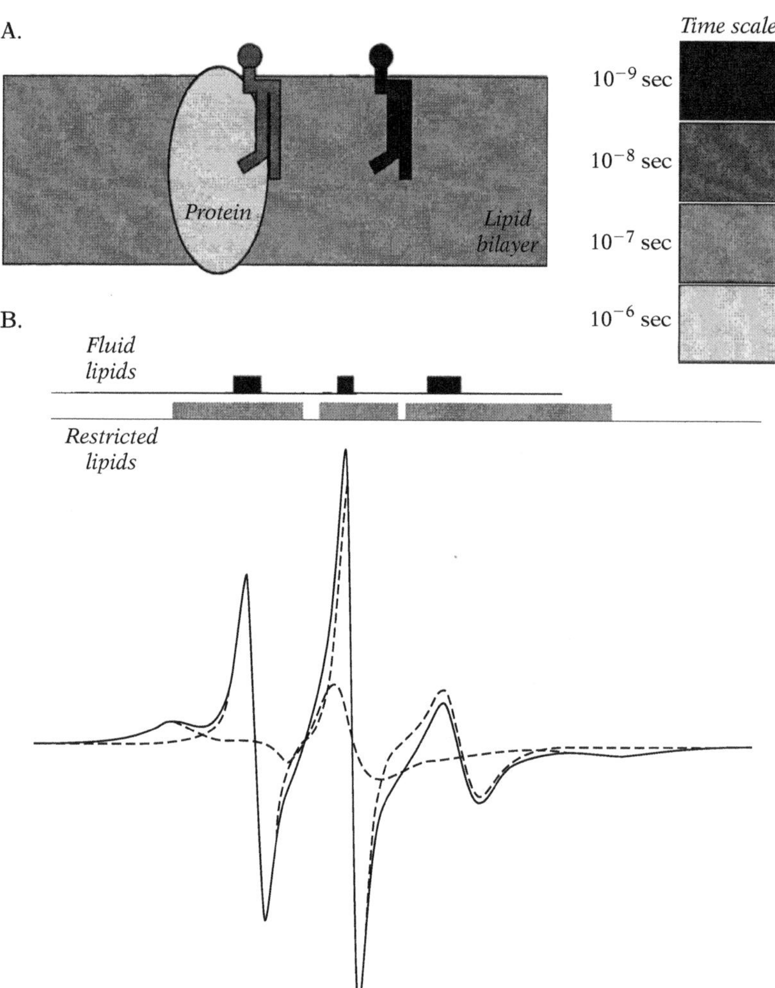

4.32 EPR detection of two populations of lipids: bulk and annular. **(A)** Diagram of components of the membrane with time scales for the rotational mobility of each. The mobility of the annular lipid is about 100-fold slower than that of the bulk lipid. **(B)** EPR spectrum of a lipid spin-labeled on C14 (see Box 4.3). The solid line is the spectrum arising from contributions of two components, annular and bulk lipids, shown by dashed lines. The spectral ranges of the two components are identified over the spectrum, with the black indicating the more fluid (bulk) lipid and the gray indicating the restricted (annular) lipid. From Marsh, D., and L. I. Horvath, *Biochim Biophys Acta*. 1998, 1376:267–296. © 1998 by Elsevier. Reprinted with permission from Elsevier.

membranes; thus the K_r for PC in this example is 1.0. When there is no selectivity, $K_r = 1$; with fairly high selectivity, K_r approaches 10. The restriction of the bilayer to two dimensions produces a high effective concentration of lipids that can further enhance the selectivity. Since K_r gives an average of affinities of a particular lipid, which may be due to several sites of quite different binding affinities, it may mask the presence of a tightly binding site for the lipid. A comparison of the lipid selectivity of different proteins shows that other proteins do not have the large variation in K_r seen with myelin proteolipid protein and some, like rhodopsin, do not discriminate at all (Figure 4.34). Thus, in general, the composition of lipids in the boundary layer appears to be quite similar to the composition in the bulk lipid bilayer.

The exchange into the annulus (on-rate) is diffusion limited (~10^8 sec^{-1}) and thus is the same for different lipids, while the off-rate reflects the specificity of interaction with the protein and can be slowed to 10^7 or even 10^6 sec^{-1}. By performing experiments at different lipid-to-protein ratios, both K_r and the fraction of spin-labeled lipid associated with protein can be determined. The stoichiometry of annular lipids for a number of different proteins has been found to correlate well with the shape of the protein. Given a helix diameter of ~1 nm and a lipid diameter of ~0.5 nm, a bitopic protein with one TM helix has ten lipids in the shell around the helix. For polytopic proteins, the number of annular lipids is proportional to the number of TM helices and dependent on whether the geometric arrangement of the helices is sandwiches or polygons. An example is the Ca^{2+}-ATPase with an estimated circumference of 14 nm, which was calculated to require a lipid shell of ~30 lipid molecules and was found by EPR measurements to have 32 annular lipids. Similar calculations may be done for oligomeric proteins by treating the geometry of the arrangement of subunits instead of the TM helices.

In integral membrane proteins that are *β*-barrels, the TM segments are more extended and thinner, with diameters of ~0.5 nm. The predicted number of annular lipids works out to be the same as the number of *β*-strands if there is no tilt as they cross the bilayer; if there is a 60° tilt, there are about twice as many annular lipids as the number of *β*-strands. The exchange rate was found to be slower for annular lipids of *β*-strands, presumably because lipids aligned along the *β*-strands

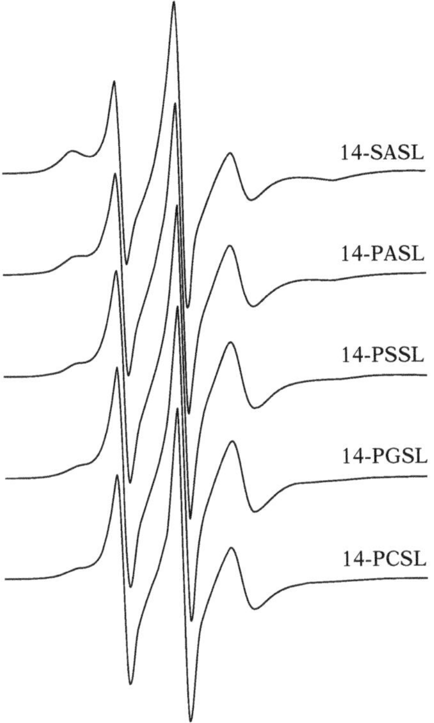

4.33 EPR spectra of different lipids reconstituted with myelin proteolipid protein in DMPC. The protein:DMPC ratio is 23:1 and the temperature is 30°C. All the lipids contain a spin label on C14. They are stearic acid (14-SASL), phosphatidic acid (14-PASL), phosphatidylserine (14-PSSL), phosphatidylglycerol (14-PGSL), and phosphatidylcholine (14-PCSL), where SL stands for the spin label in each case. The increasing relative intensity of the outer peaks arising from motionally restricted lipids indicates increasing selectivity for the protein. Redrawn from Marsh, D., and L. I. Horvath, *Biochim Biophys Acta*. 1998, 1376:267–296. © 1998 by Elsevier. Reprinted with permission from Elsevier.

are more extended and less flexible than annular lipids around α-helices. This result was determined with M13 coat protein, whose TM domains can be either α-helices or β-strands, and the exchange rate was four to five times slower for lipids associating with M13 coat in β-strand conformation than those associating with M13 coat in α-helix.

EPR experiments have addressed the detailed nature of the selectivity for annular lipids by comparing phospholipids with different headgroups and varying the ionic strength and pH, as well as by examining the importance of the glycerol backbone and the length of the acyl chains. In general, most proteins are found to prefer negatively charged lipids. In some cases, this is simply an electrostatic effect that is overcome by high ionic strengths, and in other cases the selectivity for the headgroup holds even in high ionic strength. (The examples of headgroups resolved in high-resolution structures reveal that extensive electrostatic and hydrogen-bonding interactions stabilize them in binding pockets, described in Chapter 8.) There is little or no difference when sphingomyelin and gangliosides are compared to PC, indicating the glycerol backbone is not a factor in selectivity. On the other hand, the acyl chain length is important: the free energy of association shows a linear dependence on chain length from 13 to 17 carbons.

Hydrophobic Mismatch

The importance of acyl chain length on lipid–protein interactions produced the concept of hydrophobic mismatch, which results when the nonpolar region of the bilayer is thinner or thicker than the hydrophobic thickness required by an integral membrane protein. The thickness of a bilayer is strongly influenced by its lipid

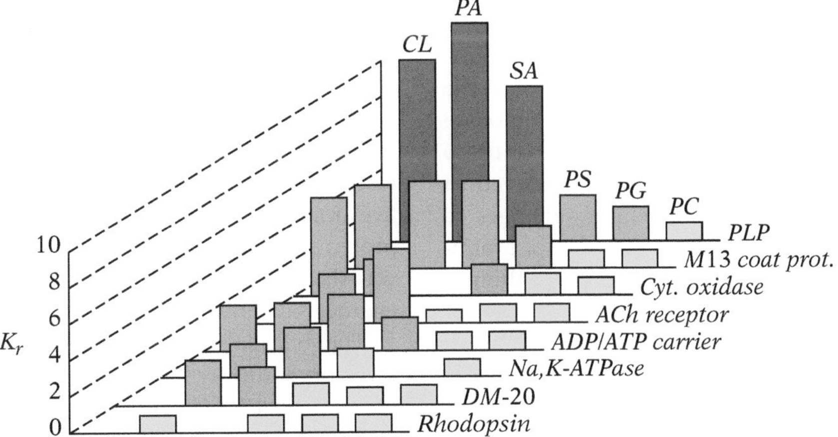

4.34 Patterns of lipid selectivity of different proteins. K_r, the relative association constant between each protein and each lipid, varies from 1 (shown in yellow) to >6 (shown in red). The data show the K_r for each protein (listed along the right edge), with each of the lipids identified at the top. Lipid selectivity increases from the front (with rhodopsin exhibiting almost no lipid selectivity) to the back (highest selectivity with PLP, the myelin proteolipid protein). Redrawn from Marsh, D., and L. I. Horvath, *Biochim Biophys Acta*. 1998, 1376:267–296. © 1998 by Elsevier. Reprinted with permission from Elsevier.

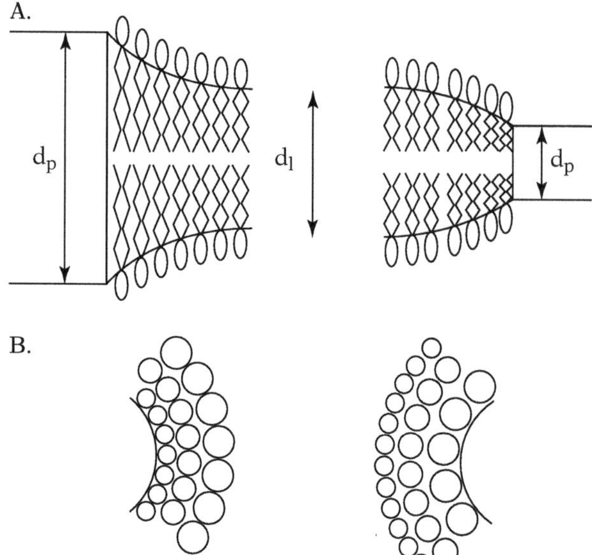

4.35 Distortion of lipid bilayers due to hydrophobic mismatch. Lipids in a bilayer can be distorted to match the hydrophobic thickness of the protein, as viewed from the side (A) and top (B). In both diagrams on the left, the hydrophobic thickness of the protein (d_p) is greater than that of the lipid bilayer (d_l), whereas in the diagrams on the right the reverse is the case. When the acyl chains stretch ($d_p > d_l$), the surface area they occupy decreases; when they compress ($d_l > d_p$), it increases. Redrawn from Lee, A. G., *Biochim Biophys Acta*. 2004, 1666:62–87. © 2004 by Elsevier. Reprinted with permission from Elsevier.

composition: for example, a PC bilayer with saturated chains is 2.5 Å wider than a bilayer with unsaturated chains of the same number of carbon atoms (see Chapter 2). If the hydrophobic regions of the protein and lipid do not match, either the lipid bilayer must stretch or compress to match the hydrophobic thickness of the protein (Figure 4.35), or the protein must change by tilting helices or rotating side chains to fit to the bilayer to avoid exposing nonpolar groups to the aqueous environment. Since proteins are more rigid than lipids, the bilayer might be expected to deform to accommodate the dimensions of their TM segments, contributing to the lateral tension of the bilayer. This is observed in cases when the thickness of lipid bilayers changes to accommodate gramicidin channels: insertion of gramicidin causes a DMPC bilayer to become 2.6 Å thinner and a DLPC bilayer to thicken by 1.3 Å. The perturbation of the membrane due to the mismatch creates a tension that contributes to its free energy. The $\Delta G°$ for bilayer deformation has been calculated to be ~1.2 kcal mol^{-1} for a large hydrophobic mismatch of 10 Å.

Not all lipid bilayers adjust to accommodate gramicidin, and similarly they do not deform to accommodate single TM helices. Synthetic peptides designed to be TM helices of different lengths have no effect on the thickness of model bilayers. Rather, NMR measurements revealed that TM helical peptides tilt with respect to the bilayer normal to match the hydrophobic thickness of the lipids. The peptides have sufficient flexibility of orientation to accommodate to the bilayer and not deform it. These findings suggest that proteins that cross the membrane with a small number of α-helices are likely to accommodate the bilayer thickness by helix tilting. However, larger proteins or proteins with less flexibility impact the lipid enough for hydrophobic mismatch to induce changes in bilayer thickness. The latter includes proteins that cross the bilayer as β-barrels (see Chapter 5), which have structural constraints that prevent them from adapting to the lipid bilayer and thus are more likely to select for lipids that provide hydrophobic matching (see Chapter 7).

A comparison of relative binding constants for PCs with acyl chains of different lengths indicates that some integral membrane proteins bind more strongly to lipid that requires no change in bilayer thickness than to lipid that requires such a change. Preference for chain length has been demonstrated with both rhodopsin and the photosynthetic reaction center. When covalently spin-labeled rhodopsin was reconstituted in PC with different-length saturated chains, it was active in DMPC (C14), segregated into protein-rich domains in DLPC (C12), and aggregated in DSPC (C18). Similarly, when the incorporation of photosynthetic reaction center into lipid bilayers was monitored by DSC, the T_m for DLPC (C12) increased by 8°C whereas that for DPPC (C16) decreased by 3°C. These differences suggested that the protein partitioned into the gel phase with the shorter acyl chains and into the liquid crystalline phase with the longer acyl chains, since the bilayer is thicker in gel phase than in fluid phase. Such findings support the idea that hydrophobic mismatch could drive integral membrane proteins to regions of the bilayer of appropriate thickness, which could be important in raft formation.

Hydrophobic mismatch could also be involved in sorting membrane proteins to different membrane compartments. For example, along the secretory pathway that carries proteins to the plasma membrane in eukaryotic cells, some proteins remain in the Golgi apparatus, where they glycosylate secreted proteins. These Golgi-resident proteins have TM domains that are typically five amino acid residues shorter than the TM domains of proteins of the plasma membrane. This length difference was shown to be critical to sorting when constructs were engineered with shorter and longer TM segments.

The bilayer thicknesses of the membranes of the secretory pathway have been determined by x-ray scattering. The ER, Golgi, basolateral, and apical plasma membranes from rat hepatocytes were treated with proteases (and puromycin and ribonuclease [RNase] as appropriate to remove ribosomes) prior to measuring their distances from P atom to P atom to determine bilayer thickness. The thickness of these membranes was expected to increase along the pathway, proportional

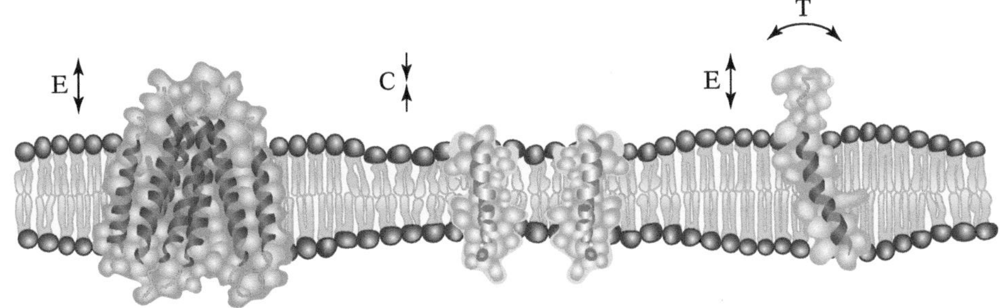

4.36 The consequence of hydrophobic mismatch in biological membranes may be a high-energy state as lipids and proteins try to compensate by extension of acyl chains (E), compression of acyl chains (C), and/or tilting of TM helices (T). Redrawn from Mitra, K., *et al.*, *Proc Natl Acad Sci USA*. 2004, 101:4083–4088. © 2004 National Academy of Sciences, U.S.A.

to their cholesterol content. While the thickness does increase from the ER to the Golgi to the apical plasma membrane, the basolateral plasma membrane is significantly thinner than the others. Since proteins targeted to the apical plasma membrane of the rat hepatocyte pass through the basolateral plasma membrane, hydrophobic mismatch must occur along the pathway. It is possible that the strain of hydrophobic mismatch puts the membrane in a high-energy state useful for vital functions such as fusion or protein insertion (Figure 4.36).

In addition to protein sorting, hydrophobic mismatch is involved in membrane protein folding (see Chapter 7). The stress induced by mismatch is likely to affect the environment in which integral membrane proteins fold and assemble, which may at least partially account for the need for specific lipids in folding certain proteins. For example, the *E. coli* transporter lactose permease (see Chapter 11) requires PE for correct folding but does not require PE for function. Thus it is proposed that the lipids have the role of chaperone in the folding process.

With this understanding of how the special environment of integral membrane proteins constrains their structure and how they interact closely and dynamically with their boundary lipids, Chapter 5 focuses on the properties of some very well-characterized proteins. The following chapters describe the kinds of functions membrane proteins carry out, the structural principles used to predict their structures, and their folding and biogenesis.

FOR FURTHER READING

Peripheral Proteins

Gerke, V., C. E. Creutz, and S. E. Moss, Annexins: linking Ca^{2+} signalling to membrane dynamics. *Nat Rev Mol Cell Biol*. 2005, 6:449–461.

Heimburg, T., and D. Marsh, Thermodynamics of the interaction of proteins with lipid membranes, in K. Merz and B. Roux (eds.), *Biological Membranes*. Cambridge, MA: Birkhauser, 1996, pp. 405–462.

Hurley, J. H., and S. Misra, Signaling and subcellular targeting by membrane-binding domains. *Annu Rev Biophys Biomol Struct*. 2000, 29:49–70.

Johnson, J. E., and R. B. Cornell, Amphitropic proteins: regulation by reversible membrane interactions. *Mol Membr Biol*. 1999, 16:217–235.

Kusters, I., and A. J. M. Driessen, SecA, a remarkable nanomachine. *Cell Mol Life Sci*. 2011, 68:2053–2066.

Mayor, S., and H. Riezman, Sorting GPI-anchored proteins. *Nat Rev Mol Cell Biol*. 2004, 5:110–119.

McLaughlin, S., and A. Aderem, The myristoyl-electrostatic switch: a modulator of reversible protein–membrane interactions. *Trends Biochem Sci*. 1995, 20:272–276.

Sardis, M. F., and A. Economou, SecA: a tale of two protomers. *Mol Microbiol*. 2010, 76:1070–1081.

Seaton, B. A., and M. F. Roberts, Peripheral membrane proteins, in K. Merz and B. Roux (eds.), *Biological Membranes*. Cambridge, MA: Birkhauser, 1996, pp. 355–403.

Curvature

Antonny, B., Mechanisms of membrane curvature sensing. *Ann Rev Biochem*. 2011, 80:101–123.

Drin, G., and B. Antonny, Amphipathic helices and membrane curvature. *FEBS Letts*. 2010, 584:1840–1847.

McMahon, H. T., and J. L. Gallop, Membrane curvature and mechanisms of dynamic cell membrane remodeling. *Nature*. 2005, 438:590–596.

Rao, Y., and V. Haucke, Membrane shaping by the Bin/amphiphysin/Rvs (BAR) domain protein superfamily. *Cell Mol Life Sci*. 2011, 68:3983–3993.

Toxins and Colicins

Falnes, P. O., and K. Sandvig, Penetration of protein toxins into cells. *Curr Opin Cell Biol*. 2000, 12:407–413.

Gouaux, E., α-hemolysin from *Staphylococcus aureus*: an archetype of β-barrel, channel-forming toxins. *J Struct Biol*. 1998, 121:110–122.

Jakes, K. S., and W. A. Cramer, Border crossings: colicins and transporters. *Ann Rev Genetics*. 2012, 46:209–231.

Young, J. A. T., and Collier, J. R., Anthrax toxin: receptor binding, internalization, pore formation, and translocation. *Annu Rev Biochem*. 2007, 76:243–265.

Zakharov, S. D., E. A. Kotova, Y. N. Antonenko, and W. A. Cramer, On the role of lipid in colicin pore formation. *Biochim Biophys Acta*. 2004, 1666:239–249.

General Features of Integral Membrane Proteins

Bowie, J. U., Membrane protein folding: how important are hydrogen bonds?. *Curr Op Str Biol*. 2011, 21:42–49.

Curran, A. R., and D. M. Engelman, Sequence motifs, polar interactions and conformational changes in helical membrane proteins. *Curr Opin Struct Biol*. 2003, 13:412–417.

Popot, J. L., and D. M. Engelman, Helical membrane protein folding, stability and evolution. *Annu Rev Biochem*. 2000, 69:881–922.

White, S. H., Biophysical dissection of membrane proteins. *Nature*. 2009, 459:344–346.

White, S. H., and G. von Heijne, Transmembrane helices before, during and after insertion. *Curr Opin Struct Biol*. 2005, 15:378–386.

White, S. H., A. S. Ladokhin, S. Jayasinghe, and K. Hristova, How membranes shape protein structure. *J Biol Chem*. 2001, 276:32395–32398.

Conserved domains for binding the membrane from the National Center for Biotechnology Information: http://www.ncbi.nlm.nih.gov/cdd/?term=membrane+binding<SITE=NcbiHome<submit.x=13<submit.y=11

Membrane proteins of known structure from the Steven White laboratory: http://blanco.biomol.uci.edu/mpstruc

Protein–Lipid Interactions

Lee, A. G., Lipid–protein interactions in biological membranes: a structural perspective. *Biochim Biophys Acta*. 2003, 1612:1–40.

Lee, A. G., How lipids affect the activities of integral membrane proteins. *Biochim Biophys Acta*. 2004, 1666:62–87.

Lee, A. G., Biological membranes: the importance of molecular detail. *TIBS*. 2011, 36: 493–500.

Marsh, D., Protein modulation of lipids, and *vice-versa*, in membranes. *Biochim Biophys Acta*. 2008, 1778:1545–1575.

Marsh, D., and T. Pali, The protein–lipid interface: perspectives from magnetic resonance and crystal structures. *Biochim Biophys Acta*. 2004, 1666:118–141.

Phillips, R., T. Ursell, P. Wiggins, and P. Sens, Emerging roles for lipids in shaping membrane protein function. *Nature*. 2009, 459:379–385.

BUNDLES AND BARRELS

5

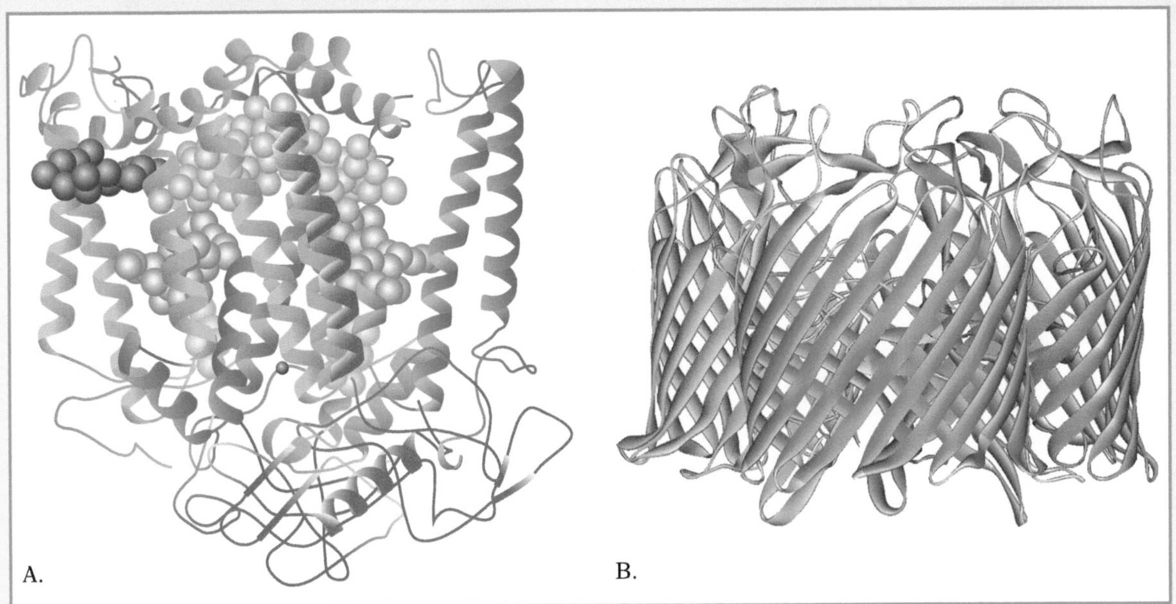

Structures of helical bundle and β-barrel membrane proteins differ in many respects, seen in the ribbon diagrams of the photosynthetic reaction center from *Rb. sphaeroides* (A) and the maltoporin trimer from *E. coli* outer membrane (B). (A) Redrawn from Jones, M. R., *et al.*, *Biochim Biophys Acta*. 2002, 1565:206–214. © 2002 by Elsevier. Reprinted with permission from Elsevier; (B) redrawn from Wimley, W. C., *Curr Opin Struct Biol*. 2003, 13:404–411. © 2003 by Elsevier. Reprinted with permission from Elsevier.

The thermodynamic arguments discussed in the previous chapter make it clear that the TM segments of proteins will utilize secondary structure to satisfy the hydrogen bond needs of the peptide backbone. While a variety of combinations of secondary structures might be imagined in polytopic membrane proteins, all known protein structures cross the bilayer with either α-helices or β-strands. This produces two major classes of integral membrane proteins: helical bundles and β-barrels. This chapter looks at the structure and function of the paradigmatic proteins in each class, then covers the surprising variation among β-barrels, and ends with a look at proteins that fit neither class.

HELICAL BUNDLES

Transmembrane (TM) α-helices have dominated the picture of membrane proteins, guided by early structural information on bacteriorhodopsin and by the first x-ray structure solved for membrane proteins, that of the photosynthetic reaction center (RC). The majority of integral membrane proteins whose high-resolution structures have been solved by x-ray crystallography exhibit the helical bundle motif (see many examples in Chapters 9 through 13). Helix–helix interactions have been analyzed in many of these, providing details of both tertiary and quaternary interactions. Identification of new integral membrane proteins in the proteome relies heavily on prediction of TM helices, as described in Chapter 6.

Bacteriorhodopsin

If a single protein dominated the thinking about structure, dynamics, and assembly of membrane proteins in the decades following 1970, that protein was bacteriorhodopsin (bR) from the purple membranes of the salt-loving bacterium *Halobacter salinarum*. From early structural

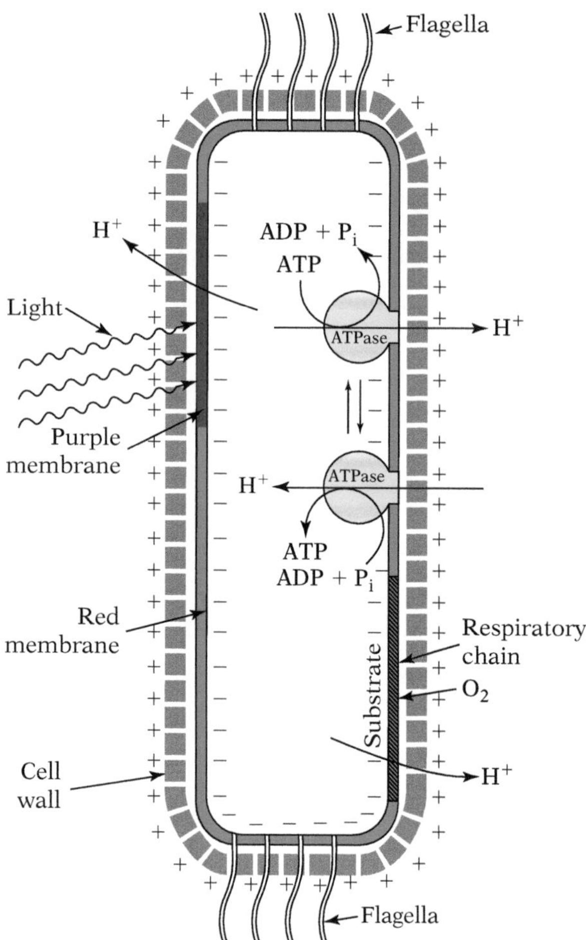

5.1 Schematic showing the different membrane domains of a halobacterium cell. The patches of purple membrane containing bacteriorhodopsin (bR) are separate from the regions of membrane containing the respiratory chain and the ATP synthase. Protons are pumped out of the cell in response either to light absorption by bR in the purple membrane or to cytosolic substrates for the electron transport chain. The ATP synthase normally uses the uptake of protons to drive the synthesis of ATP, although it can act as an ATPase and eject protons at the expense of ATP. Redrawn from Stoeckenius, W., *Sci Am*. 1976, 234:38–46. Reproduced with permission. © 1976 by Scientific American, Inc. All rights reserved.

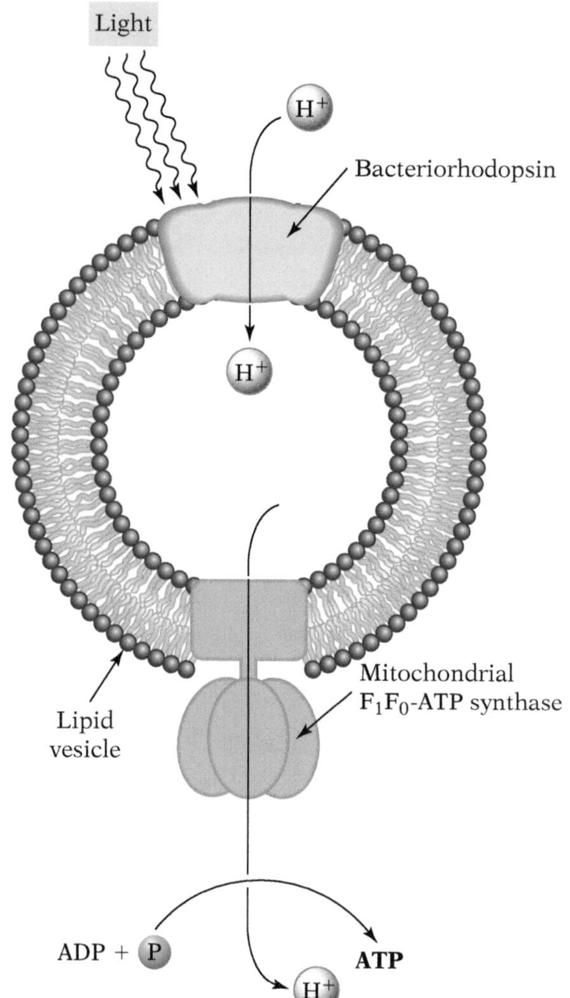

5.2 Schematic of reconstituted inside-out vesicles containing bR and ATP synthase. When the light is turned on, ATP is synthesized from ADP + Pi. When the light is turned off, ATP synthesis stops. These vesicles gave important evidence to support the chemiosmotic theory of Peter Mitchell. Redrawn from Garrett, R. H., and C. M. Grisham, *Biochemistry*, 2nd ed., Brooks/Cole, 1999, p. 897. © 1999 Wadswarth, a part of Cengage Learning, Inc. Reproduced by permission. www.cengage.com/permissions.

images and spectroscopic characterizations, bR became the paradigm for ion transport proteins and indeed for α-helical TM proteins in general. From the wealth of studies of its structure and function, a truly detailed understanding of this membrane protein has emerged.

bR is the only protein species in the discrete membrane domains called purple membranes, the light-sensitive regions of the plasma membranes of *H. salinarum* (Figure 5.1). Together with specialized lipids, this protein forms functional trimers that pack as ordered two-dimensional arrays on the bacterial cell. In photophosphorylation bR functions to pump protons out of the cell in response to the absorption of light by its chromophore, retinal, converting light energy into an electrochemical proton gradient that supports the synthesis of ATP. The ability of reconstituted vesicles containing bR and beef heart mitochondrial ATP synthase to synthesize ATP in response to light provided crucial early support for Mitchell's chemiosmotic hypothesis that the energy of an electrochemical gradient across the membrane could be used to do work (Figure 5.2).

Like rhodopsin, the light-absorbing protein in the rod outer segments of the eye's retina (see Chapter 10), bR has seven helices labeled A to G that span the membrane, and a retinal that is bound to a lysine residue in helix G via a protonated Schiff base (Figure 5.3). Whereas bR

5.3 Retinal bound to lysine 216 in bacteriorhodopsin. The Schiff base linkage (shaded) between the aldehyde of retinal and the ε-amino group of lysine 216 is protonated, as shown, before light stimulation. The all-*trans* retinal converts to 13-*cis* retinal upon absorption of a photon. Redrawn from Neutze, R., *et al.*, *Biochim Biophys Acta*. 2002, 1565:144–167. © 2002 by Elsevier. Reprinted with permission from Elsevier.

undergoes a light-induced photocycle involving conversion of the retinal from all-*trans* to 13-*cis* accompanied by conformational changes in the protein that result in proton transfer across the membrane, visual rhodopsin's activation cascade involves an 11-*cis* to all-*trans* conversion of the retinal, followed by its dissociation from the protein. In addition to bR, *H. salinarum* contains three related rhodopsins, and similar molecules are found in eubacteria and unicellular eukaryotes.

The purple membrane is composed of 75% protein and 25% lipid by weight, with ten halobacterial lipids per bR monomer. These native lipids are based on archeol (see Figure 2.6), so they differ in headgroups but not in length of acyl chains. Delipidation by treatment with a mild detergent affects the kinetics of the bR reaction; addition of halobacterial lipids but not phospholipids restores activity. When bR is crystallized (see below) the bound lipids that are retained from the membrane fit well into the grooves along the protein surface (see Figure 8.20).

bR was the first integral membrane protein whose topological organization in the membrane was elucidated. Electron diffraction of the native two-dimensional crystalline arrays of purple membrane provided early images of bR trimers, revealing the monomer structure to be seven TM α-helices arranged in an arc-like double crescent in the plane of the bilayer (Figure 5.4A, B). Models fitting the primary structure of the protein, with 70% of its 248 residues being hydrophobic, to the observed images were aided by the sensitivity of exposed loop residues to partial proteolysis *in situ* (carried out on bR in the membrane), although the precise beginning and end of each helix were uncertain for years. Such studies also showed that a few N-terminal amino acids are exposed to the exterior and the last 17 to 24 amino acids of the C terminus are accessible in the cytoplasm. The location of retinal in the center of the protein (Figure 5.4B) was determined by neutron diffraction, and the lysine to which it binds was identified by reduction

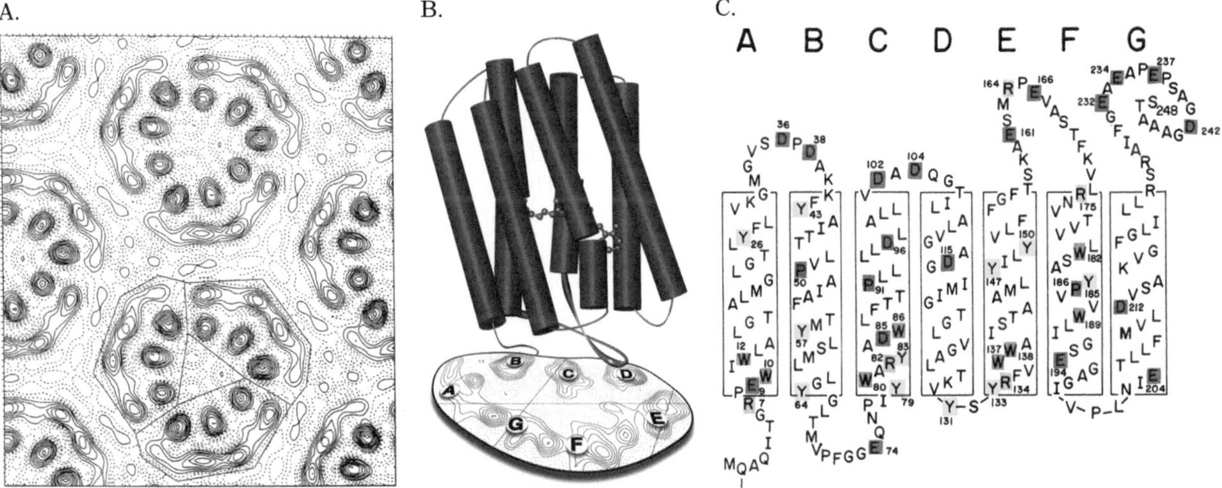

5.4 EM structure of bR. (A) The electron density profile of the two-dimensional crystalline purple membrane shows arrays of bR trimers. Each subunit of the trimer is arc-shaped with three well-resolved peaks in the inner layer and four less-resolved peaks in the outer layer. From Hayward, S. B., *et al.*, *Proc Natl Acad Sci USA*. 1978, 75:4320–4324. Reprinted with permission from the authors. (B) Seven helices are modeled to correspond to the seven peaks of a bR monomer. Based on neutron diffraction data, a retinal has been placed in the center of the protein. From Subramaniam, S., and R. Henderson, *Biochim Biophys Acta*. 2000, 1450:157–165. © 2000 by Elsevier. Reprinted with permission from Elsevier. (C) A topology model of bR shows the predicted sequence composition of the seven helices and their connecting loops. The model was adjusted periodically based on genetic mutations of targeted residues (colored boxes) until the x-ray structure was solved. From Khorana, H. G., *J Biol Chem*. 1988, 263:7439–7442. © 1988, by American Society for Biochemistry & Molecular Biology.

of the Schiff base with $NaBH_4$. Once it was clear that the retinal binds to Lys216, nearby residues were investigated by site-directed mutagenesis, which identified residues that interact with the retinal, such as Asp212 and Arg82, as well as residues crucial to proton pumping, such as Asp85 and Asp96. These results led to a model for the topology of bR that was further modified as genetic studies defined the positions of many residues (Figure 5.4C).

Extensive mutagenesis of the gene for bR provided a large collection of mutant proteins that could be studied in cell suspensions or reconstituted in lipid vesicles, with changes of pH, temperature, and salt conditions used to further characterize the protein function. In addition, a variety of retinal analogs were incorporated to observe their effects on the absorption spectrum and activity. Researchers used a number of pH-sensitive dyes to investigate the stoichiometry of proton pumping. Over many years, increasingly sophisticated instrumentation for visible and ultraviolet absorbance, fluorescence, circular dichroism, Raman, and infrared spectroscopy have been employed to follow the response of bR to light.

The primary event when bR absorbs a photon is the isomerization of retinal. This event triggers subsequent structural changes and pK_a shifts in the protein that allow deprotonation of the Schiff base, vectorial transfer of the proton to the extracellular side of the membrane, and uptake of a proton from the cytosol. These processes are accompanied by differences in the absorbance spectrum of bR, allowing detection of intermediates with lifetimes varying from a few picoseconds (ps, 10^{-12} sec) to a few milliseconds (msec, 10^{-3} sec).

The light-induced changes in bR are summarized in a photoreaction cycle, or photocycle (Figure 5.5). bR in its resting state has a λ_{max} of 570 nm (purple); when it absorbs a photon, it rapidly isomerizes to the K intermediate (λ_{max} 590 nm) and then converts to the L intermediate (λ_{max} 550 nm). The transition from L to M (λ_{max} 410 nm) occurs when the proton from the Schiff base is transferred to Asp85, the primary acceptor. At this point a structural rearrangement occurs to switch the accessibility of the Schiff base, described as M1 → M2, which is essential for vectorial proton transport by preventing reprotonation from the extracellular side that would result in a zero net effect. The next three steps are slower, each taking around 5 msec. Transfer of a proton to the Schiff base from Asp96 creates the N intermediate (λ_{max} 560 nm). With the return from 13-*cis* to all-*trans* retinal, the O intermediate (λ_{max} 640 nm) is formed and the release of a proton from Asp85 completes the cycle.

The first high-resolution structure of bR was achieved by x-ray crystallography of microcrystals prepared in bicontinuous cubic phase lipids, either monoolein or monopalmitolein. (These are not phospholipids but rather racemic mixtures of glycerol esterified to one acyl chain, either oleoyl or palmitoleoyl, on C1.) In cubic phase-

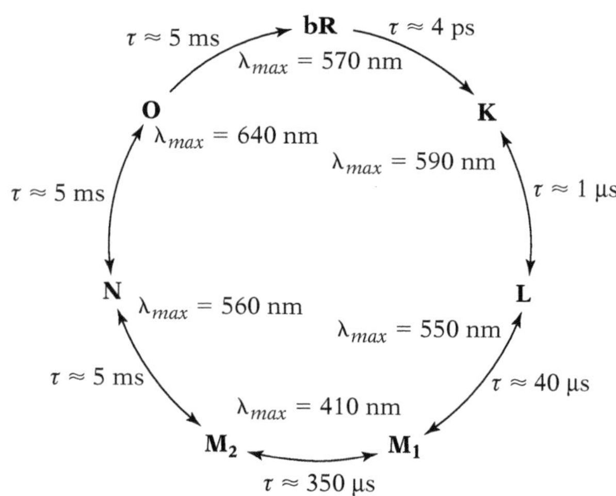

5.5 The photocycle of bR. In response to light, bR undergoes a series of transitions through intermediates K, L, M_1, M_2, N, and O, which have different lifetimes (τ) and different absorbance maxima, as shown. The photocycle is initiated by isomerization of the retinal from all-*trans* to 13-*cis* (bR → K, L), followed by transfer of a proton (L → M), the conformational change that switches the accessibility of the Schiff base from the extracellular side to the cytoplasmic side (M1 → M2), another proton transfer (M → N), and conversion of the retinal back to all-*trans* (N → O). From Neutze, R., *et al.*, *Biochim Biophys Acta*. 2002, 1565:144–167. © 2002 by Elsevier. Reprinted with permission from Elsevier.

grown crystals, bR trimers are stacked in layers that have the same orientation and lipid content as that observed in purple membrane. Furthermore, spectroscopic studies show that in the cubic phase bR undergoes the main steps of the photocycle. As expected from the electron density images, the seven TM helices cross the membrane nearly perpendicular to the plane of the bilayer and are packed closely together and connected by short loops (Figure 5.6).

The structure has now been refined to better than 2 Å resolution and provides sufficient detail to trace the proton channel, including several important water molecules (Figure 5.7A). The central cavity that contains retinal in Schiff base linkage to Lys216 is quite rigid, with the π-bulge in helix G stabilized by an H-bond from Ala215 to a water molecule. In the resting state, the positive charge on the protonated Schiff base (pK_a ~13.5) is stabilized by the nearby deprotonated carboxylate groups of Asp85 and Asp212. Polar side chains and water molecules make a clear proton path in the extracellular half of the molecule from Asp85, the proton acceptor, to the extracellular surface where the proton is released (Figure 5.7B). At the cytoplasmic surface are several acidic groups that may be involved in transferring protons from the cytoplasm, but no clear proton path connects the central cavity to them. The pK_a of Asp96, the proton donor during reprotonation of the retinal from the cytoplasmic side, is very high due to its

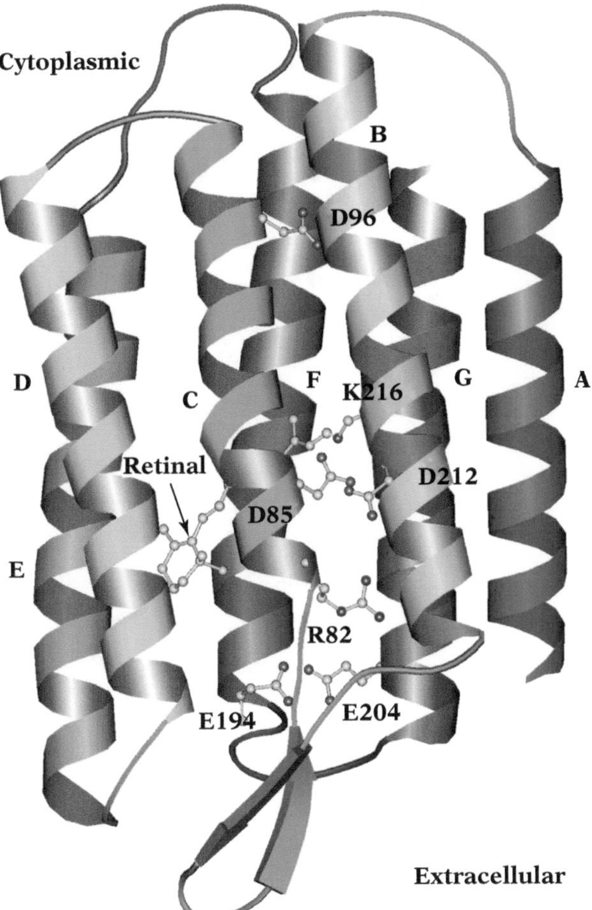

5.6 The first high-resolution structure of bR. This overview of the structure shows the seven TM helices labeled A–G, and the residues involved in proton translocation as well as the retinal. From Pebay-Peyroula, E., *et al.*, *Biochim Biophys Acta*. 2000, 1460:119–132. © 2000 by Elsevier. Reprinted with permission from Elsevier.

nonpolar environment and to its side chain H-bond to the side chain of Thr46. This part of the protein is more flexible than the extracellular half and must undergo a conformational change to open a proton path between the cytoplasmic surface and Asp96.

To relate the elegant structure of bR in its resting state to the dynamic events of its photocycle, structural information has been obtained for different intermediates in the photocycle by crystallization of mutants prevented from completing the photocycle (such as D96N, which stops at the late M state) and by "kinetic crystallography" of wild type crystals, which uses low temperatures and different wavelengths of light to trap a significant population of the molecules in the crystals in one state. By now over 20 high-resolution structures show significant movements of the cytoplasmic portions of helices F and G and the extracellular portion of helix C during the photocycle (Figure 5.8). Although there is disagreement over the details of the structural changes, together they complement the spectroscopic data to portray the steps of the cycle. Other methods, such as EPR and time-resolved wide angle x-ray scattering, continue to add insights into the dynamic mechanism of this light-driven proton pump, nature's simplest photosynthetic machine.

Photosynthetic Reaction Center

Nearly a decade before the first high-resolution x-ray structure for bR was published, the 1988 Nobel Prize in Chemistry was awarded to Hartmut Michel, Johann Deisenhofer, and Robert Huber for the elucidation of the x-ray structure of the photosynthetic RC from *Rhodopseudomonas viridis*, the first high-resolution structure achieved for integral membrane proteins. When Michel and coworkers first crystallized the RC, the gene sequences encoding its protein constituents were not even available! The beautiful structure of this multicomponent complex provided specific descriptors of its protein domains including TM helices, along with the locations of cofactors involved in light absorption and electron transfer.

In photosynthesis light energy is converted into chemical energy when the absorption of a photon drives an electron transfer that is otherwise thermodynamically unfavorable. The subsequent passage of electrons through spatially arranged carriers is coupled with the expulsion of protons, just as it is in oxidative phosphorylation, thus providing the proton gradient that drives the ATP synthase. Plants have two types of photosystems, called PSI and PSII, which differ in the electron acceptors used. The much simpler photosynthetic RCs of purple bacteria are considered the ancestors of PSII. Recent x-ray structures of both PSI and PSII reveal common structural features with the RCs in spite of their enormous size and complexity.

RCs were discovered in photosynthetic purple bacteria and characterized by biophysical techniques such as EPR (see Box 4.3) and optical spectroscopy before they proved amenable to crystallization. Soon after elucidation of the structure from *R. viridis* (renamed *Blastochloris viridis*), another high-resolution structure was obtained for the RC from *Rhodopseudomonas sphaeroides* (renamed *Rhodobacter sphaeroides*). More recently, the x-ray structure for the RC from *Thermochromatium tepidum* was solved at 2.2 Å resolution. While the cofactor–protein interactions are nearly the same in all three complexes, they represent two types of RCs. The *Rb. sphaeroides* RC is a member of group I and contains three protein subunits called L, M, and H, along with ten cofactors (see Frontispiece). The other two RCs are members of group II, and they contain an additional subunit, a *c*-type cytochrome with its four heme cofactors (Figure 5.9). The three protein subunits – L (light), M (medium), and H (heavy) – were named for their apparent molecular weights determined with SDS gel electrophoresis.

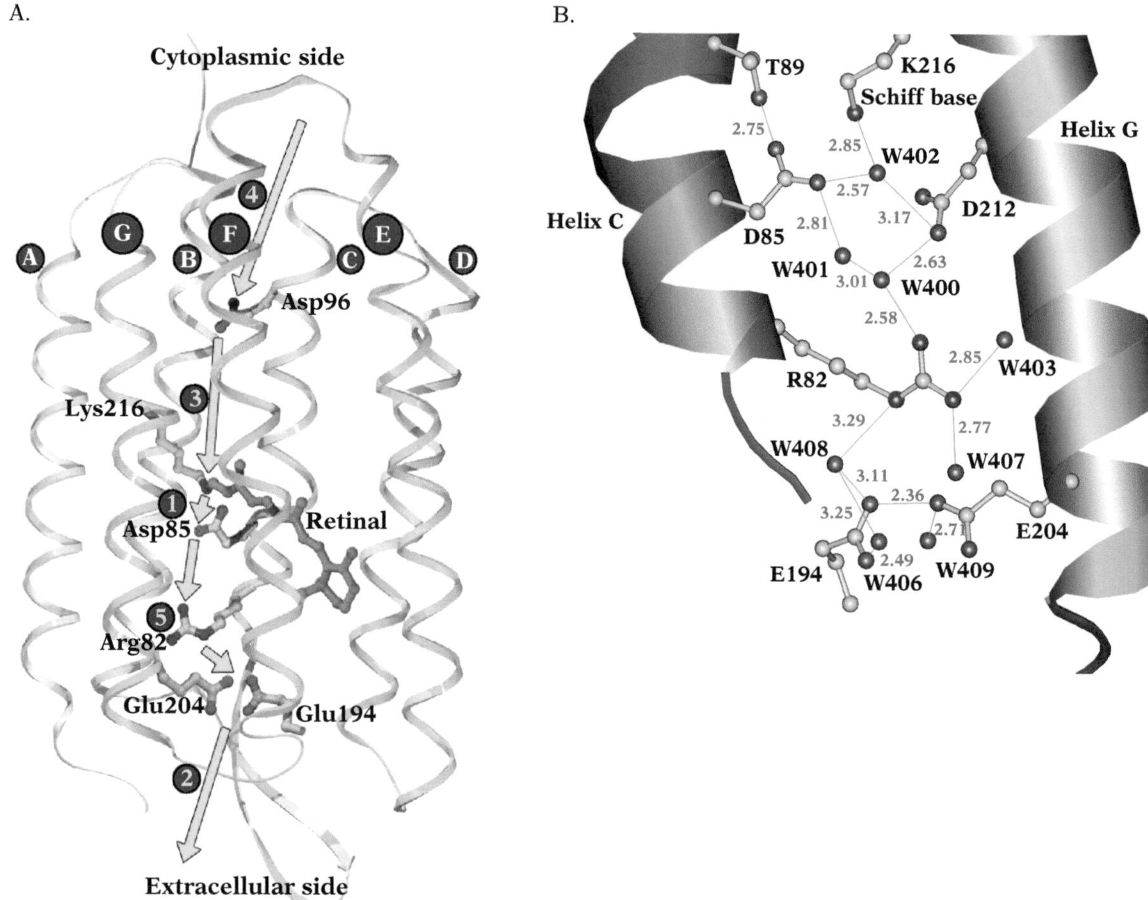

5.7 Proton path in bR. (A) The ribbon diagram for the seven TM helices, labeled A–G, is marked to show proton transfer steps indicated by arrows numbered in chronological order from 1 to 5. Step 1 is release of a proton from the Schiff base to Asp85. In step 2 a proton is released to the extracellular medium, possibly via Glu204 or Glu194. In step 3 the Schiff base is reprotonated by Asp96. Step 4 is the reprotonation of Asp96 from the cytoplasmic medium. Step 5 is the final proton transfer step from Asp85 to the group involved in proton release at the extracellular side, either Glu204 or Glu194. From Neutze, R., *et al.*, *Biochim Biophys Acta*. 2002, 1565:144–167. © 2002 by Elsevier. Reprinted with permission from Elsevier. (B) Details of the proton path on the extracellular side of the Schiff base reveal a network of hydrogen bonds between Asp85, Asp212, Arg82, Glu194, and Glu204 and discrete water molecules (red, labeled W). Interatomic distances are given in angstroms. From Pebay-Peyroula, E., *et al.*, *Biochim Biophys Acta*. 2000, 1460:119–132. © 2000 by Elsevier. Reprinted with permission from Elsevier.

The Proteins

The *B. viridis* RC has a size of ~130 Å by ~70 Å; its TM domain consists of five α-helices from L, five α-helices from M, and one α-helix from H. The remainder of the H subunit caps the structure on the cytoplasmic side, and the *c*-type cytochrome lies on the periplasmic side (Figure 5.9). The TM helices composed of 21 to 28 amino acids cross nearly perpendicular to the plane of the membrane. Three helices from each L and M subunit are nearly straight, while one is curved and one is bent more than 30° at a proline residue and ends with a 3_{10} helix (a slightly narrower helix with three amino acids per turn). The structural similarity of L and M (Figure 5.10) gives a high degree of two-fold symmetry in the TM domain in spite of only 26% sequence identity. The cytochrome with its two pairs of hemes also has two-fold symmetry, unrelated to that of L and M. Its compact structure consists of five segments: the N-terminal segment (residues C1 to C66); the first heme-binding segment (C67 to C142); a connecting segment (C143 to C225); the second heme-binding segment (C226 to C315); and the C-terminal segment (C316 to C336).

As the first available detailed structure of integral membrane proteins, the RC confirmed expectations about distribution of amino acids on its surface, yet it revealed a new concept of interior polarity. The surface of the RC complex is polar in the peripheral subunits, with a net negative charge on the periplasmic side and a net positive charge on the cytoplasmic side. The membrane-spanning surface is very hydrophobic. There are no charged amino

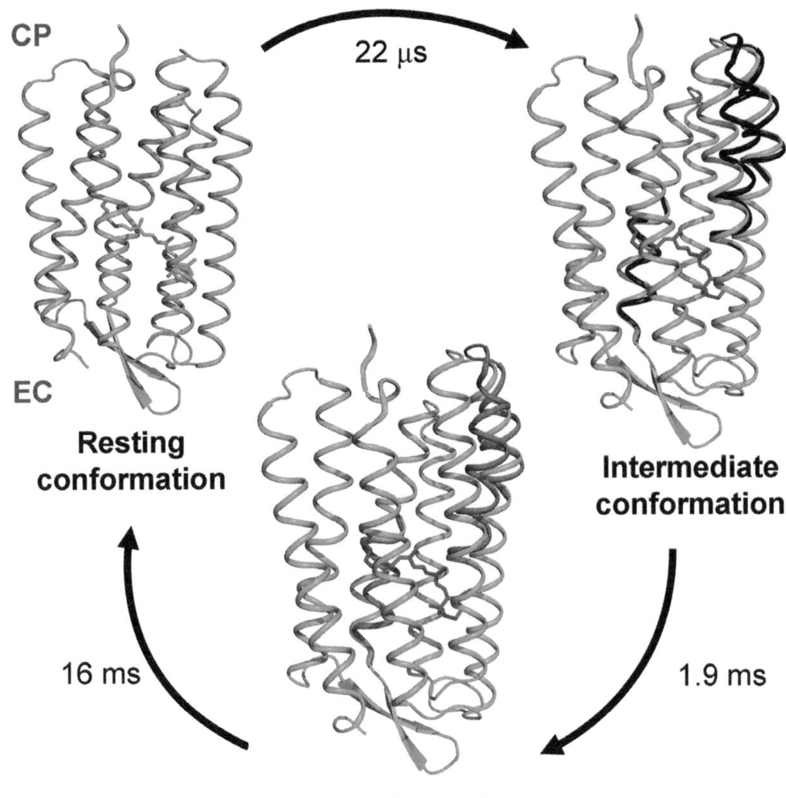

5.8 Significant conformational changes during the photocycle in bR. The ribbon diagram of the resting conformation (green) of bR is superposed with intermediate (black) and late (red) conformations to show the movements of the F, G, and C helices once the retinal is converted from all-*trans* (green) to 13-*cis* (magenta). bR is oriented with the cytoplasmic end (CP) at the top and the extracellular end (EC) at the bottom. From Andersson, M., *et al.*, *Structure*. 2009, 17:1265–1275. © 2009, Elsevier. Reprinted with permission from Elsevier.

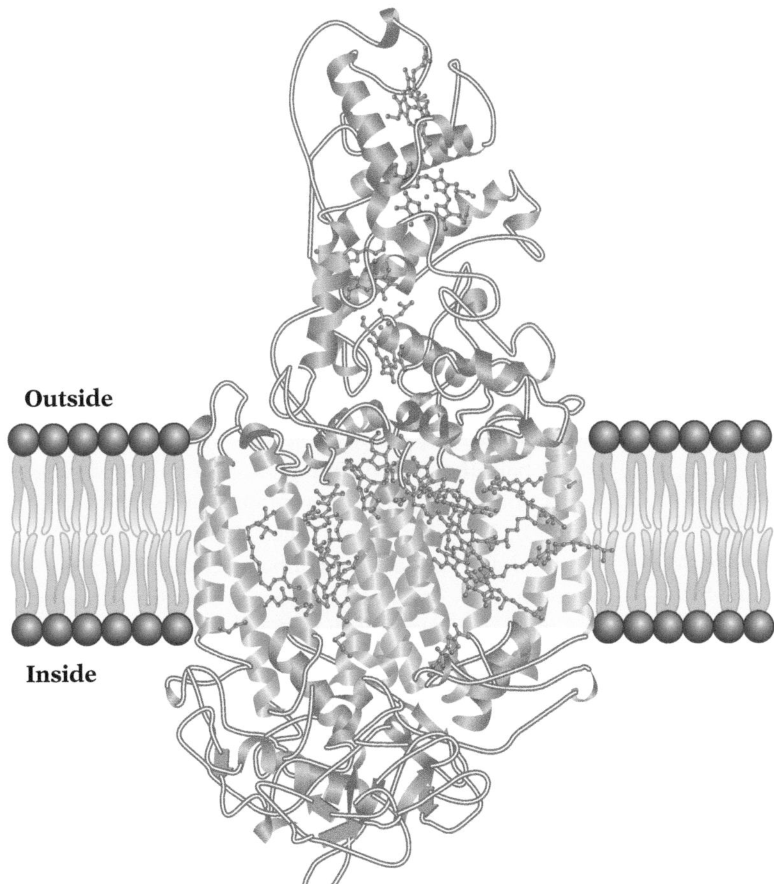

5.9 The structure of the photosynthetic reaction center from *Blastochloris viridis*, a group II reaction center. The complex contains four protein subunits, L, M, H, and a cytochrome, and 14 cofactors (red). The TM helices are highlighted in yellow. Compare it with the group I reaction center shown in the chapter frontispiece. Redrawn from Nelson, D. L., and M. M. Cox, *Lehninger Principles of Biochemistry*, 4th ed., W. H. Freeman, 2005, p. 376. © 2005 by W. H. Freeman and Company. Used with permission.

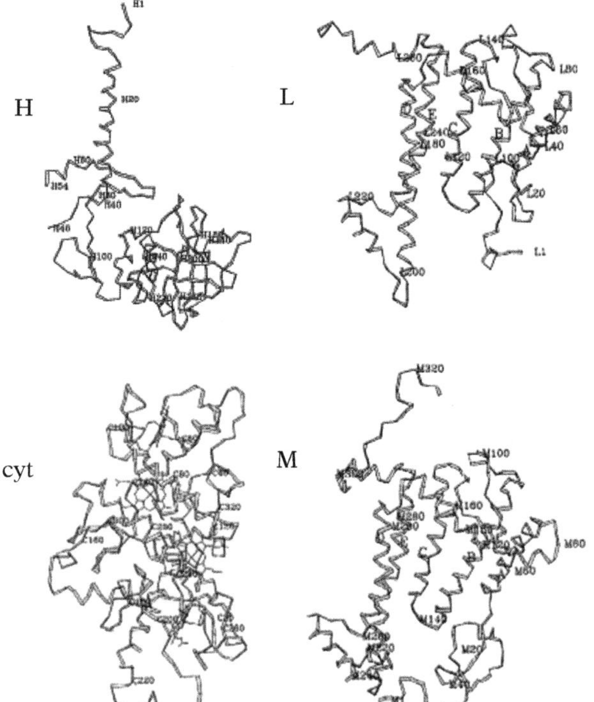

5.10 The peptide backbones of L, M, H, and cytochrome subunits of the photosynthetic reaction center from *B. viridis*. From Deisenhofer, J., *et al.*, *Nature*. 1985, 318:618–624. © 1985. Reprinted by permission of Macmillan Publishers Ltd.

acids in this 30 Å-wide band around the center of the RC, and very few water molecules associate with it. Like other membrane proteins, it has tyrosine and tryptophan residues distributed at the interfacial borders. Surprisingly, the polarity of the interior of the membrane-spanning domain is like that of the interior of soluble proteins, intermediate between the polarities of amino acids exposed to water and the hydrophobic interior of the bilayer. The ten TM helices of L and M together have 74 polar side chains. Furthermore, most of these polar residues do not appear to be involved in forming hydrogen bonds, as there are at most two hydrogen bonds between any pair of TM helices. Their predominant roles seem to be interactions with cofactors and protein subunits.

Comparison of related RC sequences indicates that residues buried in the interior of the structure are conserved more than are residues on the surface. The M, L, and H subunits of the *Rb. sphaeroides* RC have 59%, 49%, and 39% homology to subunits in *B. viridis*, respectively, and are very similar in structure. While some differences in amino acid sequence lead to differences in the interactions of the cofactors with the peptide chains, the complex has the same approximate two-fold symmetry. The site-specific mutants first available in *Rb. sphaeroides* revealed the roles of GluL212 and SerL223 for reduction and protonation of the quinone and TyrM210 for efficient electron transfer.

Lipids

Based on its size and shape, the RC is predicted by EPR to have 30–35 annular lipids (see Chapter 4). However, most of the lipids are replaced by detergent during purification. For crystallization, the RC is solubilized with LDAO (*N*,*N*-dimethyl-dodecylamine-*N*-oxide), and a few detergent molecules are included in the crystal structure. While it is often difficult to ascertain whether an acyl chain detected in the structure is from a lipid or a detergent, the x-ray structures give evidence for specific lipids (see Chapter 8), including a cardiolipin and a PE, which fit closely into hydrophobic grooves at the surface of the protein exposed to the nonpolar membrane domain.

The Cofactors

The proteins provide a scaffold for the cofactors, holding them in the same spatial arrangement in all three RCs crystallized. The RC core has ten cofactors:

- Four bacteriochlorophylls (BChl), which resemble heme except for the replacement of iron by Mg^{2+} ions, a cyclopentenone ring fused to one pyrrole ring, and different substituents off two of the pyrrole rings.
- Two bacteriopheophytins (BPh), which are BChl with two protons in place of the Mg^{2+}.
- Two quinones (one ubiquinone and one menaquinone in *B. viridis*).
- A nonheme ferrous ion.
- A carotenoid, which is a largely linear C40 polyene such as *β*-carotene.

There are two types of both BChl and BPh, a and b. The a type is found in the RC from *Rb. sphaeroides* and has either a phytyl or geranylgeranyl side chain, whereas the b type, found in the RC from *B. viridis* and *T. tepidum*, has only a phytyl side chain and has one more C=C in the side chain of ring II.

The cofactors form two symmetrically related branches within the hydrophobic environment of the closely packed TM helices (Figure 5.11). At the top, two molecules of BChl are positioned so close together that the edges of their tetrapyrrole rings overlap. Called the special pair, they receive the photon of light and release an electron in the primary event of photosynthesis. Each branch has another BChl (called the accessory BChl), a BPh, and a quinone. The nonheme iron is between the two quinones. The two branches follow the same local symmetry displayed by the L and M chains. The symmetry is not perfect, and the two branches have different electron transfer properties – in fact, electron transfer uses only the branch that associates with the L subunit. The cofactors are differentiated by groups of the protein that alter their environments. For example, the two quinones play different roles (see below) and the primary quinone is in a more hydrophobic environment than the secondary quinone. After being reduced and protonated,

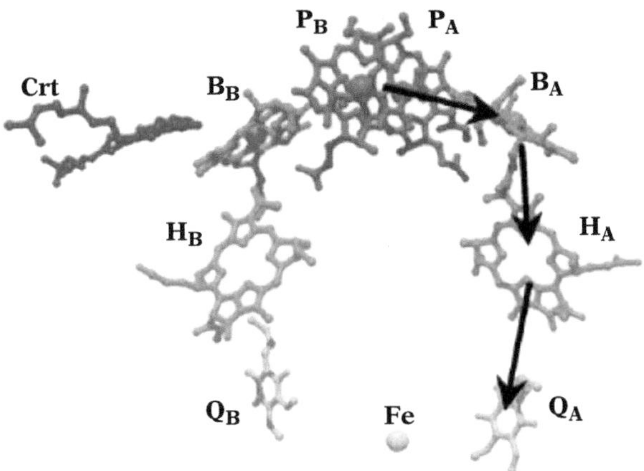

5.11 The arrangement of the RC cofactors. The tetrapyrrole rings of BChl-b, BPh-b, and the quinone headgroups follow the same local symmetry displayed by the L and M chains. The figure shows the special pair, P_A and P_B (coral); the accessory BChl molecules, B_A and B_B (rose); the two BPh, H_A and H_B (cyan); the ubiquinones, Q_A and Q_B (yellow); carotenoid, Crt (purple); and non-heme iron atom (gray). The electron path utilizes only the A half, indicated by the arrows. From Jones, M. R., *et al.*, *Biochim Biophys Acta*. 2002, 1565:206–214. © 2002 by Elsevier. Reprinted with permission from Elsevier.

the secondary quinone (now quinol) diffuses from the RC, its leaving facilitated by nearby carboxylate groups. The similarities in the arrangement of cofactors in all photosystems, including the iron–sulfur type PSI complexes, allow definition of a common motif: a dimer of tetrapyrrole molecules flanked by four monomeric tetrapyrrole molecules with an iron at the center.

The cytochrome of group II RCs has four hemes, located in pairs on the two heme-binding segments of the polypeptide, each having a helix followed by a turn and the Cys-X-Y-Cys-His sequence typical of *c*-type cytochromes. The hemes lie parallel to the axis of the helix with thioether bonds between each heme and the two cysteines (Figure 5.12). The fifth ligand to the iron is His, and for three of the four hemes the sixth ligand is a methionine residue within the helix. The reduction potentials for the hemes vary, and paradoxically the arrangement of hemes by potential does not follow the internal symmetry of the cytochrome. The closest heme to the special pair is heme-3, with the highest reduction potential (370 mV). However, in order of increasing distance from the special pair, the heme reduction potentials follow the sequence high, low, high, low.[1] Spectroscopic measurements indicate that electrons are transferred through heme-2 and heme-3 to the special pair, perhaps through heme-4, which is located between them but has a much lower reduction potential.

[1] In *B. viridis*, the reduction potentials are heme-1, –60 mV; heme-2, +300 mV; heme-3, +370 mV; and heme-4, +10 mV.

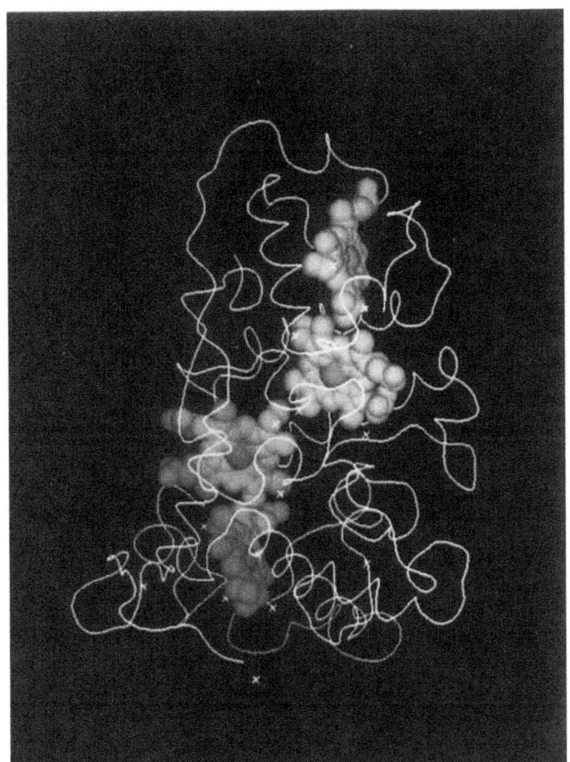

5.12 Structure of the tetraheme cytochrome subunit of the *B. viridis* reaction center. Location of the hemes is apparent when the polypeptide backbone of the cytochrome subunit is represented by a thread, while the heme groups are represented by space-filling models. The crosses mark the visible positions of the sulfur atoms in the thioether side chains for the heme bridges, as well as free cysteines. From Knaff, D. B., *et al.*, *Biochemistry*. 1991, 30:1303–1310. © 1991, American Chemical Society.

Antennae

To maximize the absorption of light, the membrane contains about 100 times more BChl molecules than RC complexes. These other chlorophylls function as antennae to collect photons and pass the energy on to the special pair, greatly increasing the efficiency of each RC. These chromophores are organized by light-harvesting proteins into complexes called LH1 and LH2. LH1 is the core complex, found in a fixed stoichiometry with the RC. LH2 is peripheral, and its synthesis depends on factors such as light intensity (Figure 5.13). LH2 absorbs light at shorter wavelengths than LH1, so it rapidly passes the energy to LH1, which passes it to the RC. Both LH1 and LH2 contain many copies of two short protein subunits called α and β (in LH2, α has 56 amino acids and β has 45 amino acids), each containing a single TM helix.

In low-resolution EM images of native *B. viridis* membrane, the RC appears to be surrounded by a ring of 15–17 LH1 molecules. A crystal structure of LH2 shows a double cylinder formed of an inner ring of its α subunits and an outer ring of its β subunits, with eight- or

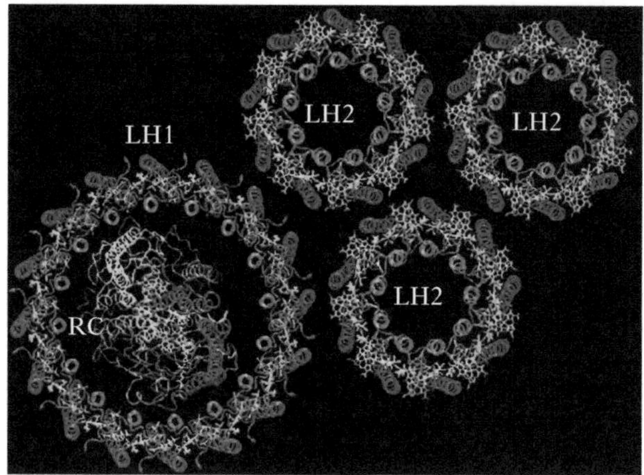

5.13 Model of the antenna system in purple photosynthetic bacteria. Viewed from above the plane of the membrane, each photosynthetic reaction center is surrounded by a light-harvesting complex called LH1. In addition, several LH2 complexes associate with it and with each other to rapidly transfer the excitation generated by absorption of light. From Voet, D., and J. Voet, *Biochemistry*, 3rd ed., John Wiley, 2004, p. 878. © 2004. Reprinted with permission from John Wiley & Sons, Inc.

nine-fold symmetry, depending on the organism of origin. A ring of nine-fold symmetry has 18 BChl molecules between the rings, arranged like a water wheel, and nine more BChl molecules between the helices of its outer wall, along with accessory pigments such as carotenoids (Figure 5.14A). However, both theoretical considerations and spectroscopic data suggest that the ring is not intact but is C-shaped. Dimers of LH1 observed at low resolution indicate that two C-shaped complexes face each other to form an S-shape (Figure 5.14B). Dimers of LH1 could be isolated only in the presence of another membrane protein called PufX, named for photosynthetic unit formation. This protein, which is required for anaerobic photosynthetic growth, associates closely in a 1:1 ratio with RC–LH1 complexes.

The Reaction Cycle

The overall reaction carried out by the photosynthetic RC is the reduction of ubiquinone by cytochrome c_2, using the energy from light to initiate the electron transfer. It occurs through the following steps (Figure 5.15):

(1) The light energy absorbed via antenna complexes LH1 and LH2 is transmitted to the RC, where it promotes a charge separation with the oxidation of the special pair of BChl and the two-electron reduction of quinone to quinol. When the special pair absorbs light, it gives a distinct optical absorption band in the near infrared. The cofactors in one branch are close enough to delocalize the electrons in a conjugated system, so electron transfer is fast – within 200 ps the electron reaches the primary quinone. The primary quinone accepts one electron and releases it to

A.

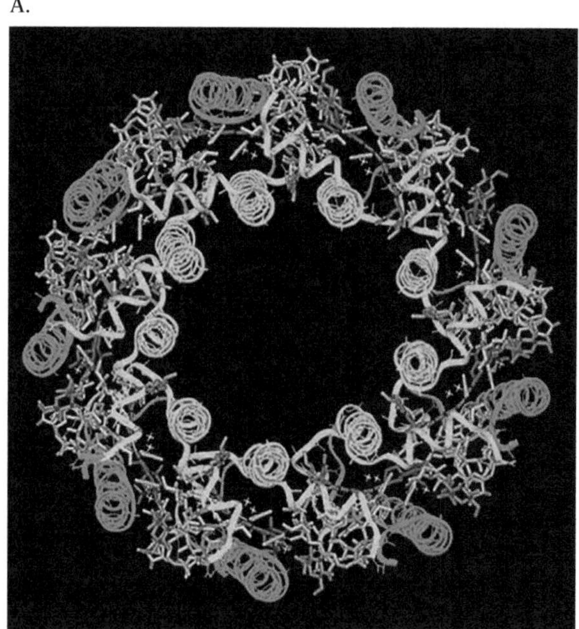

B.

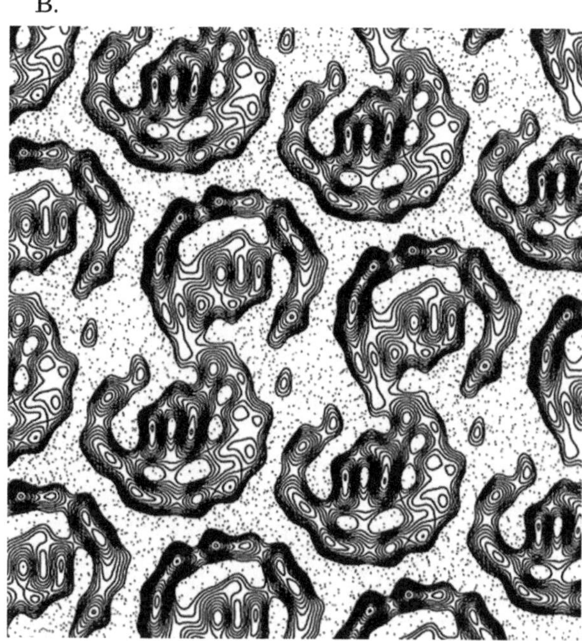

5.14 Organization of the light-harvesting complexes. (A) Crystal structure of LH2 from *Rhodopseudomonas acidophila*, viewed from the periplasmic side. The complex has nine-fold symmetry, shown with α subunits in yellow, β subunits in green, BChl in gray, and carotenoids in orange. (B) Projection map of membranes from *Rb. sphaeroides* grown under photosynthetic conditions in the presence of nitrate, in which two C-shaped LH1–RC complexes dimerize to an S shape. From Vermeglio, A., and P. Joliot, *Trends Microbiol.* 1999, 7:435–440. © 1999 by Elsevier. Reprinted with permission from Elsevier.

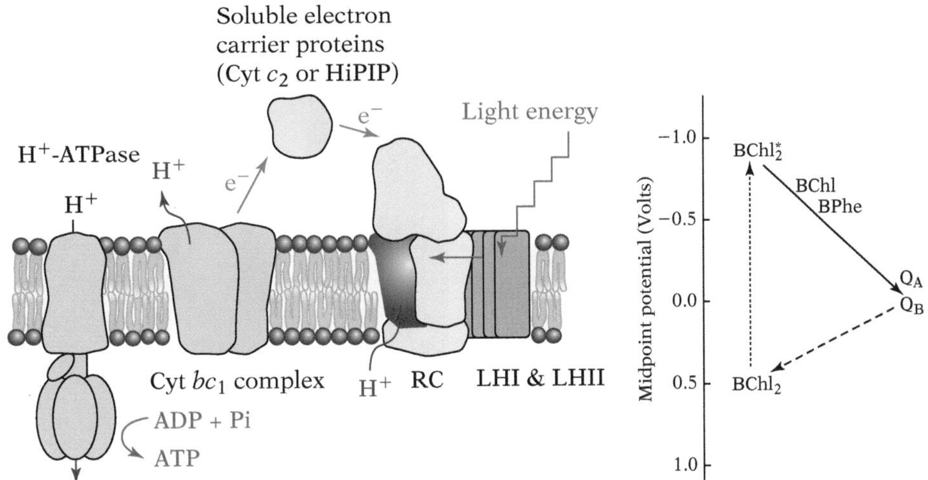

5.15 Illustration of the photosynthetic electron transfer reactions in purple bacteria. The RC (pink) accepts energy from the antenna complexes LH1 and LH2 (purple). After charge separation in the RC complex reduces the secondary quinone (Q_B) to quinol, it diffuses through the membrane to the cytochrome bc_1 complex (green), where it is oxidized back to quinone. Cytochrome bc_1 transfers the electrons to soluble electron carrier proteins (blue), either cytochrome c_2 or HiPIP, which carry them back to the RC. Cytochrome bc_1 also generates a proton gradient used by the ATP synthase (H^+-ATPase, yellow) to drive the synthesis of ATP. Redrawn from Nogi, T., and K. Miki, *J Biochem (Tokyo)*. 2001, 130:319–329. © 2001, Oxford University Press. Inset: potentials of cofactors involved in electron transfer in purple bacteria. The dotted line represents the absorption of light by the special pair of BChl, and the solid line represents the energy transfer steps in the RC to the quinones, Q_A and Q_B (shown in Figure 5.11). The dashed line represents steps that occur outside the RC. Redrawn from Allen, J. P., and J. C. Williams, *FEBS Lett*. 1998, 438:5–9. © 1998 by the Federation of European Biochemical Societies. Reprinted with permission from Elsevier.

the secondary quinone, which then accepts a second electron from a second photon event. Next the secondary quinone picks up two protons from the cytoplasm and enters the membrane's quinone pool as quinol (Q_BH_2).

(2) The quinol diffuses in the membrane to the cytochrome bc_1 complex, where it is reoxidized, with concomitant release of electrons to periplasmic electron carriers, either cytochrome c_2 or in some species (including *T. tepidum*) the abundant high-potential iron–sulfur protein (HiPIP). The cytochrome bc_1 complex is similar to complex III in the mitochondrial membrane. It releases protons into the periplasmic space, from which they reenter the cytosol via the F_1F_0-ATP synthase (see Chapter 13) to drive the synthesis of ATP.

(3) The periplasmic electron carriers bring electrons to the RC to reduce the special pair. Cytochrome c_2 is very similar to mitochondrial cytochrome c and diffuses along the surface of the membrane to the RC. The *Rb. sphaeroides* RC has been cocrystallized with cytochrome c_2, which appears to bind quite near the special pair and to reduce it directly. In group II RCs, the soluble electron carriers dock on the cytochrome subunit of the RC and reduce its hemes. The binding sites envisioned on crystal structures of the separate redox components indicate that the cytochrome c_2 interaction with the RC cytochrome is electrostatic, while that between HiPIP and the *T. tepidum* RC is hydrophobic (Figure 5.16).

Completion of the cycle takes less than 100 μs and gives a quantum yield close to one, meaning that for almost every photon absorbed by the RC, one electron is transferred. There is much more to learn about how the RC regulates the electrochemical properties of its cofactors and precisely how it takes up protons from the cytosol (when it reduces quinone) and electrons from reduced periplasmic carriers (to reduce the special pair at the end of the cycle). To address these and other questions, researchers are now exploring structural differences between RCs in the light and those adapted to darkness, between wild type and mutants, and between isolated RC and the complex consisting of RC and its partners of the photosynthetic membrane.

β-BARRELS

While the structures of bR and the photosynthetic RC came to represent the image of membrane proteins, a wholly different picture of membrane protein structure was emerging in studies of bacterial porins. Early data from circular dichroism and infrared spectroscopy indicated that these pore-forming proteins contain β-structure and little or no helical content; EM and x-ray diffraction studies suggested they span the membrane

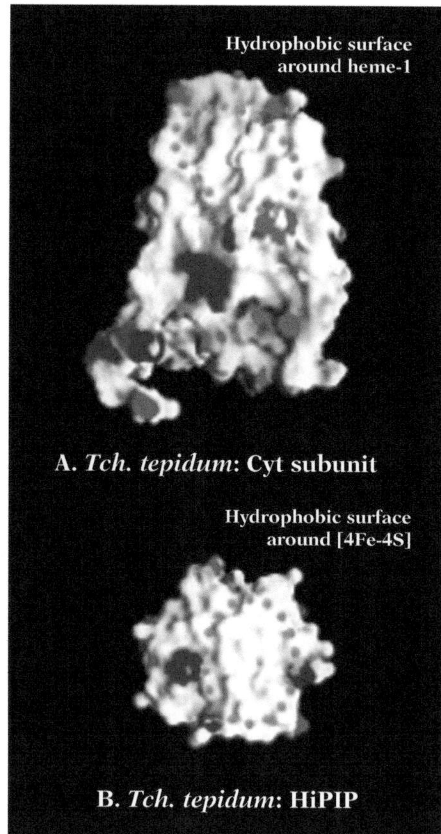

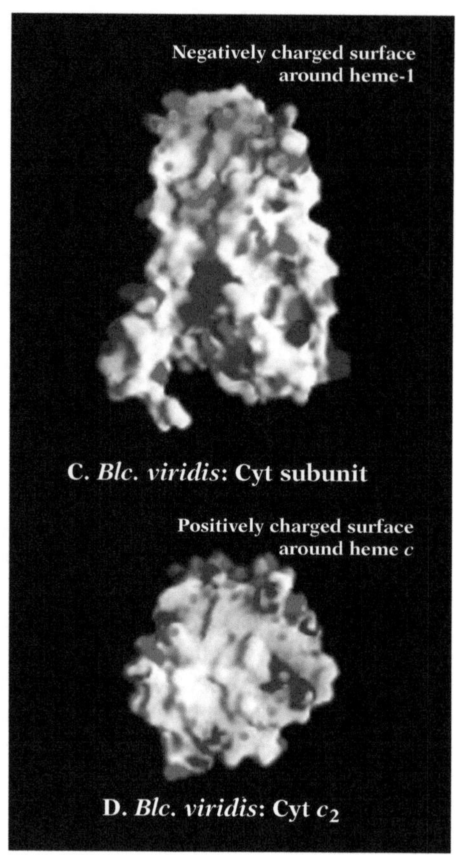

5.16 Recognition of electron carriers by the cytochrome subunit of the RC. In the case of HiPIP, the interacting surfaces on both proteins are hydrophobic ((A) and (B)), while the recognition of cytochrome c_2 is electrostatic ((C) and (D)), as the *B. viridis* cytochrome c_2 has a large basic surface and the cytochrome subunit of RC has a large acidic surface. Negatively charged surfaces are red, and positively charged surfaces are blue. The green dotted circles indicate the interacting surfaces involved in recognition. From Nogi, T., and K. Miki, *J Biochem (Tokyo)*. 2001, 130:319–329. © 2001, Oxford University Press.

as *β*-barrels. The first x-ray crystal structure of a TM *β*-barrel was that of the porin from *Rhodobacter capsulatus* (Figure 5.17) and was quickly followed by the high-resolution structures of other bacterial porins.

The cell envelope of Gram-negative bacteria consists of two membranes separated by an aqueous compartment called the periplasm and a thin layer of peptidoglycan, a net-like polymer of amino acid and sugar residues that confers structural stability (see Figure 1.1B). The *β*-barrel proteins are found in the outer membrane, while the proteins in the inner (plasma) membrane are bundles of *α*-helices. This difference is attributed to mechanisms of biogenesis, since the more hydrophilic *β*-barrel proteins are secreted into the periplasm before incorporation into the outer membrane, while the inner membrane proteins are incorporated directly into the bilayer from the translocon (see Chapter 7). New types of *β*-barrel proteins with a surprising diversity of function have been discovered and many more can be expected, since in *E. coli* alone the outer membrane is estimated to have about 100 proteins. *β*-barrel proteins are predicted to make up 2% to 3% of the proteome of Gram-negative bacteria. They also are found in mitochondria and chloroplasts of eukaryotes but not in archaea or in Gram-positive bacteria, except *Mycobacteria*.

Due to its extended peptide backbone, a *β*-strand needs only seven amino acid residues to cross the nonpolar domain of the membrane. Typically TM *β*-strands have 9–11 residues, usually tilted ~45° from perpendicular to the plane of the bilayer, with observed tilts varying from 20° to 45°. To satisfy the H-bonding of the carbonyl and amino groups along their peptide backbones within the nonpolar domain, *β*-strands must partner with other *β*-strands. By forming a circular pattern in a barrel, none of the strands are left without a partner. The interstrand H-bonding makes the structure rigid, as well as very stable: for a small *β*-barrel of eight TM strands, the H-bonding contributes a stabilization of around 40 kcal mol^{-1} (eight strands of ~10 amino acids, forming 80 H-bonds × 0.5 kcal mol^{-1} H-bond).

Among known *β*-barrel structures of membrane proteins, with one exception (VDAC, see below) the number of *β*-strands is an even number that varies from 8 to 24. In the smallest of these proteins, the center is filled with polar residues, while barrels of intermediate sizes (16–18 strands) contain aqueous pores. The larger barrels

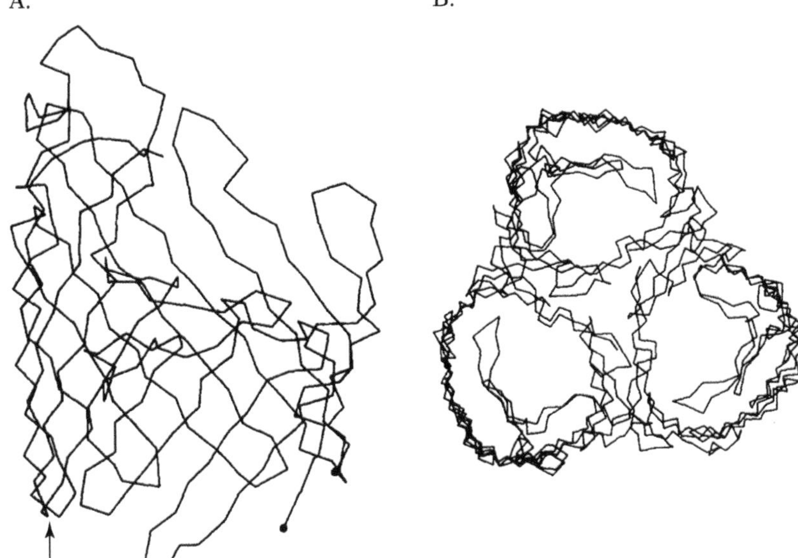

5.17 Tracing of porin from *Rhodobacter capsulatus*, the first porin with an x-ray crystal structure. (A) View from the side of the monomer, with chain ends marked by dots. (B) View of the trimer from the top. From Weiss, M. S., *et al.*, *FEBS Lett.* 1990, 267:379–382. © 1990 by the Federation of European Biochemical Societies. Reprinted with permission from Elsevier.

with 22 strands have "hatch" or plug domains that close their channels. An even number of strands allows the N and C termini to meet since the barrels are made from meandering antiparallel strands, which means all of them are hydrogen-bonded to their next neighbors along the peptide chain. With strands connected by tight turns or loops, the twisted β-sheet rolls into a cylinder. In the known structures, the periplasmic ends of the β-strands form tight turns and the other ends are loops of varying lengths, either pointing into the extracellular space or folding back inside the barrel.

In general, the β-strands are amphipathic, with polar side chains pointing to the aqueous interior of the protein alternating with nonpolar side chains contacting the hydrophobic bilayer. This means the composition of the lipid-exposed surfaces of β-barrels, like those of α-helical integral membrane proteins, is rich in aromatic and nonpolar amino acids (Phe, Tyr, Trp, Val, Leu, Ile; Figure 5.18). At the two bilayer interfaces, aromatic residues are ~40% of the lipid-exposed residues. These aromatic side chains form girdles whose intermediate polarity helps define the two nonpolar–polar interfaces of the membrane, also observed in α-helical membrane proteins. The exposed loops at the extracellular surface provide binding sites for colicins and bacteriophage, as well as recognition sites for antibodies. Interestingly, half of the strands of the small β-barrel called OmpX protrude into the external medium and nonspecifically bind proteins with exposed β-strands (see Figure 5.18A and Box 5.1). Because of this unusual feature, OmpX, which is not a porin, is called an adhesion protein. It is induced by stress and is thought to be a part of the virulence mechanism of *E. coli*.

There is much variation in quaternary structure among the known β-barrels. Several of them are monomers, while all the porins are homo-oligomers (usually trimers) with very tight interactions between the subunits. The enzyme activity of outer membrane phospholipase A (OMPLA; see Chapter 9) is regulated by its quaternary structure: it is only active as a homodimer. A mycobacterial porin forms a β-barrel similar to those of anthrax toxin protective antigen and α-hemolysin (see Chapter 4) in that each subunit contributes a β-hairpin to make the 16-stranded barrel. The rich diversity of structure and function among β-barrels continues to be uncovered, yet the paradigm for this group of membrane proteins is the porins.

Porins

The outer membranes of Gram-negative bacteria protect the cells from harmful agents while allowing nutrient uptake via porins and other transport proteins. Porins function as passive diffusion channels and allow rapid diffusion of their solutes, even at 0°C. Porins are grouped in two categories: general porins, like OmpF and OmpC, and specific porins, like PhoE (phosphoporin), LamB (maltoporin), ScrY (sucrose porin), and Tsx (the nucleoside channel). The channels of general porins do not discriminate among solutes that are hydrophilic, under ~600 Da, and not highly charged, which allows them to take up many nutrients such as mono- and disaccharides. In contrast, the specific porins have channels that are selective for their solutes, although the line of demarcation is blurred as described below for PhoE. Note that aquaporin, which is a TM channel for water molecules in plasma membranes (see Chapter 12), is not a β-barrel.

Porins were first detected by the permeability of the outer membrane of Gram-negative bacteria to hydrophilic

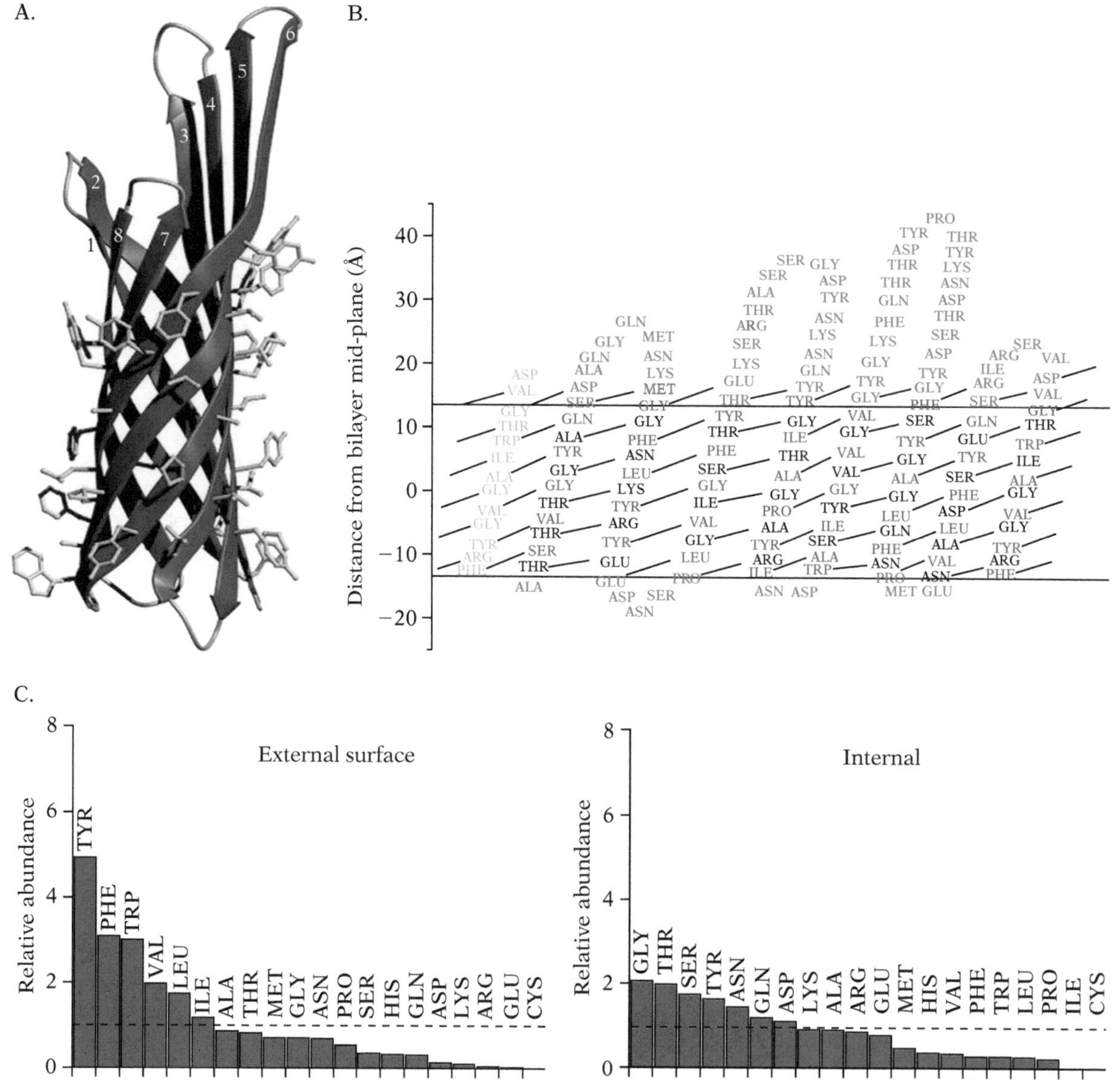

5.18 The structure of OmpX, a small, closed β-barrel protein. Because OmpX is monomeric, all of its exposed side chains go into the bilayer. Additionally, four of its eight β-strands protrude on the outside of the cell. The barrel interior is filled with polar residues that form a hydrogen-bonding network. (A) Ribbon drawing shows nonpolar side chains (yellow) on the surface of OmpX. From Schulz, G. E., *Curr Opin Struct Biol.* 2000, 10:443–447. © 2000 by Elsevier. Reprinted with permission from Elsevier. (B) Topology model shows the primary structure and distribution of amino acids in OmpX, with lipid-exposed (red), interior (black), and external loops (green). The *y*-axis gives the transbilayer location with the center of the bilayer as 0. To show the interstrand hydrogen bonds between each pair of strands, the eighth strand is repeated on the left (gray). (C) Bar graphs show the distribution of amino acid residues on the external surface (left) and in the interior (right). (B) and (C) from Wimley, W. C., *Curr Opin Struct Biol.* 2003, 13:404–411. © 2003 by Elsevier. Reprinted with permission from Elsevier.

antibiotics, which allowed them to be assayed by the rate of hydrolysis of β-lactams (penicillin and its derivatives) in the periplasm of intact cells. Channel activity of porins is evident when the purified proteins are reconstituted with lipids: porins make lipid vesicles permeable to small, hydrophilic solutes and make voltage-gated conductance channels in black films (see Chapter 3). Porins were first purified as SDS-resistant two-dimensional crystalline aggregates. They can also be solubilized by extraction with nonionic detergents, and they bind 0.6 g detergent per gram of protein (around 200 detergent monomers per protein trimer.) These remarkably stable proteins are highly resistant to proteases, chaotropic agents, and most organic solvents, in addition to many detergents.

OmpF and OmpC

Typically Gram-negative bacteria have ~10^5 copies of general porins per cell, with the number of species of porins varying in different strains. The outer membrane of *E. coli* K12 is dominated by OmpF and OmpC porins,

BOX 5.1 NMR DETERMINATION OF β-BARREL MEMBRANE PROTEIN STRUCTURE

Some of the first membrane proteins to have structures solved by solution NMR (see Box 4.2) are β-barrel membrane proteins. Structures of β-barrels are easier to solve by NMR than those of helical bundles because (1) the ^1H chemical shift dispersion is larger for β-sheets than for α-helices, especially with alternating hydrophilic/hydrophobic residues, and (2) β-barrels have higher thermal stability, so they can withstand higher temperatures for the longer times needed to get well-resolved NMR spectra. Furthermore, the two β-barrel proteins described here could be overexpressed in *E. coli* and purified in denatured form from inclusion bodies before refolding in detergent micelles (see Chapter 7), resulting in a mixed micelle particle size of 60–80 kDa. Uniform labeling of the proteins with ^2H, ^{13}C, and ^{15}N allows detection of the protein NMR signals with little or no interference from the signals of the unlabeled detergent molecules.

OmpX from *E. coli*

Triply labeled OmpX, a 148-residue protein, was solubilized from inclusion bodies with guanidinium chloride and reconstituted in dihexanoyl-PC (DHPC) micelles. Comparison of the two-dimensional COSY (Correlation Spectroscopy) and TROSY spectra (Figure 5.1.1A) shows the enhancement obtained with TROSY, which allowed the eight-stranded fold of the polypeptide backbone of OmpX to be determined from 107 Nuclear Overhauser Effect (NOE)-derived distance

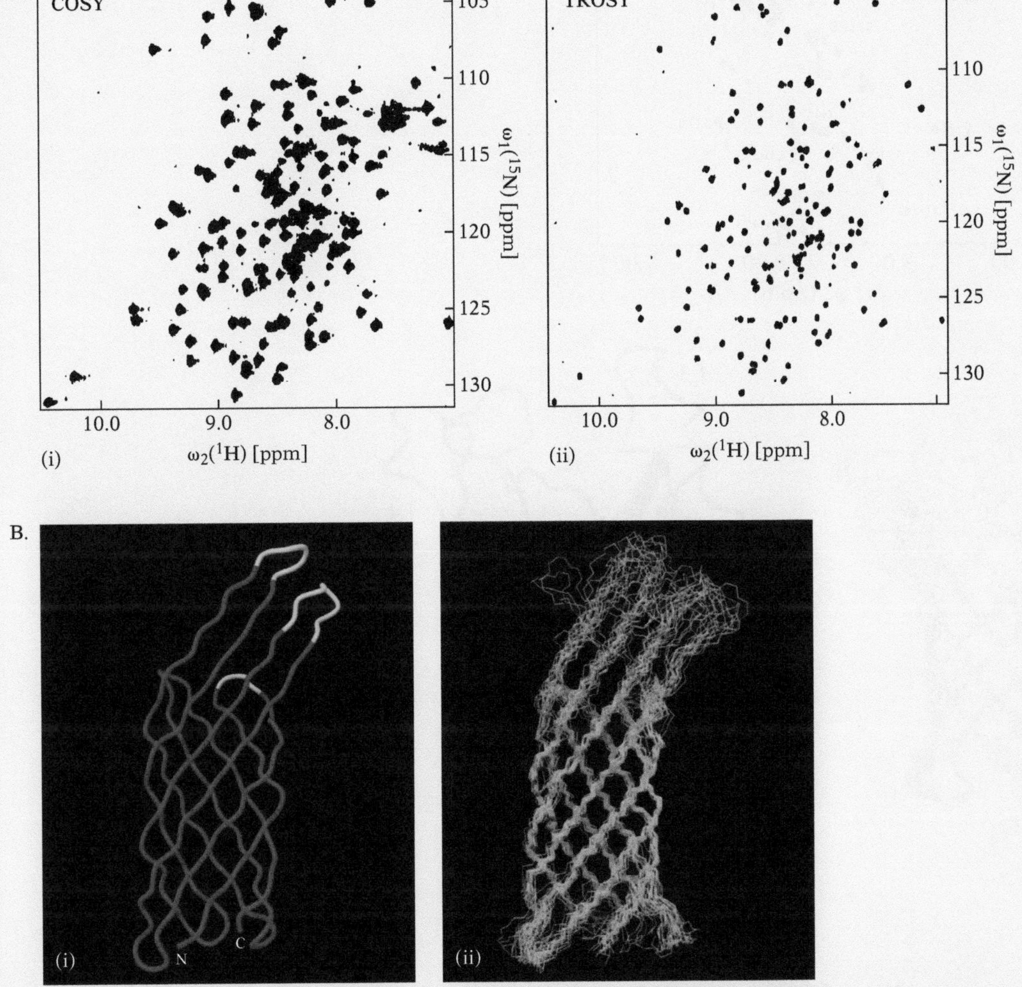

5.1.1 OmpX structure solved by solution NMR. (A) NMR spectra of OmpX in DHPC micelles obtained using ^1H ^{15}N COSY (i) and ^1H ^{15}N TROSY (ii). (B) Structure of OmpX determined by NMR. When the NMR assignments are mapped onto the x-ray structure of OmpX (i), the barrel portion is well-structured (red) and three loops are disordered (yellow). The flexibility of the loops is clear in the NMR structure represented by the superposition of 20 conformers (ii). From Fernandez C., *et al.*, *FEBS Lett.* 2001, 504:173–178. © 2001 by the Federation of European Biochemical Societies. Reprinted with permission from Elsevier.

(continued)

BOX 5.1 *(continued)*

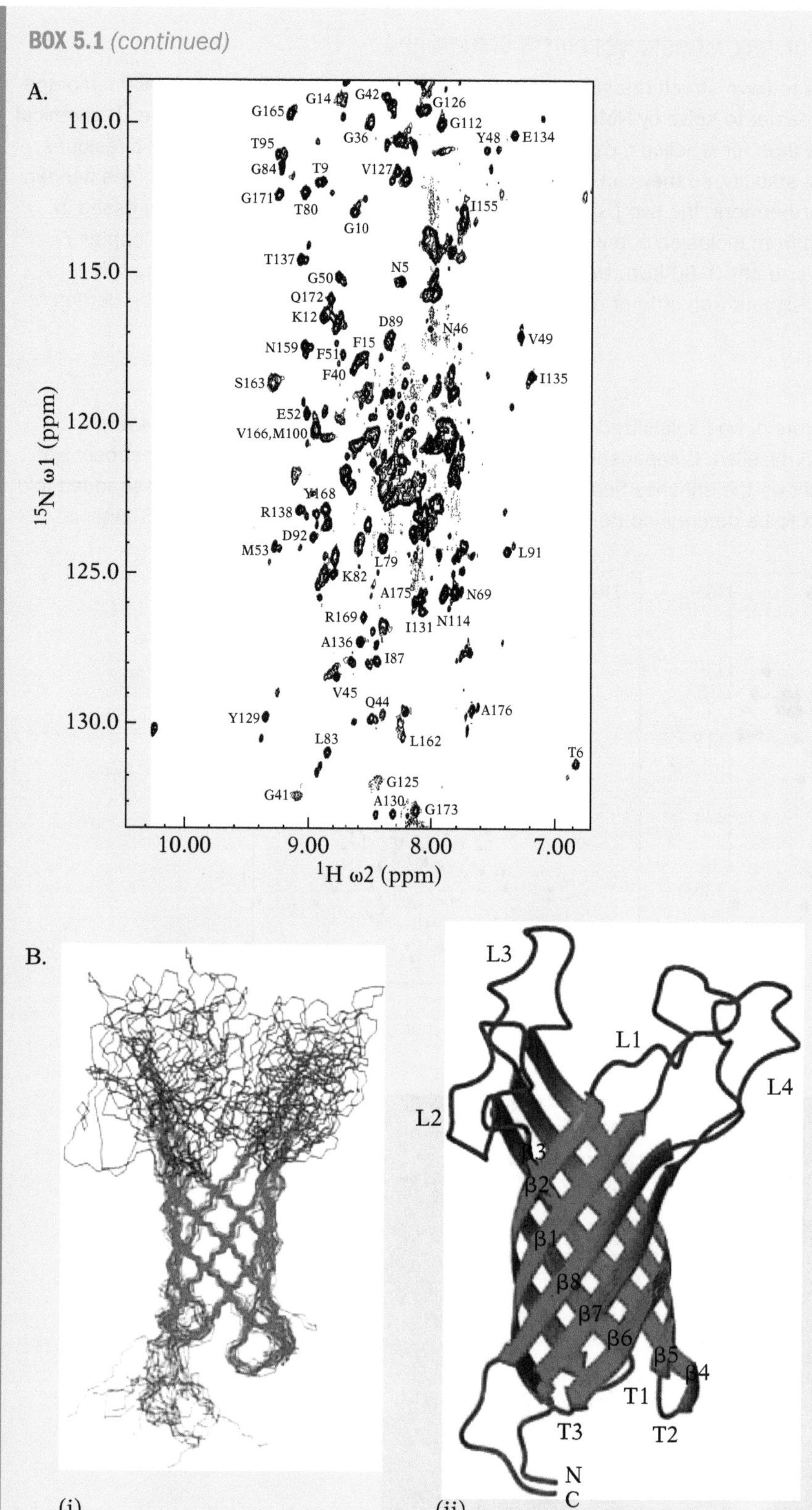

5.1.2 OmpA structure solved by NMR. (A) TROSY-based ^{1}H-^{15}N spectrum of the TM domain of OmpA in DPC micelles. (B) NMR structure of OmpA. The NMR solution structure of OmpA is represented by the superposition of the ten confomers having lowest energy (i). Again, the *β*-barrel (red) is well-defined, while the loops are disordered. Four flexible loops are resolved in the ribbon diagram based on the NMR results (ii). (C) (next page) Comparison of the NMR (red) and x-ray (blue) structures for OmpA show good agreement on the *β*-strands and considerable differences in the loops. From Arora, A., *et al.*, *Nat Struct Biol*. 2001, 8:334–338. © 2001, Nature Publishing Group. Reprinted by permission of Macmillan Publishers Ltd.

BOX 5.1 *(continued)*

C.

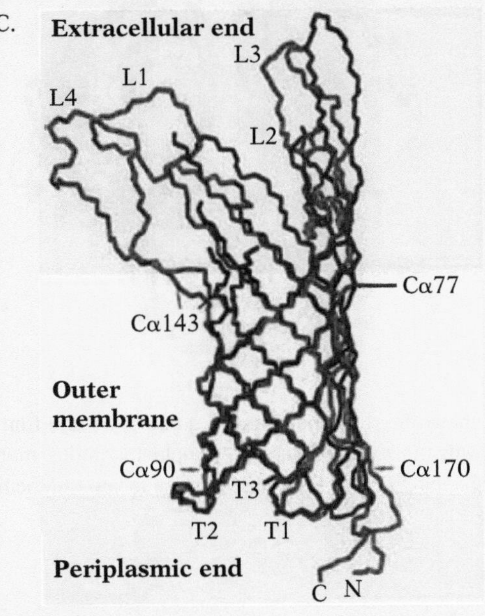

5.1.2 *(continued)*

constraints and 140 dihedral angle constraints. The structure was further refined with selective protonation of the methyl groups of Val, Leu, and Ile on a perdeuterated background, which enabled assignment of 526 NOE distance constraints. When 20 NMR conformers are superimposed, the result clearly reveals the β-barrel with its flexible loops (part ii of Figure 5.1.1B). The ribbon diagram in Figure 5.1.1B(i) maps the NMR results on the x-ray structure and distinguishes the well-structured regions (red) from the disordered loops (yellow). The NMR result is strikingly similar to the structure from x-ray crystallography and in particular confirmed the extension of β-strands on the exterior of the protein.

OmpA TM domain from *E. coli*

The β-barrel domain of OmpA (residues 172–325) was purified after denaturation in urea and refolding into dodecylphosphocholine (DPC) micelles. The protein was labeled with ^{15}N,^{13}C, and ^{2}H, and TROSY experiments were carried out at 600 and 750 MHz. The protein in a large excess of DPC gave the best NMR spectra, shown in Figure 5.1.2A with several of the assigned resonances labeled. TROSY experiments were carried out with several specific amino acid-labeled samples to aid in assignments. The backbone fold of the OmpA TM domain was initially calculated from 91 NOE distance constraints and 142 torsional angle constraints. It was refined by introducing 116 H-bond constraints between adjacent β-strands that were identified in the initial fold calculations. The structure of the eight-stranded β-barrel is well defined. Figure 5.1.2B shows ten superimposed conformers (i) and the corresponding ribbon diagram of the solution structure (ii) of OmpA. Like the OmpX structure, the NMR structure of OmpA shows much agreement with its x-ray structure, illustrated in Figure 5.1.2C, where the solution structure determined by NMR (red) is overlain with the x-ray crystal structure (blue). As NMR gives information about protein dynamics, some of the poorly defined loops are thought to have intrinsic high mobility. A study of the dynamics of the backbone from ^{15}N relaxation times indicates the H-bonded core is not completely rigid but moves on the microsecond–millisecond time scale.

Several other β-barrel membrane protein structures have been solved by NMR, including the eukaryotic voltage-dependent anion channel VDAC, the outer membrane protein OmpG, and two conformations of the outer membrane enzyme PagP.

both homotrimers of around 115 kDa whose levels are controlled by osmolarity of the growth medium and other environmental parameters. This regulation is carried out by a two-component system made up of EnvZ, the protein sensor of the environment, and OmpR, the transcriptional activator for their two genes. OmpC has been called osmoporin, because its expression is induced by high osmolarity as well as high pH and high temperature. On the other hand, higher levels of OmpF are expressed in a medium of low osmolarity. In proteoliposomes OmpF exhibits somewhat faster rates of disaccharide uptake than does OmpC; in planar bilayers OmpC is slightly more cation selective than OmpF. With 60% amino acid identity, it is not surprising that the two structures are very similar.

The first crystal structure of porin from *Rb. capsulatus* (see above) was quickly followed by the high-resolution structures of *E. coli* porins OmpF, PhoE, and LamB, and later OmpC. Now there are many structures for porins from other bacteria, along with other specific porins. These porins are homotrimers, in which each subunit forms a β-barrel (Figure 5.19). At the borders of the barrel two rings of aromatic residues interact with the interfacial regions of the lipid bilayer. The pore through each subunit is delineated by one or more of the loops that fold back into the channel, forming a constriction belt or eyelet. The OmpF barrel consists of 16 antiparallel β-strands tilted by 45°, with a salt bridge between the amino and carboxyl termini to complete the cyclic structure.

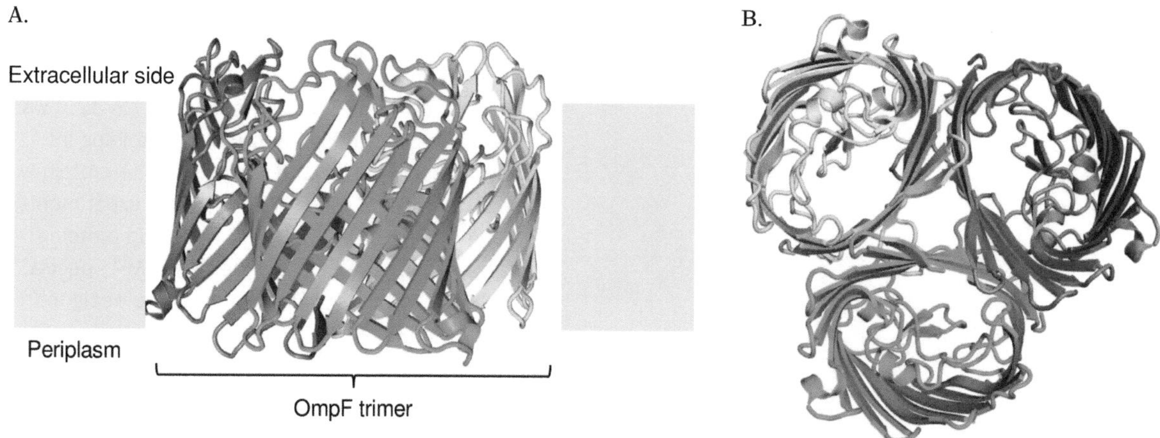

5.19 X-ray structure of OmpF porin. Ribbon diagram of the OmpF trimer is shown from the periplasmic surface (A) and from the plane of the membrane (B). In (A) the lipid bilayer (light blue) is indicated with the extracellular and periplasmic sides marked. From Yamashita, E. *et al.*, *EMBO J.* 2008, 27:2171–2180. © 2008, Nature Publishing Group. Reprinted by permission of Macmillan Publisher Ltd.

OmpF is probably the most thoroughly studied porin. Crystal structures of the wild type and several mutant OmpF proteins show how the dimensions of the channel limit the size of solutes, while the constellation of charged residues in the pore determines other transport properties. Loop 3 in OmpF, which has a highly conserved sequence motif PEFGG, constricts the pore at the middle of the barrel to 15 × 22 Å (Figure 5.20). With two acidic residues on Loop 3 across from a cluster of basic residues on the opposite wall, Loop 3 produces a local transverse electric field that accounts for the cation specificity observed in conductance studies (see Figure 3.17). Loss of five of the charges by mutations of these residues gave a larger pore size (larger radius in the crystal structure and higher uptake rates for disaccharides in the liposome-swelling assay) but lower single channel conductance. Even in the double mutant with both acidic residues mutated (D113N/E117Q), the conductance drops by 50% and the cation selectivity is removed. The role of the acidic residues in drawing cations through the constriction was illustrated by Brownian dynamics simulations.

The pore size and overall dimensions of OmpC are very like OmpF, with the same five charged residues in the eyelet of OmpC, even though its cation selectivity is slightly higher and its single channel conductance is lower. The pore lining at the extracellular side differs considerably, and external loops have different conformations, so perhaps they influence conductance through the pore.

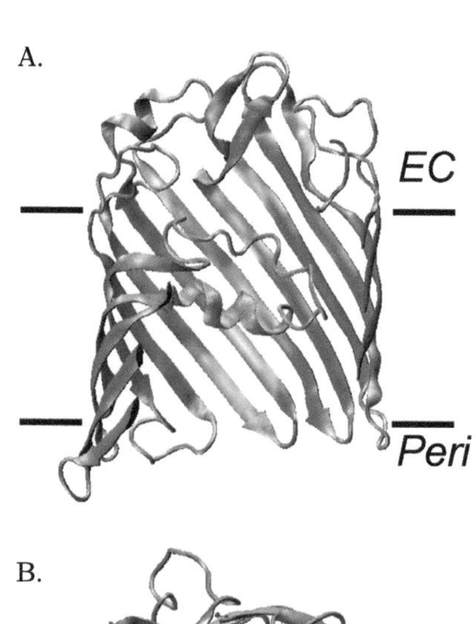

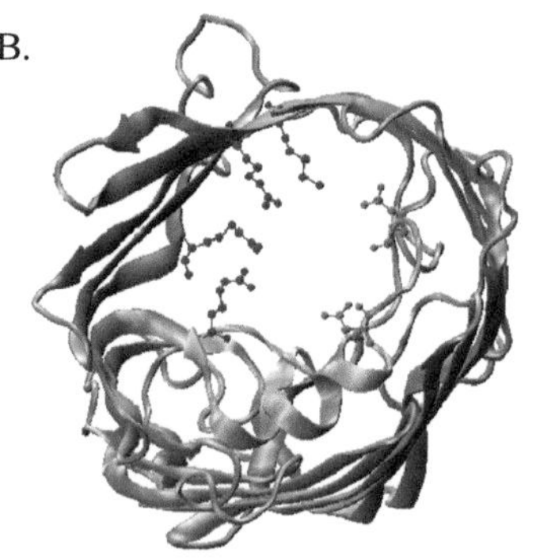

5.20 The constriction zone of OmpF porin. (A) Viewed from the plane of the membrane, the front β-strands have been removed to show the constriction of the OmpF porin monomer by Loop 3 (orange). The black lines delineate the membrane bilayer with the extracellular side (EC) on top and the periplasmic side (Peri) on the bottom. (B) A view from the periplasmic side shows the Loop 3 eyelet with two acidic residues (D113 and E117, red) across from a cluster of basic residue (K16, R42, R82, and R132, blue) in the opposite barrel wall. From Delcour, A. H., *Biochim Biophys Acta*. 2009, 1794:808–816. © 2009, Elsevier. Reprinted with permission from Elsevier.

VDAC, a Mitochondrial Porin

The outer membrane of mitochondria shares many characteristics of the outer membranes of Gram-negative bacteria, so it is not unexpected that it has a general porin called the voltage-dependent anion channel

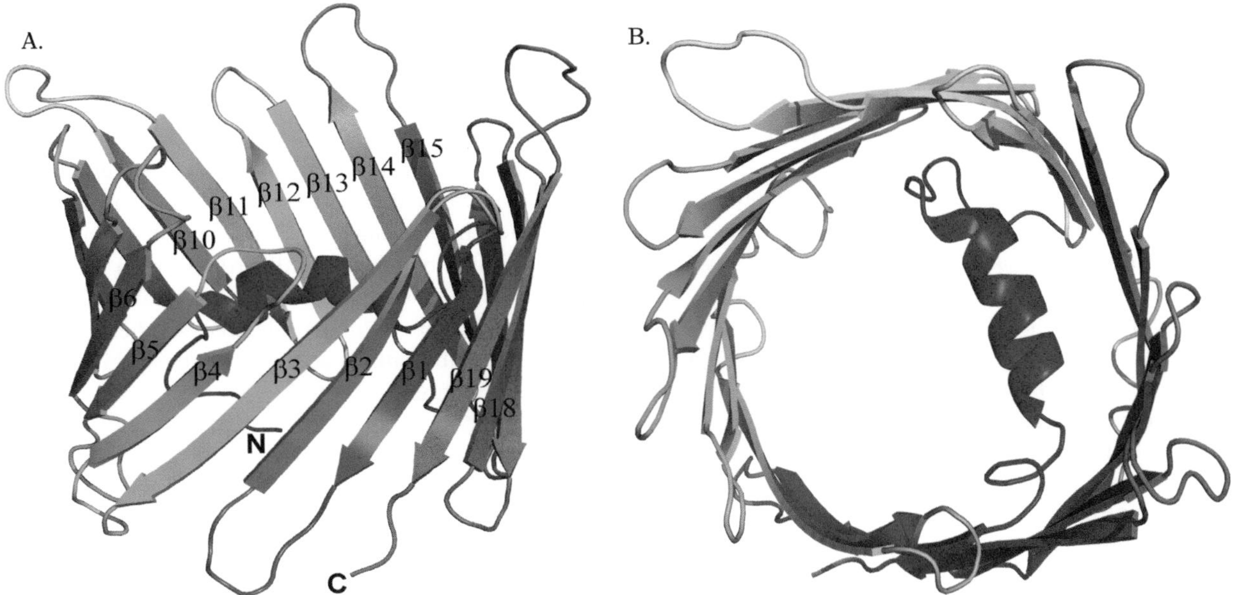

5.21 The structure of VDAC, a β-barrel with 19 β-strands. (A) The structure of VDAC from human mitochondria is shown from the plane of the membrane as a ribbon presentation (rainbow coloring from blue at the N terminus to red at the C terminus) with the β-strands numbered. Note the parallel strands β1 and β19. The residues from Tyr7 to Val17 make an α-helix inside the barrel. (B) The view from the top shows the α-helix is completely enclosed by the β-barrel. From Bayrhuber, M., *et al.*, *Proc Natl Acad Sci USA* 2008, 105:15370–15375. © 2008 National Academy of Sciences, U.S.A.

(VDAC.) The structure of VDAC from human mitochondria, solved by a combination of x-ray crystallography and NMR, is a β-barrel with 19 strands (Figure 5.21). While the first 18 β-stands are antiparallel, the last strand is parallel to the first strand. Otherwise, it shares the general architecture of the *E. coli* porins, with features such as two aromatic girdles at the bilayer perimeters. VDAC has an N-terminal α-helix that folds into the barrel to cross the channel midway within the pore, in the position of Loop 3 in OmpF and OmpC. VDACs from several species exist in different oligomerization states, from monomers to hexamers or higher oligomers. NMR relaxation rates for human VDAC in the detergent LDAO are indicative of a monomer–dimer equilibrium. VDACs share high sequence identity, suggesting this first structure will be representative of the family.

Specific Porins

Sharing many characteristics of general porins, the specific porins add the ability to discriminate among solutes. Since they carry out passive diffusion, they cannot rely on energized conformational changes to release substrates from high-affinity binding sites (see below and Chapter 11). Rather, the specific porins have low affinities for their substrates and have achieved selectivity through features of their channel architecture. Their specificity is apparent only with larger molecules because small solutes easily permeate the nonspecific channel interiors. Clues to their specific functions come from their regulation: for example, in *E. coli*, PhoE protein is induced under phosphate starvation and LamB protein is induced by growth on the carbon source maltose.

PhoE, the Phosphoporin

In spite of the very high homology between PhoE, OmpF, and OmpC, PhoE has a specific transport function. When their functions are compared in whole cells, mutants constructed to have only PhoE take up phosphate and phosphorylated compounds much more efficiently (with a nine-fold decrease in K_m of transport) than mutants with only OmpF or OmpC. Conductance studies with purified PhoE protein demonstrate a strong anion selectivity. Furthermore, polyphosphates such as ATP inhibit the flux of small ions through reconstituted PhoE channels.

Like OmpF and OmpC, the PhoE protein is a 16-stranded β-barrel, with the differences between it and OmpF confined to the loop regions and a single short turn. The constriction zone of the PhoE pore has two additional basic groups, which make the calculated electrostatic potential of the pore more strongly positive than that of OmpF. Mutation of one of these to an acidic residue (K125E) changes the ion selectivity of the PhoE pore. Interestingly, when OmpF was engineered to have the two additional lysine residues, it did not convert to an anion-specific pore, indicating that other residues in PhoE must contribute to its selectivity. Even though it lacks a well-defined specific binding site, PhoE facilitates the transport of phosphorylated compounds into cells, and its blockage by polyphosphates is very similar to the inhibition of the LamB pore by maltodextrins.

LamB, the Maltoporin

LamB protein is named for its role as the receptor for phage λ. It is required for growth of *E. coli* in chemostats with limiting maltose as the sole carbon source, and was shown to be specific for maltose and maltodextrins by comparing rates of sugar uptake in the liposome-swelling assay (see Figure 3.26). The weak affinity for maltodextrins (K_ds in the low mM range) can be quantitated by their inhibition of glucose uptake as well as by binding to immobilized starch. Maltodextrins also block conductance through the LamB channel when reconstituted in black films.

Overall, the high-resolution structure of LamB protein shows similar architecture to the other porins: it is a homotrimer in which each monomer has 18 strands in the β-barrel (see Frontispiece). Three of the loops fold in to constrict the channel to a minimum diameter of 5 Å. Six aromatic residues line one side of the channel, forming the "greasy slide," a hydrophobic path through the otherwise aqueous pore (Figure 5.22A). When soaked into the maltoporin crystals, maltotriose lies lengthwise in the channel with the hydrophobic sides of its pyranose rings along the greasy slide. Sucrose soaked into the crystals gets stuck above the channel constriction, which explains its very low rate of uptake into liposomes. Interestingly, a sucrose-specific porin, called SrcY, is homologous to LamB protein, with a few strategic differences in the residues revealed in its high-resolution structure. Thus the selectivity of these pores for their sugar substrates is achieved by the specificity built into their channels.

Other β-barrel Transporters

Other specific transporters in the outer membrane of *E. coli* have similarly selective pores. One example is the nucleoside transporter, Tsx, named because it is the receptor for T6 phage. The Tsx transporter is a 12-stranded monomer with several distinct nucleotide-binding sites along its narrow channel (Figure 5.22B).

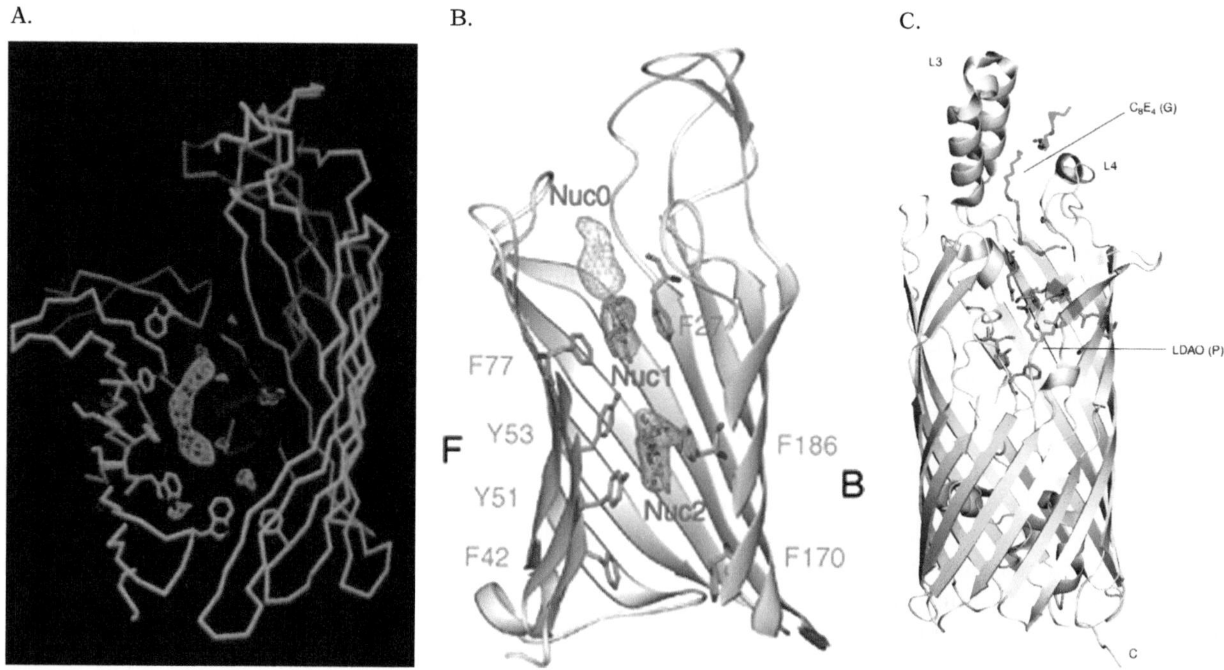

5.22 β-barrels with substrate specificity. (A) When crystals of LamB protein are soaked in maltotriose, electron density corresponding to the substrate (cyan) is detected at the "greasy slide" made up of six aromatic residues: Trp74 from the adjacent monomer, Tyr41, Tyr6, Trp420, Trp358, and Phe227 (stick side chains, yellow). In this cut-away view of the LamB monomer, Loop 1 has been removed, and Loops 3 (red) and 6 (green) can be seen folding into the barrel. From Schirmer, T., *et al.*, *Science* 1995, 267:512–514. © 1995, American Association for the Advancement of Science. Reprinted with permission of AAAs. (B) Crystals of Tsx protein that have been soaked in thymidine exhibit electron density corresponding to three molecules of the nucleoside (pink) within the barrel along a greasy slide (stick models, blue) similar to that in LamB protein, as seen in this cut-away view. From Ye, J. and B. van den Berg, *EMBO J.* 2004, 23:3187–3195. © 2004 Nature Publishing Group. Reprinted by permission of Macmillan Publishers Ltd. (C) The crystal structure of FadL shows a hydrophobic groove and pocket which are occupied by detergent molecules and are presumably on the path for fatty acid transport. Two molecules of C_8E_4 (green stick) are in the groove and an LDAO molecule (green stick) is in the pocket. Positively charged residues (Arg157 and Lys317, blue) are within hydrogen-bonding distance of the LDAO headgroup, while hydrophobic residues (salmon) are close to its alkyl chain. Note that the N-terminal residues 1–42 make a hatch domain (light brown) From van den Berg, B., *Curr Opin Struct Biol.* 2005, 15:401–407. © 2005, Elsevier. Reprinted with permission from Elsevier.

The pore has a greasy slide like that in LamB protein, with substrate sites in it consisting of pairs of aromatic residues flanked by ionizable residues.

Transport of hydrophobic compounds across the outer membrane is accomplished by a family of proteins including FadL, the long chain fatty acid transporter from *E. coli*. These transporters provide long-chain fatty acids for phospholipid biosynthesis and for catabolism as an energy source. The x-ray structure of FadL shows a very long (~50 Å) 14-stranded β-barrel that lacks an interior channel since there is a plug – called a hatch domain – that consists of three short helices near the N terminus (Figure 5.22C). The barrel protrudes far into the extracellular space and has a hydrophobic groove that points toward a prominent hydrophobic pocket. Occupied by detergent molecules in the x-ray structure, these are thought to be low-affinity and high-affinity binding sites for substrates. A second crystal structure shows a conformational change that allows the detergent in the pocket to move toward the periplasm, although it does not show the movement of the hatch required for it to be released. A similar need to remove a plug domain is observed in the iron receptors (see below); however, FadL differs in that it does not bind TonB for energy input.

Iron Receptors

Receptors involved in iron transport are β-barrel proteins with high affinities for their substrates. Iron is abundant but unavailable in the environment due to its insolubility as ferric hydroxides. To solubilize this essential mineral, microorganisms synthesize and secrete siderophores, iron chelators of 500–1500 Da with extremely high affinities for ferric ions. The enteric bacteria, including *E. coli*, synthesize and transport an iron chelator called enterobactin and also have transport systems for siderophores secreted by other microorganisms, such as ferrichrome. These transport systems consist of outer membrane receptors, periplasmic binding proteins, and inner membrane transport proteins. The iron receptors in the outer membrane are also used by colicins and antibiotics to enter the cell.

Unlike the porins, the outer membrane iron receptors bind their substrates with high affinities (K_d ~0.1 μM) and pump them into the periplasm at the expense of energy. The energy is provided via a complex of three proteins, TonB, ExbB, and ExbD, anchored in the inner membrane. In *E. coli* the iron receptors, along with BtuB, the receptor for vitamin B_{12}, interact with TonB at a conserved sequence of five amino acids near the N terminus called the "TonB box" (see Chapter 11). Although these interactions have been thoroughly studied, the mechanism of energy delivery is unknown.

The first high-resolution structures of iron receptors, the *E. coli* receptors for enterobactin (FepA) and ferrichrome (FhuA), show a common architecture consisting of two domains, a C-terminal domain that makes a 22-stranded β-barrel, and a globular N-terminal domain of ~150 residues that fills the interior, making a plug (Figure 5.23). The barrel has a diameter of ~40 Å and

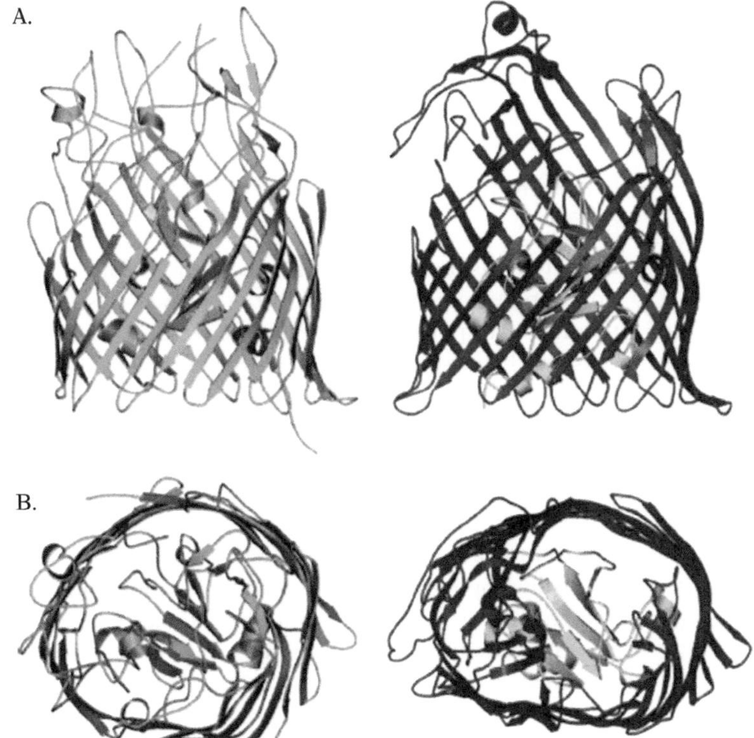

5.23 Structures of FepA and FhuA proteins involved in iron transport. **(A)** Side views with the extracellular surface on top and the periplasmic end on the bottom. **(B)** Views from the external solvent show the blockage of the interior. The FhuA β-barrel is blue and plug domain is yellow; the FepA β-barrel is green and plug domain is orange. From Ferguson, A. D., and J. Deisenhofer, *Biochim Biophys Acta*. 2002, 1565:318–381. © 2002 by Elsevier. Reprinted with permission from Elsevier.

extends beyond the bilayer on the outside. Some of the 11 loops on the external membrane surface are unusually long, up to 37 residues in FepA. The plug domain, a four-stranded β-sheet and interspersed α-helices and loops, has two loops that extend 20 Å beyond the outer membrane interface to frame a pocket with the binding site for the siderophores. The largest difference between the two receptors is the nature of this site, which is tailored for siderophores that differ in composition and charge. Comparison of the x-ray structures of FhuA in the presence and absence of siderophores reveal small conformational changes in the N-terminal domain upon ligand binding, insufficient to open a transport channel.

There are now structures of a dozen TonB-dependent transporters, including other iron and colicin receptors, with architecture similar to FepA and FhuA and customized ligand-binding sites. In addition, structures have been solved for iron receptors in complex with the periplasmic domain of TonB (see Chapter 11). The understanding of iron transport in microorganisms, especially iron receptors in pathogenic bacteria, is important from a medical standpoint because acquisition of iron is critical for invading microbes, which extract iron from host proteins such as transferrin and hemoglobin.

Some pathogenic bacteria gain a competitive advantage by scavenging heme from their environment. *Serratia marcescens* excretes a heme-binding protein HasA to capture heme and then binds the HasA hemophore with an outer membrane receptor, HasR. HasA forms a tight complex with HasR, allowing heme to transfer to a binding site in HasR. A TonB-type of energy transducer is required to release heme into the periplasm. The high-resolution structure of a HasA–HasR complex shows the HasR structure is similar to other TonB-dependent transporters with a C-terminal β-barrel and an N-terminal plug domain (Figure 5.24). HasR has long extracellular loops that bind HasA.

From the porins to the heme receptor, outer membrane transporters have much in common. The β-barrel structure is also used by outer membrane enzymes, including two that are discussed in Chapter 9, OMPLA and OmpT. While the first β-barrels characterized, the porins, consist of β-strands of fairly uniform lengths, others are distorted by varied strand lengths that make the barrel very asymmetric, as seen in OmpX and FadL. In addition, many have small regions of α-helix, for example the short appendage to the barrel in Tsx, the constriction across the barrel in VDAC, and the mixed α/β plug domains in the iron receptors FepA, FhuA, and HasR. Even the amazing TolC that spans the periplasm (described in Chapter 11) forms a classic β-barrel to cross the outer membrane. Until recently it could be said that all outer membrane proteins shared this basic architecture.

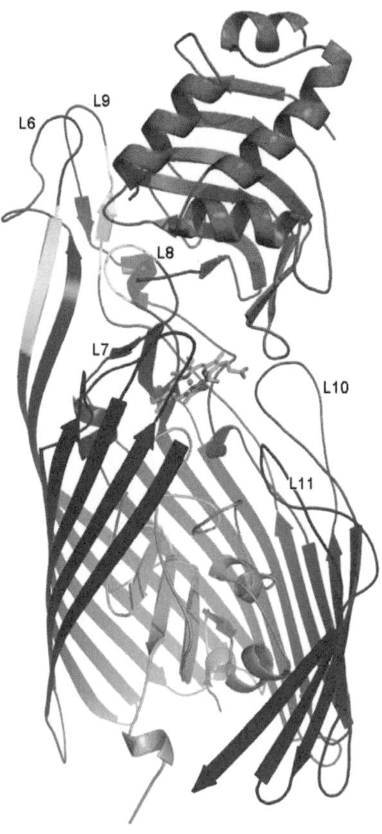

5.24 X-ray structure of the ternary complex HasA–HasR–Heme. Ribbon representations of the hemophore HasA (red) and its receptor HasR (blue, with the plug domain in orange) show the structure after the heme (green wire model) has passed into HasR. The first five strands and loops L1–L3 of HasR are omitted to allow a view into the barrel interior. The portions of the extracellular loops (Loops 6–11, labeled) and long β-strands that are essential for HasA binding are highlighted (yellow). From Krieg, S., *et al.*, *Proc Natl Acad Sci USA* 2009, 106:1045–1050.

Outer Membrane Secretory Proteins: Not β-barrels

Export across the outer membrane is required for secretion of polymers synthesized in the periplasm, such as the capsular polysaccharide that envelopes bacterial cells and helps them to evade the host immune system. The high-resolution structure of Wza, which exports capsular polysaccharide, is the first crystal structure of an outer membrane secretory protein. It shows a layered, elongated structure that ends in a unique α-helical barrel (Figure 5.25). Wza is an octomer, in which each subunit contributes an amphipathic α-helix to form a barrel with a hydrophobic exterior and a hydrophilic interior. The channel in Wza has an internal diameter of 17 Å, allowing extended polysaccharides to pass. The lower domains of Wza have hydrophilic exteriors and protrude into the periplasm to interact with the complex

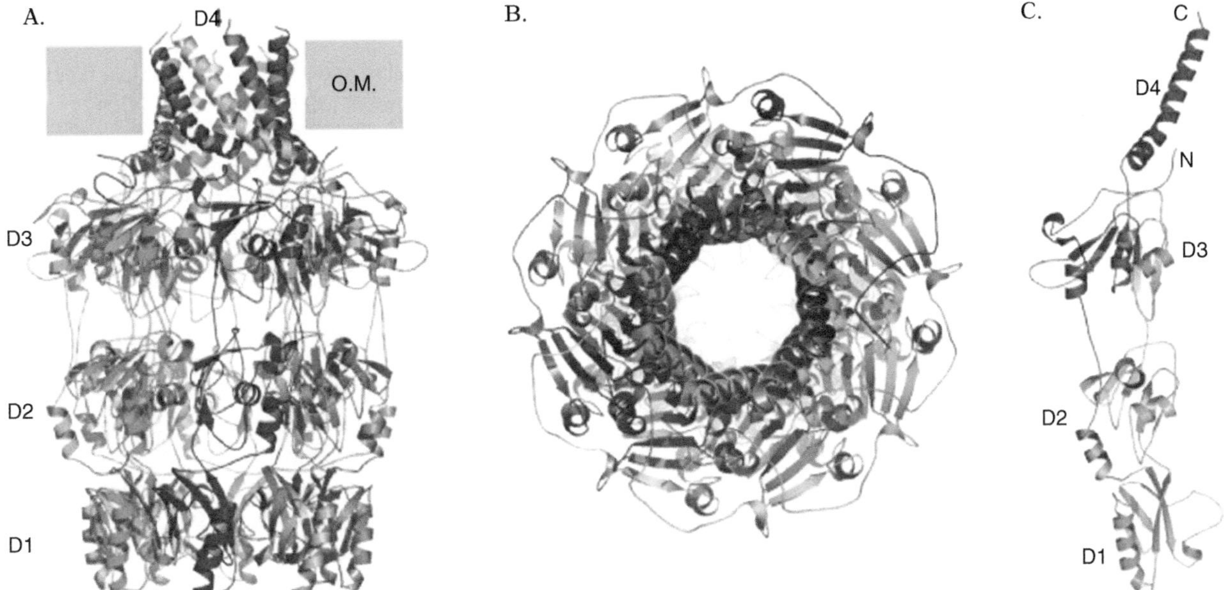

5.25 X-ray structure of the secretory protein Wza. The structure of the Wza octomer (ribbon representation with each subunit a different color) is viewed from the plane of the membrane (A) and the extracellular side looking down through the helical barrel (B). The four segments of Wza are numbered D1–D4. (C) A side view of the monomer (rainbow colored with the N terminus blue and the C terminus red) shows the extended C-terminal region that makes the helical barrel. From Collins, R. F., and J. P. Derrick, *Trends Microbiol.* 2007, 15:96–100. © 2007, Elsevier. Reprinted with permission from Elsevier.

of proteins that assemble the polymer on the outside of the inner membrane. Since EM images of other outer membrane secretory proteins, including those for the fibers that make the bacterial pili, appear very similar to Wza, it is likely that this unusual structure will be seen again. With its α-helical barrel, Wza clearly does not fit into the standard two classes of integral membrane proteins.

CONCLUSION

The two large classes of integral membrane proteins, bundles of α-helices and β-barrels, show very different characteristics. Paradigms for each are drawn from early examples: bacteriorhodopsin and photosynthetic reaction center, and the porins. Many more examples of helical membrane proteins are described in later chapters. Although fewer in number, β-barrel proteins show surprising diversity. Tremendous progress has produced many elegant structures of β-barrel membrane proteins, along with enhanced understanding of their transport mechanisms. Even so, it is likely the known examples represent a small fraction of the proteins in this class of membrane proteins. If the genomic analyses are correct, there are many more β-barrel proteins to be characterized. The next chapter looks at bioinformatics techniques used to predict and analyze membrane proteins and shows how they are grouped in families.

FOR FURTHER READING

Bacteriorhodopsin Reviews

Andersson, M., E. Malmerberg, S. Westenhoff, *et al.*, Structural dynamics of light-driven proton pumps. *Structure.* 2009, 17:1265–1275.

Haupts, U., J. Tittor, and D. Oesterhelt, Closing in on bacteriorhodopsin. *Ann Rev Biophys Biomol Struct.* 1999, 28:367–399.

Hirai, T., S. Subramaniam, and J. K. Lanyi, Structural snapshots of conformational changes in a seven-helix membrane protein: lessons from bacteriorhodopsin. *Curr Opin Struct Biol.* 2009, 19:433–439.

Lanyi, J. K., and H. Luecke, Bacteriorhodopsin. *Curr Opin Struct Biol.* 2001, 11:415–419.

Neutze, R., E. Pebay-Peyroula, K. Edman, A. Royant, J. Navarro, and E. M. Landau, Bacteriorhodopsin: a high resolution structural view of vectorial proton transport. *Biochim Biophys Acta.* 2002, 1565: 144–167.

Seminal Papers

Henderson, R., and P. N. T. Unwin, Three-dimensional model of purple membrane obtained by electron microscopy. *Nature.* 1975, 257:28–32.

Oesterhelt, D., and W. Stoeckenius, Functions of a new photoreceptor membrane. *Proc Natl Acad Sci USA*. 1973, 70:2853–2857.

Pebay-Peyroula, E., G. Rummel, J. P. Rosenbusch, and E. M. Landau, X-ray structure of bacteriorhodopsin at 2.5 angstroms from microcrystals grown in lipidic cubic phases. *Science*. 1997, 277:1676–1681.

Winget, G. D., N. Kanner, and E. Racker, Formation of ATP by the adenosine triphosphatase complex from spinach chloroplasts reconstituted together with bacteriorhodopsin. *Biochim Biophys Acta*. 1977, 460:490–499.

Photosynthetic Reaction Centers Reviews

Deisenhofer, J., and H. Michel, Structures of bacterial photosynthetic reaction centers. *Annu Rev Cell Biol*. 1991, 7:1–23.

Nogi, T., and K. Miki, Structural basis of bacterial photosynthetic reaction centers. *J Biochem (Tokyo)*. 2001, 130:319–329.

Vermeglio, A., and P. Joliot, The photosynthetic apparatus of *Rhodobacter sphaeroides*. *Trends Microbiol*. 1999, 7:435–440.

Seminal Papers

Deisenhofer, J., O. Epp, K. Miki, R. Huber, and H. Michel, Structure of the protein subunits in the photosynthetic reaction centre of *Rhodopseudomonas viridis* at 3Å resolution. *Nature*. 1985, 318: 618–624.

Deisenhofer, J., O. Epp, I. Sinning, and H. Michel, Crystallographic refinement at 2.3 Å resolution and refined model of the photosynthetic reaction centre from *Rhodopseudomonas viridis*. *J Mol Biol*. 1995, 246:429–457.

β-Barrels

Bishop, R. E., Structural biology of membrane-intrinsic β-barrel enzymes: sentinels of the bacterial outer membrane. *Biochim Biophys Acta*. 2008, 1778:1881–1896.

Buchanan, S. K., β-barrel proteins from bacterial outer membranes: structure, function and refolding. *Curr Opin Struct Biol*. 1999, 9:455–461.

Fairman, J. W., N. Noinaj, and S. K. Buchanan, The structural biology of β-barrel membrane proteins: a summary of recent reports. *Curr Opin Struct Biol*. 2011, 21:523–531.

Schulz, G. E., β-barrel membrane proteins. *Curr Opin Struct Biol*. 2000, 10:443–447.

Wimley, W. C., The versatile β-barrel membrane protein. *Curr Opin Struct Biol*. 2003, 13:404–411.

Porins

Achouak, W., T. Heulin, and J.-M. Pages, Multiple facets of bacterial porins. *FEMS Microbiol Lett*. 2001, 199:1–7.

Delcour, A. H., Solute uptake through general porins. *Frontiers Biosci*. 2003, 8:1055–1071.

Delcour, A. H., Outer membrane permeability and antibiotic resistance. *Biochim Biophys Acta*. 2009, 1794:808–816.

Dutzler, R., Y.-F. Wang, P. J. Rizkallah, J. P. Rosenbusch, and T. Schirmer, Crystal structures of various maltooligosaccharides bound to maltoporin reveal a specific sugar translocation pathway. *Structure*. 1996, 4:127–134.

Nikaido, H., Porins and specific channels of bacterial outer membranes. *Mol Microbiol*. 1992, 6:435–442.

Schirmer, T., General and specific porins from bacterial outer membranes. *J Struct Biol*. 1998, 121:101–109.

Iron Receptors

Chimento, D. P., R. J. Kadner, and M. C. Wiener, Comparative structural analysis of TonB-dependent outer membrane transporters: implications for the transport cycle. *Proteins*. 2005, 59:240–251.

Clarke, T. E., L. W. Tari, and H. J. Vogel, Structural biology of bacterial iron uptake systems. *Curr Top Med Chem*. 2001, 1:7–30.

Krewulak, K. D., and H. J. Vogel, TonB or not TonB: is that the question?. *Biochem Cell Biol*. 2011, 89:87–97.

Létoffé, S., G. Heuck, P. Delepelaire, N. Lange, and C. Wandersman, Bacteria capture iron from heme by keeping tetrapyrrol skeleton intact. *Proc Natl Acad Sci USA*. 2009, 106:11719–11724.

Noinaj, N., M. Guillier, T. J. Barnard, and S. K. Buchanan, TonB-dependent transporters: regulation, structure and function. *Ann Rev Microbiol*. 2010, 64:43–60.

Outer Membrane Secretory Proteins

Collins, R. F., and J. P. Derrick, Wza: a new structural paradigm for outer membrane proteins?. *Trends Microbiol*. 2007, 15:96–100.

Cuthbertson, L., I. L. Mainprize, J. H. Naismith, and C. Whitfield, Pivotal roles of the outer membrane polysaccharide export and polysaccharide copolymerase protein families in export of lipopolysaccharides in Gram-negative bacteria. *Microbiol Mol Biol Rev*. 2009, 73:155–177.

First Crystal Structures of Proteins Discussed in Chapter 5

Baslé, A., G. Rummel, P. Storici, J. P. Rosenbusch, and T. Schirmer, Crystal structure of osmoporin OmpC from *E. coli* at 2.0 Å. *J Mol Biol*. 2006, 362:933–942.

Bayrhuber, M., T. Meins, M. Habeck, *et al.*, Structure of the human voltage-dependent anion channel. *Proc Natl Acad Sci USA*. 2008, 105:15370–15375.

Buchanan, S. K., B. S. Smith, L. Venkatramani, *et al.*, Crystal structure of the outer membrane active transporter FepA from *Escherichia coli*. *Nat Struct Biol*. 1999, 6:56–63.

Cowan S. W., T. Schirmer, G. Rummel, *et al.*, Crystal structures explain functional properties of two *E. coli* porins. *Nature*. 1992, 358:727–733.

Deisenhofer, J., O. Epp, K. Miki, R. Huber, and H. Michel, Structure of the protein subunits in the photosynthetic reaction centre of *Rhodopseudomonas viridis* at 3Å resolution. *Nature*. 1985, 318:618–624.

Dong, C., K. Beis, J. Nesper, *et al.*, Wza the translocon for *E. coli* capsular polysaccharides defines a new class of membrane protein. *Nature*. 2006, 444:226–229.

Ferguson, A. D., E. Hofmann, J. W. Coulton, K. Diederichs, and W. Welte, Siderophore-mediated iron transport: crystal structure of FhuA with bound lipopolysaccharide. *Science*. 1998, 282:2215–2220.

Forst, D., W. Welte, T. Wacker, and K. Diederichs, Structure of the sucrose-specific porin ScrY from *Salmonella typhimurium* and its complex with sucrose. *Nat Struct Biol*. 1998, 5:37–45.

Kreusch, A., A. Neubüser, E. Schiltz, J. Weckesser, and G. E. Schulz, Structure of the membrane channel porin from *Rhodopseudomonas blastica* at 2.0 Å resolution. *Protein Sci*. 1994, 3:58–63.

Krieg, S., F. Huche, K. Diederichs, *et al.*, Heme uptake across the outer membrane as revealed by crystal structures of the receptor-hemophore complex. *Proc Natl Acad Sci USA*. 2009, 106:1045–1050.

Pebay-Peyroula, E., G. Rummel, J. P. Rosenbusch, and E. M. Landau, X-ray structure of bacteriorhodopsin at 2.5 angstroms from microcrystals grown in lipidic cubic phases. *Science*. 1997, 277:1676–1681.

Schirmer, T., T. A. Keller, Y. F. Wang, and J. P. Rosenbusch, Structural basis for sugar translocation through maltoporin channels at 3.1Å resolution. *Science*. 1995, 267:512–514.

van den Berg, B., P. N. Black, W. M. Clemons, and T. A. Rapoport, Crystal structure of the long chain fatty acid transporter FadL. *Science*. 2004, 304:1506–1509.

Weiss, M. S., A. Kreusch, E. Schiltz, *et al.*, The structure of porin from *Rhodobacter capsulatus* at 1.8 Å resolution. *FEBS Lett*. 1991, 280:379–382.

Ye, J., and B. van den Berg, Crystal structure of the bacterial nucleoside transporter Tsx. *EMBO J*. 2004, 23:3187–3195.

FUNCTIONS AND FAMILIES

6

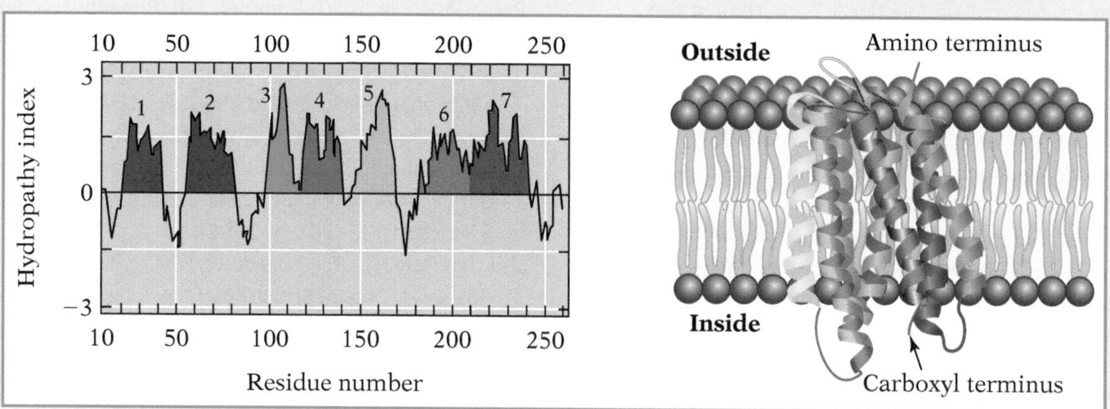

A hydropathy plot predicts bilayer-spanning regions of membrane proteins like bacteriorhodopsin, shown here with its TM helices colored to match the corresponding hydrophobic peaks on the plot. From Nelson, D. L., and M. M. Cox, *Lehninger Principles of Biochemistry*, 4th ed., W. H. Freeman, 2005, pp. 375–376. © 2005 by W. H. Freeman and Company. Used with permission.

The functions of most membrane proteins are traditionally described as enzymes, transporters and channels, and receptors, although the demarcations of these groups are blurred. For example, ATPases that are ion channels and other active transport proteins ("permeases") are studied as enzymes as well as transporters. Some receptors involved in signaling are also tyrosine kinases; other receptors are gated ion channels. Membrane proteins can also be classified using data from bioinformatics, which describe families of proteins, functions of homologous domains, and evolutionary relationships. This chapter utilizes both approaches to look at the roles of membrane proteins before turning to tools for prediction of membrane protein structure and genomic analysis of membrane proteins. It starts by describing the general characteristics of membrane enzymes, transporters, and receptors, giving specific examples and briefly identifying their protein families.

MEMBRANE ENZYMES

The enzymes found in membranes carry out diverse catalytic functions. In addition to solute transport and signaling, membrane-bound enzymes participate in electron transport chains and other redox reactions, as well as the metabolism of membrane components such as phospholipids and sterols. Many membrane enzymes require specific lipids or particular types of lipids for activity (see Chapter 4). In addition, soluble enzymes may bind to the membrane periphery for catalysis involving substrates in the membrane, for efficient access to substrates that are passed along a series of bound enzymes to avoid dilution in the cytoplasm, or for regulation by modulation of their activities, as described in Chapter 4. In all these cases, the heterogeneity and dimensionality of the membrane affect the activity of the enzymes.

Whether membrane enzymes require lipids for their activity or catalyze reactions involving membrane-bound substrates, they are restricted to the available lipid or substrate near them in the membrane. This means the rate of enzyme activity (velocity) depends on the concentration of available lipid or substrate in the bilayer in the vicinity of the enzyme, not the concentration in the bulk solution. For kinetic treatment of these cases, the amount of substrate available to the enzyme is not determined by its bulk concentration but by its concentration in the lipid bilayer, so it is described with surface concentration terms, either mole fraction ([substrate]/{[substrate] + [total lipid]}) or mole percent (mole fraction × 100). Then the classical Michaelis–Menten equation is applicable, with surface concentration units

BOX 6.1 SURFACE DILUTION EFFECTS

A description of the actions of lipid-dependent enzymes must consider both three-dimensional bulk interactions that occur in solution and two-dimensional surface interactions in the bilayer. Thus a kinetic model for surface dilution effects takes into account the concentrations of the required lipid in both phases, whether the enzyme is reconstituted into micelles or liposomes. The detergent–lipid mixed micelle is particularly amenable for study of surface dilution kinetics because it allows both the bulk concentration and the surface concentration of a lipid substrate to be varied. Then the surface concentration of the lipid is expressed in terms of its mole fraction, [lipid] / ([lipid] + [detergent]). With a fixed number of lipids at several different micelle concentrations, the bulk concentration of lipids does not vary, but the lipid:micelle ratio changes. This is illustrated by the lipid-dependent binding of a phorbol ester by PKC in mixed detergent–lipid micelles (see Figure 6.1.1). PKC requires PS, as described in Chapter 4. When the level of Triton X-100 is increased as the amount of PS is held constant, the activity drops: the required PS is not as available to activate the enzyme. If the ratio of PS to Triton X-100 is held constant as the concentration of Triton X-100 increases, the activity stays the same because the mole percent of PS is maintained. Thus PKC shows a loss of activity at high detergent concentration (without addition of PS) because the surface concentration of lipid has been diluted.

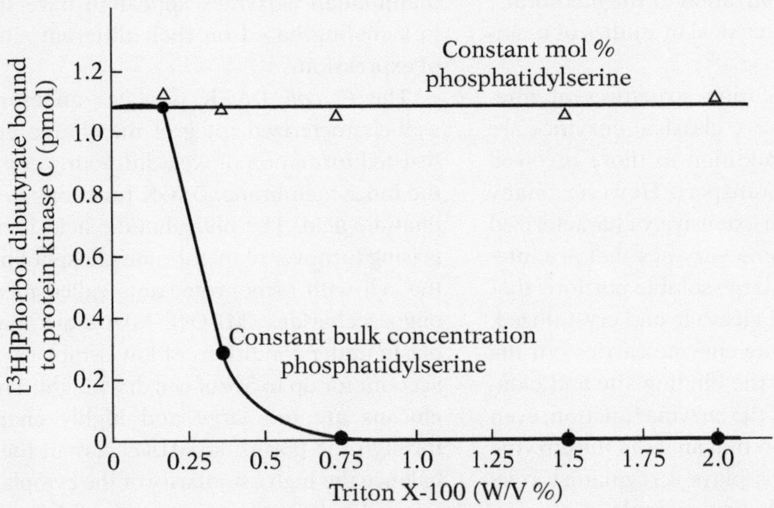

6.1.1 Surface dilution effect on the lipid-dependent enzyme, PKC. Redrawn from Gennis, R. B., *Biomembranes: Molecular Structure and Function*, Springer-Verlag, 1989, p. 228, © 1989 by the American Society for Biochemistry & Molecular Biology.

The surface dilution kinetic model was developed for cobra venom PLA_2 (also described Chapter 4) in Triton X-100/PC mixed micelles. In the case of such peripheral proteins, the analysis includes the binding step that takes place in the three-dimensional bulk solution and results in restricting the interactions to the two-dimensional surface of the membrane. Thus the first step is the binding of the soluble enzyme to the micelles, followed by binding the substrate in the bilayer:

Bulk Step Surface Step

$$E + A \underset{k_{-1}}{\overset{k_1}{\rightleftharpoons}} EA \qquad EA + B \underset{k_{-2}}{\overset{k_2}{\rightleftharpoons}} EAB \underset{k_{-3}}{\overset{k_3}{\rightleftharpoons}} EA + Q,$$

where E is the enzyme, A is the mixed micelle, EA is the enzyme–mixed micelle complex, B is the substrate, EAB is the catalytic complex, and Q is the product. (Note that the equation for the first step applies whether the enzyme binds nonspecifically, in which case A is the sum of the molar concentrations of the lipid and the detergent, or specifically to a phospholipid species, in which case A is the molar concentration of that lipid.) Once bound, the association between EA and B is a function of their surface concentrations, expressed in units of mole fraction or mole percent. For a water-insoluble integral membrane enzyme, the protein is delivered to the assay as a detergent–protein mixed micelle, which is likely to fuse with lipid micelles. In this case, E represents the concentration of the enzyme–detergent complex.

(continued)

BOX 6.1 *(continued)*

The kinetic equation becomes

$$v = \frac{V_{max}[A][B]}{K_s{}^A K_m{}^B + K_m{}^B[A] + [A][B]},$$

where the dissociation constant, $K_s{}^A = k_{-1}/k_1$, and the interfacial Michaelis constant, $K_m{}^B = (k_{-2}+k_3)/k_2$, are expressed in surface concentration units.

used for the substrate concentration (see Box 6.1). When an enzyme that loses activity upon dilution of the lipid is solubilized and reconstituted into micelles or liposomes, the amount of lipid remaining will affect its activity. For this reason, the extent of separation of lipid and protein components during solubilization of the membrane components (see Chapter 3) is critical in studies of membrane enzymes.

A relatively few high-resolution structures of integral membrane proteins that are classical enzymes are available (see Chapter 9) in addition to those involved in energy transduction and transport. However, many membrane enzymes have been extensively characterized biochemically. In addition, some enzymes that are integral membrane proteins have large soluble portions that can be removed by proteolytic cleavage and crystallized. When the soluble portion of the enzyme carries out the catalysis, its structure reveals the binding site and catalytic groups to give a picture of the enzyme function, even though it is missing the portion that anchors the enzyme in the membrane and perhaps plays a regulatory role. Diacylglycerol kinase (DAGK) is an example of a membrane enzyme that was thoroughly investigated long before its high-resolution structure was known. Some of the P450 cytochromes provide examples of membrane enzymes whose soluble portions have been crystallized and their structure solved. Both of these examples are enzymes that occur in mammals in numerous isoforms, different forms of the enzymes that are encoded by different genes. Isoforms, also called isozymes, are catalytically and structurally similar and are typically located in different tissues of the organism, where they respond to different regulators.

Diacylglycerol Kinase

Diacylglycerol kinase carries out the reaction

$$\text{Diacylglycerol} + \text{MgATP} \rightarrow \text{Phosphatidic acid} + \text{MgADP}$$

with Michaelis–Menten kinetics and rates limited by substrate diffusion. Both the substrate and product of the DAGK reaction are lipid-soluble allosteric effectors and second messengers in signal transduction in mammals, which have ten isoforms of DAGK. Localized to the cytosol or the nucleus, the mammalian DAGKs are peripheral proteins that dock on the membrane to access their substrate. Two of the isozymes are activated by both PE and cholesterol and inhibited by sphingomyelin when reconstituted in large unilamellar vesicles. The mammalian isozymes appear to have specialized roles in signaling based on their different sites and patterns of expression.

The *E. coli* DAGK provides an example of a very well-characterized integral membrane enzyme that has resisted formation of well-diffracting crystals. Located in the inner membrane, DAGK functions to replenish phosphatidic acid. The phosphatidic acid is needed in a surprising turnover of membrane phospholipid that provides the cell with osmoprotectants called membrane-derived oligosaccharides (MDOs). MDOs are made in the periplasm under conditions of low osmolarity, when they can account for up to 5% of cell dry weight. These hydrophilic glucans are too large and highly charged to diffuse through the porins, so MDOs stay in the periplasm and balance the high osmolarity of the cytoplasm. MDOs contain 6–12 glucose units joined by β-1,2 and β-1,6 linkages that are variously substituted with *sn*-1-phosphoglycerol, phosphoethanolamine, and O-succinyl ester residues. The phosphoglycerol and phosphoethanolamine are enzymatically added from PG and PE, respectively, leaving diacylglycerol (DAG). A second source of DAG is the utilization of phosphoethanolamine from PE in the biosynthesis of lipopolysaccharide. Too much DAG could affect the membrane by driving formation of nonbilayer phases. It is the job of DAGK to phosphorylate the DAG to return it to the phospholipid pool in the membrane. The activity of DAGK in *E. coli* is determined by the rate of TM flip-flop supplying diacylglycerol from the outer to the inner leaflet of the plasma membrane.

The smallest known kinase, *E. coli* DAGK, is a homotrimer of 13-kDa subunits (121 amino acids) with three active sites. Purified DAGK has good activity when reconstituted in lipid–detergent mixed micelles and phospholipid vesicles, as well as in bicelles, amphipols, and cubic phase monolein. Formation of a DAGK–DAG–MgATP ternary complex does not depend on order of substrate addition, supporting a random equilibrium kinetic mechanism with direct transfer of the phosphoryl group

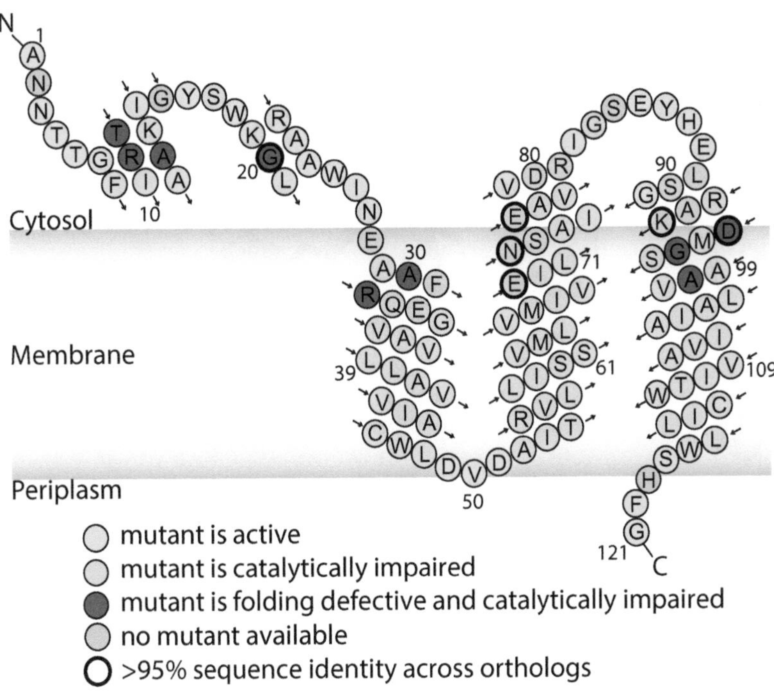

6.1 Topology of DAGK, showing sites critical for catalysis and/or folding. Mutants were generated by replacing each residue in a cysteineless mutant form of DAGK with cysteine. Residues are represented by spheres that are colored to show their roles: Blue have at least 20% wild type activity, yellow fold but have less than 20% wild type activity, magenta do not fold (hence inactive). Highly conserved residues have a bold outline. From Van Horn, W. D., *et al.*, *Science*. 2009, 324:1726–1729. © 2009, American Association for the Advancement of Science. Reprinted with permission from AAAS.

from ATP to DAG. The K_M for DAG corresponds to the concentration of DAG in the membrane of cells grown in low osmolarity when the MDO cycle is active.

In native membranes DAGK is unusually stable, active even after a few minutes at 100°C, and mutants have been engineered to increase its stability even more. The effects of mutations that alter every residue in the sequence show that most critical residues of DAGK for both folding and catalysis are located in or next to the active site, along with several critical residues clustered near the N terminus. Patterns of disulfide bond formation between single-cysteine mutants allowed prediction of intramolecular distances. These results, along with spectroscopic studies and topology predictions, indicate the DAGK monomer contains three TM helices plus two short amphipathic helices at its N terminus (Figure 6.1).

Despite years of effort, no highly diffracting crystals of DAGK suitable for x-ray crystallography have been obtained. In a major accomplishment, the backbone structure of DAGK in dodecyl PC micelles was solved by solution NMR, employing structural constraints from paramagnetic relaxation enhancement-derived long-range distances and residual dipolar coupling-based orientational restraints, in addition to distances derived from biochemical disulfide mapping. DAGK was one of the first polytopic α-helical membrane proteins to be solved by NMR (see Box 4.2). The nine TM helices (three from each subunit) are arranged around a central core of the three TM2 helices in a left-handed parallel bundle (Figure 6.2). TM3 is domain-swapped, meaning it interacts with TM1 and TM2 from the adjacent subunit. The entire molecule can be viewed as three overlapping four-helix bundles, each containing helices from all three subunits (Figure 6.2C, D). The interface between each four-helix bundle contains a cavity bound by a portico: TM1 and TM3 are pillars and the loop between TM2 and TM3 forms the cap. The spatial limits of the portico determine the substrate specificity, restricting the head group of the lipid binding in the cavity to glycerol without restricting the composition of the acyl chains. (A similar lipid binding site in a portico is seen in OMPLA, see Chapter 9). Titrations of DAGK with its substrates DAG and MgATP, product (phosphatidic acid), and a nonhydrolyzable ATP monitored by NMR show no global conformational change upon binding. Given the structure lacks side chain positions, it is not possible to see a well-defined nucleotide-binding site, and the N terminus is too dynamic to appear.

The *E. coli* DAGK is clearly unlike the water-soluble DAGKs, and it lacks the structural and functional motifs found in kinases in general. The only other member of the family is undecaprenol kinase (UDPK), which plays an analogous role in Gram-positive bacteria. UDPK phosphorylates undecaprenol, another lipid carrier of hydrophilic molecules needed for biosynthesis – in this case, of peptidoglycan in the cell wall. The *dgkA* gene that encodes both DAGK and UDPK is widely distributed in prokaryotes but rarely found in archea and eukaryotes, and the two enzymes have no significant homology with any other proteins. The two enzymes present a unique solution to the need for catalysis within membranes.

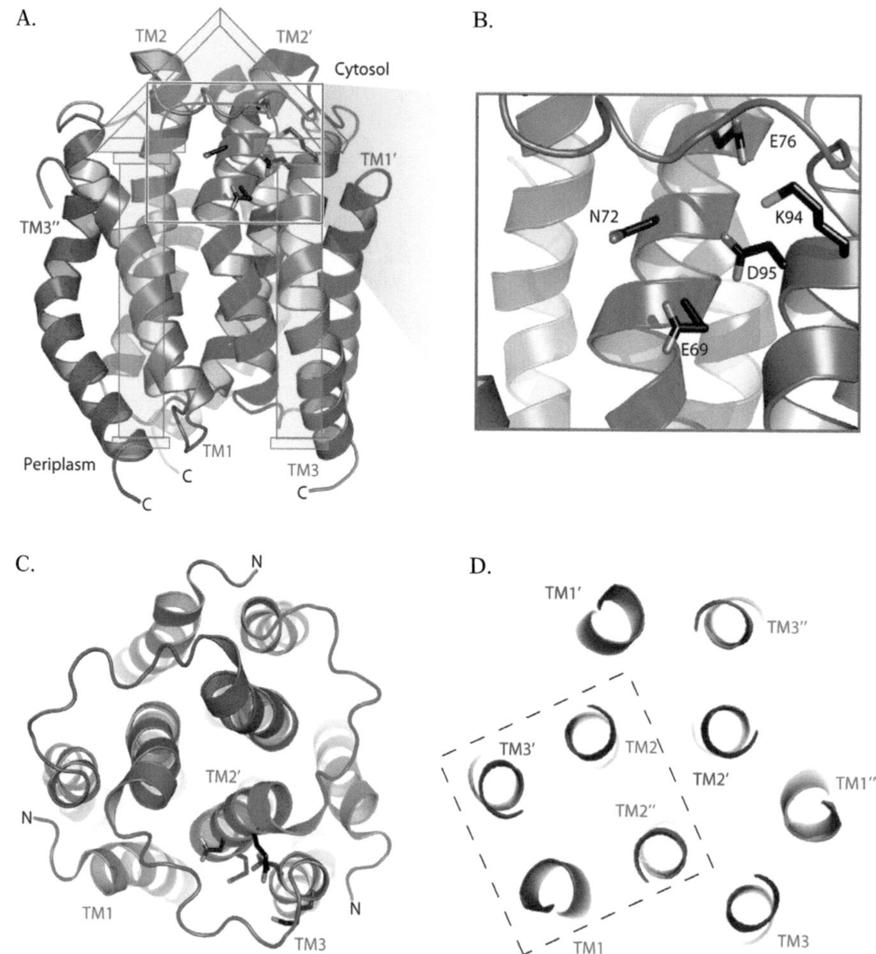

6.2 Backbone structure of DAGK solved by solution NMR. Twenty residues at the N terminus are missing. The three subunits are colored red, blue, and green. (A) Ribbon diagram of the trimer from the plane of the membrane highlighted to show a single portico-like active site. (B) Close-up of that active site, with the highly conserved residues indicated in black stick representation. (Additional highly conserved residues are found on the missing N terminus.) (C) Cytosolic view of the structure, again with active site residues of one site in stick representation. (D) View of clipped TM helices to show the topology, with one four-helix bundle highlighted with a dashed square. Note the domain swapping: each TM2 interacts with TM1 and TM3 from the adjacent subunit. From Van Horn, W. D., and Sanders, C. R., *Ann Rev Biophysics*. 2012, 41:81–101. © 2012 by Annual Reviews. Reprinted with permission from the Annual Review of Biophysics, www.annualreviews.org.

Presenilin, an intra-membrane protease

Remarkably, intra-membrane proteases or I-CLiPs (intramembrane-cleaving proteases) are able to hydrolyze peptide bonds in the hydrophobic milieu of the membrane utilizing the catalytic mechanisms of soluble proteases. Four families of intra-membrane proteases have been identified: rhomboid, signal peptide peptidase, site 2 protease, and presenilin. Rhomboid proteases are serine proteases that are important in embryogenesis in eukaryotes. A bacterial rhomboid, GlpG, has been crystallized and is described in Chapter 9. Signal peptide peptidase (SPP) is an aspartyl protease that plays a role in the biogenesis of membrane and secreted proteins by digesting the signal peptides removed during their export from the cytosol (see Chapter 7). Site protease 2 (SP2) is a metalloprotease that releases a transcription factor from the sterol regulatory element-binding protein involved in control of cholesterol and FA synthesis. Both SPP and SP2 require precleavage of their substrates, which is how they are regulated.

Presenilin is an aspartyl protease included in the γ-secretase complex that carries out the final step in the release of the Aβ peptide associated with Alzheimer's disease (AD). In fact, the enzyme is named for its role in the formation of senile plaques that accumulate in the brain, causing dementia in AD patients. Because it affects millions of people worldwide, the formation of Aβ is the focus of much research.

Aβ is produced from amyloid precursor protein (APP), a bitopic membrane protein expressed at high levels in the brain, through proteolytic steps carried out

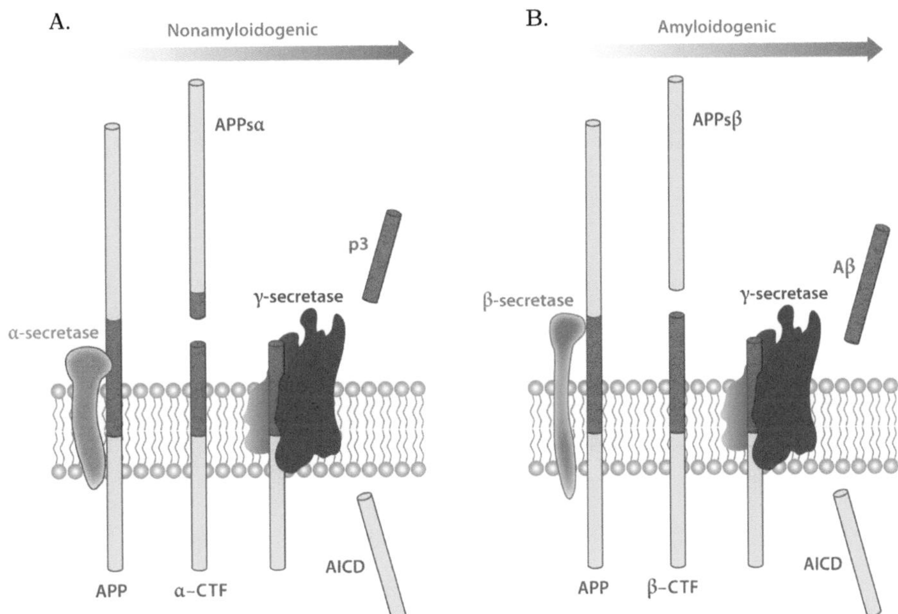

6.3 Role of membrane proteases in Aβ production. The sequential cleavage of the amyloid precursor protein (APP) occurs by two pathways. (A) Nonamyloidogenic processing involves cleavage by α-secretase within the Aβ peptide region (red) to create a soluble fragment (APPsα) and a membrane-bound C-terminal fragment (α-CTF), followed by γ-secretase cleavage of the α-CTF to release a small peptide (p3) and an intracellular C-terminal fragment (AICD). (B) Amyloidogenic processing starts with cleavage by β-secretase to release a soluble fragment (APPsβ) from the C-terminal fragment (β-CTF, also called C99) in the membrane. Then γ-secretase cleaves off the Aβ peptide and again releases AICD. From O'Brien, R. J., and P. C. Wong, *Ann Rev Neurosci.* 2011, 34:185–204. © 2011 by Annual Reviews. Reprinted with permission from the Annual Review of Neuroscience, www.annualreviews.org.

by secretases in the membrane. In spite of intense investigation, the physiological function of APP is not known. In mice, deletion of APP is not deleterious, and overexpression of APP is not harmful. APP metabolism is complicated by alternate pathways of proteolysis, only some of which generate Aβ (Figure 6.3). When APP is cleaved first by α-secretase and then γ-secretase (usually on the cell surface), it releases a peptide that does not form toxic amyloid fibrils. When it is cleaved by β-secretase and then γ-secretase (usually in endosomes formed from clathrin-coated pits) Aβ is produced and released to the outside of the cell, where it may aggregate and result in fibril formation. Therefore, segregation of APP into lipid rafts that are more likely to undergo endocytosis favors the amyloidogenic pathway.

Neuronal β-secretase (also called BACE1) is an aspartyl protease that cleaves APP outside the membrane at one of two sites, numbered +1 and +11 based on the sequence of Aβ. This cleavage generates two C-terminal fragments C99 and C83. C99 is the substrate for γ-secretase, which can cleave it at one of several residues in the γ-site within the TM segment (Figure 6.4). With several cleavage sites for γ-secretase, Aβ of different lengths are observed, of which $A\beta_{1-40}$ (Aβ40) and $A\beta_{1-42}$ (Aβ42) are the most common. Several lines of experiments indicate that Aβ42 is more toxic than Aβ40. Clearly the ratio of different proteolytic products is critical in the development of amyloid disease, which is why most of the mutations in APP that result in early onset AD cluster around the cleavage sites.

γ-secretase is a multiprotein complex composed of presinilin (PS1 or PS2), nicastrin (Nct), anterior pharynx defective-1 (Aph-1), and presenilin enhancer-2 (Pen-2) in a ratio of 1:1:1:1. PS1 and PS2 are isoforms whose expression varies in different tissues; they are the catalytic subunits with two key Asp residues in the active site. Nct has a single TM segment and a large glycosylated extracellular domain and assists in substrate selection. Aph-1 has seven TM segments and promotes complex formation and trafficking. Pen-2 has two TM segments and triggers endoproteolysis that is involved in regulation.

The active site of presinilin is similar to that in SPP, with the Asp residues in the same motif, YDX_nLGhGD, where X is any amino acid and h is a hydrophobic residue. Interestingly, they are oriented oppositely in the membrane, allowing them to cleave oppositely oriented TM helices in their substrates: presenilin cleaves type I proteins (with their C terminus in the cytosol), while SPP cleaves type II proteins. The low-resolution structure of the γ-secretase complex indicates it has an interior aqueous cavity.

Because presinilin has other substrates that are involved in cell growth and differentiation, inhibition of

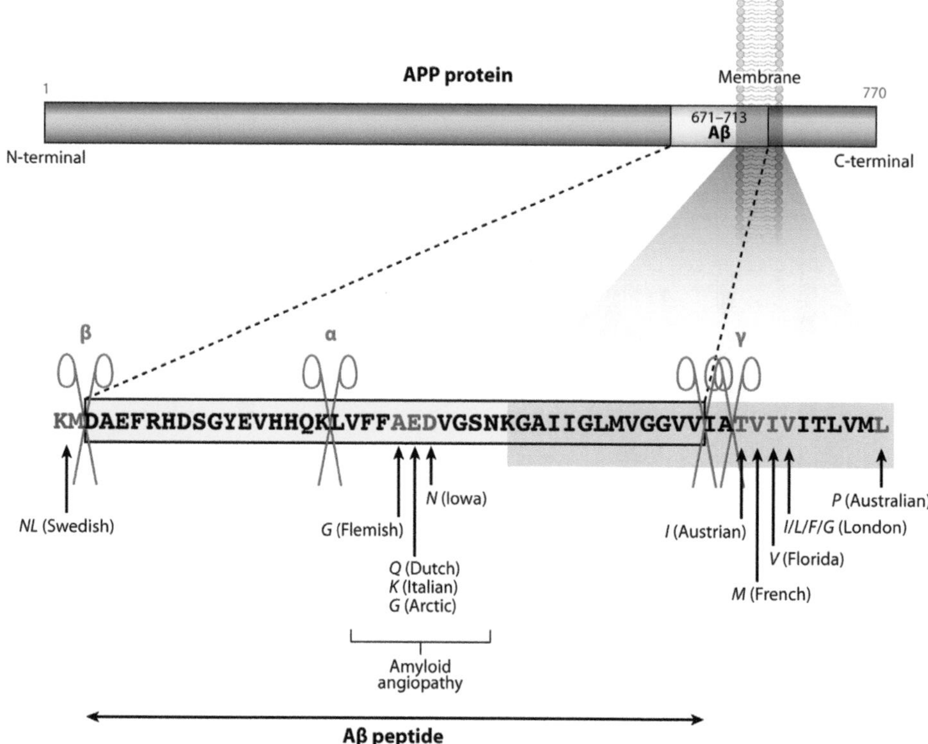

6.4 Cleavage sites in amyloid precursor protein. The different fragments of the amyloid precursor protein result from different cleavage sites for α-, β-, and γ-secretase (red scissors). Clustered near the cleavage sites are sites of familial mutations (residues in green at black arrows) that affect the N- and C-terminal regions of Aβ. An additional cluster is adjacent to a critical turn region in Aβ that is related to amyloid formation. From Straub, J. E., and Thirumalai, D., *Ann Rev Phys Chem*. 2011, 62:437–463. © 2011 by Annual Reviews. Reprinted with permission from the Annual Review of Physical Chemistry, www.annualreviews.org.

γ-secretase is not a feasible approach for drug therapy. New Alzheimer's drugs are designed to modulate the enzyme to decrease cleavage of C99 while allowing it to accept the other substrates. Another strategy for therapy targets cholesterol: Elevated levels of cholesterol are correlated with amyloid formation, and amyloid plaques in AD patients are highly enriched in cholesterol. It has been shown by NMR that cholesterol binds APP, which favors partitioning into rafts and thus pushes more of it into the amyloidogenic pathway.

P450 Cytochromes

P450 cytochromes are a ubiquitous superfamily of heme-containing monooxygenases that are named for their absorption band at 450 nm. They are involved in metabolism of an unusually wide range of endogenous and exogenous substances. They participate in the metabolism of steroids, bile acids, fatty acids, eicosanoids, and fat-soluble vitamins, and they convert lipophilic xenobiotics (foreign compounds) to more polar compounds for further metabolism and excretion. P450 cytochromes catalyze hydroxylation of an organic substrate, RH, to R–OH with the incorporation of one oxygen atom of O_2, while reducing the other oxygen atom to H_2O. Their source of reducing power is NAD(P)H, with electrons either donated directly from a flavin-containing reductase (class II) or shuttled to the P450 by small soluble electron carrier proteins (class I; Figure 6.5). Some self-sufficient P450s in bacteria have been found to contain heme, flavin, and iron–sulfur centers in one polypeptide.

In eukaryotes, most P450 cytochromes are bound to the mitochondrial inner membrane or to the ER. The P450s in the ER are integral membrane proteins, each bound by a single N-terminal TM domain. Truncation of the N-terminal domain is not sufficient to express soluble protein for crystallization and must be accompanied by several point mutations to disrupt a peripheral membrane-binding site. Even so, detergent is needed to prevent aggregation. High-resolution structures of such constructs show the large soluble domain is a triangular prism shape with β-sheets along part of one side and α-helices forming the rest (see Figure 6.5). Structures of these P450 catalytic domains in the presence and absence of substrates reveal that dramatic conformational flexibility must be needed for binding such a variety of compounds. In order to react with partners that provide electrons to P450s, such as cytochrome b5, the TM segments of both are required. The first

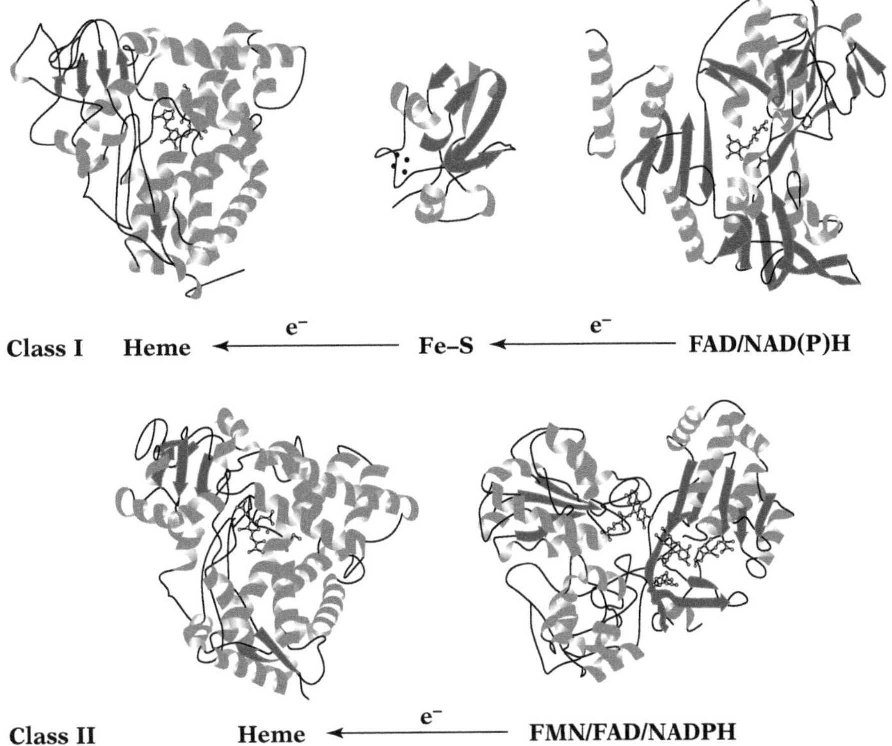

6.5 Examples of the two classes of P450 cytochromes. The classes are distinguished by their redox partners. Class I is represented by the P450 system from *Pseudomonas putida* with P450cam (with heme), putidaredoxin (with Fe–S), and putidaredoxin reductase (with FAD/NAD(P)H). Class II is represented by P450BM3 from *Bacillus megaterium* (with heme) and cytochrome P450 reductase from rat liver (with FMN/FAD/NADPH). The high-resolution structures for P450cam and P450BM3 have been solved for proteins lacking their TM segments. From Li, H., and T. L. Poulos, *Curr Top Med Chem*. 2004, 4:1789–1802. © 2004. Reprinted by permission of Bentham Science Publishers Ltd.

structure of a full length cytochrome P450, the lanosterol 14αdemethylase (ScErg11p) from yeast, reveals an acute angle between the catalytic domain and the TM helix that brings the edge of the catalytic domain into the lipid bilayer.

About half of the 57 P450 cytochromes in the human genome metabolize endogenous compounds, while many of the rest metabolize drugs and other xenobiotics. Some plants have around 300 or more P450s. The genes are classified into families based on sequence identity: to the root symbol CYP is added a number for the family (one of more than 200 groups with >40% sequence identity), followed by a letter for subfamilies (having >55% identity), followed by a number for the gene. For example, sterol 27-hydroxylase is CYP27A and vitamin D 24-hydroxylase is CYP27B, because their amino acid sequences are >40% identical (placing them both in CYP27) but <55% identical. If there were another P450 cytochrome in the CYP27A class, then it would be CYP27A2. The presence of many gene clusters, each containing up to 15 CYP genes, in most genomes suggests that the diversity of P450 cytochromes arose from many gene duplications in addition to likely gene amplifications and lateral transfers.

TRANSPORT PROTEINS

Transport of molecules across the bilayer is obviously an important function of membrane proteins and utilizes a variety of mechanisms. These mechanisms are defined and classified by both the stoichiometry and the energetics of the transport process. The definitions are essential for understanding the systematic classification of all membrane transport proteins. Uniport is the movement of one molecule at a time across the membrane, and cotransport is the tightly coupled movement of more than one substrate. Symport is the tightly coupled transport of two different molecules in the same direction, and antiport is the tightly coupled transport of two different molecules in opposite directions. When symport and antiport involve transport of ions, they may be electroneutral – resulting in no net transfer of charge – or electrogenic – creating a charge separation across the membrane.

Proteins involved in transport carry out either active or passive transport. Active transporters enable a cell to accumulate solutes against electrical and/or concentration gradients by making use of energy sources to "pump" them thermodynamically "uphill," whereas passive transporters allow the "downhill" flow of solutes across the membrane until their electrical and/or concentration gradients are dissipated. Like the specific porins described in the last chapter, passive transport proteins provide saturable pathways that are susceptible to competitive inhibition. The classic example is the glucose transporter in erythrocytes, a glycoprotein predicted to have 12 TM α-helices (Figure 6.6). Unlike the porins, the channel of the glucose transporter does not leak small molecules or ions, so it is described as a "gated pore" that opens alternatively to one side of the membrane or the other, depending on its conformation (Figure 6.7). This alternating access mechanism for transport was proposed over 50 years ago and is now the dominant model for many active transporters as well (see Chapter 11).

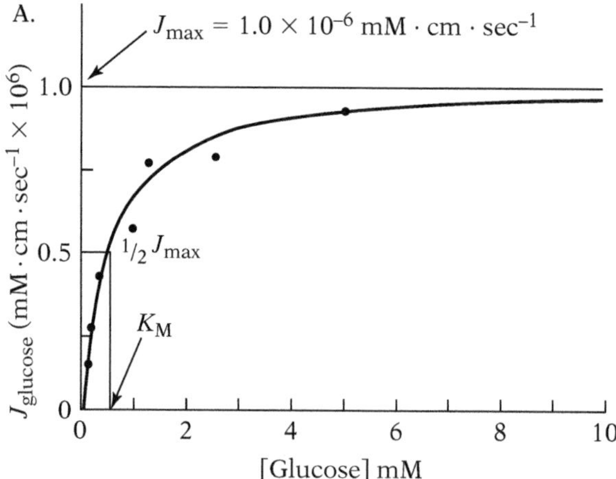

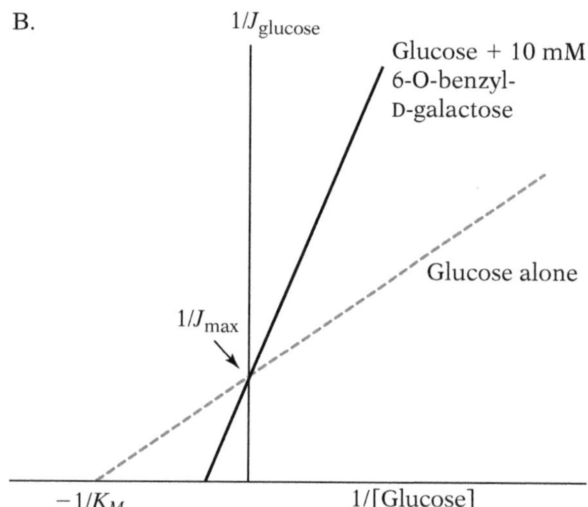

6.6 Glucose transport into erythrocytes. (A) A plot of the flux for glucose uptake by erythrocytes at 5°C as a function of the external concentration of glucose shows the passive uptake of glucose follows Michaelis–Menten kinetics, with the flux (J, in units of mM cm sec^{-1}) in place of the velocity of a reaction. The saturable curve indicates that the flux utilizes a carrier. (B) The flux of glucose can be inhibited by 6-O-benzyl-D-galactose, and the kinetic analysis indicates it is competitive inhibition, providing evidence for a binding site on the carrier. Redrawn from Voet, D., and J. Voet, *Biochemistry*, 3rd ed., John Wiley, 2004, pp. 728–729. © 2004. Reprinted with permission from John Wiley & Sons, Inc.

Active transport is further divided into primary processes, which are directly driven by hydrolysis of ATP (or another exergonic chemical reaction), and secondary processes, which carry out symport or antiport coupled to ion gradients made by primary active transporters like ATPases. These distinctions are part of the functional basis for classification of all transporters into hierarchies of families and superfamilies.

Transport Classification System

The transport classification (TC) system endorsed by the International Union for Biochemistry and Molecular Biology describes five broad classes: channels and pores, carriers or porters, primary active transporters, group translocators, and TM electron carriers, in addition to lists of accessory factors and incompletely characterized transport systems (Table 6.1). About 400 families of transport proteins are now included in the TC system, divided as follows:

- Class 1: The channels and pores are proteins (and peptides) that allow the relatively free flow of solute through the membrane. This class is divided into five subclasses: α-helical protein channels; β-barrel porins (see Chapter 5); channel-forming toxins, including colicins, diphtheria toxin, and others (see Chapter 4); nonribosomally synthesized channels, such as gramicidin and alamethicin (see Chapter 4); and holins, which function in export of enzymes that digest bacterial cell walls in an early step of cell lysis.
- Class 2: Carriers form complexes by binding their solutes before transporting them across the bilayer by the secondary processes of symport or antiport. For this reason, class 2 is called electrochemical potential-driven transporters. A very large family of porters in this class is called the major facilitator superfamily (MFS) and is exemplified by the lactose permease (see Chapter 11). Class 2 also contains the ionophores like valinomycin and nigericin, which are nonribosomally synthesized ion carriers, as well as the TonB family of proteins involved in transferring energy to the outer membrane of Gram-negative bacteria (see Chapter 11).
- Class 3: The class of primary active transporters includes the ATPases and the ATP-binding cassette transporters described next. Other examples of primary active transporters are those driven by oxidoreduction reactions and those driven by light, including the photosynthetic reaction center (RC; see Chapter 5).

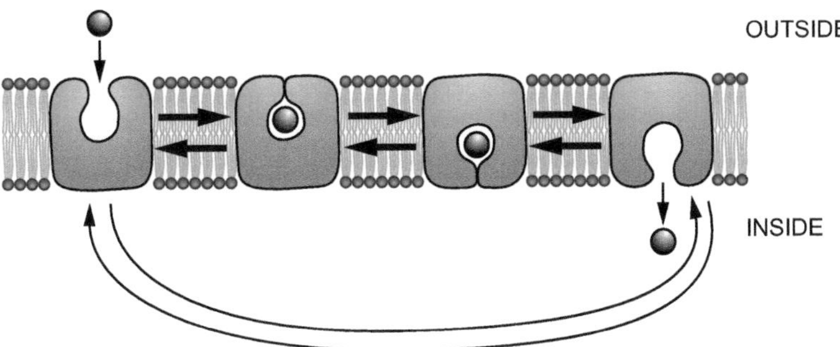

6.7 The model for conformational changes involved in transporting glucose via a gated pore. Step 1, glucose binds from outside; step 2, a conformational change closes the gate to the outside; step 3, conformational change to inward facing; step 4, gate opens to inside, releasing glucose to the cytoplasm. All the steps are reversible, as indicated, so uptake is driven by the concentration gradient across the membrane.

- Class 4: The group translocators provide a special mechanism for the phosphorylation of sugars as they are transported into bacteria, described below.
- Class 5: The TM electron transfer carriers in the membrane include two-electron carriers, such as the disulfide bond oxidoreductases (DsbB and DsbD in *E. coli*) as well as one-electron carriers such as NADPH oxidase. Often these redox proteins are not considered transport proteins.

Cells have multiple transport systems to take up many nutrients, and the families of uptake systems employed for any nutrient type are not limited to one class. For example, the uptake of sugars is carried out by nine families of ABC transporters, 20 families of secondary carriers, seven families of porins, and six families of group translocation systems. In addition to descriptions of transport functions, the TC system derives evolutionary relatedness from sequence information obtained using the tools of bioinformatics described below that allow investigators to compare primary sequences, predict topologies, and analyze genomes.

TABLE 6.1 THE TRANSPORTER CLASSIFICATION SYSTEM[a]

Class		Subclass
1 Channels/pores	1a	α-helical protein channels
	1b	β-barrel protein porins
	1c	Toxin channels
	1d	Nonribosomally synthesized channels
	1e	Holins
2 Electrochemical potential-driven transporters	2a	Protein porters
	2b	Nonribosomally synthesized porters
	2c	Ion gradient-driven energizers
3 Primary active transporters	3a	P~P-bond hydrolysis-driven systems
	3b	Decarboxylation-driven systems
	3c	Methyl transfer-driven systems
	3d	Oxidoreduction-driven systems
	3e	Light absorption-driven systems
4 Group translocators	4a	Phosphotransfer-driven systems
5 Transmembrane electron carriers	5a	Two-electron transfer carriers
	5b	One-electron transfer carriers

[a] The five main classes are listed on the left, with major subclasses provided on the right.

Source: Saier, M., *et al.*, The Transporter Classification Database: recent advances. *Nucl Acid Res.* 2009, 37:D274–D278. © 2009, Oxford University Press.

Superfamilies of ATPases

The TC system has two superfamilies of ATPases, one for the P-type ATPases that are found in plasma membranes and include the Na^+, K^+ ATPase and the Ca^{2+}, pump (see Chapter 9). The other superfamily consists of three other types of ATPases: F-type, such as the mitochondrial and bacterial ATP synthases (see Chapter 13); A-type, which transports anions such as arsenate and is mainly found in archaea; and V-type, which maintains the low pH of vacuoles in plant cells and lysosomes, endosomes, the Golgi, and secretory vesicles of animal cells. Both the subunit compositions and their protein sequences indicate that the A-type is more similar to the V-type ATPases than to the F-type.

The Na^+, K^+ ATPase, called the Na^+-K^+ pump, transports two K^+ in and three Na^+ out for every ATP hydrolyzed, establishing gradients that are critical to osmotic balance in all animal cells and to the electrical excitability of nerve cells, as well as driving the uptake of sugars and amino acids. The high-resolution structure of the Na^+, K^+ ATPase reveals details of its architecture, which consists of a catalytic α subunit, a bitopic β subunit, and a tissue-specific γ subunit, and its mechanism (see Chapter 9). ATP phosphorylates an Asp residue in the highly conserved sequence DKTG only in the presence of Na^+; the aspartyl-phosphate thus produced is hydrolyzed only in the presence of K^+, giving strong evidence for two conformational states. Thus the mechanism of coupling active transport with ATP hydrolysis involves shifting from the phosphorylated form with high affinity for K^+ and low affinity for Na^+ to the dephosphorylated form with high affinity for Na^+ and low affinity for K^+. This electrogenic process helps create the typical TM potential of –50 to –70 mV (inside negative) across the plasma membrane of most cells. The Na^+, K^+ ATPase is inhibited by digitalis in the well-known treatment for congestive heart failure.

ABC Transporter Superfamily

An important class of ATP-dependent transporters is the superfamily of ABC transporters, named for their *A*TP-*b*inding *c*assettes, which are nucleotide-binding domains. ABC transporters are found in large numbers in animals, plants, and microorganisms; the genome of *E. coli* has 80, while the human genome has 49. Some ABC transporters function in uptake and others in efflux. Some are very specific for their substrates while others are quite promiscuous. The first ones to be identified functioned in uptake of amino acids, peptides, sugars, and vitamin B_{12}; the high-resolution structures of the maltose transporter and the vitamin B_{12} transporter are described in Chapter 11. ABC transporters also export cell wall polysaccharides, participate in cytochrome *c* biogenesis, and secrete cellulases, proteinases, and toxins. The first ABC exporter to have a crystal structure was the drug efflux protein Sav1866 from *Staphylococcus aureus* (see Chapter 11).

All ABC transporters have two nucleotide-binding domains and two membrane domains, which are separate subunits in archaea and eubacteria and are usually fused together in higher organisms (Figure 6.8). In addition, the ABC transporters that function in uptake have a substrate-specific domain or protein on the external side (the periplasmic binding proteins in Gram-negative bacteria). The membrane domains typically have 12 TM helices in either a single peptide or two subunits.

The nucleotide-binding domains of ABC transporters are highly conserved, whether they are separate subunits or are fused to the TM domains. Several of them have been crystallized either as aqueous proteins or water-soluble fragments of transporters, and they may be seen in x-ray structures that have been solved for intact transporters (see Chapter 11). As expected, the various NBDs show strong structural similarities.

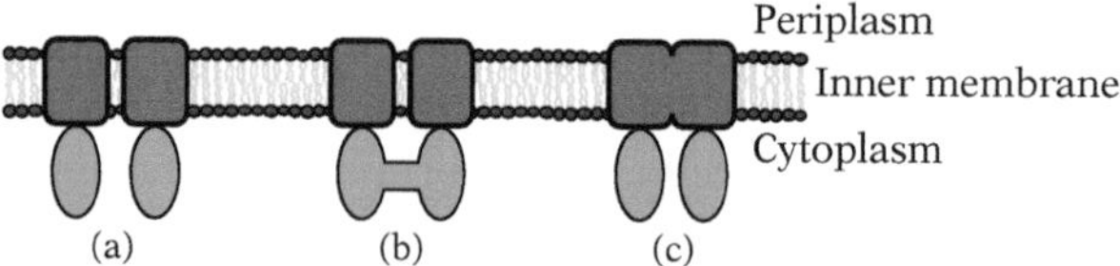

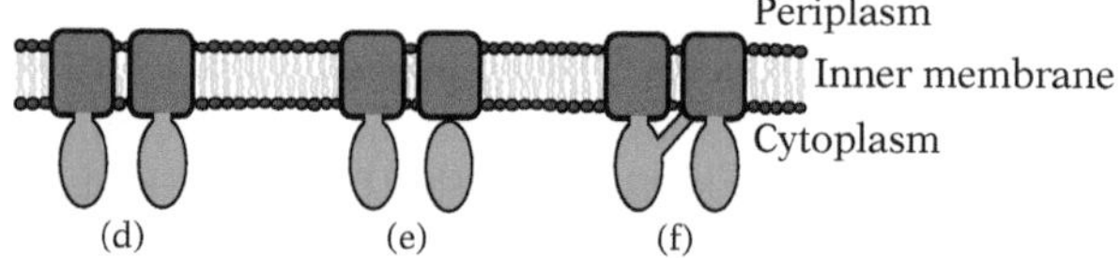

6.8 Organization of the structures of ABC transporters. The subunit composition of the ABC transporter system varies in different organisms. The two nucleotide-binding domains (NBDs) (pink) and two TM domains (red) may be four separate subunits or fused into one, two, or three proteins, as shown. Redrawn from Higgins, C. F., *Res Microbiol*. 2001, 152:205–210. © 2001 by Elsevier. Reprinted with permission from Elsevier.

Each NBD contains an ATPase subdomain with similarities to the F_1 ATPase (see Chapter 13) and a helical subdomain specific to the ABC transporters (Figure 6.9A). The ATPase subdomain contains typical Walker A and Walker B motifs found in proteins that hydrolyze ATP and has a β-sheet region that positions the base and ribose moieties of the nucleotide. The helical subdomain contains the signature motif of ABC transporters, LSGGQ. In the dimer, the Walker A motif from one subunit binds the same ATP as the LSGGQ from the other subunit, thus the two NBDs make an "ATP sandwich." Each round of ATP hydrolysis involves three conformations of the NBDs: a nucleotide-free conformation, an ATP-binding conformation, and a more open conformation after the hydrolysis reaction that are displayed in the MalK protein, the NBD of the maltose transport system, crystallized in all three conformations (Figure 6.9B). The movements of the NBD are transmitted to the TM regions of the membrane transporter to drive active transport.

The substrate-specific components of ABC transporter systems in the periplasm of Gram-negative bacteria are obligatory to the transport of their substrates. High-resolution structures of several of the periplasmic binding proteins with and without bound substrate indicate large conformational changes typically occur upon binding. Each protein has two lobes that close over the substrate when it binds, giving a very high substrate affinity (micromolar K_d). For some transport systems, mutants of the inner membrane transporters have been characterized in which delivery of substrate by these soluble binding proteins is no longer obligatory.

Many of the ABC transporters in humans have medical significance. The CFTR protein, the cystic fibrosis transmembrane conductance regulator, is a Cl^- channel that is defective in most cases of cystic fibrosis (see Chapter 7). In humans the P-glycoprotein that confers multidrug resistance to cancer cells is in this class (see Chapter 11), as is TAP, the transporter for antigen processing in the immune system (see Chapter 14).

Group Translocation

A different mechanism of coupling to an exergonic reaction is utilized for group translocation of sugars in bacteria, in which the sugar substrates are phosphorylated during the transport process. This unusual group of transport systems constitutes class 4 in the TC system. The best-characterized example is the uptake of glucose, fructose, mannose, and other sugars by the phosphoenolpyruvate-dependent phospho-transferase system (PEP PTS) of *E. coli* (Figure 6.10). The two cytoplasmic proteins involved in transfer of the phosphoryl group from PEP, EI (enzyme I), and HPr (histidine-containing phosphocarrier protein) are used in transport of all the sugar substrates of this system. Uptake

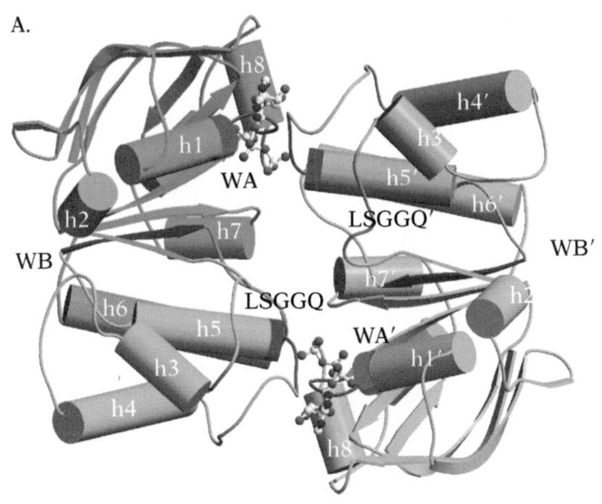

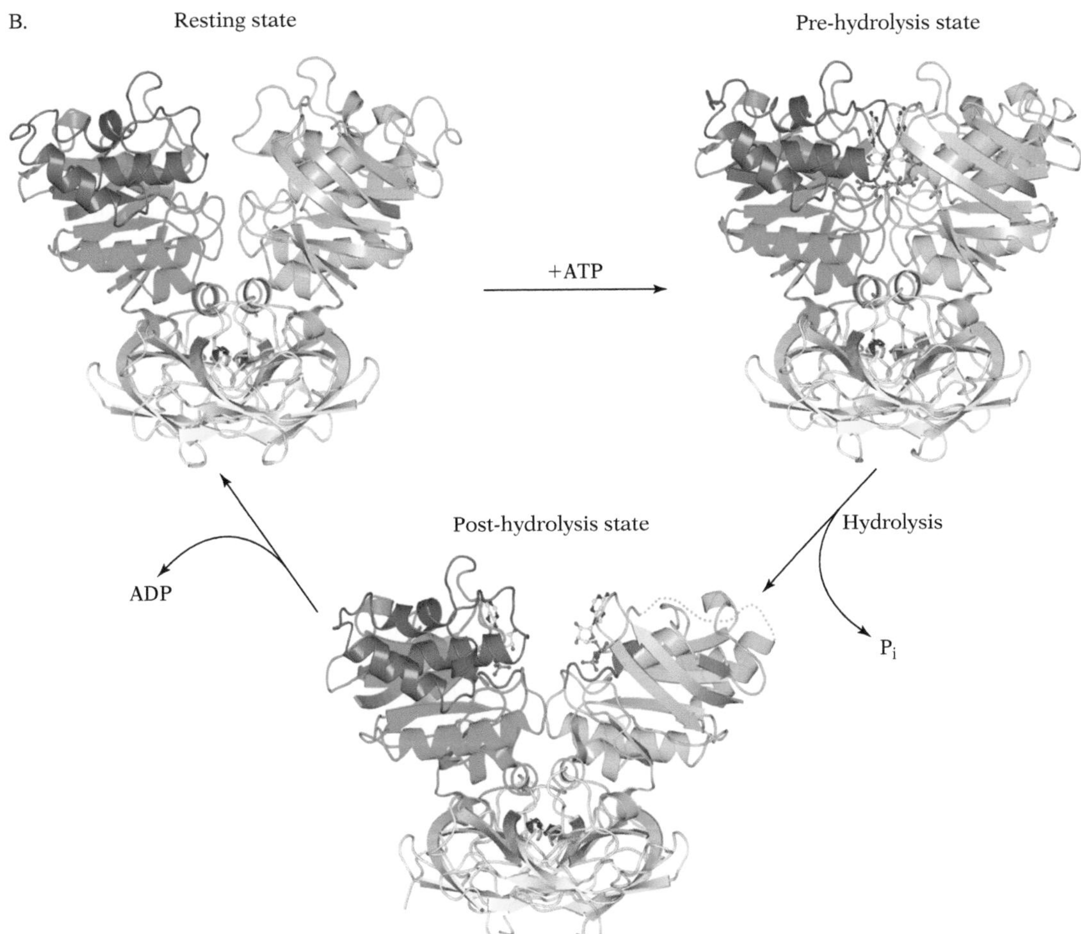

6.9 The MalK dimer, an example of an NBD of an ABC transporter. (A) The x-ray structure of the NBDs of the MalK dimer shows the ATP-binding site. Each NBD has two subdomains, an ATPase subdomain (green) and a helical subdomain (cyan). The conserved segments shown are the Walker A motif (WA, red), Walker B motif (WB, blue), LSGGQ motif (magenta), and the Q-loop (yellow) that connects the ATP-binding region to the helical subdomain, as well as to the TM portions of the ABC transporter. The bound ATP is shown in a ball-and-stick model. The regulatory domain of MalK has been omitted for clarity. From Davidson, A. L., and J. Chen, *Annu Rev Biochem*. 2004, 73:241–268. © 2004 by Annual Reviews. Reprinted with permission from the Annual Review of Biochemistry, www.annualreviews.org. (B) The conformation of the MalK dimer varies from the resting state (with the two NBDs separated from each other), the ATP-bound state (closed), and the post-hydrolysis state (open). Each subunit has an NBD (green and blue or cyan) and a regulatory domain (yellow). The Walker A motif (red) shows where the nucleotide (ball-and-stick model) binds. From Lu, G., *et al.*, *Proc Natl Acad Sci USA*. 2005, 102:17969–17974. © 2005 by National Academy of Sciences, U.S.A.

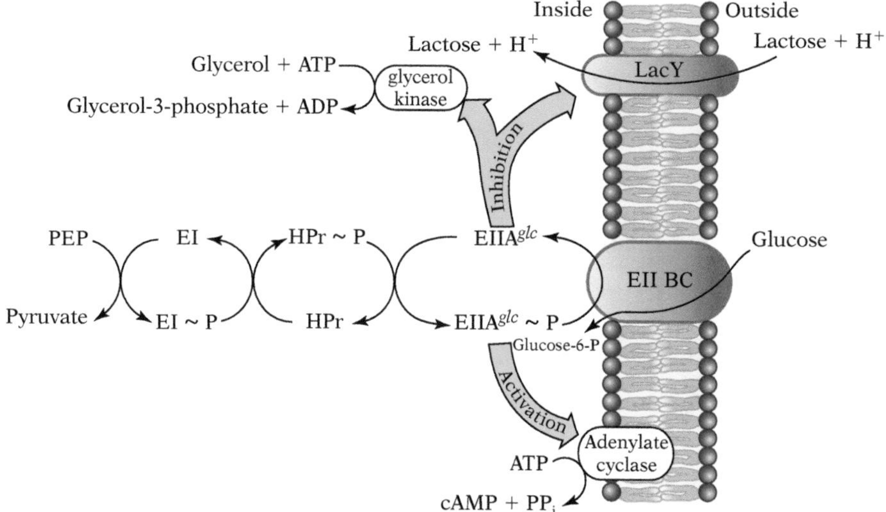

6.10 The PEP PTS for glucose in *E. coli*. Group translocation utilizes a combination of shared and specialized proteins, as exemplified by the PTS for glucose. EI and HPr are the shared cytoplasmic proteins utilized for transfer of phosphoryl groups. $EIIA^{glc}$ is a specialized cytoplasmic protein that phosphorylates the glucose as it enters the cytoplasm and is sensitive to catabolite repression. $EIIBC^{glc}$ is the TM glucose transporter. In the cytoplasm EI is phosphorylated by PEP; the phosphoryl group is transferred to HPr, and then to a specific EIIA. Then EIIA phosphorylates the membrane protein(s), in this case EIIBC, which then phosphorylate the substrate as it enters. Variations in the number of EII proteins are seen with different substrates. LacY is shown transporting lactose, which plays an inhibitory role in uptake of glucose by the PEP PTS. Redrawn from Voet, D., and J. Voet, *Biochemistry*, 3rd ed., John Wiley, 2004, p. 745. © 2004. Reprinted with permission from John Wiley & Sons, Inc.

of each sugar uses a specific TM component called EII, which in some cases has one or more separate cytoplasmic subunits, called EIIA and EIIB. For example, the TM component of the PTS for glucose is $EIIBC^{glc}$, which transports and phosphorylates glucose and transfers it to the cytosolic component. The cytosolic component, $EIIA^{glc}$, is sensitive to catabolite repression. Thus transfer of the high-energy phosphoryl group in PEP to glucose drives its uptake as glucose-6-phosphate. Many components of the PEP PTS have been characterized thoroughly, and several of the soluble proteins have high-resolution structures. Although the EI and HPr components are common elements that are shared by transport systems for different substrates, different forms of them are usually present, encoded by more than one gene. For example, the PTS families are represented in the *E. coli* genome by five genes for EI homologs and six for HPr homologs. There are 21 EII complexes, including seven for the fructose family and seven for the glucose family, in addition to some regulatory proteins.

Symporters

Secondary active transport that uses ion gradients typically involves cotransport of the substrate and the ion. The well-characterized lactose permease from *E. coli* (see Chapter 11) carries out symport of protons and lactose, driven by the proton gradient across the membrane (Figure 6.11). Bacterial proton symporters are also used to accumulate arabinose, ribose, and some amino acids, while Na^+ symport is used in *E. coli* for uptake of melibiose, glutamate, and other amino acids, as well as in the intestinal epithelium for uptake of glucose and some amino acids (Figure 6.12).

When symport uses Na^+, it is driven by the Na^+ gradient created by the Na^+, K^+ ATPase. The energy used can be calculated from the chemical potential and electrical potential of the ion gradient. For example, for the Na^+ gradient,

$$\Delta G = nRT\ln([Na^+]_{in}/[Na^+]_{out}) + n\mathfrak{F}\Delta E.$$

In intestinal Na^+–glucose symport, $n = 2$ for both the concentration and electrical terms, as two Na^+ ions enter per glucose. Using typical values for ΔE (–50 mV), $[Na^+]_{in}$ (12 mM) and $[Na^+]_{out}$ (145 mM), the ΔG for glucose transport is –22.5 kJ, enough to pump glucose inside until its concentration is 9000 times the external concentration.

Antiporters

Antiport is important for ion exchange across many biological membranes. For example, the Na^+/H^+antiporter in the plasma membrane exchanges extracellular Na^+ for intracellular H^+ and is activated in response to mitogenic agents such as epidermal growth factor. The slight increase in cytoplasmic pH produced by this exchanger may activate other enzymes in the mitogenic response.

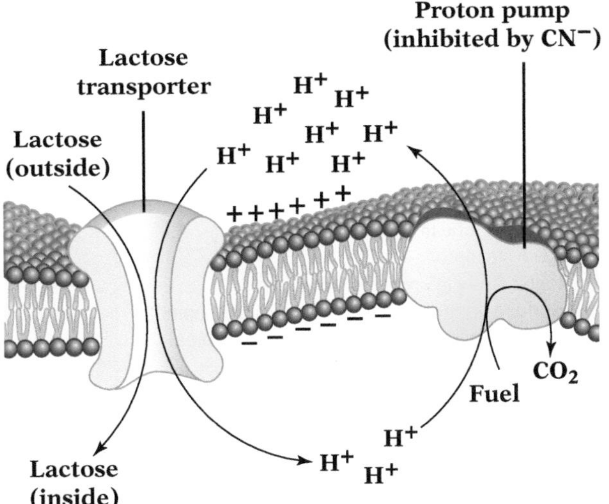

6.11 Symport of lactose and protons in *E. coli*. The proton gradient made by the respiratory chain or other proton pumps is used to drive the uptake of lactose by the lactose permease. When cells are treated with CN^- to inhibit the energy-yielding oxidation pathways that support the proton pump, active transport of lactose is abolished. Under this condition, the lactose permease carries out passive transport. Redrawn from Nelson, D. L., and M. M. Cox, *Lehninger Principles of Biochemistry*, 4th ed., W. H. Freeman, 2005, p. 404. © 2005 by W. H. Freeman and Company. Used with permission.

A prominent antiporter in the erythrocyte membrane is the anion exchanger called Band 3, which is about 25% of the total protein in that membrane. It carries out a one-to-one exchange of Cl^- and bicarbonate (HCO_3^-) to facilitate the removal of CO_2 from respiring cells. (Carbonic anhydrase in the erythrocyte converts waste CO_2 to HCO_3^-, which then circulates to the lungs where it reenters erythrocytes to be converted back to CO_2 and exhaled.) Since the exchange of anions is obligatory, the net flow of transport depends on the sum of the chloride and bicarbonate concentration gradients. The membrane portion of the Band 3 protein is predicted to have 12 or 14 TM helices, and the cytoplasmic portion interacts with cytoskeletal components while the extracytoplasmic portion carries the carbohydrate antigenic determinants for several blood groups. Besides the anion exchanger in the erythrocyte, humans have two other anion exchangers that are closely related to Band 3. Similar exchangers are also found in plants and microorganisms.

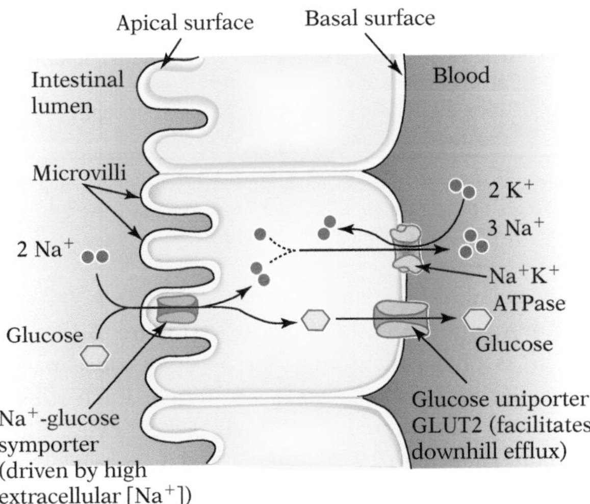

6.12 Symport of glucose and Na^+ ions in the intestinal epithelium. The Na^+ gradient made by the Na^+, K^+ ATPase provides the driving force for glucose uptake from the intestinal lumen. A passive glucose uniporter exports glucose to the bloodstream. Redrawn from Nelson, D. L., and M. M. Cox, *Lehninger Principles of Biochemistry*, 4th ed., W. H. Freeman, 2005, p. 405. © 2005 by W. H. Freeman and Company. Used with permission.

Not all antiporters are electroneutral; for example, the ADP/ATP carrier, the most abundant protein in the mitochondrial inner membrane, exchanges one ATP^{4-} (exiting) for one ADP^{3-} (entering). This electrogenic exchange is driven by the inside-negative potential across the mitochondrial membrane and utilizes one-third of the energy stored in the electrochemical proton gradient generated by the respiratory chain! To meet the energy needs of the cell, it facilitates the exchange of adenine nucleotides with a turnover number of 500 per minute. Two natural inhibitors bind to alternate sides of the transporter, indicating that it has two conformations that allow its adenine nucleotide binding site to face in or out. The high-resolution structure of the ADP/ATP translocator in the presence of one of these inhibitors shows a bundle of six TM α-helices forming a basket with the inhibitor inside (see Chapter 11).

The ADP/ATP carrier belongs to the mitochondrial carrier family (MCF), whose functions ensure that metabolites (such as pyruvate, tricarboxylates and dicarboxylates of the Krebs cycle, carnitine, and citrulline) as well as nucleotides and reducing power are exchanged between the cytosol and the mitochondrial matrix. The family also includes uncoupling protein of brown adipose tissue, which carries protons back into the matrix, allowing respiration to generate heat, and is involved in type II diabetes. The transporters in this family are in the inner membrane of mitochondria and share 20% sequence identity and a common three-fold symmetry in their structure. New members of the family are readily identified by the MCF signature, which consists of three copies of the sequence PX[D,E]XX[K,R], along with the three-fold sequence repeats that generate the overall fold. Mitochondria in different eukaryotes contain from 35 to 55 different mitochondrial carriers; the human genome encodes 48.

Ion Channels

Most eukaryotic membranes contain selective ion channels that are unlike the ion pumps driven by the hydrolysis of ATP. They are characterized by ion specificity, high fluxes – up to 10^8 ions sec^{-1} – and gating (opening and closing) in response to ligands or voltage detected by patch clamp measurements (see Chapter 3).

High-resolution structures for a number of ion channels reveal their basis of selectivity and transport mechanisms as described in Chapter 12. Ion channels that are involved in the passage of signals in nerves and muscle cells are also receptors for agonists and antagonists that control their activity.

MEMBRANE RECEPTORS

Membrane receptors are integral proteins that function in information transfer by triggering a response to their binding of ligands. Receptors have many diverse functions, listed in Table 6.2. Many receptors are involved in cell surface interactions, while some trigger endocytosis or recycling of membrane components. Among the receptors with enzyme activity are many receptors involved in signaling, such as the insulin receptor. The insulin receptor is an $\alpha_2\beta_2$ tetramer that responds to insulin by autophosphorylation of three Tyr residues on each β subunit, activating its protein kinase activity to start a signal cascade.

TABLE 6.2 FAMILIES OF SOME REPRESENTATIVE RECEPTORS IN EUKARYOTES

Immunoglobulin superfamily – many diverse functions
- T cell receptor (α, β subunits)
- Major histocompatibility complex II MHC (α, β subunits)
- Lymphocyte function – associated antigen-3 (LFA-3)
- CD2 (T cell LFA-2)
- Immunoglobulin A (IgA)/IgM receptor
- IgG F_c receptor
- High-affinity IgE receptor (α subunit)
- Surface immunoglobulins (heavy, light chains)
- N-CAMs (neuronal cell adhesion molecules)
- Myelin-associated glycoprotein

Integrins – bind to extracellular matrix and adhesion proteins
- Fibronectin receptors
- Vitronectin receptors
- Platelet glycoprotein complex (IIb/IIIa)
- Leukocyte adhesion proteins (LFA-1, Mac 1)
- T cell very late antigens (VLA family)

Mitogen/growth factor receptors with tyrosine kinase activity – stimulate cell growth
- Epidermal growth factor (EGF) receptor
- Platelet-derived growth factor (PDGF) receptor
- Insulin receptor
- Insulin-like growth factor-1 (IGF-1) receptor
- Colony stimulating factor-1 (CSF-1) receptor

Neurotransmitter receptors/ion channels – receptor-operated channels
- Nicotinic acetylcholine receptor (nAChR)
- γ-aminobutyric acid (GABA) receptor
- Glycine receptor

Receptors that activate G proteins
- β-adrenergic receptors (β_1, β_2)
- α-adrenergic receptors (α_1, α_2)
- Opsins (rhodopsin)
- Muscarinic acetylcholine receptors (M1, M2)

Miscellaneous
- (A) Asialoglycoprotein receptors
  - Low-affinity (lymphocyte) IgE receptor (unknown function)
- (B) Insulin-like growth factor-2 (IGF-2) receptor
  - Cation-independent mannose-6 phosphate receptor
  - Cation-dependent mannose-6-phosphate receptor
  - Cation-dependent mannose-6-phosphate receptor (extracytoplasmic domain)

Source: Gennis, R. B., *Biomembranes: Molecular Structure and Function*, Springer-Verlag, 1989, p. 324. © 1989 by Robert B. Gennis. Reprinted by permission of the author.

Nicotinic Acetylcholine Receptor

One of the most thoroughly studied receptors is the nicotinic acetylcholine receptor (nAChR), a neurotransmitter receptor in the Cys-loop superfamily. It differs greatly from the muscarinic acetylcholine receptor, a GPCR (see below). The Cys-loop superfamily, named for the signature 13-residue loop linked by a disulfide bond, includes excitatory receptors, such as those for acetylcholine and serotonin, and inhibitory receptors, such as those for γ-aminobutyric acid and glycine. The first high-resolution x-ray structure representing this family is that of the glutamate-gated chloride channel from *Caenorhabditis elegans* described in Chapter 10.

The nAChR is expressed throughout the nervous system and affects a wide variety of physiological functions. It is perhaps best known for its role mediating voluntary movement in skeletal muscle, which is affected by disease-causing mutations. When nAChR binds acetylcholine, it opens a channel for cations (Na^+, Ca^{2+}, and K^+), triggering depolarization of the membrane that causes the muscle fiber to contract. After several milliseconds the channel closes and acetylcholine is released.

Knowledge of the structure of the nAChR from two-dimensional EM has been complemented by an x-ray structure of an acetylcholine-binding protein from snail, which has homologous ligand-binding domains. The detailed structural model provides much insight into how the receptor functions (Figure 6.13). The nAChR is a heteropentamer, composed of $\alpha_2\beta\delta\varepsilon$ in adult muscle (ε is replaced by γ in the fetus). Each subunit comprises three domains: a largely β-sheet extracellular domain, an α-helical TM domain, and an α-helical cytoplasmic domain. Acetylcholine binds to two sites at the interfaces of each α subunit with either ε or β in pockets lined by aromatic residues and capped by a mobile loop, Loop C.

The TM domain of each highly homologous subunit has four TM helices (M1–M4), of which M2 lines the channel. Reversible gating of the channel by constriction and dilation at the narrowest point allows control of ion

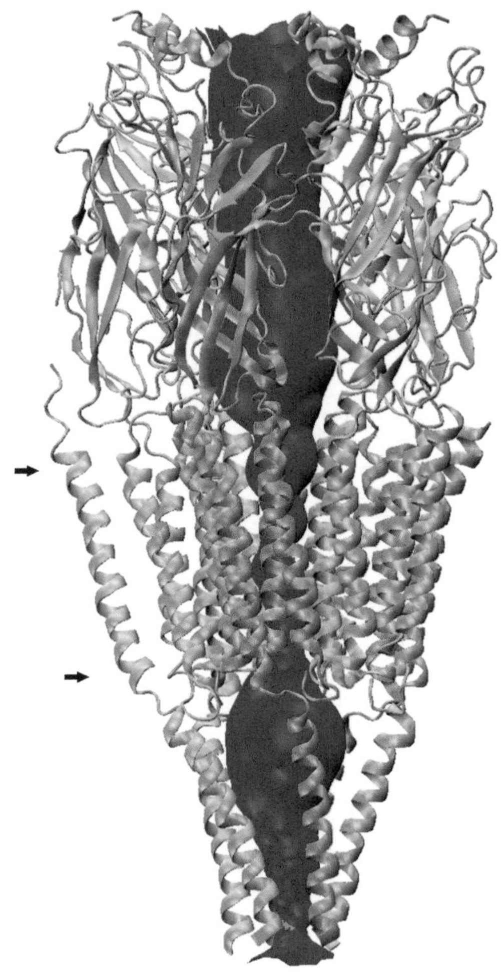

6.13 Structural model of the acetylcholine receptor from the *Torpedo* electric organ. The hetero-pentamer is viewed from the plane of the membrane as a ribbon diagram highlighting the α-subunits (gold; other subunits, cyan). The channel is visualized as the outer surface of the central vestibule (dark blue). The membrane boundaries (arrows) correspond to the interfaces of the three domains. From Sine, S. M., and A. G. Engel, *Nature*. 2006, 440:448–455. © 2006, Nature Publishing Group. Reprinted with permission of Macmillan Publishers Ltd.

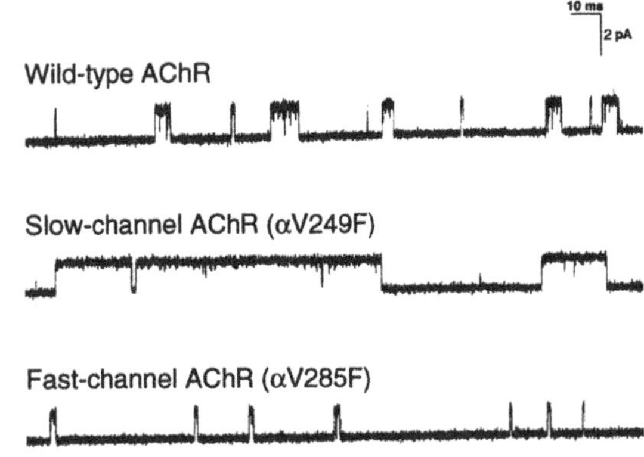

6.14 Activity of the nAChR revealed in single-channel currents from wild type and congenital myasthenic syndrome. Mutations associated with congenital myasthenic syndrome cause either fast-channel or slow-channel activity, shown here with single-channel currents measured for representative mutant receptors and the wild type expressed in human embryonic kidney fibroblasts. Both phenotypes cause muscle weakness but by different mechanisms. From Sine, S. M., and A. G. Engel, *Nature*. 2006, 440:448–455. © 2006, Nature Publishing Group. Reprinted by permission of Macmillan Publishers Ltd.

flux. The open state of nAChR has a much greater affinity for acetylcholine than the closed state has. Mutant studies have probed the coupling of ligand binding to channel opening, which appears to involve the loops that connect the β-sheet domain with the α-helical channel. Congenital myasthenic syndromes that cause severe neuromuscular disorder are caused by mutations in muscle nAChR that affect either gating or agonist binding and produce either slow channels or fast channels (Figure 6.14).

The Cys-loop neurotransmitter receptors are considered part of an even larger superfamily, the ligand-gated ion channels (LGICs). Together these superfamilies have over 200 entries in the LGIC database.

G Protein-Coupled Receptors

The largest and most diverse group of receptors involved in signaling is the G protein-coupled receptors (GPCRs). The majority of these receptors respond to ligand binding by activating heterotrimeric guanine nucleotide binding proteins (G proteins) on the cytoplasmic surface of the membrane, while a few open ion channels directly. The G proteins transmit and amplify the signals by triggering changes in the concentrations of second messengers, such as cAMP (Figure 6.15). GPCRs respond to many intercellular messenger molecules as well as sensory messages. The more than 1000 GPCRs in humans include receptors for nucleotides, neurotransmitters (acetylcholine, dopamine, histamine, and serotonin), prostaglandins and other eicosanoids, and many hormones, as well as olfactory and gustatory receptors. With a common protein fold, the GPCRs are called serpentine receptors because their seven TM α-helices "snake" across the membrane. The first member of the GPCRs to have its structure solved at high resolution is rhodopsin (described in Chapter 10). Intense interest has led to numerous high-resolution structures of GPCRs, which are expected to aid in drug design, as approximately half of modern pharmaceuticals are targeted at GPCRs. The GPCRs form a superfamily that is divided into six families with no statistically significant sequence similarity between them. Their seven-TM topology was attributed to convergent evolution until some of the bioinformatics tools described next enabled searches for distant homology and new members of the superfamilies.

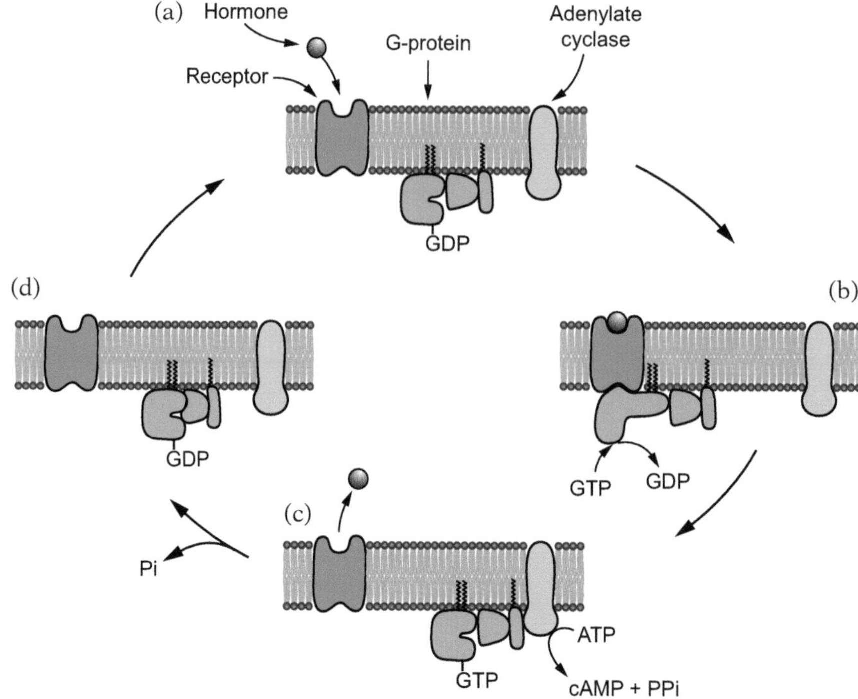

6.15 Interaction of a receptor with a G protein to stimulate the production of cAMP. The activation/deactivation cycle for hormonally stimulated adenylate cyclase (AC) starts with inactive AC and with guanosine diphosphate (GDP) bound to the G protein. When hormone binds to its receptor (a), the hormone–receptor complex stimulates the G protein to exchange GDP for guanosine triphosphate (GTP; b). The GTP–G protein complex binds to AC, activating it to produce cAMP (c). When the G protein catalyzes hydrolysis of GTP to GDP, it dissociates from AC, inactivating it (d). Modified from Voet, D., and J. Voet, *Biochemistry*, 3rd ed., John Wiley, 2004, p. 674. © 2004. Reprinted with permission from John Wiley & Sons, Inc.

BIOINFORMATICS TOOLS FOR MEMBRANE PROTEIN FAMILIES

Because purification and crystallization of membrane proteins are complicated by the presence of lipids and detergent (see Chapters 3 and 8), the number of high-resolution structures of membrane proteins, while growing rapidly, is still a small proportion of those predicted in the genome. With over 400 structures of membrane proteins in the Protein Data Bank (see Box 6.2), analysis of the wealth of genomic information available relies on methods for interpretation and comparison of the approximately 540,000 identified protein sequences in Swiss-Prot, up to one-third of which could be membrane proteins. The planar dimensionality and the hydrophobicity of the bilayer simplify the application of bioinformatics to membrane proteins to give topology models describing the number and orientation of TM segments. Classification of membrane proteins into families has relied on both sequence homologies and common topologies or folds. An early advance was the construction of hydrophobicity plots to identify potential TM α-helices based on occurrences of ~18 or more contiguous nonpolar amino acids. The problem then became how best to assign values for the hydrophobicity of the different residues, with numerous scales sometimes offering contrasting results. Application of various statistical tools brought large improvements in prediction methods, and methods were added to identify β-barrels. The rest of this chapter looks at these developments in some detail.

Structural predictions of integral membrane proteins have evolved from fairly simple computations using physical data to the increasingly sophisticated algorithms that are the tools of bioinformatics today. At the heart of the searches for structural similarities that are the basis for determining families and superfamilies is the ability to predict TM segments from the primary structure data.

Predicting TM Segments

With the vast majority of membrane proteins expected to be bundles of α-helical TM segments, the prediction of integral membrane protein structure has focused on recognizing those portions of polypeptides with favorable free energy changes for transfer from the aqueous solvent to the nonpolar milieu of the membrane. The transfer free energy change (ΔG_{tr}, see Chapter 1) for partitioning of each amino acid between ethanol and water was the basis of the first scale for hydrophobicity used to predict which sequences of amino acids were likely to

BOX 6.2 BIOINFORMATICS BASICS

Bioinformatics develops statistical methods and computational models for analysis of biological data to identify and predict the composition and structure of biological molecules. The rapid growth of sequence data and the huge numbers of structures solved by x-ray crystallography or NMR created the need for a central repository for structure information, the Protein Data Bank (PDB) (http://www.pdb.org/). In use since 1971, the PDB provides atomic coordinates, chemical and biochemical features, details of structure determination, and features derived from the structures of over 80,000 proteins, with more added yearly. Many of the PDB entries are redundant, because the same structure was determined by different groups or resulted when the sequence contained a point mutation, so the Astral compendium (http://astral.berkeley.edu) provides tools to provide nonredundant data based on unique protein domains. Swiss-Prot (http://expasy.org) is a widely used annotated protein sequence databank that contains over 500,000 entries for sequences assigned to proteins based on homologies or known identities. It provides a description of the proteins' features, such as disulfide bonds, binding sites, and secondary structure elements (including TM α-helices), along with references and links to other databases. It also points out conflicts between data provided by different references.

Many bioinformatics tools use the structural information of the PDB to display images that allow visualization of the structures, to simulate their dynamic interactions, to classify proteins into families and superfamilies, and to predict three-dimensional structures based on sequence homologies. Two programs in use for pairwise sequence alignments are BLAST, a Basic Local Alignment Search Tool (http://www.ncbi.nlm.nih.gov/BLAST), and FASTA (http://www.ebi.ac.uk/Tools/sss/fasta), which search a chosen database for sequences that align with the query sequence. A more powerful search can be carried out with the new program PSI-BLAST (at the same URL as BLAST), an iterative process that extends the search based on the profile first generated by BLAST.

Created to probe evolutionary relationships, SCOP, the Structural Classification of Proteins (http://scop.mrc-lmb.cam.ac.uk/scop), is a database that describes the three-dimensional structure of proteins. SCOP organizes proteins by family, superfamily, fold, and class, based on common structural domains (see Figure 6.2.1). A similar database, CATH, for Class, Architecture, Topology, and Homologous superfamily (www.cathdb.info), takes a slightly different approach to classify folds into 30 major protein architectures, such as α-bundle, β-barrel, $\alpha\beta$-propeller, and so on. There are over 2,000 superfamilies of proteins identified in CATH and SCOP. Other tools probe the interactions between proteins (see, for example, DIP, the Database of Interacting Proteins; http://dip.doe-mbi.ucla.edu) and between proteins and their ligands or their solvent. These tools were developed for soluble proteins; bioinformatics of membrane proteins relies more on specialized algorithms described later in the chapter.

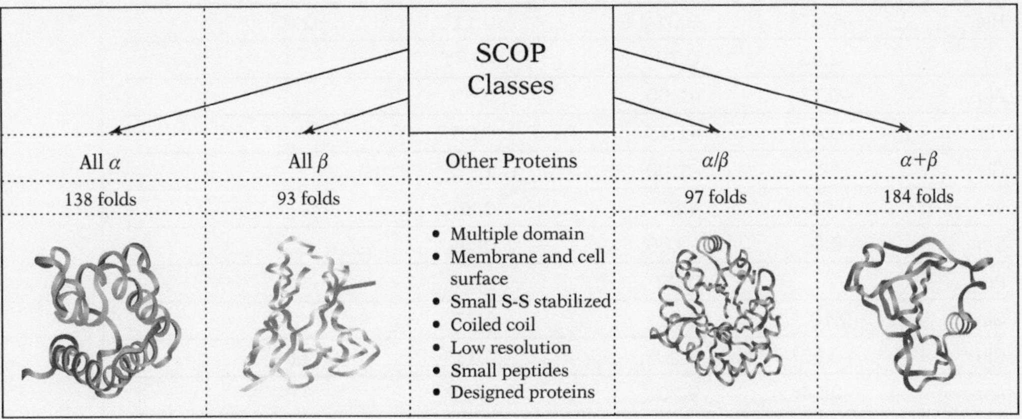

6.2.1 SCOP classifies proteins into four classes based on common folds. From Reddy B. V. B., and P. E. Bourne, P. E. Bourne and H. Weissig (eds.), *Structural Bioinformatics*, Wiley-Liss, 2003, pp. 239–248. © 2003, John Wiley and Sons.

form a TM helix. To emphasize the polarity of the side chain, each ΔG_{tr} was normalized to that of glycine to subtract the contributions of the amino group, the carboxyl group, and the α carbon:

$$\Delta G_{tr} = RT\ln(\chi^{w}/\chi^{o}) - RT\ln(\chi_{gly}{}^{w}/\chi_{gly}{}^{o}),$$

where χ is the mole fraction of the amino acid in the aqueous phase (w) or the organic phase (o) at equilibrium.

Numerous other hydrophobicity scales have been developed to model partitioning into the membrane interior based on partitioning into other organic solvents, vapor pressures of side chain analogs, or distributions of amino acids in the interior of soluble proteins. Two of the most frequently used are the Goldman–Engelman–Steitz (GES) and the Wimley–White (WW) scales (Table 6.3). The GES scale is based on calculated estimates for the ΔG_{tr} of amino acids in α-helical peptides arrived at by combining terms for the appropriate hydrophobic and hydrophilic contributions. To the data from partitioning experiments the hydrophobic component adds a function for the surface area of the amino acid side chain in an α-helix, computed based on geometric considerations. The hydrophilic component considers the pK_as for charged groups and the energy required to produce an uncharged species by protonation or deprotonation. In contrast, the WW scale is an empirical scale based on water/octanol partitioning of small random-coil peptides, so it provides an empirical measure that includes the contribution from the peptide backbone. It is augmented by varying the charged states of Asp, Glu, and His residues to allow for formation of salt bridges. For example, neutralizing Asp115 in bR improved detection of helix D as a TM segment (see below). These biophysical hydrophobicity scales can now be compared with more biological hydrophobicity scales: one that assesses the effect of amino acid substitutions on the unfolding of a membrane protein in liposomes and one that describes the tendency for each amino acid to be inserted in the membrane as part of a TM helix during biogenesis (see Chapter 7).

Hydrophobicity Plots

A simple yet powerful advance was the creation of hydrophobicity plots, also called hydropathy plots, to display

TABLE 6.3 HYDROPHOBICITY SCALES FOR PARTITIONING OF THE AMINO ACIDS INTO NONPOLAR PHASES, AS DETERMINED BY NOZAKI AND TANFORD; GOLDMAN, ENGELMAN, AND STEITZ (GES); AND WIMLEY AND WHITE (WW)[a]

Amino acid	Nozaki and Tanford	Goldman-Engelman-Steitz	Wimley-White	Wimley-White with charged side chains
Phe	−2.65	−3.70	−1.71	
Cys	–	−2.00	−0.02	
Ile	−2.96	−3.11	−1.12	
Met	−1.30	−3.39	−0.67	
Leu	−2.41	−2.80	−1.25	
Val	−1.68	−2.61	−0.46	
Trp	−3.01	−1.90	−2.09	
His	−0.67[b]	3.01	0.11[b]	2.33
Tyr	−2.53	0.70	−0.71	
Ala	−0.73	−1.60	0.50	
Thr	−0.44	−1.20	0.25	
Gly	–	−1.00	1.15	
Asn	0.01	4.80	0.85	
Ser	−0.04	−0.60	0.46	
Pro	–	0.20	0.14	
Gln	0.10	4.11	0.77	
Glu	−0.55[b]	8.20	0.11[b]	3.63
Asp	−0.54[b]	9.20	0.43[b]	3.64
Arg	−0.73	12.3		1.81
Lys	−1.50	8.80		2.80

[a] The ΔG_{tr} is given for the transfer of single amino acids from water to ethanol (Nozaki and Tanford), from water to an α-helix in the membrane (GES), and for the transfer from water to octanol of the amino acid in a peptide (WW). Note that Wimey and White compare the un-ionized and ionized states of Asp, Glu, and His. They also determined the ΔG_{tr} for residues linked in salt bridges, such as Arg^+… Asp^- (not shown).

[b] Values for the un-ionized state.

Sources: Jones, M. N., and D. Chapman, *Micelles, Monolayers, and Biomembranes*, Wiley-Liss, 1995, p. 16 (Tanford and GES); Jayasinghe, S., *et al.*, Energetics, stability, and prediction of transmembrane helices. *J Mol Biol.* 2001, 312:927–934.

the distribution of hydrophobic segments along the linear peptide chain and thus predict its two-dimensional topology. The original Kyte–Doolittle algorithm moves a sliding window along the sequence of amino acids and calculates the mean ΔG_{tr} at each position. To correspond to the length of bilayer-spanning helices, the size of the window should be around 18 amino acids. When plotted with the primary structure along the *x*-axis and hydrophobicity along the *y*-axis, with positive values (for hydrophobic residues) above the line, the peaks correspond to predicted TM segments. This method became widely used, with results depending heavily on the hydrophobicity scale used (Figure 6.16). Hydrophobicity plots give a satisfying picture of the predicted fold of many integral membrane proteins (see Box 6.3), yet they are quite imprecise at the end points of the TM helices. The typical bands of aromatic Trp and Tyr residues at the boundaries of nonpolar and interfacial regions (see Chapter 5) can help to define the ends of the TM region. Even then, hydrophobicity plots are unable to establish the orientation of the TM helices.

Orientation of Membrane Proteins

Establishing the correct orientation of an integral membrane protein has long been a challenge for membrane biochemists, and many biochemical techniques have been used to label the external portions of membrane proteins, with varying degrees of success. Chemical modification with impermeant reagents may be invalidated by findings that the reagent can permeate the membrane after all. Protease sensitivity is limited by the extensive protease resistance of many membrane proteins, once they are fully folded and inserted into the bilayer. Epitopes recognized by antibodies do identify external binding sites as long as the antibody-binding data give unambiguous results.

A successful, though laborious, genetic approach to determining orientation was developed in *E. coli* using gene fusions with reporter enzymes inserted into predicted loop regions of the membrane protein. Three reporter enzymes that each require a particular location in the cell are commonly used. Alkaline phosphatase (PhoA) is normally exported to the periplasm, where essential disulfide bonds are formed, so if it is expressed in the cytoplasm (on a cytoplasmic domain of a membrane protein) it gives little or no activity. β-lactamase (Bla) is also effective only in the periplasm, because its substrate is ampicillin, which inhibits peptidoglycan formation; thus only when this reporter is exported to the periplasm does it confer ampicillin resistance to the cell. The third reporter enzyme, β-galactosidase (LacZ), is active only in the cytoplasm because attempts to translocate it across the inner membrane leave it embedded in the membrane, which renders it inactive. When normalized to their expression levels, the enzyme activities obtained for different inserts can be correlated with the sites of localization and used to map the results as a topology model (Figure 6.17).

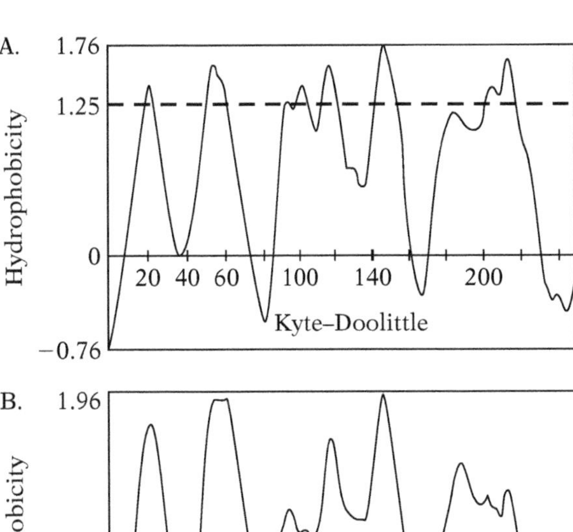

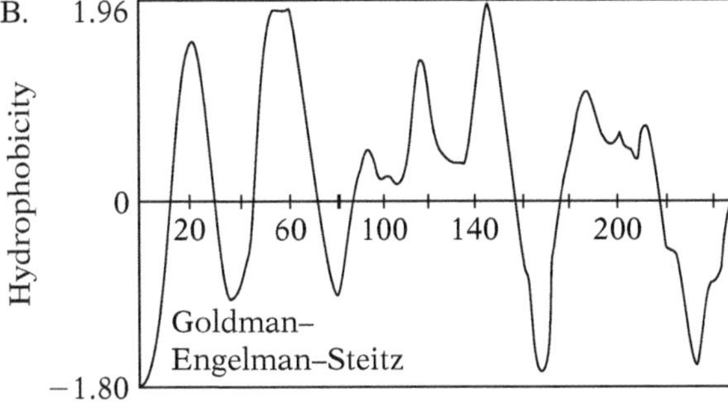

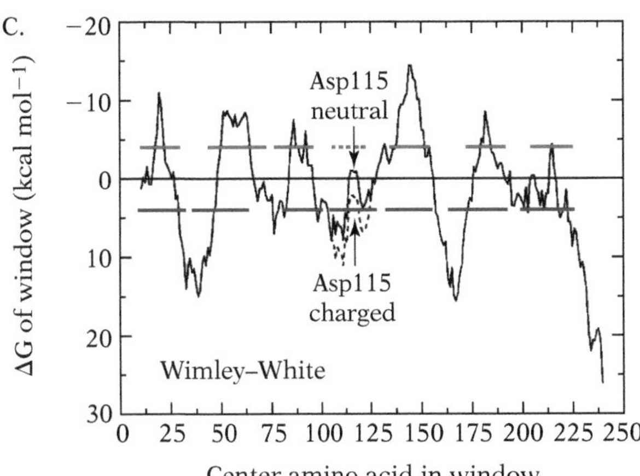

6.16 Hydrophobicity plots for bR using different hydrophobicity scales. (A) Kyte–Doolittle; (B) Goldman–Engelman–Steitz (GES); (C) Wimley–White (WW) with varying charge on Asp115. The agreement between the WW plot and the structure of bR is indicated by the lines for identified (red) and known (blue) TM helices in (C), and is illustrated in the frontispiece to this chapter. (A) and (B) redrawn from Gennis, R. B., *Biomembranes: Molecular Structure and Function*, Springer-Verlag, 1989, p. 125. © 1989 by Robert B. Gennis. Reprinted by permission of the author; (C) from Jayasinghe, S., *et al.*, *J Mol Biol.* 2001, 312:927–934. © 2001 by Elsevier. Reprinted with permission from Elsevier.

While the gene fusion approach works well for bacterial proteins, relatively few eukaryotic membrane proteins have been cloned into *E. coli* for expression and testing with these markers. Identification of glycosylation

> **BOX 6.3 MAKING AND TESTING HYDROPHOBICITY PLOTS**
>
> Membrane Protein Explorer (MPEx, http://blanco.biomol.uci.edu/mpex) is a tool for the simple generation of hydrophobicity plots using the WW scale and a window size set initially to 19. It allows the user to choose a protein sequence from the PDB database or to choose a protein of known topology. The site divides the latter proteins into three groups: "3D-helix" contains helical membrane proteins whose structures are known from x-ray crystallography or NMR; "1D-helix" contains proteins with evidence from gene fusions and other techniques that supports their predicted topology; and "3D-other" contains *β*-barrels and monotopic proteins. Alternative functions allow for analysis of *β*-barrels and for use of the translocon-based biological partitioning scale (see Chapter 7). It is easy and fun to generate and compare hydrophobicity plots for known integral membrane proteins. However, MPEx predicts TM segments in a large fraction (up to 43%) of soluble proteins chosen from the PDB because it picks up segments that are simply buried in the hydrophobic interior of a globular protein, as well as cleavable signal sequences on secreted proteins (see Chapter 7). As a research tool, it must be combined with other methods to improve the success of topology predictions (see text).
>
> *Source*: Snider, C., *et al.*, MPEx: A tool for exploring membrane proteins. *Protein Sci.* 2009, 18:2624–2628.

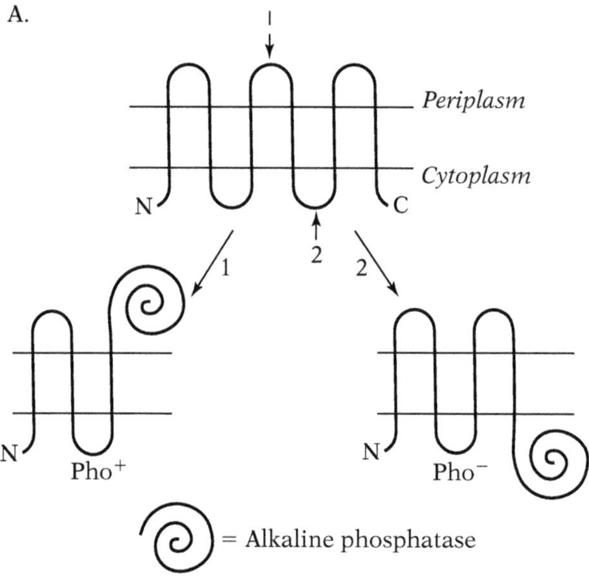

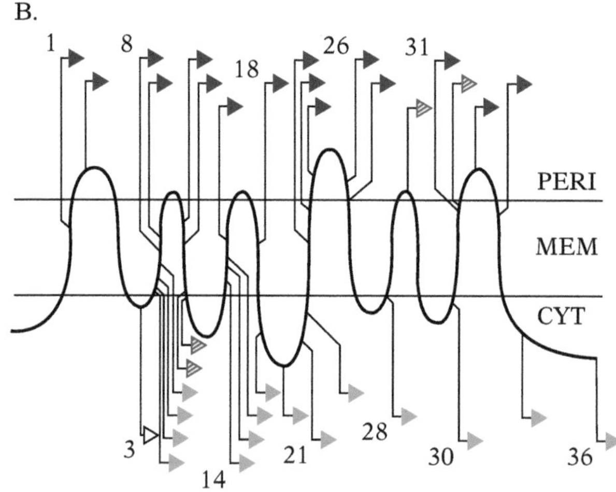

6.17 Analysis of membrane protein topology using alkaline phosphatase fusions. (A) Diagram that demonstrates the method: When the fusion puts alkaline phosphatase in the periplasm, as at site 1, it is active, and when it puts it in the cytoplasm, as at site 2, it is inactive. From Manoil, C., and J. Beckwith, *Trends Genet*. 1988, 4:223–226. © 1998 by Elsevier. Reprinted with permission from Elsevier. (B) Application of the method to the LacY protein. The sites of fusions are labeled by the level of alkaline phosphatase activity they produced: high (dark blue arrows), >190 units; medium (hatched green arrows), 46–99 units; and low (light blue arrows), <35 units. The ends of some TM segments are adjusted to maximize their hydrophobicity. Redrawn from Calamia J., and C. Manoil, *Proc Natl Acad Sci USA*. 1990, 87:4937–4941. © 1990 National Academy of Sciences, U.S.A.

sites has long been used to indicate which loops or domains are exported from the cytoplasm in eukaryotes, since the sugar chains are always extracellular. Another strategy to localize proteins in eukaryotes uses fusions to green fluorescent protein, which can reveal the fusion protein's compartmentalization if the resolution of the fluorescence microscope is high enough.

The Positive-Inside Rule

Using the topological data on *E. coli* inner membrane proteins plus data from the proteins in the photosynthetic RC, a statistical analysis of the amino acids of internal and external loops revealed a prevalence of basic residues in the cytoplasmic loops. In general the nontranslocated loops of membrane proteins have two to four times more Lys and Arg residues than found in translocated domains. This characteristic is the basis of the von Heijne positive-inside rule and makes it possible to predict the orientation of the TM helices from the amino acid sequences. Specifically, a loop with several basic residues within 30 residues from the end of a TM helix will be an internal loop, probably due to restricted insertion of helical hairpins carrying highly positively charged sequences.

The positive-inside rule was tested by engineering the charge distribution in the *E. coli* inner membrane

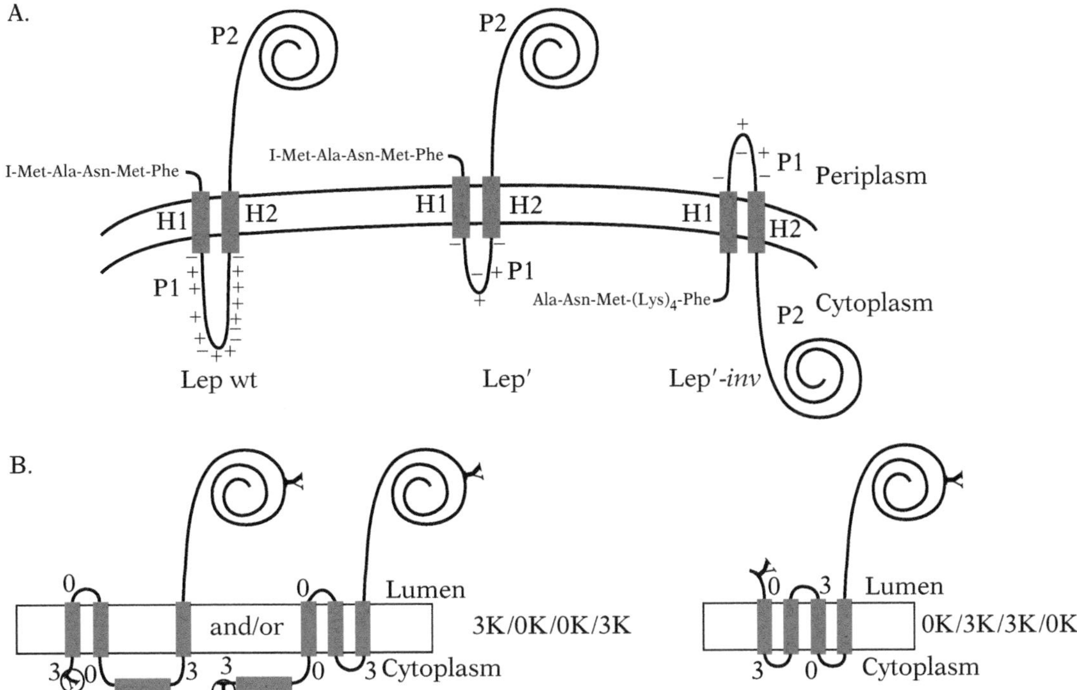

6.18 Demonstration of the positive-inside rule with leader peptidase (Lep). (A) Mutations that affect charge distribution in the loops can change the topology of Lep. At left is wild type Lep in its normal orientation with P1 in the cytoplasm. The middle construct is a mutant with a shortened P1, still in normal orientation. The right illustrates that the addition of four basic residues to the N-terminal end gives an inverted orientation. From von Heijne, G., *Annu Rev Biophys Biomol Struct*. 1994, 23:167–192. © 1994 by Annual Reviews. Reprinted with permission from the Annual Review of Biophysics and Biomolecular Structure, www.annualreviews.org. (B) A variety of Lep constructs have been engineered from duplicates of the *lep* gene to have four TM segments and a varying number of lysine residues in soluble portions, as indicated. Some of them, such as the 3K/0K/0K/3K construct, are "frustrated," meaning that regions that would normally span the membrane cannot do so. Redrawn from Gafvelin, G., *et al.*, *J Biol Chem*. 1997, 272:6119–6127. © 1997 by American Society for Biochemistry & Molecular Biology.

protein leader peptidase (Lep). Lep has two TM helices (H1 and H2) separated by a cytoplasmic domain (P1), with a short N terminus and a large C terminus (P2), both in the periplasm. The small P1 domain has nine positively charged residues. When basic residues were redistributed to put more in the N terminus than the P1 domain, the orientation of Lep flipped to the inverted topology (Figure 6.18A). More tests of engineered proteins with at least four TM segments show that altering the charge distribution can invert or "frustrate" the topology of the protein (Figure 6.18B). An example of the positive-inside rule at work during evolution is the pair of highly homologous proteins called RnfA and RnfE, membrane proteins involved in nitrogen fixation in *Rhodobacter capsulatus* that differ in orientation across the membrane and exhibit opposite distributions of Lys and Arg residues (Figure 6.19).

Analysis of the amino acid distributions in integral membrane proteins from over 100 genomes (see below) indicates that the positive-inside rule is universal. Combining the positive-inside rule with hydrophobicity plots significantly improves the power of topology predictions, as carried out by TopPred, an algorithm that calculates a standard hydrophobicity profile and ranks the possible topologies according to the positive-inside rule. Structural predictions also benefit from knowledge of which side of the membrane the N or C terminus is on. Another approach is the use of gene fusions to rapidly identify the localization of C termini using PhoA as a periplasmic marker and green fluorescent protein as a cytoplasmic marker, since it does not fold correctly in the *E. coli* periplasm. This approach enabled a global topology analysis of over 700 inner membrane proteins in *E. coli*.

Inverted Repeats

Repeated structural elements that are common in transporters have proven to be useful in predicting alternate conformations once one structure is available. Such structural repeats consisting of two or more groups of TM helices are recognized in the fold of many secondary transporters (see Chapter 11). They likely arise from gene duplication but are often difficult to detect in sequence analyses due to sequence divergence during evolution. However, they may be recognized in the topology once the three-dimensional architecture is known

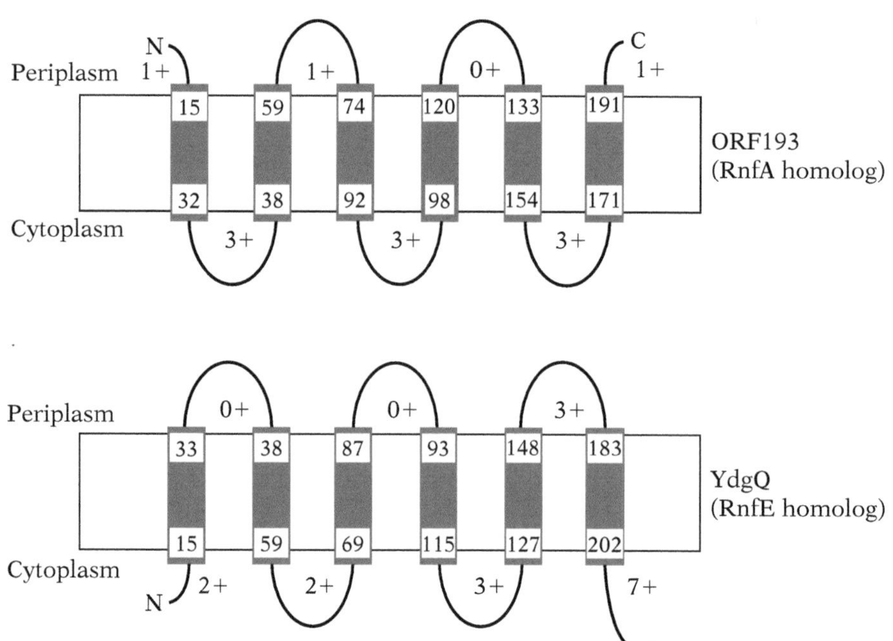

6.19 Topology of the homologous proteins RnfA and RnfE from *R. capsulatus*. The *E. coli* homologs, ORF193 (RnfA) and YdgQ (RnfE), were tested by PhoA fusion analysis. The opposite orientations of the two proteins are consistent with the positive-inside rule, indicated by the number of Lys + Arg residues in each loop and tail (with the number of + charges shown). Redrawn from von Heijne, G., *Q Rev Biophys*. 1999, 32:285–307. © Cambridge University Press 1999.

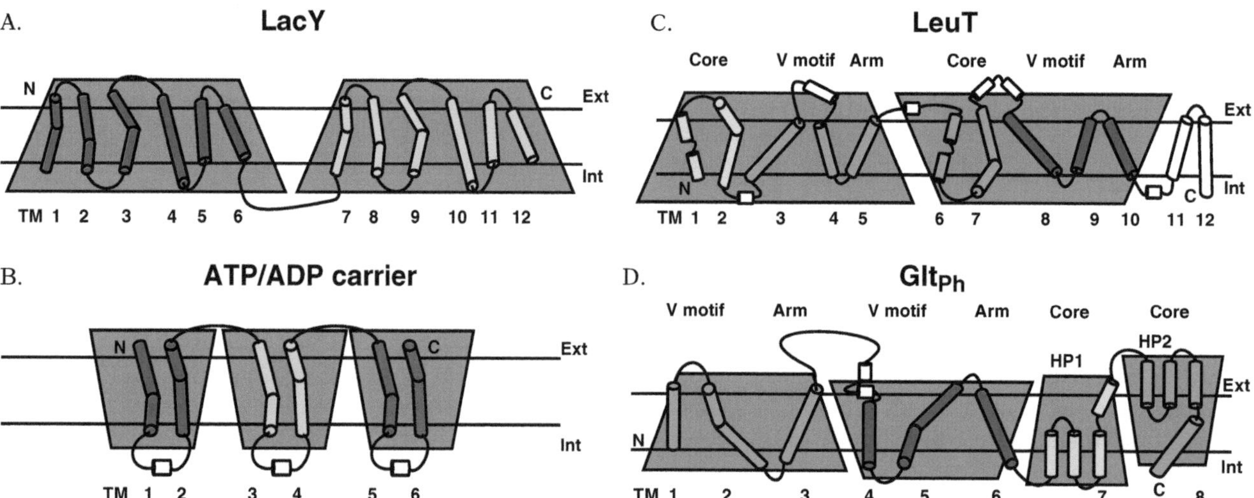

6.20 Structural repeats in transporters. The membrane topologies of many transporters show parallel and inverted structural repeats. Examples shown are for transporters described in detail in Chapter 11: LacY (A), the ATP–ADP carrier (B), LeuT (C), and GltPh (D). The structural repeats are highlighted by trapezoids (shaded gray) to emphasize their relative orientation, parallel (in (A) and (B)) or inverted (in (C) and (D)), with TM helices (rods) colored to show their relationships. TM helices that are not part of the symmetrical repeats are white, and nonhelical regions in the middle of core TMs are shown as lines. From Boudker, O., and Verdon, G., *Trends in Pharmacol Sci*. 2010, 31:418–426. © 2010, Elsevier.

(Figure 6.20). The repeats are frequently "inverted" because they have an antiparallel orientation related by a pseudo-symmetry axis perpendicular to the plane of the membrane. In other words, after a rotation of approximately 180° of one half of the repeat, the structures are closely aligned when superimposed.

The symmetry of the structural repeats observed in transporters is likely related to the alternating access model of transport described earlier. Often a transporter, such as LacY, energetically favors one conformation (open to the cytosol in the case of LacY) so much that it is difficult to crystallize in the other conformation. To

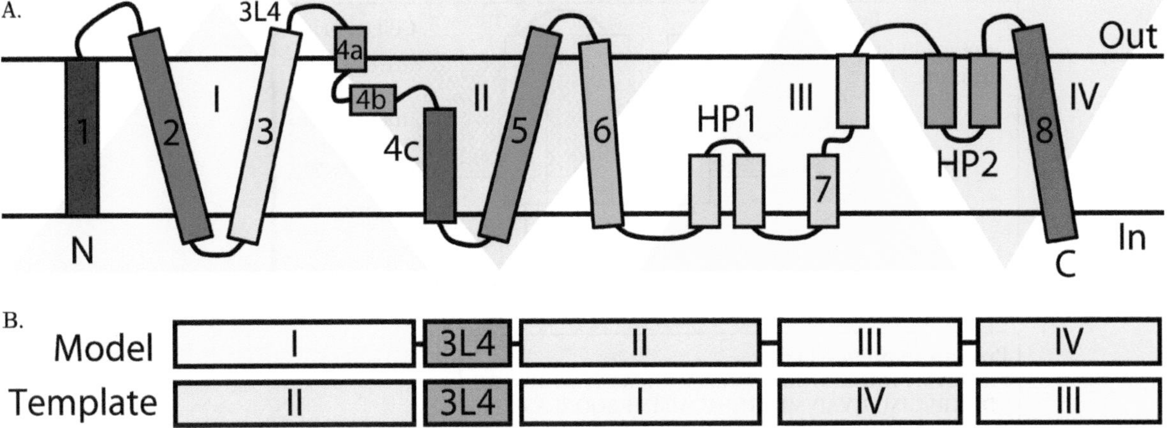

6.21 Detection of inverted repeats in GltPh. The structure of GltPh, first crystallized in an outward-facing conformation, shows the relationship between the topology (A) and the sequence alignment for inverted repeats (B). The four repeat segments, I–IV, are colored the same in (A) and (B) to illustrate domain swapping. For example, segment I (blue) uses the structure of segment II (green) as a template, which flips its orientation for the prediction. With the four segments flipped and the conformation of the loop between TM3 and TM4 (3L4, gray) based on the x-ray structure, the template provides the predicted inward-facing structure, later validated by the x-ray structure of GltPh in the inward-facing conformation. From Chrisman, T. J., *et al.*, *Proc Natl Acad Sci USA*, 2009, 106:20752–20757.

predict the other conformation (LacY open to the periplasm) the conformation of the structural repeats are swapped *in silico*, generating a model of the alternate transport state (Figure 6.21). This approach has been validated with examples in which the crystal structure of the second state was then obtained, such as GltPh, a bacterial sodium/aspartate symporter related to glutamate transporters in the brain (see Chapter 11).

Genomic Analysis of Membrane Proteins

Identification of membrane proteins in genomes requires fast, accurate methods designed to start with the primary structure of a protein derived from a genetic sequence and predict its topology from the locations of TM segments (Figure 6.22). A number of algorithms carry out sophisticated statistical analyses (Table 6.4). Some, for example TMHMM and HMMTOP, use hidden Markov models, while MEMSAT uses dynamic programming to optimally "thread" a polypeptide chain through a set of topology models, and PHD builds a neural network for predicting secondary structure from multiple sequence alignments (see Box 6.4). As these methods were applied, a number of difficulties became evident and newer predictors were developed.

One challenge is the need to distinguish between signal peptides and TM segments, since both contain stretches of hydrophobic amino acids. Signal peptides occur at the N terminus of many secreted and membrane proteins and are usually removed by membrane peptidases (see Chapter 7). Independent programs designed to determine the presence of a signal peptide, such as SignalP (an artificial neural network) and SignalP-HMM, can be run prior to performing the topology prediction. To avoid this extra step, combination programs such as Phobius (using HMM) were developed to predict both TM segments and signal peptides.

Since most predictors define a TM segment as a stretch of 15–25 amino acids, another kind of problem emerges from the irregularities of helical bundle proteins in actual membrane proteins. Such irregularities include wide variations in lengths of TM helices, kinks in helices, and reentrant loops that vary in secondary structure. OCTOPUS is a combination of artificial neural networks and HMM models that attempts to identify reentrant regions and interface helices along with TM segments. SPOCTOPUS adds the prediction of signal peptides.

Another difficulty arises when, after testing the algorithms with known data sets, they are less accurate with other proteins. It is a common experience to apply several algorithms to a protein of interest and get varying results. In a comparison with a test set of 60 bacterial integral membrane proteins with known topology, the highest fraction (around three-quarters) was correctly predicted with two algorithms using hidden Markov model, TMHMM, and HMMTOP, while MEMSAT and TopPred had less success and the lowest fraction (around one-half) were correctly predicted with the neural network predictor, PHD. The best results were achieved by carrying out analyses by all five and searching for consensus: a very high rate of correct predictions occurs when all five or four of the five agree. Therefore algorithms were developed to compare the results of different methods: CONPRED compares the results of nine, while TOPCONS compares the results of five algorithms.

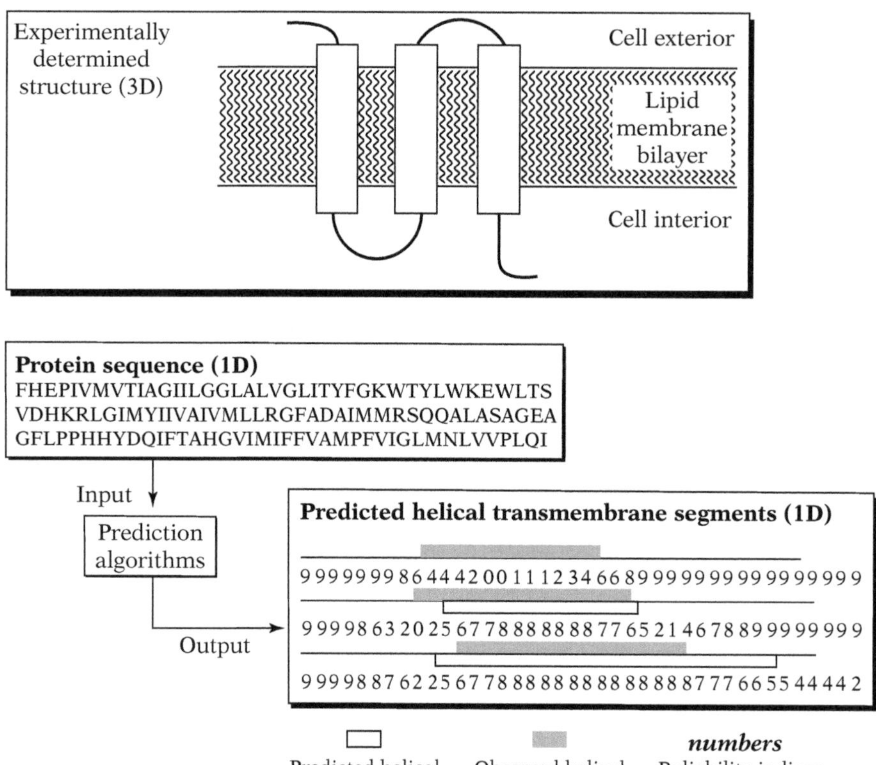

6.22 Strategy for predicting topology from protein sequence. Algorithms such as TMHMM, PHD, and MEMSAT all derive topology information from the primary structure of proteins. The example shows a partial sequence of cytochrome *o* ubiquinol oxidase subunit 1 (cyob), drawn to show the three TM helices determined experimentally (top) and the analysis by PHDhtm (bottom). The bars in the last panel allow comparison between predicted (white) and observed (shaded) TM segments, and the numbers below the line give the reliability of the prediction for each residue on a scale of 0 to 9. The low reliability values for the top segment illustrate how TM segments can be underpredicted. Redrawn from Rost, B., *et al.*, *Protein Sci.* 1995, 4:521–533. © 1995 The Protein Society.

As expected, these methods give the highest accuracy in comparisons with the individual algorithms.

Since the five topology predictors incorporated into TOPCONS require a PSI-BLAST search against a sequence database to incorporate evolutionary information, the process is relatively slow: running a database of 101 proteins took 4483 sec, compared to 2–26 sec for the individual methods. In order to speed the process to scan whole genomes, a combination of topology predictors that do not involve the PSI-BLAST search was constituted as TOPCONS – single. This program took only 64 sec for the database of 101 proteins and could analyze the complete human genome (~21,000 sequences) in about one hour.

The strengths and weaknesses of the methods in Table 6.4 are revealed when they are compared using different data sets. For example, TOPCONS, SCAMPI, and MEMSAT3 perform the best (~80% correct predictions), testing sets of experimentally verified topologies, and they do even better when the location of the C terminus is known. On the other hand, Phobius and related programs, SPOCTOPUS, HMMTOP, and TMHMM, more successfully predict identified extracellular glycosylation sites in over 700 proteins. Polyphobius is the best predictor of GPCRs in a set of 1300 proteins suspected to be GPCRs. Finally Phobius, PolyPhobius, Philius, and SCAMPI are most successful at distinguishing between membrane and nonmembrane proteins when tested with sets of cytosolic and secreted proteins. Even so, when applied to genomes they perform far worse than is typically reported (99%) for small test sets. A recommended strategy is to filter nonmembrane proteins with PolyPhobius and then predict the topology of the remaining proteins with TOPCONS.

A different approach is the application of ROSETTA to membrane proteins. ROSETTA is a protein-folding algorithm that predicts three-dimensional structures from *ab initio* calculations. To adapt ROSETTA to membrane proteins requires embedding the protein chain into a model membrane in order to maximize the

TABLE 6.4 MEMBRANE TOPOLOGY PREDICTORS IN CURRENT USE

Method	Release year	Algorithm	URL	MSA	SP	Constrained
HMM-TM	2006	HMM	http://biophysics.biol.uoa.gr/HMM-TM	–	–	+
HMMTOP	2001	HMM	www.enzim.hu/hmmtop	*	–	+
MEMSAT1	1994	ANN + grammar	http://single.topcons.net	–	–	–
MEMSAT3	2007	ANN + grammar	http://bioinf.cs.ucl.ac.uk/psipred	+	–	–
MEMSAT-SVM	2009	SVM + model	http://bioinf.cs.ucl.ac.uk/psipred	+	+	–
OCTOPUS	2008	ANN + HMM	http://octopus.cbr.su.se	+	–	+
Philius	2008	DBN	www.yeast-ctermrc.org/philius	–	+	–
Phobius	2004	HMM	http://phobius.sbc.su.se	–	+	+
PolyPhobius	2005	HMM	http://phobius.sbc.su.se/poly.html	+	+	–
PRO	2004	HMM	http://topcons.net	+	–	+
PRODIV	2004	HMM	http://topcons.net	+	–	+
SCAMPI	2008	Hydrophobicity + model	http://topcons.net	+	–	+
SCAMPI – single	2008	Hydrophobicity + model	http://scampi.cbr.su.se	–	–	+
SPOCTOPUS	2008	ANN + HMM	http://octopus.cbr.su.se	+	+	–
TMHMM	2001	HMM	www.cbs.dtu.dk/services/TMHMM	–	–	–
TOPCONS	2009	HMM – consensus	http://topcons.net	+	–	+
TOPCONS – single	2011	HMM – consensus	http://single.topcons.net	–	–	–
TopPred2	1994	Hydrophobicity profiles	http://mobyle.pasteur.fr	–	–	–

MSA: the predictor uses multiple sequence alignments as input; *: the original version of HMMTOP is capable of utilizing homologous sequences to improve predictions, however we benchmark the single-sequence version; SP: the predictor also predicts signal peptides; Constrained: the predictor allows constrained predictions, i.e., it allows using (experimentally derived) topology information as input.
Source: Tsirigos, K. D., *et al.*, A guideline to proteome-wide α-helical membrane protein topology predictions. *Proteomics*. 2012, 12:2282–2294. © 2011 by Annual Reviews. Reprinted with permission from the Annual Review of Physical Chemistry www.annualreviews.org.

exposure of hydrophobic residues to the hydrophobic region of the membrane and minimize their exposure to the hydrophilic environment outside. Fragments of the protein are built up from libraries of helix pairs and evaluated as additional helices are added to a growing chain. ROSETTA has been found to work well for families of related structures, such as the GPCRs. While *ab initio* methods are computationally expensive, they have the advantage of mimicking aspects of how membrane proteins fold in cells.

Analysis of 26 genomes for membrane proteins produced a total of 637 families of polytopic TM domains, based on a combination of TM helices predicted using TMHMM and those annotated in Swiss-Prot. The domains were classified based on homologies with known families or characterized using sequence alignment, hydrophobicity plots with the GES scale, and identification of consensus sequences for TM segments (Figure 6.23). When the families are sorted by the number of TM helices, the number of families with domains of a given number of TM helices decreases as the number of helices increases (Figure 6.24). Within this trend, the plot shows the highest occurrences of two- and four-TM helices, a slight excess of seven-TM helices, and a major spike at 12-TM helices due to the prevalence of this topology among transporters and channels.

Interestingly, the total number of membrane protein domains in the genome is roughly proportional to the number of open reading frames in all of the genomes except that of the nematode worm *C. elegans*, which has a huge number (754) of seven-TM chemoreceptors. In fact, the genome of *C. elegans* conforms to the general picture when its three large families of chemoreceptors are removed from the total. Clearly this organism that lacks sight and hearing has finely developed chemosensation to find its food!

Approximately half of the membrane proteins identified by genome analysis have an even number of TM helices with the N_{in}-C_{in} topology. The other three combinations of N- and C-terminal locations are about equally represented in the other half. The high proportion of proteins with both N and C termini inside is attributed to the mechanism of biogenesis, because this topology results from insertion of helical hairpins (see Chapter 7).

BOX 6.4 STATISTICAL METHODS FOR TM PREDICTION

While most of the methods listed in Table 6.4 employ hydrophobicity analysis and the positive-inside rule to find TM domains, some are greatly enhanced by the computational power of the statistical methods they employ. These methods give improved accuracy over the simple sliding-window approach that is limited to analysis of the linear sequence of the protein of interest. In particular, they can be based on multiple sequence alignment since there is now a sufficient databank of membrane proteins of known topology to be used in the search for other membrane proteins. Two successful approaches are neural networks and hidden Markov models.

Neural Networks

The program PHD uses neural networks to score "preferences" for TM helices generated from multiple sequence alignments. Neural networks compute relationships between units (amino acid residues) in layers, described as the input layer, one or more hidden layers, and the output layer, that together define the network architecture (Figure 6.4.1). For multiple protein sequences, the input is a sequence profile of a protein family, with each sequence position represented by the amino acid residue frequencies derived from multiple sequence alignments in the database. The hidden layer is set up to determine the state of each unit as it depends on the states of the units in the previous layer to which it is connected and the weights of the connections. The output layer has the results of the analysis.

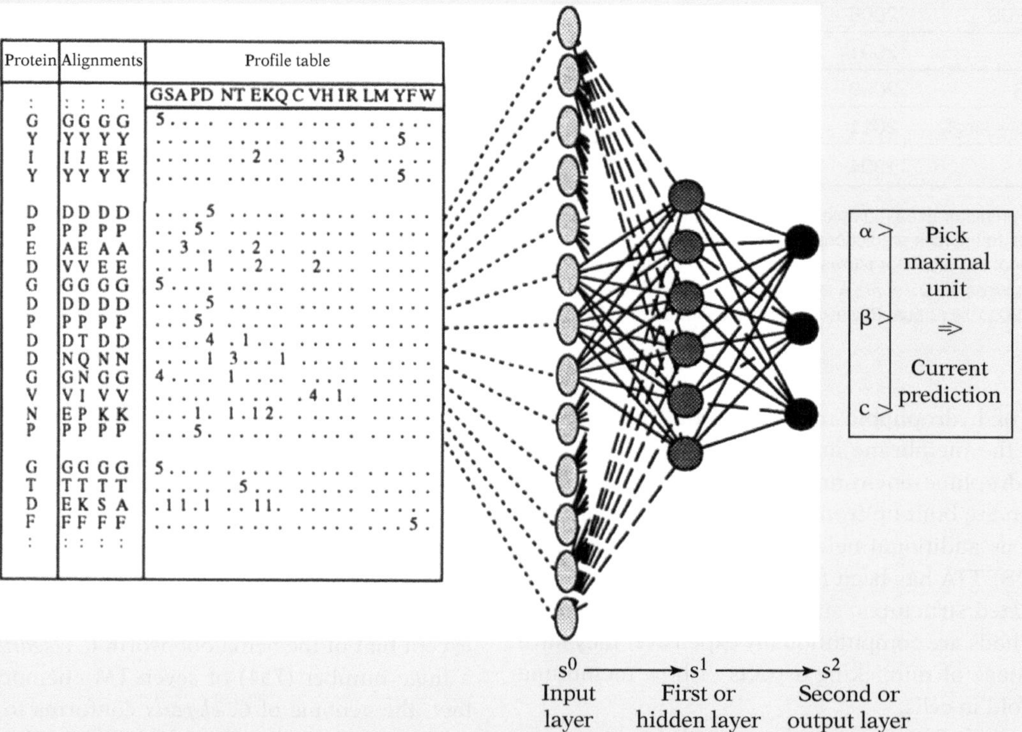

6.4.1 Neural networks compute relationships between residues in layers to make predictions based on aligned sequences. From Rost, B., and C. Sander, *Proc Natl Acad Sci USA*. 1993, 90:7558–7562. © 1993 National Academy of Sciences, U.S.A.

PHD analyzes protein sequences with two levels of neural network analysis, so that the output from the first level is the input for the second. For the first level, the local input is a weighted term including the frequency of occurrence of each amino acid at that position in the multiple alignments, plus the number of insertions and deletions in the alignment for that residue. The output of the first level indicates whether or not the segment is predicted to be a TM helix. This is the local input for the second level. The global input for both levels describes characteristics of the protein outside the window of 13 residues and includes its amino acid composition and length, plus the distance (number of residues) from the first residue in the window of 13 adjacent residues to the N terminus and the distance from the last residue in the window to the C terminus. The output is expressed in units that signify TM helix and not-TM helix.

BOX 6.4 *(continued)*

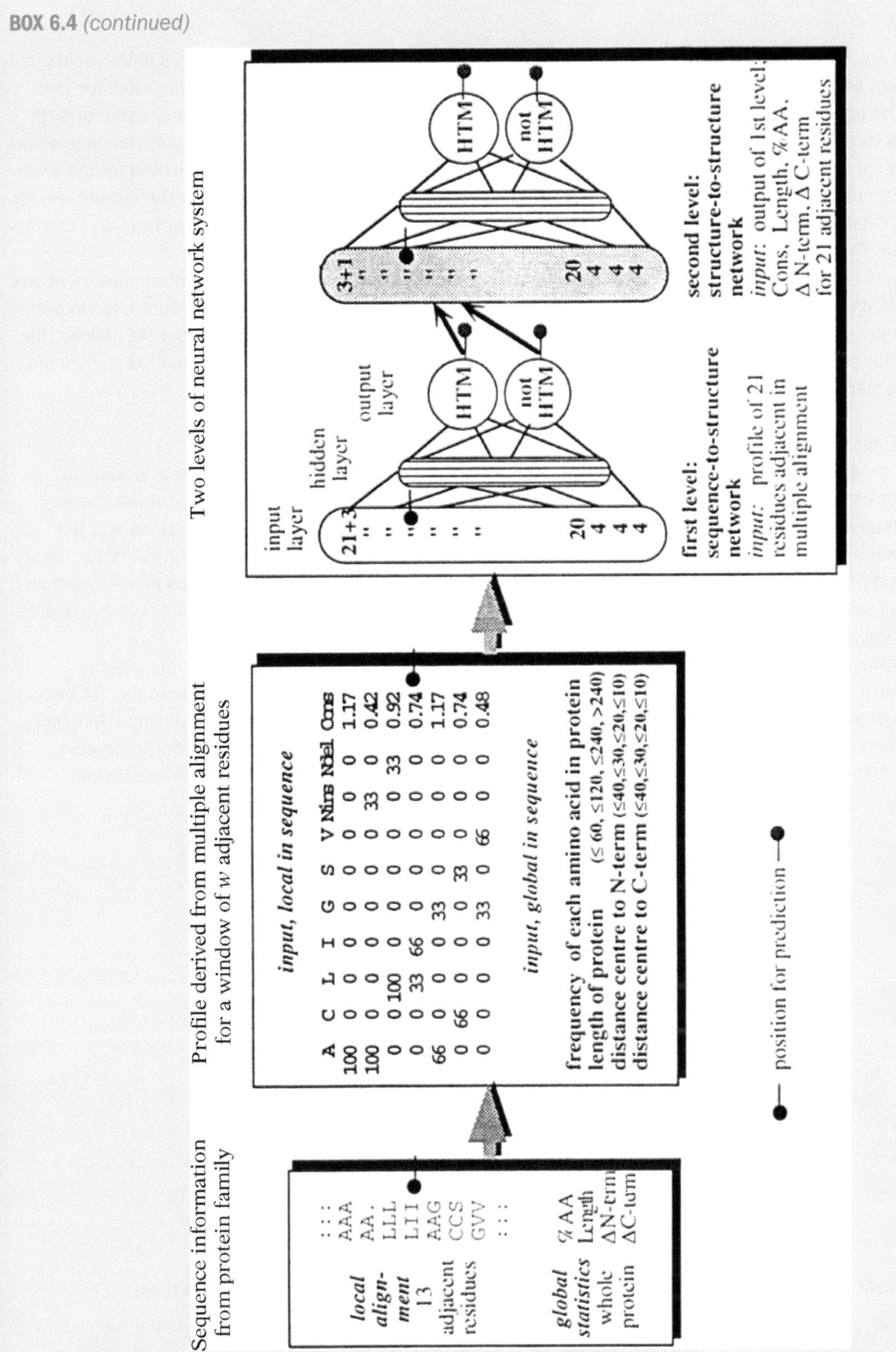

6.4.2 PHD uses two levels of neural networks to score structure preferences generated from multiple sequence alignments. From Rost, B., *et al.*, *Protein Sci.* 1995, 4:521–533. © 1995 The Protein Society.

(continued)

BOX 6.4 *(continued)*

As shown in Figure 6.4.2, the first level of PHD analyzes the protein sequence in windows of 13 amino acids (only seven are shown) and produces the output code for HTM (helix TM) or NotHTM, which becomes the input for the second level. The box on the left shows the data from sequence information from the protein family used for both local and global factors. The center box shows how at each position in the alignment the amino acid frequencies are compiled, the number of insertions and deletions are counted, and a conservation weight is computed for the local input. Global information gives the amino acid composition, length of the protein, and position of the current window from the N and C termini. All of this was codified and fed into the neural network input for the first level and then to the second level, shown in the box on the right.

When PHD was tested with an initial data set of 69 membrane proteins with experimentally determined locations of TM segments, its accuracy was >95%. Assessments of its accuracy when tested on a data set not used in setting up the analysis are considerably lower. One weakness of the system is the step it uses to filter out TM helices that are too long (>35) or too short (<17). If too long, they are split in the middle into two helices; if too short, they are deleted or elongated.

Hidden Markov Models

Hidden Markov models (HMMs) apply statistical profiles to describe a series of states connected by transition probabilities. For proteins, each state corresponds to the columns of a multiple sequence alignment, which are intermediate steps in the algorithm that the user does not see. A matrix describes the possible states and the transitions between the states, and an algorithm is employed for a "random walk" through the states, that is, one that derives each possibility from the previous state. For sequence alignment, the HMM generates profiles compiled of high scores (if sequence is highly conserved), low scores (if sequence is weakly conserved), and negative scores (if sequence is unconserved) and identifies sequences with the highest scores.

In HMMs for membrane protein topology, structural states are defined to describe portions of the protein. HMMTOP uses five structural states: inside loop (I), inside tail (i, the region of a loop that is close to the TM helix), TM helix (h), outside tail (o), and outside loop (O) (Figure 6.4.3). TMHMM further divides the residues in TM helices according to whether they are in the center of a helix (helix core) or on one end of a helix (helix cap). Thus seven states are considered in TMHMM: helix core, helix caps, short loop on inside, short and long loop on outside,

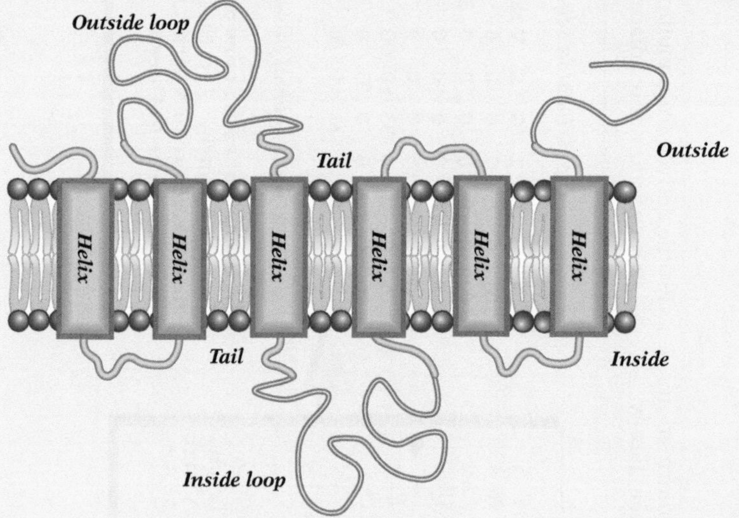

Amino acid seq:	MGDVCDTEFGILVA . . . SVALRPRKHGRWIV. . . FWVDNGTEQ . . . PEHMTKLHMM . . .
State seq:	o ooooooooohhhhh. . . hhhhi i i i i i i hhh. . . hh hoooo OO . . . OOOooooohhh . . .

Topology: *Tail* / *Out* — *Helix* — *Tail-Tail* / *Short loop* — *Helix* — *Tail - Loop - Tail* / *Long loop*

6.4.3 Structural states defined for HMMTOP, where thick lines represent tails while thin lines are loops. Redrawn from Tusnady, G. E., and I. Simon, *J Mol Biol*. 1998, 283:489–506. © 1998 by Elsevier. Reprinted with permission from Elsevier.

BOX 6.4 *(continued)*

6.4.4 The layout of the hidden Markov model. Redrawn from Krogh, A., *et al.*, *J Mol Biol.* 2001, 305:567–580.

and globular regions. Each state has a probability distribution over the 20 amino acids, based on the data set with known topologies.

The overall layout of the HMM is a function of the different structural states (Figure 6.4.4). Arrows show the possible connections between the different states, which are limited by the constraints of the protein structure as depicted in (A), with boxes corresponding to one or more states in the model. The connectivity of the different states varies. For the inside and outside loops and helix caps the connectivity is shown in (B), and for the helix core it is shown in (C). This model allows the helix core to be between 5 and 25 residues, which makes the entire length of TM helices (including caps) between 15 and 35 residues.

The program follows rules of "grammar" that state a helix must be followed by a loop, and inside and outside loops must alternate. Then it calculates probabilities for the sequences of these states. This is depicted in the architecture of HMMTOP (Figure 6.4.5), which shows states within the same transition

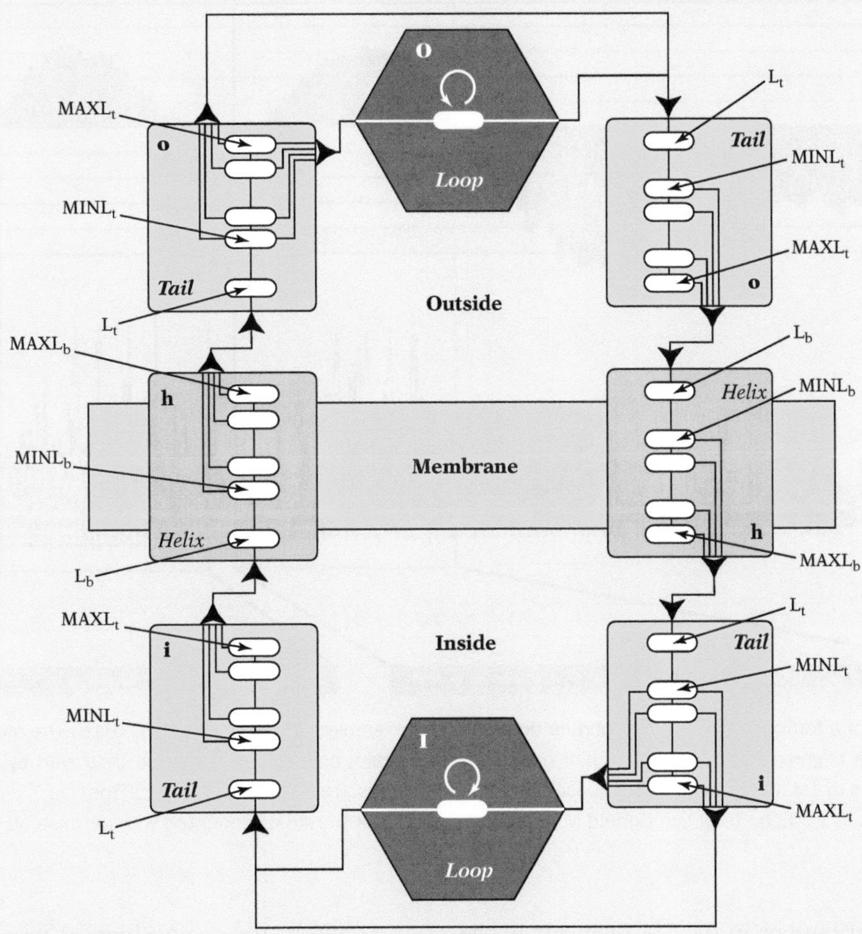

6.4.5 The architecture of HMMTOP. Redrawn from Tusnady, G. E., and I. Simon, *J Mol Biol.* 1998, 283:489–506. © 1998 by Elsevier. Reprinted with permission from Elsevier.

(continued)

BOX 6.4 *(continued)*

matrices (blue = helix states, yellow = tail states, red = loop states). The rectangular areas represent fixed-length states, which include the helices whose lengths need to be sufficient to cross the nonpolar domain of the bilayer and the tails that are defined to be the short segments of loops adjacent to ends of helices. The hexagonal areas represent the non-fixed-length states, which are the loops that are allowed to be any length. By considering fixed-length states and non-fixed-length states, the methods allow realistic length constraints on TM helices.

While the current programs using these methods (listed in Table 6.4) are very powerful, new computational approaches can be expected to make them even more useful in the near future.

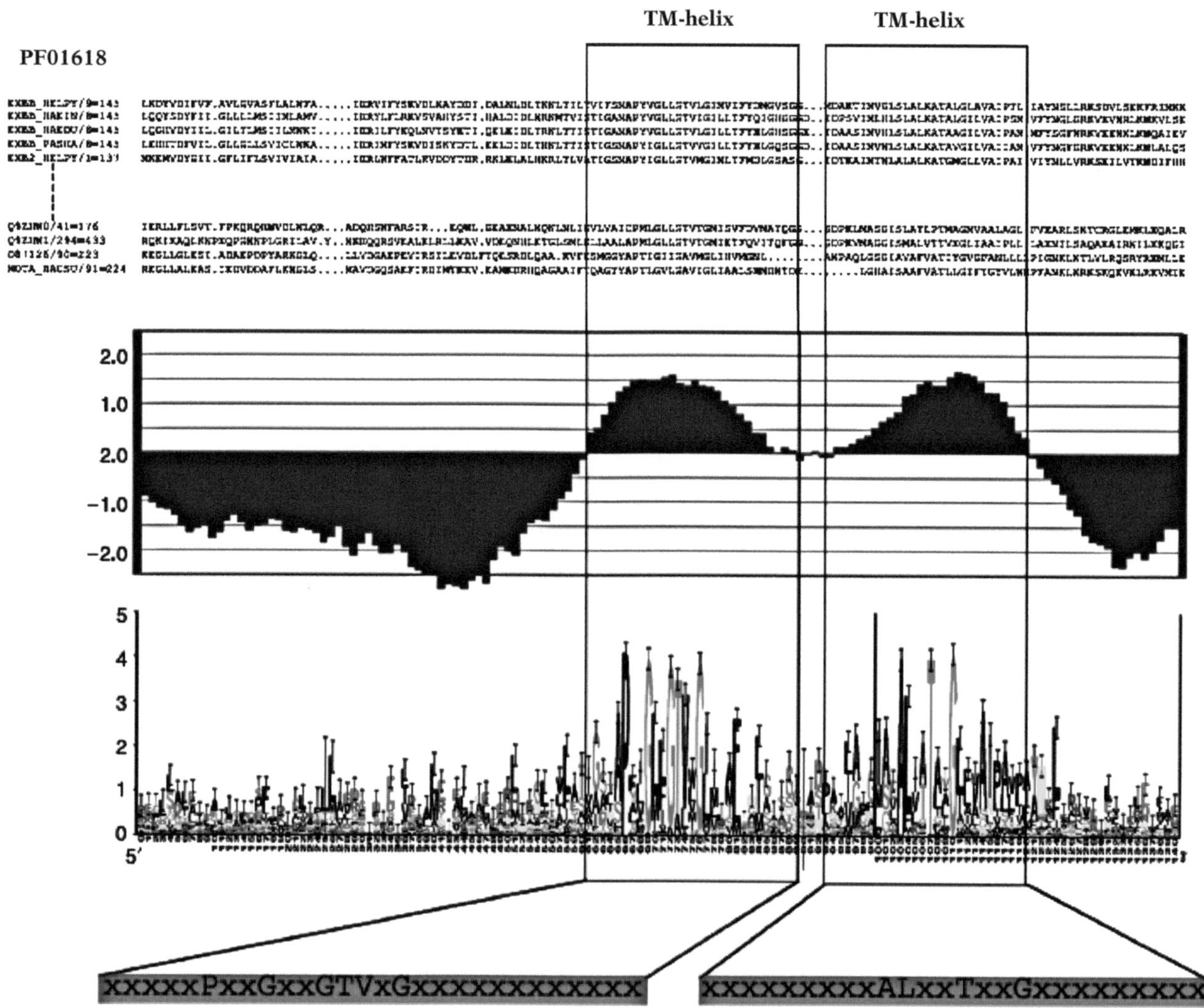

6.23 Classification of a family of polytopic membrane domains. The example shown is family PF01618. The steps involved are (top to bottom): sequence alignment, hydrophobicity plot based on GES scale, consensus sequence displayed by sequence logo, and consensus sequences of TM helices, where the nonconserved amino acids are represented by "x." From Liu Y, *et al.*, *Genome Biology*, 2002, 3:54.1–54.12. © 2002 by Yang Lui, Donald M. Engelman, and Mark Gerstein. Reprinted with permission from the authors.

Amino acid distributions in the TM segments of the putative membrane proteins in the 26 genomes show the expected high amounts of nonpolar amino acids, along with the polar residues Ser and Thr that participate in hydrogen bonding (see below; Figure 6.25A).

Similarly, the composition of nearly 50 000 TM segments annotated in the Swiss-Prot database (see Box 6.2) reveals that the six amino acids Leu, Ile, Val, Phe, Ala, and Gly make up two-thirds of TM residues. When the sequences are aligned to determine conserved residues

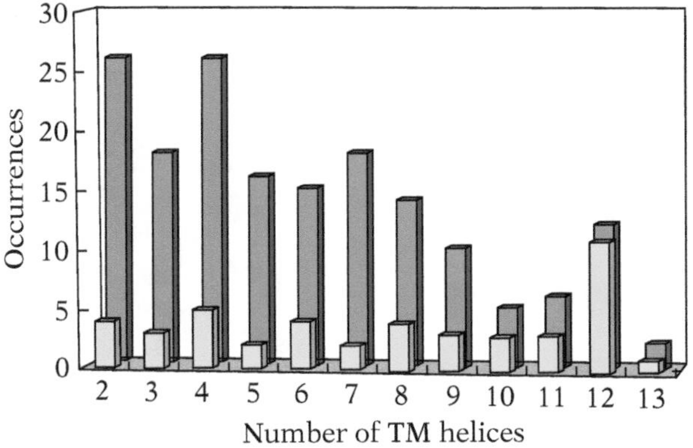

6.24 The number of TM helices in families of polytopic membrane domains. The number of families of polytopic membrane domains is plotted on the *y*-axis as a function of the number of TM helices, plotted on the *x*-axis. The green bars are the numbers from all studied families from the Pfam-A database. The yellow bars are the numbers from families from the Pfam-A database that are annotated as transporters and channels. Redrawn from Liu, Y., *et al.*, *Genome Biology*, 2002, 3:54.1–54.12. © 2002 by Yang Lui, Donald M. Engelman, and Mark Gerstein. Reprinted with permission from the authors.

(see Figure 6.23), the nonpolar amino acids are not prevalent in conserved positions, presumably because they are quite interchangeable. Interestingly, the amino acids that are prevalent at highly conserved positions in the helices are Gly, Pro, and Tyr (Figure 6.25B). The proline residues form kinks in the helices, which are conserved even after mutation of the Pro residues, while tyrosine residues play a special role near the interfaces due to their electronic properties (see Chapter 4). The positions of glycine residues are often highly conserved in soluble proteins because they occur at positions where the proteins do not accommodate larger side chains. This is also the case in TM helices, where Gly is commonly observed where two helices are in close contact. The GxxxG motif, first observed in glycophorin dimers, places both Gly residues on the same side of the α-helix (see Figure 4.31).

Genomic analysis for pair motifs shows a very high presence of GxxxG and GxxxxxxG pairs, as well as similar motifs with Ala or Ser replacing Gly residues. The resulting location of small side chains is expected to be important for helix–helix interactions in a broad range of membrane proteins.

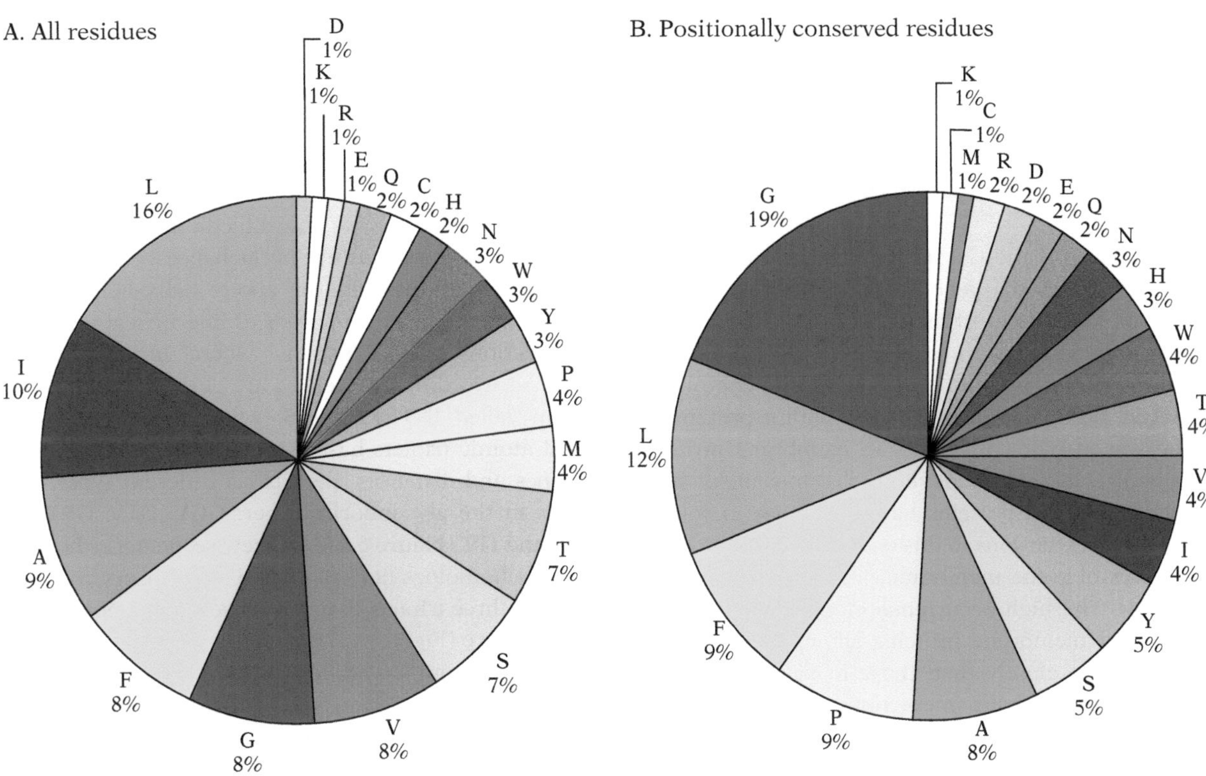

6.25 Amino acid distributions in TM helices. Amino acid distributions were determined for the TM segments from the 168 families from the Pfam-A database that have more than 20 members. **(A)** A pie diagram of the amino acid compositions of TM helices shows their high content of nonpolar residues, along with glycine, serine, and threonine. **(B)** Consensus sequences identified as shown in Figure 6.23 allow comparison of the amino acid residues in conserved positions of the TM helices. The diagram of the compositions of these positionally conserved residues shows that the prevalence of three amino acids (Gly, Pro, and Tyr) has increased significantly, indicating they are the ones whose positions are highly conserved. Redrawn from Liu, Y., *et al.*, *Genome Biology*, 2002, 3:54.1–54.12. © 2002 by Yang Lui, Donald M. Engelman, and Mark Gerstein. Reprinted with permission from the authors.

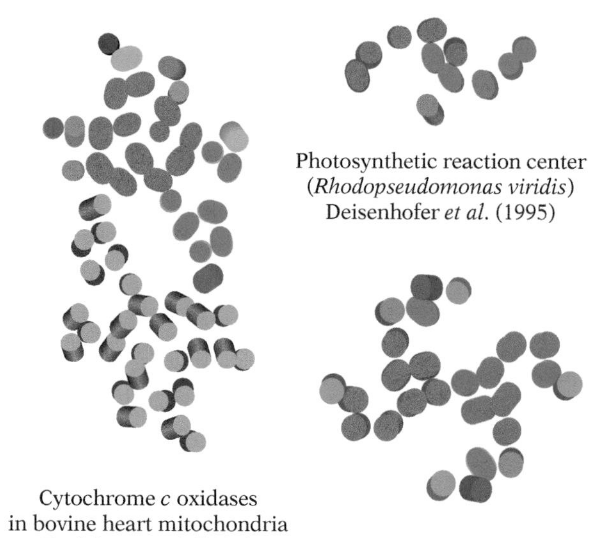

6.26 Helix interactions viewed from the membrane midplane. Positions of five-residue sections at the middle of TM helices are shown for photosynthetic reaction center, cytochrome c oxidase, and cytochrome bc_1 complex, as labeled. The subunits are colored differently. The gray-scale image of cytochrome c oxidase shows that the subunit composition cannot be inferred from the relationships of the helices. Redrawn from Liu, Y., *et al.*, *Proc Natl Acad Sci USA*. 2004, 101:3495–3497. © 2004 National Academy of Sciences, U.S.A.

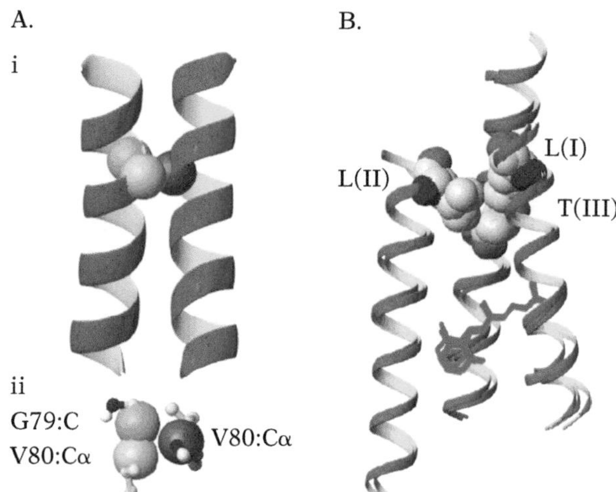

6.27 Three-body contacts, or triplets, in helix–helix interactions. (A) The TM helices (residues 73–91) of a glycophorin dimer with one of the triplets illustrated in space-filling representation. The atoms involved, C from Gly79 and Cα from Val80 on both chains, are shown in space-filling representation (i), and viewed from the top (ii). (Orange, C from G79, chain A; green, Cα from V80, chain A; blue, Cα from V80, chain B.) (B) A conserved triplet in the archael rhodopsins is shown by superposition of helices C, E, and F from bacteriorhodopsin, halorhodopsin, and sensory rhodopsin II. The residues in the triplet are two Leu and a Thr, corresponding to L97, L152, and T178 in bR, again in space-filling representation. The retinal is drawn in green. From Adamian, L., *et al.*, *J Mol Biol*. 2003, 327:251–272. © 2003 by Elsevier. Reprinted with permission from Elsevier.

HELIX–HELIX INTERACTIONS

The prevalence of the GxxxG motif in polytopic membrane proteins in the genome reflects both the close packing of TM helices observed in many helix-bundle proteins and the importance of protein–protein interactions in oligomeric membrane proteins and complexes. To analyze helical packing patterns within proteins, a comparison of pairs of helices in membrane proteins and pairs in soluble proteins shows that most helix–helix pairs in membrane proteins have homologs in soluble proteins. The exceptions to this correlation are the irregular helices of some membrane proteins (described in Chapter 4). The high occurrence of GxxxG and GxxxA in helices of membrane proteins allows the helices to approach more closely than those in soluble proteins, forming "knob-into-hole" interactions, as well as to interact over longer distances, that is, over the width of the bilayer. The information on helix pairs within proteins probably also applies to helix–helix interactions between subunits because the TM helices from different subunits often align as much as TM helices within one subunit. This can be seen with oligomeric proteins, where a cross-section through the middle of the membrane shows the extensive interaction of helices from different subunits (Figure 6.26): the gray-scale image of cytochrome *c* oxidase in the figure shows how difficult it is to assign subunits from the relationships of the helices!

Additional information about helix–helix interactions comes from the analysis of interhelical three-body interactions to find "triplets" where three atoms from at least two different helices are closely packed. A program called INTERFACE-3, which computes interhelical atomic triplets based on algorithms for geometric shapes and distances, detects six different types of triplets in the glycophorin dimer: GGV, GTV, GVV, ILL, IIT, and ITT (Figure 6.27A). Likewise, bacteriorhodopsin and its homologs halorhodopsin and sensory rhodopsin II have three triplets that are conserved in sequence and in structure (Figure 6.27B).

Furthermore, an additional 13 triplets are conserved structurally but not formed by identical residues, allowing conservative mutations that suggest that the triplet interaction has been conserved to maintain the orientation of the helices in the three highly homologous proteins. Comparison with a set of soluble α-helical proteins identified triplets that are unique to membrane proteins, such as AGF, AGG, GLL, and GFF. Such triplets are often found in regions of the closest contact between helices, and are therefore often correlated with GxxxG-type motifs.

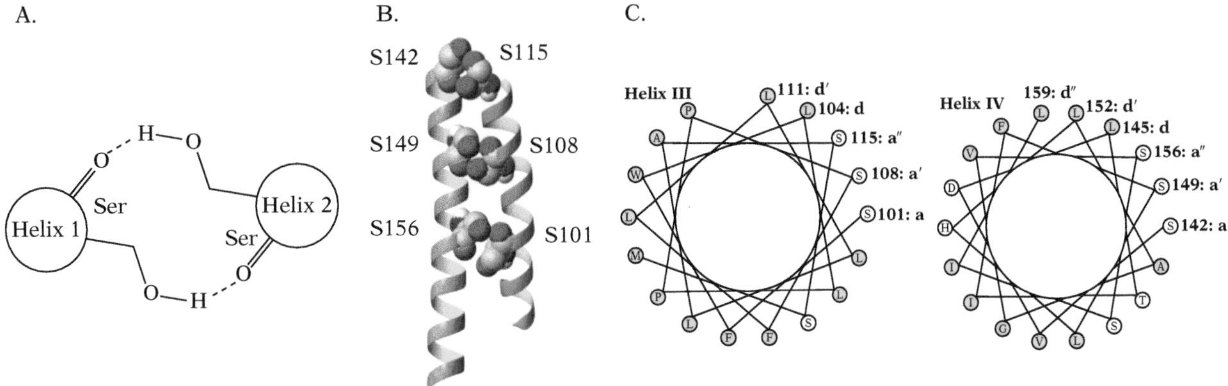

6.28 The serine zipper motif in helix–helix interactions. (A) Schematic representation of hydrogen bonding between two serine residues on two adjacent helices. (B) An example of a serine zipper in bovine cytochrome c oxidase. The two helices shown are helices III and IV, and the hydrogen-bonded serine residues are S101–S156, S108–S149, and S115–S142. (C) The pairs of serines can be predicted from helical wheels of helices III and IV from subunit 1 of bovine cytochrome c oxidase. A helical wheel designates residues *i* and *i* + 3 as a and d, and shows how they fall on the same side of the helix. Nonpolar residues are shaded pink. Three Ser–Ser pairs and two Leu–Leu pairs form a zipper at different a and d positions in this example. (A) and (B) redrawn from Adamian, L., and J. Liang, *Proteins*. 2002, 47:209–218. © 2002 Wiley-Liss, Inc., (C) from Adamian, L., *et al.*, *J Mol Biol*. 2003, 327:251–272. © 2003 by Elsevier. Reprinted with permission from Elsevier.

Close contact between TM helices is also stabilized by two types of hydrogen bonding. The hydrogen bonds observed in glycophorin dimers are relatively weak bonds because they use the Cα proton as the donor. Each of these hydrogen bonds contributes less than 1 kcal mol^{-1} to the stability of the dimer. The second type of hydrogen bond makes use of polar residues in TM segments; such bonds are detected in high-resolution structures of polytopic membrane proteins such as bacteriorhodopsin. Interactions between polar residues account for around 4% of all atomic interhelical contacts in membrane proteins, as well as in soluble proteins. In contrast to soluble proteins, where interacting ionized residues typically form salt bridges, the types of polar interactions are more varied in TM segments of membrane proteins, which also have H-bonds between ionizable and polar residues (the most common are D–Y and Y–R) and between polar nonionizable residues (the most common are Q–S and S–S).

How prevalent are these interhelical hydrogen bonds? In a data set based on the high-resolution structures of 13 membrane proteins, 134 unique TM helices form nearly 300 helical pairs, 53% of which are connected by at least one H-bond. In the 299 interhelical H-bonds identified, almost half involve Ser, Tyr, Thr, and His. Hydrogen bonds between two side chains and hydrogen bonds between a side chain and a backbone carbonyl oxygen occur at the same frequency. The majority of these helical pairs have one H-bond between two amino acids from two neighboring helices. However, two other types of H-bonding patterns emerge. One type is a "serine zipper" motif, named for its similarity to a leucine zipper (Figure 6.28). A search for homologous serine zippers by PSI-BLAST identified more than 100 sequences with highly conserved Ser residues in positions 7, 14, and 21 of one helix and 1, 8, and 15 of the other. A three-body analysis also found triplets of two serine residues with a leucine. The other type of H-bonding pattern is called a polar clamp because the side chain of an amino acid at a given position *i* that is capable of forming at least two hydrogen bonds (i.e., E, K, N, Q, R, S, T) is "clamped" by H-bonding to two other residues, one at position *i* + 1 or *i* + 4 and one on the other helix (Figure 6.29). The polar clamp shown in rhodopsin, involving residues T160 and W161 (both on helix IV), and N78 (on helix II), is highly conserved in the GPCR family.

When the numbers of intermolecular and intramolecular helix–helix interactions are estimated for membrane proteins in whole genomes, the totals are in the millions. Clearly an understanding of them will deepen as the number of high-resolution structures increases and will provide in turn a more complete insight for predicting membrane protein structure. Furthermore, helix–helix interactions are a critical step in the assembly of polytopic membrane proteins, as described in the next chapter. Since the nonpolar residues are quite nonspecific in their interactions, the often conserved hydrogen bonds between polar residues in TM segments must be crucial in helix alignments.

PROTEOMICS OF MEMBRANE PROTEINS

With thousands of membrane proteins predicted by genomic analysis, many questions are raised about their identities and interactions. Using TMHMM predictions combined with PhoA/GFP fusions to localize the C termini, a global topology analysis of *E. coli* membrane

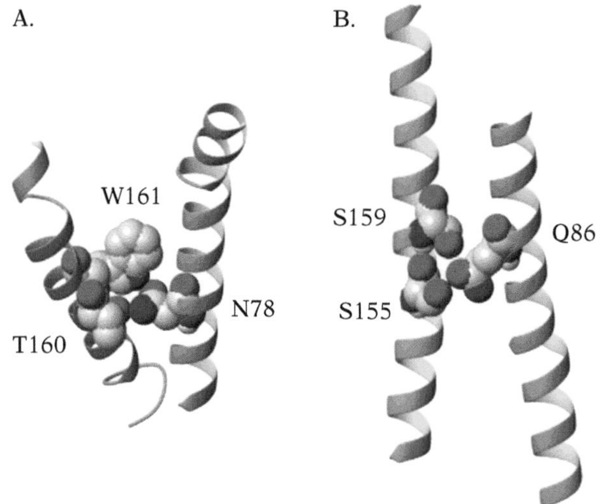

6.29 Polar clamp hydrogen bonding in helix–helix interactions. (A) Three residues in rhodopsin, W161, T160, and N78, form a clamp between helix IV and helix II. The side chain of N78 (on helix II) makes two hydrogen bonds: its O atom is hydrogen-bonded to the N atom from W161, and one of its amide hydrogen atoms is hydrogen bonded to the oxygen of T160. (B) A polar clamp in subunit I of cytochrome *c* oxidase from *Thermus thermophilus* involves residues S155 and S159 from helix α4 and Q86 from helix α2. From Adamian, L., and J. Liang, *Proteins*. 2002, 47:209–218. © 2002 Wiley-Liss, Inc.

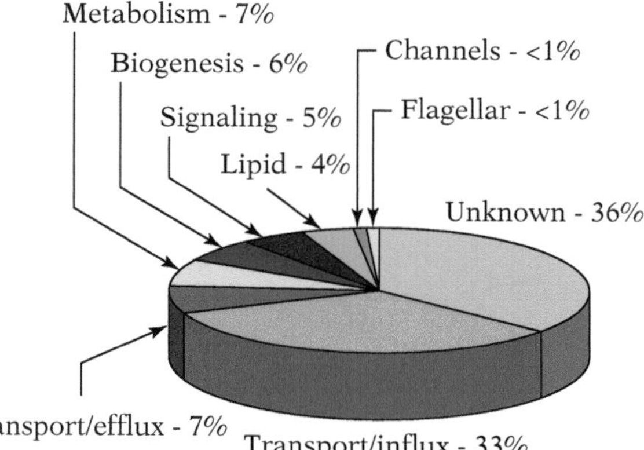

6.30 Functional categorization of the *E. coli* inner membrane proteome. The 737 proteins identified in the inner membrane (including 601 with localized termini) are assigned to different functional categories. Redrawn from Daley, D. O., *et al.*, *Science*. 2005, 308:1321–1323. © 2005, American Association for the Advancement of Science. Reprinted with permission from AAAS.

proteins detected 601 inner membrane proteins. When these proteins are sorted by their known or predicted functions, 40% of them are involved in transport, while nearly 20% are involved in metabolism, biogenesis, and signaling (Figure 6.30). Over one-third are "orphans" with unknown functions.

The functions of some orphan membrane proteins can be ascertained from their association with known proteins in complexes. (The yeast two-hybrid analysis, an important proteomic tool for identifying protein–protein interactions in complexes, does not work with membrane proteins because loss of function in the fusion proteins generated can result from loss of correct compartmentalization, topology, or orientation in the membrane.) Complexes of membrane proteins can be detected by two-dimensional gel electrophoresis using native gel electrophoresis for the first dimension, combined with mass spectrometry to identify the proteins. To carry out this procedure with Gram-negative bacteria such as *E. coli*, the inner and outer membrane fractions were first separated by gradient centrifugation and solubilized in a mild detergent, such as 0.5% *n*-dodecyl-β-D-maltoside. When native gel electrophoresis was carried out in Coomassie blue, complexes involving 44 inner membrane proteins and 12 outer membrane proteins were isolated. The results enabled roles to be assigned to six orphan proteins of unknown function, in addition to identifying a number of known proteins. A majority of the inner membrane proteins in complexes are members of respiratory chains (e.g., succinate dehydrogenase, F_1F_0-ATP synthase, and cytochrome bo_3 ubiquinol oxidase). Others are involved in biogenesis (including the SecYEG translocon; see Chapter 7) and some are transporters (including the intact PTS systems for mannose and galactitol, as well as ABC transporters for maltose and glutamine). The trimeric porins in the outer membrane were also identified, as the method also recognizes homo-oligomers.

PREDICTING β-BARRELS

The prediction methods described above for genomic analysis utilize algorithms that identify TM α-helices and are not useful for recognition of β-barrel membrane proteins. One way to illustrate this is to apply the hydropathy analysis of MPtopo to the β-barrel proteins in group 3 of the program (see Box 6.3). Now MPtopo also performs a β-barrel analysis, using one of several methods available. These prediction algorithms make use of the main characteristics of β-barrels (an even number of β-strands, antiparallel strands, periplasmic N and C termini, aliphatic residues on the surface except for the girdle of aromatic residues) along with their relatedness (amino acid composition and secondary structure element alignments). When applied to a sample outer membrane protein of known structure, MPtopo accurately identifies the β-strands (Figure 6.31). Other algorithms called PRED-TMBB, TMB-HMM, TMBETAPRED-RBF, and TMBpro use computational approaches discussed above, including statistical propensities, nearest neighbor methods, neural networks, and hidden Markov models. A new algorithm to predict

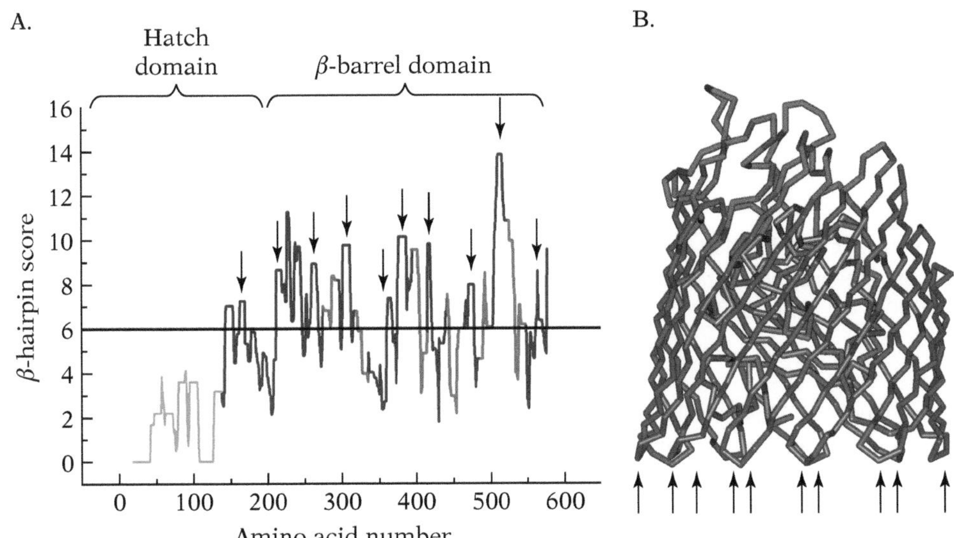

6.31 Prediction of the topology of BtuB, a β-barrel protein. The correlation between predicted (A) and actual (B) regions of the *E. coli* outer membrane protein BtuB is emphasized with color coding. The arrows show the peaks that correspond to the β-strands in the ribbon diagram of the x-ray structure. From Wimley, W. C., *Curr Opin Struct Biol*. 2003, 13:404–411. © 2003 by Elsevier. Reprinted with permission from Elsevier.

β-barrels based on OCTOPUS and MEMSAT-3 is called BOCTOPUS. BOCTOPUS uses a combination of support vector machines and HMM and correctly predicted 30 out of 36 β-barrel proteins. The "Z-coordinate predictor" ZPRED can then construct a three-dimensional model fairly accurately, except in the case of flattened (elliptical) β-barrels such as OMPLA.

Application of these methods to genomes of different Gram-negative bacteria indicates that 2% to 3% of the genomes encode β-barrel proteins, which corresponds to <10% of integral membrane proteins. These results signify how many β-barrel proteins remain to be characterized: one-third of the approximately 100 β-barrels they identify in the *E. coli* genome are unknown proteins. There are now 35 families of β-barrels, including 29 in Gram-negative bacteria. The others are in mitochondria, chloroplasts, and the acid-fast Gram-positive bacteria, such as mycobacteria. There is no significant sequence similarity between members of different families, in spite of frequent resemblances in the fold.

Equipped with increasingly sophisticated computational tools, bioinformatics of membrane proteins provides a wealth of information about families of membrane proteins and their characteristics, especially for the large class of helical bundle proteins. It provides patterns and frameworks that clarify much of what is known about membrane proteins, and by viewing data on hundreds of predicted membrane proteins with unknown structures and functions, it stimulates the desire to know more about the many important roles these proteins play. The expansive reach of genomics will make bioinformatics important for many years to come.

FOR FURTHER READING

Enzymes

Carman, G. M., R. A. Deems, and E. A. Dennis, Lipid signaling enzymes and surface dilution kinetics. *J Biol Chem*. 1995, 270:18711–18714.

DAGK

Van Horn, W. D., and Sanders, C. R., Prokaryotic diacylglycerol kinase and undecaprenol kinase. *Ann. Rev. Biophysics*. 2012, 41:81–101.

Van Horn, W. D., H.-J. Kim, C. D. Ellis, *et al.*, Solution nuclear magnetic resonance structure of membrane-integral diacylglycerol kinase. *Science*. 2009, 324:1726–1729.

Proteases

Beel, A. J., and Sanders, C. R., Substrate specificity of γ-secretase and other intramembrane proteases. *Cell Mol Life Sci*. 2008, 65:1311–1334.

Erez, E., D. Fass, E. Bibi, How intramembrane proteases bury hydrolytic reactions in the membrane. *Nature*. 2009: 459:371–378.

O'Brien, R. J., and P. C. Wong, Amyloid precursor protein processing and Alzheimer's disease. *Ann Rev Neurosci*. 2011, 34:185–204.

Straub, J. E., and D. Thirumalai, Toward a molecular theory of early and late events in monomer to amyloid fibril formation. *Ann Rev Phys Chem*. 2011, 62:437–463.

Cytochromes P450

Li, H., and T. L. Poulos, Crystallization of cytochromes P450 and substrate–enzyme interactions. *Curr Top Med Chem*. 2004, 4:1789–1802.

Transporters

Abramson, J., and Wright, E. M., Structure and function of Na^+-symporters with inverted repeats. *Curr Opin Struct Biol*. 2009, 19: 425–432.

Boudker, O., and G. Verdon, Structural perspectives on secondary active transporters. *Trends Pharm Sci*. 2010, 31:418–426.

Busch, W., and M. H.Saier, Jr., The transporter classification (TC) system. *Crit Rev Biochem Mol Biol*. 2002, 37:287–337.

Gruber, G., H. Wieczorek, W. R. Harvey, and V. Müller, Structure–function relationships of A-, F- and V-ATPases. *J Exp Biol*. 2001, 204:2597–2605.

Law, C. J., P. C. Maloney, and D. N. Wang, Ins and outs of major facilitator superfamily antiporters. *Ann Rev Microbiol*. 2008, 62:289–305.

Locher, K. P., Structure and mechanism of ATP-binding cassette transporters. *Phil Trans R Soc B*. 2009, 364:239–245. (See also http://nutrigene.4t.com/humanabc.htm)

Pebay-Peyroula, E., and G. Brandolin, Nucleotide exchange in mitochondria: insight at a molecular level. *Curr Opin Struct Biol*. 2004, 14: 420–425.

Tchieu, J. H., V. Norris, J. S. Edwards, and M. H. Saier, Jr., The complete phosphotransferase system in *Escherichia coli*. *J Mol Microbiol Biotechnol*. 2001, 3:329–346.

Ziegler, C., E. Bremer, and R. Krämer, The BCCT family of carriers: from physiology to crystal structure. *Molec Microbiol*. 2010, 78: 13–34.

Transporter Classification Database from the Saier Lab Bioinformatics Group: www.tcdb.org

The Ligand-Gated Ion Channel Database from European Bioinformatics: www.ebi.ac.uk/compneur-srv/LGICdb/LGICdb.php

Receptors

Nicotinic acetylcholine receptor

Chen, L., In pursuit of the high resolution structure of nicotinic acetylcholine receptors. *J Physiol*. 2010, 588:557–564.

Sine, S. M., and A. G. Engel, Recent advances in Cys-loop receptor structure and function. *Nature*. 2006, 440:448–455.

Sixma, T. K., and A. B. Smit, Acetylcholine binding protein (AchBP)... pentameric ligand-gated ion channels. *Annu Rev Biophys Biomol Struct*. 2003, 32:311–334.

Yakel, J. L., Gating of nicotinic ACh receptors: latest insights into ligand binding and function. *J Physiol*. 2010, 588:597–602.

GPCRs

Katritch, V., V. Cherezov, and R. C. Stevens, Structure-function of the G protein-coupled receptor superfamily. *Ann Rev Pharmacol Toxicol*. 2013, 53:531–556.

Lebon, G., T. Warne, C. G. Tate, Agonist-bound structures of G protein-coupled receptors. *Curr Opin Struct Biol*. 2012, 22:1–9.

Rosenbaum, D. M., S. G. F. Rasmussen, and B. K. Kobilka., The structure and function of G-protein-coupled receptors. *Nature*. 2009, 459:357–363.

Information system on G protein-coupled receptors: www.gpcr.org/7tm

Transmembrane proteins in the PDB (Protein Data Bank): http://pdbtm.enzim.hu

Predictions and Genomic Analysis

Daley, D. O., M. Rapp, E. Granseth, K. Melén, D. Drew, and G. von Heijne, Global topology analysis of the *Escherichia coli* inner membrane proteome. *Science*. 2005, 308:1321–1323.

Elofsson, A., and von Heijne, G., Membrane protein structure: prediction versus reality. *Ann Rev Biochem*. 2007, 76:125–140.

Forrest, L. R., R. Krämer, and C. Ziegler, The structural basis of secondary active transport mechanisms. *Biochim Biophys Acta*. 2011, 1807:167–188.

Hayat, S. and A. Elofsson, BOCTOPUS: improved topology prediction of transmembrane β barrel proteins. *Bioinformatics*. 2012, 28:516–522.

Kall, L., A. Krogh, and E. L. Sonnhammer, A combined transmembrane topology and signal peptide prediction method. *J Mol Biol*. 2004, 338:1027–1036.

Lehnert, U., Y. Xia, T. E. Royce, *et al.*, Computational analysis of membrane proteins: genomic occurrence, structure prediction and helix interactions. *Q Rev Biophys*. 2004, 37:121–146.

Tsirigos, K. D., A. Hennerdal, L. Kall, and A. Elofsson, A guideline to proteome-wide α-helical membrane protein topology prediction. *Proteomics*. 2012, 12: 2282–2294.

Tusnady, G. E., and I. Simon, Principles governing amino acid composition of integral membrane proteins: application to topology prediction. *J Mol Biol*. 1998, 283:489–506.

Yarov-Yarovoy, V., J. Schonbrun, and D. Baker, Multipass membrane protein structure prediction using Rosetta. *Proteins*. 2006, 62:1010–1025.

Protein–Protein Interactions

Adamian, L., R. Jackups, Jr., T. A. Binkowski, and J. Liang, Higher-order interhelical spatial interactions in membrane proteins. *J Mol Biol*. 2003, 327:251–272.

Stenberg, F., P. Chovanec, S. L. Maslen, *et al.*, Protein complexes of the *Escherichia coli* cell envelope. *J Biol Chem*. 2005, 280:34409–34419.

PROTEIN FOLDING AND BIOGENESIS

7

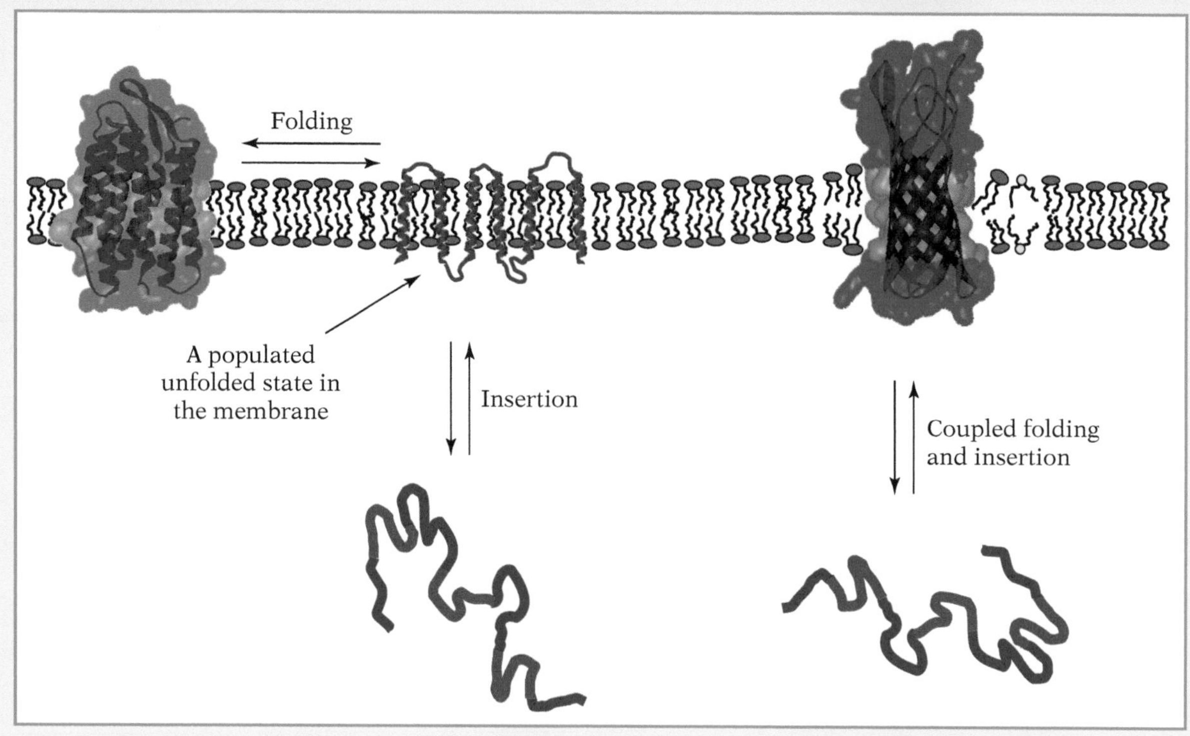

Folding and insertion of helical bundle and β-barrel membrane proteins utilize different mechanisms, according to results from *in vitro* folding studies. From Bowie, J. U., *Proc Natl Acad Sci USA*. 2003, 101:3995–3996. © 1972 National Academy of Sciences, U.S.A.

With an appreciation of the structural characteristics and functional diversity of membrane proteins, the question of their biogenesis arises. How does a nascent peptide, a chain of amino acids emerging from the ribosome, fold into a three-dimensional structure and insert into the membrane bilayer? Evidence suggests that folding and insertion are coupled processes in cells. However, given the nature and complexity of these processes, different approaches are taken to study folding and insertion *in vitro*.

Protein folding studies give information about the thermodynamic forces that drive folding and about its kinetic pathways, identifying transient but detectable intermediates. Recently these techniques have been applied to purified membrane proteins of both the α-helical and β-barrel classes. While the folding mechanisms determined by *in vitro* studies of membrane proteins may differ significantly from mechanisms of their biogenesis in the cell, these studies do provide insights into the necessary steps, as well as their stability and their lipid requirements. Thermodynamic analysis of the *in vitro* folding process gives insights into the evolution of the complex machinery used to assemble membrane proteins in cells. Finally, the *in vitro* studies provide valuable practical information on refolding techniques that are applicable to other membrane proteins that have been denatured during purification.

Insertion of nascent proteins into the membrane involves their translocation out of the cytoplasm by the same export machinery used to secrete proteins. There are strong similarities between prokaryotic and eukaryotic protein translocation systems in the recognition of the signal for export, structure of the main translocation apparatus, and mechanisms for insertion into the bilayer. Progress in defining and characterizing the components of these systems is leading to a detailed picture of the processes involved in export and integration of proteins. This chapter views how membrane proteins fold *in vitro* before turning to the complex cellular process of translocation and integration of nascent proteins into the membrane. It ends with a look at how misfolding of membrane proteins can lead to diseases in humans.

PROTEIN FOLDING

Before discussing folding studies of membrane proteins, it is useful to review some principles established with protein folding studies using purified small soluble proteins, with which both theory and methodology for *in vitro* folding studies were developed. Typically the protein of interest is unfolded with high concentrations of a denaturant such as urea or guanidinium hydrochloride, whose removal by dilution or dialysis triggers refolding. Temperature and extremes of pH can also be used to cause denaturation, although frequently it is not reversible in these cases. Of course, solution conditions including temperature, pH, and ionic strength are critical in all folding studies.

A standard assay follows the changes in intensity and wavelength of the intrinsic fluorescence of the protein: As it unfolds, the aromatic side chains move to more polar environments and their fluorescence emissions shift to longer wavelengths. A complementary method monitors the loss of secondary structure in the protein using circular dichroism, a method of spectroscopy that detects secondary structure in a peptide backbone by following the differential absorption of circularly polarized light. Other techniques include monitoring changes with ultraviolet and visible absorption, FTIR spectroscopy, FRET, Raman resonance spectroscopy, NMR, and even AFM. Both kinetic and equilibrium measurements are informative: Kinetic experiments reveal intermediates in the folding pathway, and equilibrium measurements give thermodynamic constants if the process is reversible.

An intact peptide chain can fold to a vast set of different configurational isomers, whose relative free energies are determined by the interactions within the fold and the surfaces they present to the environment. A few configurations have a relatively low free energy, corresponding to multiple minima in the energy landscape that depicts how the energy of the system (comprising all configurations of the protein) depends on the positions and orientations of all its atoms during the folding process. The native protein samples this conformational space by dynamic motions, with small, fast fluctuations between configurations.

This dynamical complexity is overlooked in the elementary form of the folding reaction, which views it as a two-state process: U ↔ N, where U is the unfolded protein and N is the native (folded) protein. For small, one-domain globular proteins, the unfolding process

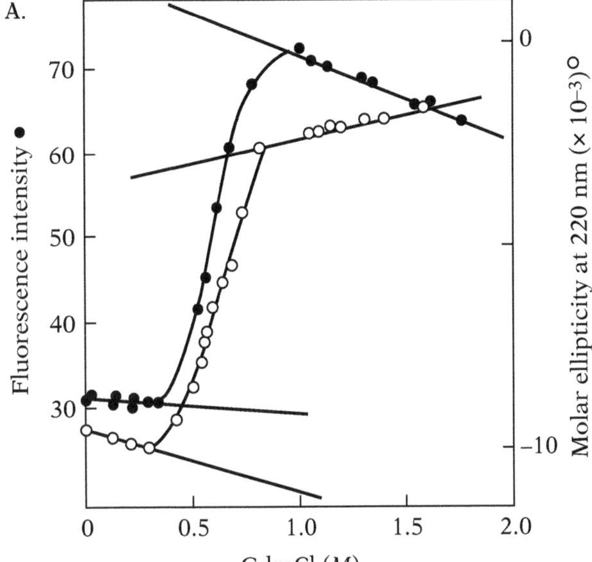

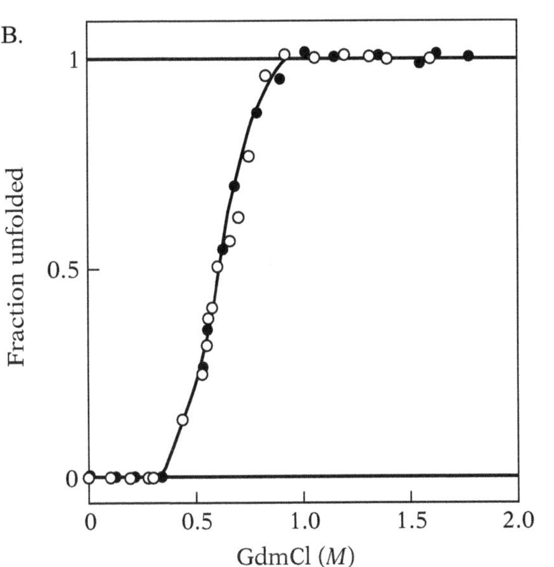

7.1 Folding studies of the soluble enzyme phosphoglycerate kinase. The data for unfolding phosphoglycerate kinase show a two-state mechanism. (A) Unfolding as a function of guanidinium chloride (GdmCl) concentration is monitored by fluorescence (●) and circular dichroism (○). Samples are allowed to go to equilibrium. The steepness of both curves indicates that unfolding is a highly cooperative process. (B) When the fraction unfolded is plotted as a function of the concentration of GdmCl, the two methods give the same unfolding curve, consistent with a two-state mechanism. Redrawn from Creighton, T. E., *Proteins: Structures and Molecular Properties*, 2nd ed., W. H. Freeman, 1992, p. 288. © 1992 by W. H. Freeman and Company. Used with permission.

appears to be a first-order transition, which typically exhibits cooperativity (Figure 7.1). When this reaction is fully reversible, it can go to equilibrium, at which point

$$\Delta G_{folding} = G_N - G_U = -RT\ln K_{eq},$$

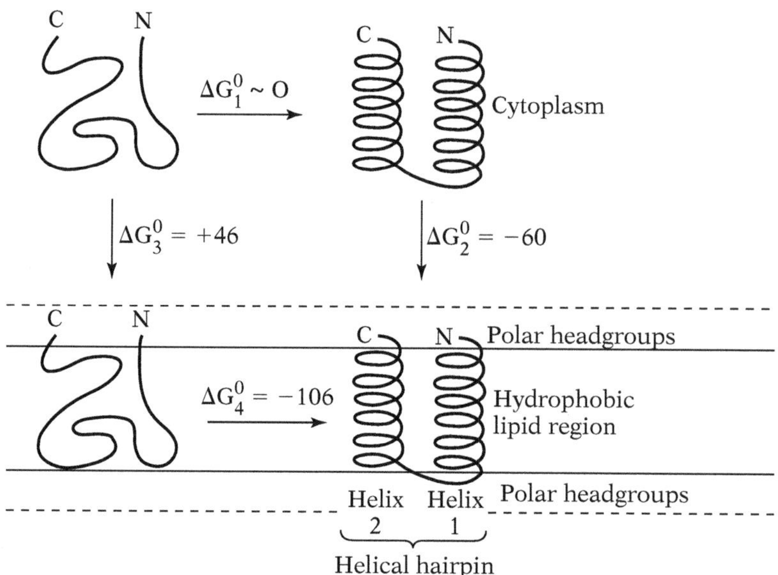

7.2 The helical hairpin hypothesis describes the insertion of a hairpin structure composed of two helices into the nonpolar interior of the bilayer. It is driven by the free energy arising from burying hydrophobic helical surfaces, estimated to be –60 kcal mol^{-1} for a pair of membrane-spanning helices. The alternate pathway of inserting a random coil prior to folding in the bilayer, with a free energy change of +46 kcal mol^{-1}, is so unfavorable that the insertion of an unfolded peptide effectively cannot occur. Redrawn from Engelman, D. M., and T. A. Steitz, *Cell*. 1981, 23: 411–422. © 1981 by Elsevier. Reprinted with permission from Elsevier.

where K_{eq} = [N]/[U]. K_{eq} is also equal to $(1 - \alpha)/\alpha$, with α defined as the average fraction of unfolding, so it is directly obtained from equilibrium folding or unfolding data.

In most unfolding studies, the environment of the protein is simply aqueous solvent; for membrane proteins, it is complicated by the presence of either detergent or lipids due to the need for an amphipathic system that provides a nonpolar domain to solubilize the native protein. However, once such a system has been defined, it can be exploited to give additional information, such as demonstrating the importance of bilayer elasticity by obtaining thermodynamic parameters in the presence of different lipids, as shown below.

Folding α-helical Membrane Proteins

Because integral membrane proteins require the special environment of the lipid bilayer (as described in Chapter 4), their *in vitro* folding process includes their insertion into a lipid bilayer, micelle, or other model membrane. An important early proposal for the mechanism of spontaneous membrane insertion was the helical hairpin hypothesis. It described how two hydrophobic α-helices connected by a hairpin turn can more favorably penetrate the nonpolar region of the bilayer as a pair of helices than as unfolded peptides (Figure 7.2). The next important development built on the recognition that TM helices of a polytopic membrane protein could independently insert into the membrane as a first step in assembling the helical bundle, the premise of the widely accepted two-stage model for folding membrane proteins (Figure 7.3). In stage I, the hydrophobic α-helical TM segments partition into the bilayer, and then in stage II, they assemble by packing together. For some proteins, there is a third stage that involves the binding of prosthetic groups, the folding of loops, the assembly of oligomers, or even the movement of other segments into the bilayer region of the protein, all of which can occur once the bundle of helices creates an environment that is both more organized and more polar than the bilayer itself.

Stage I of the two-stage model describes the insertion of each individual α-helix as driven by the hydrophobic effect and stabilized by hydrogen bonding along the backbone. The prediction that TM segments can insert independently is supported by the partitioning behavior of peptides that result from either gene splicing that cuts polytopic membrane proteins into separate TM pieces or synthesis of peptides that correspond to single TM domains of membrane proteins. Furthermore, the TM segments observed in known structures of membrane proteins fit quite well with those predicted based on hydrophobicity algorithms, supporting the idea that they each could interact with the lipid nonpolar domain before clustering together to decrease protein–lipid contacts.

For thermodynamic analysis, stage I can be further separated into three stages: partitioning, folding, and insertion (Figure 7.4). As described in Chapter 4, the hydrophobic side chains provide more than enough free energy for partitioning, even considering the changes in solvation of the protein (or peptide), the perturbation of the lipid by the protein, and the loss of degrees of freedom of the protein. Folding to α-helix is likely to be induced by partitioning, when the ΔG of partitioning the folded peptide is more negative than the ΔG of partitioning the unfolded peptide (see Box 7.1). Insertion of the helix to span the bilayer restricts both its rotational degrees of freedom and its space, in addition to ordering the nearby acyl chains. These entropic costs are also compensated by the hydrophobic effect.

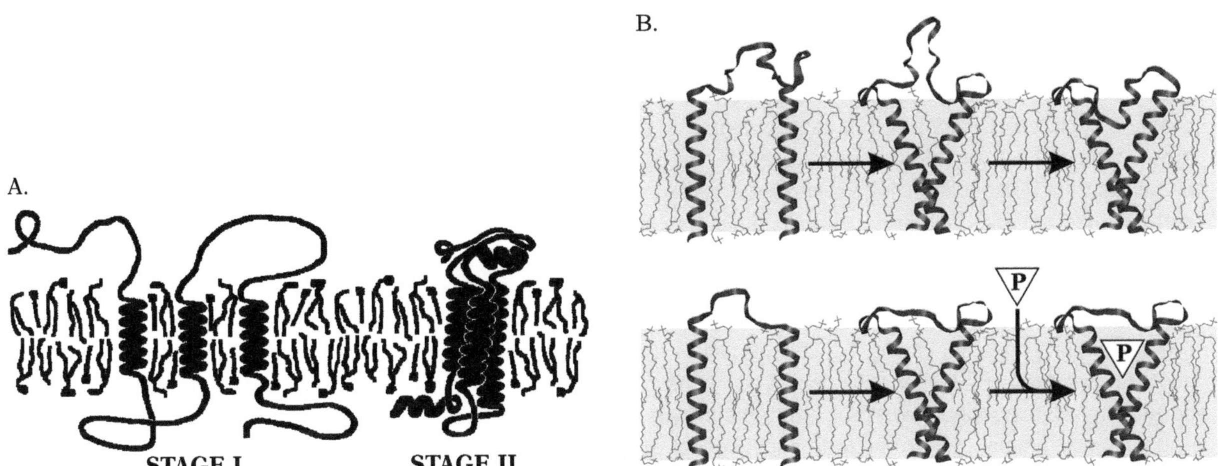

7.3 The two-stage model for folding helical membrane proteins. (A) The two-stage model for folding helical membrane proteins consists of (I) helix insertion and (II) helix interaction. In stage I, the stability of individual helices is postulated to allow them to insert as stable domains in the lipid bilayer. In stage II, side-to-side helix association results in a folded and functional protein. From Popot, J.-L., and D. M. Engelman, *Annu Rev Biochem*. 2000, 69:881–922. © 2000 by Annual Reviews. Reprinted with permission from the Annual Review of Biochemistry, www.annualreviews.org. (B) A third stage for some proteins involves additional folding steps, such as loop rearrangement (top), or addition of prosthetic groups (P, bottom). From Engelman, D. M., *et al.*, *FEBS Lett*. 2003, 555:122–125. © 2003 by the Federation of European Biochemical Societies. Reprinted with permission from Elsevier.

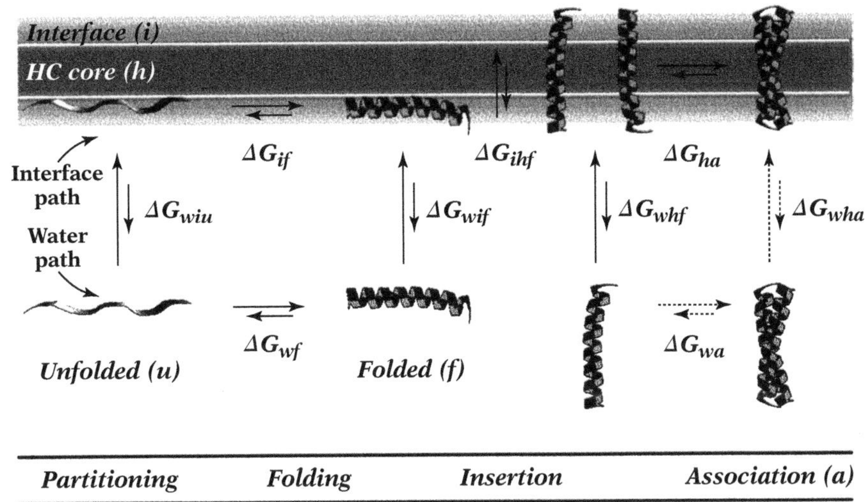

7.4 The four-step model for the thermodynamics of folding helical membrane proteins breaks stage I into three steps: partitioning, folding, and insertion. All four steps are represented occurring in the aqueous phase as well as at the bilayer, with the possibility of partitioning at each step. The process can follow either the ΔG_{wiu} or ΔG_{wif} step, or a combination of both. The ΔG symbols indicate the standard transfer free energies, with subscripts indicating the steps: *w*, water; *i*, interface; *h*, hydrocarbon core of the bilayer; *u*, unfolded; *f*, folded; and *a*, association. For example, ΔG_{wif} is the standard free energy of transfer from water to interface of a folded peptide. This allows the overall standard transfer free energy to be described by a sum of contributions from the various steps, which may be obtained from different types of experiments. From White, S. H., and W. C. Wimley, *Annu Rev Biophys Biomol Struct*. 1999, 28:319–365. © 1999 by Annual Reviews. Reprinted with permission from the Annual Review of Biophysics and Biomolecular Structure, www.annualreviews.org.

Since stage I is determined by interactions relating to the hydrophobic effect, other factors must drive the assembly of helices in stage II. These factors include intrinsic forces such as packing, electrostatic effects, and interactions among the loops between helices, as well as interactions with prosthetic groups or with components at the surface of the membrane. Hydrogen-bonding side chains (Asp, Asn, Glu, Gln, Ser, Thr, Tyr, His) are often important in helix–helix interactions (see Chapters 4 and 6). As helix packing increases helix–helix interactions, it

BOX 7.1 ENERGETICS OF FOLDING AND INSERTING A HYDROPHOBIC α-HELIX INTO THE BILAYER

While numerous peptide studies have been carried out, it is not feasible to obtain values of all the free energy changes depicted in Figure 7.4 from experiments, for several reasons. First, most peptides that are sufficiently hydrophobic to partition into the membrane will aggregate in water and possibly in the interfacial region of the lipid bilayer as well. Without changing experimental conditions, most such peptides will either partition to the interface or perhaps insert across the bilayer, and it is difficult to control the insertion step. Finally, peptides that form a helix in the bilayer will not unfold in the bilayer.

The last of these reasons is illustrated by considering the energetics associated with folding and insertion of a peptide of 20 hydrophobic amino acids. When the peptide forms an α-helix in aqueous solution, the net number of hydrogen bonds does not change significantly, since there is an exchange of some of the hydrogen bonds with the water for hydrogen bonds along the helical backbone. When the peptide partitions into the bilayer, the favorable partitioning of its nonpolar side chains into the bilayer gives an estimated free energy change on the order of -30 kcal mol^{-1}, mainly due to the hydrophobic effect discussed in Chapter 1. This is the favorable ΔG for inserting the hydrophobic helix into the bilayer. In contrast, if the unfolded peptide transfers from water to the nonpolar milieu of the bilayer, it will dehydrate and the loss of hydrogen bonds is estimated to cost about 40 kcal mol^{-1}.

Inserting the backbone of the helix into the nonpolar region of the membrane is unfavorable, since the hydrogen-bonded peptide bonds are hydrophilic. The cost of moving each peptide bond from the water to the bilayer, based on the ΔG_{tr} obtained for partitioning an internally hydrogen-bonded peptide bond from water into an alkane, is $+2.1$ kcal mol^{-1}. Including the partitioning of the αC of each residue brings that cost down to $+1.15$ kcal mol^{-1}, which can be viewed as the per residue cost of transferring a polyglycine α-helix into the bilayer. It means the full cost of dehydrating the backbone of a 20-residue α-helix is 23 kcal mol^{-1}.

What is the cost if the peptide is not folded into a helix? The ΔG_{tr} for a peptide bond that is not internally hydrogen bonded to partition into alkane is $+6.4$ kcal mol^{-1}. Again, partitioning of the αC is favorable, -0.95 kcal mol^{-1}, so the cost of partitioning a residue of an unfolded peptide into the nonpolar milieu of the bilayer is about $+5.4$ kcal mol^{-1}.

For a helix in the bilayer to unfold, the per residue cost is therefore $+5.4$ kcal mol^{-1} -1.15 kcal mol^{-1} (because the cost of dehydration has already been paid), which comes to ~4.2 kcal mol^{-1}. This means the cost to unfold a 20-residue helix in the bilayer is 84 kcal mol^{-1}!

Clearly, the energetics depend on the amino acid composition of the TM helix. In the example of glycophorin, the membrane-spanning helix has 19 hydrophobic residues. The ΔG for its insertion is estimated to be -36 kcal mol^{-1}. Since the cost of dehydrating the helix backbone is $+26$ kcal mol^{-1}, the net energy favoring insertion is -10 kcal mol^{-1}.

also increases lipid–lipid interactions while decreasing helix–lipid interactions. This effect lessens the entropic cost of helix–lipid interactions due to the packing of relatively "soft" lipids against the rigid, relatively "hard" helices.

Helix packing is sufficiently close for helix–helix interactions to be dominated by van der Waals forces. The tight packing between two helices typically involves "knobs" formed by branched residues such as valine and isoleucine fitting into "holes" formed by glycine residues. This was originally seen in the dimerization of glycophorin A, a protein with only one TM helix, as described in Chapters 4 and 6 (Figure 7.5). Similar knob-into-hole packing has been observed in many multispanning (polytopic) membrane proteins. Dimerization motifs for helix–helix interactions have been identified in many membrane proteins, but as they are absent from helix pairs in others, sequence motifs are not sufficient to explain helix packing.

While insertion and packing of separate TM helices are emphasized in the two-stage model, the loops between helices can also be critical. This was demonstrated in experiments that tested the ability of bacteriorhodopsin (bR, described in Chapter 5) expressed as two fragments to assemble into stable, functional molecules in lipid bilayers. When its seven helices (A to G, see Figure 5.6) were combined as two fragments, such as AB + CDEFG, bR usually assembled as a stable protein of the correct structure. However, some fragment combinations, such as ABC + DEFG, produced less stable proteins. Therefore the connecting loops between helices C and D, and also between E and F, while not indispensable for assembly, are essential for the stability of the native protein. A different approach tested the contribution of each loop by genetically replacing each with a linker of Gly–Gly–Ser of the same length, and showed that four loops (AB, CD, EF, and FG) contribute to the resistance of bR to SDS. Furthermore, only helices A, B, D, E, and also C when

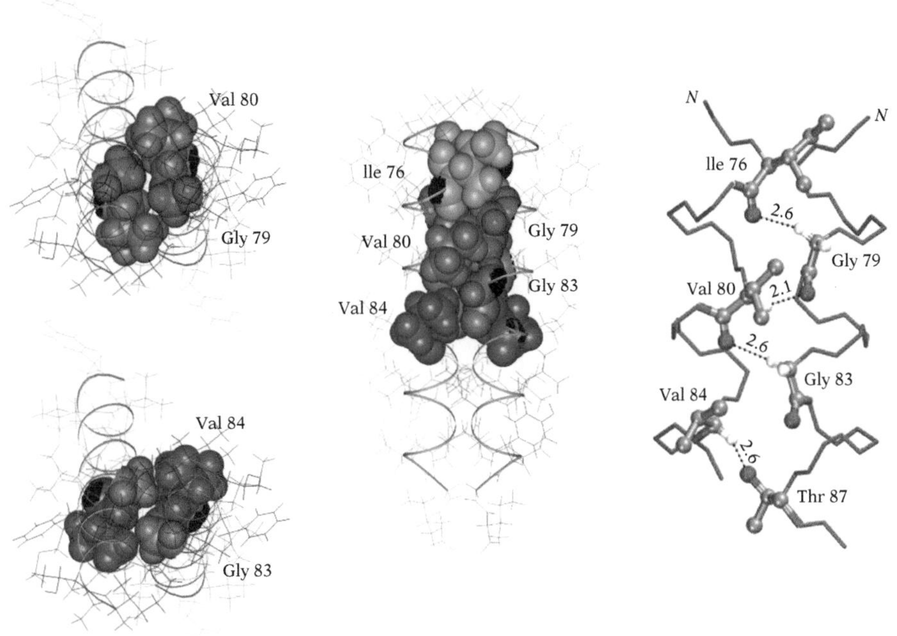

7.5 "Knobs-into-holes" packing is observed at helix–helix interfaces in many helical membrane proteins. Glycophorin A dimerizes when the TM helices from two molecules associate, bringing the "knobs" of branched amino acid side chains into "holes" due to glycine residues. Viewed from the top (left), the close contacts between pairs of valine and glycine residues are evident. Viewed from the side (center), the proximity of such pairs allows the two helices to interact tightly. The third drawing (right) shows the close contacts between the two TM strands, with interatomic distances given in angstroms. From Smith, S. O., *et al.*, *Biochemistry*. 2001, 40:6553–6558. © 2001 by American Chemical Society.

its aspartate residues are protonated, are able to insert independently into phospholipid vesicles, as expected by the two-stage model. A great deal has now been learned about bR folding *in vitro*.

bR Folding Studies

bR was the first integral membrane protein to be fully unfolded and then refolded to regain its activity. Early studies demonstrated its complete unfolding requires formic acid or anhydrous trifluoroacetic acid; when transferred into SDS, it regains about half its helical content, and then when added to retinal and mixed micelles (e.g., cholate or CHAPS and lipid), it refolds to the native structure. Since bR sufficiently unfolds in SDS to lose its native structure and its chromophore, kinetic studies can follow the folding and insertion of SDS-unfolded bR into micelles (or bilayers) using circular dichroism, fluorescence, or absorption spectroscopy. The protein in the absence of retinal is called bacterio-opsin (bO). Binding of retinal to bO is marked by the return of the absorbance maximum at ~560 nm, as well as by its quenching of intrinsic fluorescence.

Stopped-flow absorption spectroscopy and circular dichroism reveal a rate-limiting, pH-dependent, and lipid-dependent step that produces an intermediate with most of the normal helical content. This could be attributed to the slow folding of some of the helices or to incomplete formation of all seven helices. The simplest pathway consistent with the kinetic data from folding SDS-denatured bR is:

$$\text{bO} \rightleftharpoons \text{I}_1 \rightleftharpoons \text{I}_2 \xrightleftharpoons{\text{R}} \text{I}_\text{R} \Rightarrow \text{bR}.$$

The first intermediate, I_1, is accompanied by changes in the surrounding lipid/detergent solvent. The second intermediate, I_2, has native-like secondary structure but lacks some tertiary contacts of the native protein. Formation of I_2 is rate-limiting in apoprotein formation, and this folding event must occur before retinal can bind. I_2 has a binding pocket that enables it to bind retinal noncovalently, forming the intermediate Ir. Formation of the Schiff base between retinal and Lys216 converts I_R to bR. The intermediate I_R has at least two forms distinguished by their absorption maxima at 380 and 440 nm, which are populated differently at different values of pH, suggesting there are parallel folding pathways to the same native structure. Since the intermediates in bO folding are different from those in bR unfolding, pseudo two-state equilibria are attained by unfolding bR in SDS to I_R (which has ~56% of the α-helix content of folded bR and binds retinal) and then

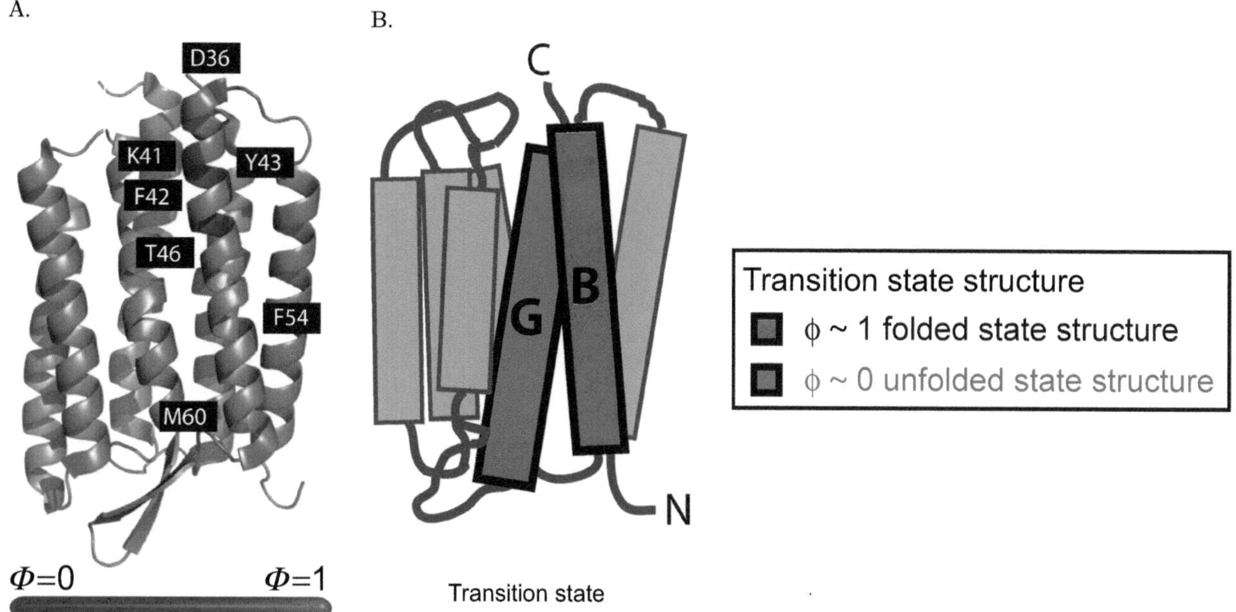

7.6 φ value analysis of the transition state in bacteriorhodopsin folding. (A) A ribbon representation of bR (gray, with the cytoplasmic side at the top) shows the residues of helix B mutated to Ala for φ value analysis. Most of the mutations give $\varphi > 0.8$ (deep red color), meaning they are sites that are folded in the transition state; Y43A and T46A give intermediate values that are harder to interpret. From Curnow, P., and P. J. Booth, *Proc Natl Acad Sci USA*, 2009, 106:773–778. (B) Schematic representation of the φ value data for helix B and helix G shows that in the transition state helix B is largely folded (red), while helix G is more like the unfolded state (blue). From Booth, P. J., *Curr Opin Struct Biol*. 2012, 22:465–475. © 2012, Elsevier. Reprinted with permission from Elsevier.

refolding. This system gives good agreement between the apparent equilibrium denaturation curves and the kinetic measurements.

The nature of the folding intermediate can be probed with a powerful method developed to characterize transition states of soluble proteins. This method is called Phi (φ) value analysis and involves measuring the changes in the free energy of intermediate, transition, and folded states of a protein due to single mutations. With conditions established for carrying out equilibrium folding reactions and studying the folding kinetics, the free energy of folding ΔG_f and the activation energy $\Delta G^{\dagger}$ are determined for mutant and wild type proteins. The difference in free energies between the mutant and wild type is the $\Delta\Delta G$ for each. The φ value is the ratio of the differences $\Delta\Delta G^{\dagger}/\Delta\Delta G_f$ and gives the change in transition state energy relative to the change in the folded state. When $\varphi = 1$, the mutation has induced the same change in both states, so the transition state has the same structure as the folded state at the site of that mutation. When $\varphi = 0$, the structure at the site of the mutation is the same in the transition state as in the unfolded state. With site-directed mutations in regions of the protein structure, φ value analysis reports whether that region is folded or unfolded at the transition state.

To apply φ value analysis to bR, each residue of a single TM helix was mutated to Ala and its $\Delta\Delta G^{\dagger}$ and $\Delta\Delta G_f$ determined from refolding studies in DMPC–CHAPS mixed micelles (Figure 7.6). Quite different results are obtained for the Ala scans of helix B and helix G: In the transition state, helix B is mainly folded while helix G is mainly unfolded. This result obtained from *in vitro* folding studies fits well with the mechanism of folding in the cell (see below): Since the TM helices insert in order of synthesis, the C-terminal helix G is last to be synthesized and last to be folded.

A detergent-free system for bR refolding provides the ability to investigate the effect on folding of the state of the bilayer lipids, in particular the effect of lateral pressures that derives from the tendency of each leaflet to curve away from the plane of the bilayer (see Figure 2.18). This curvature stress confers bending rigidity to the bilayer, decreasing its elasticity. Such stress can be introduced into reconstituted bilayers by mixing bilayer-forming lipids with a nonbilayer former (e.g., PC and PE) and can be altered by varying acyl chain composition; for example, it is reduced with a shorter chain length that decreases crowding (pressure) near the middle of the bilayer.

The yields of bR formed from bO added to retinal and different lipid vesicles are clearly affected by curvature stress of the bilayer (Figure 7.7). The vesicles are made of DPoPC (dipalmitoleoyl PC) under conditions that give 70% refolding. The yield increases with addition of

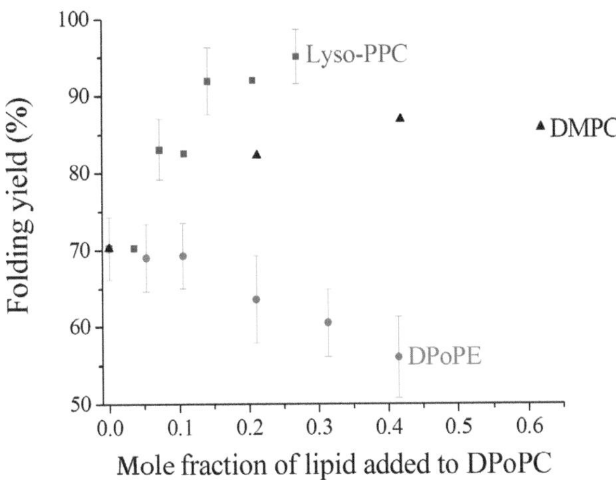

7.7 Effect of bilayer curvature stress on folding of bacteriorhodopsin. Three different lipids are added to DPoPC: DpoPE (red circles), which increases curvature stress; DMPC (black triangles), which decreases curvature stress; and lyso-PPC (blue squares), which relaxes the bilayer still further. The folding yields are determined from the area under the purple absorption band of folded functional bR. From Allen, S. J., *et al.*, *J Mol Biol*. 2004, 342:1293–1304. © 2004 by Elsevier. Reprinted with permission from Elsevier.

DMPC, and even more with addition of a lysophospholipid having one acyl chain removed to reduce the lateral pressure. The yield decreases with the addition of PE with unsaturated acyl chains, which increases the lateral pressure in the center of the bilayer. In bicelles consisting of different proportions of DHPC (dihexanoyl PC, chain length of six) and DMPC (chain length of 14), the bending rigidity of the bilayer is calculated to increase two-fold as the mole fraction of DMPC increases from 0.3 to 0.7. In refolding experiments in this system the rate of bR folding decreased ten-fold over the same range, revealing a clear effect of lateral pressure of the bilayer on kinetics of bR folding. These results suggest that bO has more difficulty inserting into stressed, rigid bilayers as it refolds to bR.

From φ value analysis of its folding transition to its refolding in different conditions of bilayer stress, *in vitro* folding studies with bR produce a sophisticated understanding of the interplay of the protein with the membrane during the folding process. The elegant folding studies of bR provide a paradigm for other helical membrane proteins.

Folding Studies of β-barrel Membrane Proteins

How do β-barrels fit the model for folding and insertion of membrane proteins? They are fundamentally unlike bundles of α-helices, because each helix can be formed independently with hydrogen bonds along the helix axis while β-barrels have hydrogen bonds between neighboring strands, even between the strands closest to the N and C termini. A single β-strand is not a stable structure; rather, formation of at least three to five strands is required for stability in all β-structures, including domains of soluble proteins. Furthermore, geometric considerations make it difficult to envision a portion of a barrel folding first, so all the strands of a β-barrel can be expected to form at roughly the same time. Since the β-barrel proteins are found in outer membranes, they fold and insert from an aqueous environment, such as the periplasm of Gram-negative bacteria.

The first and most extensive *in vitro* folding studies of a β-barrel membrane protein have been done with the β-barrel domain of OmpA ($OmpA_{BB}$), an eight-stranded monomer that can be unfolded in urea and refolded upon dilution into a suspension of PL vesicles. OmpA is an abundant protein in the outer membrane of *E. coli* that exhibits mobility shifts on SDS polyacrylamide gel electrophoresis upon folding. In the structure solved with NMR (see Box 5.1), the positions of its five tryptophan residues are oriented such that four of them are near one end of the barrel while the fifth is close to the opposite end. During assembly of the β-barrel, four Trp residues must translocate to the outer leaflet of the bilayer, crossing its midsection, while the fifth remains in the inner leaflet. Using a series of $OmpA_{BB}$ mutants with all but one of the Trp residues deleted, the locations of the remaining Trp could be determined during the folding process with a set of "spectroscopic rulers," lipids carrying a quenching bromine atom at different positions on their acyl chains (Figure 7.8A). As predicted, during insertion four Trp residues cross the bilayer while the fifth does not. The time course of insertion suggests a model for β-barrel assembly from a flattened disc composed of β-strands at the interface to an inserted "molten globule" as the β-strands penetrate the bilayer in a relatively slow temperature-dependent step.

To further analyze the association of neighboring β-strands during the insertion of $OmpA_{BB}$, single Cys residues were introduced into strands adjacent to the position of the single Trp residues in the mutants (Figure 7.8B). To measure proximity between them during folding, the Cys was reacted with a nitroxide spin label that quenches fluorescence only at short ranges. Fluorescence was monitored during folding from urea into DOPC SUVs over a period of 60 min. Pairs of Trp and Cys on the *trans*-side of β-strands 1, 2, and 3 reached close proximity sooner than pairs on β-strands 1 and 8 (which come together to seal the barrel), while the last pairs to come close were on the *cis*-side (periplasmic side) of β-strands 1 and 2. The observed folding intermediates are consistent with simultaneous β-strand association and insertion into the bilayer.

Similar results are obtained with FRET experiments designed to follow folding by placing an acceptor molecule (a naphthalene derivative) in various locations

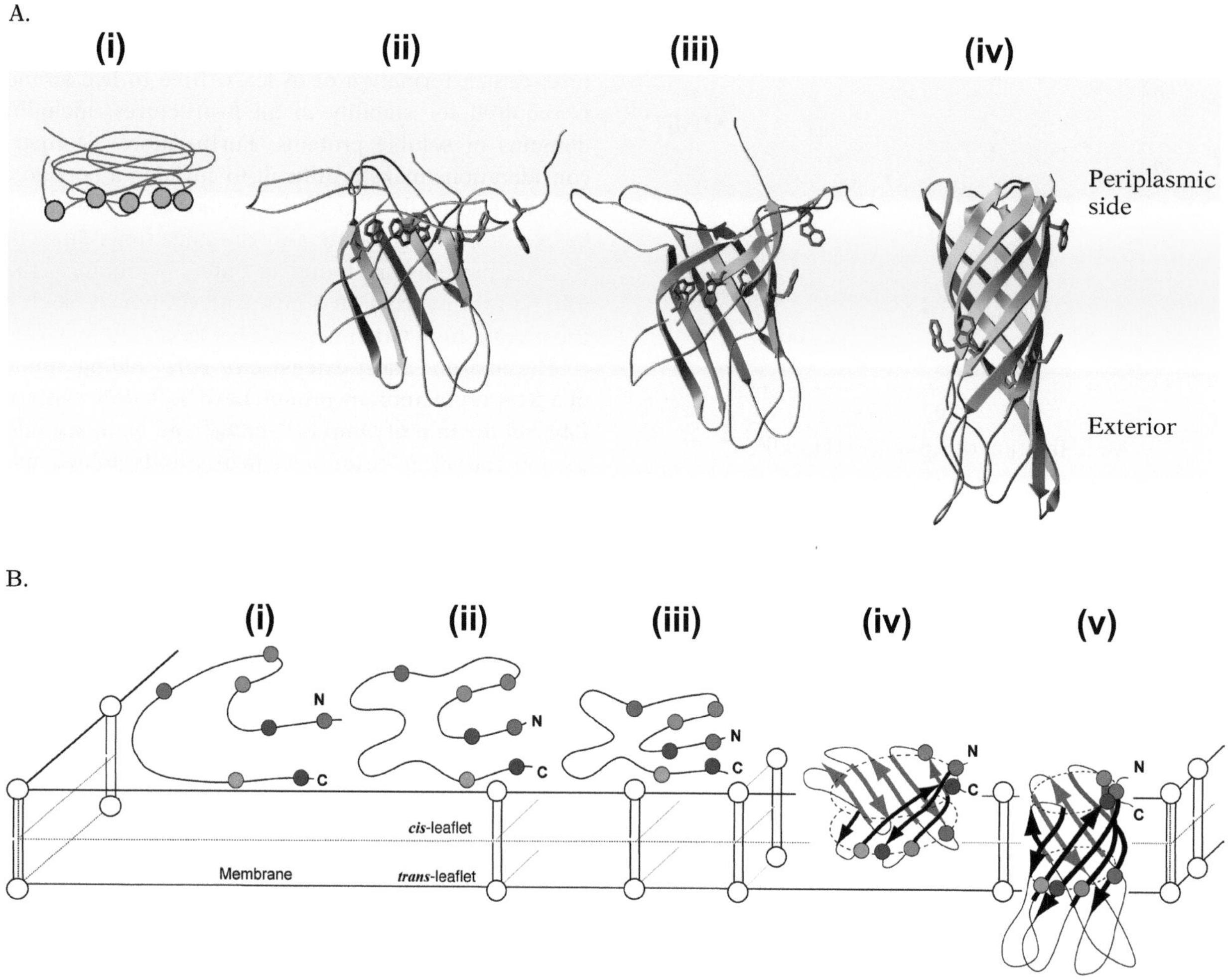

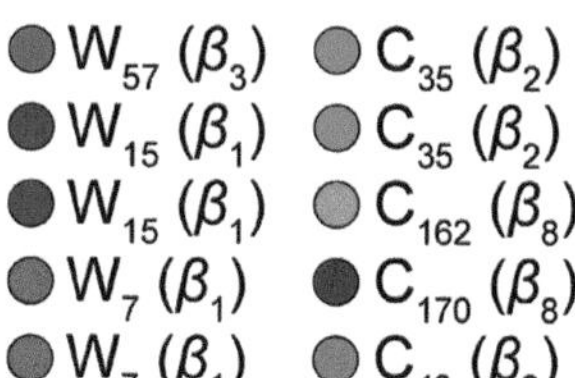

7.8 Folding studies of the TM barrel domain of OmpA. (A) Schematic drawing of four stages in the insertion of OmpA, with the locations of the Trp residues at each stage determined from fluorescence quenching with lipids carrying Br at different positions. In stage (i), each circle represents a Trp residue; in (ii)–(iv) the Trp side chains are stick figures (purple). From Otzen, D. E., and K. K. Andersen, *Arch Biochem Biophys*. 2013, 531: 34–43. © 2013, Elsevier. Reprinted with permission from Elsevier. (B) Schematic drawing to show when associations of neighboring β-strands during folding and insertion of $OmpA_{BB}$ produce quenching of individual Trp residues due to nitroxyl labels on the indicated Cys residues (circles colored according to key). Strands β_1, β_2, and β_3 are contiguous, while β_1 and β_8 only become neighbors when the barrel is formed. Closing of the barrel is required for insertion into the *trans* leaflet of the bilayer. Different intermediates can be trapped at different temperatures, with temperatures >20°C required to reach the final folded stage. From Kleinschmidt, J. A., *et al.*, *J Mol Biol*. 2011, 407:316–332. © 2011, Elsevier. Reprinted with permission from Elsevier.

relative to single Trp residue donor (Figure 7.9). FRET spectra show decreasing distances between donor and acceptor as $OmpA_{BB}$ folds from urea into DPPC over a period of 15–60 min for all donor–acceptor pairs except for one, where they are on the same strand and loop together in the unfolded state. The data support a model of initial insertion of a compact pore, followed by slower traversing of the bilayer.

The full length OmpA protein ($OmpA_{FL}$) consists of $OmpA_{BB}$ (171 residues) and a periplasmic domain, $OmpA_{PER}$ (154 residues). Since $OmpA_{PER}$ lacks Trp residues, CD spectroscopy is used to monitor unfolding and refolding of $OmpA_{PER}$ and to investigate its effect on $OmpA_{BB}$. When $OmpA_{FL}$ is unfolded in urea, aggregation competes with refolding upon dilution. However, reversible unfolding and refolding of $OmpA_{FL}$ in octyl

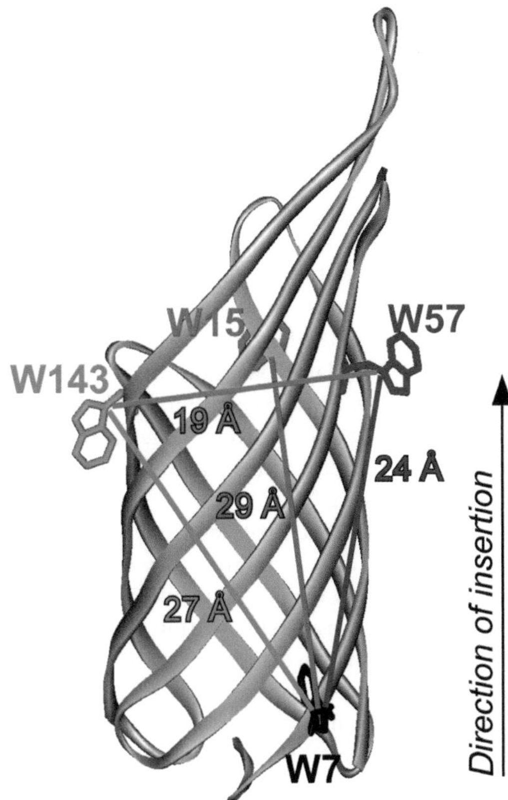

7.9 Residues in $OmpA_{BB}$ used in FRET experiments. Single Trp mutants of $OmpA_{BB}$ were used in FRET experiments, with Trp15, Trp57, and Trp143 as donors. The acceptor was a cysteine-linked naphthalene derivative placed at W7C or W57C. The distances between *β*-carbons of the donor and acceptor in the folded protein are given. Note that W7 and W15 are on the same strand. During folding the *β*-barrel forms as the strands insert into the membrane (direction of arrow), which brings the FRET pairs close enough for fluorescence energy transfer. From Kang, G., *et al.*, *Biochim Biophys Acta*. 2012, 1818:154–161. © 2012, Elsevier. Reprinted with permission from Elsevier.

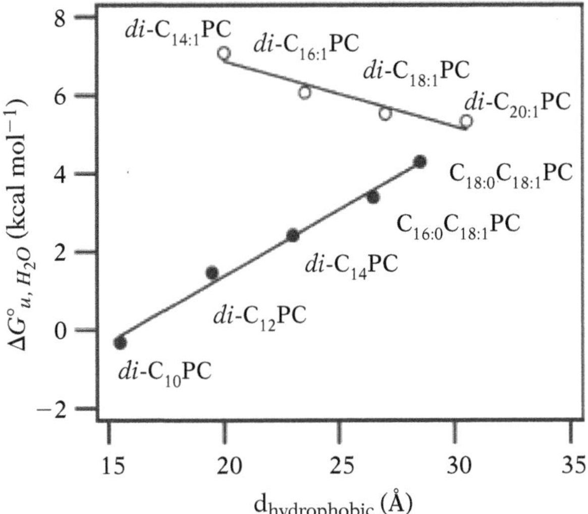

7.10 Folding studies of the TM barrel domain of OmpA reveal the effects of bilayer stresses. The graph shows the free energy of unfolding (ΔG°_u, in kcal mol^{-1}) as a function of the hydrophobic thickness of the PC bilayer. For saturated and monounsaturated chains (filled circles), the ΔG°_u increases linearly with chain length (thickness), while for *cis* double-unsaturated acyl chains (open circles), it decreases linearly with chain length. From Hong, H., and L. K. Tamm, *Proc Natl Acad Sci USA*. 2004, 101:4065–4070. © 2004 by National Academy of Sciences, U.S.A.

maltoside micelles was accomplished with guanidinium chloride and monitored on SDS polyacrylamide gel electrophoresis (PAGE), and produced kinetic results that indicate the existence of a folding intermediate.

Folding of $OmpA_{BB}$ is also affected by stresses in the bilayer created by variations in lipid compositions that introduce curvature or increase lateral pressure. Refolding was observed both by fluorescence and electrophoretic mobility. The reference bilayer was POPC with 7.5 mol% POPG, which gives two-state folding with a free energy of unfolding of 3.4 kcal mol^{-1}. The ΔG° for unfolding was linearly proportional to chain lengths when saturated and monounsaturated acyl chains were used, and inversely proportional to chain lengths when double-unsaturated acyl chains with more curvature stress were used (Figure 7.10). A possible explanation for the increasing folding rates with increased curvature is that the more curved surfaces bring the *β*-strands closer together when they adsorb on the surface prior to insertion. Refolding also increased with increased lateral pressure due to addition of PE. In spite of their different mechanisms for folding, the studies with both bR and OmpA make it clear that folding and stability of membrane proteins are affected by the material characteristics of the bilayer, in particular the bending rigidity resulting from curvature stress (Figure 7.11).

Other Folding Studies

Classical *in vitro* folding studies in lipids or detergents have now been carried out with many other membrane proteins, including the enzyme DAGK, the lactose permease LacY, the potassium channel KcsA, and the *β*-barrels OMPLA and PagP. Some studies have employed different strategies to gain different types of information about folding membrane proteins. For example, the stability of various mutants of DAGK was assessed with thermodynamic analysis of refolding the partially denatured protein. When DAGK is reversibly denatured in SDS, it retains most of its *α*-helical content and the refolding that takes place upon the removal of SDS is sufficient to differentiate the stability of different mutants.

To study the TM portion of the KcsA potassium channel (see Chapter 9), a novel approach combined partial

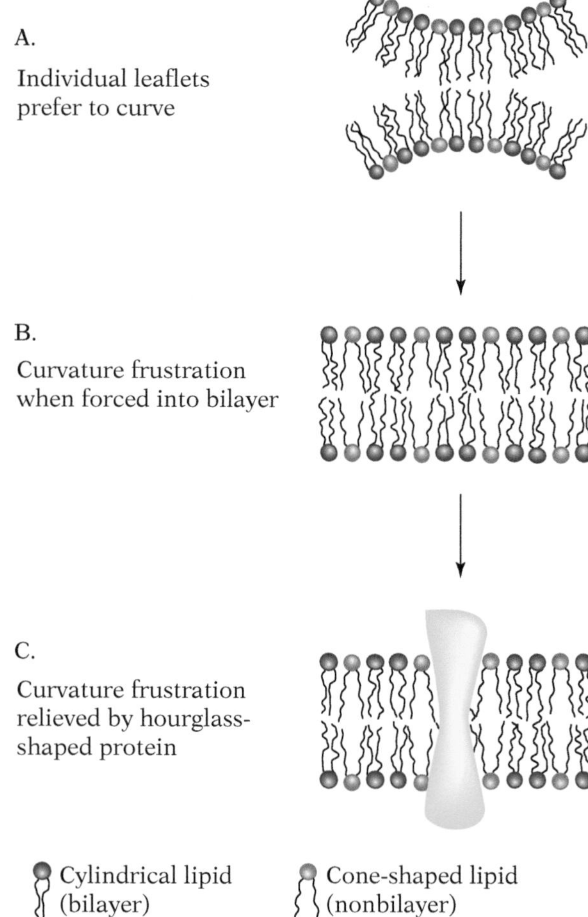

7.11 Effect of protein insertion on bilayer curvature stress. Bilayer stress resulting from incorporation of a nonbilayer lipid (cone shaped) can be relieved by insertion of a protein with an hourglass shape. Redrawn from Bowie, J., *Nature*. 2005, 438:581–589. © 2005. Reprinted by permission of Macmillan Publishers Ltd.

de novo synthesis with folding. The semisynthetic TM protein was constructed by fusion of a recombinant peptide (residues 1–73) with a synthetic peptide (residues 74–125). The refolding of this construct was then used to define the lipid requirement for folding.

Whole Protein Hydrophobicity Scale

Folding studies of the β-barrel enzyme outer membrane phospholipase A (OMPLA, see Chapter 9) provide quantitative assessment of the effect of substituting different amino acids into a TM segment that is the basis for a new hydrophobicity scale (see Chapter 6). The spontaneous, reversible unfolding of OMPLA into DLPC LUVs at pH 3.8 is followed by Trp fluorescence, and refolding is verified by enzyme assay and protease protection. OMPLA is the host for guest substitution of every other amino acid at Ala210, a position in the middle of the β-barrel exposed to lipid (Figure 7.12). By subtracting the free energy of unfolding the OMPLA host from the free energy of unfolding the host–guest, the difference ($\Delta\Delta G_{w,l}$) gives the energy of partitioning of the guest side chain from an aqueous environment into the membrane relative to the alanine side chain in the wild type sequence. Comparison of $\Delta\Delta G_{w,l}$ for every guest is the basis of the whole protein hydrophobicity scale (Figure 7.12B). Since low pH is required for reversibility, the acidic residues are less charged than at neutral pH and partition more readily. However, the scale is the first to represent the thermodynamics of water to bilayer partitioning in the context of a TM segment of a native protein. Further, by varying the position of the guest residue, the energetics of side chain partitioning can be determined as a function of depth in the membrane (Figure 7.12C).

Folding of membrane proteins can be a very important step of their purification after they are harvested in denatured form from insoluble inclusion bodies, which are lipid-bounded storage sites within the cytoplasm of bacterial cells that are overproducing the proteins from high-expression vectors. An example of a membrane protein that was refolded after it was obtained in inclusion bodies is the enzyme OMPLA (described in Chapter 9). The denatured OMPLA was solubilized in 8 M urea and diluted into Triton X-100, which produced a mixture of folded and unfolded OMPLA that was resolved by ion-exchange chromatography to recover the native enzyme. The choice of detergent is critical, as refolding studies under different conditions revealed a strong preference for Triton X-100 in the case of OMPLA.

Finally, the use of cell-free systems to synthesize membrane proteins reveals conditions needed for folding and insertion of the protein of interest. Using *in vitro* transcription/translation systems in the presence of lipid vesicles or micelles, combinations of lipids and detergents are tested. For example, in pioneering studies of the *E. coli* outer membrane protein PhoE (see Chapter 5), folding and insertion require lipopolysaccharide. In similar studies of the *E. coli* lactose permease (see Chapter 11), protein folding took place only with inside-out vesicles, mimicking the vectorial insertion of the lactose permease into the inner membrane. The common problems caused by expressing high amounts of eukaryotic membrane proteins in bacteria because their targeting signals may not be recognized or their overexpression is toxic has led to increasing use of cell-free systems for eukaryotic proteins. In these cases information gleaned from *in vitro* folding may be relevant to folding into cell membranes.

BIOGENESIS OF MEMBRANE PROTEINS

Folding and insertion of membrane proteins are just two elements of the very complex process needed for the assembly of proteins into cell membranes. The

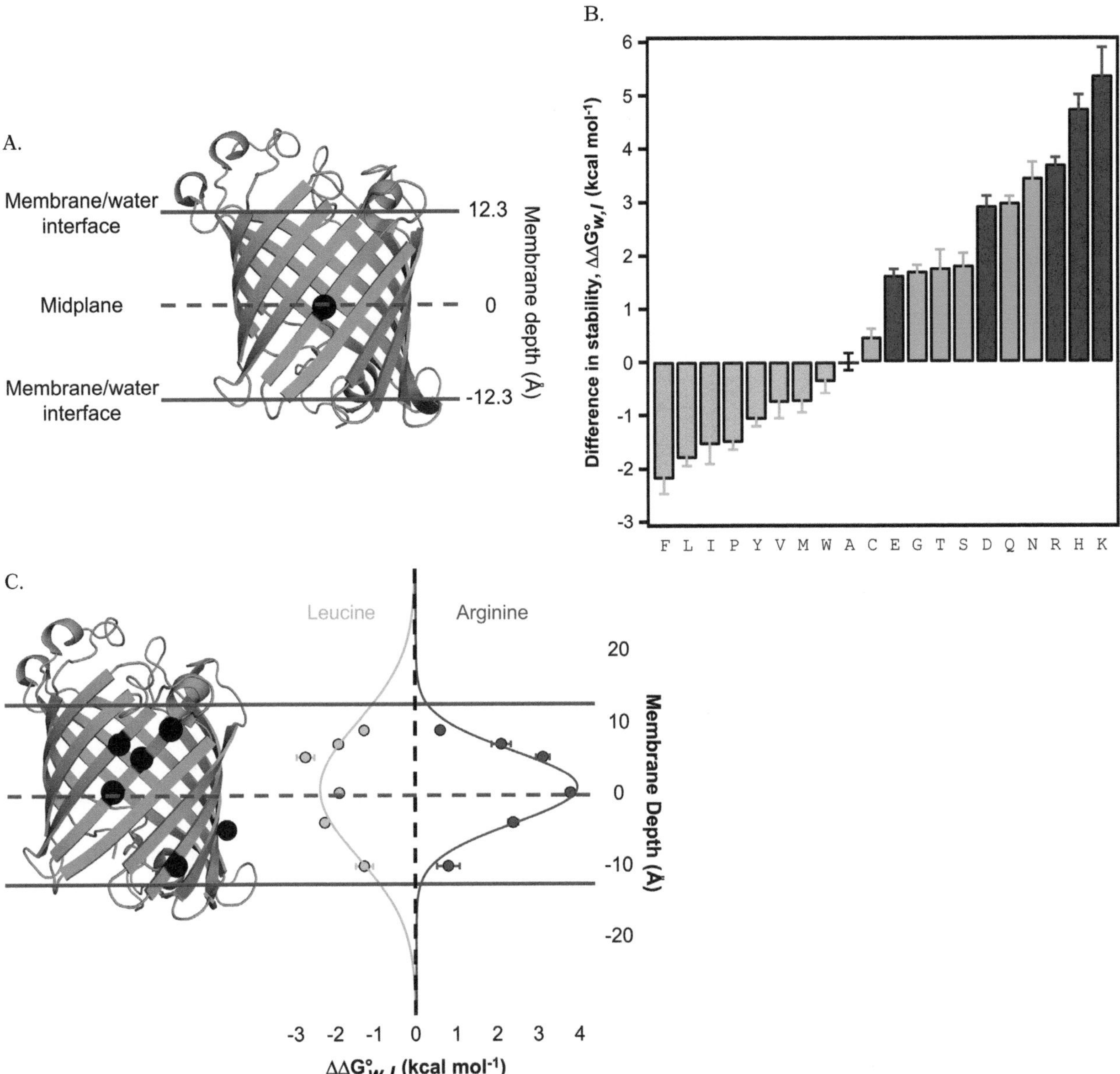

7.12 Folding of OMPLA as a host–guest system for measuring the membrane insertion energies of lipid-exposed amino acid residues. (A) Ribbon diagram of OMPLA showing the Ala210 site for guest substitutions (black sphere) at midplane (dashed blue line) of the transmembrane region. (B) The whole-protein hydrophobicity scale determined from the OMPLA host–guest system with each amino acid variant at position 210. The free energy of unfolding ($\Delta\Delta G_{w,l}$) of each amino acid variant is compared to the wild type OMPLA, with error bars for the standard errors of the mean from individual titration experiments. Residues are colored for types: F, I, L, M, P, V, W, and Y (orange); C and G (gray); N, Q, S, and T (teal); D, E, H, K, and R (blue). (C) Use of other guest sites shows how the energetics of side chain partitioning varies by depth in the membrane. The positions used are at residues 120, 164, 210, 212, 214, and 223 (black spheres). The $\Delta\Delta G_{w,l}$ of leucine and arginine variants (relative to alanine variants) are shown aligned with the OMPLA image, along with normal distributions fit to the data. From Moon, C. P., and K. G. Fleming, *Proc Natl Acad Sci USA*. 2011, 108:10174–10177.

first steps of that process are shared by membrane and secreted proteins and utilize a complex apparatus called a translocon that has been conserved through eukaryotic, bacterial, and archael kingdoms. In later steps integral membrane proteins are differentiated from secreted proteins by a process that allows their lateral integration into the lipid bilayer and determines their topology. The process of integration and topogenesis (the genesis of topology) will be examined after describing the process for export of nascent (newly synthesized) proteins from the cytosol and the partners involved in the translocation step.

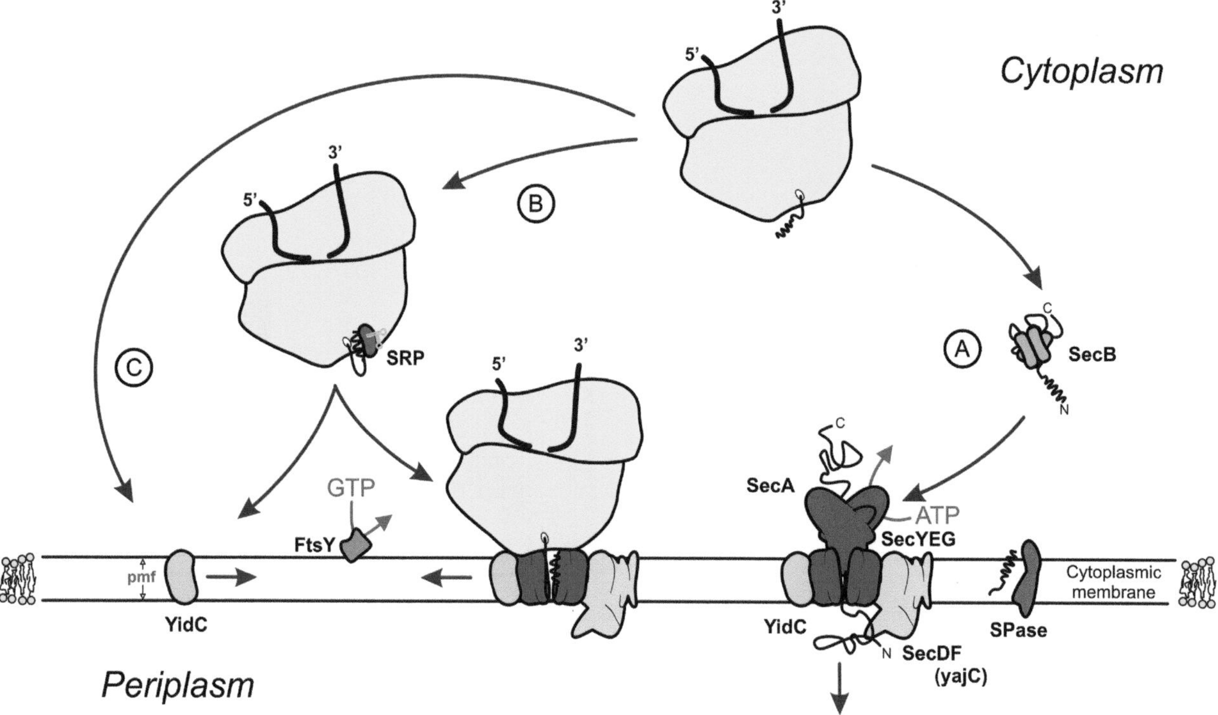

7.13 Diagram illustrating posttranslational translocation and cotranslational translocation in bacteria. The Sec translocon (blue) spans the cytoplasmic membrane (CM) and consists of SecY, SecE, and SecG. SecA (red) is the peripherally associated motor protein. (A) Proteins synthesized at the ribosome (light yellow) that bind SecB (light blue) are brought to SecA at the translocon. Their posttranslational translocation may be aided by SecDF (gray) and or by YidC (yellow), and their signal peptide cleaved by signal peptidase (green). (B) Cotranslational translocation begins with the nascent chain recognized by SRP (pink), which brings it to the SRP receptor, FtsY (purple), and targets it to the translocon. (C) A subset of proteins can be inserted by YidC alone. The three pathways use different energy sources: SecA hydrolyzes ATP, SRP/FtsY and the ribosome hydrolyze GTP, and YidC uses the proton motive force (pmf). From du Plesis, D. J. F., *et al.*, *Biochim Biophys Acta*. 2011, 1808:851–865. © 2011, Elsevier. Reprinted with permission from Elsevier.

Export from the Cytoplasm

Membrane proteins are targeted for their destinations as they exit the ribosomes. In eukaryotes, they may enter the ER and end up in the plasma membrane or endocytic organelles (the ER pathway), or they may follow other pathways for mitochondrial, chloroplast, and nuclear membranes. The destinations in prokaryotes are the plasma membrane and, in Gram-negative bacteria, the outer membrane. Once targeted for export, the processes of translation and translocation may be coupled (cotranslational translocation) or sequential (posttranslational translocation), each of which involve many specialized proteins (Figure 7.13).

The classical signal sequence, the first recognized, is located at the N terminus of the nascent peptide and is cleaved during export. (For other types of signals, see "Topogenesis in Membrane Proteins" below.) The signal sequence consists of ~20 amino acids, with a few polar and basic residues followed by a stretch of 7–15 primarily hydrophobic residues before the polar cleavage site at the beginning of the mature protein. When cleavage is blocked, the larger size of the precursor protein is readily recognized by SDS-PAGE (see Box 7.2). Although not all N-terminal signals are cleaved, this size difference between precursor and mature forms of the protein has been used to follow export in thousands of experiments.

Signal sequences are highly degenerate and tolerate many mutations as long as the nonpolar region is sufficiently hydrophobic and long enough (but not so long as to become a TM segment). This stretch of hydrophobic amino acids binds to a hydrophobic groove on a chaperone or piloting protein, which recognizes the signal as it emerges from the ribosome, protects the nascent chain from aggregation, and targets it to the membrane. This role is often carried out by a ribonucleoprotein complex called the signal recognition particle (SRP). The SRPs in prokaryotes and eukaryotes have homologous core components, as do their receptors in the membrane (Figure 7.14). Both SRP and its receptor are GTPases, and when the SRP is docked, the catalytic sites on both SRP and the SRP receptor come together to form a unique catalytic chamber

BOX 7.2 EVIDENCE FOR CLEAVABLE SIGNAL SEQUENCES INVOLVED IN PROTEIN TRANSLOCATION

The protease protection assay distinguishes between translocated membrane proteins and secreted proteins and provides evidence that both can have cleavable signal sequences. The experiment uses SDS polyacrylamide gel electrophoresis (PAGE) to distinguish products of *in vitro* protein synthesis in the presence or absence of membrane vesicles, with and without added protease. When translation of a protein with a classical N-terminal cleavable signal sequence is carried out *in vitro* in the absence of membranes, a precursor containing the signal is produced. The precursor is not protected by membranes, so it is digested by added protease (sample 1 in Figure 7.2.1). When it is carried out in the presence of membrane vesicles, the signal inserts into the membrane, along with any TM segments of the protein (samples 2 and 3). Signal peptidase on the membrane cleaves the signal from both secreted and inserted proteins, resulting in mature forms that migrate faster on SDS-PAGE as shown. To determine whether the protein is integrated into the membrane or secreted across it, a nonspecific protease such as proteinase K is added. If the protein is soluble and outside the vesicles, it is fully digested (sample 1); if it is integrated into the membrane, the external portion is digested, resulting in a smaller protein or fragments (sample 2); and if it is fully transported into the vesicle (secreted), it is protected from the protease and retains its size (unaltered mobility on the gel, sample 3). Many applications of this basic protocol have revealed the nature and role of signal sequences on hundreds of proteins.

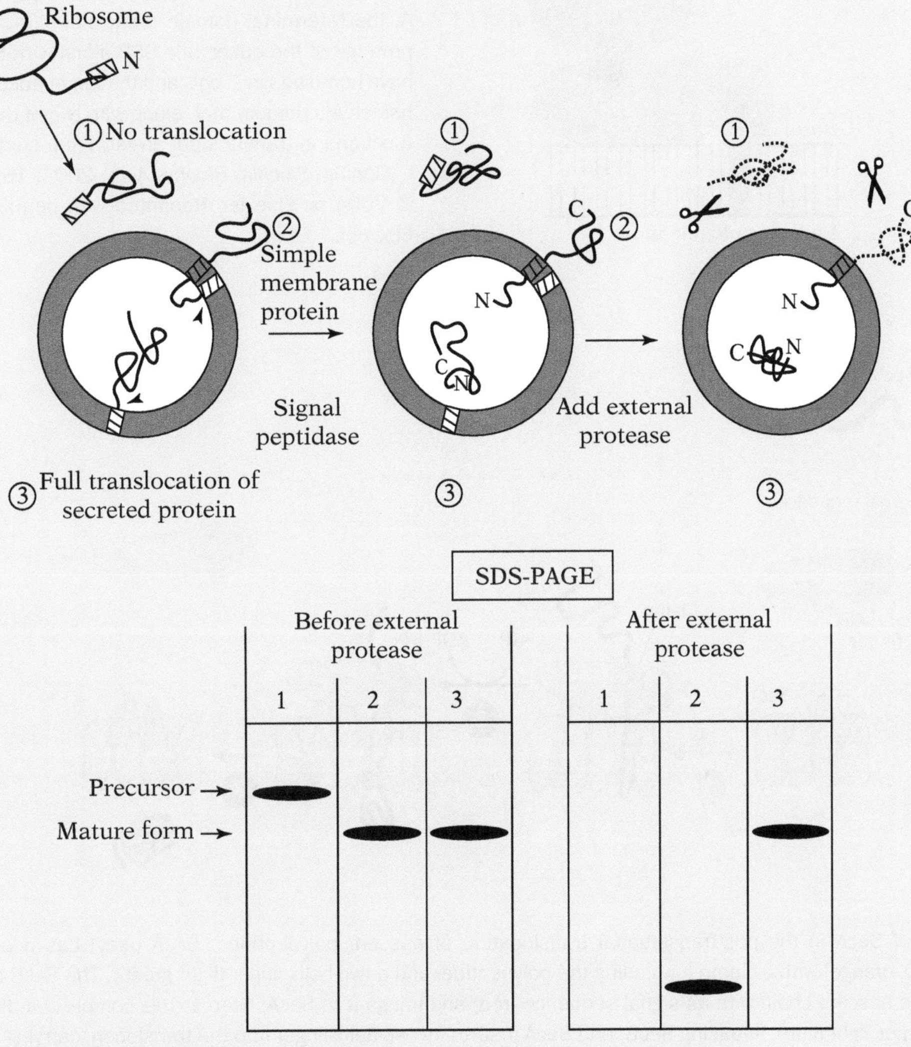

7.2.1 *In vitro* assay for protein insertion into membrane vesicles, utilizing protease sensitivity (top) and analysis by SDS-PAGE. Reprinted with permission from Arnold Driessen, based on unpublished figure.

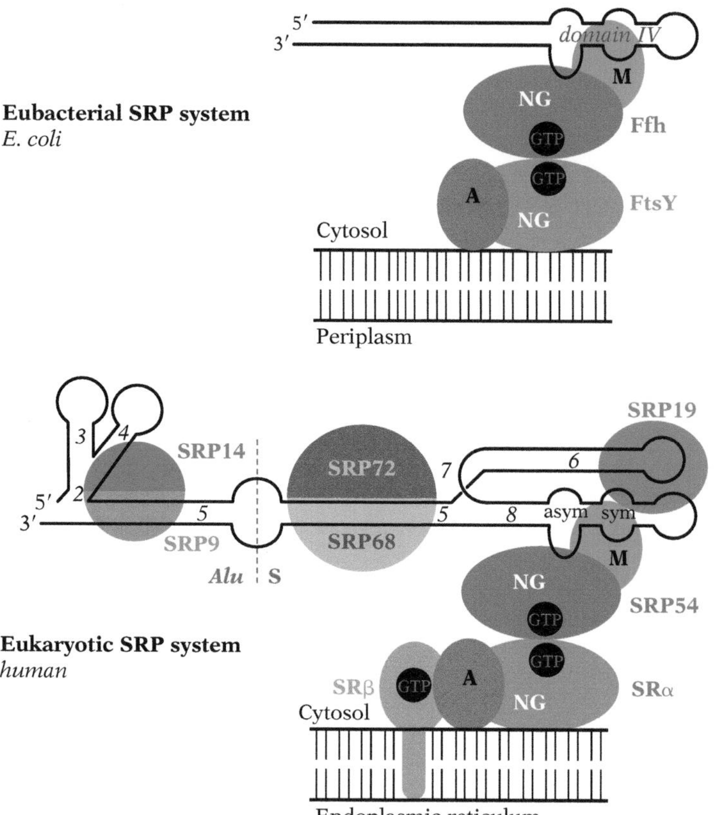

7.14 The core components of the *E. coli* and human signal recognition particle (SRP) and its receptor are highly homologous. Both SRP54 in eukaryotes and the Ffh protein of the *E. coli* SRP have three domains. The M domain is methionine-rich and binds RNA (lavender). The N domain is the α-helical domain, and the G domain is a nucleotide-binding site with GTPase activity (drawn together in purple). The receptor, which is FtsY in *E. coli* and SRα in eukaryotes, also has similar domains, homologous N and G (light pink) and A, the N-terminal domain (dark pink). The remaining proteins of the eukaryotic SRP along with SRβ do not have homologs in *E. coli*, and the larger eukaryotic RNA has an Alu domain that, along with two of the proteins, functions in translational arrest. From Luirink, J., and I. Sinning, *Biochim Biophys Acta*. 2004, 1694:17–35. © 2004 by Elsevier. Reprinted with permission from Elsevier.

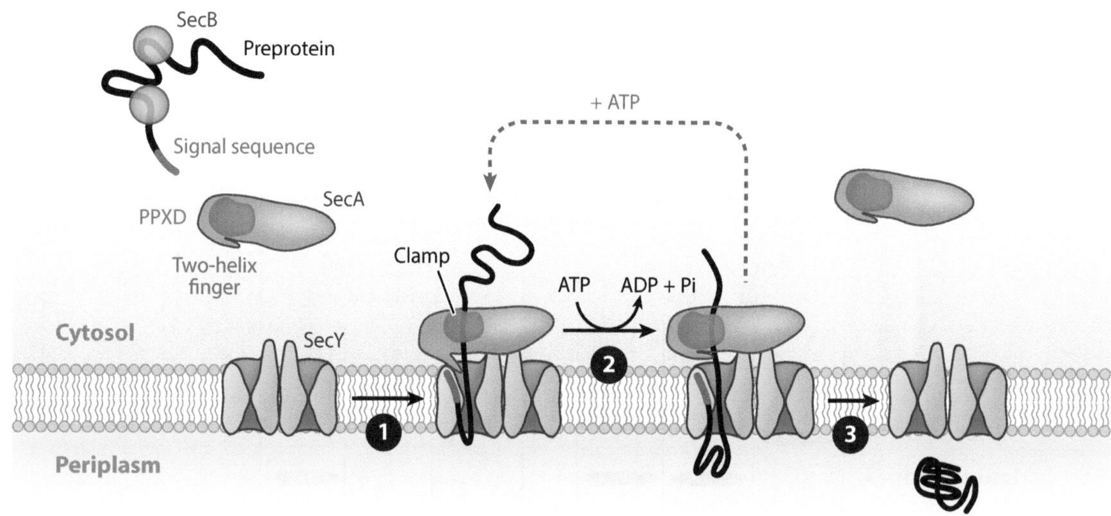

7.15 Model for the role of SecA in the posttranslational translocation of nascent polypeptides. SecA (gray) has a polypeptide crosslinking domain (PPXD, orange) with a clamp for binding the polypeptide, and a two-helix finger (light green). The SecB chaperone (lavender circles) binds the nascent chain with its signal sequence (red) and brings it to SecA. Step 1: this complex binds the SecY translocon (shown as a dimer, light blue), releasing SecB, and SecA inserts its two-helix finger into the translocon, carrying the bound precursor. Step 2: when SecA hydrolyzes the ATP, it pushes the polypeptide into the channel, then withdraws the two-helix finger from the translocon, while the peptide remains clamped within the channel. SecA binds ATP and another segment of the peptide chain to repeat the cycle (dotted line). Step 3: after sufficient repetitions, the peptide chain is fully translocated and SecA diffuses into the cytoplasm. From Park, E., and T. A. Rapoport, *Ann Rev Biophys*. 2012, 41:21–40. © 2012 by Annual Reviews. Reprinted with permission from the Annual Review of Biophysics, www.annualreviews.org.

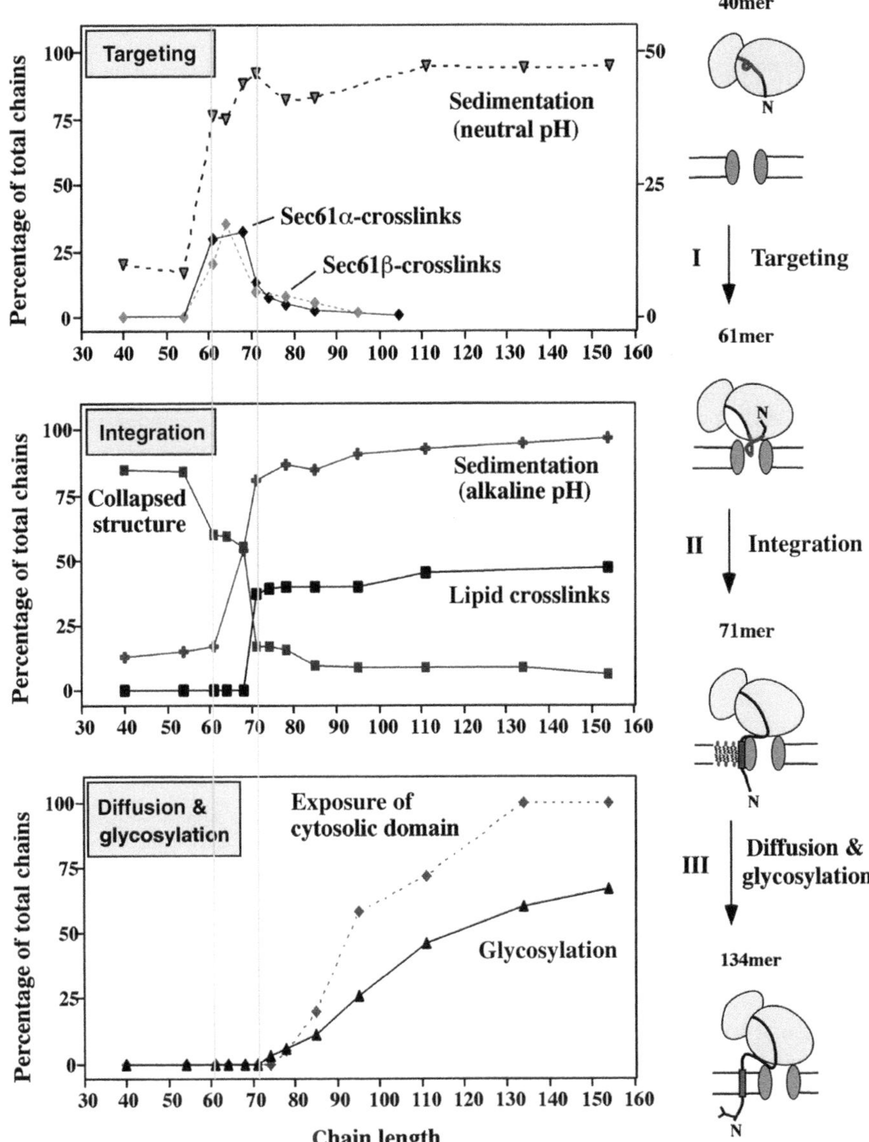

7.16 Crosslinking results indicate three stages in the biosynthesis of the SAI protein, a signal-anchor protein. Data were obtained as described in Box 7.3 and were quantitated to show the percentage of total chains that had each characteristic plotted (*y*-axis) as a function of the length of the truncated chain (*x*-axis). (Note: the top plot also gives a scale for the percentage sedimented at neutral pH, right axis.) The top graph shows crosslinking to Sec61, which requires a minimum of about 60 amino acids. The middle graph shows that over 70 residues are required before the nascent chain sediments at alkaline pH, at which length it is integrated into the membrane. At the same length, it can be crosslinked to lipid. The bottom graph shows that glycosylation occurs on chains longer than 85 residues, indicating they have reached the outside, while the proportion that is exposed to protease after translocation continues to grow. These stages are diagrammed along the side of the graphs. From Heinrich, S., *et al.*, *Cell*. 2000, 102:233–244. © 2000 by Elsevier. Reprinted with permission from Elsevier.

that binds two GTPs. Once bound, they stimulate each other's GTPase activity. The hydrolysis of GTP by the complex releases SRP, which enables the ribosome to dock on the translocon and resume translation to carry out cotranslational translocation. In *E. coli*, SRP and its receptor FtsY are required for integration of very hydrophobic proteins into the inner membrane.

A different set of chaperones are used for posttranslational translocation (export that follows the synthesis of the entire peptide chain) of less hydrophobic proteins, including those destined for the outer membrane. In *E. coli* (and other bacteria) the signal sequence is recognized by SecB, which brings it to SecA. SecB is a tetrameric chaperone that prevents newly synthesized proteins from folding or aggregating in the cytosol by binding to exposed hydrophobic surfaces without an expenditure of ATP. SecB has well-defined binding sites for SecA and pilots the nascent peptide to the SecA protein. Transfer of the nascent peptide chains from SecB to SecA releases SecB and allows the initiation of translocation. SecA is the motor, a cytosolic ATPase that provides the energy for translocation. Highly conserved in bacteria, it functions as a dimer that binds to the membrane and can insert into the translocon (see Chapter 4). A groove called a clamp between the nucleotide binding domains and the peptide binding domains is hypothesized to hold the peptide while a two-helix finger pushes it through the translocon as SecA hydrolyzes ATP in repetitive cycles (Figure 7.15). An x-ray structure of SecA with the translocon is shown in Figure 7.20, below.

For posttranslational translocation in yeast (and probably other eukaryotes) several chaperones can bind the chain as it emerges from the ribosome in the cytosol to prevent aggregation prior to export. In the ER lumen an essential chaperone called Bip captures the chain during export and prevents it from sliding back into the cytosol.

BOX 7.3 CROSSLINKING TRACES NASCENT PEPTIDES THROUGH THE TRANSLOCON INTO THE BILAYER

To investigate the path of nascent membrane proteins through the translocon, full-length or truncated proteins are engineered with sites for incorporation of chemical crosslinking agents. Translation intermediates of various lengths are generated using polymerase chain reaction (PCR) with different downstream primers to amplify portions of the gene before transcription to produce a set of truncated mRNAs. *In vitro* translation is carried out in the presence of ^{35}S-methionine and is stopped with cycloheximide, which inhibits the peptidyl transferase activity of the ribosome.

Many different experiments have employed crosslinking to study the fate of translocated proteins. For the protocol described here, the truncated mRNAs are designed to allow the incorporation of a photoactivatable derivative at a unique position in the TM segment of the protein to be studied by incorporating a UAG stop codon at the position of interest. A suppressor tRNA carrying the anticodon CUA will bind to the UAG stop codon and incorporate its amino acid into the growing chain, instead of letting translation terminate. When the suppressor tRNA carries trifluoromethyl-diazinyl phenylalanine, it incorporates this UV-sensitive crosslinking agent in that position (Figure 7.3.1). Only when irradiated will it react, forming a covalent bond with other molecules in its immediate environment. In addition, chemical crosslinking was carried out with bis-maleimidohexane, a homobifunctional sulfhydryl-reactive agent (Figure 7.3.1). It can react with two cysteines, linking one in the protein substrate to another within ~15 Å. In this case, the targeted cysteine is in the Sec61β subunit.

7.3.1 Chemical structures of the crosslinking agents used: trifluoromethyl-diazinyl phenylalanyl-tRNA (left), which is added during translation to incorporate a photoreactive derivative of Phe in the TM segment, and bis-maleimidohexane (right), a reagent for homobifunctional sulfhydryl-reactive crosslinking with a spacer arm of 13 Å).

The substrate protein for these experiments is a signal-anchor protein engineered for glycosylation at the N terminus as well as for crosslinking as described (Figure 7.3.2A). The TM domain from residues 19 to 41 (shaded gray) contains a site for incorporation of the photoreactive phenylalanine derivative (*, residue 28 labeled "stop codon") as well as the cysteine (C) residue at position 35 for the chemical crosslinking. Below it the positions marked with (++) indicate where two arginine residues were inserted in experiments designed to test whether charges in the TM segment disrupted translocation and lipid partitioning (and they do).

In vitro translation is carried out in the presence or absence of rough (i.e., not highly purified) microsomal membranes, and proteinase K is added to see at what point the translation intermediates are no longer protected from digestion across the membrane (see Box 7.2). The extent of glycosylation indicates whether the N terminus extends into the lumen of the microsomes. The results in Figure 7.3.2B are autoradiographs from experiments in which radiolabeled chains are synthesized from different length messages in the presence or absence of rough microsomes and in the presence and absence of proteinase K, as indicated. Samples are analyzed by SDS-PAGE prior to autoradiography. The ^{35}S-labeled products show that at least 61 amino acids (the 61mer) are needed to see some protection from proteolysis by the addition of microsomal membranes (bands marked with arrows), indicating the ribosome-nascent chain complex has reached the translocon at that length. By the length of 85 amino acids (85mer), glycosylation can take place, producing slower-migrating bands (marked with stars), which indicates that the chain is emerging on the other side of the membrane. (Open arrows and open stars mark nonproteolyzed nonglycosylated and glycosylated chains, respectively; closed arrows and closed stars mark membrane-protected nonglycosylated and glycosylated chains, respectively, in the 85mer to the 154mer.)

When samples are irradiated after translation and then sedimented at either neutral or alkaline pH, crosslinks to the translocon (Sec61α and Sec61β) and to lipid are observed (Figure 7.3.3). Sedimentation at neutral pH that does not occur at alkaline pH indicates that a nascent peptide chain is associated at the membrane periphery and not yet integrated across it, as observed with the 61mer and 68mer in Figure 7.3.3A. These two nascent chains are also able to crosslink to the Sec61α of the translocon (marked with closed diamonds, lanes 14 and 22; see also Figure 7.3.3B). Chemical crosslinking to Sec61β appears with the 71mer, the 78mer, and the 85mer (closed circles). Membrane integration begins with the 71mer, with clear crosslinking to lipid (open diamonds; see also

BOX 7.3 *(continued)*

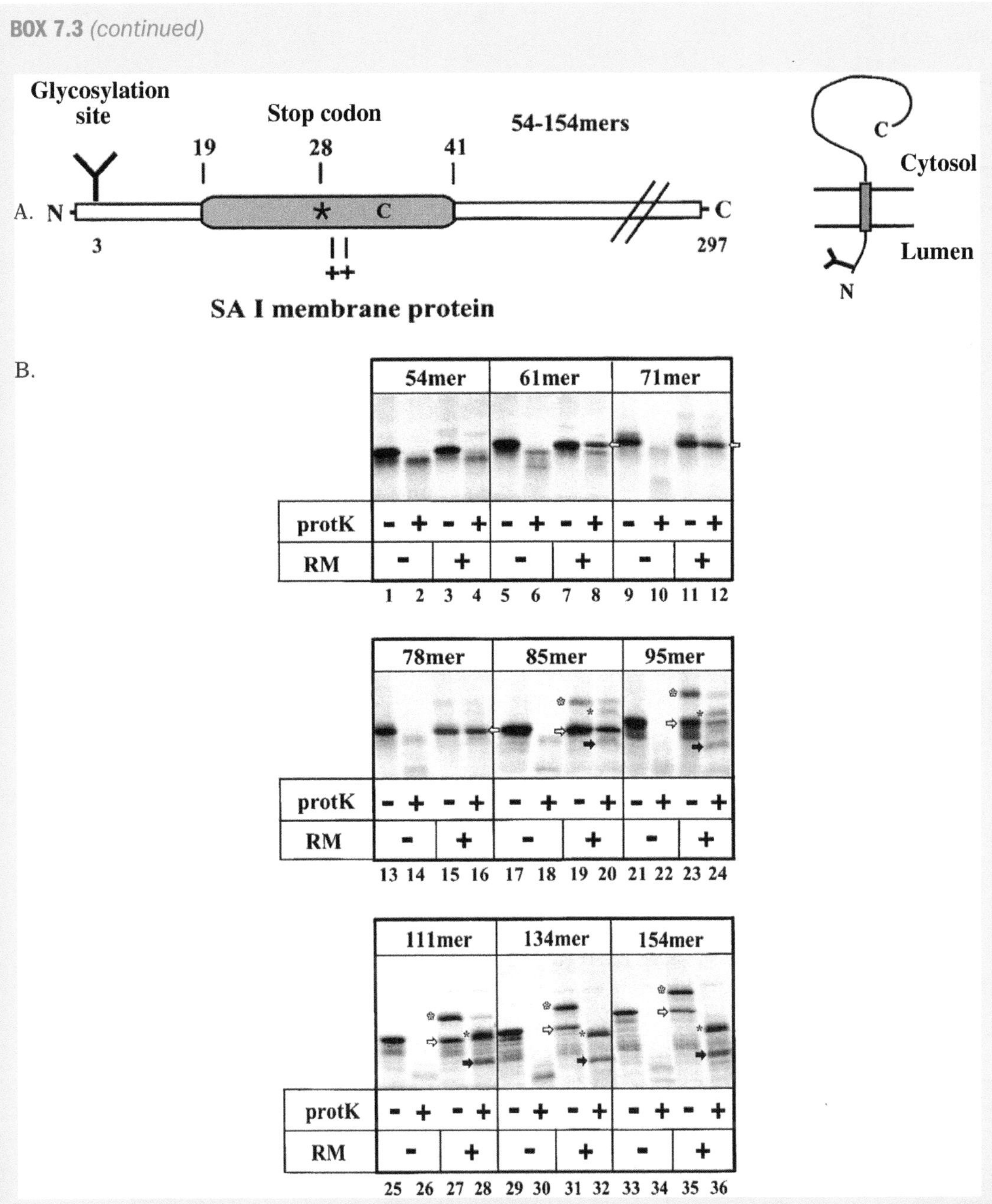

7.3.2 (A) The signal-anchor membrane-protein used in photocrosslinking experiments. (B) Protection assay of radiolabeled chains of different lengths. Synthesis of the signal-anchor nascent chains of different lengths in the presence and absence of rough microsomes (RM) was followed with and without treatment with proteinase K (protK) by SDS-PAGE and autoradiography. From Heinrich, S., *et al.*, *Cell*. 2000, 102:233–244. © 2000 by Elsevier. Reprinted with permission from Elsevier.

Figure 7.3.3C), while the 71mer does not crosslink to Sec61α. As the length increases, the lipid crosslinks remain, indicating the TM segment remains in the lipid bilayer. Figure 7.3.3B shows the results when the samples are immunoprecipitated with antibodies to Sec61α and verifies that the peptides of 61 and 68 residues are the only ones that remain in its close vicinity. Figure 7.3.3C shows what happens when the crosslinked 71mer is treated with phospholipase A and verifies that the faster-migrating band is the result of crosslinking to lipid.

When all the data are plotted according to the length of the translation intermediate, they give a sequential picture of contacts, indicating that the nascent TM segment is first exposed to the cytosol, and then enters the translocon and inserts into the lipid bilayer; finally, the N terminus becomes glycosylated as the C terminus diffuses away from the translocon into the cytosol (see Figure 7.16).

(continued)

BOX 7.3 *(continued)*

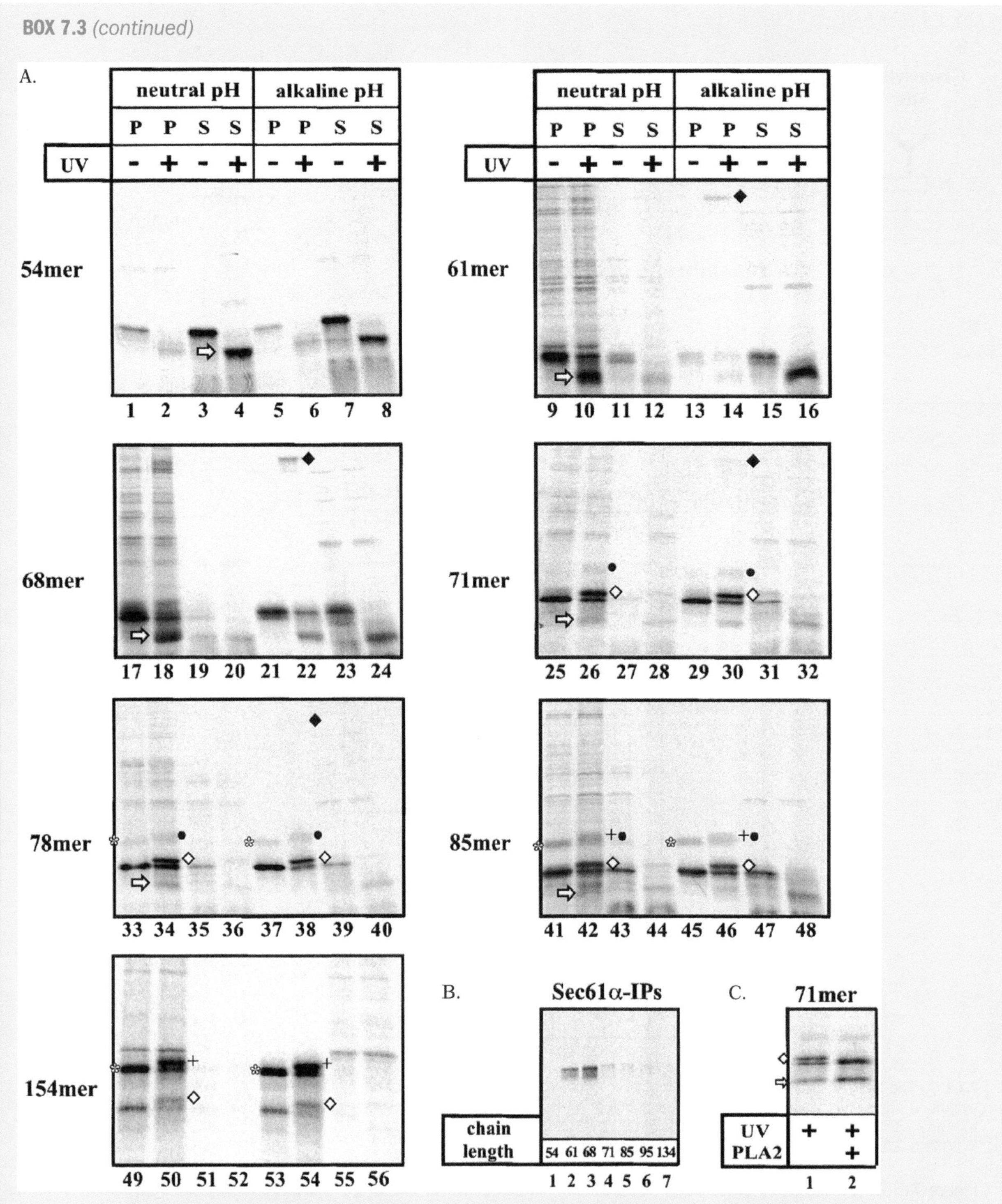

7.3.3 Chain length requirement for membrane integration. (A) Chains of different lengths were crosslinked by exposure to ultraviolet irradiation and the membranes were sedimented at alkaline or neutral pH. Pellets (P) and supernatant (S) were analyzed. (B) Immunoprecipitation with antibodies to Sec61α showed which chains interacted with the translocon. (C) Treatment with phospholipase A_2 showed which chains interacted with lipids. From Heinrich, S., *et al.*, *Cell*. 2000, 102: 233–244. © 2000 by Elsevier. Reprinted with permission from Elsevier.

In addition to the purification and characterization of the many components of the translocation process, its detailed mechanism has been probed with sophisticated studies designed to follow the progress of a nascent peptide. The first recognition of a nascent peptide destined for the membrane (or secretion) occurs in the exit tunnel of the ribosome. Fluorescence and crosslinking experiments indicate that two proteins of the large subunit make contact with helical TM segments of nascent membrane proteins and not with signal sequence segments of nascent secretory proteins. In cotranslational translocation in *E. coli*, one of the proteins at the ribosome exit, L23, makes contact with SRP and then with the translocon when the ribosome docks at the membrane (see Figure 7.13).

Once the SRP guides the ribosome translating a peptide to the translocon, or SecB/SecA chaperone a fully synthesized protein to it, the protein threads into the translocon and either crosses to the periplasm or the ER lumen or inserts into the lipid bilayer. To track nascent TM segments as they move through the translocon and into the lipid bilayer, photoactivated crosslinking groups are incorporated into nascent peptides at known locations and irradiated at different times (see Box 7.3). The nascent peptides carrying TM segments were observed to crosslink initially to cytosolic components, then to subunits of the translocon, and then to lipid (Figure 7.16). The data suggest that the translocon has a dual pathway, a prediction borne out by structural studies.

The Translocon

Remarkably, the insertion of nascent membrane proteins into either ER membranes or bacterial plasma membranes utilizes the same translocation apparatus that exports most proteins bound for extracellular destinations. Therefore the translocon is a protein-conducting channel that functions in two dimensions, allowing proteins to go into the membrane or to cross it into the lumen of the ER or the periplasm of Gram-negative bacteria. It does this while maintaining the permeability barrier of the membrane even though its channel is large enough to accommodate at least one α-helix.

The translocon is a passive pore that associates with various partners to provide the driving force for translocation (see above). A universally conserved heterotrimer, it is called the Sec61 translocon in eukaryotes, with subunits named Sec61 α, β, and γ, and the SecY translocon in bacteria, with subunits SecY, E, and G (see Table 7.1). Two of the three proteins are essential for viability, while the third (β or SecG) is not. SecG stimulates translocation as well as the ATPase activity of SecA in *E. coli*. Additional protein partners in *E. coli* include the YidC protein involved in insertion of hydrophobic TM segments into the bilayer and the SecDF heterodimer that facilitates protein translocation (see below).

TABLE 7.1 COMPONENTS OF THE PROTEIN TRANSLOCATION APPARATUS IN BACTERIA AND MAMMALIAN CELLS[a]

Function	Bacterial system: *E. coli*	Mammalian system: ER
Translocon core	SecYEG	Sec61$\alpha\beta\gamma$
Homologs	SecY	Sec61α
	SecE	Sec61γ
	SecG	Sec61β
Piloting complex	Ffh + 4.5 S RNA	SRP = 6 proteins + 7S RNA
Pilot receptor	FtsY	SRP receptor
Driving force	SecA + ATP; pmf	Ribosome + GTP

[a] The bacterial system is represented by that of *E. coli*, and the mammalian system is that found in the ER.
Source: Based on Wickner, W., and R. Schekman, *Science*. 2005, 310:1452–1456, and earlier papers.

The first images of translocons from canine ER, yeast, and *E. coli* were obtained by cryo-EM of two-dimensional crystals and showed large structures, suggestive of dimers or larger assemblies. The first high-resolution structure of the translocon from *Methanococcus jannaschii* shows a visible pore in the Sec heterotrimer, illustrating that it can function as a unit. A popular hypothesis is that dimers of translocons associate when contacted by a ribosome; however, a high-resolution EM of Sec61 in complex with a ribosome with an active synthesis of a translocated peptide shows that a monomer is sufficient (Figure 7.17).

The translocon structure

The first crystal structure of a translocon, SecY$\alpha\beta\gamma$ from the archaea *M. jannaschii*, greatly advanced understanding of the structure of the heterotrimeric translocon and suggested how it could move proteins both across the membrane and laterally into the bilayer. The α subunit has ten TM helices positioned in the membrane to form a sort of barrel with a rectangular shape when viewed from the cytosol and a pseudo-symmetry between two halves formed by TM1–5 and TM6–10 (Figure 7.18). The loop between TM5 and TM6 connecting the two halves is proposed to act as a hinge at the back side of the rectangle, with the γ subunit wrapping around to hold them together. Many of the helices are tilted up to 35° from the bilayer normal, contributing to the funnel shape of the central cavity. Not all the helices span the bilayer fully, and in particular TM2a extends only halfway through the membrane and is only partially hydrophobic. The nonessential β subunit, a single TM at the edge of the complex, makes limited contact with the α subunit near its N terminus, close to the C terminus of the γ subunit. The γ subunit makes a band along two sides of the rectangle. It consists of two helices: an amphipathic helix

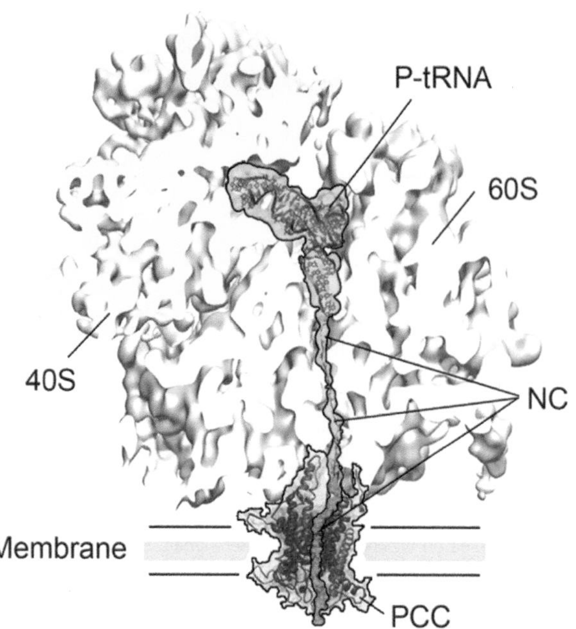

7.17 Cryo-EM image of Sec61 translocon with a eukaryotic ribosome actively translating a polypeptide chain. The surface representation of a eukaryotic ribosome (40S, yellow, 60S gray) docked on the translocon or protein-conducting channel (PCC, multicolored TM helices in interior) is visualized by high-resolution cryo-EM. The nascent chain (NC, green) extends from the P-tRNA (green) at the elongation site of the ribosome through the ribosome and through the translocon. The ribosome is interacting with a single Sec complex (rather than a dimer). From Becker, T., *et al.*, *Science*. 2009, 326:1369–1373. © 2009, American Association for the Advancement of Science. Reprinted with permission from AAAS.

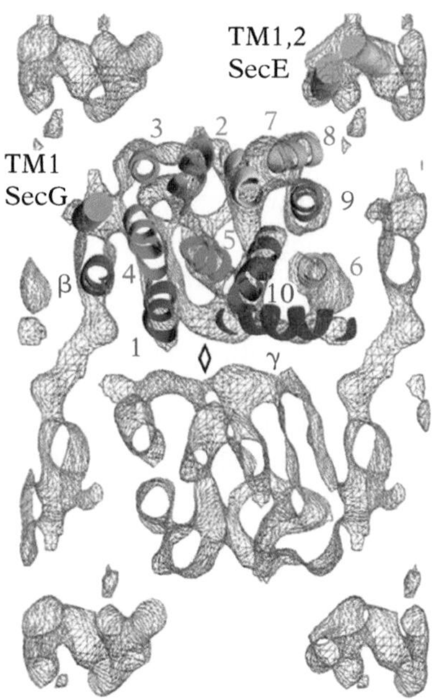

7.19 Superposition of the x-ray structure of the *M. jannaschii* translocon on the cryo-EM structure of the *E. coli* translocon. The *M. jannaschii* x-ray structure (numbered helices, colored as in previous figure) was visually docked onto the electron density map of the *E. coli* SecY complex from cryo-EM (light blue). The labeled gray cylinders represent TM helices in *E. coli* that have no correspondence in *M. jannaschii*. The diamond indicates the axis of two-fold symmetry in the *E. coli* complex. From van den Berg, B., *et al.*, *Nature*. 2004, 427:36–44. © 2004, Nature Publishing Group. Reprinted by permission of Macmillan Publishers Ltd.

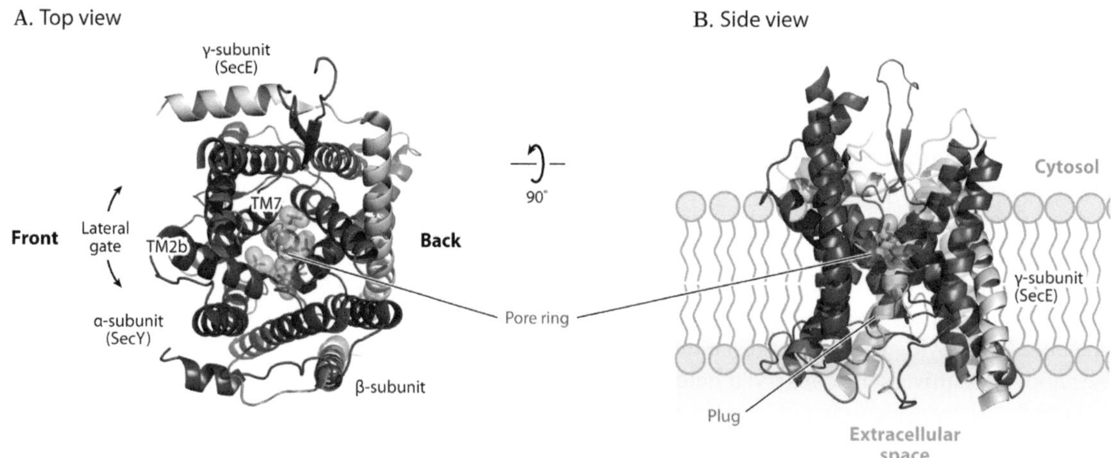

7.18 Crystal structure of the *Methanococcus jannaschii* SecY channel. (A) When the SecY channel is viewed from the cytosol, a pseudosymmetry is seen between TM1–5 (blue) and TM6–10 (red) of the α subunit (SecY in prokaryotes). The β subunit (SecG in prokaryotes, gray) is seen at the bottom, and the γ subunit (SecE in prokaryotes, beige) across the back and top. Across from the back is the lateral gate, between TM2b and TM7. The pore ring consists of bulky nonpolar residues (transparent spheres with side chains, green). (B) In the view from the plane of the membrane, the hourglass shape shows the constriction at the pore ring. In addition, the plug helix (yellow) can be seen below the pore ring. From Park, E., and T. A. Rapoport, *Ann Rev Biophys*. 2012, 41:21–40. © 2012 by Annual Reviews. Reprinted with permission from the Annual Review of Biophysics, www.annualreviews.org.

that lies on the cytoplasmic surface and a long, curved helix that runs along the back side of the α subunit. The *M. jannaschii* structure can be superimposed on the EM structure of *E. coli* translocon (SecYE) at 8 Å resolution with a good match of the TM α-helices (Figure 7.19). It also agrees very well with several bacterial translocons that have been crystallized more recently.

This first high-resolution structure provided a number of insights into the function of the translocon. The channel is shaped like an hourglass, with funnel shapes above and below its central constriction (see Figure 7.18B). The opening of the constriction is <5 Å in diameter and is lined by a ring of bulky hydrophobic side chains. Crosslinking experiments that engineered cysteine residues in SecY showed the constriction is the only region that bonded to a translocating polypeptide, giving strong evidence that this is the pore. However, the small diameter of the constriction means that other TM helices would have to shift to open the channel to 12–14 Å to allow passage of an extended peptide or an α-helical peptide; this may happen dynamically as a peptide is passing through the channel to accommodate the peptide without a leak. TM2a is proposed to be a plug that closes the channel. Indeed, when the TM2a segment in *E. coli* is locked by an engineered disulfide bond to the γ (SecE) subunit, the channel stays open. The plug is displaced (but not enough to open the channel) in the structure of SecY with SecA present.

In a crystal structure of a complex of SecY and SecA from *Thermotoga maritima* with ADP plus beryllium fluoride present to mimic ATP, the SecA lies above the plane of the membrane with its clamp just over the SecY channel (Figure 7.20). Also the two-helix finger of SecA is inserted into the cytoplasmic side of the channel in SecY, in a position that might allow it to drag the peptide substrate into the pore.

TM Insertion

The x-ray structure of the *M. jannaschii* translocon also suggested how a TM segment of the nascent protein could be released into the lipid bilayer, with a hinge movement between TM5 and TM6 at the back that would open the front to allow the peptide access to the bilayer. An x-ray structure of the translocon from *Pyrococcus furiosus* containing its two essential subunits, SecY and SecE, in a different crystallographic packing shows the same general architecture as the *M. jannaschii* translocon except for a lateral exit portal opposite the hinge (Figure 7.21). The seal of the central ring of hydrophobic residues is compromised where a cavity large enough for a helix opens along TM7. This portal

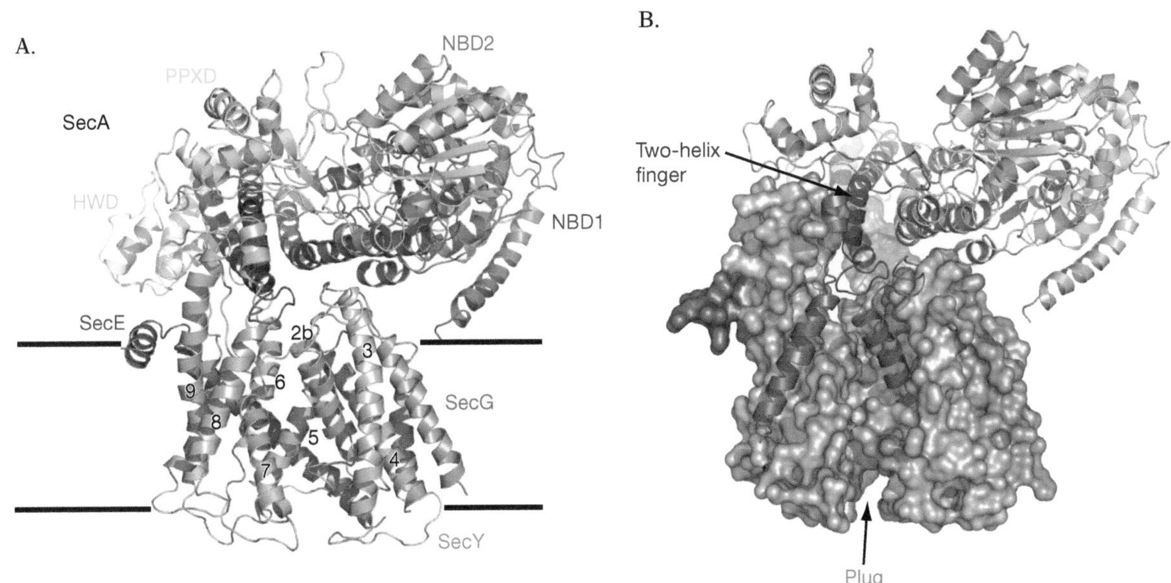

7.20 X-ray structure of a complex of SecA and SecY from *Thermotoga maritima*. (A) A ribbon representation of the SecA–SecYEG complex (SecY, gray, SecE red, SecG, green) is viewed from the plane of the membrane (solid lines). The domain structure of SecA is highlighted, with two parts of the nucleotide-binding domain, NBD1 (blue) and NBD2 (cyan), the polypeptide-crosslinking domain (PPXD, yellow) and the helical wing domain (HWD, gray). (B) A ribbon representation of SecA with a space-filling representation of the translocon highlights the two-helix finger of SecA (red) that is inside the channel of the translocon (SecYEG, colored as in (A)) as well as the plug residues (orange). Ribbon representation highlights TM2b, TM8, and the tip of the 6–7 loop in SecY. From Zimmer, J., *et al.*, *Nature*. 2008, 455:936–943. © 2008, Nature Publishing Group. Reprinted with permission of Macmillan Publisher Ltd.

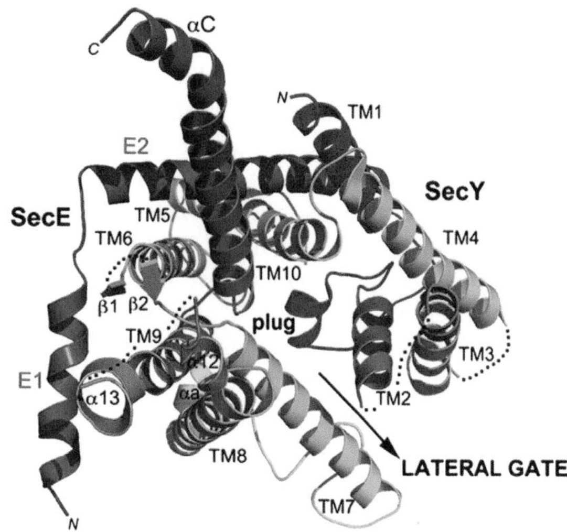

7.21 X-ray structure of the translocon from *Pyrococcus furiosus* with the lateral gate open. A ribbon representation of the SecYE from *P. furiosus* is viewed from the cytoplasm. The helices are numbered E1 and E2 from SecE (red) and TM1–TM10 from SecY (rainbow coloring, with N terminus blue and C terminus red). The TM2a plug occludes the channel in the center, and the lateral gate is open along TM7 (direction of arrow) dividing the two halves of SecY. From Egea, P. F., and R. M. Stroud, *Proc Natl Acad Sci USA*. 2010, 107:17182–17187.

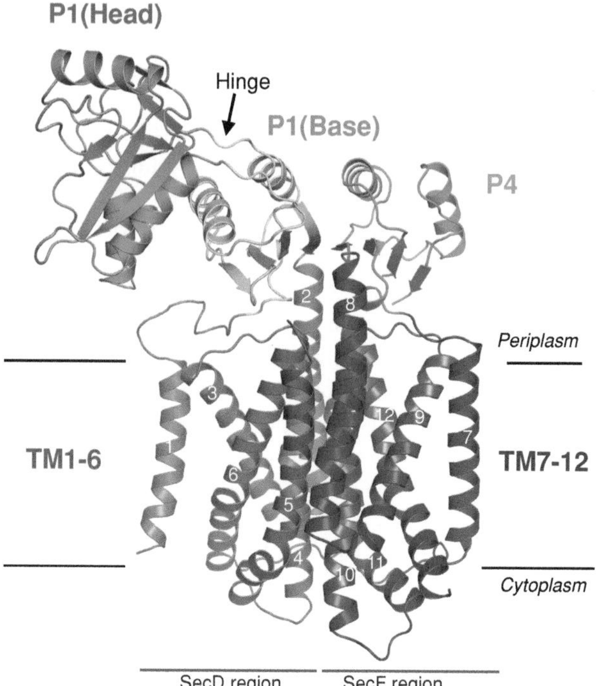

7.22 X-ray structure of SecDF from *Thermus thermophilus*. A ribbon representation of the SecDF is viewed from the plane of the membrane. In *T. thermophilus* SecDF is composed of a single peptide chain, containing 12 TM helices that correspond to SecD (TM1–6, red) and SecF (TM7–12, blue) in other bacteria. In addition it has two major periplasmic domains, P1 (orange and green) and P4 (cyan). Two conformations of the periplasmic domains are observed in different crystals, which are related by a conformational movement at the hinge in P1 that moves the P1 head over the P1 base. From Tsukazaki, T., *et al.*, *Nature*. 2011, 474:235–238. © 2011, Nature Publishing Group. Reprinted by permission of Macmillan Publishers Ltd.

spans the membrane, providing a lateral gate for peptide substrates to contact the lipid bilayer while the plug blocks the opening to the *trans* side of the membrane. If the peptide is sufficiently nonpolar, it will partition into the lipid as a TM helical segment.

Insertion of bacterial membrane proteins can also utilize YidC along with or independent of the SecY translocon. YidC has homologs in mitochondria (Oxa1) and in chloroplasts (Alb3). YidC was discovered for its role in inserting small membrane proteins that were categorized as Sec-independent proteins, such as the M13 phage coat protein. YidC is a polytopic membrane protein with five TM helices and a periplasmic domain with a β fold. Both YidC and the Sec translocon are required for the insertion of certain proteins, such as subunit a of the ATP synthase. While photo-crosslinking studies show these substrates interact with SecY before YidC, the precise role of YidC is not known.

Yet another site for integration of membrane proteins is the SecDF complex that uses the proton gradient (or in some bacteria, a sodium ion gradient) to enhance protein export. Its role was clarified with studies using a substrate protein, proOmpA, blocked during translocation by an engineered disulfide bond creating a 59-residue loop. When the loop was released by reduction with dithiothreitol, translocation via SecDF could resume in the absence of ATP, driven by the proton motive force. The crystal structure of SecDF from *Thermus thermophilus* shows a 12-TM membrane domain that conducts protons, along with two periplasmic domains that undergo conformational changes when they interact with substrate proteins (Figure 7.22). This heterodimer is a member of the RND family of transporters, with a proton channel similar to AcrB (see Chapter 11). It may be associated with YajC, a small bitopic membrane protein whose function is unknown.

An additional participant in membrane protein integration in eukaryotic cells is called TRAM (translocating chain-associated membrane protein). TRAM is required for translocation of some proteins and could be reconstituted with the Sec61 translocon and SRP to transport them in liposomes. Data from crosslinking experiments of the sort described in Box 7.3 show that TRAM specifically interacts with signal sequences during their passage through the Sec61 channel and associates with TM segments as they leave the translocon and integrate into the lipid bilayer. TRAM may guide the integration of TM segments to provide the correct topology.

Numerous other proteins that play important roles in the translocation of nascent polypeptides include enzymes and chaperones. Cleavable N-terminal signal sequences are removed by signal peptidase (also called leader peptidase, Lep), which is anchored with its active site on the noncytoplasmic side of the membrane. An *E. coli* chaperone called trigger factor (TF) competes with the SRP for emerging signal peptides to sort nascent peptides between the SecB/A pathway and the SRP pathway. Other *E. coli* chaperones in the periplasm, such as Skp and SurA, prevent aggregation of outer membrane proteins secreted by the translocon. Also in the periplasm, the Dsb system oxidizes cysteine residues to form disulfide bonds, and other enzymes make bonds to lipid or lipopolysaccharide. Similarly, covalent modification of the nascent proteins to bind sugars and/or lipid is a critical part of the maturation process in the ER lumen of eukaryotes. Clearly, the translocon is at the epicenter of a complex molecular machine that carries out many dynamic processes.

Other translocons or protein export complexes are much less well understood than the well-characterized Sec system. The cytoplasmic membrane of bacteria and the thylakoid membrane in plants have a poorly understood system called the twin arginine translocation (TAT) pathway that transports folded proteins across the membrane. Precursor proteins are recognized by their cleavable signal sequences with twin arginine residues in the motif S/T-RRXFLK. The protein components TatA, TatB, and TatC form homo- and hetero-oligomeric complexes, with pore formation induced by the folded Tat precursor proteins. The structure of the pore and the mechanism of transport are unknown.

Another interesting protein transport system is the assembly complex for insertion of β-barrel proteins into the outer membrane, called the BAM complex (previously Omp85). The complex in *E. coli* consists of a pore-forming protein BamA and four lipoproteins, BamB, BamC, BamD, and BamE, of which BamA and BamD are essential for outer membrane biogenesis. BamA has a large N-terminal periplasmic domain and a C-terminal β-barrel domain in the membrane. The periplasmic portion of BamA has five smaller domains called POTRA (polypeptide-transport associated) domains that are thought to bind unfolded outer membrane protein precursors in the periplasm: They recognize a motif in the C-terminal β-strand that ends with an obligatory aromatic residue as the C terminus. A recent crystal structure shows BamB to be a β-propellar, but how it and the other lipoproteins interact with the BamA pore is not known. Nor is it understood how a β-barrel pore can enable other β-barrels to fold and insert into the membrane. Similar systems are found in the outer membranes of chloroplasts (Toc75) and mitochondria (the SAM complex), where they are one of the systems called translocators (transport complexes) in the specialized mechanisms for protein import into chloroplasts and mitochondria (see Box 7.4).

Biological Hydrophobicity Scale

The mechanism of integration of proteins into the membrane by opening the lateral gate in the translocon signifies that when nascent TM segments are exposed to the lipid as well as the aqueous channel of the translocon they partition into the bilayer due to their hydrophobicity. Thus the nonpolar character of these segments allows their prediction from the primary sequence based on hydrophobicity scales. As shown in Table 6.3, traditional biophysical scales for hydrophobicity give values of the free energy for partitioning of each amino acid into organic solvents representing the lipid phase.

A biological hydrophobicity scale encompassing the role of the translocon is derived from an ingenious set of experiments that determined the apparent free energy (ΔG_{app}) of inserting each amino acid, as a part of a TM helix, into the cell membrane. The amino acid is placed in a test segment engineered to follow the two TM segments of leader peptidase (see Figure 6.18) so that when synthesized in the cell, it is either inserted into the membrane or exported past it; in the latter case, it is glycosylated, which allows quantification (Figure 7.23A). The test segment consists of leucine and alanine residues flanked by GGPG linkers, with the amino acid of interest in the center position.

As expected, the nonpolar amino acids have $\Delta G_{app} < 0$, promoting membrane insertion, while the polar and charged residues have $\Delta G_{app} > 0$ (Figure 7.23B). The biological scale for membrane insertion correlates very well with the White–Wimley scale derived earlier, supporting the postulate that during translocation the segment is directly exposed to the bilayer. Interestingly, some amino acids give different values when they are positioned away from the center of the segment: notably, positions closer to the ends of the segment are more favorable than the center for Trp and Tyr, as well as Lys and Asn. Structural data show a prevalence of Trp and Tyr in aromatic girdles of integral membrane proteins at the interfacial regions of the bilayer, while "snorkeling" of Lys and Arg residues enables their basic groups to move to the more polar interface (see Chapter 4).

During the insertion of multispanning membrane proteins, TM segments that alternate with hydrophilic segments have stop transfer sequences, which are regions of the peptide sufficiently nonpolar to partition into the membrane. The presence of more than one stop transfer sequence in polytopic proteins allows each hydrophobic TM segment to be inserted sequentially as each one reaches the translocon. An engineered protein that contains four copies of a stop transfer sequence inserts each as a TM segment, provided the loops between them are of sufficient length (Figure 7.24). However, some

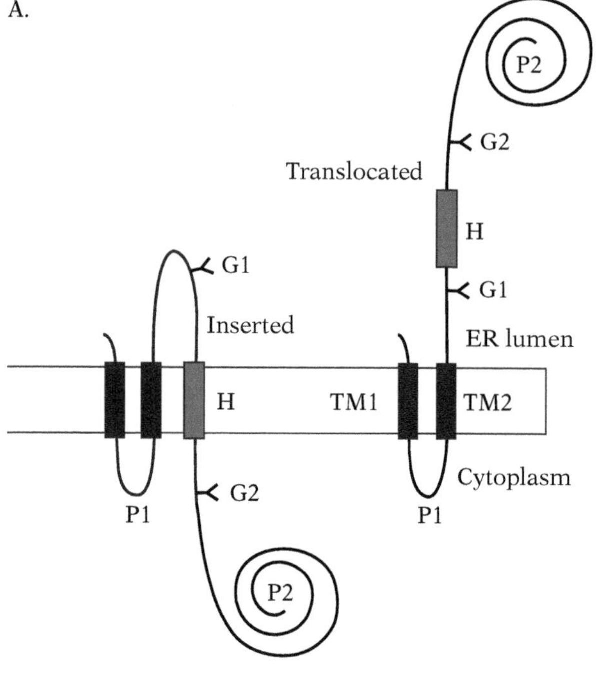

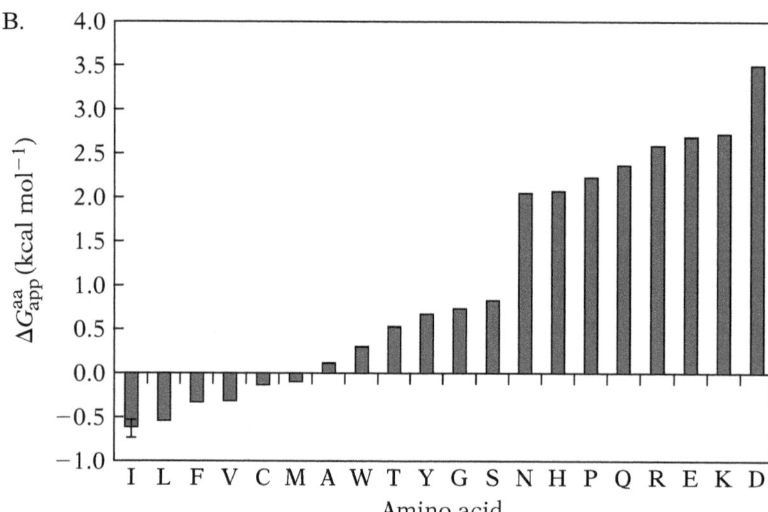

7.23 Biological hydrophobicity scale for insertion of TM segments. (A) A test segment is engineered in leader peptidase (Lep) constructs that will be either inserted into the membrane or translocated across it. The test segment, H, consists of a flanking sequence GGPG, followed by 19 leucine residues, then the flanking sequence GPGG, with each amino acid to be tested inserted in its center. Two sites for *N*-glycosylation are also inserted, as indicated by G1 and G2. Each construct was expressed in BHK cells that were immunoprecipitated with Lep antiserum after labeling with ^{35}S-methionine. The bands corresponding to singly and doubly glycosylated proteins were quantified on a phosphorimager after separation by SDS-PAGE. (B) The ΔG_{app} for membrane insertion of each amino acid placed in the center of a TM segment is calculated from the apparent equilibrium constant of the singly glycosylated and doubly glycosyated Lep molecules obtained with the contructs described in (A). The bar indicates the standard deviation in the ΔG_{app} for Ile; the standard deviation for each of the others is similar. Redrawn from Hessa, T., *et al.*, *Nature*. 2005, 433:377–381. © 2005. Reprinted by permission of Macmillan Publishers Ltd.

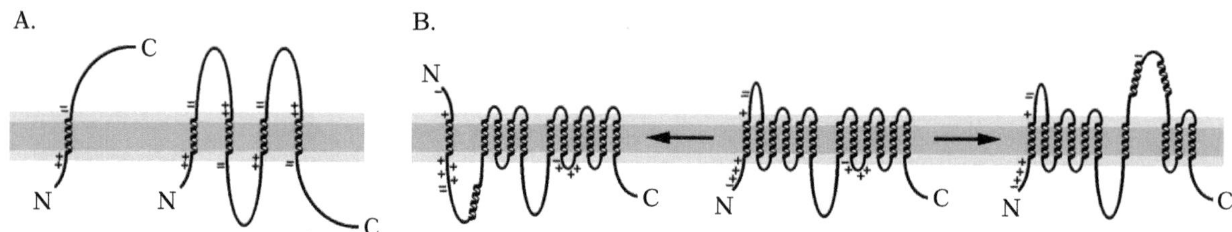

7.24 Functional topogenic determinants throughout the polypeptide. Downstream portions of polytopic membrane proteins may influence topology, depending on loop size. (A) Insertion of the bitopic ASGP receptor follows the positive-inside rule. However, in an engineered protein consisting of the four copies of its signal-anchor sequence with flanking charges, separated from each other by ~100 residues, the TM segments insert sequentially in spite of their flanking charges. It appears that the long loops between TM segments allow insertion to override the positive-inside rule. (B) A natural protein with 12 TM segments with short loops between them, Glut1, inserts according to the positive-inside rule (wild type, in center). Perturbing its topogenic determinants by introduction of different charges disturbs the local topology but not the other TM segments (right and left). From Goder., V., and W. Spiess, *FEBS Lett*. 2001, 504:87–93. © 2001 by the Federation of European Biochemical Societies. Reprinted with permission from Elsevier.

BOX 7.4 IMPORT OF MITOCHONDRIAL PROTEINS

Nearly all (98%) of the >1000 mitochondrial proteins are synthesized in the cytosol and imported into the mitochondria, destined either for the matrix, the intermembrane space, the inner membrane, or the outer membrane. The imported proteins are classified in two major groups: precursor proteins with cleavable amino-terminal signals, called presequences; and hydrophobic membrane proteins with targeting signals dispersed throughout their primary structures. In addition, the proteins of the outer membrane and the intermembrane space constitute special classes.

All imported proteins are transported across the outer membrane of the mitochondria via the translocase of the outer membrane (TOM). In addition, the outer membrane has a complex for the assembly of *β*-barrel proteins into the outer membrane, after they have been taken into the intermembrane space via the TOM complex (Figure 7.4.1). This SAM complex is similar to the BAM complex in bacterial outer membranes. Two more translocators are located in the mitochondrial inner membrane: the TIM23 complex transports cleavable preproteins, while the TIM22 complex inserts hydrophobic proteins into the inner membrane (Figure 7.4.2). The components of these complexes are proteins that act as receptors, form pores, and/or exhibit chaperone functions and are well characterized in yeast. These complexes appear to accumulate at adhesion sites where the inner and outer membranes are 18–20 nm apart, which could allow protein complexes in the two membranes to cooperate during translocation.

The TOM Complex

Translocation across the outer membrane is a passive process mediated by receptors (Tom20, Tom22, and Tom70) and a Tom40 complex. The presequences of cleavable precursor proteins are positively charged and can form amphipathic helices that fit into a hydrophobic groove of Tom20. They are transferred to Tom22 by its recognition of the opposite, charged surface of the helices. Tom70 is a receptor for hydrophobic precursors and forms oligomers upon binding them, which could help prevent aggregation. The structures of the soluble receptor domains of Tom70 and the related Tom71 have been determined by x-ray crystallography, and the structure of Tom20 is known from both NMR and x-ray studies. The main channel is formed by Tom40, a *β*-barrel with a pore of ~20 Å diameter – large enough to accommodate two *α*-helical polypeptide segments but not a folded protein. In the membrane, Tom40 functions with the smaller proteins Tom5, Tom6, and Tom7, which may help stabilize several Tom40 dimers to form

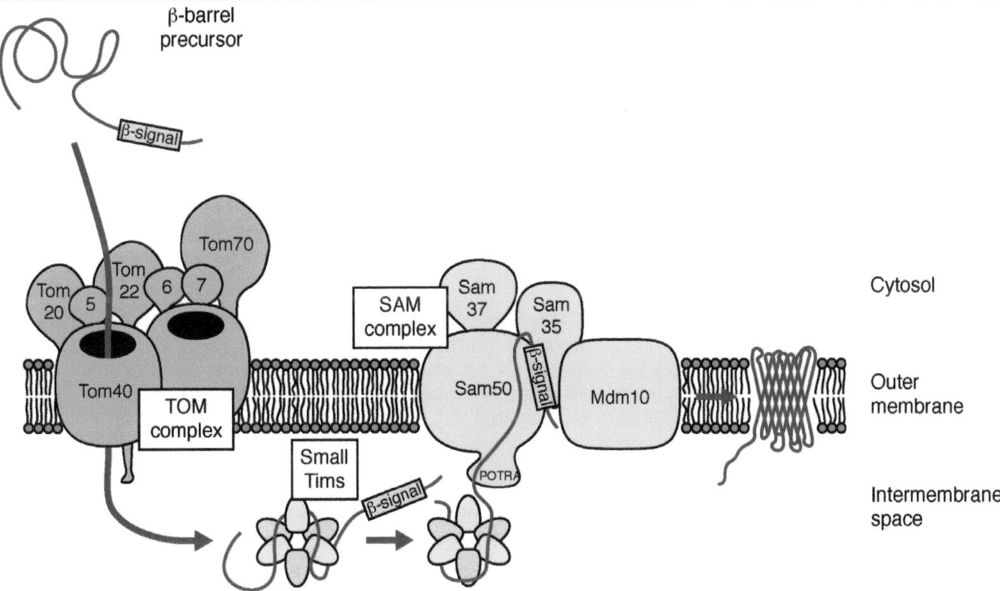

7.4.1 Biogenesis of mitochondrial outer membrane proteins. The two translocons in the mitochondrial outer membrane are the TOM complex and the SAM complex. The TOM complex is used by all mitochondrial proteins that are synthesized in the cytosol, including those destined for the outer membrane. The precursors of *β*-barrel proteins use the TOM complex to enter the intermembrane space, where they are chaperoned by the small Tim proteins. Then they are folded and inserted into the outer membrane by the SAM complex, and perhaps assembled with the help of morphology maintenance proteins, such as Mdm10. From Becker, T., *et al.*, *Curr Opin Cell Biol.* 2009, 21:484–493. © 2009, Elsevier. Reproduced with permission from Elsevier.

(continued)

BOX 7.4 *(continued)*

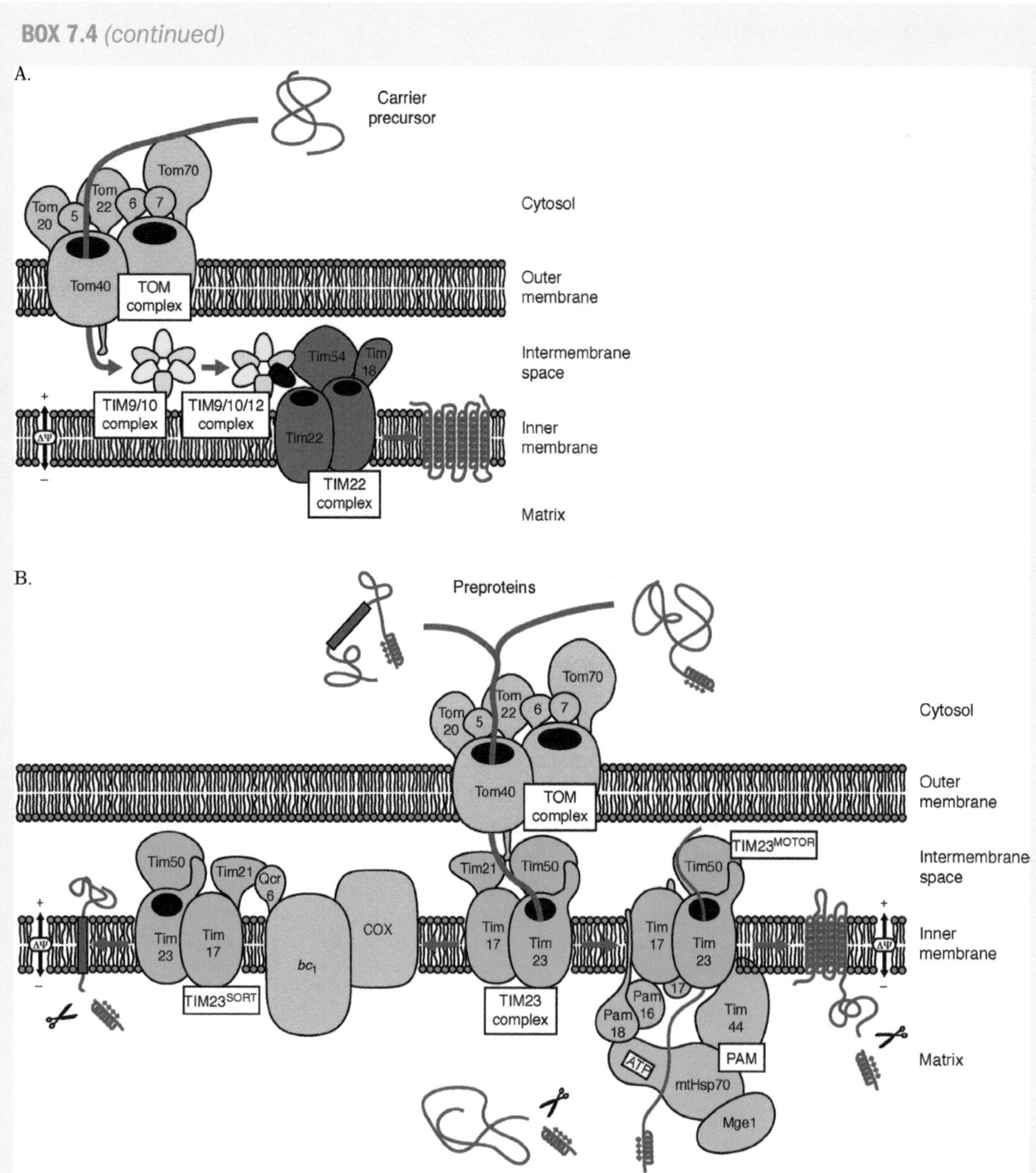

7.4.2 Biogenesis of mitochondrial inner membrane proteins. After transport across the outer membrane via the TOM complex, inner membrane proteins use either the TIM22 complex (A) or the TIM23 complex (B). (A) The TIM22 complex is the carrier pathway used by metabolite carrier proteins that are generally very hydrophobic. They are chaperoned in the intermembrane space by the Small Tim proteins prior to insertion into the inner membrane. (B) The proteins with presequences that are destined for the inner membrane or for the matrix use the TIM23 complex, along with its multi-subunit motor, the PAM complex. From Becker, T., *et al.*, *Curr Opin Cell Biol*. 2009, 21:848–893. © 2009, Elsevier. Reproduced with permission from Elsevier.

a large, dynamic complex containing several pores. Interestingly, in biogenesis the import of precursors to the Tom proteins utilizes the TOM apparatus, before they are inserted into the membrane by SAM.

Assembly of β-barrel proteins into the mitochondrial membrane utilizes a highly conserved complex with a poorly understood mechanism. The last β-strand of these proteins contains a β-signal that is crucial for targeting them to the SAM complex. Sam50 (analogous to BamA in bacteria) has been shown to bind Tom40 (as a substrate) before it inserts into the membrane. Two peripheral proteins are involved in the SAM complex on the cytosolic side: Sam35 interacts with the β-signal sequence, and Sam37 helps release the substrate proteins. The small Tim proteins, which are soluble hexameric α-propellers composed of Tim9/10 or Tim8/13, chaperone the unfolded substrates in the intermembrane space. The morphology maintenance protein, Mdm10, associates with the SAM complex and appears to help Tom40 assemble with the other Tom proteins to form the TOM complex.

BOX 7.4 *(continued)*

TIM Complexes

The pathways for two other classes of imported proteins diverge after translocation via TOM (see Figure 7.4.2). Translocation of presequence-containing proteins across the inner membrane uses the TIM23 complex, which consists of Tim23, Tim17, Tim21, and Tim50 in the inner membrane. The TIM23 complex may be sufficient to insert single-spanning proteins. To insert polytopic proteins the TIM23 complex interacts with a peripheral motor, the PAM complex anchored by the Pam18 membrane protein. In the PAM complex, Tim44 provides a binding site for mitochondrial Hsp70 (mHsp70 or Ssc1p). The mitochondrial Hsp70 hydrolyzes ATP in repeated cycles of binding of the transported peptides in the matrix to prevent their return. It works with a nucleotide exchange factor, Mge1, that promotes ADP release. The structures of the soluble domains of Pam18 and Tim21, as well as the peripheral proteins Tim44 and Pam16 are known from x-ray crystallography. When reconstituted in liposomes, purified TIM23 forms a voltage-activated cation-selective channel that is activated by presequence peptides and a membrane potential ($\Delta\Psi$, negative inside), while inhibited by presequence peptides alone. While this suggests that $\Delta\Psi$ is required for channel opening, translocation across the membrane also uses hydrolysis of ATP. The mitochondrial processing peptidase (MPP) cleaves the amino-terminal presequence in the matrix. TIM23 also interacts with complexes III and IV of the respiratory chain (see Chapter 13).

The TIM22 complex is a translocator responsible for insertion of hydrophobic inner membrane proteins, including the metabolite carrier proteins such as the ATP–ADP carrier (described in Chapter 12), as well as the Tim proteins themselves. It utilizes the Tim9–Tim10 complex in the intermembrane space, the peripheral protein Tim12, and three integral membrane proteins, Tim22, Tim54, and Tim18, of which only Tim22 is essential for viability. Tim22 contains a double channel and uses the membrane potential to drive insertion; when purified and reconstituted, it forms a gated channel with two pore sizes, the larger of which allows insertion of tightly packed loops.

The mitochondrial translocators are clearly fruitful areas for research.

polytopic proteins appear to exhibit cooperation between adjacent TM segments, which implies more than one can be in the translocon simultaneously. Although the channels observed in high-resolution structures of the translocon are only wide enough for the passage of an unfolded polypeptide or a single α-helix, it is possible that alternate conformations of the translocon involving dimers or higher oligomers allow more than one helical peptide segment to fit in the channel. This would explain the evidence that some TM segments can influence the insertion of neighboring TM domains in some polytopic proteins (see below).

Topogenesis in Membrane Proteins

In discussing the signals that determine the topology of integral membrane proteins, which have been studied extensively in the mammalian ER and also in *E. coli*, it is useful to refer to the outside compartment (lumen or periplasm) as the exocytoplasmic side of the membrane. Thus for bitopic membrane proteins, type I are oriented N_{exo}/C_{cyt} and type II are N_{cyt}/C_{exo}. Multispanning or polytopic proteins, called type III membrane proteins, may have an even or an odd number of TM segments and thus have their N and C termini either on the same side of the membrane or on opposite sides with either orientation. The first signal occurring in the sequence of a polytopic protein is treated much as the signal of a bitopic protein, and insertion of the remaining TM segments can be assumed to follow in sequence, even though there are other factors influencing insertion, as will be discussed.

Signals that govern topogenesis are classified as (1) signal sequences, (2) signal-anchor sequences, and (3) reverse signal-anchor sequences. The presence of one of these sequences determines the orientation of the first TM segment of the protein (Figure 7.25). Insertion of the later TM segments of polytopic membrane proteins is initiated by stop transfer sequences, which have sufficient length to span the membrane and can insert stably into the bilayer.

Signal sequences have already been discussed for their role in targeting the nascent chain to the membrane. Not all signal sequences are cleaved by signal peptidase. Those that are cleaved are released and degraded by signal peptide peptidase. After cleavage, which makes a new N terminus outside the membrane, the rest of the peptide is fully exported as a secreted protein unless it is followed by a stop transfer sequence to create a TM segment (see *a* and *b* of Figure 7.25).

A stop transfer sequence that can insert stably in the membrane is called a signal-anchor sequence. Signal-anchor sequences remain uncleaved and have enough nonpolar residues to span the hydrophobic domain of the bilayer. When it is the only TM segment and is not preceded by a cleavable signal sequence, a signal-anchor inserts to create a type II membrane protein (*c* in Figure 7.25).

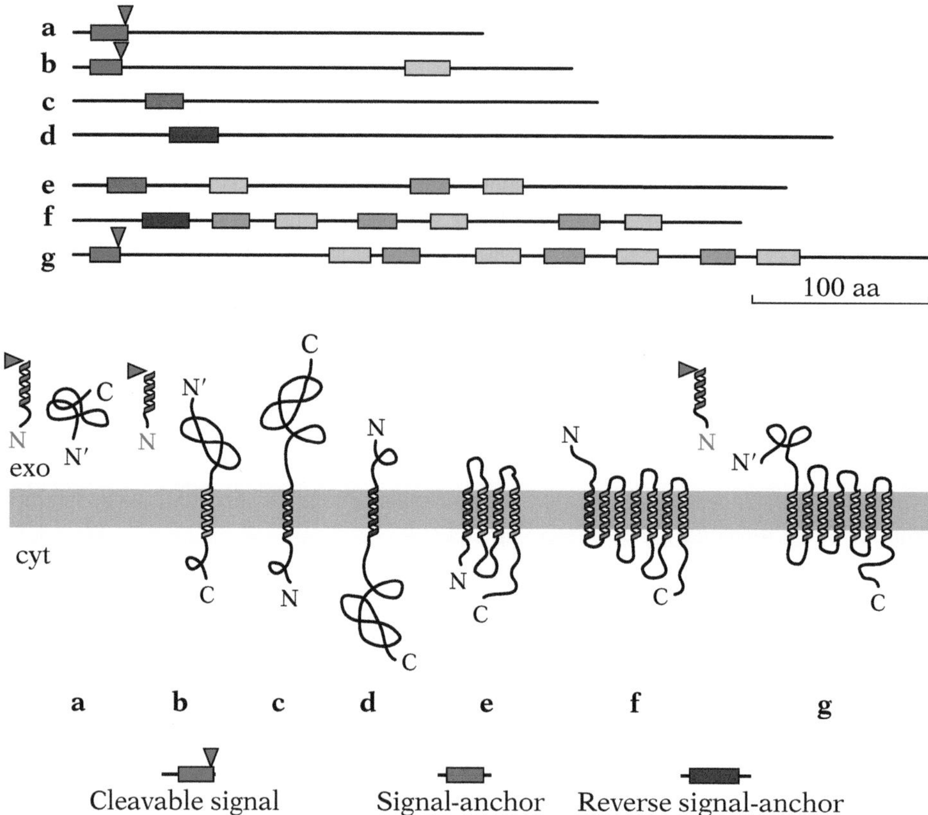

7.25 Three types of signals initiate topogenesis. Cleavable signal sequences (green, with arrowheads for cleavage sites) are translocated, leading the N terminus to the outside before they are cleaved. After cleavage by signal peptidase, they are digested by signal peptide peptidase outside the membrane (not shown). The protein is released (a) unless the cleaved signal is followed by another TM segment (b). Uncleaved signal-anchor sequences (red) induce translocation of the rest of the peptide while they remain TM (c). Reverse signal-anchors (blue) insert to become TM as the N terminus is translocated (d). Thus for bitopic proteins, both a cleaved signal and a reverse signal-anchor produce the N_{exo}/C_{cyt} orientation (b and d), while a signal-anchor produces the N_{cyt}/C_{exo} orientation (c). For polytopic proteins (e, f, and g), additional TM segments insert in alternating orientations (light red for N_{cyt}/C_{exo} and light blue for N_{exo}/C_{cyt}). The sequences shown represent a secreted protein (a, prolactin), two type I membrane proteins (b, asialoglycoprotein receptor; d, cation-dependent mannose-6-phosphate receptor), a type II membrane protein (c, synaptotagmin I), and three type III proteins (e, gap junction protein; f, vasopressin receptor V2; g, glucagons receptor). From Higy, M., *et al.*, *Biochemistry*. 2004, 43:12716–12722. © 2004 by American Chemical Society.

Reverse signal-anchors are like signal-anchors except that they create type I membrane proteins (*d* in Figure 7.25). They are differentiated from signal-anchors not by obvious physical characteristics but by more subtle factors that affect their function, allowing them to insert with an N_{exo}/C_{cyt} orientation. The factors that influence the orientation of the signal-anchor sequence as it emerges from the translocon are just being elucidated.

The topologies that result from the different signals indicate that a TM segment can insert into the bilayer in either orientation and can thus translocate its C terminus or its N terminus (Figure 7.26). Current evidence points to five factors that influence the orientation of signals within the translocon.

(1) The charges flanking the hydrophobic sections of signals greatly influence the resulting topology. As described in the last chapter, the positive-inside rule, discovered in bacteria where positive residues are more abundant in cytoplasmic loops than in periplasmic loops, holds for all organisms. In eukaryotes the charge difference between the flanking segments of a signal's hydrophobic core, described as Δ(N–C), results in the more positive flanking sequence preferring the cytoplasmic side. Therefore if the signal can change directions while in the translocon, it will orient to allow the more positive flanking residues to stay on the more negative side of the membrane and perhaps to interact with residues of the translocon on the cytoplasmic side.

(2) Because a large globular folded domain will not get through the translocon, an N-terminal domain that folds into a large globular domain will not be translocated, even if the flanking charges suggest it should. Evidence for this comes from many

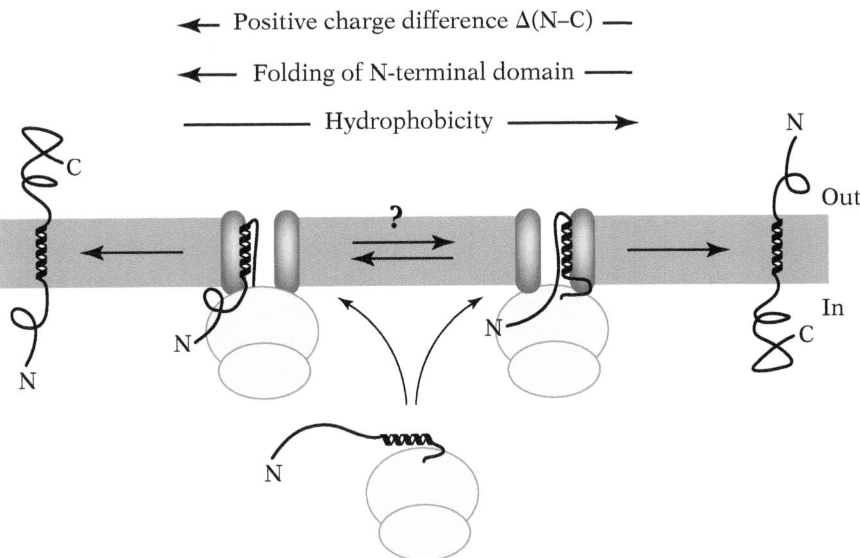

7.26 Several factors affect the orientation of signals in the translocon. Signals translocate either their C or their N terminus once they enter the translocon, depending on factors such as the difference in positive charges, the hydrophobicity, and the folding of the N-terminal domain. These factors are thought to affect the orientation of signal-anchor and reverse signal-anchor sequences as they engage with the translocon. The resulting orientation determines whether the C terminus or the N terminus emerges. From Goder, V., and W. Spiess, *FEBS Lett.* 2001, 504:87–93. © 2001 by the Federation of European Biochemical Societies. Reprinted with permission from Elsevier.

studies with fusions and truncations that attach or eliminate rapidly folding globular domains. However, natural proteins that have extensive globular domains are probably protected from folding in the cytoplasm by chaperones.

(3) Another factor that can override the positive-inside rule is hydrophobicity, specifically the length and composition of the nonpolar sequence of the signal. When artificial signals are made up of leucine residues, orientation is a function of length: Leu_7 drives mostly C-terminal translocation, resulting in N_{in}/C_{out}, while Leu_{22} and Leu_{25} produce N-terminal translocation, resulting in N_{exo}/C_{cyt}. When amino acids are ranked for their ability to promote N-terminal translocation, the best are the most hydrophobic, which also have the highest propensity to form α-helices in nonpolar environments.

(4) An additional factor for multispanning membrane proteins derives from cooperation with the rest of the molecule. Charge mutations designed to invert the topology of polytopic proteins can instead affect a region of the protein without inverting it, as observed with the human glucose transporter, Glut1, a 12-TM helix bundle (Figure 7.24). One or two hydrophobic segments can be "frustrated" enough by the mutation to remain outside the membrane, rather than perturbing the rest of the topology. The short loops between many TM segments suggest they insert as helical hairpins, preserving the expected topology, if the translocon accommodates two helices.

(5) Finally, all glycosylated proteins have sites for glycosylation on the outside. In the ER, an oligosaccharyl transferase associates with the translocon and glycosylates nascent chains as they emerge. An engineered site that becomes glycosylated will force that portion of the polypeptide to end up on the outside and prevent reorientation of that segment.

Several observations, such as the reverse orientation of type II proteins, raise questions of how nascent chains emerge from the ribosome and from the translocon. At what point do TM helices form? Proteolysis after pulse-chase labeling, FRET results, and cryoEM images make it clear that helices form in the last ~30 Å of the ribosome exit tunnel (Figure 7.27). Furthermore, a TM peptide sequence that is helical inside the exit tunnel stays helical when it emerges if it is directly inserted into the membrane. (In contrast, peptide sequences from soluble proteins have negligible helicity both inside and upon emerging from the exit tunnel.)

Can two or more TM helices interact in the translocon? The topology of some polytopic proteins is dependent on the order of TM segments in the sequence, suggesting that "strong" TM domains can help integrate adjacent "weak" TM domains. The interactions between strong and weak helices seem to take place in the translocon when the two helices are close enough in the sequence to enter it together, because in experiments that prevent the interaction

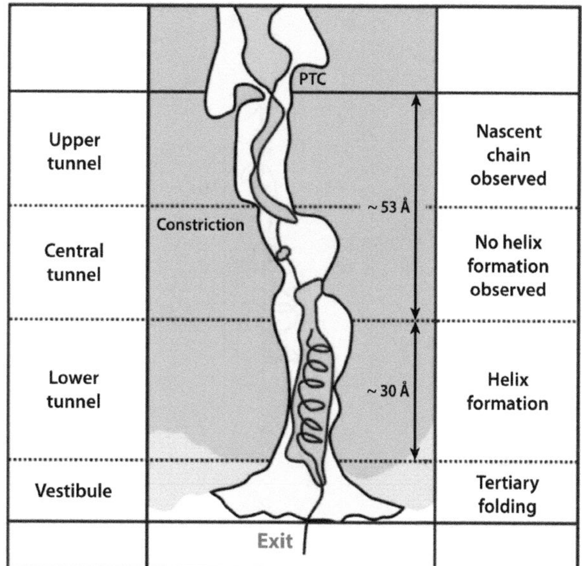

7.27 Protein folding in the ribosome exit tunnel. A profile of the exit tunnel of the eukaryotic ribosome was constructed based on cryo-EM maps of different ribosome-bound nascent chains that are observed with different peptidyl tRNAs. Helix formation is observed in about 30 Å near the exit. From Fedyukina, D. V., and S. Cavagnero, *Ann Rev Biophys*. 2011, 40:337–359. © 2011 by Annual Reviews. Reprinted with permission from the Annual Review of Biophysics, www.annualreviews.org.

by engineering a longer distance between them, the "weak" helix no longer inserts into the bilayer. Therefore mechanisms may exist that allow simultaneous insertion of more than one helix from the translocon (Figure 7.28). Such mechanisms are not ruled out by the small channels observed in x-ray structures of the translocon because they may utilize dynamic changes within the translocon or between interacting translocons.

Finally, a number of polytopic proteins can insert into the membrane with alternative topologies, including the EmrE multidrug resistance protein (see Chapter 11), the prion protein, PrP, and the ion channel involved in cystic fibrosis, CFTR (see below). The mechanisms for their biogenesis are important subjects for research, especially for proteins whose altered topology results in disease.

MISFOLDING DISEASES

Since correct insertion of membrane proteins determines their topology (and therefore their fold) and mistakes cannot be remedied simply by refolding due to the barrier to reorienting TM segments across the bilayer, errors in this process can be costly. In an

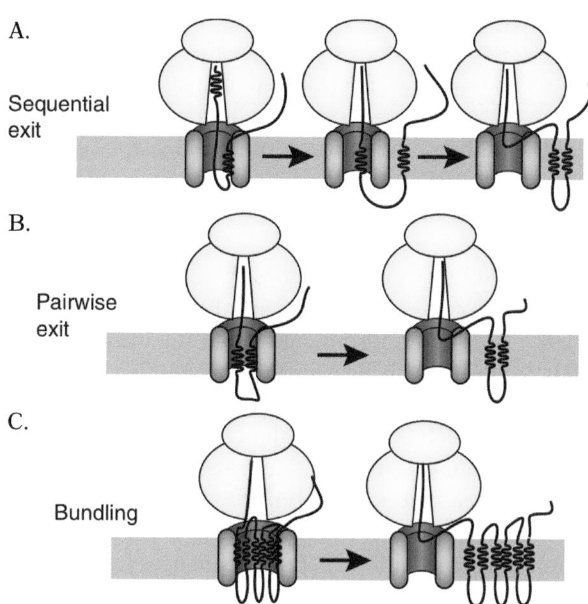

7.28 Models for mechanisms of TM integration. (A) The conventional model is for sequential integration of each TM helix independently from the translocon into the lipid bilayer. (B) A second model allows TM helices to transfer from the translocon as a pair. (C) It is possible that larger groups of helices collect in the translocon and transfer together into the lipid environment. From Skach, W. R., *Nat Struct Mol Biol*. 2009, 16:606–611. © 2009, Nature Publishing Group. Reprinted by permission of Macmillan Publishers Ltd.

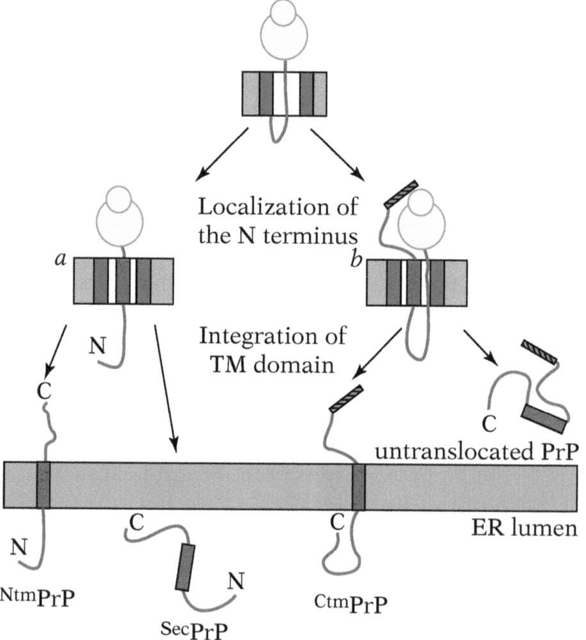

7.29 The three topologies of the prion protein, PrP. The TM segment of PrP can insert in the bilayer in both orientations, N_{ext}/C_{cyt} (^{Ntm}PrP) and N_{cyt}/C_{ext} (^{Ctm}PrP). In addition, it can be secreted into the lumen in a soluble form (^{Sec}PrP), which can then become attached to a GPI-anchor. There is evidence for a role of ^{Ctm}PrP in the neurodegenerative scrapie diseases. Redrawn from Ott, C. M., and V. R. Lingappa, *Biochemistry*. 2004, 43:11973–11982. © 2004 by American Chemical Society.

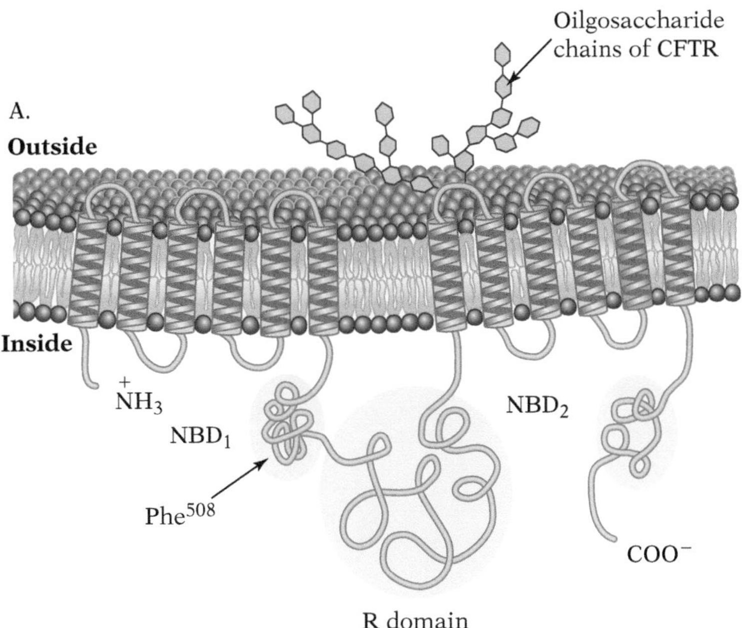

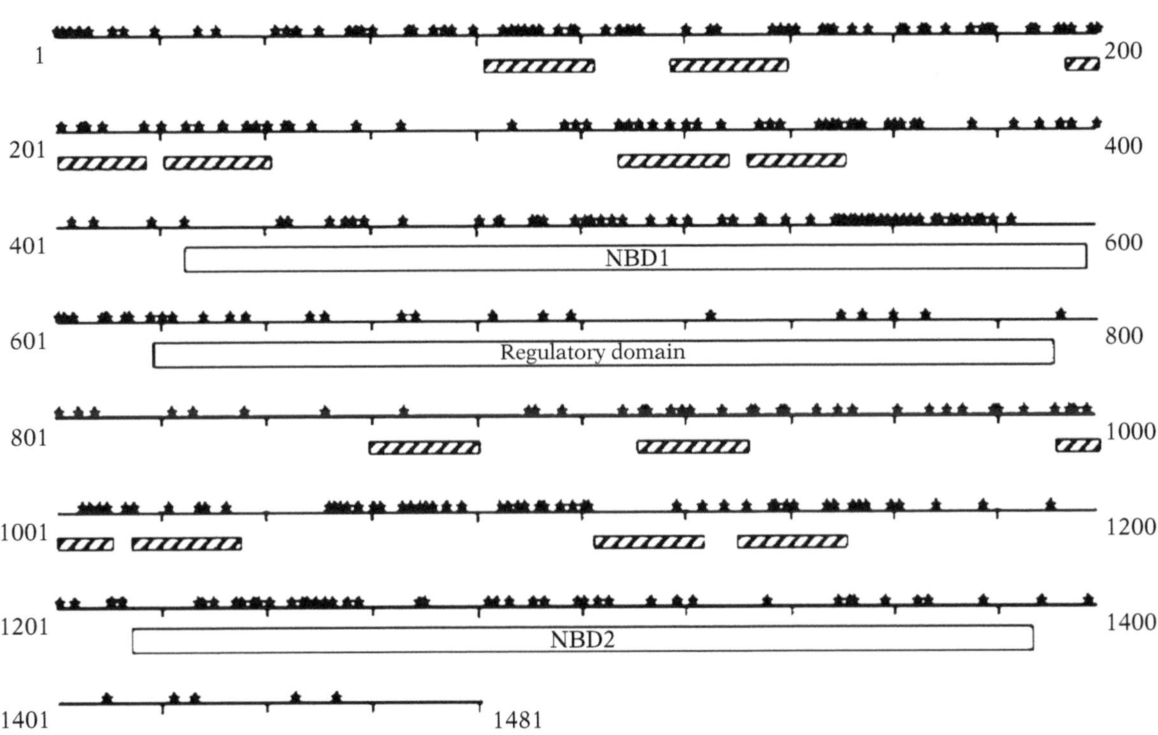

7.30 The CFTR protein and cystic fibrosis disease. **(A)** The topology of the cystic fibrosis TM conductance regulator (CFTR) is shown with 12 TM helices and three cytoplasmic domains. Two of the cytoplasmic domains are nucleotide-binding domains, NBD1 and NBD2, and the third is a regulatory domain (R domain) in which the protein is phosphorylated by cAMP-dependent protein kinase. The position of the most common mutation in cystic fibrosis patients, Phe508, is indicated. Redrawn from Nelson, D. L., and M. M. Cox, *Lehninger Principles of Biochemistry*, 4th ed., W. H. Freeman, 2005, p. 403. © 2005 by W. H. Freeman and Company. Used with permission. **(B)** The locations (starred sites) of other mutations in the CFTR sequence that result in cystic fibrosis disease. The hatched bars below the sequence correspond to TM segments of the protein. From Sanders, C. R., and J. K. Myers, *Annu Rev Biophys Biomol Struct*. 2004, 33:25–51. © 2004 by Annual Reviews. Reprinted with permission from the Annual Review of Biophysics and Biomolecular Structure, www.annualreviews.org.

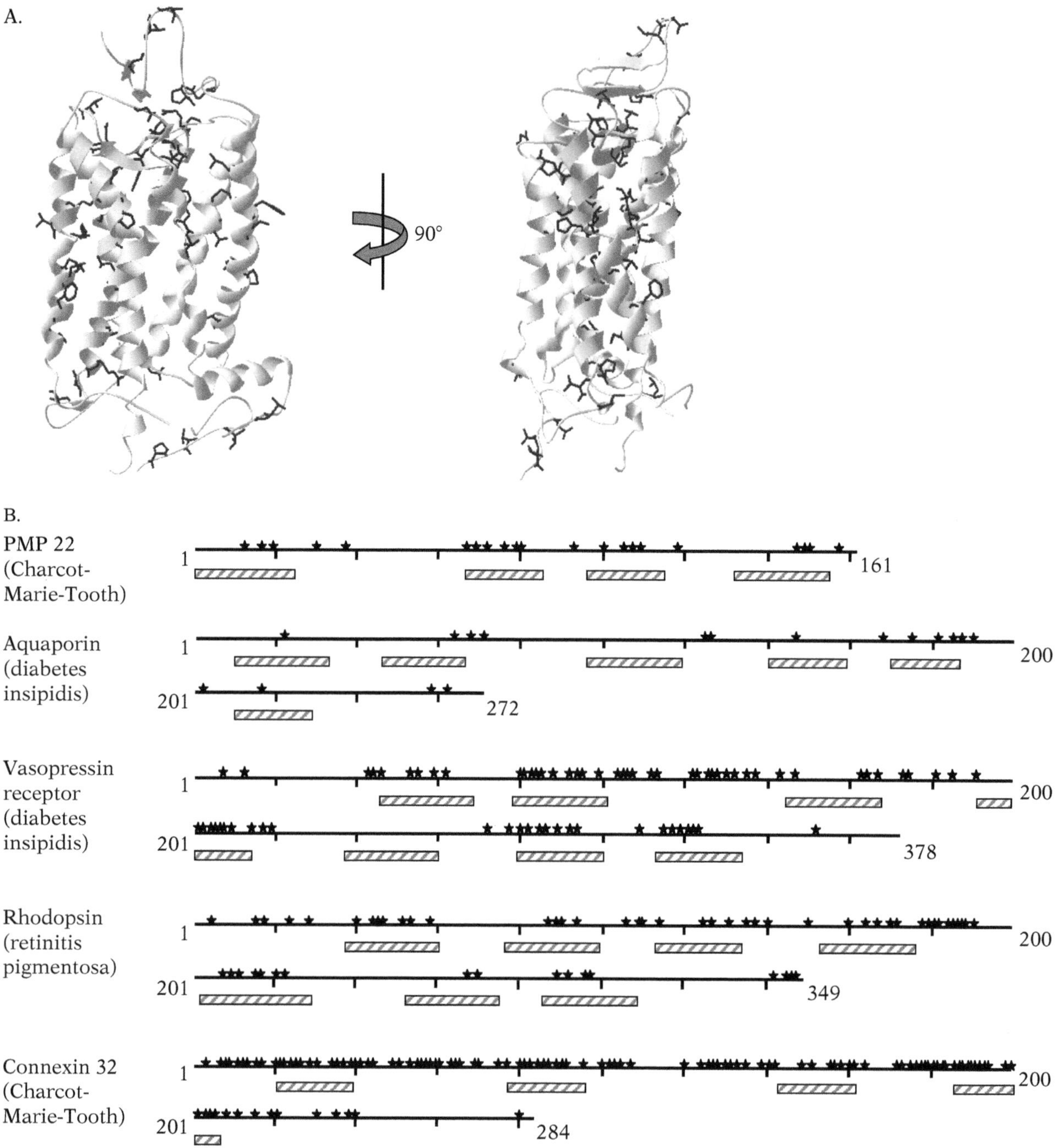

7.31 Sites in membrane proteins of mutations linked to diseases. The sites for which mutations have been linked to disease are well distributed throughout the sequences of numerous membrane proteins. (A) Ribbon diagram of the structure of rhodopsin showing the locations of missense mutations implicated in retinitis pigmentosa. The affected side chains (shown in black) are distributed all over the structure. (B) The sequences of rhodopsin and other membrane proteins linked to diseases, with the positions of mutations (asterisks) and of TM segments (hatched bars) shown. From Sanders, C. R., and J. K. Myers, *Annu Rev Biophys Biomol Struct*. 2004, 33:25–51. © 2004 by Annual Reviews. Reprinted with permission from the Annual Review of Biophysics and Biomolecular Structure, www.annualreviews.org.

intriguing case, when the prion protein, PrP, is inserted with its N terminus in the ER lumen, it does not result in disease; however, when it is inserted with its N terminus in the cytoplasm, it can lead to neurodegeneration, apparently by a mechanism involving aggregation (Figure 7.29). In general, damage can result from either accumulation of misfolded proteins or from loss of needed functions of the affected proteins.

In humans the misfolding of membrane proteins is known to cause several heritable diseases, such as cystic fibrosis and retinitis pigmentosa. Many of the 1000-point mutations causing cystic fibrosis result in the misfolding of the cystic fibrosis TM conductance regulator (CFTR), and even the wild type CFTR assembles with a low efficiency under some conditions. Almost one-third of the 55 genes linked to disorders of the human retina encode integral membrane proteins, with many of the disorders caused by missense mutations. Over 100 different rhodopsin mutants produce retinitis pigmentosa.

Cystic fibrosis is a relatively common hereditary disease that results in severe obstruction of the respiratory and gastrointestinal tracts of people with two defective copies of the gene for CFTR. The CFTR protein (Figure 7.30A) functions as a Cl^- ion channel that is activated by cAMP-protein kinase. Loss of the channel activity in epithelial cells allows the lungs to be clogged with debris and become subject to frequent bacterial infections that damage the lungs, reduce respiratory efficiency, and shorten the afflicted person's lifetime. In 70% of cystic fibrosis cases, a deletion of Phe508 from CFTR results in misfolding of the protein, which results in its degradation. In addition, many other disease-related mutations are distributed throughout the protein sequence (Figure 7.30B).

Like CFTR, the mutations in rhodopsin that lead to retinal disease are located in many different parts of the protein structure (Figure 7.31A). From the variety of domains affected by mutations, it is evident that the mutations do not target an active site, or even a couple of localized regions. In fact, this pattern of widely distributed disease-causing mutations is true in five other membrane proteins associated with diseases (Figure 7.31B). It strongly suggests that these are misfolding mutations, because a large variety of single changes in virtually any part of a protein are not expected to critically impair function. They could, instead be crucial to folding by tipping the delicate energetic balance between correct folding and misassembly. The high frequency of misfolding of these membrane proteins provides new targets for drug development, as ligands that stabilize the native state could act as chemical chaperones that rescue the misfolded membrane proteins.

FOR FURTHER READING

Seminal Papers

Egea, P. F., and R. M. Stroud, Lateral opening of a translocon upon entry of protein suggests the mechanism of insertion into membranes. *Proc Natl Acad Sci USA*. 2010, 107:17182–17187.

Engelman, D. M., and T. A. Steitz, The spontaneous insertion of proteins into and across membranes: the helical hairpin hypothesis. *Cell*. 1981, 23:411–422.

Gruner, S. M., Instrinsic curvature hypothesis for biomembrane lipid composition. *Proc Natl Acad Sci USA*. 1985, 82:3665–3669.

Hartl, F. U., S. Lecker, E. Schiebel, J. P. Hendrick, and W. Wickner, The binding cascade of SecB to SecA to SecY/E mediates preprotein targeting to the *E. coli* plasma membrane. *Cell*. 1990, 63:269–279.

Heinrich, S. U., W. Mothes, J. Brunner, and T. A. Rapoport, The Sec61p complex mediates the integration of a membrane protein by allowing lipid partitioning of the transmembrane domain. *Cell*. 2000, 102:233–244.

Mitra, K., C. Schaffitzel, T. Shaikh, *et al.*, Structure of the *E. coli* protein-conducting channel bound to a translating ribosome. *Nature*. 2005, 438: 318–324.

Randall, L. L., Translocation of domains of nascent periplasmic proteins across the cytoplasmic membrane is independent of elongation. *Cell*. 1983, 33:231–240.

Simon, S. M., and G. Blobel, A protein-conducting channel in the endoplasmic reticulum. *Cell*. 1991, 65:371–380.

van den Berg, B., W. M. Clemons, I. Collinson, *et al.*, X-ray structure of a protein-conducting channel. *Nature*. 2004, 427:36–44.

Zimmer J., Y. Nam, and T. Rapoport, Structure of a complex of the ATPase SecA and the protein-translocation channel. *Nature*. 2008, 455: 936–943.

Selected Reviews

Folding

Booth, P. J., A successful change of circumstance: a transition state for membrane protein folding. *Curr Opin Struct Biol*. 2012, 22:1–7.

Bowie, J., Solving the membrane protein folding problem. *Nature*. 2005, 438:581–589.

Kleinschmidt, J. H., Folding kinetics of the outer membrane proteins OmpA and FomA into phospholipid bilayers. *Chem Phys Lipids*. 2006, 141:30–47.

Popot, J.-L., and D. M. Engelman, Helical membrane protein folding, stability and evolution. *Annu Rev Biochem*. 2000, 69:881–922.

Sanders, C. R., and J. K. Myers, Disease-related misassembly of membrane proteins. *Annu Rev Biophys Biomol Struct*. 2004, 33:25–51.

White, S. H., and W. C. Wimley, Membrane protein folding and stability: physical principles. *Annu Rev Biophys Biomol Struct*. 1999, 28:319–365.

Biogenesis

Calo, D., and J. Eichler, Crossing the membrane in archaea, the third domain of life. *Biochim Biophys Acta*. 2011, 1808:885–891.

Dalbey, R. E., P. Wang, and A. Kuhn, Assembly of bacterial inner membrane proteins. *Ann Rev Biochem*. 2011, 80:161–187.

du Plesis, D. J. F., N. Nouwen, and A. Driessen, The Sec translocase. *Biochim Biophys Acta*. 2011, 1808:851–865.

Higy, M., T. Junne, and M. Spiess, Topogenesis of membrane proteins at the endoplasmic reticulum. *Biochemistry*. 2004, 43:12716–12722.

Park, E., and T. A. Rapoport, Mechanisms of Sec61/SecY-mediated protein translocation across membranes. *Ann Rev Biophys*. 2012, 41:21–40.

Sardis, M. F., and A. Economou, SecA: a tale of two protomers. *Molec Microbiol*. 2010, 76:1070–1081.

Skach, W. R., Cellular mechanisms of membrane protein folding. *Nat Struct Mol Biol*. 2009, 16: 606–611.

White, S. H., and G. von Heijne, Transmembrane helices before, during and after insertion. *Curr Opin Struct Biol*. 2005, 15:378–386.

Bacterial Outer Membrane Biogenesis

Hagan, C. L., T. J. Silhavy, and D. Kahne, β-barrel membrane protein assembly by the Bam complex. *Ann Rev Biochem*. 2011, 80:189–210.

Knowles, T. J., A. Scott-Tucker, M. Overduin, and I. R. Henderson, Membrane protein architects: the role of the BAM complex in membrane protein assembly. *Nat Rev Micro*. 2009, 7:206–214.

Mitochondrial Biogenesis

Becker, T., M. Gebert, N. Pfanner, and M. van der Laan, Biogenesis of mitochondrial membrane proteins. *Curr Opin Cell Biol*. 2009, 21:848–893.

Chacinska, A., C. M. Koehler, D. Milenkovic, T. Lithgow, and N. Pfanner, Importing mitochondrial proteins: machineries and mechanism. *Cell*. 2009, 138:628–644.

Endo, T., K. Yamano, and S. Kawano, Structural insight into the mitochondrial protein import system. *Biochim Biophys Acta*. 2011, 1808: 955–970.

DIFFRACTION AND SIMULATION

8

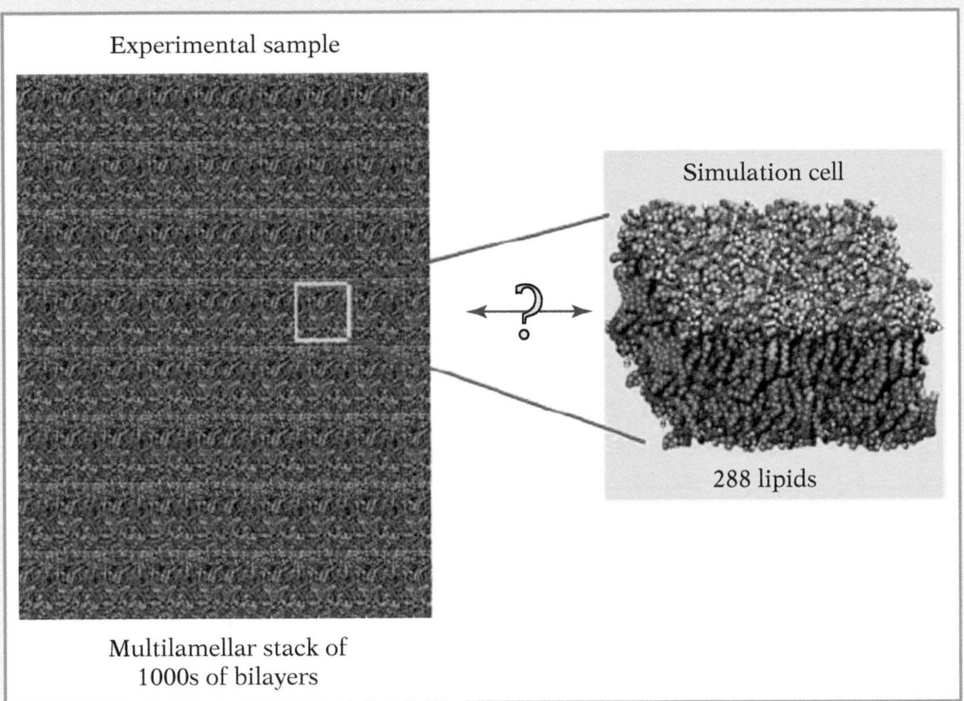

Diffraction studies focus a beam of x-rays or neutrons on multilamellar stacks containing thousands of lipid bilayers and thus differ greatly in scale from simulations of a bilayer containing only a few hundred lipid molecules. From Benz, R. W., *et al.*, *Biophys J.* 2005, 88:805–817. © 2005, Elsevier. Reproduced with permission from Elsevier.

Important tools for structural determination of membrane components include x-ray and neutron diffraction techniques. While the most familiar use of x-ray diffraction is the solution of crystal structures providing high-resolution structures of proteins and lipids in crystalline arrays, other diffraction techniques can provide structural information on membranes or reconstituted systems with lipids in the fluid phase. Such membrane diffraction studies give information that is one-dimensional, normal to the bilayer plane, because of the liquid nature of the acyl chains. The constant motions of lipids in the L_α phase (see Chapter 2) introduce several types of disorder (Figure 8.1) that prevent precise delineation of their structure at atomic resolution and invite description of their dynamic properties by sophisticated computer modeling. Today the interplay between diffraction techniques and simulation methods contributes even more to understanding the structure of the fluid membrane. This chapter describes diffraction and simulation methods as they apply to the lipid bilayer and then looks at the lipids that are resolved in crystal structures of membrane proteins. It will close by looking at some of the new techniques for crystallography of membrane proteins that are allowing solution of many more high-resolution structures. The following chapters illustrate how x-ray crystallography of membrane proteins is providing insights toward detailed understanding of their structures and functions.

BACK TO THE BILAYER

A starting point to depict the lipids in a bilayer is a view of the static structures obtained from x-ray crystallography of several pure lipids in crystalline phase (Figure 8.2). Analysis of such single-crystal structures

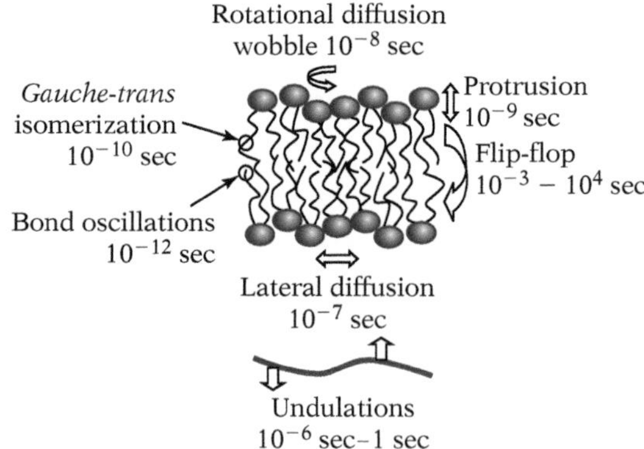

8.1 Disorder in lipid bilayer is due to several kinds of lipid motions, shown here with their approximate correlation times. Redrawn from Gawrisch, K., in P. Yeagle (ed.), *The Structure of Biological Membranes*, 2nd ed., CRC Press, 2005, p. 149. © Taylor & Francis Group LLC Books, 2004.

8.2 Single-crystal structures of four phospholipids. The structures of DMPC, DLPA, DMPG, and $DLPEM_2$ (dilauroylglycero-1-phosphate-*N*,*N*-dimethylethanolamine, or DLPC lacking one methyl group) are shown as labeled. The crystal structures were all determined with DL lipids. The four differ in both chain stacking and the orientation of the glycerol group. Redrawn from Hauser, H., *et al.*, *Biochemistry*. 1988, 27:1966.

describes the conformations of lipids in L_c phase, giving their dimensions, torsion angles, and tilt angles relative to the bilayer normal, along with the positions of their headgroups. In addition, several high-resolution structures of membrane proteins have a few well-delineated lipids (see below), although these must be viewed as boundary lipids with constraints on their mobility from the close interactions with the proteins.

Because the biologically relevant fluid lipid bilayer is a two-dimensional liquid crystal with lateral motions of its molecular components and constant fluctuations of its acyl chains, its structure is much less amenable to analysis than that of less dynamic lipid phases. A simple parameter can be difficult to pin down: for example, experimental measurements of the interfacial area of DPPC in L_α phase from diffraction and NMR studies give results varying from 56 Å^2 to 72 Å^2 per monomer. Recent simulations employing different values suggest that the appropriate value for the area is 69 Å^2, which exemplifies how contributions from both experimental and computational approaches can refine knowledge of fundamental membrane characteristics such as bilayer thickness and distributions of components.

A number of methods have been used to determine such general structural characteristics for a well-hydrated lipid bilayer. X-ray diffraction studies give a measure of D, the thickness of the bilayer, as described below. Flotation experiments can determine V_L, the volume of a lipid, by finding what mixture of H_2O and D_2O matches its density. Using data that correlate relative humidity with the activity of water determined either gravimetrically or isotopically, the number of water molecules (n_w) is calculated from the percentage humidity, and the volume of a water molecule (V_W) is known. Thus the area can be calculated from the volume and the thickness:

$$AD = 2(V_L + n_w V_W),$$

since the thickness is for two lipid layers. For fully hydrated DPPC at 50°C, these methods give D = 67.2 Å, V_l = 1230 Å^3, and A = 71.2 Å^2.

LIQUID CRYSTALLOGRAPHY

The techniques called liquid crystallography use data derived by diffraction from oriented multilamellar arrays of lipids (see Frontispiece and Box 8.1). Since the disorder inherent in the fluid bilayer defies structural delineation at atomic resolution, the positions of atoms in bilayer lipids are properly described by broad statistical distribution functions. Typically presented as one-dimensional projections along the axis normal to the bilayer surface, the structure derived from diffraction data gives the time-averaged probability distributions of the groups that make up the lipid (Figure 8.3). Examples of profiles of DOPC bilayers obtained with

BOX 8.1 X-RAY AND NEUTRON SCATTERING

In diffraction experiments beams of x-rays or neutrons (or electrons or light) are directed at ordered samples and reflect off components in the sample. X-rays are scattered by electrons, while neutrons are scattered by nuclei. The diffracted waves interfere with each other and many cancel each other out, while some are added together to produce a diffraction pattern that can be analyzed mathematically.

The radiation from an x-ray beam penetrates enough to act with atoms in many layers of the sample. The distance between layers is represented by the variable *d*, and θ is the tilt angle of reflection, as shown in Figure 8.1.1. Thus $d \sin \theta$ is the difference in path length between two waves. The mathematical relationship between the diffraction pattern and the object that produced the scattering is called a Fourier transform. It is a computation that converts an intensity signal (from the electron density) into a series of numbers that characterize the relative amplitude and phase components of the signal as a function of frequency (the diffraction pattern): The Fourier transform of a function f(x)* is defined as

$$F(S) = \int_{-\infty}^{\infty} f(x) e^{-i2\pi sx}\, dx,$$

which has the reversible property

$$f(x) = \int_{-\infty}^{\infty} F(S)_e^{-i2\pi sx}\, dx$$

(see Figure 8.1.2). Many results of Fourier transforms can be seen at www.ysbl.york.ac.uk/~cowtan/fourier/fourier.html.

These principles apply to the more familiar technique, x-ray crystallography, in which the object is a crystal with the particles in a very ordered array, as well as to diffraction studies of liquid-crystalline lipid bilayers. Since the observed diffraction is the result of time-averaged positions of the diffracting particles, it is affected by thermal motion, which is much larger in the bilayers than in crystalline materials.

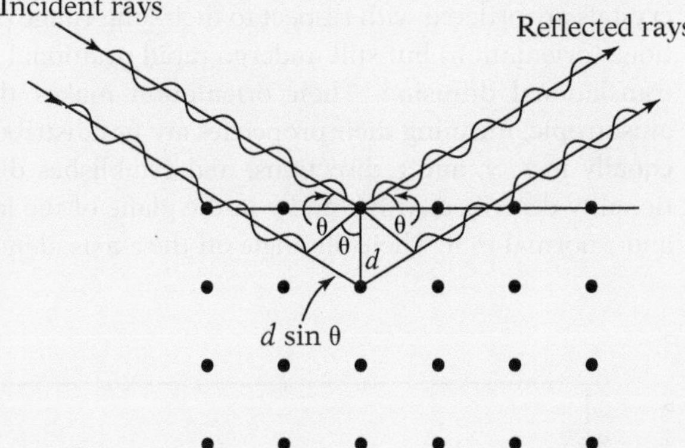

8.1.1 Reflection of x-rays from different layers of atoms, with *d* the distance between layers and θ the tilt angle of reflection. From Chang, R., *Physical Chemistry for the Chemical and Biological Sciences*, University Science Books, 2000, p. 835.

The diffraction pattern may be analyzed by Bragg's law, which is based on the fact that constructive interference occurs only when the difference in distance of the x-rays reflected from adjacent planes is equal to the wavelength of the x-ray beam. Therefore $2d \sin \theta = \lambda$, where θ is the reflection angle, *d* is the distance between the layers, and λ is

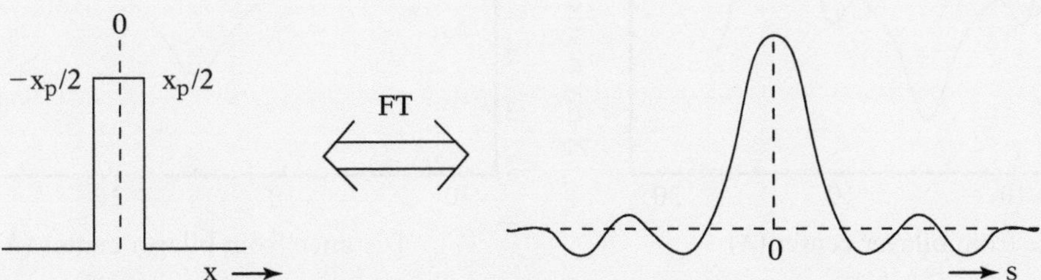

8.1.2 Fourier transform (FT) pair that relates a step function of width x_p to a sin y/y function when $x_p s/2 = n\pi$, as an example of the Fourier transform, F(S), of a function f(x). Redrawn from Campbell, I. D., and R. A. Dwek, *Biological Spectroscopy*, Benjamin Cummings, 1984, pp. 359–360.

(continued)

BOX 8.1 *(continued)*

the wavelength. A stack of fluid lipid bilayers can produce five to ten sharp Bragg reflections, analyzed as structure functions whose Fourier transform yields the one-dimensional structure profile across the bilayer. When x-rays are scattered by electrons, the profile is an electron density distribution; when neutrons are scattered by nuclei, the profile is called a scattering length distribution. Both can be generalized as probability distributions.

The energy of a neutron beam is very weak, compared with x-rays, but neutron diffraction has been utilized very productively with selectively deuterated molecules incorporated into the bilayer. Since protons and deuterons scatter neutrons very differently, it is easy to pinpoint the locations of the deuterated groups. In addition, nondeuterated lipid can be suspended in D_2O to locate the water molecules in the bilayer with neutron diffraction.

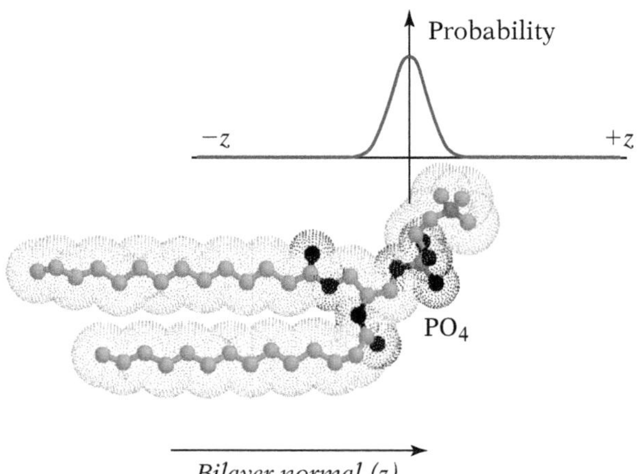

8.3 Illustration of the correspondence between a peak in the time-averaged probability distribution in the *z*-axis and the group that it represents. The ball-and-stick model shows the structure of DMPC with shading to illustrate its space-filling dimensions lacking the hydrogen atoms. The peak represents the phosphate group. From Wiener, M. C., and S. H. White, *Biophys J.* 1991, 59:162–173, 174–185. © 1991, Elsevier. Reprinted with permission from Elsevier.

x-ray electron density and neutron scattering show how simple the data appear (Figure 8.4); however, they contain the information needed for liquid crystallography to describe a fully resolved bilayer structure profile. Two distinct approaches are taken for liquid crystallography: direct analysis with liquid crystal theory and the application of the refinement methods of crystallography to these lamellar diffraction patterns.

Liquid Crystal Theory

The L_α phase of the fluid bilayer denotes a liquid crystalline material that is intermediate between the liquid and solid states of matter and can be treated like smectic liquid crystals (Figure 8.5). The constituents of liquid crystals are ordered with respect to their long-range positions (orientation) but still undergo rapid rotational and translational diffusion. Their orientation makes them anisotropic, meaning their properties are not distributed equally in x, y, and z directions, and establishes directionality described with x and y in the plane of the layer and z normal to it. Their tilt angle off the z-axis, denoted

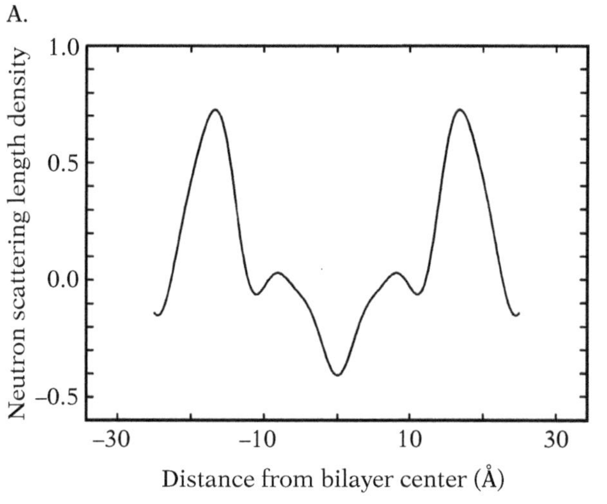

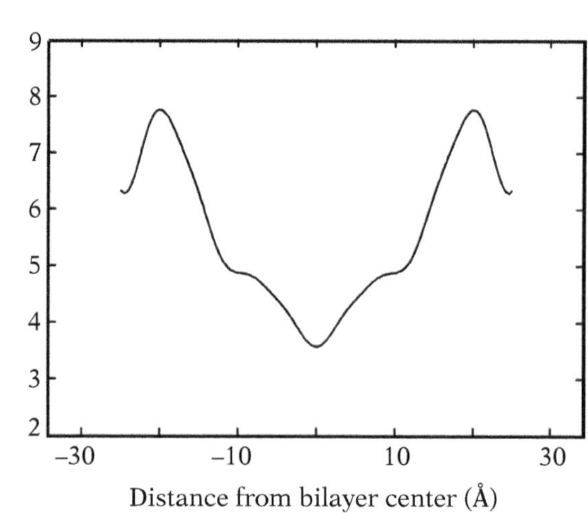

8.4 Examples of diffraction data for DOPC bilayers at 23°C and 66% relative humidity presented as eight-order scattering-length density profiles. **(A)** Neutron diffraction profile, whose peaks correspond to the carbonyl groups. **(B)** X-ray diffraction profile, whose peaks correspond to the phosphate groups. From Wiener, M. C., and S. H. White, *Biophys J.* 1992, 61:434–447. © 1992, Elsevier. Reprinted with permission from Elsevier.

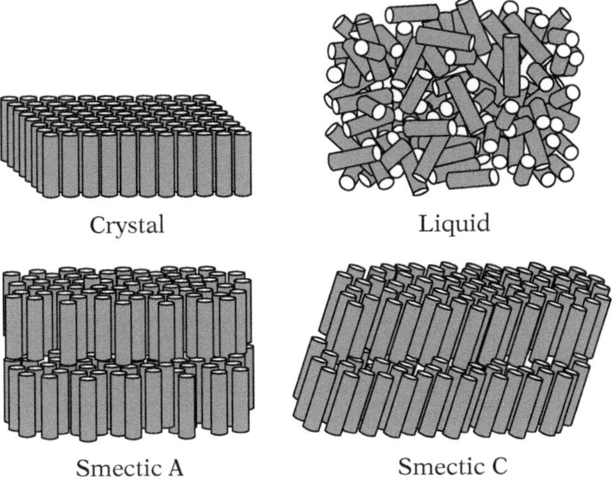

8.5 Crystals, smectic liquid crystals, and liquids. Smectic liquid crystals are ordered in layers consisting of rod-like molecules oriented with their long axes perpendicular to the plane of the layers. They may also be at an angle from the normal, as shown. The layers are free to slide over one another, giving the structural properties of a two-dimensional solid. Redrawn from Chang, R., *Physical Chemistry for the Chemical and Biological Sciences*, University Science Books, 2000, p. 884. © 2000. University Science Books. Reprinted with permission.

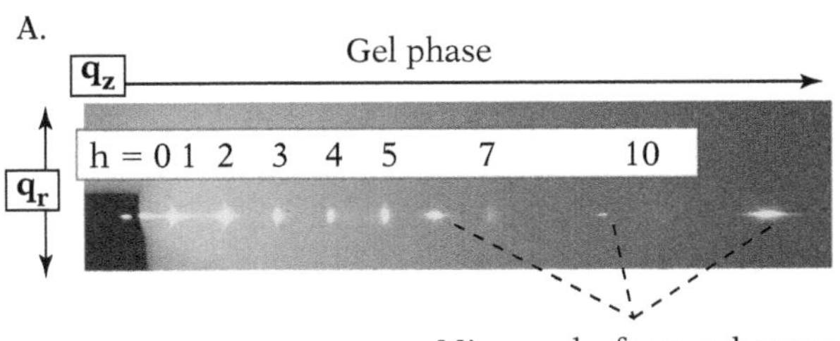

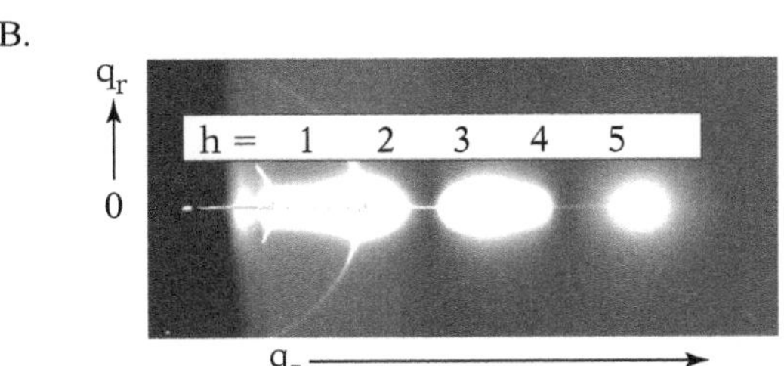

8.6 Bragg peaks for x-ray diffraction of phospholipid bilayers. **(A)** X-ray scattering data from a sample of DMPC at 10°C (gel phase) show peaks (h) up to the seventh order. The variable q represents the scattering in two dimensions, z and r. **(B)** X-ray scattering data from fully hydrated DOPC in fluid phase is much more diffuse, with overlap between Bragg orders 1 and 2. From Tristram-Nagle, S., and J. F. Nagle, *Chem Phys Lipids*. 2004, 127:3–14. © 2004 by Elsevier. Reprinted with permission from Elsevier.

θ, can vary significantly. Their degree of orientational order is described by the function $(3\cos^2\theta - 1)/2$, which varies from 0 (totally random, i.e., isotropic) to 1 (totally aligned, so $\theta = 0°$ and $\cos\theta = 1$). The order parameter, S, of a liquid crystal is the average of this function: $S = \langle(3\cos^2\theta - 1) / 2\rangle$.

X-ray diffraction of a smectic liquid crystal often produces strong first- and fourth-order scattering peaks, but the peaks are very diffuse in x-ray scattering data from fluid lipid bilayers (Figure 8.6). The peaks decrease in intensity as the order number increases until diffuse scattering overcomes the ability to detect the peak. Liquid crystal theory provides a mathematical treatment of the scattering that yields electron density profiles for samples with diffuse scattering that extends beyond the fifth Bragg order. In addition, the analysis of diffuse scatter also provides elasticity information, so there is actually more information in the diffuse scatter than in the Bragg peaks. Using this technique, very accurate fluid phase structures and bending moduli have been obtained on numerous pure lipids and on biomimetic membranes with cholesterol, bioflavinoids, and peptides as additives.

Joint Refinement of X-ray and Neutron Diffraction Data

Since the number of observable diffraction orders is limited when x-ray or neutron diffraction data are collected on lipid bilayers, a powerful advance in determining bilayer structure combines the information from both. The joint refinement of x-ray and neutron diffraction data is effective because the two kinds of data report on different regions of the bilayer. X-rays interact with electrons, so they scatter most strongly from the phosphate group, while neutrons interact with nuclei and scatter most strongly from the carbonyl groups (see Figure 8.4). Thus to more fully describe bilayer structure, both sets of data are combined, after they are scaled mathematically to establish agreement on the positions of the groups. With this joint refinement of x-ray and neutron diffraction data, the positions of eight groups have been defined for a phospholipid molecule plus associated waters (Figure 8.7). The result is the structure of a DOPC bilayer in L_α phase (Figure 8.8). Each peak represents the time-averaged distribution of a principal structural group of the lipid projected onto the *z*-axis normal to the bilayer plane. Each peak is a Gaussian distribution whose area represents the number of structural groups the peak represents. The apex of each peak gives the position of the group in the bilayer normal, and the half-width of the peak gives the thermal motion.

A number of features of the bilayer are evident in the structure profile obtained for DOPC. For example, there is no water in the internal hydrocarbon core. Also the two interfacial layers (each about 15 Å thick) together make up about half the total thickness of the bilayer. Furthermore, this method of presenting the

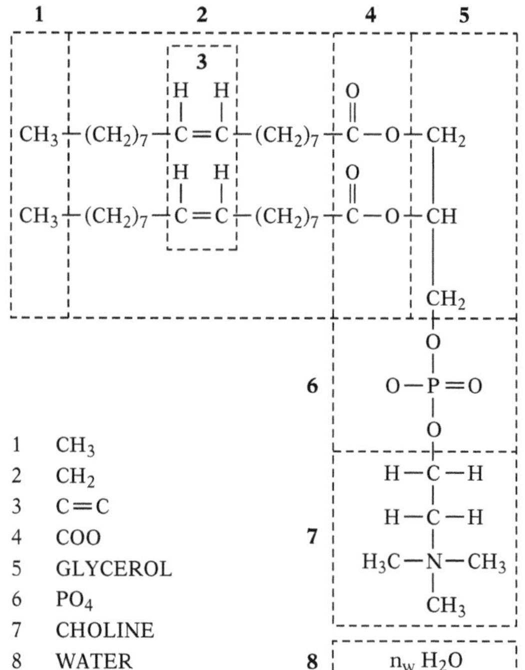

8.7 A phospholipid such as DOPC is divided into submolecular groups to be used in the structure determination. Redrawn from White, S. H., and M. C. Wiener, in K. M. Merz, Jr., and B. Roux (eds.), *Biological Membranes: A Molecular Perspective from Computation and Experiment*, Birkhauser, 1996.

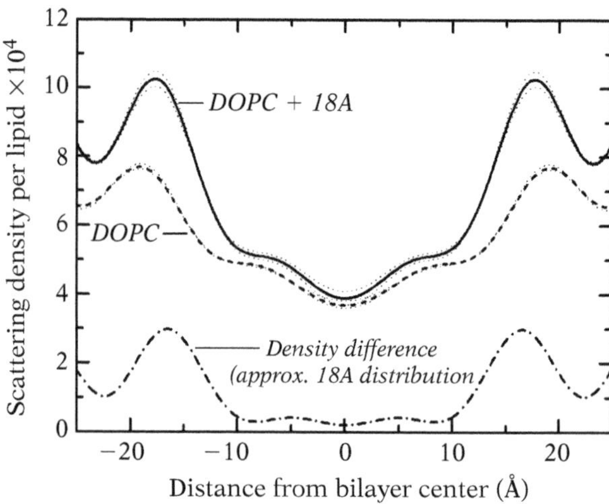

8.9 The scattering density profiles of a DOPC bilayer with and without 5 mol % Ac-18A-NH_2 (18A) expressed on a per lipid basis The dashed curve gives DOPC alone; the solid curve gives DOPC + Ac-18A-NH_2; and the dot-dash curve gives the difference showing the localization of the peptide. From Hristova, K., *et al.*, *J Mol Biol*. 1999, 290:99–117.© 1999 by Elsevier. Reprinted with permission from Elsevier.

structure of the bilayer can convey other features of membranes. For example, when a peptide consisting of 18 alanine residues with neutralized N- and C-terminal groups (Ac-18A-NH_2) inserts into a DOPC bilayer, it is detected entirely in the headgroup region, indicating it is not hydrophobic enough to insert as a TM helix (Figure 8.9).

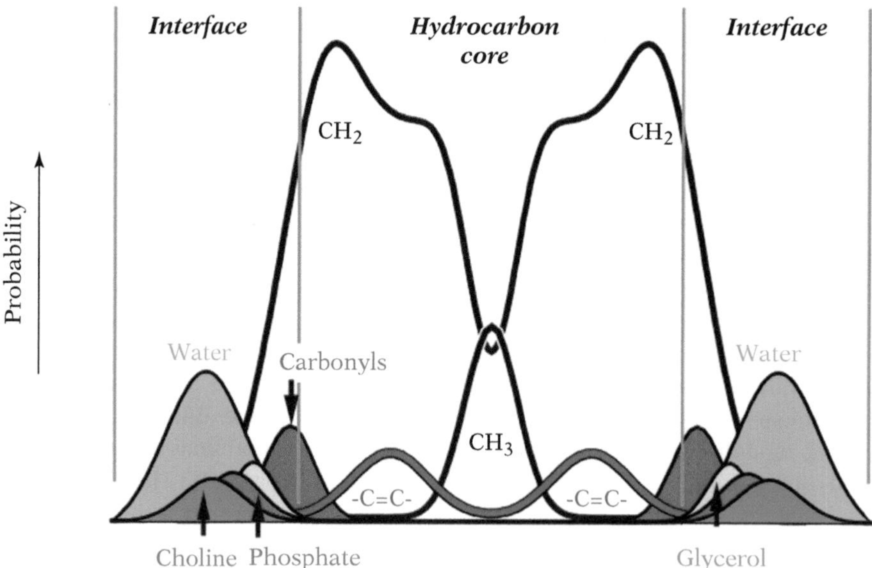

8.8 The structure of a DOPC bilayer determined by the joint refinement of x-ray and neutron diffraction data. The peaks correspond to the methyl groups (CH_3), methylene groups (CH_2), double bonds (C=C), carbonyl groups, glycerol, phosphate, choline, and water. The bilayer is divided into two interfacial regions (defined by the penetration of water) and a hydrocarbon core, as labeled at the top. From White, S. H., *et al.*, *Biochim Biophys Acta*. 2003, 555:116–121. © 2003 by the Federation of European Biochemical Societies. Reprinted with permission from Elsevier.

MODELING THE BILAYER

Simulations of Lipid Bilayers

For over three decades, advances in computer hardware and software, including computer graphics, have made simulations of lipid bilayers more realistic and more predictive. To look at the resulting molecular models as more than pretty pictures requires some appreciation of the theoretical steps that go into generating a simulation. A static model requires specification of all the atoms in the system, which can be done with the familiar Cartesian coordinates x, y, and z. Therefore a system of N particles (atoms) can be described by 3N coordinates, which becomes formidable when the system contains more than a few hundred atoms. The system is also dynamic, so each atom has a velocity and direction and therefore is constantly changing positions. Furthermore, the atoms are part of molecules, so they have interactions due to bonds specified by lengths, angles, and rotations, as well as nonbonded interactions, such as electrostatic and van der Waals interactions. By specifying all these factors in a model of the system, well-established algorithms calculate the potential energy of the system as a function of its atomic coordinates.

The changes in potential energy of all the atoms in the system are considered as movements on a multidimensional surface called the energy surface (Figure 8.10). Stable structures correspond to minima on the energy surface. Since any movement away from a minimum describes a configuration with a higher energy, algorithms make small changes in the coordinates and determine whether the energy increases or decreases. In a complex energy surface with more than one minimum, the one with the lowest energy is called the global minimum, which usually (but not always) corresponds to the state observed in a biological system.

When experimentalists measure some property of a system, they usually detect an average of that property over a large (often macroscopic) number of molecules and over the time it takes to make their measurement. If the property is called A, the experiment produces the time average, A_{ave}. However, computers can rapidly examine a large number of replications of the system and produce an ensemble average, denoted $\langle A \rangle$, the average value of property A over all replications of the ensemble generated by the simulation. The ensemble is defined by the variables held constant as the replications are generated: The canonical ensemble has a constant number of particles, volume, and temperature, and is therefore referred to as NVT. It can be useful to work with other ensembles, such as NPT (holding pressure constant instead of volume) or NVE (holding energy constant instead of temperature). Simulations allow calculation of thermodynamic properties, such as heat capacity, and kinetic properties, such as order parameters. When possible, comparison of the calculated results with results from experiments is used to evaluate the accuracy of the simulation.

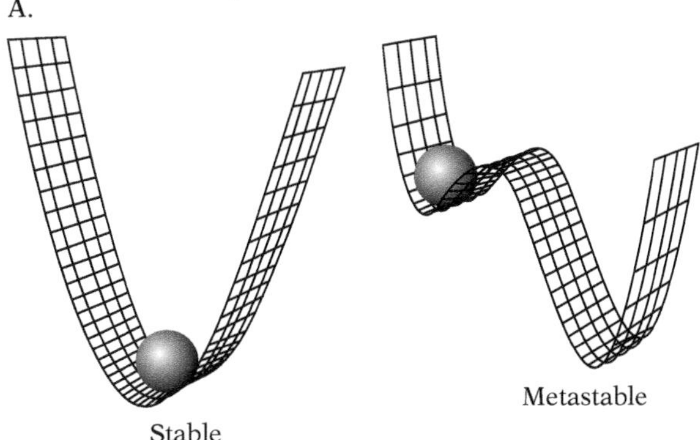

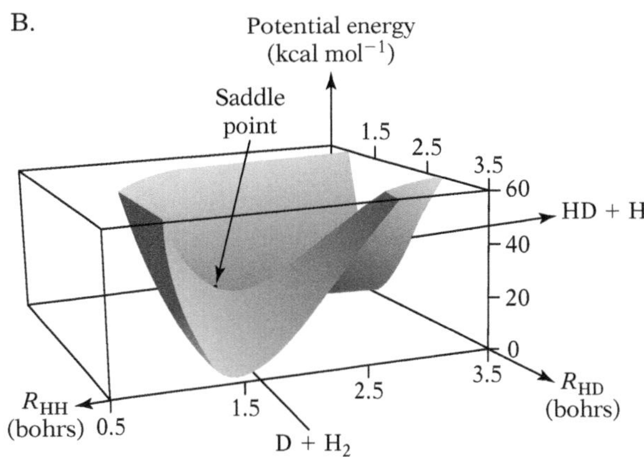

8.10 Energy surfaces depict potential energy with analogy to forces on a ball. (A) In one dimension, a ball will roll downhill until it reaches equilibrium at the bottom. A minimum corresponds to a stable or metastable state. In the metastable state, it is at a local minimum but not at the global minimum. (B) In three dimensions, the energy surface can depict the transition states of a reaction, as shown here for the reaction D + H_2 → HD + D, or it can depict the various conformational states of a molecule or macromolecular system. Redrawn from Dill, K. A., and S. Bromberg, *Molecular Driving Forces*, Garland Science, 2003, pp. 29, 349. © 2003 by Ken A. Dill, Sarina Bromberg, Dirk Stigter. Reproduced by permission of Taylor & Francis, a division of Informa plc.

Computer simulations in frequent use for lipid bilayers employ molecular dynamics (MD) and, to a lesser extent, Monte Carlo (MC) methods.

Molecular Dynamics

The first MD simulation of a lipid bilayer by van der Ploerg and Berendsen in 1982 consisted of two leaflets of 16 decanoate molecules each and lacked waters or lipid headgroups. Advances in computational abilities greatly increased the size of bilayer simulations. By 1996, a

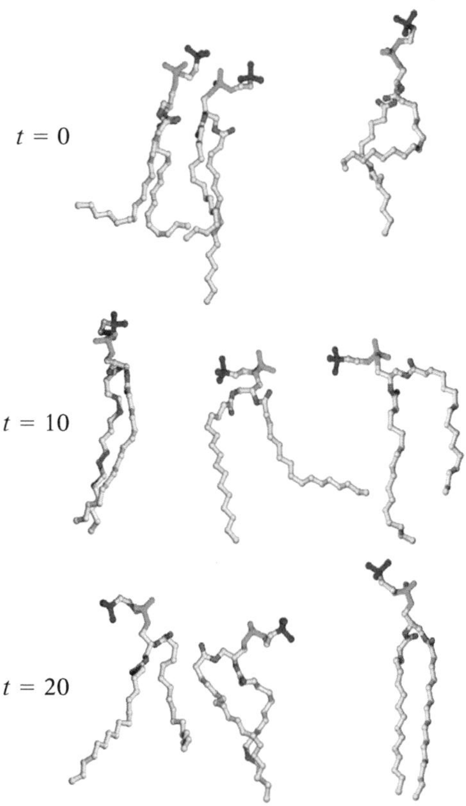

8.11 Mobility of PL molecules in a fluid bilayer. The rapid rotation around C–C bonds of the acyl chains in L_α phase phospholipids results in striking chain mobility, illustrated in snapshots of three individual lipid molecules from the molecular dynamics simulation of a DPPC bilayer shown at 0, 10, and 20 ns. Kindly provided by S. E. Feller and R. W. Pastor.

simulation of 17 000 atoms, representing 72 phospholipid molecules and 2511 water molecules, could be run for a duration of ~10 ns. A major quantitative result from MD simulations is the very rapid rate (20 ns^{-1}) of isomerization of dihedral angles along the acyl chains in L_α phase phospholipids (Figure 8.11), which increases the mobility of acyl chains compared with the standard picture of the fluid lipid bilayer. Today longer and more complex simulations allow representation of the bilayer with more than one kind of constituent (for example, causing curvature), simulation of membrane proteins in different environments, and computational analysis of dynamic actions.

Molecular dynamics calculates time averages of properties based on the dynamics of the system. Sequential determination of sets of atomic positions at very short time steps (1–10 femtoseconds) are derived using Newton's equations of motions (see Box 8.2). From the initial coordinates all intramolecular and intermolecular interactions of the atoms are computed to determine where the atoms will move and, with many repetitions, to generate their trajectories and thus predict the future state of the system. Because in a fluid bilayer the force on one atom depends on its position relative to many other atoms, solution of Newton's equations for all the atoms becomes very computationally expensive.

BOX 8.2 MOLECULAR DYNAMICS CALCULATIONS

Newton's laws of motion state that (1) a body in motion continues to move in a straight line at constant velocity unless a force acts on it; (2) force equals mass times acceleration ($F = ma$), which is the rate of change of momentum; and (3) to every action, there is an equal and opposite reaction. The trajectory for a motion of a particle is obtained by solving the differential equation

$$d^2x_i/dt^2 = Fx_i/m_i,$$

where m_i is the mass of the ith particle, x_i is the coordinate along which it moves, and Fx_i is the force on the particle in the x direction. Similarly, the trajectories in the other two coordinates are Fy_i/m_i and Fz_i/m_i.

The forces on each atom are calculated from the derivatives of the potential energy function:

$$Fx = dU/dx,\ Fy = dU/dy, \text{ and } Fz = dU/dz,$$

with Fx as the x component of force, Fy as the y component of force, and Fz as the z component of force.

U is the potential energy of the system. For a simulation, U is equal to $\langle E \rangle$, which is the ensemble average of the energies of states generated during the course of the simulation. The total energy is calculated as:

$$E_{tot} = \sum_{bonds} k_b(r-r_0)^2 + \sum_{angle} k_\alpha(\alpha-\alpha_0)^2 + \sum_{UB} k_{1\text{-}3}(r^{1\text{-}3}-r_0^{1\text{-}3})^2 + \sum_{improper} k_\gamma(\gamma-\gamma_0)^2 + \sum_{dihedrals} v[\cos(n\iota-\iota_0)+1] + \sum_{nonbonded}\left\{\frac{q_i q_j}{4\pi\varepsilon_0 r_{ij}} + \varepsilon\left[\left(\frac{\sigma}{r_{ij}}\right)^{12} - \left(\frac{\sigma}{r_{ij}}\right)^{6}\right]\right\}.$$

This form of the energy function is used in a program called CHARMM. The algorithm performs calculations using CHARMM or other available programs, such as AMBER and GROMOS.

More detailed introductions to the mathematics of computer simulations may be found in *Molecular Modelling: Principles and Applications*, by Andrew R. Leach, second edition, Prentice Hall, 2001, and *Molecular Modeling*, by H. D. Höltje, W. Sippl, D. Rognan, and G. Folkers, third edition, Wiley-VCH, 2008.

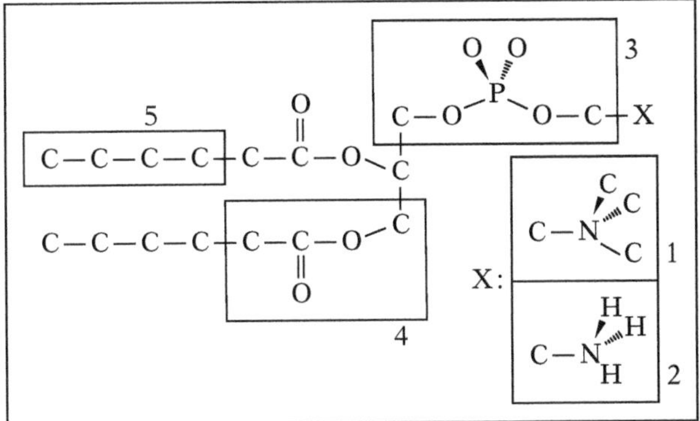

8.12 The portions of a phospholipid molecule (either PC or PE) that are used to determine model compounds for the parameters in an MD simulation: (1) trimethanolammonium moiety of the PC headgroup, modeled by tetramethylammonium or choline; (2) an ammonium ion from the PE headgroup, for which parameters are taken from the protonated ε-amino group of lysine or *ab initio* calculations for ethanolammonium; (3) phosphate group, like those in the nucleic acid parameter set; (4) ester bond to the acyl chains, modeled by methyl acetate, methyl propionate and ethyl acetate; (5) aliphatic chain, for which parameters are taken from aliphatic amino acids in proteins. Redrawn from Schlenkrich, M., *et al.*, in K. M. Merz, Jr., and B. Roux (eds.), *Biological Membranes: A Molecular Perspective from Computation and Experiment*, Birkhauser, 1996, p. 36. © 1996 by Springer-Verlag. With kind permission of Springer Science + Business Media B. V.

The steps in setting up an MD simulation include the following:

(1) Specify initial conditions, giving the positions and velocities of the particles and the interparticle forces. Impose boundary conditions.
(2) Describe the potential energy function and algorithm to be used.
(3) Determine the simulation time period, which depends on computer power.

To get starting parameters for a simulation, the phospholipid molecule has been divided into portions that are similar to groups on other macromolecules or to small molecules (Figure 8.12). Some of these portions correspond to small model compounds whose geometries and interaction energies have been determined with *ab initio* calculations. Alternatively, the starting point can be derived from x-ray structures of lipids in the L_c phase. Initial velocities can be assigned to the atoms using Maxwell–Boltzmann distributions at the temperature of interest. Now that several MD simulations of lipids are available, they provide the starting point for others. To avoid the unrealistic effect of atoms hitting the wall at the boundary of the simulation, the boundary is considered permeable. Since the number of particles in the system is held constant, when a particle leaves the system an identical particle enters from the opposite side.

Starting from the initial set of coordinates and velocities, the forces on each atom are calculated from the derivatives of the potential energy function. The potential energy function, U, is differentiated into a number of components or parameters, as shown in Box 8.2. They include intramolecular parameters, such as bond length, bond angle, vibrational modes, and torsional potential (for rotation along the C–C axis), and intermolecular forces such as van der Waals interactions and electrostatic interactions. Each parameter is described as the mathematical sum of the differences between instantaneous and equilibrium values. For example, the harmonic (meaning symmetrical) function for bond length (l), is

$$U(l) = \sum_{\text{BONDS}} k_b (l - l_0)^2,$$

where l_0 is an equilibrium bond length taken from a simple model compound of known geometry – for example, for an alkyl chain – and k_b is the vibrational frequency for that same compound.

The bond angle energy and vibrational energy of the system can each be represented by a similar quadratic function, while the equation for torsional potential can describe two local minima for *gauche* conformations and a global minimum for the *trans* position, or it can describe a double bond having *cis* and *trans* states. The van der Waals interaction is modeled to account for the attractions between atoms and the sizes of atoms and is limited by the short-range nature of these interactions. Its value is based on experimental data such as heats of vaporization and densities. For electrostatic interactions the partial charge of each atom is needed to include the charge component of headgroup–headgroup, headgroup–solvent, and solvent–solvent interactions. These assignments may be based on *ab initio* calculations or may be calculated with both short-range and long-range summations.

Each simulation has two phases, equilibration phase and production phase. For a simulation of a lipid bilayer, equilibration can take a relatively long time. When little or no change occurs, the system is assumed to have reached equilibrium. The longer the production phase, the more likely macroscopic properties will emerge. With times typically in the microsecond range, MD can now be used to simulate bilayer phenomena such as curvature and domain formation.

By the late 1990s an MD simulation of a DPPC bilayer demonstrated just how mobile the acyl chains could be in the fluid phase (Figure 8.13). Four simulations with constant particle number, pressure, temperature, and area (an NPAT ensemble) were carried out to determine the best value for the surface area per DPPC molecule, which turned out to be 62.9 Å^2. That simulation has provided a starting point for numerous other simulations and allowed comparisons of different lipids, varying acyl chains and/or headgroups. For example, when bilayers

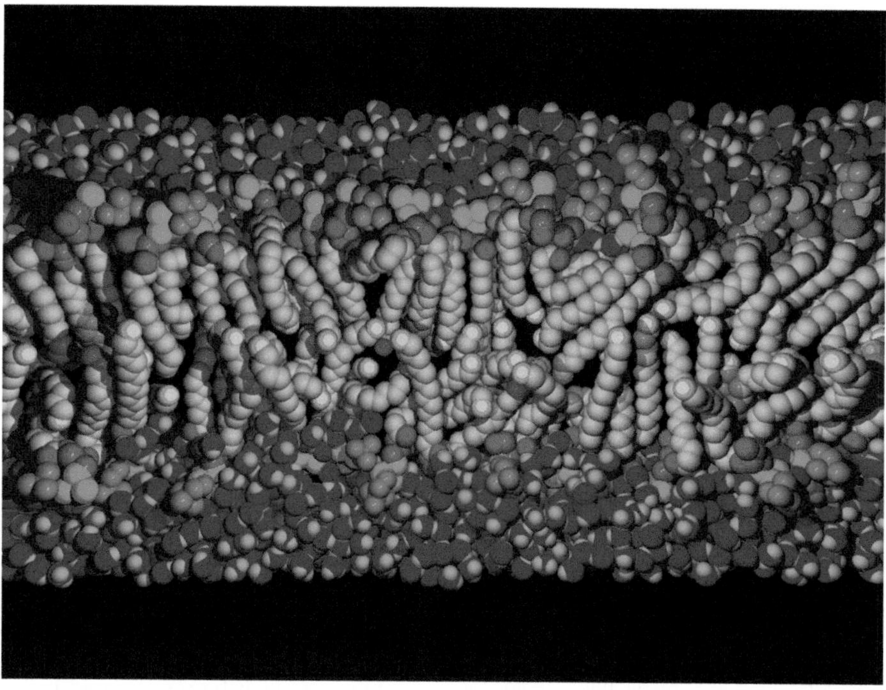

8.13 MD simulation of a DPPC bilayer. This view of a DPPC bilayer is from an 800-ps trajectory with A = 62.9 Å^2/lipid. The atoms and atom groups are colored as follows: yellow, chain terminal methyl; gray, chain methylene; red, carbonyl and ester oxygen; brown, glycerol carbon; green, phosphate; pink, choline; dark blue, water oxygen; and light blue, water hydrogen. From Feller, S. E., *et al.*, *Langmuir*. 1997, 13:6555–6561. © 1997, American Chemical Society.

of PE are compared with PC, the much more extensive hydrogen bond network among the PE headgroups dominates the dynamics of the interfacial region.

The extreme mobility of the acyl chains in L_α-phase DPPC could make it surprising that kinks made by *cis* double bonds in unsaturated chains (Figure 2.1B) increase the disorder of the acyl chains enough to significantly lower melting transitions of unsaturated fatty acids compared with saturated fatty acids (see Chapter 2). The dynamic effects of unsaturated acyl chains can be appreciated in MD simulations of bilayers containing acyl chains with *cis* and *trans* double bonds. The effects of *trans* double bonds are interesting in view of the link between consumption of *trans* fatty acids in foods and heart disease. In a systematic study, a number of 18-carbon hydrocarbons were compared, including stearoyl (C18:0), oleoyl (C18:1Δ9*cis*), and elaidoyl (C18:1 Δ9*trans*), in both same-chain lipids (e.g., DOPC, DEPC) and mixed-chain lipids (e.g., SOPC). Since the double bond is between C9 and C10, chain packing of lipids was analyzed from the MD simulations by representing each chain as a pair of vectors stretching from C2 to C9 and from C10 to C17. The ensemble-averaged measure of the relative orientations of the chain segments, shown as probability distributions (Figure 8.14), indicate that for both stearoyl and elaidoyl chains the most likely state is a parallel orientation (maximum probability at 1), while a much higher

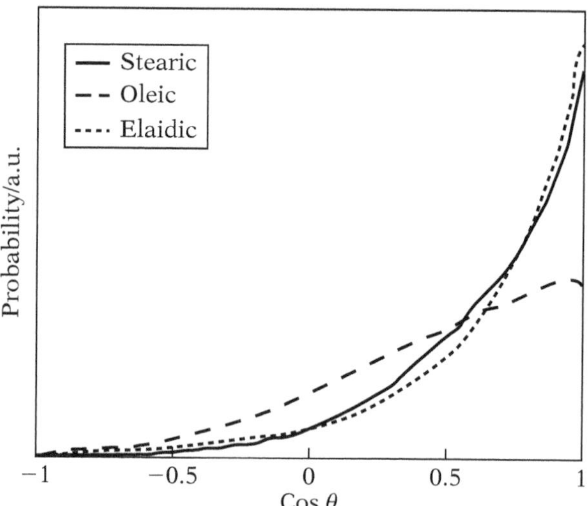

8.14 Probability distribution of a kink in the acyl chain. The presence of a kink at the carbon–carbon double bond is indicated by the probability distribution function for the angle made between vectors representing the upper and lower halves of the acyl chain on C1 of PC. When cos $\theta = 1$, the two halves are parallel, as is dominant for both stearic (no double bond) and elaidic (with a *trans* double bond) but not oleic (with a *cis* double bond), which has a fairly broad distribution of angles. Redrawn from Roach, C., *et al.*, *Biochemistry*. 2004, 43:6344–6351. © 2004 by American Chemical Society.

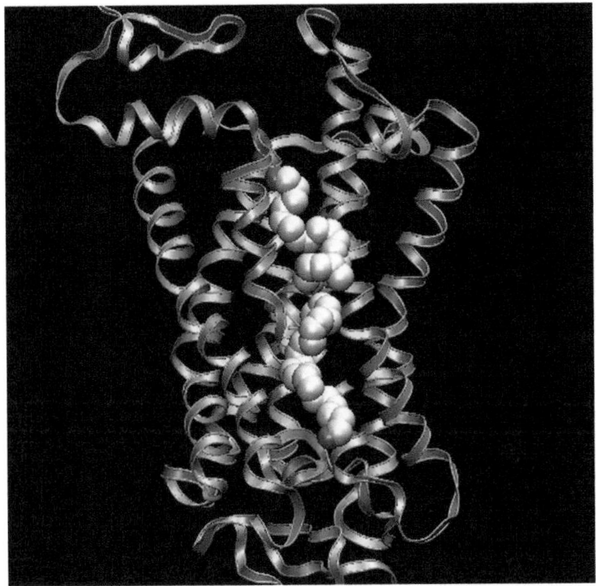

8.15 Snapshot of rhodopsin interacting with two molecules of 1-stearoyl-2-docosahexaenoyl-PC in a simulated bilayer. Note how the polyunsaturated chains (spheres) bend in conforming to the surface of the protein (ribbon diagram). From Feller, S. E., and K. Gawrisch, *Curr Opin Struct Biol.* 2005, 15:416–422. © 2005 by Elsevier. Reprinted with permission from Elsevier.

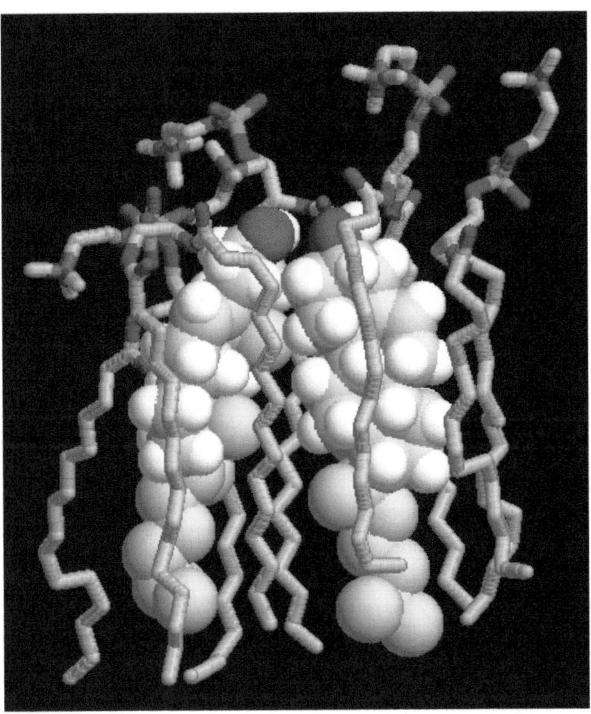

8.16 Simulation of contacts between DPPC and cholesterol. The acyl chains of DPPC (sticks) coat the surface of cholesterol (space-filling models) in a system of DPPC:cholesterol at a 7:1 ratio. The hydroxyl groups of the cholesterol molecules are hydrogen bonded to different lipids, but are also quite close to each other. From Chiu, S. W., *et al.*, *Biophys J.* 2002, 83:1842–1853. © 2002, Elsevier. Reprinted with permission from Elsevier.

distribution of bent conformations (kinks) is observed for oleoyl chains. Experimental measurements of fluidity and lateral mobility also reveal similarities between the saturated lipids and lipids with *trans* double bonds, suggesting that replacement of *cis* unsaturated fatty acids by *trans* fatty acids depletes the lipids of the disorder that is needed in the membrane.

MD simulations have also examined the role of polyunsaturated fatty acids such as the beneficial omega-3 fatty acids (see Chapter 2). In contrast to early views that suggested polyunsaturated fatty acids increased the rigidity of the bilayer, MD simulations along with NMR studies indicate that the extra flexibility and rapid conformational fluctuations of these hydrocarbon chains increase the softness of the bilayer. The most prevalent omega-3 fatty acid, docosahexaenoic acid (DHA, C22:6), is found in high concentrations (up to 50 mol%) in membranes of the nervous system. It is the dominant fatty acid in the rod cell membrane, where it is required for rhodopsin activation (see Chapter 10). DHA could be affecting rhodopsin indirectly by modulating membrane elasticity or activating it directly, since MD simulation indicates it interacts closely with helices on the surface of rhodopsin (Figure 8.15).

Simulations of DPPC or DMPC and cholesterol have been done in several labs and at several different concentrations of cholesterol. The condensing effect of cholesterol is evident in the decreased lipid surface area above approximately 10% cholesterol and can be attributed to close contact with acyl chains that "wets" the cholesterol, allowing it to come closer together (Figure 8.16). The L_o phase appears when cholesterol is ~12% to 50%; the lower concentration suggests that one cholesterol molecule can affect eight to nine PC molecules.

With vast increases in computing capabilities, MD simulations on the order of msec allow studies of membrane proteins in bilayers. One example is a simulation of the mechanism of voltage sensing in potassium channels (see Chapter 12) that addresses how charged residues move past a hydrophobic group in response to a voltage gradient (Figure 8.17).

Monte Carlo

Because the fully atomistic simulations are limited on both time and length scales, coarse-grained approaches are often used to model cooperative, large-scale behavior. In contrast to MD simulations where the successive configurations of the system are connected in time, in Monte Carlo (MC) simulations small random changes in conformation are generated and each is compared only with its predecessor to determine whether it represents a state of lower potential energy. By calculating the potential energy after making random small changes,

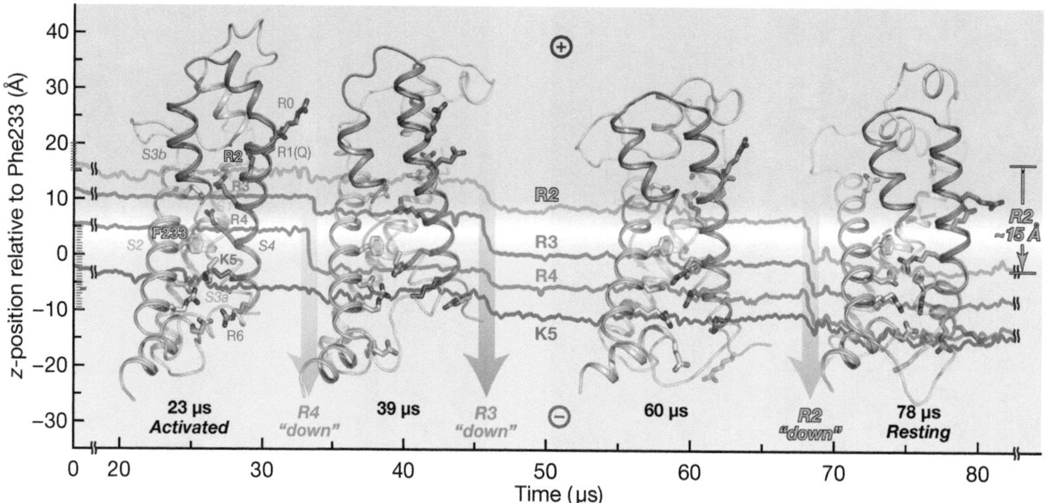

8.17 MD simulation of voltage gating. All atom simulation of a voltage-gated potassium channel shows how its voltage sensor switches between activated and deactivated states. The S4 helix (red) of the voltage sensor domain moves to allow a succession of basic residues (R2, orange; R3, green; R4, cyan; and K5, purple) to move past a hydrophobic residue (F233, yellow) as a current is applied across the membrane (positive on top and negative below). The different states are observed over a time scale of ~100 μs. From Jensen, M. Ø., *et al.*, *Science*. 2012, 336:229–233. © 2012, American Association for the Advancement of Science. Reprinted with permission from AAAS.

such as in the rotation of a bond, and assigning a higher acceptance probability for the new configuration if the potential energy has decreased and a lower probability if it has increased, the simulation progresses toward configurations of lowest energy (greatest stability) without a kinetic input. Traditional MC simulations use NVT ensembles, although like MD, it can use others, such as NTP.

For an MC simulation of a PL molecule, the hydrogen atoms may be omitted, reducing the lipid to the equivalent of three chains, the two acyl chains plus a chain corresponding to the phosphate and headgroup (Figure 8.18). The flexibility of each of these chains gives an enormous number of degrees of freedom that contribute to the conformation of the molecule. For example, DMPC in L_α phase has around 15–20 degrees of freedom per molecule, mainly associated with the acyl chains. Initial MC simulations applied lattice or polymer methodology to analyze the conformations of the acyl chains. A configurational-bias MC method is used in combination with MD to give more complete equilibration and sampling. For this method, the MD simulation stops at random times, and around 100 MC configurations are generated – for example, for each position of a randomly chosen acyl chain – to push the system closer to equilibration or energy minima.

Simulations are providing increasingly complex glimpses into intricate membrane processes. As an important way to study dynamics, they complement spectroscopic methods (see Box 8.3). They must be constantly informed by empirical results, and they make predictions used to design further experiments.

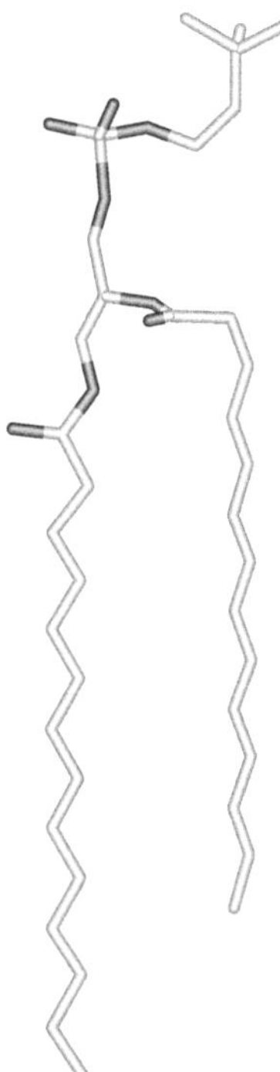

8.18 Stick model of DMPC used in Monte Carlo simulations. The model shows three flexible chains: the upper chain is the polar headgroup, and the two lower chains are the hydrophobic fatty acyl chains. All have large degrees of freedom, providing large numbers of random conformations during the simulation. From Scott, H. L., in K. M. Merz, Jr., and B. Roux (eds.), *Biological Membranes: A Molecular Perspective from Computation and Experiment*, Birkhauser, 1996. © 1996 by Springer-Verlag. With kind permission of Springer Science + Business Media B. V.

BOX 8.3 STUDYING DYNAMICS IN SOLUTION

MD simulations take as their starting points detailed coordinates from static x-ray structures and make predictions that can be tested empirically with techniques capable of exploring dynamics. These techniques utilize applications of NMR (see Box 4.2), EPR (see Box 4.3), fluorescence (see Box 2.1), circular dichroism, and other spectroscopic techniques. Measurements can be done in solutions subjected to varying conditions and monitored with time, in contrast to crystallography, which requires making new crystals under the different conditions or soaking added ligands into the crystals. Experiments in solution allow direct measurement of the effects on protein conformation of adding substrates, ions, inhibitors, and other effectors. Changes in pH altering the concentration of a very important ligand, the proton, can titrate groups critical to enzyme activity, affect ligand binding, and provide counterions for transport. Temperature changes can be used to affect rates of conformational changes. Other physical changes trigger some specific processes – such as turning on and off the light for light-sensitive proteins like bR and applying a voltage for voltage-sensitive channels and enzymes.

A robust method to study conformational change that is seeing increasing utilization is fluorine NMR. Incorporation of a ^{19}F-amino acid (during cell culture) allows specific labeling without significant perturbation of structure because the size of the fluorine and hydrogen are so similar. Because ^{19}F is highly sensitive to its environment, particular shifts in its spectra indicate regions of the protein that move to new surroundings. Spectral changes observed over time in ^{19}F-labeled proteins give rates of exchange between different conformations.

Other powerful techniques require site-specific incorporation of labels that report on distance changes within the protein. Typically, cysteine residues are engineered at sites expected to move during conformational changes and reacted to link to spectroscopically active tags. For many years these have been used to determine solvent exposure of particular residues. FRET pairs have been used to monitor the opening and closing of lobes of transporters during the transport mechanism. FRET experiments require a fluorescent donor and a suitable energy acceptor to come within a few nanometers to allow the energy transfer to occur (see Box 2.1). A similar approach that uses EPR is double electron–electron resonance (DEER). Distances of 1.5 to 8 nm can be very precisely measured between nitroxide labels placed on pairs of Cys replacements. Changes in distance distributions are indicative of conformational changes. These approaches can also be applied to studying subunit–subunit interactions, giving information about complex formation. Future applications of these and other even more sophisticated biophysical techniques, such as electron-nuclear double resonance (ENDOR), 2D IR, and fluorescence correlation spectroscopy, can be expected to make important contributions to the dynamic characterization of membrane components.

LIPIDS OBSERVED IN X-RAY STRUCTURES OF MEMBRANE PROTEINS

Both membrane diffraction and bilayer simulations enhance the appreciation of the environment of membrane proteins by providing insight into the nature of bulk lipids in the membrane. In contrast, observations of lipids in the x-ray structures of membrane proteins can show how specific lipid–protein interactions affect the structure of individual lipid molecules. Overall, these lipids exhibit much more varied conformations than the bulk lipids of the bilayer (Figure 8.19), sometimes even displaying energetically disfavored eclipsed angles. The unusual configurations of these lipids are most likely stabilized by strong electrostatic and van der Waals interactions with specific groups or regions of the proteins, which probably also accounts for their adherence to the protein during the purification and crystallization procedures. It is possible, however, that they are forced into those configurations during crystallization, as discussed below.

In the analysis of the x-ray data, lipids are modeled to fit electron-dense regions observed in clefts or at the edges of the crystallized proteins. Unless the resolution is very high, better than ~2 Å, the lipid is probably not well defined. Often only portions of the lipid are clearly resolved; for example, the relatively fixed glycerol and top part of the acyl chains of a phospholipid may be clear while the ends of the chains are lost. Headgroups are frequently not well resolved (as in Figure 8.19), which suggests they are bound with lower affinity or less specificity than the rest of the molecule. In the cases where the headgroups are complete in the x-ray structure, they usually deviate from the conformation observed in L_c phase lipids.

As the number of high-resolution structures of membrane proteins increases, the data on lipid configuration grow (Table 8.1). These lipids fall into three categories: annular lipids in the first shell of lipid surrounding the TM regions of the protein; nonannular lipids in crevices between subunits; and integral lipids in unusual positions within the proteins (Table 8.2). Annular lipids mediate between integral membrane proteins and the bilayer, as they fit tightly in the grooves and clefts on the protein to cover its rough surface. The crystal structure

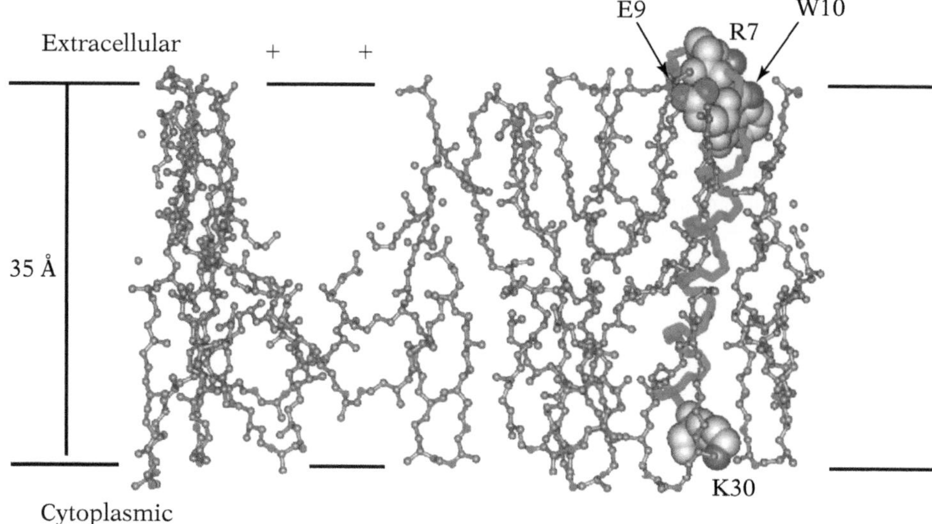

8.19 Lipid molecules in the high-resolution structure of bacteriorhodopsin. The identified lipids are varied in conformation. Lipids on the outer surface of the bR trimer are modeled as 2,3-diphytanyl-*sn* propane. Oxygen atoms are red. The green helix is helix A with Arg7, Glu9, and Trp10 at the extracellular end and Lys30 at the cytoplasmic end in space-filling representation. From Lee, A. G., *Biochim Biophys Acta*. 2003, 1612:1–40. © 2003 by Elsevier. Reprinted with permission from Elsevier.

of bR, which is unusual in showing a nearly complete shell of annular lipids, clearly demonstrates the complementarity between the protein surface and bound lipid molecules (Figure 8.20). Some tightly bound lipids on the surface are essential to the function of the proteins, such as the cardiolipin in the photosynthetic RC (Figure 8.21).

In contrast to annular lipids, which are readily exchanged with bulk lipids in the bilayer, nonannular lipids are more tightly bound in compartments defined by protein molecules, between TM helices or between subunits of either oligomeric membrane proteins or large membrane complexes. A nonannular lipid in bR is the haloarchaeal glycolipid STGA (3-HSO3-Galpβ1-6Manpα1-2Glcpα1-archaeol), which binds in the central compartment of the trimer and therefore is separated from the bulk lipid (Figure 8.22).

Integral lipids reside within a membrane protein complex and may be involved in its function and/or assembly. In contrast to both other groups, integral lipids may not even be aligned normal to the bilayer as they fit into internal sites. For example, the yeast cytochrome bc_1 complex has an internal lipid, assigned as PI, whose acyl chains extend into a hydrophobic cleft that is parallel to the membrane plane and whose phosphate group is submerged 10 Å below the interfacial zone of phosphodiester groups (Figure 8.23). Of the six PE molecules identified in the structure of cytochrome *c* oxidase from *Rb. sphaeroides*, four are between subunits and are responsible for binding subunit IV in the complex.

Analysis of specific lipid-binding sites on crystal structures of membrane proteins reveals evolutionarily conserved residues (Figure 8.24). This information can define a lipid-binding motif; for example sites in G protein-coupled receptors (see Chapter 10) interact with cholesterol by aromatic stacking interactions in a basin stabilized by inter-helical hydrogen bonding. A comparison of crystal structures of aquaporins (see Chapter 12) from three very different sources finds seven annular lipids in the same grooves in each, although the lipids in their native membranes are very different.

Obviously lipids observed in static crystals of membrane proteins may be very dissimilar from the lipids in the native membrane. First of all, after detergent solubilization during protein purification, detergent micelles cover the nonpolar regions of membrane proteins (see Figure 3.10) and most of the lipid is lost. Often addition of nonnative lipids is used to enhance crystallization of proteins in detergent. Crystallization in cubic phase lipids (see below) can also replenish the lipid. Indeed, the number of lipids bound to bR when crystallized in cubic phase matches the stoichiometry of the purple membrane (see Chapter 5). Second, to be detected by crystallography, the lipids must be "frozen" in place, having lost both their dynamic fluctuations and fast exchange rates (~10^8 sec^{-1} for annular lipids according to EPR studies; see Chapter 4). The unusual lipid configurations observed in annular lipids could result from crystal packing, for example, when flexible acyl chains are forced to adapt to the irregular protein surface during the dehydration step. Finally, important elastic properties of the membrane (such as lateral pressure) are obviously lost in crystals and could be essential to understanding key roles of the lipids. Nonetheless, the information provided by high-resolution structures has enriched the view of membrane lipids by expanding notions of their possible configurations.

TABLE 8.1 LIPIDS ASSOCIATED WITH INTEGRAL MEMBRANE PROTEINS IN THE PROTEIN DATABASE

Lipid	Protein	Resolution (nm)	PDB file
DSPC	*Paracoccus denitrificans* CO	0.30	1QLE
DSPE	*Rb. sphaeroides* CO	0.23	1M56
PLPC	Bovine CO	0.18	1V54
SAPE	Bovine CO	0.18	1V54
PVPG	Bovine CO	0.18	1V54
$Acyl_4CL$	Bovine CO	0.18	1V54
$Acyl_2PE$	*Saccharomyces cerevisiae* CR	0.23	1KB9
$Acyl_2PC$	*S. cerevisiae* CR	0.23	1KB9
$Acyl_2PI$	*S. cerevisiae* CR	0.23	1KB9
$Acyl_4CL$	*S. cerevisiae* CR	0.23	1KB9
$Acyl_2PE$	Chicken CR	0.316	1BCC
$Acyl_2PC$	Bovine AAC	0.22	1OKC
$Acyl_4CL$	Bovine AAC	0.22	1OKC
$Acyl_2PC$	*Rb. sphaeroides* RC	0.255	1M3X
(Glc Gal) $acyl_2Gro$	*Rb. sphaeroides* RC	0.255	1M3X
$Acyl_4CL$	*Rb. sphaeroides* RC	0.255	1M3X
$Acyl_4CL$	*Rb. sphaeroides* RC	0.21	1QOV
$Acyl_4CL$	*Rb. sphaeroides* RC	0.27	1E14
DPPE	*Thermochromatium tepidum* RC	0.22	1EYS
DPPG	*Synechococcus elongatus* PS I	0.25	1JBO
Galactosyl S_2Gro	*S. elongatus* PS I	0.25	1JBO
DOPC	*Mastigocladus laminosus* cyt b_6f	0.30	1UM3
$Acyl_4CL$	*E. coli* Fdh-N	0.16	1KQF
$Acyl_4CL$	*E. coli* Sdh	0.26	1NEK
$Acyl_2PE$	*E. coli* Sdh	0.26	1NEK
$Acyl_2Gro$	*Streptomyces lividans* KcsA	0.20	1K4C
Phy_2Gro	*Halobacterium salinarum* bR	0.27	1BRR
Triglycosyl Phy_2Gro	*H. salinarum* bR	0.27	1BRR
Phy_2Gro	*H. salinarum* bR	0.155	1C3W
Phy_2Gro	*H. salinarum* bR	0.19	1QHJ
Phy_2Ptd	*H. salinarum* bR	0.25	1QM8
(Triglycosyl) CL	*H. salinarum* bR	0.25	1QM8
Phy_2PG-*P*	*H. salinarum* bR	0.25	1QM8
Phy_2PG	*H. salinarum* bR	0.25	1QM8
LPS	*E. coli* FhuA	0.25	1QFG

PL abbreviations as in Appendix II with acyl chains S, stearoyl; P, palmitoyl; Phy, phytanyl; L, linoleoyl; V, vaccenoyl; A, arachidonoyl; and unidentified acyl chains where indicated. Fragments of PLs abbreviated Ptd, phosphatidyl; Gro, glycerol. Other abbreviations: Glc, glucose; Gal, galactose; LPS, lipopolysaccharide; CO, cytochrome *c* oxidase; CR, cytochrome *c* reductase (cytochrome bc_1 complex); AAC, ADP/ATP carrier; RC, photosynthetic reaction center; PS, photosystem; cyt b_6f, cytochrome b_6f complex; Fdh-N, nitrate-induced formate reductase; Sdh, succinate dehydrogenase; KcsA, pH-gated potassium channel; bR, bacteriorhodopsin; FhuA, Fe-siderophore transporter.

Source: Marsh, D., and T. Pali, *Biochim Biophys Acta*. 2004, 1666:118–141. © 2004 by Elsevier. Reprinted with permission from Elsevier.

TABLE 8.2 NUMBERS OF ANNULAR AND NONANNULAR LIPIDS OBSERVED IN X-RAY STRUCTURES OF INTEGRAL MEMBRANE PROTEINS

			Nonannular lipids[a]	
Protein	PDB code	Annular lipids	Between helices	Between subunits
Bacteriorhodopsin	1QHJ	6		2
Rhodopsin	1GZM	1		
Bacterial photosynthetic reaction centers				
Rb. sphaeroides	1QOV			1
Rb. sphaeroides	1M3X	1?	1?	1
Tch. tepidum	1EYS			1
Photosystem 1 from *S. elongatus*	1JB0	1	2	1
Light-harvesting complex from spinach	1RWT			2
Cytochrome *c* oxidase from *P. denitrificans*	1QLE		1	1
Cytochrome bc_1 from *S. cerevisiae*	1KB9		1	4
Cytochrome b_6f from *Chlamydomonas reinhardtii*	1Q90			2
Succinate dehydrogenase from *E. coli*	1NEK	1		1
Nitrate reductase	1Q16			1
ADP/ATP carrier from mitochondria	1OKC	7		
Potassium channel KcsA	1K4C			1

[a] Nonannular lipids are classified as being located either between transmembrane α-helices within a monomer or between subunits in a multimeric complex.

Source: Lee, A. G., *Biochim Biophys Acta*. 2004, 1666:62–87. © 2004 by Elsevier. Reprinted with permission from Elsevier.

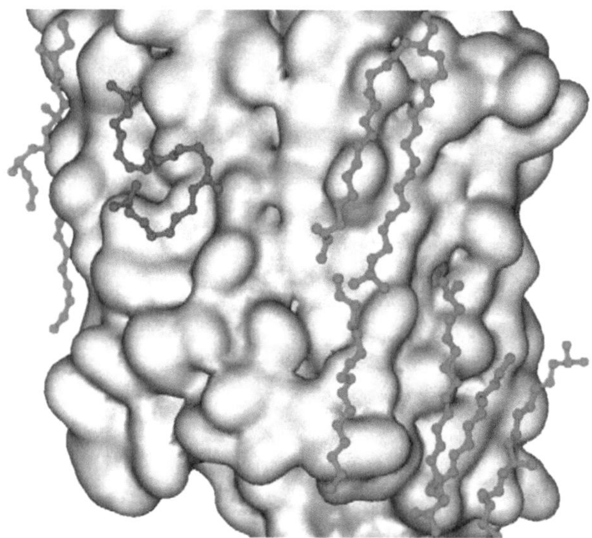

8.20 Annular lipid molecules on the surface of bacteriorhodopsin. The high-resolution structure of bR is shown with portions of lipids bound to the surface. Lipids are green, and a squalene molecule is red. From Lee, A. G., *Biochim Biophys Acta*. 2003, 1612:1–40. © 2003 by Elsevier. Reprinted with permission from Elsevier.

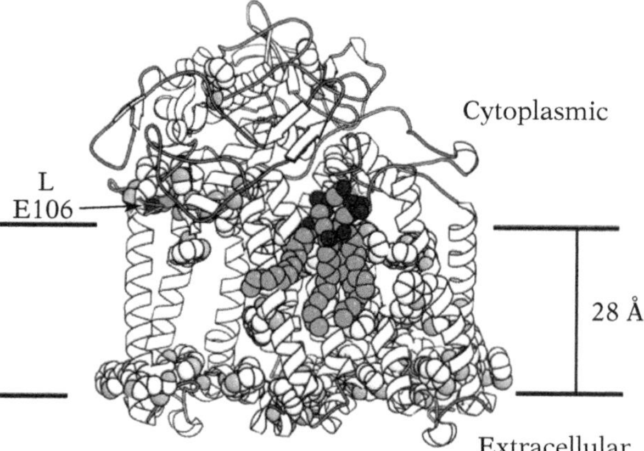

8.21 An essential annular lipid. A molecule of cardiolipin that is required for function binds to the hydrophobic surface of the photosynthetic reaction center of *Rb. sphaeroides*. In a view from the plane of the membrane, the cardiolipin (space-filling representation) is located in a depression among three of the TM helices. Trp residues (space-filling representation) define the interfacial region and the position of Glu106 is indicated to help define the cytoplasmic surface. From Lee, A. G., *Biochim Biophys Acta*. 2003, 1612:1–40. © 2003 by Elsevier. Reprinted with permission from Elsevier.

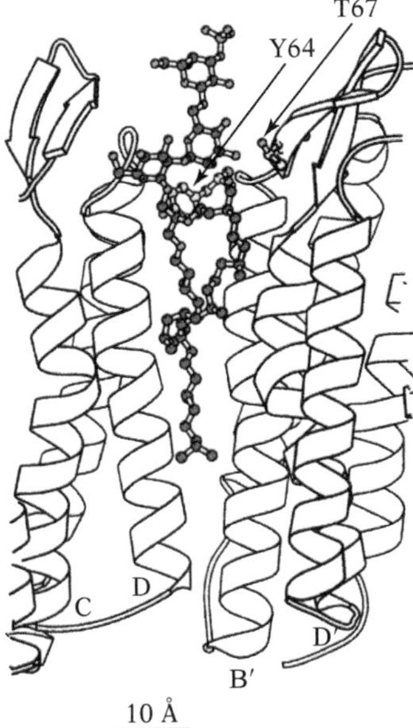

8.22 A lipid in the central cavity of the bacteriorhodopsin trimer. A nonannular lipid molecule, STGA (3-HSO_3-Galpβ1-6Manpα1-2Glcpα1-archaeol) is observed in the central cavity of the bR trimer. Viewed from the trimer interior, this STGA molecule lies among helices from two neighboring bR monomers, with hydrogen bonds to Tyr64 and Thr67 (in ball-and-stick models) on one monomer. From Lee, A. G., *Biochim Biophys Acta*. 2003, 1612:1–40. © 2003 by Elsevier. Reprinted with permission from Elsevier.

CRYSTALLOGRAPHY OF MEMBRANE PROTEINS

High-resolution crystal structures of membrane proteins provide exciting revelations of detailed structure–function relationships, as shown in the following chapters. Some of these proteins were notoriously difficult to crystallize. More than 1 g of purified LacY protein was required for crystallization attempts as more than 1000 crystals were examined at a synchrotron over a ten-year period before its structure was determined. Even as the rate of acquisition of these structures became exponential (see Figure 9.1), obtaining each new crystal structure was a feat that required overcoming many barriers. Often multiple homologs of the protein of interest were screened to select the one most suitable for crystallization.

Determining the conditions to crystallize a particular membrane protein is still largely empirical, although new systematic approaches are increasing the efficiency of the process. Many labs are tackling the challenges of improving expression systems for the proteins of interest and identifying ways to stabilize them for growth of well-ordered large crystals. Some are setting up pipelines, high throughput systems to crystallize larger numbers of membrane proteins, often screening for those from eukaryotic sources that have a high probability of success.

A.

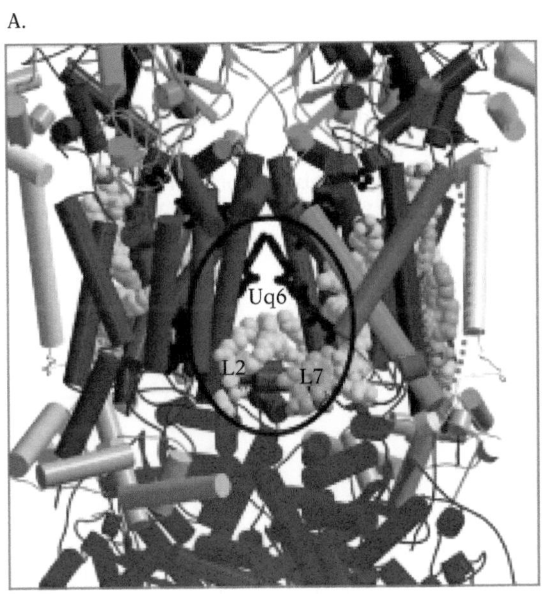

B.

8.23 Internal phospholipids in the yeast cytochrome bc_1 complex. (A) The tightly bound phospholipids of the cytochrome bc_1 complex from yeast include lipids in two internal cavities, circled on the structure (black circle and brown broken circle). The lipids, including those in the center numbered L2 and L7, are shown in space-filling representation (yellow, except for cardiolipin, which is cyan). Cofactors, including ubiquinone 6 (Uq6), are shown as ball-and-stick models (black), and the helices are colored according to subunits: cytochrome *b* (red), cytochrome c_1 (black), Rieske protein (green), Qcr6p (cyan), Qcr7p (mid-gray), Qcr8p (white), and Qcr9p (magenta). (B) For context, the homodimeric complex is viewed from the side, with the intermembrane space at the top and the matrix at the bottom (same color scheme), and with yellow and red bars at the sides delineating the plane of the membrane. For a thorough discussion of this complex, see Chapter 13. From Palsdottir, H., and C. Hunte, *Biochim Biophys Acta* 2004, 1666: 2–18. © 2004, Elsevier. Reprinted with permission from Elsevier.

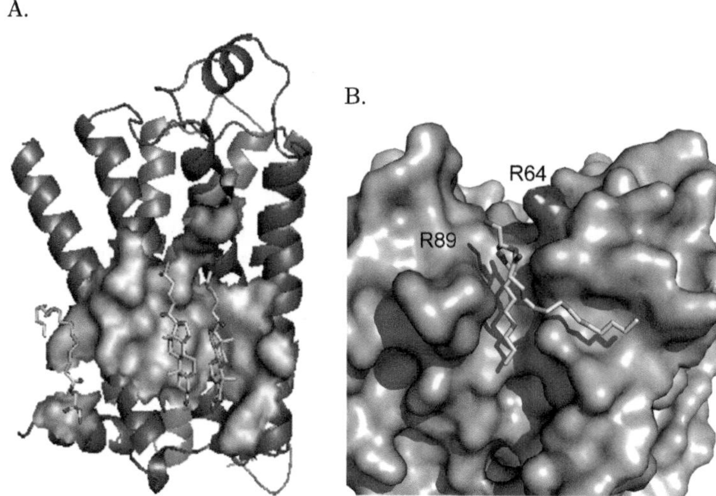

8.24 Conservation of residues at lipid-binding sites seen in crystal structures of membrane proteins. (A) Recent x-ray structures of the human β_2-adrenergic receptor (see Chapter 10) show binding of two molecules of cholesterol (sticks) at the same binding site (surface contours) in a shallow surface depression of the protein. The cholesterol-binding sites share five of the same residues (pink) that are highly conserved among GPCRs, while other residues at the sites are not conserved (gold). The acyl chain of an additional phospholipid shares the site. (B) The lipid binding site of the KcsA potassium channel (see Chapter 12) has 9 out of 13 conserved residues (same color scheme as in A). Arg64 (blue) is proposed to interact with the negatively charged phosphate group in PG. (A) and (B) From Adamian, L., *et al.*, *Biochim Biophys Acta*, 2011, 1808: 1092–1102. © 2011, Elsevier. Reprinted with permission from Elsevier.

When the success comes, the pathway is often not as straightforward as a summary indicates (Figure 8.25).

The excitement of seeing a new x-ray structure, usually visualized as a complex ribbon diagram, can eclipse the uncertainties inherent in its determination. One obvious issue is the limit of resolution in the observed electron density (discussed below). Another issue is how complete the structure is. Sometimes portions of the protein are removed to facilitate its crystallization. Even when present in the molecule, the ends of polypeptide chains, as well as some internal loops, may be disordered enough in the crystal to go undetected by x-ray diffraction. The degree of disorder is indicated by the B factor, also called the temperature factor or the Debye–Waller factor (Figure 8.26). The B factor describes the degree to which the electron density is spread out. A higher B factor indicates a higher degree of uncertainty in fitting a model of the structure to the electron density, which can be due to higher mobility of the molecule in the region or can result from an error in the model. It is typical for membrane proteins to be quite ordered in the plane of the membrane and to have more disordered loops (with higher B factors) in the solvent-exposed loops (see Figure 8.26).

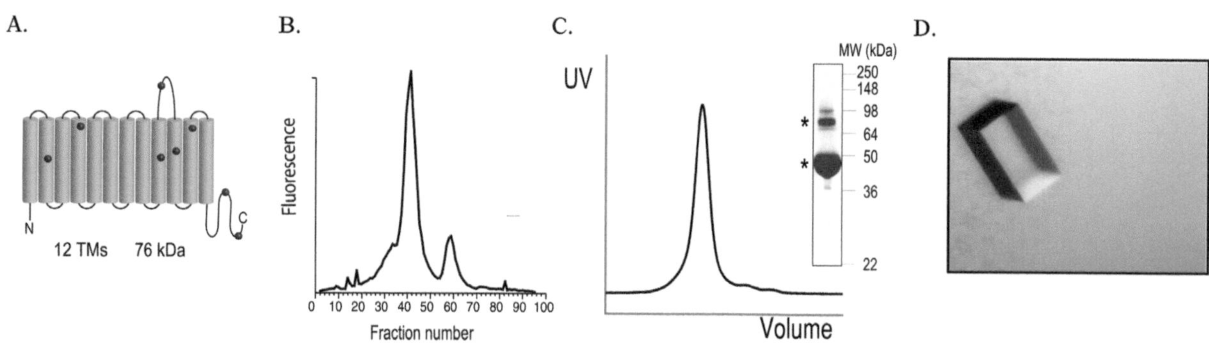

8.25 Successful procedure for obtaining crystals of the rice anion exchanger. (A) The experiment starts with a topology prediction for the rice anion exchanger PT-2, which has eight cysteine residues (red spheres). (B) PT-2 is solubilized in the detergent dodecyl-β-maltopyranoside (DDM) and chromatographed by size exclusion with a fluorescence detector (FSEC). (C) PT-2 is purified by further size exclusion chromatography (SEC) with a UV detector, and the pooled peak analyzed by SDS-PAGE as well as mass spectrometry (not shown). (D) Crystal of PT-2 in detergent DDM. From Sonoda, Y., *et al.*, *Structure*. 2011, 19:17–25. © 2011, Elsevier. Reprinted with permission from Elsevier.

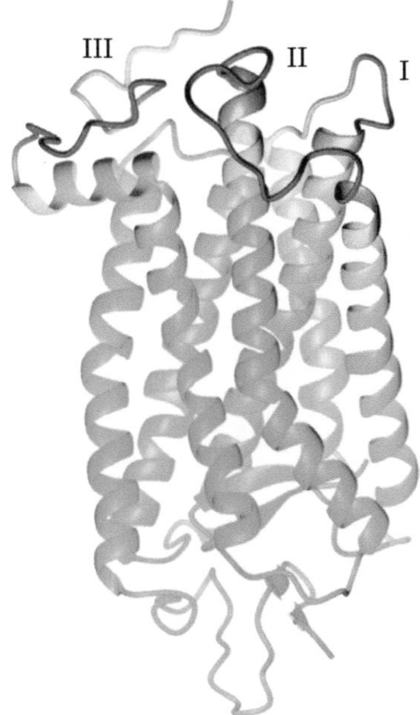

8.26 Uncertainties in fitting structure to electron densities limit the resolution of structures produced by x-ray crystallography, as indicated by the B factors represented on this ribbon diagram of rhodopsin. The TM helices have a low B factor (green), while the external loops and some of the regions toward the loops (labeled I, II, and III) have higher B factors (yellow to red, with red representing the highest uncertainty). Loop II is incomplete because of ambiguity in the electron density. From Palczewsk, K., *Annu Rev Biochem*. 2006, 75:743–767. © 2006 by Annual Reviews. Reprinted with permission from the Annual Review of Biochemistry, www.annualreviews.org.

Technologies have advanced on several fronts. The choice of detergent is critical, and often the detergent used for purification is not appropriate for crystallization due to polydispersity or large micellar size. Optimization of conditions typically employs multiple trials with different detergents in microtiter plates. Analysis of the detergents used in successful crystallizations of helical membrane proteins puts the alkyl maltopryanosides (dodecyl maltoside, DDM, and decyl maltoside, DM) at the top, followed closely by the alkyl glucosides (octyl glucoside, OG, and nonyl glucoside, NG). For reasons that are not clear, OG is more effective for channels, as well as for the β-barrel proteins. Another comparison of detergents measured the stability of several transport and channel proteins in six different detergents using an assay for unfolding, and again DDM performs the best (Figure 8.27). Stability in detergent is important for obtaining well-ordered crystals, as shown by a correlation of the stability in LDAO with the resolution of the resulting crystal structures.

Since the detergent micelle covers the hydrophobic region, crystal formation depends on contacts between the exposed polar ends of the protein. Membrane proteins with small hydrophilic portions may lack sufficient exposed regions for protein–protein contacts to form a crystal lattice. In this case crystallization is aided by fusion partners, small hydrophilic proteins or domains that can be linked to the protein to enlarge the hydrophilic regions and create space for detergent micelles. It is often convenient to use a Fab or Fv fragment of an antibody. This method was first used to crystallize cytochrome oxidase and has been used successfully

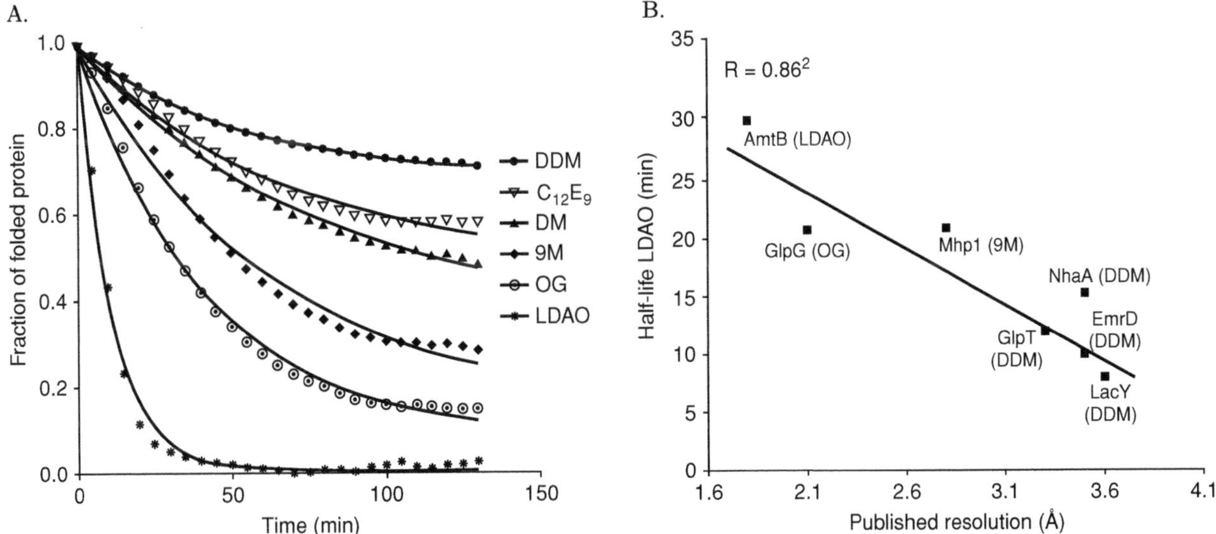

8.27 Choice of detergent for protein stability and success in crystallization. (A) Membrane protein stability is assayed by measuring unfolding at 40°C for 130 min in dodecylmaltoside (DDM, filled circle), decylmaltoside (DM, filled triangle), nonylmaltoside (9M, filled diamond), LDAO (asterisk), $C_{12}E_9$ (open triangle), and octyl glucoside (OG, open circle.) (B) Stability judged by unfolding rates in LDAO correlates to the published resolution of the membrane proteins studied. The detergent used for crystallization is listed in brackets beside each protein. From Sonoda, Y., *et al.*, *Structure*. 2011, 19:17–25. © 2011, Elsevier. Reprinted with permission from Elsevier.

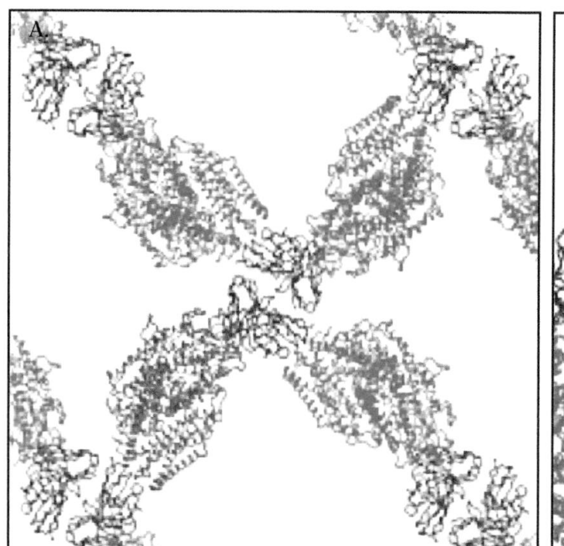

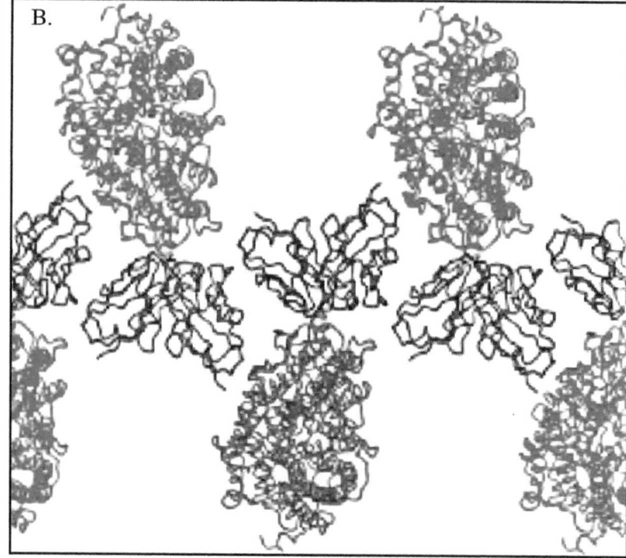

8.28 Crystallization of cytochrome oxidase from *Paracoccus denitrificans* mediated by Fv antibody fragments. Crystal lattices of the four-subunit cytochrome oxidase (COX) in (A) and the two-subunit COX in (B) both show crystal contacts involving the antibody fragments. Colors indicate the pairs, so a blue COX is bound to a red Fv and a green COX is bound to a magenta Fv. From Hunte, C., and H. Michel, *Curr Opin Struct Biol*. 2002, 12:503–508. © 2002 by Elsevier. Reprinted with permission from Elsevier.

for other components of the respiratory chain (Figure 8.28). The antigen-binding site of the antibody gives high specificity for the protein of interest. Another advantage of this approach is that it enables the protein to be purified, employing affinity chromatography with a tag fused to the antibody fragment. Because the relatively large antibody fragment (56 kDa or 28 kDa) may alter portions of the membrane protein structure, a single-chain antibody produced in camels (only 15 kDa) has been used, giving success in trapping an active GPCR complex (see Chapter 10).

A powerful method for crystallization avoids the conventional limitations of detergents by its unusual path for crystal formation in a system of lipids in cubic phase (see Chapter 2). The lipidic cubic phase (LCP) usually consists of monoolein, the racemic monooleoyl glycerol, and has viscoelastic properties that mimic the native membrane. The cubic phase allows protein diffusion throughout the sample to feed the crystal nuclei, forming a crystalline array of stacked two-dimensional sheets (Figure 8.29). Pioneered for crystallization of bR (see Chapter 5), LCP crystallization has now been successful with many different membrane proteins. A simple method of preparing the LCP with two coupled syringes – one containing the protein in detergent and the other containing the lipid – popularized the method, now automated in a commercial robotic system.

Many of the strategies described above have the goal of increased stability of the protein so it can form large, well-ordered crystals. However, the growth of large protein crystals is not a requirement for a promising new method called femtosecond nanocrystallography, which collects room-temperature crystallography data from hundreds of thousands of nanocrystals injected across a pulsed x-ray beam (Figure 8.30). Using 70 fs (or 70×10^{-15} sec) pulses from a high-energy x-ray free-electron laser, scattering data are recorded from tiny crystals–sometimes

8.29 Cartoon of crystallization in lipidic cubic phase. The integral membrane protein (represented by β_2AR-T4L, blue and green) in detergent (pink) is added to a bicontinuous cubic phase (tan), which consists of curved bilayers. When precipitants are added to shift the equilibrium away from the cubic phase, the protein molecules diffuse from the cubic phase into a lamellar lattice of the advancing crystal face. In this example, the protein co-crystallizes with cholesterol (purple). The components are drawn to scale with a lipid bilayer of ~40 Å. From Caffrey, M. and V. Cherezov, *Nat Protoc*. 2009, 4:706–731. © 2009, Nature Publishing Group. Reprinted by permission of Macmillan Publishers Ltd.

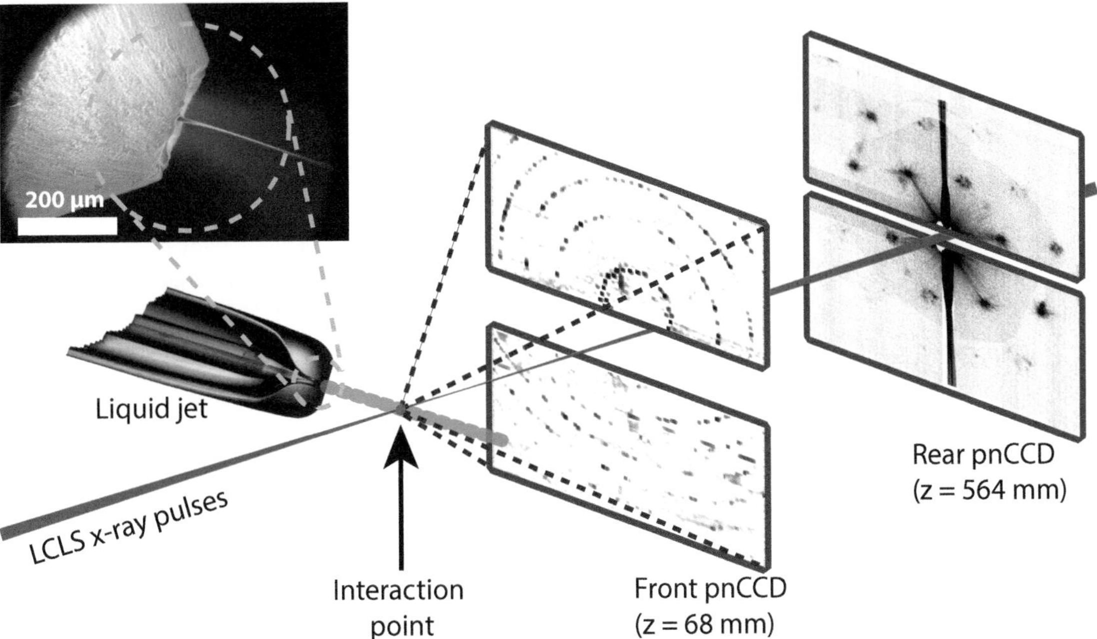

8.30 Femtosecond nanocrystallography. Nanocrystals flow in their buffer solution in a gas-focused, 4-µm-diameter jet at a velocity of 10 m s^{-1} perpendicular to the pulsed x-ray free-electron laser (FEL) beam that is focused on the jet. Two pairs of high-frame-rate detectors (pnCCDs) record low- and high-angle diffraction from single x-ray FEL pulses at a rate of 30 Hz. Crystals arrive at random times and orientations in the beam, and the probability of hitting one is proportional to the crystal concentration. The source of the powerful FEL beam is the Linac Coherent Light Source (LCIS) at Stanford University. The inset shows a scanning electron micrograph of the nozzle, flowing jet and focusing gas. From Chapman, H.N., *et al.*, *Nature*. 2011, 470:73–81. © 2011, Nature Publishing Group. Reprinted by permission of Macmillan Publishers Ltd.

containing as few as several hundred unit cells – in liquid suspension in the short time before they are destroyed by photoelectron damage, giving "diffraction before destruction." Diffraction patterns are generated at a fast rate (currently 7200 per min) and the integrated Bragg peak intensities are calculated through orientational alignment and averaging of the diffraction patterns. Even dynamic studies are possible; for example, time-resolved data for light-stimulated photosystems are obtained with the addition of an optical laser to the system.

Further progress in the field of membrane protein crystallography is needed to obtain more x-ray structures of membrane proteins with better than 2 Å resolution, as needed to provide many details in the positions of amino acid side chains and small molecules such as lipids and water (Figure 8.31). It is also important to obtain more than one crystal form of the protein to eliminate artifacts arising from the crystallization contacts. Finally, there is much interest in multiple structures showing the effects of bound substrate or ligands to represent active states or conformational changes during the function of the protein. The many structures of Ca^{2+} ATPase show the power of solving the structure with different ligands bound (Chapter 9).

A MULTIDISCIPLINARY APPROACH

Even the snapshots of the Ca^{2+} ATPase that give views of its mechanistic cycle (see Chapter 9) do not provide a moving, dynamic portrayal of its action. The capability of x-ray crystallography is limited to providing static, ordered structures captured in an environment far from the fluid and complex native membrane. It is satisfying that most of the structures of membrane proteins that have been solved by both x-ray crystallography and NMR (see Boxes 4.2 and 5.1) show few discrepancies. These two methods can give the highest resolution needed for mechanistic studies and drug design (see Figure 8.31). Understanding the complexity of membrane proteins requires a multidisciplinary approach that combines various computational and experimental technologies, including dynamic spectroscopic methods (see Box 8.3).

The interplay between simulations and crystallography will continue to enhance the understanding of membrane proteins and serves as a reminder that the x-ray structures of membrane proteins, including those described in the next chapters, should be viewed in the rich complexity of their dynamic lipid environment, the fluid mosaic membrane.

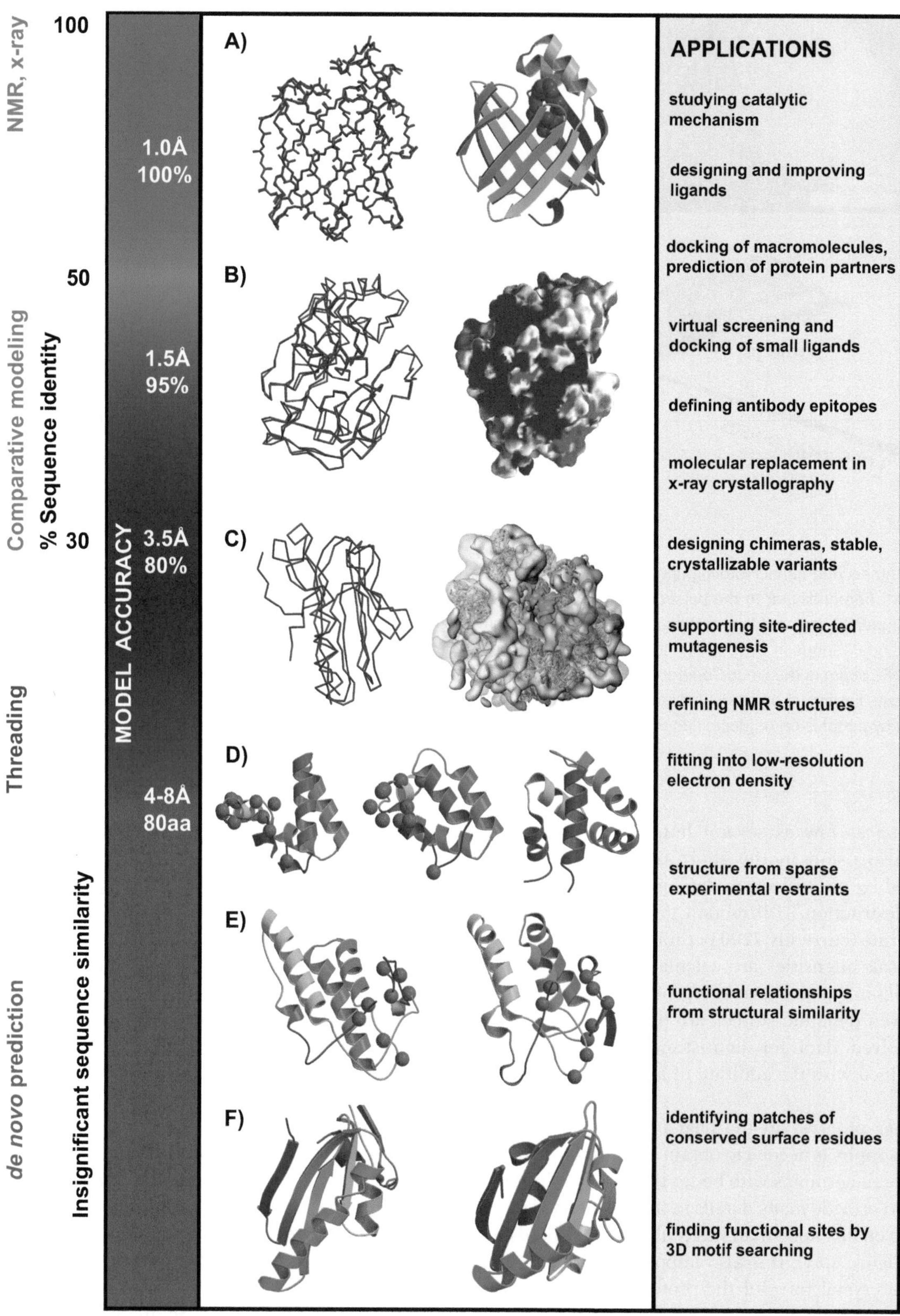

8.31 Resolution in structure determination. The different ranges of model accuracy from various methods (left column) allow different applications (right column). Backbone structures predicted with Rosetta (red) compared to actual structures (blue) increase in accuracy as the resolution increases from better than 4 Å (C, B, A). From Baker, D., and A. Sali, *Science*. 2001, 294:93–96. © 2001, American Association for the Advancement of Science. Reprinted with premission from AAAS.

FOR FURTHER READING

Membrane Diffraction

Tristram-Nagle, S., and J. F. Nagle, Lipid bilayers: thermodynamics, structure, fluctuations and interactions. *Chem Phys Lipids*. 2004, 127:3–14.

Tristram-Nagle, S., R. Chan, E. Kooijman, *et al.*, HIV Fusion peptide penetrates, disorders, and softens T-cell membrane mimics. *J Mol Biol*. 2010, 402:139–153. See also www.cmu.edu/biolphys/jfstn.

White, S. H., and M. C. Wiener, The liquid-crystallographic structure of fluid lipid bilayer membranes, in K. M. Merz, Jr., and B. Roux (eds.), *Biological Membranes, A Molecular Perspective from Computation and Experiment*. Boston, MA: Birkhauser, 1996.

Wiener, M. C., and S. H. White, Fluid bilayer structure determined by the combined use of x-ray and neutron diffraction. *Biophys J*. 1991, 59:162–173.

Molecular Modeling

Feller, S. E., Molecular dynamics simulations as a complement to nuclear magnetic resonance and x-ray diffraction measurements. *Methods Mol Biol*. 2007, 400:89–102.

Gumbart, J., Y. Wang, A. Aksimentiev, E. Tajkhorshid, and K. Schulten, Molecular dynamics simulations of proteins in lipid bilayers. *Curr Opin Struct Biol*. 2005, 15:423–431.

Khalili-Araghi, F., J. Gumbart, P. C. Wen, *et al.*, Molecular dynamics simulations of membrane channels and transporters. *Curr Opin Struct Biol*. 2009, 19:128–137.

Schlenkrich, M., J. Brickmann, A. D. MacKerell Jr., and M. Karplus, An empirical potential energy function for phospholipids, in K. M. Merz, Jr., and B. Roux (eds.), *Biological Membranes: A Molecular Perspective from Computation and Experiment*, Boston, MA: Birkhauser, 1996, pp. 31–81.

Scott, H. L., Modeling the lipid component of membranes. *Curr Opin Struct Biol*. 2002, 12: 495–502.

Lipids Viewed in Membrane Protein Structures

Adamian, L., H. Naveed, and J. Liang, Lipid-binding surfaces of membrane proteins: evidence from evolutionary and structural analysis. *Biochim Biophys Acta*. 2011, 1808:1092–1102.

Lee, A. G., Lipid–protein interactions in biological membranes: a structural perspective. *Biochim Biophys Acta*. 2003, 1612:1–40.

Lee, A. G., Biological membranes: the importance of molecular detail. *Trends Biochem Sci*. 2011, 36:493–500.

Marsh, D., and T. Pali, The protein–lipid interface: perspectives from magnetic resonance and crystal structures. *Biochim Biophys Acta*. 2004, 1666:118–141.

Marsh, D., Protein modulation of lipids and vice versa in membranes. *Biochim Biophys Acta*. 2008, 1778:1545–1575.

Palsdottir, H., and C. Hunte, Lipids in membrane protein structures. *Biochim Biophys Acta*. 2004, 1666:2–18.

Crystallography of Membrane Proteins

Cherezov, V., Lipidic cubic phase technologies for membrane protein structural studies. *Curr Opin Struct Biol*. 2011, 21:559–566.

Chun, E., A. A. Thompson, W. Liu, *et al.*, Fusion partner toolchest for the stabilization and crystallization of G protein-coupled receptors. *Structure*. 2012, 20:967–976.

Fromme, P., and J. C. H. Spence, Femtosecond nanocrystallography using x-ray lasers for membrane protein structure determination. *Curr Opin Struct Biol*. 2011, 21:509–516.

Hunte, C., and H. Michel, Crystallisation of membrane proteins mediated by antibody fragments. *Curr Opin Struct Biol*. 2002, 12: 503–508.

Landau, E. M., and J. P. Rosenbusch, Lipidic cubic phases: a novel concept for the crystallization of membrane proteins. *Proc Natl Acad Sci USA*. 1996, 93:14532–14535.

Sonoda, Y., S. Newstead, N. Hu, *et al.*, Benchmarking membrane protein detergent stability for improving throughput of high-resolution x-ray structures. *Structure*. 2011, 19:17–25.

Vergis, J. M., M. D. Purdy, and M. C. Wiener, A high-throughput differential filtration assay to screen and select detergents for membrane proteins. *Anal. Biochem*. 2010, 407:1–11.

White, S. H., Biophysical dissection of membrane proteins. *Nature*. 2009, 459:344–346.

MEMBRANE ENZYMES

9

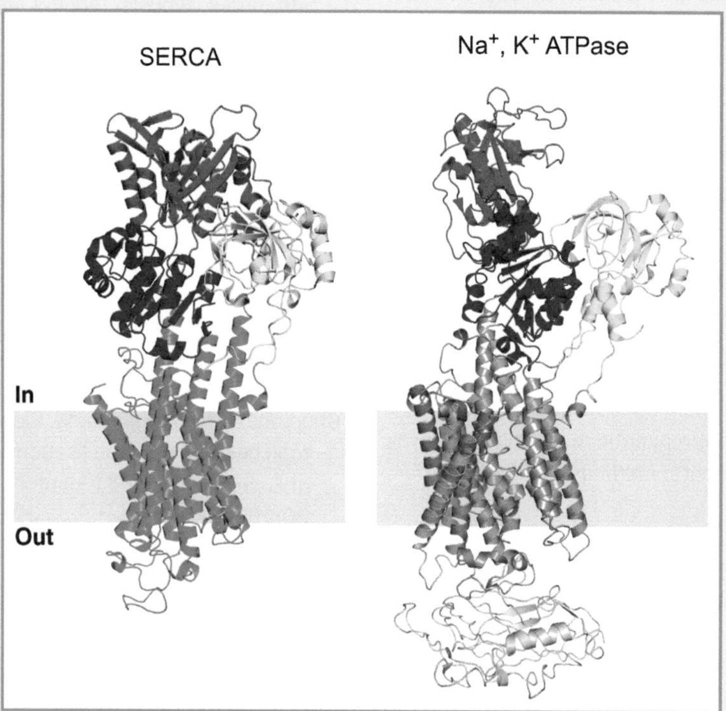

High-resolution structures show similarities of two P-type ATPases, the Ca^{2+} ATPase from sarcoplasmic reticulum (SERCA) and the Na^+, K^+ ATPase from the plasma membrane, shown here in ribbon diagrams. They share a common overall domain organization, with three cytoplasmic domains (colored to show correspondence) involved in the hydrolysis of ATP. In addition to the main catalytic (α) subunit, the Na^+, K^+ ATPase has a β subunit (tan) with a single TM helix and a luminal domain and a γ subunit (cyan). SERCA is the first membrane enzyme to have high-resolution structures of numerous intermediate steps in its reaction cycle. From Bublitz, M., *et al.*, *Curr Opin Struct Biol.* 2010, 20:431–439. © 2010, Elsevier. Reprinted with Permission from Elsevier.

An understanding of their lipid environment, structural constraints, and biogenesis lays the foundation for a survey of membrane protein structures. Recent progress in determining the structures of membrane proteins has produced an exponential increase in the total number of unique structures since the first structure was solved in 1985 (Figure 9.1). While restricted by topological considerations and largely confined to the classes of helical bundles and β-barrels, membrane protein structures show varied architecture employing elegant designs. The remaining chapters showcase a gallery of high-resolution structures selected to be representative of the different well-characterized membrane proteins.

The availability of structures has provided great insight into how membrane proteins function, igniting fresh excitement in the field. Most of these first structures are representative of many others in their families; occasionally one appears to give a unique solution to a particular need. This chapter looks at examples of membrane enzymes that are not part of extensive macromolecular machines; therefore, their structures tell much about how they carry out their functions. Chapter 10 presents

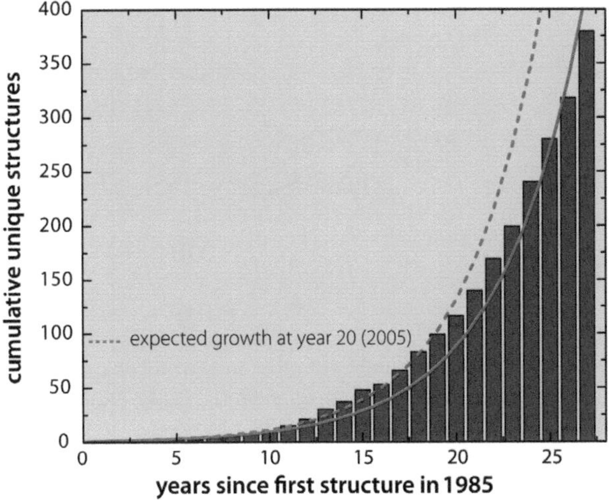

9.1 Progress in determining membrane protein structures at high resolution. The cumulative number of unique high resolution structures of membrane proteins has shown exponential growth when plotted against the number of years since the first structure was reported in 1985, although the increase has slowed in the last five years. This growth approximates the exponential growth in numbers of soluble protein structures that occurred 25 years earlier. From White, S. H., http://blanco.biomol.uci.edu/mpstruc/listAll/list. © 2009, Nature Publishing Group. Reprinted by permission of Macmillan Publishers.

the exciting progress made in structural biology of membrane receptors. Chapters 11 and 12 describe structures of transporters and channels that provide amazing specificity in the passage of small molecules and ions across membranes. Chapter 13 views some of the large and complex assemblies of membrane proteins that interact to carry out electron transport, generate a proton gradient, and harvest the energy of that gradient.

The number of structures of membrane enzymes solved in their entirety is still not large (see Chapter 6). The high-resolution structures of membrane enzymes that are available portray a variety of often surprising arrangements for carrying out catalysis in the membrane environment. Furthermore, they may differ from soluble enzymes in having overlapping functions, typically as membrane receptors, ion channels, or transporters. This chapter covers a broad range of membrane enzymes, from a monotopic enzyme to catalytic β-barrels, electron carriers, and large ATPases that undergo substantial conformational changes during their reaction cycles. Prostaglandin H_2 synthase (PGHS) is a membrane protein that sits in one leaflet of the membrane to have ready access to its substrates. The outer membrane phospholipase A (OMPLA) is a β-barrel that traps its phospholipid substrate when it dimerizes. Another family of β-barrels including OmpT carry out proteolysis of extracellular substrates. Very different proteases such as rhomboid hydrolyze peptide bonds of TM segments in the membrane. Fumarate reductase creates a "redox loop" for the generation of a proton motive force under anaerobic conditions. Finally the structures of P-type ATPases that pump calcium ions and that exchange sodium for potassium ions give insight into how the hydrolysis of ATP is tightly coupled to ion transport.

PROSTAGLANDIN H_2 SYNTHASE

In vertebrates from humans to fish, two isoforms of PGHS, also called cyclooxygenase (COX),[1] carry out the committed step in prostaglandin synthesis. Their main catalytic function is the conversion of arachidonic acid to prostaglandin H_2 (PGH_2), using two molecules of oxygen (Figure 9.2). These enzymes are monotopic integral membrane proteins: They bind to the luminal leaflet of the ER membrane, as well as to nuclear membranes, and they require detergent solubilization to release them from the membrane. Implicated in thrombosis, inflammation, neurological disorders, and cancer, they receive a great deal of attention as the targets of nonsteroidal anti-inflammatory drugs (NSAIDs) such as aspirin, acetaminophen, and ibuprofen, as well as newer drugs such as rofecoxib (Vioxx) and celecoxib (Celebrex) that specifically target the second isoform.

The active enzymes are homodimers with subunit molecular weights of 70 kDa. The sequences of PGHS-1 and PGHS-2 have 60–65% identity and only minor differences in structure, but they differ in expression and function. PGHS-1 is expressed constitutively in a wide range of tissues, while expression of PGHS-2 is induced by inflammatory and proliferative stimuli and it is mainly found in nervous system, immune, and renal cells. Due to catalytic controls, PGHS-2 is active in cells where PGHS-1 is latent (see Box 9.1). PGHS-2 also accepts a wider range of substrates, which allows for design of inhibitors that specifically target this isoform (see below).

Ovine PGHS-1 (from rams) was the first monotopic protein for which a high-resolution structure was obtained. The crystal structure for the murine PGHS-2 (from mice) showed a peptide backbone that was superimposable to PGHS-1. In the following description features of PGHS (and residue numbers) are drawn from ovine PGHS-1. The crystallized protein lacks its signal sequence, which has been cleaved from the N terminus as is typical for proteins targeted to the ER, and lacks the C-terminal end, which is not resolved in the x-ray structure. Bound oligosaccharides are observed: Glycosylation is necessary for folding but not for enzyme activity once folded.

[1] The enzyme is called PGHS here, rather than the more familiar name COX, because it contains two active sites, COX and peroxidase (POX), that are discussed in more detail below.

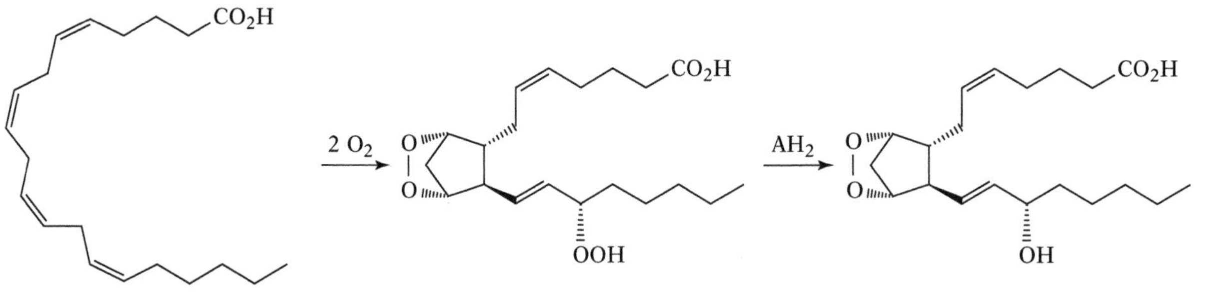

9.2 The main reaction carried out by prostaglandin H_2 synthase. The enzymes accept a number of other fatty acid substrates, and the second isoform accepts neutral derivatives of arachidonate. AA, arachidonic acid; PGG_2, prostaglandin G_2; PGH_2, prostaglandin H_2. Redrawn from Rouzer, C. A., and L. J. Marnett, *Biochem Biophys Res Commun*. 2005, 338:34–44.

BOX 9.1 MECHANISM OF ACTION OF PROSTAGLANDIN H_2 SYNTHASE

The crystal structures, along with data from mutant studies, spectroscopy, and EPR, provide insight into the mechanism of the reaction. The mechanism at the COX site is a controlled free radical chain reaction that can be divided into four stages (see Figure 9.1.1). Entry of arachidonic acid into the substrate channel where it interacts with Arg120 positions the *proS* hydrogen on carbon 13 next to a free radical on Tyr385. Abstraction of this hydrogen creates an arachidonyl radical centered at carbon 13. A rearrangement of the radical to carbon 11 is followed by attack by O_2 to form an 11-peroxy radical. The 11-peroxy radical attacks carbon 9 to form the endoperoxide with isomerization of the radical to carbon 8, and ring closure occurs between carbon 8 and carbon 12, producing the bicyclic peroxide. This is hypothesized to change the configuration of the substrate, repositioning the acyl tail to allow attack by the second oxygen on carbon 15 to generate a 15-hydroperoxyl radical. Abstraction of a hydrogen atom from Tyr385 produces PGG_2 and regenerates the tyrosyl radical for the next round of catalysis.

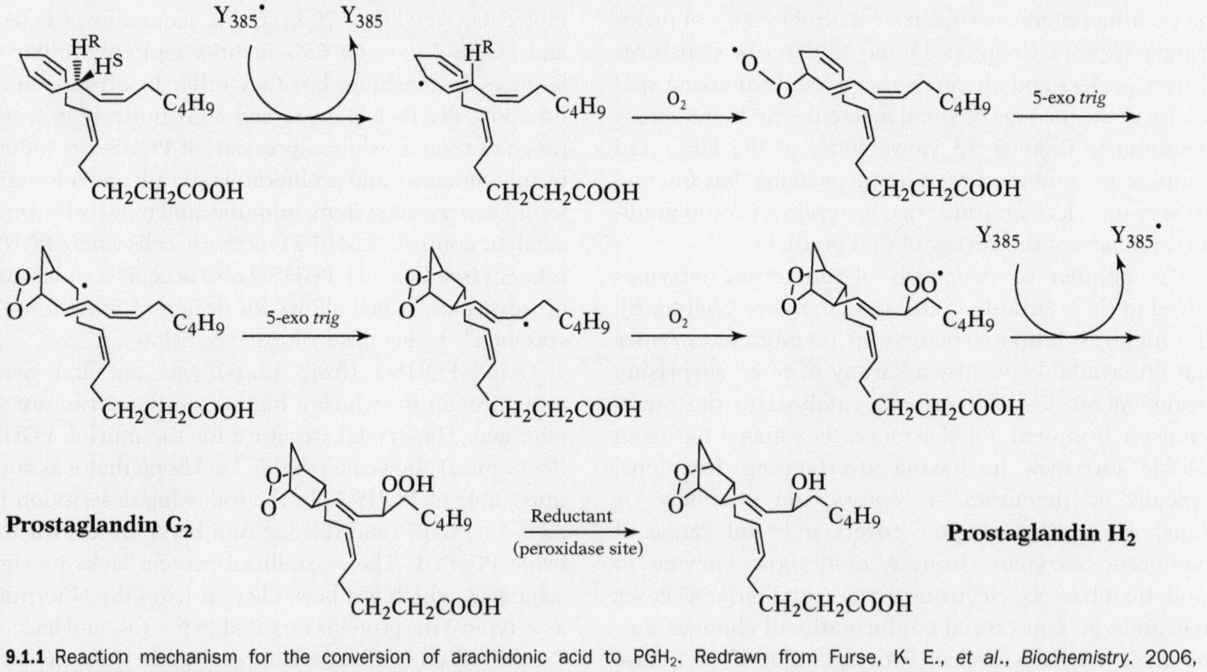

9.1.1 Reaction mechanism for the conversion of arachidonic acid to PGH_2. Redrawn from Furse, K. E., *et al.*, *Biochemistry*. 2006, 45:3189–3205. © 2006 by American Chemical Society.

The free radical at Tyr385 is generated by a redox reaction at the POX site that involves a hydroperoxide to produce a ferryl-oxo derivative of the heme. Intramolecular migration of the electron transfers the radical from the heme to Tyr385. PGG_2, the product of the COX reaction, serves as the substrate for the peroxidase at the POX site, generating PGH_2. The initial need for peroxide to generate the free radical provides a way to differentiate between the PGHS isoforms catalytically. Activation of PGHS-1 by peroxide is much less efficient than activation of PGHS-2. Under many conditions, glutathione levels and cellular glutathione peroxidase keep the peroxide concentrations below those needed for PGHS-1. Therefore even though its expression is constitutive, PGHS-1 is catalytically latent in cells under many conditions.

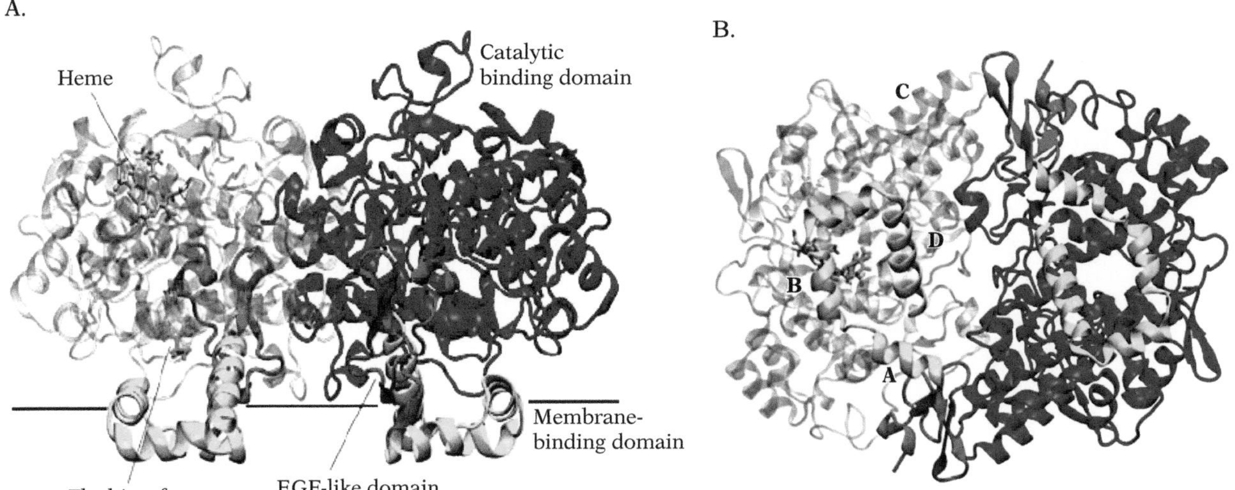

9.3 (A) The structure of the PGHS dimer viewed from the side and colored to indicate the different domains: EGF domains (red), membrane-binding domains (MBDs, yellow), and catalytic domains (blue and gray). In the monomer on the left, the heme is pink and a bound flurbiprofen (an NSAID) is green. The plane of the membrane is indicated by the line at the bottom. (B) The side of the PGHS dimer that faces the membrane interior is viewed from the bottom. The four α-helices of the MBD are labeled A, B, C, and D in the monomer on the left. From Fowler, P. W., and P. V. Coveney, *Biophys J.* 2006, 91:401. © 2006, Elsevier. Reprinted with permission from Elsevier.

Each subunit of the dimer has three domains: A domain similar to epidermal growth factor (EGF, residues 33–72); a membrane-binding domain (MBD, residues 73–116); and a catalytic domain (residues 117–586; Figure 9.3A). The interface in the dimer involves the EGF-like and catalytic domains, with patches of polar, electrostatic, and hydrophobic contacts between the subunits. The role of the EGF-like domain is unclear, although it is a feature of many cell surface proteins. It contains three highly conserved interlocking disulfide bonds, with a fourth disulfide linking it to the catalytic domain.

The membrane-binding domain consists of four amphipathic α-helices, three of which lie roughly in the same plane, while the fourth angles away from them into the catalytic domain (Figure 9.3B). The hydrophobic and aromatic residues along the surface interact with the lipid bilayer and contribute to a strong interaction with the membrane (ΔG for binding of -37 kcal mol^{-1}). MD simulations suggest that once inserted, PGHS also forms hydrogen bonds with lipid phosphate groups.

The large catalytic domain has two active sites for the two steps of the reaction: the COX site that converts arachidonic acid to the hydroperoxy endoperoxide, prostaglandin G_2 (PGG_2), and a heme-containing peroxidase (POX) site that reduces PGG_2 to the hydroxyl endoperoxide, PGH_2 (Figure 9.4). The catalytic domain has direct homology to members of the myeloperoxidase family, which makes PGHS the only membrane protein in the superfamily of heme-dependent peroxidases.

At the POX site, the heme is unusually open to solvent. It lies in a shallow cleft on the surface opposite

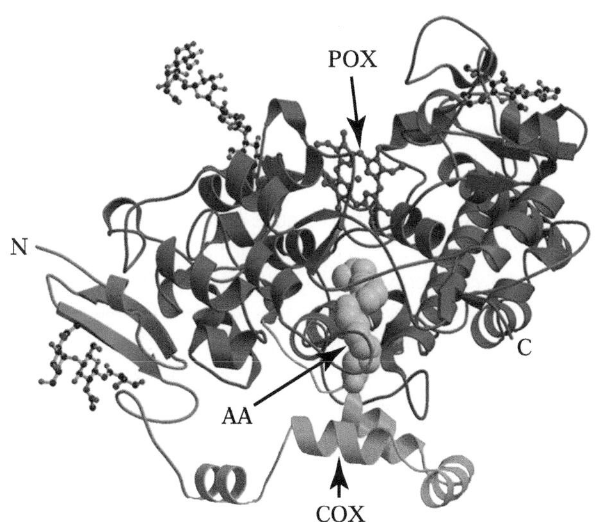

9.4 The x-ray structure of the PGHS monomer. Each monomer has three domains. The catalytic domain (blue) has two active sites, the POX site (top, at the heme) and the COX site (bottom), where arachidonic acid (yellow space-filling model) is bound. The membrane-binding domain (orange) is below the arachidonic acid, and the epidermal growth factor domain (green) is on the side that becomes the subunit interface in the dimer. From Garavito, R. M., and A. M. Mulichak, *Annu Rev Biophys Biomol Struct.* 2003, 32:183–206. © 2003 by Annual Reviews. Reprinted with permission from the Annual Review of Biophysics and Biomolecular Structure, www.annualreviews.org.

A.

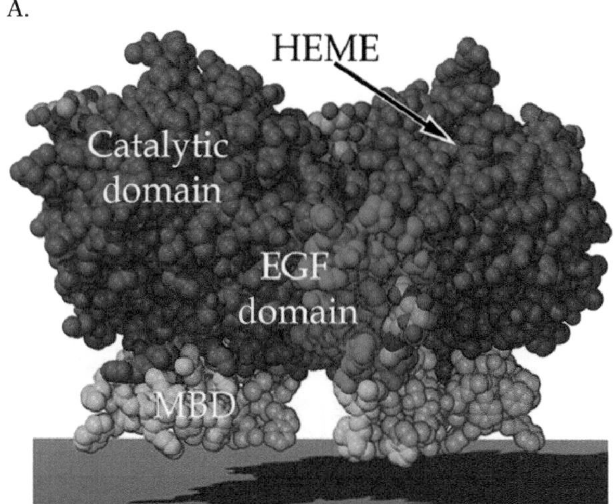

B.

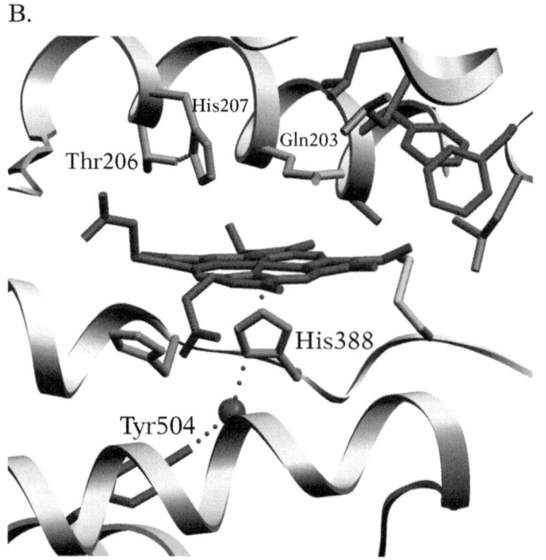

9.5 Heme in the POX site of the catalytic domain of PGHS. (A) The heme is in a cleft that is quite exposed on the surface of the catalytic domain. (B) A close-up of the heme-binding site of PGHS-1 shows the proximal (His388) and distal (His207) ligands, along with other amino acids that define the heme-binding pocket. From Garavito, R. M., and A. M. Mulichak, *Annu Rev Biophys Biomol Struct*. 2003, 32:183–206. © 2003 by Annual Reviews. Reprinted with permission from the Annual Review of Biophysics and Biomolecular Structure, www.annualreviews.org.

the MBD, coordinated by proximal (His388) and distal (His207) axial ligands (Figure 9.5). Although the groups around the heme are not highly conserved, the ferric/ferrous midpoint potential is quite similar in the two isoforms of PGHS. Below the heme is a long channel for binding long-chain fatty acid substrates (Figure 9.6A). The COX catalytic center encompasses half the channel, from Arg120 to Tyr385. Arachidonic acid binds in an extended L-shape, with its carboxylate liganded by the guanidinium group of Arg120 (Figure 9.6B). Carbons 7 through 14 weave around the side chain of Ser530 (the residue acetylated by aspirin) and carbon 13 is pointed toward the phenolic oxygen of Tyr385. The rest of the acyl chain (carbons 14–20) binds in a hydrophobic groove above Ser530 but is not resolved in crystal structures. There is room for movement in the channel that enables a few minor products to be synthesized from arachidonic acid and allows several other fatty acids to bind as substrates, forming other bioactive products. In PGHS-2 substitution of a valine for Ile523 creates an additional pocket that allows fatty acyl derivatives to bind, such as 2-arachidonyl-glycerol and arachidonyl-ethanolamine. The products of these two reactions are both endogenous ligands for the cannabinoid receptors that mediate the effects of the active component of marijuana. Clearly the flexibility at the active site has important biological consequences of interest.

Many NSAIDs act as competitive inhibitors of PGHS. High-resolution structures solved with NSAIDs bound to the enzyme show how they prevent substrate binding by occupying the upper part of the COX channel between Arg120 and Tyr385 (Figure 9.7). The interactions between the drugs and the enzyme are hydrophobic, with the exception of interactions of the acidic NSAIDs with Arg120 and the potential of forming a hydrogen bond with Ser530. The extra binding pocket in the active site of PGHS-2 has been exploited to design specific inhibitors that do not fit the channel of PGHS-1 (commonly known as COX-2 inhibitors). For example, flurbiprofen interacts with Arg120 and fills a portion of the substrate channel in PGHS-1, while the phenylsulfonamide group of SC-588 fits into the pocket of the channel in PGHS-2. Other NSAIDs follow binding with covalent modification of the enzyme. When aspirin acetylates Ser530 it blocks the binding of arachidonic acid to PGHS-1, thereby inhibiting the enzyme completely. Acetylation has a different effect on PGHS-2, in which it does not inhibit substrate binding but results in formation of a different product, presumably because it affects the alignment of the substrate in the channel. As these differences are likely to have significant biological consequences, developing new inhibitors to help minimize side effects ensures continuing interest in these enzymes.

OMPLA

The outer membrane phospholipase A is an unusual membrane enzyme for a number of reasons. It is one of the few integral membrane proteins in the outer membrane of Gram-negative bacteria that has enzyme activity. It belongs to a large family of lipolytic enzymes (enzymes that catalyze the hydrolysis of lipids and phospholipids), yet it has no sequence homology with the water-soluble members of the family. Like most other outer membrane proteins, this 31-kDa protein has a β-barrel structure. Most unusual is the fact that its active site is on the external surface of the β-barrel.

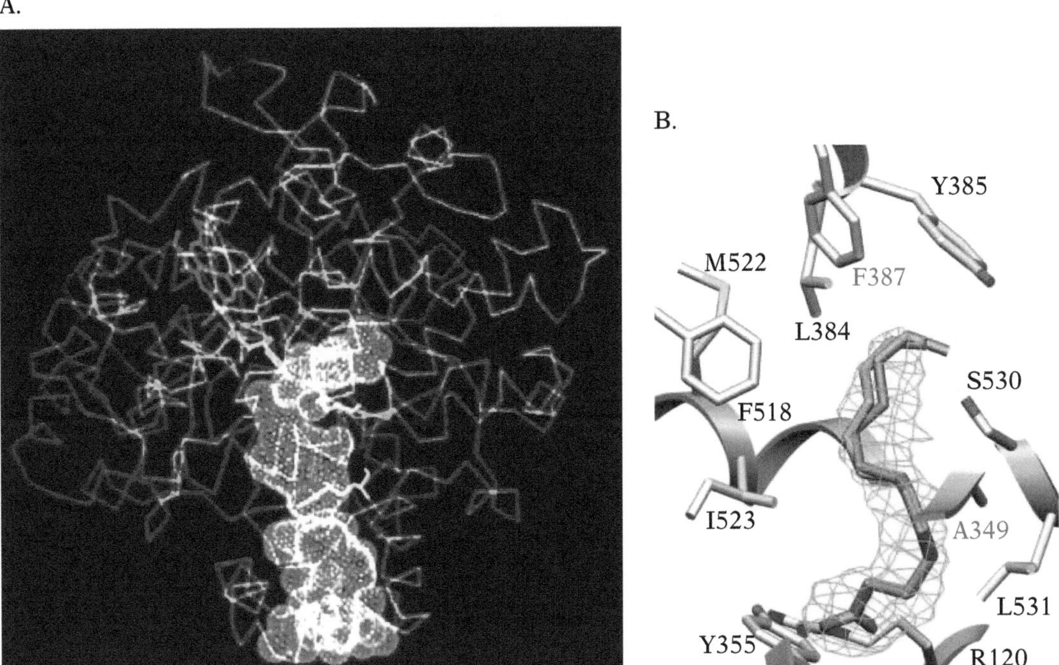

9.6 (A) Substrate access channel of the COX site of the catalytic domain of PGHS. The α-C tracing of the PGHS-1 monomer (blue) is shown with the substrate access channel highlighted (pink). Active site residues are Arg120, Tyr385, and Ser530 (yellow), and the heme is also indicated (red). From Rouzer, C. A., and L. J. Marnett, *Biochem Biophys Res Commun*. 2005, 338:34–44. © 2005 by Elsevier. Reprinted with permission from Elsevier. (B) A portion of arachidonic acid in the substrate channel of a mutant PGHS-1. The first 12 carbon atoms of arachidonic acid are modeled (pink) into the electron density (light green) and compared with a simulated structure (blue). The structure shows that the mutations, V349A and W387F, do not significantly affect the structure of the active site. The side chains that contact the substrate are identified and shown with stick representations (gray carbon atoms, red oxygens, dark blue nitrogens, and yellow sulfur). From Harman, C. A., *et al.*, *J Biol Chem*. 2004, 279:42929–42935. © 2004, by the American Society for Biochemistry & Molecular Biology.

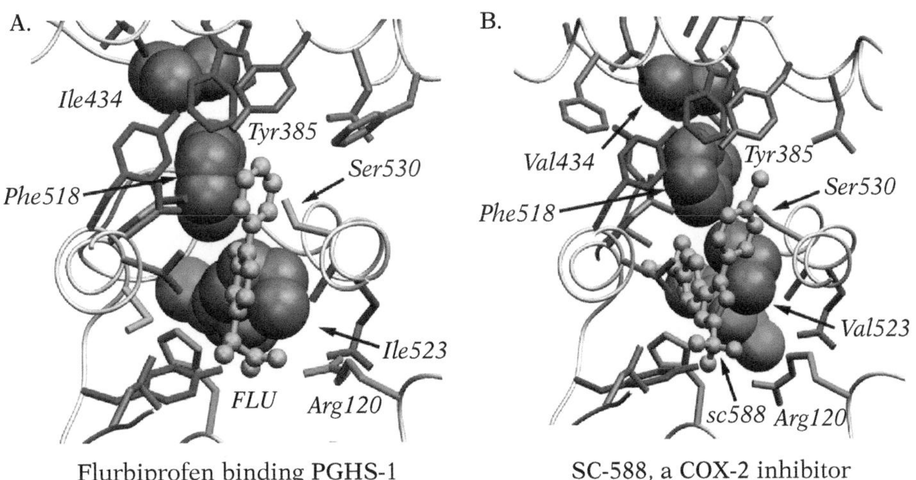

9.7 Binding of NSAIDs to PGHS as revealed by crystal structures. The amino acid side chains that define the binding sites are shown, with those critical to the differences between the two isoforms in space-filling models: isoleucine residues at positions 434 and 523 in PGHS-1 are replaced with valine residues in PGHS-2, allowing a shift in Phe518 (all three copper-colored) that gives access to a more polar side pocket between Ser530 and Tyr385 for specific COX-2 inhibitors. (A) Flurbiprofen (yellow-green) binds in the substrate channel of PGHS-1. (B) SC-588 binds in the substrate channel of PGHS-2, filling the side pocket with the phenylsulfonamide group that prevents its binding to the active site of PGHS-1. From Garavito, R. M., and A. M. Mulichak, *Annu Rev Biophys Biomol Struct*. 2003, 32:183–206. © 2003 by Annual Reviews. Reprinted with permission from the Annual Review of Biophysics and Biomolecular Structure, www.annualreviews.org.

The enzyme activity of OMPLA is dependent on calcium. It has phospholipase A_1 and A_2 activities, which denote the ability to cleave the acyl chain from either carbon 1 or carbon 2 of phospholipids. In addition, it can cleave the acyl chain from a lysophospholipid. Its broad substrate specificity shows a minimum requirement of a polar headgroup esterified to an acyl chain of at least 14 carbon atoms.

With such broad specificity, OMPLA must be strictly regulated to avoid significant losses of membrane phospholipids. The monomer of OMPLA is inactive, while the dimer is active. Activation of the enzyme is triggered by perturbations of the integrity of the outer membrane from events such as heat shock, EDTA treatment and spheroplast formation, phage-induced lysis, and colicin release. It is likely that such disruptions affect the normal lipid asymmetry of the outer membrane, which consists of an outer leaflet of lipopolysaccharide and an inner leaflet of phospholipids. The resulting nonbilayer structures allow phospholipids access to the active site of OMPLA, which promotes dimerization.

What then is its physiological role? OMPLA is constitutively expressed. In *E. coli* OMPLA is involved in release of bacteriocins, and in pathogenic bacteria it is involved in virulence, most likely because an increase in lysophospholipid content of the outer membrane gives increased invasive capacity. In addition, it has been proposed to enable cells to tolerate organic solvents by increasing phospholipid turnover to allow incorporation of *trans* fatty acids produced by a periplasmic *cis–trans* isomerase, which makes the membrane less permeable.

The *E. coli* gene for OMPLA, called *pldA*, codes for 289 amino acids, the first 20 of which are a typical signal sequence that is removed during export from the cytosol (see Chapter 7). Comparison of the *E. coli* gene sequence with the OMPLA genes from 15 other Gram-negative bacteria reveals 20 strictly conserved residues and an additional 18 that are highly conserved. One conserved region contains the catalytic triad, which consists of histidine, serine, and asparagine residues (Asn156-His142-Ser144 in *E. coli*; Figure 9.8). Thus OMPLA is a serine hydrolase, and site-directed mutagenesis of Ser144 abolishes its activity except in the case of S144C, which has 1% the activity of the wild type. Similarly, H142G has an activity four orders of magnitude lower than the wild type. Asn156 is less strictly conserved (being replaced by Asp or Gln); its role is to orient the histidine residue, in the same manner of an Asp residue in classical serine proteases. In the mechanism, Ser144 performs a nucleophilic attack on the carbonyl carbon of the ester bond in the substrate, forming a tetrahedral intermediate. Cleavage of the bond generates a free lysophospholipid, which can diffuse away, and an acyl–enzyme intermediate, which is subsequently cleaved by hydrolysis, releasing a fatty acid.

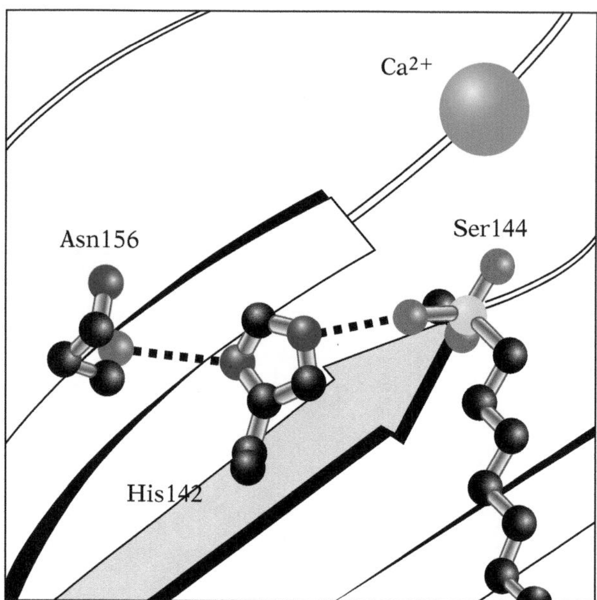

9.8 The catalytic triad of OMPLA. Serine, histidine, and asparagine residues at the active site of OMPLA, with a hexadecylsulfonyl inhibitor covalently attached to Ser144, are shown from the crystal structure of OMPLA of *E. coli*. The dashed lines represent hydrogen bonds. Note the proximity of the calcium ion. Redrawn from Kingma, R. L., *et al.*, *Biochemistry*. 2000, 39:10017–10022. © 2004 by American Chemical Society.

The x-ray structure of OMPLA, solved with a resolution of 2.6 Å, shows a *β*-barrel composed of 12 *β*-strands with polar loops on the outside end of the barrel and short turns on the inside end (Figure 9.9). The TM *β*-strands of OMPLA are amphipathic, giving it a polar interior and a hydrophobic surface, with interfacial regions rich in aromatic residues. The active site residues His142 and Ser144 are located at the exterior of the barrel in the outer leaflet side of the membrane.

The OMPLA monomer is flattened on one side, and dimers form by the association of their flat sides (Figure 9.10A). The dimer interface is hydrophobic, with a patch of four leucine residues from each monomer forming knobs that fit into holes on the opposite subunit (Figure 9.10B). Embedded in the hydrophobic region is a strictly conserved glutamine residue, Gln94, which makes a double hydrogen bond with the same residue in the other monomer. When cysteine residues are inserted into the flat side with the mutation H26C, the disulfide bond crosslinks OMPLA to produce a covalent dimer.

Reversible dimerization is triggered by addition of Ca^{2+} *in vitro*, and binding of the inhibitor hexadecanesulfonyl fluoride stabilizes the dimer. The high-resolution structure of the dimer binding the inhibitory hexadecanesulfonyl chain shows very little difference from the structure of the monomer. The dimer has two deep clefts that extend 25 Å from the active sites along the subunit interface (Figure 9.10C). The 16 carbon atoms of the

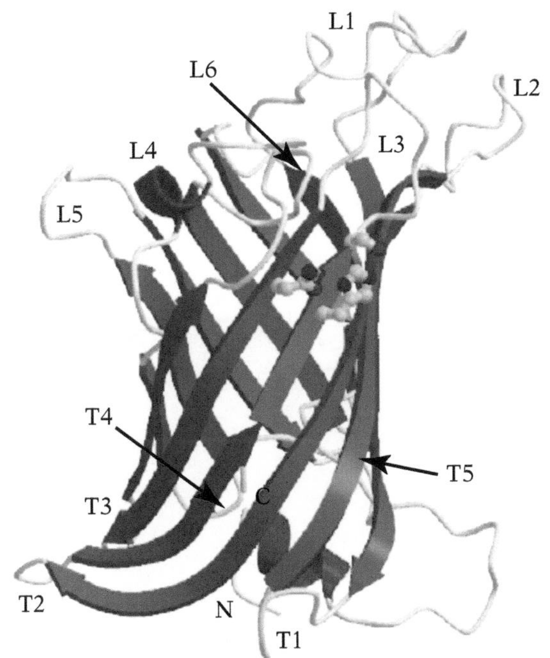

9.9 Structure of a monomer of OMPLA. The monomer has 12 β-strands (blue), and two small α-helices (red). The active site residues on the external surface are shown as ball-and-stick models. From Snijder, H. J., *et al.*, *Nature*. 1999, 401:717–721. © 1999. Reprinted by permission of Macmillan Publishers Ltd.

inhibitor fit into these clefts and interact with both monomers. Thus the substrate binding pocket is only formed in the dimeric enzyme.

Calcium is abundant in the bacterial outer membrane, and each monomer has two Ca^{2+}-binding sites, one with ten-fold higher affinity than the other. However, in the dimer, both have high affinity (K_d ~50 μM). The Ca^{2+}-binding site seen in the structure of the monomer is around 10 Å from the active site, between loops L3 and L4, with two aspartate side chains as ligands (Asp149 and Asp184). The second Ca^{2+}-binding site in the dimer is located at the active site and is the catalytic calcium site (Figure 9.11). The binding involves the side chain of Ser152 and one main-chain carbonyl oxygen atom from each monomer, with three water molecules in the binding site. The effect of the catalytic calcium ion is to polarize the two water molecules, forming an oxyanion hole that stabilizes the negatively charged intermediates during the reaction.

Thus three factors contribute to the inactivity of the monomer: (1) the absence of a substrate-binding pocket; (2) the lack of the oxyanion stabilization that results from Ca^{2+} binding the catalytic site; and (3) the physical separation of the active site (outer leaflet) from the substrate (PL in the inner leaflet). Perturbations of the outer membrane can make substrate available; then substrate binding triggers dimer formation and generates an active complex. Such a perturbation might be triggered by the colicin release protein that presents phospholipids in the outer membrane, allowing dimerization to activate OMPLA. In this case, the

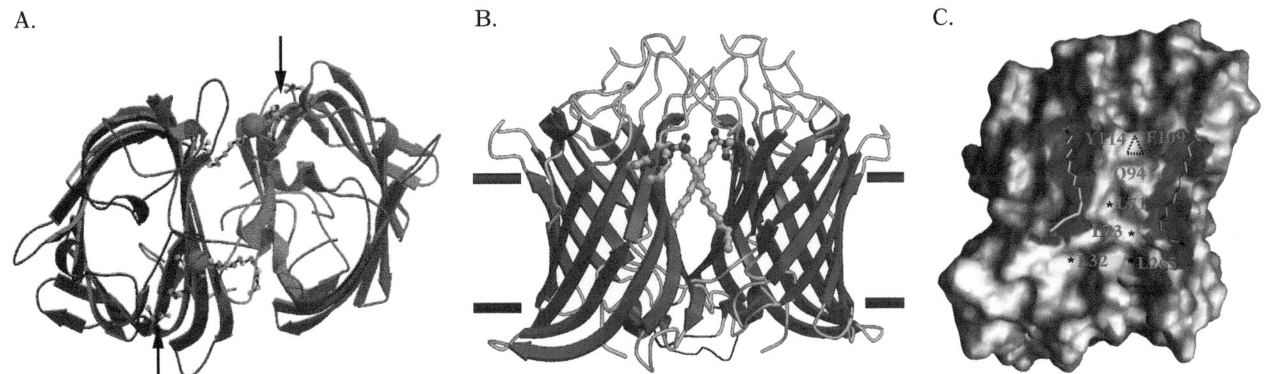

9.10 The OMPLA dimer. **(A)** One side of the OMPLA monomer is flattened, and this side interacts with another subunit to form the dimer, as evident in a view of the crystal structure of the OMPLA dimer from the top, looking down on the β-barrels. The arrows point to the active sites, with the active site residues and the hexadecanesulfonyl inhibitor shown in ball-and-stick and the calcium ions represented by large spheres. From Snijder, H. J., and B. W. Dykstra, *Biochim Biophys Acta*. 2000, 1488:91–101. © 2000 by Elsevier. Reprinted with permission from Elsevier. **(B)** The structure of the OMPLA dimer inhibited with hexadecanesulfonyl fluoride is shown (colored as in Figure 9.9). Two inhibitor molecules (ball-and-stick models) occupy the active site between the subunits. The black lines indicate the polar–nonpolar interface of the membrane. **(C)** The surface representation of the subunit interface on one monomer shows the clefts for substrate binding by indicating the surface area within 1.5 Å of the van der Waals surface of the inhibitors (blue). Residues involved in the dimer interaction are labeled, with asterisks marking the knob-and-hole pattern and a dashed triangle to label the hydrophilic cavity. From Snijder, H. J., *et al.*, *Nature*. 1999, 401:717–721. © 1999. Reprinted by permission of Macmillan Publishers Ltd.

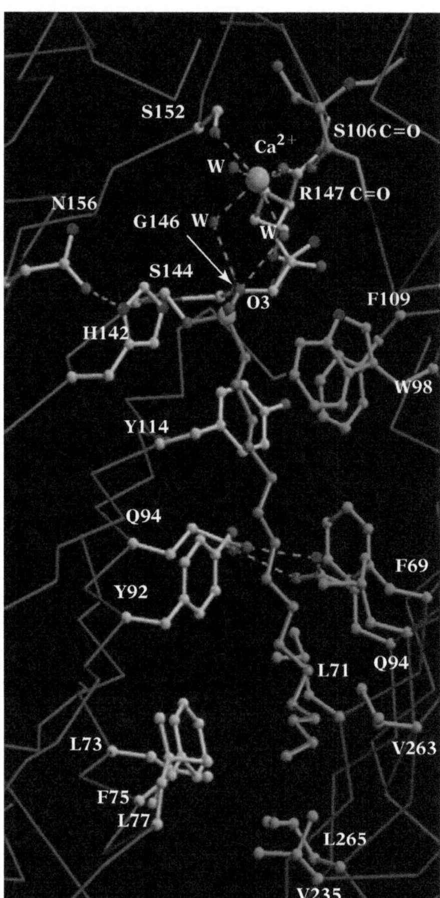

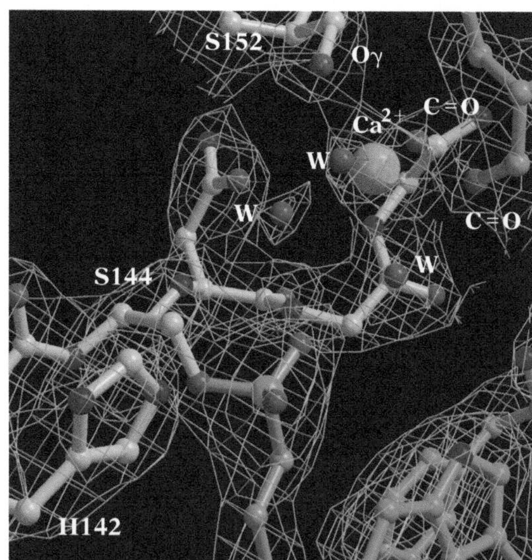

9.11 View of the active site of the OMPLA dimer inhibited with hexadecanesulfonyl. (A) The binding site at the active site of OMPLA is formed by residues on the outsides of both β-barrel monomers (yellow and green). The calcium (white sphere) is liganded by the carbonyl groups from Ser106 and Arg147 along with three water molecules; none of the calcium ligands is charged. Below it, the inhibitor hexadecanesulfonyl (purple) occupies the substrate-binding pocket formed between the two monomers. The catalytic residues from the subunit on the right, Asn156–His283–Ser144, are clearly visible, with an arrow indicating the sulfonyl oxygen that occupies the oxyanion hole. (B) A close-up view of the polar end of the site shows electron densities of the active site residues, water molecules, and Ca^{2+}. From Snijder, H. J., *et al.*, *Nature*. 1999, 401:717–721. © 1999. Reprinted by permission of Macmillan Publishers Ltd.

job of OMPLA is to hydrolyze PLs to increase the permeability of the outer membrane so the colicins may be secreted. Clearly the structure of this unusual dimer of β-barrels held the key to understanding its regulation as well as its function.

MEMBRANE PROTEASES

Proteases in the membranes bring surprises in their structures and mechanisms. Structurally unique are a class of β-barrel proteases from enterobacterial outer membranes, which cleave their substrates outside the cell. In contrast, intramembrane proteases are helical integral membrane proteins that cleave peptide bonds within transmembrane segments. These types of proteases use different approaches to capture the scissile bonds (the bonds to be cleaved) of their substrates for hydrolysis within the membrane.

Omptins

The β-barrel proteases are named Omptins for the first identified outer membrane protease, OmpT of *E. coli*. Omptins from a dozen enteric bacteria have 50–75% sequence identity, and yet they play very different physiological roles, from housekeeping (removal of unfolded proteins) to pathogenesis – even determining the virulence of plague infections by *Yersinia pestis*.

Reminiscent of OmpX (see Chapter 5), the ten β-strands of Omptins protrude far into the extracellular space, where they form the active site. The first

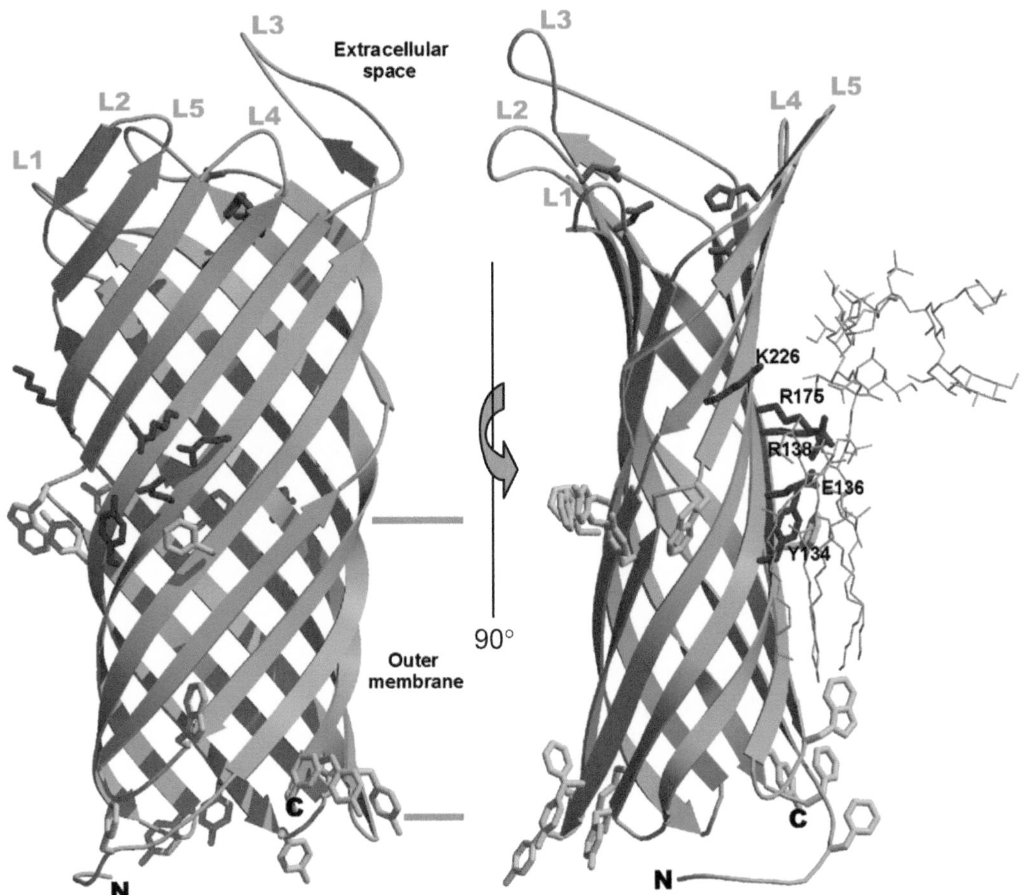

9.12 Structure of OmpT from *E. coli*. The OmpT β-barrel is viewed from two angles with the extracellular space on top and the external loops labeled L1 to L5. The position of the membrane is delineated by the aromatic girdles (yellow side chains). Also shown are residues proposed to be involved in catalysis (Asp83, Asp85, Asp210, and His212, red) and in binding LPS (Lys226, Arg175, Arg138, Glu136, and Tyr134, purple). The LPS (gray) is modeled based on that observed in a structure of FhuA. From Vandeputte-Rutten, L., *et al.*, *EMBO J.* 2001, 20:5033–5039. © 2001, Nature Publishing Group. Reprinted by permission of Macmillan Publishers Ltd.

structure of OmpT revealed a partially squashed β-barrel with catalytic residues in a large, external groove and a binding site for lipopolysaccharide (LPS) on the side (Figure 9.12). LPS binding is absolutely required for protease activity. The catalytic groove is negatively charged and contains 18 highly conserved residues. Proteolysis is carried out by a His212–Asp210 dyad on one side of the groove across from a pair of Asp residues, Asp83 and Asp85, whose likely role is activation of a water molecule to perform the nucleophilic attack on the scissile peptide bond. Specificity of OmpT for a cleavage site between basic residues could be explained by a deep negatively charged pocket which binds either Lys or Arg of the substrate at the bond to be cleaved and a shallow, hydrophobic site which binds a small hydrophobic residue (Ile, Val, or Ala) two residues away in the substrate.

A high-resolution structure of Pla, the Omptin from *Yersinia* whose primary structure is 50% identical to OmpT, reveals water molecules at the active site, supporting a role for a nucleophilic water molecule in the catalysis (Figure 9.13A). It also shows density corresponding to acyl chains of the bound LPS, proposed to push together the two sides of the active site (Figure 9.13B). The requirement for LPS is the only known regulation for Omptins and prevents them from digesting cytoplasmic or periplasmic proteins.

Intramembrane proteases

Proteases that cleave other proteins within the membrane are called I-CLiPs (intramembrane-cleaving proteases, see Chapter 6), and they release substrate domains into the cytoplasm, lumen, or outside the cell for a wide variety of functions. The common feature of their substrates is a single TM helix (type I and type II membrane proteins), and most have a helix-breaking segment that allows the protease access to the scissile

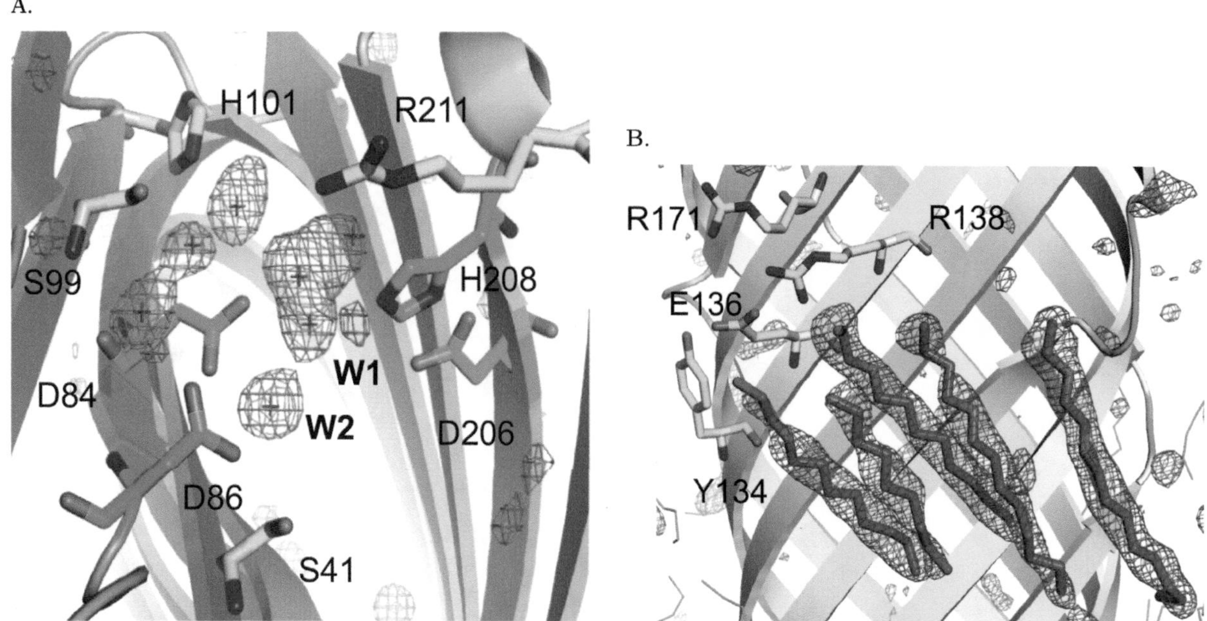

9.13 Features of the high-resolution structure of Pla from *Yersinia pestis*. (A) A view of the active site from the extracellular side shows water molecules (red crosses with blue mesh for Fo-Fc densities) among the active site residues (green stick models) and nearby important residues (yellow stick models). The nucleophilic water molecule that participates in catalysis is W1, and W2 is the water molecule between Asp84 and Asp86. (B) A view of the side of the Pla β-barrel shows the residues (yellow stick models) in the putative LPS-binding site, with bound acyl chains (magenta) fit to the electron densities (blue mesh). Both from Eren, E., *et al.*, *Structure*. 2010, 18:809–816. © 2010, Elsevier. Reprinted with permission from Elsevier.

peptide bond. Such proteases, found in all kingdoms of life, fall into three mechanistic classes: serine proteases, aspartyl proteases, and metalloproteases. As they bear no structural resemblance to soluble proteases of these types, intramembrane proteases achieved their similar mechanisms of hydrolysis by convergent evolution.

Serine proteases make up the largest family and differ from the others in that they do not require substrate precleavage and they release the cleaved portions of their substrates to the exterior of the cell. Rhomboid protease, the best characterized member of this family, is described fully below. Metalloproteases and aspartyl proteases do require precleavage of their substrates, which generates the required single TM span. Membrane metalloproteases have at least four TM helices and the HExxHx$_n$DG metal-binding motif. Some are induced by protein misfolding in the mammalian ER and the bacterial periplasm. Site protease 2 (SP2) is a metalloprotease that cleaves the sterol regulatory element-binding protein (after it is cleaved by SP1) to release a transcription factor for genes involved in cholesterol and FA synthesis. The structure of SP2 from *Methanocaldococcus jannaschii* shows that residues from two TM helices coordinate zinc at a catalytic site near the cytosol, with a channel for water entry.

The two classes of aspartyl proteases, represented by signal peptide peptidase (SPP) and γ-secretase, both have nine TM segments and two catalytic Asp residues in a YDx$_n$LGhGD motif. Bacterial SPP clears the cleaved signal peptides produced in the cytoplasmic membrane during protein export (see Chapter 7). Homologous proteases exist in eukaryotes, including humans, but their function is unknown. γ-secretase, a large complex of four proteins, carries out the final step in the release of the Aβ peptide associated with Alzheimer's disease (see Chapter 6). The requirement for precleavage is the main mechanism of regulation of these proteases. I-CLiPs are also regulated by compartmentalization.

Rhomboid protease

A rhomboid protease was first discovered in *Drosophila* embryogenesis, where it functions to cleave precursors of epidermal growth factor (such as Spitz) from the membrane, releasing them to act as transcription factors in neighboring cells (Figure 9.14). The several dozen rhomboid proteins now identified share less than 20% identity and play different roles in different organisms, from regulating apoptosis in mitochondria to facilitating invasion by pathogens such as malaria parasites. Rhomboids do not require cofactors or an energy source; some are stimulated by a particular lipid when reconstituted in detergents, while others are not. Rhomboids are not sensitive to the typical serine protease inhibitors, with the exception of isocoumarins. Their substrate specificity depends on helix-disruptors, in particular a Gly–Ala pair in the TM strand, while the portions of the substrate

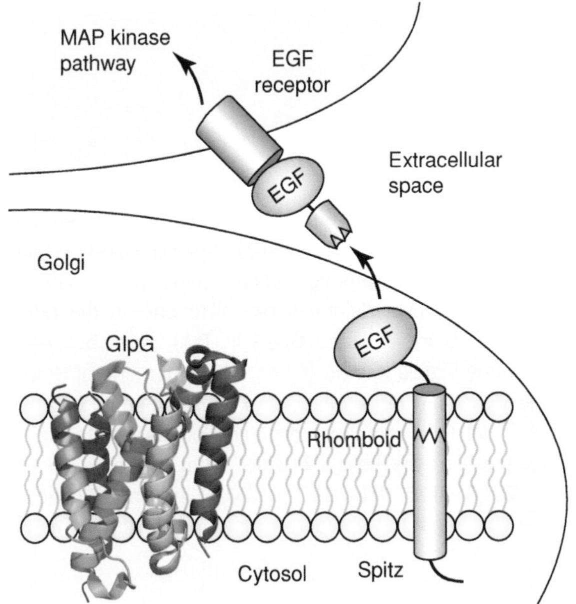

9.14 Rhomboid function in *Drosophila*. Cartoon illustration shows the role of rhomboid protease in releasing the epidermal growth factor (EGF) from its precursor Spitz. The product is secreted to activate EGF receptor on neighboring cells. Rhomboid is represented by the structure of GlpG from *E. coli*. From Erez, E. *et al.*, *Nature*. 2009, 459:371–378. © 2009, Nature Publishing Group. Reprinted by permission of Macmillan Publishers Ltd.

outside the membrane play no role in the reaction, as demonstrated with chimeras of Spitz and artificial substrates.

Numerous x-ray structures of the bacterial rhomboid GlpG (named for its gene in the Glycerol 3P operon) from *E. coli* and *Haemophilus influenza*, including one with inhibitor present, make it the best described rhomboid even though its function and natural substrate are unknown. Most are structures of GlpG in detergent, one is in bicelles of DMPC plus CHAPSO, and all have the TM core lacking the N-terminal domain. The eukaryotic rhomboids differ from GlpG in that they have an additional TM segment and longer external domains. Rhomboids have 20 highly conserved residues that reside in the active site or in specific structural elements described below.

Structure of the bacterial rhomboid GlpG

GlpG is a compact bundle of six TM helices in a unique arrangement: five TM segments encircle a central TM helix that does not span the entire membrane bilayer, leaving open a water-filled cavity on the periplasmic side (Figure 9.15). The active site Ser residue is at the top of the central helix, TM4, and can be covered by the flexible loop from TM5 to TM6 (L5), making a "cap." Adding asymmetry is a protruding loop L1 between TM1 and TM2, which is aligned with the upper leaflet of the membrane as if to position the molecule in the bilayer. L1 includes two short helices in a hairpin extension that is highly stabilized by hydrogen bonds and contains the motif E/QxWRxxS/T. Hydrogen bonded to L1 is the first residue of the signature motif of rhomboids, GySG (where y is one of several permitted residues and is Leu in *E. coli* GlpG) that includes Ser201, the catalytic serine, in TM4. A universally conserved motif AHxxGxxxG in TM6 contains the catalytic histidine residue, His254. Close proximity between the GxxxG in TM6 and TM4 brings together the catalytic Ser and His residues. A fourth structural element is the sharp turn made by GxxxExxxG at the cytosolic end of TM2. Hydrogen bonds from it to both TM1 and TM3 bring them into a V-shape with L1 in the gap between them (Figure 9.16).

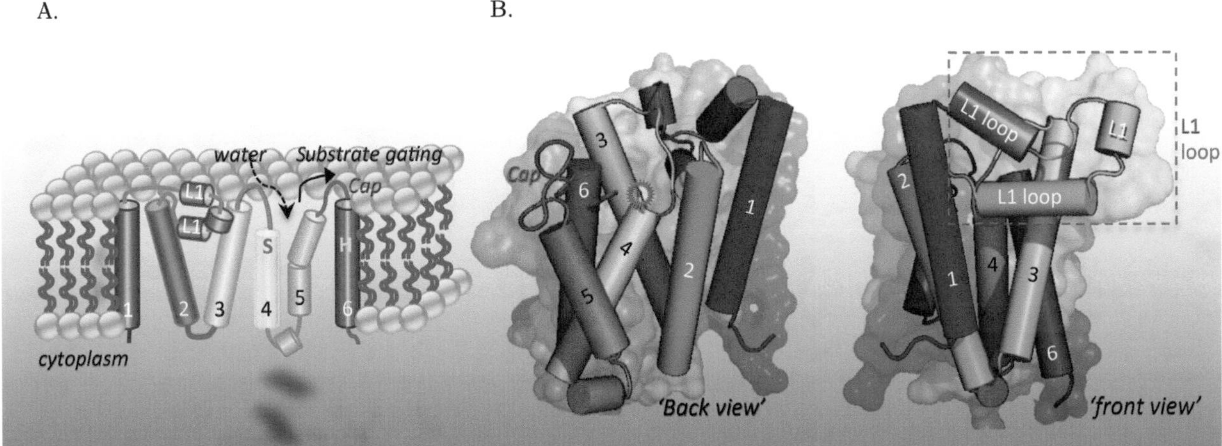

9.15 Structure of the bacterial rhomboid GlpG. GlpG is a bundle of α-helices (rainbow-colored rods), shown with the catalytic Ser denoted S (red) on TM4 (yellow) and the helical hairpin of L1 (light blue rods) on the surface of the bilayer. L5 is designated the Cap (red). **(A)** Topology and **(B)** the x-ray structure. From Urban, S. *Biochem J*. 2010, 425:501–512. © 2010 by the Biochemical Society. Reprinted with permission from Portland Press Ltd.

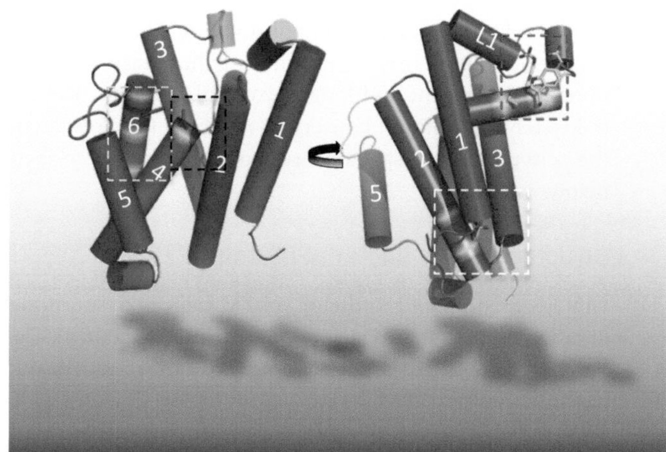

9.16 Motifs in the structure of GlpG. Four structural elements (yellow bands within dashed boxes) stabilize the interactions among helices (blue rods) and Loop 1 (L1) of GlpG. Visible in the structure on the left are AHxxGxxxG in TM6 and GySG in TM4, with the catalytic residues highlighted (red). From the other side of GlpG (on the right) can be seen ExWRxxT in L1 and GxxxExxxG in TM2. All four contribute hydrogen bonds to stabilize the structure of the helical bundle. From Urban, S. *Biochem J.* 2010, 425:501–512. © 2010 by the Biochemical Society. Reprinted with permission from Portland Press Ltd.

In the active site of GlpG the nucleophilic Ser201 is hydrogen bonded to His254 in a catalytic dyad, with the imidazole ring oriented by stacking interactions to Tyr205 (Figure 9.17A). These residues lie at the bottom of a funnel-shaped cavity lined by hydrophilic residues that leads to a quite large opening (10 Å diameter). In the crystal structure of GlpG bound to an isocoumarin inhibitor, both catalytic groups are covalently bound to the inhibitor (Figure 9.17B). The structure of this alkylated enzyme reveals the position of the oxyanion hole that stabilizes the acyl enzyme intermediate: four residues make hydrogen bonds to the benzoyl carbonyl oxygen that mimics a peptide carbonyl (Figure 9.17C). The L5 "cap" over the active site is lifted away when the inhibitor binds, with slight displacement of TM5 and TM6, but the conformational changes are not large enough to give access to the TM segment of a peptide substrate. Gating is thought to involve TM5 and L5, which is supported by mutagenesis studies. However, the GlpG–isocoumarin complex gives insight into the basis of substrate specificity, defining a shallow pocket at the amino side of the scissile bond where small amino acids bind and a hydrophobic cavity in the second position at the carbonyl side.

The unusual shape of GlpG leads to interesting protein–lipid interactions. In the crystals from bicelles a ring of annular lipids completely surrounds GlpG, with close fit of the acyl chains into grooves on the protein surface (Figure 9.18). Other lipids fill cavities around L1. The bilayer thickness also varies, with lipid headgroups at different depths around the protein. Thinning of the membrane bilayer is very clear in MD simulations, which also show water penetrating to the top of TM4 (Figure 9.19).

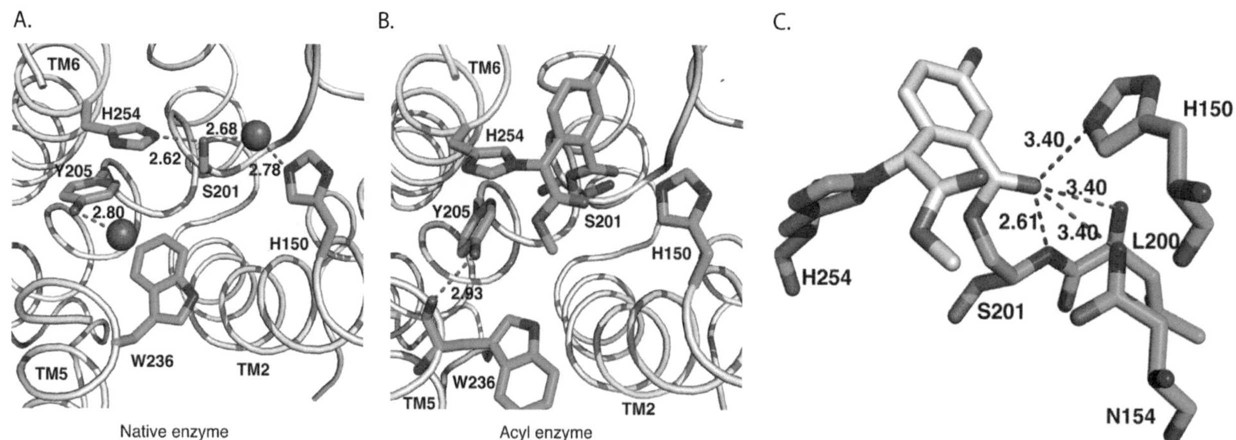

9.17 The active site of GlpG in the native enzyme and acyl enzyme (with inhibitor). (A) Ser201 and His254 form a hydrogen bond in the active site in the native enzyme, and His254 is stacked over Tyr205. Additional hydrogen bonds involve two water molecules (red spheres). (B) Covalent bonds form between both Ser201 and His254 and the inhibitor, dichloroisocoumarin. The water molecules are displaced and Tyr205 tilts toward Trp236. (C) The oxyanion hole is revealed with four hydrogen bonds from the benzoyl carbonyl oxygen of the inhibitor, with the strongest to the main chain carbonyl of Ser201. The main chain amide of Leu200, the side chains of Asp 154 and His150 are also hydrogen bond donors. From Vinothkumar, *et al.*, *EMBO J.* 2010, 29:3797–3809. © 2010, Nature Publishing Group. Reprinted by permission of Macmillan Publishers Ltd.

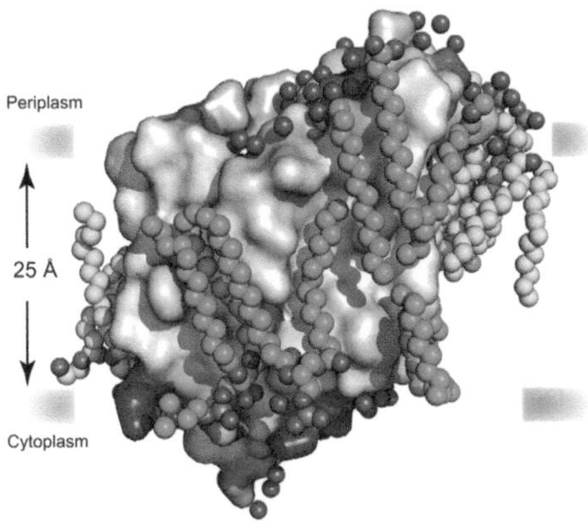

9.18 Lipid–protein interactions in GlpG crystallized in bicelles. The annular lipids seen in the x-ray structure of GlpG (surface representation with positively charged residues, blue; negatively charged residues, red; polar residues, light blue; and the rest gray) are seen as acyl chains from lipids on the same molecule (green spheres) and those from the adjacent molecules in the crystal (yellow spheres). Water molecules (blue spheres) and phosphodiester groups (red and orange) mark the boundary of the lipid bilayer. From Vinothkumar, K. R., *J Mol Biol*. 2011, 407:232–247. © 2011, Elsevier. Reprinted with permission from Elsevier.

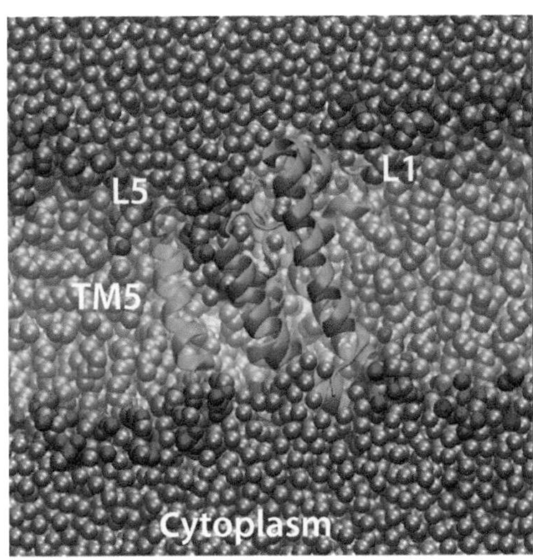

9.19 Membrane thinning around GlpG. An all-atom molecular dynamic simulation of one GlpG molecule, ~500 POPE molecules and 30000 water molecules shows the effect of GlpG on the lipid bilayer, as well as the access to water at the active site. The bilayer is thinned by ~4 Å and the protein is tilted by ~12°. A cut-away view of one simulation of GlpG (green, with TM5, L1 and L5 labeled) in a POPE bilayer (alkyl carbons, cyan; phosphorous, orange; oxygen, red; nitrogen, blue) reveals the deep penetration of bulk water molecules (pink). From Bondar, A.-N., *et al.*, *Structure*. 2009, 17:395–405. © 2009, Elsevier. Reprinted with permission from Elsevier.

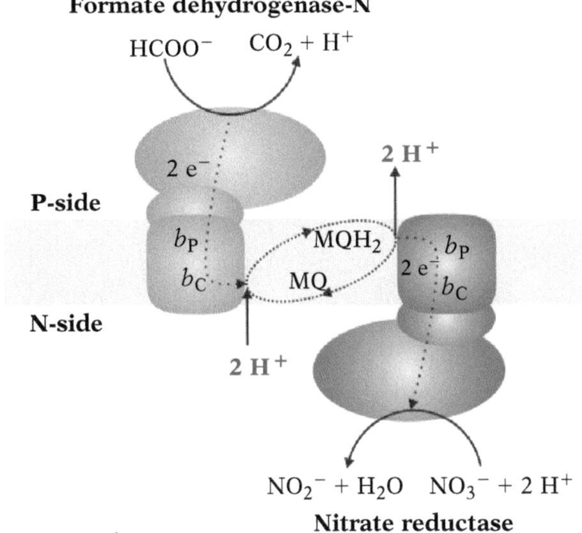

9.20 Diagram of the nitrate respiratory pathway. Formate dehydrogenase-N and nitrate reductase form a redox loop in which formate is oxidized and nitrate reduced with electron transfers that are accompanied by the transfer of protons across the membrane. From Jormakka, M., *et al.*, *FEBS Lett.* 2003, 545:25–30. © 2003 by the Federation of European Biochemical Societies. Reprinted with permission from Elsevier.

The thinning of the bilayer may help position the peptide substrate at the active site. Membrane thinning, along with a flotation device to position the protein, likely contributes to the unexpected ability of GlpG to carry out peptide hydrolysis within the membrane.

FORMATE DEHYDROGENASE

A very different membrane enzyme is formate dehydrogenase. Many prokaryotes can live anaerobically and carry out oxidative phosphorylation with a variety of terminal electron acceptors, including nitrogen. The presence of nitrate under anaerobic conditions induces the nitrate respiratory pathway, which consists of two membrane enzymes, formate dehydrogenase and nitrate reductase (Figure 9.20). Both enzymes are heterotrimers with numerous cofactors involved in electron transfers that enable proton transfer across the membrane to generate a proton motive force that can be used for the synthesis of ATP by F_1F_0-ATP synthase (see Chapter 13).

During anaerobic respiration, bacteria can produce formate from acetyl-coenzyme A derived from pyruvate. In the reaction carried out by formate dehydrogenase, formate is oxidized to CO_2 and H^+ on the periplasmic side of the cytoplasmic membrane, releasing two electrons that are transferred across the membrane to a lipid-soluble quinone called menaquinone (MQ). The reduced MQ picks up two protons from the cytosol to form menaquinol (MQH_2). The redox potentials for the

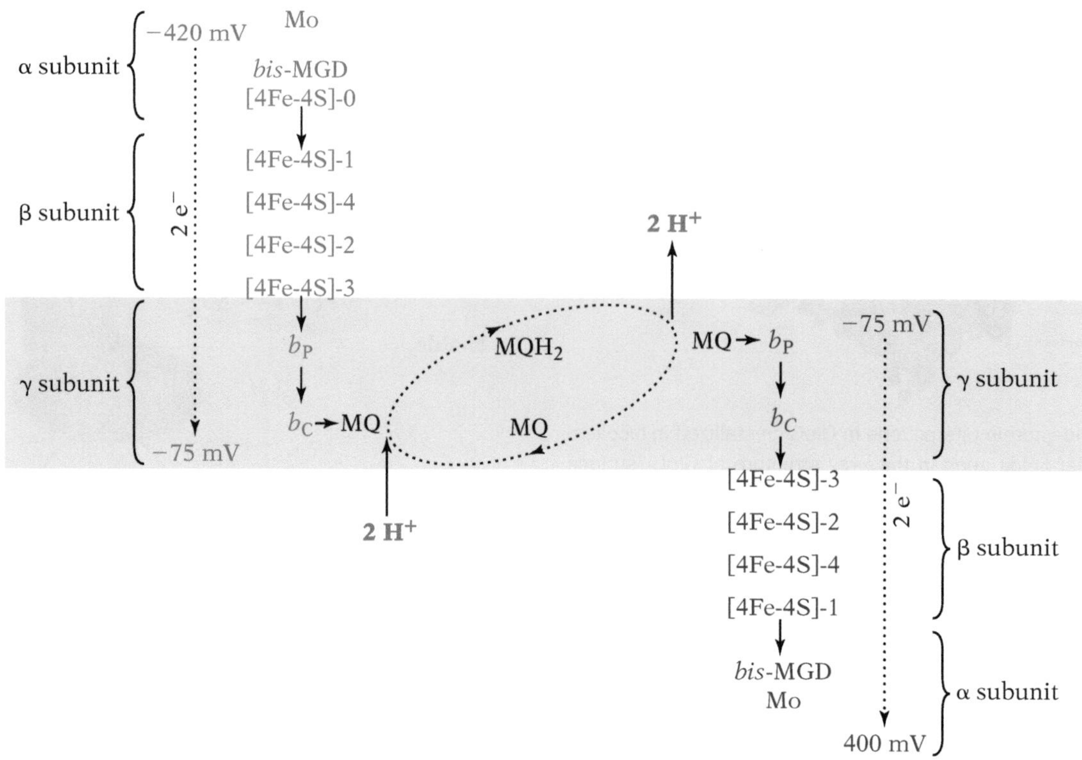

9.21 Redox potentials of the electron carriers in the nitrate respiratory pathway. Electron carriers of FDH-N are given in red, and those of nitrate reductase are given in blue. The highly exergonic reaction allows the electrons to be transferred against the membrane potential. From Jormakka, M., *et al.*, *FEBS Lett.* 2003, 545:25–30. © 2003 by the Federation of European Biochemical Societies. Reprinted with permission from Elsevier.

electron carriers indicate this is a highly exergonic reaction (Figure 9.21).

The MQH_2 then diffuses through the membrane to nitrate reductase, which oxidizes it, releasing its two protons to the periplasm and transferring its two electrons via a series of cofactors to the nitrate reduction site on the cytoplasmic side. There nitrate is reduced to nitrite, consuming two more protons from the cytosol. The transfer of electrons from the outside to the inside of the membrane, followed by their transport back to the outside accompanied by protons, was called a "redox loop" in Mitchell's original chemiosmotic theory.

The high-resolution structure of formate dehydrogenase-N (FDH-N) from *E. coli* provides insight into the steps involved in generating a proton motive force from the electron transfers of the nitrate respiratory pathway. This enzyme is a 510-kDa dimer of a heterotrimer that consists of three subunits called α, β, and γ (Figure 9.22). The catalytic α subunit is a large periplasmic polypeptide composed of 982 amino acids, including an intrinsic selenium-cysteine (SeCys) residue, a [4Fe-4S] cluster, and a molybdopterin guanine dinucleotide (MGD) cofactor. The β subunit (289 amino acids) has one TM helix near its C terminus, which is on the cytoplasmic side of the membrane, while its N terminus is on the periplasmic side. It coordinates four [4Fe-4S] clusters positioned in two symmetrical domains (Figure 9.23). The γ subunit (216 amino acids) is a polytopic protein with four TM helices and both N and C termini on the cytoplasmic side (Figure 9.24). This subunit is a cytochrome *b* that contains two heme *b* groups, one coordinated by His residues of TMs II and IV near the periplasmic side and the other coordinated by His residues of TMs I and IV near the cytoplasmic side of the membrane. It also has the site for MQ reduction.

Based on the linear path of the 11 redox centers in an $\alpha\beta\gamma$ heterotrimer, as well as the close distances between them, it appears that the functional unit of FDH-N is the heterotrimer (Figure 9.25). The electron path spans almost 90 Å across the enzyme. The sequence is initiated by the oxidation of formate by the α subunit, in a reaction like that of the soluble formate dehydrogenases, whose catalytic subunits are superimposable with the FDH-N α subunit. The Mo at the active site is at the end of a funnel lined with positively charged residues. The Mo is coordinated by *cis*-thiolate groups of the two MGD cofactors (Figure 9.25 inset), the SeCys, and a hydroxide ion, and directly receives electrons from the substrate. Positioned very close to the substrate, the SeCys may be involved in proton removal and transfer to a nearby His side chain.

The location of the quinone-binding site was determined by soaking the crystals with the inhibitor 2-heptyl-4-hydroxyquinoline-*N*-oxide (HOQNO) and was

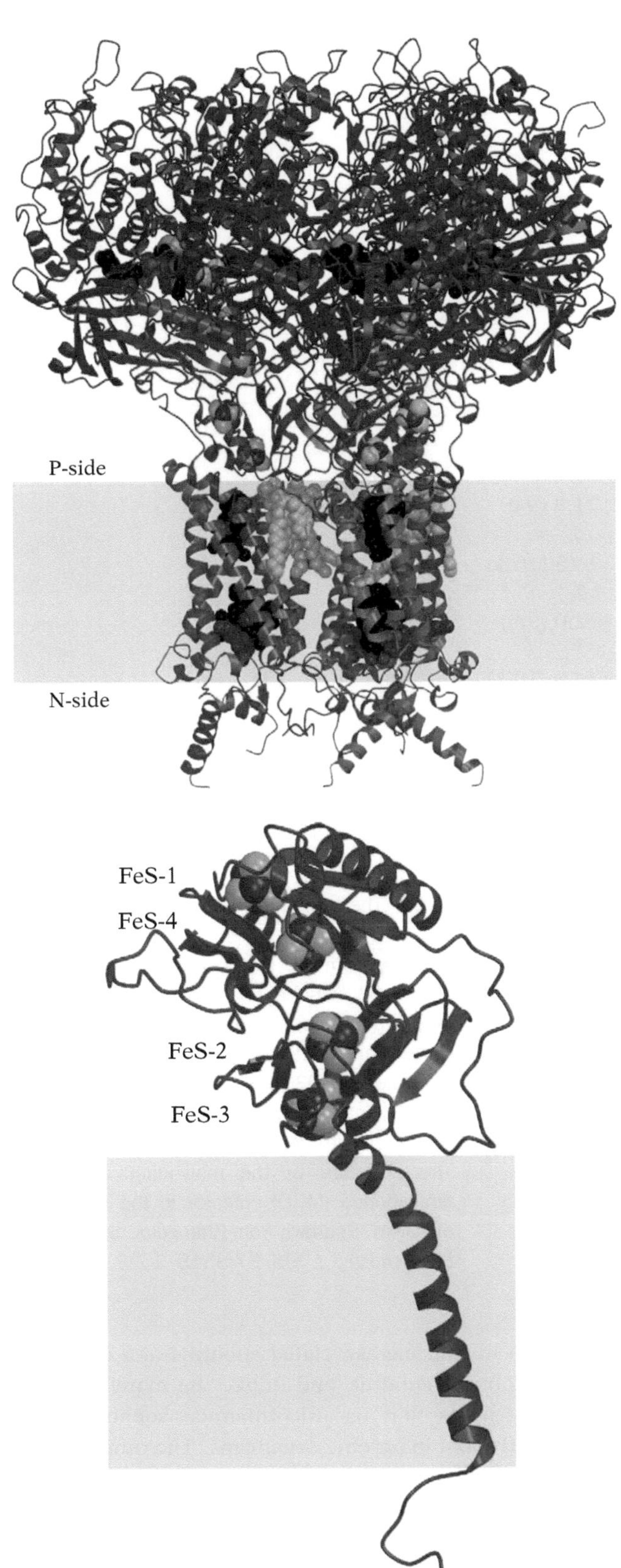

9.22 High-resolution structure of FDH-N. The mushroom-shaped heterotrimer is viewed from the side with the catalytic α subunit (dark blue), the β subunit (red), and the γ subunit (light blue). Electrons are transferred from the active site in the periplasm with its Mo atom (green) to five [4Fe–4S] clusters (gray/yellow) to two heme *b* groups (black) and then to the quinone (purple). A single cardiolipin (yellow-green) is also visible in the trimer interface. The P-side is the periplasmic side of the membrane and the N-side is the cytoplasmic side of the membrane. From Jormakka, M., *et al.*, *Curr Opin Struct Biol.* 2003, 13:418–423. © 2003 by Elsevier. Reprinted with permission from Elsevier.

9.23 Structure of the β subunit of FDH-N. Each of the two symmetric domains outside the membrane contains two [4Fe–4S] clusters (colored as in Figure 9.22). From Jormakka, M., *et al.*, *Curr Opin Struct Biol.* 2003, 13:418–423. © 2003 by Elsevier. Reprinted with permission from Elsevier.

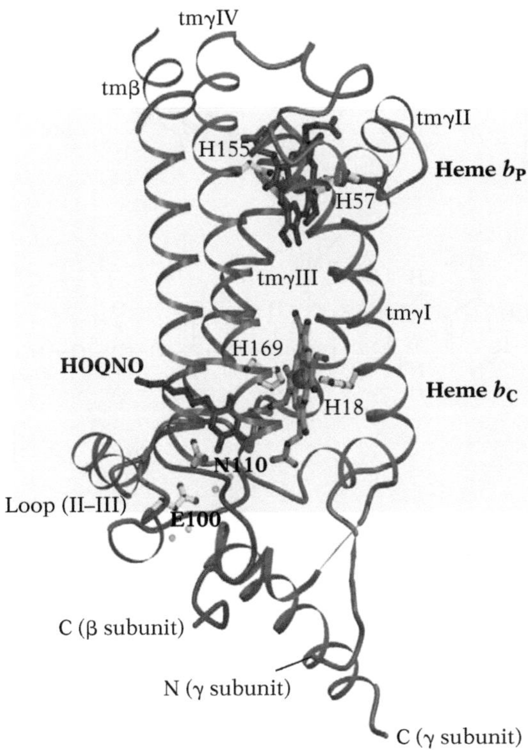

9.24 Structure of the integral membrane domain of FDH-N. The γ subunit has four TM helices numbered tmγI through tmγIV and the TM helix of the β subunit (tmβ; red) is also shown. Histidine residues that coordinate heme *b* and residues involved in binding MQ are labeled. From Jormakka M., *et al.*, *Curr Opin Struct Biol.* 2003, 13:418–423. © 2003 by Elsevier. Reprinted with permission from Elsevier.

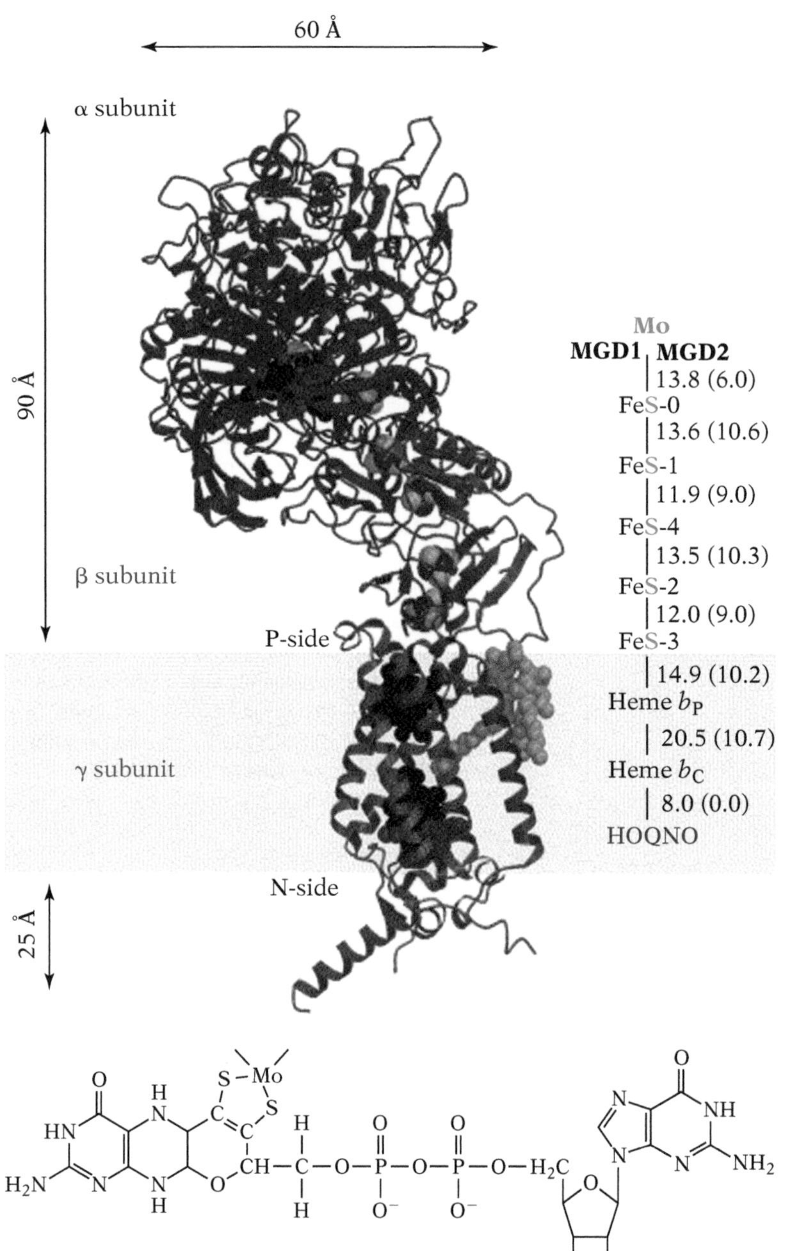

9.25 The electron path in an $\alpha\beta\gamma$ protomer of FDH-N. The linear electron transfer pathway from Mo to the quinone-binding site is shown with center-to-center and edge-to-edge (parentheses) distances between the redox centers. From Jormakka, M., *et al.*, *FEBS Lett.* 2003, 545:25–30. © 2003 by the Federation of European Biochemical Societies. Reprinted with permission from Elsevier. *Inset*: The structure of the molybdenum guanine dinucleotide (MGD) cofactor in the α subunit of FDH-N. Redrawn from Khangulov, S. V., *et al.*, *Biochemistry.* 1998, 37:3518–3528.

found to be near the heme *b* on the cytoplasmic side of the membrane. This establishes a mechanism for proton translocation, since earlier EPR studies showed that the MQH_2-binding site on nitrate reductase is close to its heme *b* on the periplasmic side. A group of waters near the FDH-N quinone-binding site suggests a proton path from the cytosol to the bound MQ (Figure 9.26). Thus when two electrons are transferred from the formate oxidation site in the periplasm, two protons are taken up from the cytoplasm. When they are translocated across the membrane (as MQH_2), they are released to the periplasm from the quinone-binding site in nitrate reductase.

The overall structure of FDH-N is like that of other membrane-bound respiratory enzymes, which generally have two membrane-associated subunits and one integral membrane subunit, and unlike the many formate dehydrogenases that use nicotinamide adenine dinucleotide (NAD^+) in aerobic organisms. The other NAD^+-independent formate dehydrogenases differ in the wide variety of redox centers they utilize. However, in *E. coli*, FDH-N is similar to formate dehydrogenase-O, another membrane-bound heterotrimer, and its α subunit is highly homologous to the catalytic subunit of the soluble FDH-H described above. The high-resolution structure of FDH-N not only gives insight into the group of formate dehydrogenases with SeCys at their active sites, it provides a model for how a redox loop generates a proton gradient.

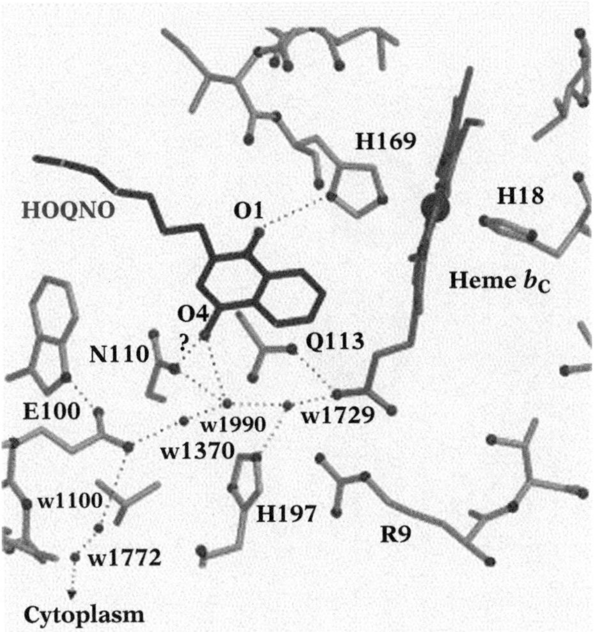

9.26 Menaquinone-binding site in FDH-N. The inhibitor HOQNO (dark blue) binds very close to the heme *b* (red) at the cytoplasmic side. A possible proton pathway from the cytoplasm to the bound MQ is indicated by the location of several waters. From Jormakka, M., *et al.*, *FEBS Lett.* 2003, 545:25–30. © 2003 by the Federation of European Biochemical Societies. Reprinted with permission from Elsevier.

TABLE 9.1 SUBFAMILIES OF P-TYPE ATPASES

Family	Specificity	Main functions (distribution)
P_{Ia}	K^+	Turgor pressure (bacteria and archaea only)
P_{Ib}	Cu^+, Cu^{2+}, Ag^+, Cd^{2+}, Zn^{2+}, Pb^{2+}, Co^{2+}	Detoxification, trace element homeostasis (all kingdoms)
P_{IIa}	Ca^{2+}, Mn^{2+}	Signaling, muscle relaxation (all kingdoms)
P_{IIb}	Ca^{2+}	Signaling, Ca^{2+} transport across plasma membrane (eukaryotes only)
P_{IIc}	Na^+/K^+, H^+/K^+	Plasma membrane potential, kidney function, stomach acidification (fungi, invertebrates and vertebrates)
P_{IId}	Na^+, Ca^{2+}	Unknown (fungi and invertebrates only)
P_{IIIa}	H^+	pH homeostasis, plasma membrane potential (bacteria, fungi, invertebrates and vertebrates)
P_{IIIb}	Mg^{2+}	Unknown (bacteria only)
P_{IV}	Phospholipids	Lipid transport (eukaryotes only)
P_V	Unknown	Unknown (eukaryotes only)

Source: Based on Bublitz, M., In and out of the cation pumps: P-type ATPase structure revisited. *Curr Opin Struct Biol.* 2010, 20:431–439. © 2010, Elsevier. Reprinted with permission from Elsevier.

P-TYPE ATPASES

An important class of membrane enzymes is the ATPases, which use the energy from the hydrolysis of ATP to drive transport. Organisms of all types expend as much as 75% of their total ATP to pump in nutrients and pump out toxins and to create and maintain ion gradients. The ABC transporters described in Chapter 11 and the F_0F_1 ATPase (the ATP synthase) described in Chapter 13 are important examples of these transporters. To examine them as enzymes, this chapter focuses on the structural characterizations of P-type ATPases that offer mechanistic insight into their reaction cycle and how they couple the hydrolysis of ATP to transport.

With substrates varying from ions to phospholipids, P-type ATPases perform vital functions in prokaryotes and eukaryotes (Table 9.1). Their name refers to the phosphorylated enzyme intermediate formed during the reaction when the γ-phosphate group of ATP is transferred to an aspartate residue at the beginning of the conserved motif DKTGT. There are over 300 known members of this superfamily, including 36 in humans, grouped in five subclasses P_I–P_V. They share a common structure consisting of 6–12 TM helices and three cytoplasmic domains called the phosphorylation (P) domain, the nucleotide-binding (N) domain, and the actuator (A) domain, which moves to dephosphorylate the phosphorylated enzyme intermediate (Figure 9.27). Some P-type ATPases are hetero-oligomers composed of a catalytic α subunit having this structure, a smaller β subunit with a single TM segment, and frequently an additional third modulator protein called the γ subunit or FXYD protein (see Frontispiece).

The functions of P-type ATPases require coupling the energy from ATP hydrolysis on the cytoplasmic side to a conformational change that leads to translocation of specific substrates, usually ions, through the membrane. This mechanism involves conversion between two main conformations, E1 and E2, that have different affinities and access channels. The elegant structures of SERCA, the Ca^{2+} ATPase from sarcoplasmic reticulum (SR), and of the Na^+, K^+ ATPase provide a framework for understanding the mechanism of many other P-type ATPases.

Ca^{2+} ATPases

Ca^{2+} ATPases enable organisms to keep the intracellular concentration of calcium low (~10^{-7} M) to avoid formation of insoluble calcium phosphates and interference with protein and RNA function. Calcium release is an important signal in many tissues, especially nerves and

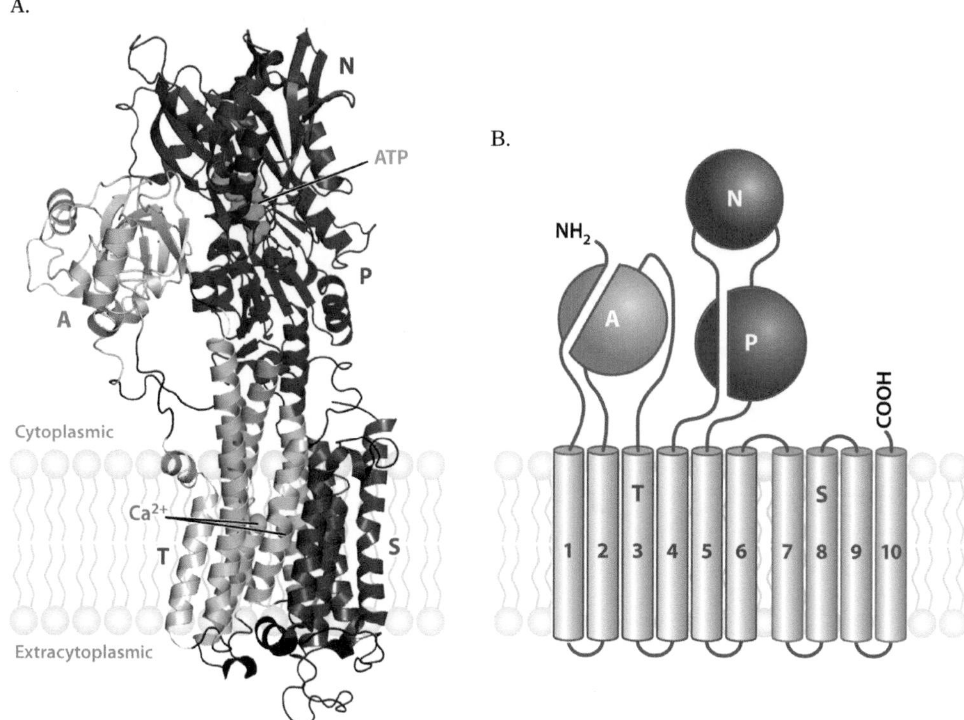

9.27 Architecture of the P-type ATPases. (A) A ribbon diagram based on the x-ray structure of the Ca^{2+} ATPase from sarcoplasmic reticulum (SERCA) shows the domain organization of P-type ATPases. There are three cytoplasmic domains, A, P, and N (gold, blue, and red) and two membrane domains, T and S (cream and gray). ATP (green space-filling model) binds to the N domain in a crevice next to the P domain. The transported Ca^{2+} ions (cyan spheres) bind in the middle of the T domain. (B) A schematic diagram of the SERCA structure shows the cytoplasmic and TM domains (colored to match (A)). The peptide chain forms the A domain at the N terminus along with an extension of the first cytoplasmic loop, and inserts the N domain into the P domain in the second cytoplasmic loop. The first six TM helices form the T domain and the latter four form the S domain in the membrane-spanning region. From Palmgren, M. G., and P. Nissan, *Ann Rev Biophys*. 2011, 40:243–266. © 2011 by Annual Reviews. Reprinted with permission from the Annual Review of Biophysics, www.annualreview.org.

muscles. For example, in muscle contraction a nerve impulse triggers the release of millimolar levels of Ca^{2+} from the SR. The released Ca^{2+} binds troponin, triggering a conformational change that exposes myosin binding sites to bind actin in thin filaments. To terminate the muscle contraction, the Ca^{2+} ATPase in the SR membrane (SERCA) performs the reuptake of calcium into the SR, helping to restore its steep concentration gradient.

SERCA is the best characterized P-type ATPase. Early work indicated SERCA has two conformations, E1 with a high-affinity Ca^{2+}-binding site exposed on the cytoplasmic side and E2 with a low-affinity Ca^{2+}-binding site exposed to the lumen. For each ATP hydrolyzed the Ca^{2+} ATPase exports two Ca^{2+} ions and imports two or three H^+. (Because the SR membrane is permeable to protons, SERCA does not create a proton gradient.) The events of the transport cycle are summarized as follows (Figure 9.28). E1 binds two Ca^{2+} ions and ATP from the cytoplasm. Phosphorylation at Asp351 leads to formation of a high-energy intermediate with occlusion of

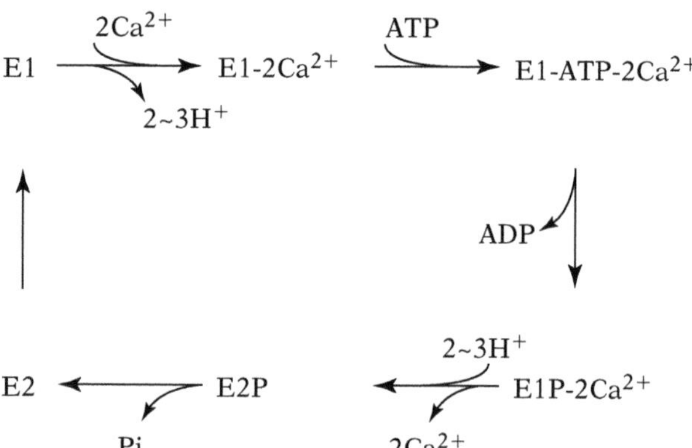

9.28 The E1–E2 reaction scheme for the Ca^{2+} ATPase. The E1 form of the enzyme binds two Ca^{2+} ions from the cytoplasm, along with ATP bound to Mg^{2+} (not shown) before the phosphorylation reaction. Phosphorylated enzyme (E1P-$2Ca^{2+}$) then exchanges Ca^{2+} ions for H^+ on the outside as it converts to E2P. E2P is dephosphorylated and then converts back to E1. This scheme is simplified because both the phosphorylation step and the dephosphorylation step are reversible, and also either the ATP or the Ca^{2+} can bind first. From Toyoshima, C. *et al.*, *Ann Rev Biochem*. 2004, 73:269–292.

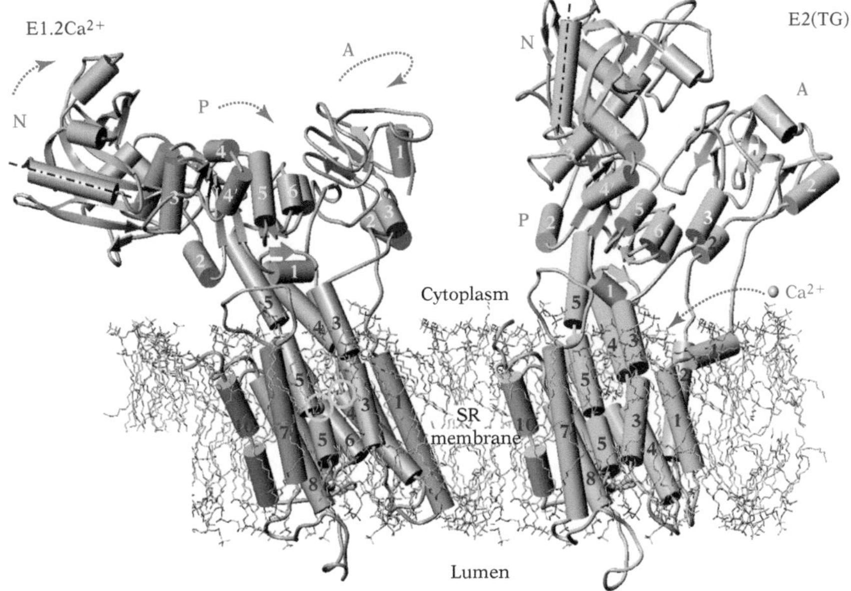

9.29 The x-ray structure of SERCA in the presence and absence of calcium ions. The Ca^{2+}-free form of the enzyme was crystallized bound to an inhibitor, thapsigargin (TG). In the structure of E1.2Ca^{2+} (purple) the three cytoplasmic domains are splayed open, with the ATP-binding site available. In the absence of Ca^{2+}, the E2(TG) structure (green) shows the N, P, and A domains are gathered into a compact headpiece. The bilayer shown is an MD simulation of DOPC. From Toyoshima *et al.*, *FEBS Letts.* 2003, 555:106–110. © 2003 by the Federation of European Biochemical Societies. Reprinted with permission from Elsevier.

the Ca^{2+} ions. Conformational change to the low-energy E2P intermediate allows release of Ca^{2+} to the lumen, in exchange for protons. Dephosphorylation of E2P to E2 allows the enzyme to convert back to E1 and release protons into the cytoplasm, where it picks up two more Ca^{2+} to repeat the cycle. Pumping of Ca^{2+} ions by SERCA involves reciprocal opening of two half-channels, so the transport mechanism fits the alternating access model (see Chapter 11).

The Ca^{2+} ATPase from rabbit skeletal muscle SR, the SERCA1a isoform (simply called SERCA in the following discussion), is a single polypeptide of 994 residues (Mr 110 kDa) that form ten TM helices and an extensive cytoplasmic headpiece. The TM helices form two domains, a transport domain of six helices and a support domain of four helices. The cytoplasmic headpiece consists of three domains: the P domain where phosphorylation takes place, the N domain that binds nucleotides and participates in the phosphorylation reaction, and the A domain that dephosphorylates and actuates the gating mechanism (see below). With over 20 crystal structures of SERCA bound to different ligands, there are now high-resolution structures representing every major stage in its catalytic cycle, making it possible to envision the conformational changes involved in coupling ATP hydrolysis with Ca^{2+} transport. To follow these changes in some detail, the roles of the different domains and the features of the ion-binding and nucleotide-binding sites must be clear.

The cytoplasmic domains. The highly mobile **A domain** rotates at each main step in the cycle, allowing it to move between the P and N domains and to dock its TGES motif where ADP leaves the N domain in E2P. Supporting these movements are three linkers to TM helices, A-M1, M2-A, and A-M3, whose lengths are critical: for example, deletion of a single residue in A-M1 can block the E1P to E2P transition. Moreover, release of the strain on the A-M3 link caused by rotation of A is a driving force for opening of the passage for ions. The large **N domain** connects to the two halves of the P domain with two β-like strands that allow big domain movements. The N domain has a mostly hydrophobic binding site for adenosine, allowing aromatic interactions between Phe487 and the adenine ring. It also contains Arg560 that bridges ATP and the P domain. The highly conserved phosphorylation site DKTG in the **P domain** starts with Asp351, which gets phosphorylated, at the C terminus of a typical Rossmann fold.[2] Other critical residues include Lys684, which binds the γ-phosphate of ATP, and Thr353 and Asp703 that coordinate the Mg associated with the nucleotide. The P domain is wedge-shaped with a flat top surface that allows rotation of the A domain upon it. Clamped to the TM domain by the integration of M5, the P domain is brought underneath the TGES loop of the A domain when M5 bends towards M1 to form the compact (E2) state (Figure 9.29).

[2] A Rossmann fold is a common protein structural motif for binding nucleotides that usually consists of two units containing three parallel β-strands linked by two α-helices. The Rossmann fold in SERCA has seven β-strands.

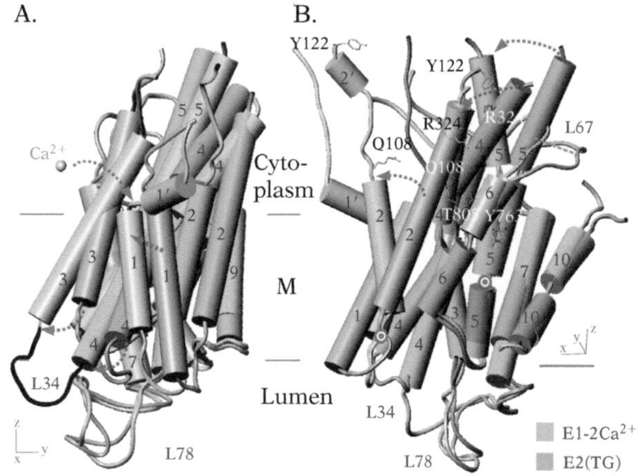

9.30 Rearrangement of TM helices when SERCA binds calcium. The TM helices of SERCA (numbered) are superimposed in positions observed in E1.2Ca^{2+} (lavender) and E2(TG) (green). Both (A) and (B) are viewed from the side, with a 90° rotation in the viewpoints such that the view in (B) is the back of the view in Figure 9.29. Helices M8 and M9 are removed in (B) to show the movements of M5, M2, and M4. The double circles (red and white) show pivot positions of M2 and M5. The red arrows indicate the direction of movements during the change from E1.2Ca^{2+} to E2(TG) and the blue arrow shows the proposed pathway for entry of the first Ca^{2+} ion. From Toyoshima *et al.*, *Ann Rev Biochem*. 2004, 73:269–292. © 2004 by Annual Reviews. Reprinted with permission from the Annual Review of Biochemistry, www.annualreviews.org.

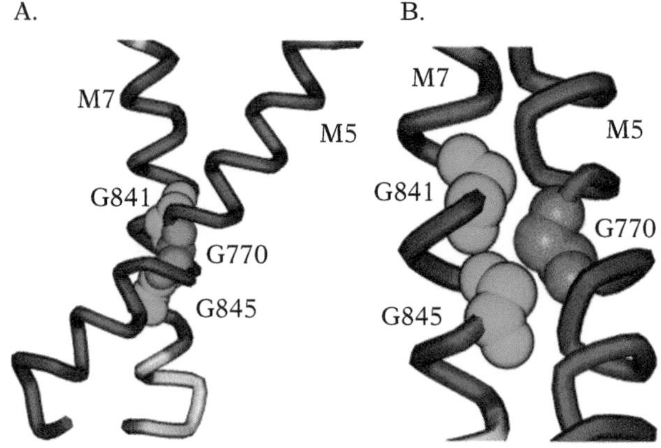

9.31 Close contact between TM helices M5 and M7 in SERCA. Gly841 and Gly845 in the GxxxG motif (green space-filling models) in M7 interacts with Gly770 (blue space-filling model) at the pivot point in M5. From Lee, A. G., *Biochim Biophys Acta*. 2002, 1565:246–266. © 2002 by Elsevier. Reprinted with permission from Elsevier.

The TM domains. The first six of the ten TM helices in SERCA (M1–M6) form an ion transport domain found in all P-type ATPases that moves considerably during the reaction cycle, while the remaining four (M7–M10) remain stationary as a support domain (Figure 9.30). M2–M5 are extra-long helices that extend into the cytoplasm. The 60-Å long M5 is the "spine" of the molecule, which bends and straightens at Gly770 in close contact with a GxxxG motif in M7 (Figure 9.31). The cytoplasmic end of M5 is integrated into the Rossmann fold on the P domain and held by hydrogen bonds with the long, rigid loop between M6 and M7. Further stabilization comes from the rigid V-shape formed by M1 and M2 that transmits the movements of the A domain to the TM domain. M1 is deeply embedded in the lipid bilayer in E1, whereas it is shifted in E2 and bent so its amphipathic end lies on the cytoplasmic membrane surface. M4 and M6 have Pro residues in the middle where they are partly unwound. M4–M6 and M8 contain residues in the two Ca^{2+}-binding sites in the TM domain near the cytoplasmic surface.

The Ca^{2+} binding sites. SERCA binds calcium with high affinity. Binding is cooperative, as the first Ca^{2+} ion enters through an incompletely formed site II (Ca2) to reach site I (Ca1). Surrounded by M5, M6, and M8, Ca1 is formed by side chain oxygen atoms from Asn768 and Glu771 in M5, Thr799 and Asp800 in M6, and Glu908 in M8, plus two water molecules (Figure 9.32). Closer to the cytoplasmic surface, Ca2 consists of three main chain carbonyl groups from the partly unwound M4, plus the side chains of Asn796 and Asp800 in M6 and both oxygen atoms of the carboxyl group of Glu309 in M4. Asp800, the only residue in both sites, is not in position for Ca2 until after the first Ca^{2+} ion binds. Because Glu309 caps

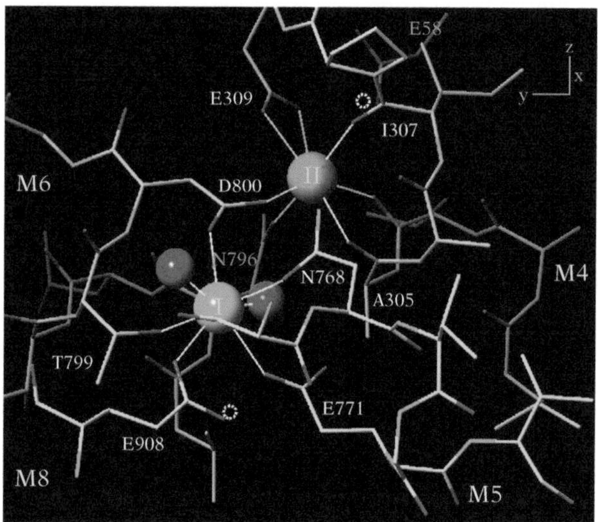

9.32 The calcium-binding sites in SERCA. There are two high-affinity sites that bind Ca^{2+} (blue spheres), each with seven oxygen ligands. Site I (Ca1) is formed by side chain oxygen atoms from Asn768, Glu771, Thr799, Asp800, and Glu908, plus two water molecules (red spheres), while site II (Ca2) consists of three main chain carbonyl groups from the partly unwound M4, plus the side chains of Asn796, Asp800, and Glu309. From Toyoshima, C. *et al.*, *Ann Rev Biochem*. 2004, 73:269–292. © 2004 by Annual Reviews. Reprinted with permission from the Annual Review of Biochemistry, www.annualreviews.org.

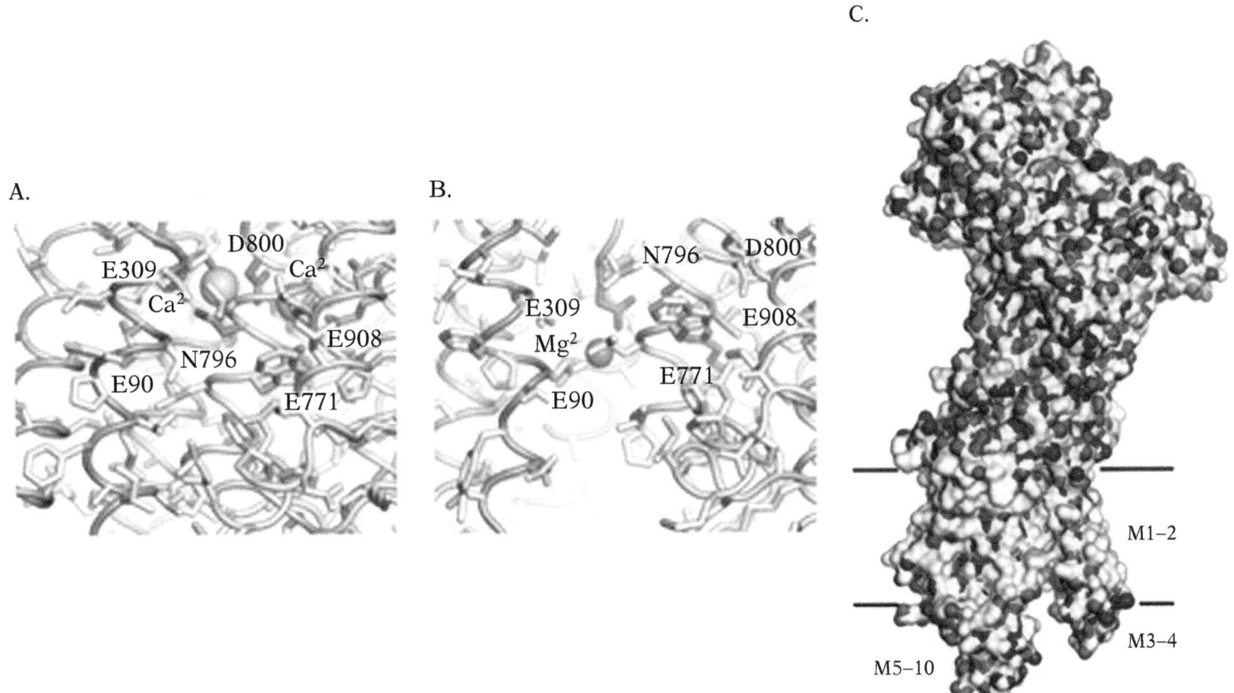

9.33 Opening of the luminal gate in SERCA. (A) Occluded Ca^{2+}-binding sites in the $Ca_2E1{\sim}P$ form of SERCA, showing a backbone trace with full stick representation and the Ca^{2+}-binding residues (carbon, green; oxygen, red; nitrogen, blue) and Ca^{2+} ions (light blue spheres). (B) Luminal exit pathway and exposure of Ca^{2+}-ligating residues in the E2-P form, obtained as a E2-BeF_3^- structure, viewed from a similar orientation and with the same representation as in (A), in addition to a Mg^{2+} ion (gray-blue sphere). (C) Space-filling molecule of the E2-P form of the SERCA molecule showing the exit pathway to the lumen. Carbon, white; sulfur, yellow; nitrogen, blue; oxygen, red. Membrane localization is based on the well-defined boundaries of apolar surfaces. From Olesen, C., *et al.*, *Nature*. 2007, 450:1036–1042. © 2007, Nature Publishing Group. Reprinted by permission of Macmillan Publishers Ltd.

Ca2, it is considered the gating residue when the calcium ions become occluded from the cytoplasm.

In the absence of Ca^{2+} ions the carboxyl groups at the Ca1 and Ca2 sites bind protons due to their elevated pK_as in the membrane environment. When the Ca^{2+} entry path exposes the sites to water, the now deprotonated carboxyl groups gain high affinity for Ca^{2+}. The entry for Ca^{2+} ions from the cytoplasm appears to go from the M1 kink and along M2 to the binding sites. Ca^{2+}-binding is required for productive binding of ATP because it straightens the M5 helix, moving the P domain away from the A domain and allowing ATP to reach the phosphorylation site. In phosphorylated SERCA the Ca^{2+}-binding sites are occluded mainly by residues of M1. Ca^{2+}-binding sites are distorted in the drastic movement of M1–M6 during the E1P to E2P transition: M5 bends towards M1; M4 moves toward the lumen, shifting Glu309 by 6 Å; and M6 rotates, moving Asp800 and Asn796 away from their positions in the sites (Figure 9.33). At the same time a passage to the lumen opens to allow Ca^{2+} release and entry of protons (see below and Figure 9.33C).

ATP binding and hydrolysis. ATP binding is only productive after the Ca^{2+} ions bind and an Mg^{2+}-binding site opens near Asp351, which assures tight coupling between ion transport and phosphorylation. ATP binds near a hinge between P and N domains and crosslinks them as its β-phosphate positions Arg560 for a salt bridge with Asp627. The γ-phosphate and associated Mg^{2+} ion bind to sites on the P domain, and extensive hydrogen bonds form as the P domain bends, tilting the A domain by 30°. The electrostatic strain between the γ-phosphoryl group and Asp351 is relieved by the Mg^{2+} ion and by Lys684, allowing in-line nucleophilic attack to transfer the phosphoryl group to the aspartate. This covalent transfer severs the bridge between P and N, allowing the transition to E2P with a 120° rotation of the A domain. Now the E2P state has a low affinity for nucleotide, so ADP leaves and the TGES loop of the A domain takes its place. Closure of the exit pathway for ions repositions Glu183 of the TGES loop to allow water to enter the catalytic site for the dephosphorylation of Asp351.

Numerous analogs of ATP, ADP, and inorganic phosphate have been used to trap SERCA in different stages of the reaction (Figure 9.34). Metallo-fluorides as phosphate analogs produce different effects; for example, AlF_4^- mimics phosphoryl transfer from ATP to Asp351 or in dephosphorylation, yielding E1-P-ADP or E2-P_i; BeF_3^- represents the covalent phosphoenzyme, E2P; and

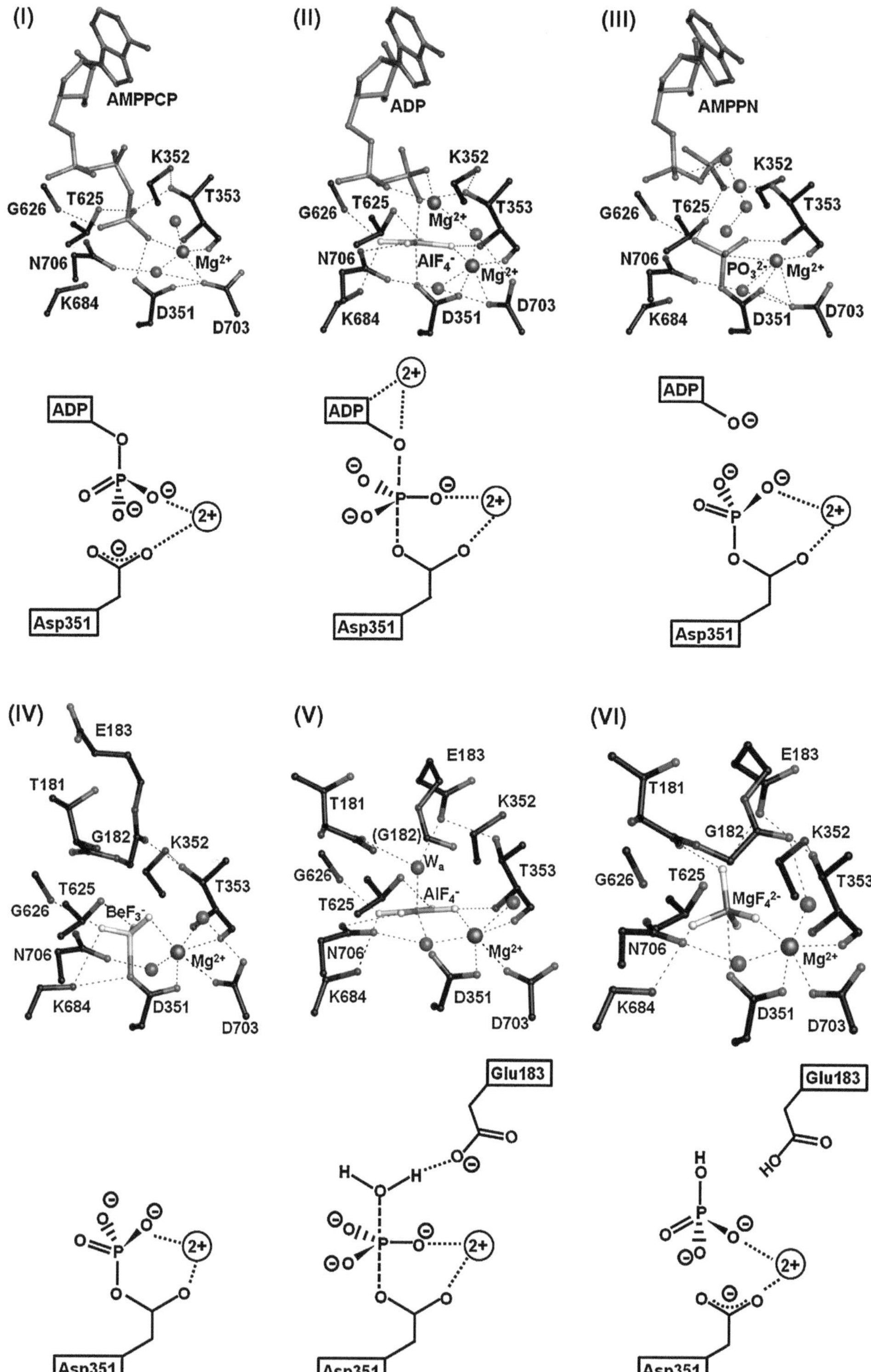

9.34 Reaction cycle at the phosphorylation site in SERCA. Snapshots from different structures representing different stages in the phosphorylation reaction, as depicted below each structure. (I) substrate bound state Ca_2E1–ATP, mimicked by AMPPCP; (II) AlF_4^-–ADP complex mimicking the calcium-occluded transition state of phosphorylation [Ca_2]E1-P-ADP; (III) phosphoryl aspartate state [Ca_2]E1P:ADP mimicked with AMP phosphoramidon; (IV) BeF_3^- complex mimicking the phosphoenzyme intermediate E2P; (V) AlF_4^- complex mimicking the transition state of dephosphorylation [$H_{2\text{-}3}$]E2-P; (VI) MgF_4^{2-} complex mimicking the phosphate product complex [$H_{2\text{-}3}$]E2:Pi. Mg^{2+} is coordinated close to the γ-phosphate, decreasing electrostatic repulsion and protecting against attack by water molecules. Upon dephosphorylation, Glu183 acts as a general base to activate the attacking water molecule (Wa). From Bublitz, M. *et al.*, *Curr Opin Struct Biol.* 2010, 20:431–439. © 2010, Elsevier. Reprinted with permission from Elsevier.

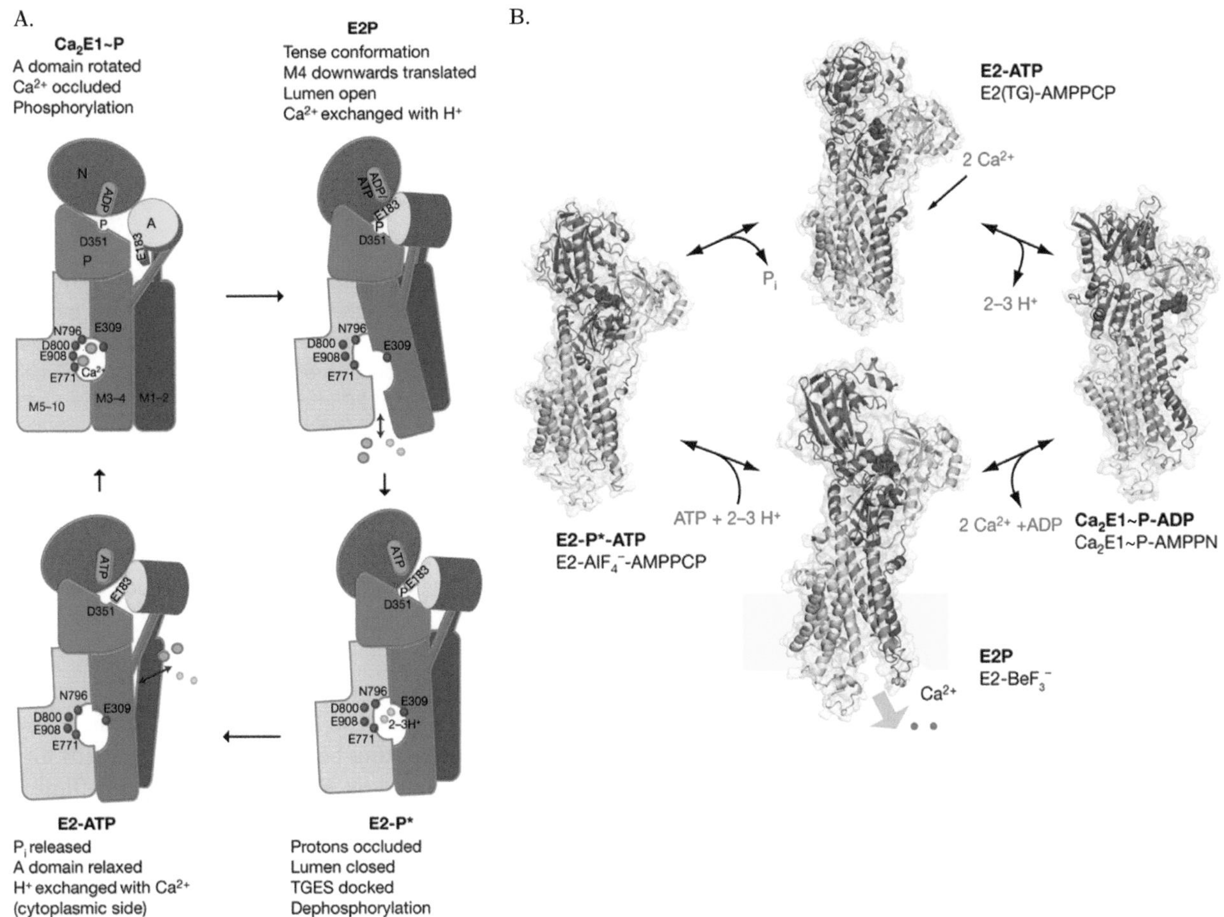

9.35 Summary of major conformational changes of SERCA during the reaction cycle. (A) Schematic representation of the major features of SERCA showing the rotations of the A domain dragging M1–M2 and the changes in the position of the P domain and N domain pushing M3–M4 in an outward and downward direction. A domain, yellow; N domain, red; P domain, blue; helix M1–2, purple; M3–4, green; M5–10, wheat; Ca^{2+} ions and protons, green and gray spheres, respectively. (B) New structures of SERCA showing key states of the reaction cycle. Structures of $Ca_2E1{\sim}P$-AMPPN, E2-BeF_3^- and E2-AlF_4^- complexes and (slightly larger) the E2-BeF_3^- complex are depicted by gray, transparent surfaces with ribbon diagrams colored as in (A) (A domain, yellow; N domain, red; P domain, blue; TM segments M1–2, purple; M3–4 green) except that M5–6 is wheat and M7–10 is gray. The TGES motif is represented by pink space-filling, residues 309, 771, and 796 as sticks, and bound Ca^{2+} ions as gray spheres. Cation- and nucleotide-exchange reactions are indicated. From Olesen, C., *et al.*, *Nature*. 2007, 450:1036–1042. © 2007, Nature Publishing Group. Reprinted by permission of Macmillan Publishers Ltd.

MgF_4^{2-} represents the trapped product state, E2-P_i. In a tour-de-force, the structure of the high-energy intermediate E1~P:ADP was obtained with AMPPNP, which is such a slow substrate that the cleaved AMPPN (an ADP analog) and phosphorylated enzyme are observed in the crystal structure.

Conformational changes of SERCA. Hydrolysis of ATP is coupled to the uptake of Ca^{2+} ions by dramatic conformational changes involving concerted movements of TM helices and rotation and tilting of the A domain. The huge conformational changes between E1 and E2 forms of SERCA were already evident in the first two crystal structures of the E1.2Ca^{2+} and E2 forms of the enzyme, solved at 2.6 Å and 3.1 Å, respectively (see Figure 9.29). (The Ca^{2+}-free form, E2, was first crystallized bound to an inhibitor, thapsigargin (TG); a newer crystal structure of the E2 state lacking inhibitors shows a nearly identical structure to E2:TG.) These two structures showed that major conformational changes allow the widely separated A, N, and P domains in E1.2Ca^{2+} to convert to the compact headpiece of E2:TG. In this transition the N domain tilts away from the P domain, allowing the A domain to rotate and tilt upwards into the crevice between them (Figure 9.35). The movements of A and P domains pull on the connected TM helices M1–M3 and M4 and M5, opening the luminal calcium exit channel.

Additional structures of SERCA bound to different substrate analogs show how its conformational changes are key to transitions in the catalytic cycle and the transport of calcium (Figure 9.35B). In the absence of calcium SERCA binds ATP but does not hydrolyze it because the formation of the phosphorylation site results from movements of the cytoplasmic domains triggered by Ca^{2+}-binding – namely the rotation of the A domain as the M5 helix straightens and brings the N domain towards the P domain. Further movements of the A domain accompany the kinase activity, which releases the N domain from the P domain, and allow the TGES loop of the A domain to replace the leaving ADP. This movement of the A domain pulls M1–M2 and pushes M3–M4 in the TM domain, disrupting the Ca^{2+}-binding sites and opening an exit pathway. Once protons have replaced the Ca^{2+} ions, the opening to the lumen is resealed with another rotation of the A domain. Upon dephosphorylation the N domain is moved away from the P domain by the bending of M5, and an opening to the cytoplasm facilitates exchange of protons with Ca^{2+} ions.

These enormous conformational changes have been visualized in animations of the SERCA reaction cycle. The change in height of SERCA between E1 and E2 states is sufficient to be detected in solid-supported SR membranes by single molecule scanning AFM. Why does pumping ions involve such large conformational changes? If the conformational changes were small, thermal motions might be sufficient to trigger enough change to allow ions to leak through. Such thermal fluctuations are seen in all-atom MD simulations of SERCA that can now portray phosphorylation, dephosphorylation, and passage of Ca^{2+} ions for more than 100 ns.

Structural biology – supported by many solution studies including binding and activity of substrate analogs, characterization of mutants, protection from proteolysis, and spectroscopic studies – provides an intimate picture of the mechanism of SERCA1a. This calcium pump represents the family of Ca^{2+}-ATPases that includes SERCA 1, 2, and 3 coded by three distinct genes. The different isoforms (e.g., SERCA1a, 1b, 2b, etc.) result from alternate splicing and occur in different tissues. Humans have 12 isoforms, with 2b the prime target of cardiotonic steroids in the heart muscle. While they are expected to be very similar in structure and mechanism, the different isoforms have different turnover rates and different responses. The detailed knowledge of SERCA1a is therefore of wide application to other Ca^{2+}-ATPases as well as to other members of the P-type ATPase family.

Na^+, K^+ ATPase

A second important and well-characterized member of the P_{II} family is the Na^+, K^+ ATPase in the plasma membrane of animal cells. The Na^+, K^+ ATPase is a hetero-oligomer, with an α subunit homologous to SERCA and a β subunit that extends in the extracellular space, plus a tissue-specific γ subunit called FXYD. It drives the electrogenic exchange of 2 K^+ for 3 Na^+ with the hydrolysis of ATP, alternating between an E1 form with higher affinity for Na^+ and an E2 form with higher affinity for K^+ (Figure 9.36). In this cycle proposed by Post and Albers, Na^+ binding is required for phosphorylation of the enzyme, leading to formation of the occluded Na_3-E1P-ADP state; then Na^+ release and K^+ binding to modulated sites stimulate dephosphorylation and leads to the K_2-E2 state. The Na^+, K^+ ATPase generates large electrochemical gradients that are required for electrical excitability and secondary transport systems that govern the uptake of nutrients, neurotransmitters, and ions and extrusion of toxins, and control of cell volume and pH.

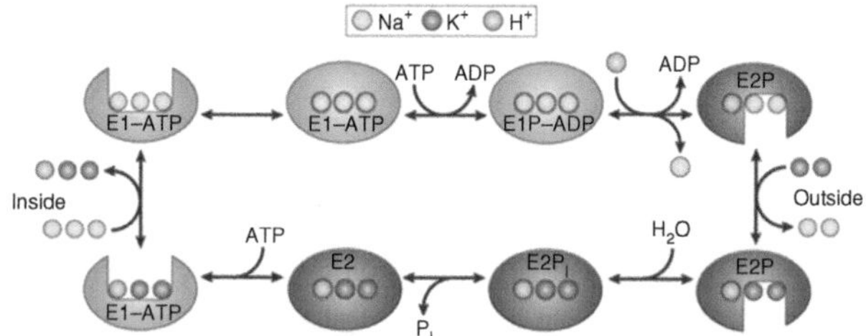

9.36 Post–Albers reaction cycle for the Na^+, K^+ ATPase. As described in the text, the E1 conformation (blue) binds ATP and three Na^+ ions from inside the cell, leading to an occluded state. Phosphorylation of the enzyme and loss of ADP occurs with transition to the E2P state (lavender) with reduced affinity for Na^+. The Na^+ ions exit and K^+ ions enter, along with a proton at the vacant Na3 site. E2P dephosphorylation occurs in another occluded state, allowing ATP binding, release of K^+ inside, and transition to E1 to repeat the cycle. From Morth, J. P., *et al.*, *Nature Reviews Mol Cell Biol*. 2011, 12:60–70. © 2011, Nature Publishing Group. Reprinted by permission of Macmillan Publisher Ltd.

The ATPase activity responsible for maintaining the Na^+ and K^+ gradients in nerves was proposed to reside in a membrane protein as early as 1957 by Jens C. Skou, who was awarded the Nobel Prize in Chemistry in 1997 for the first discovery of an ion-transporting enzyme, or ion pump. Low-resolution structures of the Na^+, K^+ ATPase became available in the 1980s. In recent years the first x-ray structure of the Na^+, K^+ ATPase from pig kidney was solved at 3.5 Å resolution, followed by the structure from shark at 2.4 Å resolution. Both structures capture the enzyme in the K_2-E2-Pi state, including rubidium ions mapping the K^+-binding sites and MgF_4^{2-} mimicking the bound phosphate. The overall architecture of the α subunit of Na^+, K^+ ATPase shows striking similarity to SERCA (Figure 9.37A and Frontispiece). Even the ion-binding sites are similar, with most of the same amino acids, in spite of the larger size of K^+ ions.

The α subunit of the Na^+, K^+ ATPase has the now familiar A, P, and N cytoplasmic domains, along with ten TM helices (Figure 9.37B). As in SERCA, M1–M6 form the transport core, with M7–M10 providing structural

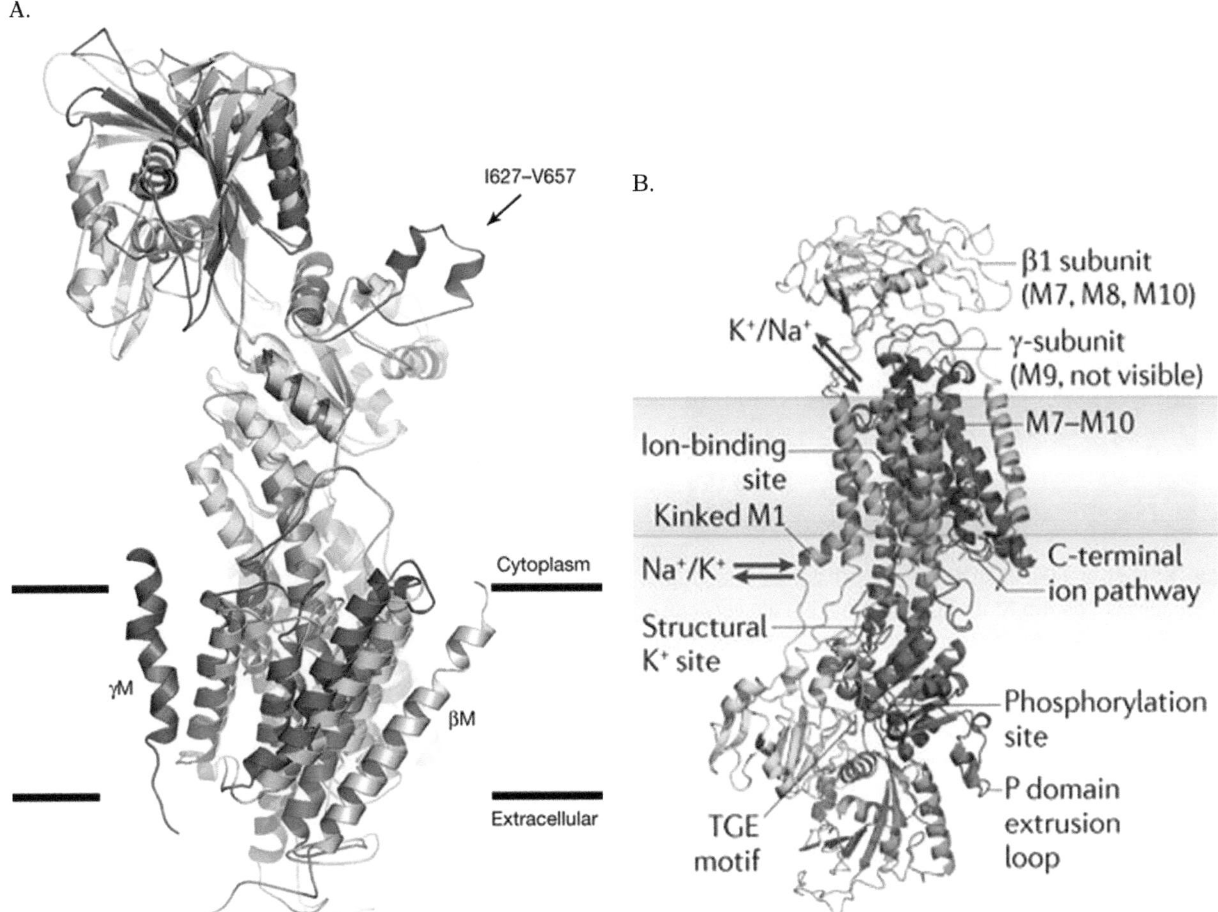

9.37 Crystal structure of the Na^+, K^+ ATPase. (A) Superposition of the Na^+, K^+ ATPase and SERCA structures. By aligning the A and P domains in the ribbon diagrams for Na^+, K^+ ATPase (blue) and SERCA (yellow), the strong similarities in the two structures are apparent. (In this view the A domain is not very visible.) SERCA lacks the β (wheat, TM strand only) and γ (red) subunits. The arrow indicates a protrusion of the P domain unique to the Na^+, K^+ ATPase. From Morth, J. P., *et al.*, *Nature*. 2007, 450:1043–1049. © 2007, Nature Publishing Group. Reprinted by permission of Macmillan Publisher Ltd. (B) Features of the structure of the Na^+, K^+ ATPase. The ribbon diagram of the Na^+, K^+ ATPase is colored to show the domains (A, yellow; N, pink; P, blue; M1–6, gray; M7–10, lavender; β subunit, wheat; γ subunit, green) and labeled to show the features involved in ATP hydrolysis and ion transport. The structural K^+ site is the location of a bound K^+ ion that is not transported. Note the orientation with the cytoplasmic domains at the bottom is opposite that in A. From Morth, J. P., *et al.*, *Nature Reviews Mol Cell Biol*. 2011, 12:60–70. © 2011, Nature Publishing Group. Reprinted by permission of Macmillan Publishers Ltd.

support. Other similarities to SERCA include the phosphorylation site Asp369 of the DKTGT motif in the P domain, unwinding of M4 and M6 in the middle of the bilayer at the region of ion binding with a glutamate residue in M4 (Glu327) capping the ion binding sites, and stabilization of the 90° kink in M1 by FGGF residues 90–93 at the entry from the cytoplasm (residue numbering from pig enzyme). The binding of substrate ions (in this case, three Na^+ ions) triggers formation of an Mg^{2+}-binding site, which allows an Mg^{2+} ion along with Lys691 to offset the electrostatic repulsion at the phosphorylation site and allow transfer of the γ-phosphate to Asp369. One main difference is the distances between the domains: The N domain is farther from the A domain with only an essential salt bridge (Glu223–Arg551) between them below the ATP-binding site. The N domain is also farther from the P domain. In the Na^+, K^+ ATPase a cholesterol molecule is seen shielding an unwound portion of M7 at the cytoplasmic side of the exterior, between the α and β subunits. Other significant differences include the kink in M7 at Gly848 and the protrusion loop formed by the C terminus that could interact with regulatory proteins. The structural similarities suggest that the Na^+, K^+ ATPase undergoes conformational changes analogous to those in SERCA.

Ion-binding sites. Two of the ion-binding sites of the Na^+, K^+ ATPase observed directly with Rb^+ ions in both pig and shark crystals overlap with the Ca^{2+} sites in SERCA (Figure 9.38). Located only 4 Å apart in a shared and occluded cavity between M4, M5, and M6, they are

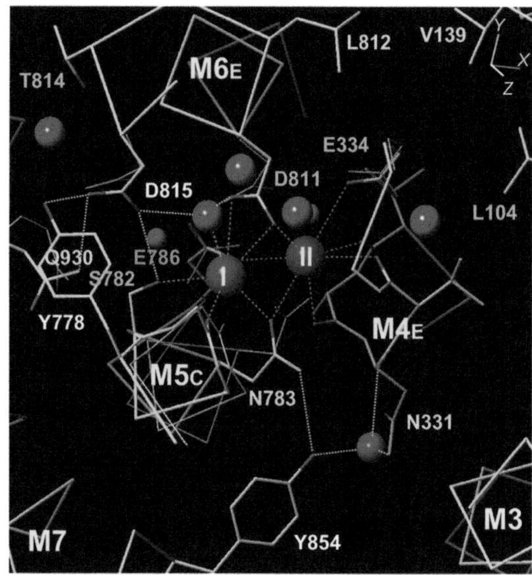

9.38 K^+ binding sites in shark Na^+, K^+ ATPase as viewed from the cytoplasm. The two K^+ ions (purple spheres) are very close, with no oxygen ligands or waters (red spheres) between them. The amino acids involved in coordinating the ions are labeled with the residue numbers for the shark enzyme, which is offset by seven from that given in the text for the pig enzyme. Thus sAsp815 = pAsp808, sAsp811 = pAsp804, sGlu334 = pGlu327, sAsn783 = pAsn776, sSer782 = pSer775, sGlu786 = pGlu779, and sGln930 = pGln923. From Shinoda, T., *et al.*, *Nature*. 2009, 459:446–450. © 2009, Nature Publishing Group. Reprinted by Permission of Macmillan Publishers Ltd.

the sites for the two K^+ ions and presumably for two of the three Na^+ ions. Na1 involves side chains of Thr807 from M6, Asn776 and Glu779 from M5, and Gln923 from M9. The side chain of Asp808 is shared by Na1 and Na2. Na2 utilizes the side chain of Glu327 and the carbonyls of Ala323 and Val325 from M4, plus the side chains of Asp804 and Asp808 from M6. The only difference from the SERCA ion-binding sites are Asp804 and Gln923 that correspond to Asn and Glu residues in SERCA, illustrating that different orientations of the same residues can give different ion selectivity.

Two possible locations for the third Na^+ site called Na3a and Na3b involve residues in M7–M10, in addition to M5. Each have acidic residues, Asp926 and Glu954, with pK_as ≥7 that are likely protonated in the x-ray structure, supporting a transition from H_2EK_2 to $HENa_3$ during the catalytic cycle. The Na3b site resides at the end of a buried channel from the cytoplasm termed the C-terminal ion pathway (see Figure 9.37B). In the x-ray structures this channel is plugged by the C terminus, whose final five residues are essential for high Na^+ affinity, but not K^+ affinity (Figure 9.39). The C terminus may be part of a switch that gives access to this ion pathway in response to voltage, since Na^+ binding is voltage sensitive and the presence of multiple Arg residues (positions 933, 934, 998, 1003, 1004, and 1005) is suggestive of a voltage sensor like that in K^+ channels (see Chapter 12). Interestingly, two of these Arg residues are among target

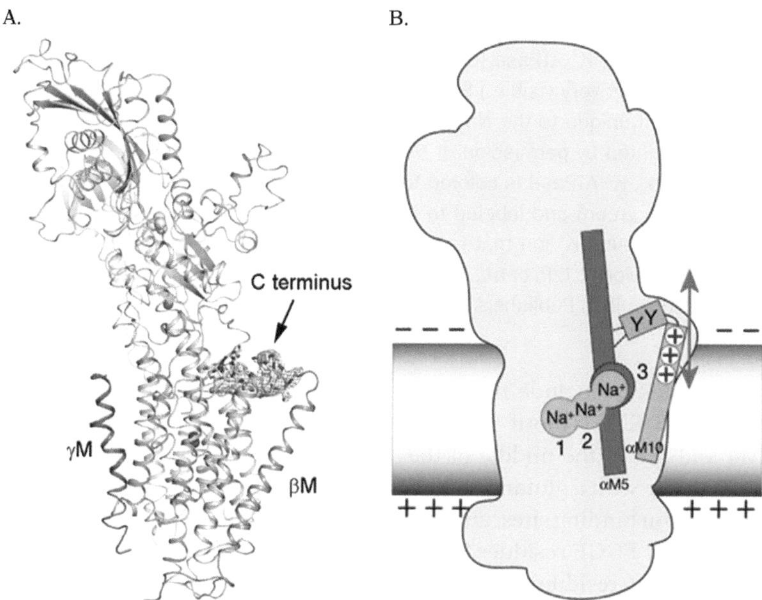

9.39 The role of the C terminus in the voltage sensitivity of the Na^+, K^+ ATPase. (A) The crystal structure of the E2P enzyme is shown as a ribbon diagram with the cytoplasmic side facing up. The C terminus (represented by electron density mesh with acidic and basic residues colored red and blue, respectively) is positioned over the C-terminal ion pathway, where it must move for access to the Na3 sites. From JBC comment, *JBC*. 2009, 284:18715–18725. © 2009 by the American Society for Biochemistry and Molecular Biology. (B) Cartoon of the proposed C-terminal switch, where the cluster of Arg residues are expected to move in response to changes in the voltage. From Morth, J. P., *et al.*, *Nature*. 2007, 450:1043–1049. © 2007, Nature Publishing Group. Reprinted by permission of Macmillan Publishers Ltd.

residues that by mutation are associated with familial hemiplegic migraine 2 (see below).

Other subunits. The β subunit is required to traffic the Na^+, K^+ ATPase to the plasma membrane and also affects its affinity for K^+. It is a 45-kDa protein with a short cytoplasmic tail, one TM helix, and a larger and highly glycosylated domain outside the membrane, where it interacts with extracellular loops of the α subunit (see Figure 9.37B). The β TM helix crosses the membrane at a tilt of 45° and interacts with M7 of the α subunit at a highly conserved YXXYF motif (where X is any hydrophobic residue). This interaction sandwiches the bound cholesterol and stabilizes the unwound part of M7, which may contribute to the effect of β on K^+ binding. The β subunit also interacts with some regulatory proteins that affect Na^+, K^+ ATPase activity.

In some cell types a γ subunit associates with the Na^+, K^+ ATPase to regulate its activity in a tissue-specific way. The γ subunit is from a family of small regulatory proteins called FXYD for their extracellular signature sequence (X is typically Tyr, but may also be Thr, Glu, or His). They share a highly conserved core of 35 amino acids in and around a single TM helix, with a basic cytoplasmic C terminus and an acidic extracellular N terminus. The structures of isolated FXYD proteins have been solved by NMR; in the Na^+, K^+ ATPase x-ray structures the TM helix of γ clearly interacts with M9 of the α subunit (Figure 9.40). Presence of a FXYD protein finely tunes the activity of Na^+, K^+ ATPase, mainly by effects on the Na^+ affinity, but also effects on K^+ affinity and transport rates. The seven FXYD proteins in humans interact with four isoforms of the α subunit along with three isoforms of the β subunit, all with various tissue distributions. FXYD1, called phospholemman and found in the heart, is a chief substrate for cAMP-dependent protein kinases A and C. FXYD6 and 7 are expressed in brain tissue, and FXYD3 and 5 are upregulated in several cancers.

Medical significance of Na^+, K^+ ATPase. It is difficult to overestimate the importance of the sodium–potassium pump to human health. In addition to the role of FXYD expression in certain cancers, the Na^+, K^+ ATPase malfunctions in cardiovascular, neurological, renal, and metabolic diseases. While the direct mechanisms are unknown, the Na^+, K^+ ATPase is strongly inhibited by ouabain, a glycosidic steroid that belongs to a class of hormones called cardiotonic steroids. Endogenous cardiotonic steroids affect renal sodium transport and blood pressure, regulation of cell growth, differentiation, apoptosis, and control of some brain functions. Ouabain and the related compound digitoxin (brand names are, for example, *Crystodigin*, *digoxin*, and *Lanoxin*) have long been used to treat heart failure. By inhibiting the Na^+, K^+ ATPase and thus decreasing Na^+ gradients, ouabain treatment decreases the activity of the Na^+/Ca^{2+} exchanger, leaving the concentration of Ca^{2+} higher in the heart muscle cell to allow stronger contractions. Consistent with mutational data, the binding of ouabain to the Na^+, K^+ ATPase has recently been observed in a high-affinity site with the lactone ring buried in a cavity in the TM domain and the glycosidic moiety facing the extracellular medium (Figure 9.41). The enzyme conformation is altered only by a slight rotation of the A domain and shifts in M1–M4 that close off access to ion-binding sites.

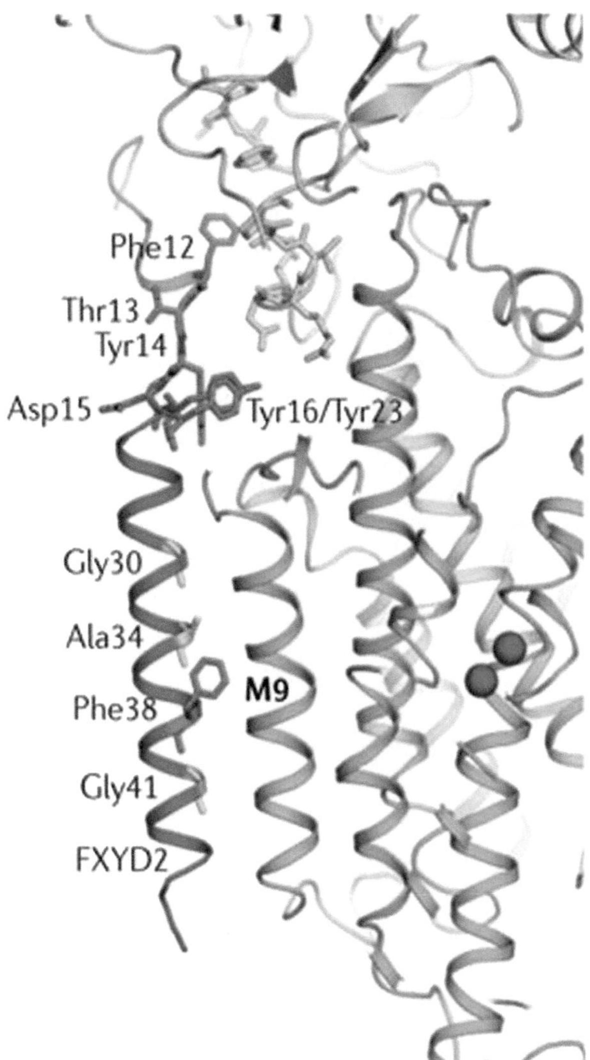

9.40 Detailed view of the interaction of FXYD2 with the α subunit in the shark Na^+, K^+ ATPase. The FXYD sequence starting at Phe12 (green portion of the ribbon) is near the extracellular end (upper left). Other highly conserved residues interact with M9 of the α subunit, whose exposed surface is similar in all human isoforms, and with the cytoplasmic portion of the β subunit (wheat). From Morth, J. P., *et al.*, *Nature Reviews Mol Cell Biol.* 2011, 12:60–70. © 2011, Nature Publishing Group. Reprinted by permission of Macmillan Publishers Ltd.

Additional details about the importance of the Na^+, K^+ ATPase in brain function come from structure–function

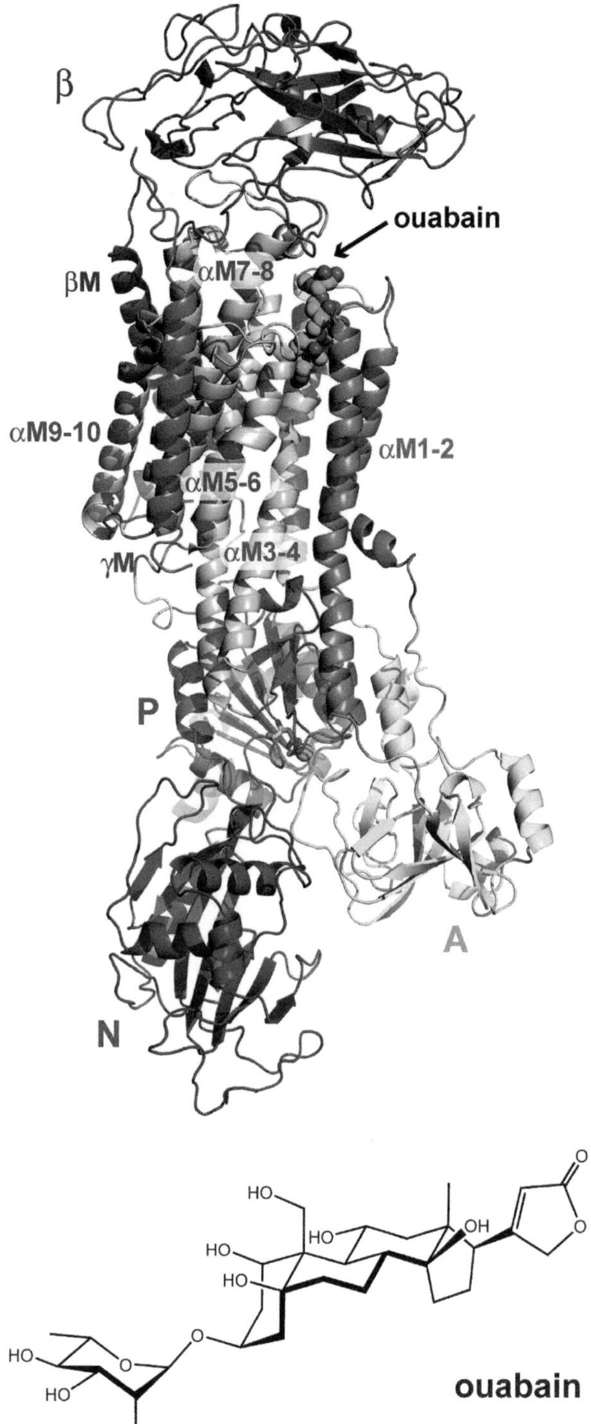

9.41 Binding of ouabain to the high-affinity site of the Na^+, K^+ ATPase. The cardiotonic steroid ouabain binds a phosphorylated form of the enzyme (E2P) with high affinity (Kd ~3 nM). A crystal structure of the complex solved at 4.6-Å resolution shows the ouabain (green and red spheres) inserted into the TM region between M1–2 and M3–4. This site partially overlaps with the low-affinity site observed in earlier crystal structures. The structure of ouabain is shown. Orientation shows cytoplasmic domains at the bottom. From Yatime, L., *et al.*, *J Struct Biol.* 2011, 174:296–306. © 2011, Elsevier. Reprinted with permission from Elsevier.

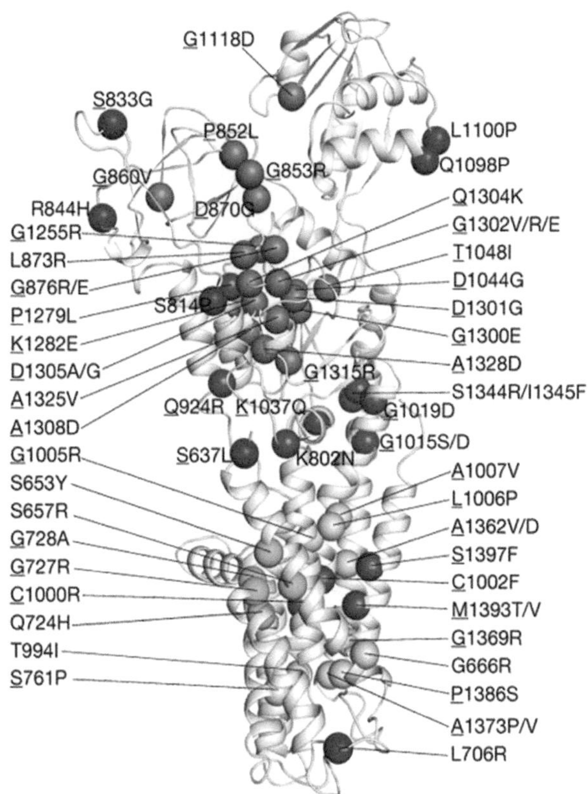

9.42 Distribution of human ATPA7 missense mutations associated with Menkes' disease. Mutated residues are represented by spheres on the ribbon diagram for CopA, the Cu^+-ATPase from *L. pneumofila*, and colored to show their likely functional roles (phosphorylation/dephosphorylation, green; domain interfaces/ATP coupling, blue; putative copper entry and exit pathways, wheat; Cu1 and Cu2 binding sites, dark brown; GG kink in amphipathic helix, cyan; surface exposed sites, purple). Underlined residues are conserved and residue numbers refer to human ATP7A. From Gourdon, P., *et al.*, *Nature*. 2011, 475:59–64. © 2011, Nature Publishing Group. Reprinted by permission of Macmillan Publishers Ltd.

studies in mutants. Two of the over 50 mutations in the gene for the Na^+, K^+ ATPase that cause severe migraines have been shown to affect the C-terminal ion pathway. Mutations that cause a severe form of dystonia with Parkinsonism map to the same region of the enzyme and appear to affect binding of the third Na^+ ion. The powerful combination of structural biology with electrophysiology and genetics will likely elucidate the nature of these and other neurological disorders.

Other P-type ATPases

The P-type ATPase superfamily has many other important members. The crucial ion gradients established by the Na^+, K^+ ATPase in animals are established in other eukaryotes (plants and fungi) by H^+-ATPases, which belong to the P_{III} subfamily. An x-ray structure of a proton pump, the H^+-ATPase from *Arabidopsis thaliana* in complex with an ATP analog, shows marked similarities to the P_{II} ATPases described above. In addition it has a large, water-filled cavity in the TM domain, with an Asp–Asn pair that serves as a gateway to the proton transport pathway and an Arg residue that stimulates proton release. Such H^+-ATPases are capable of supporting membrane potentials of 200–300 mV.

ATPases in the P_I subfamily include those that pump out heavy metals such as Cu^+, Zn^{2+}, and Co^{2+} to avoid their toxic effects. The crystal structure of CopA, a Cu^+-ATPase from *Legionella pneumophila* that has high sequence identity with the human Cu^+-ATPases, exhibits the architecture typical of P-type ATPases, although the three cytosolic domains are smaller. In addition to two Cu^+-binding sites positioned in the TM domain analogous to the Ca^{2+}-binding sites in SERCA, CopA has two additional TM helices at the N terminus that may provide a platform for copper-chaperones to dock. Also at the N terminus, one or more non-essential heavy-metal binding domains are found in tandem. Defects in the human Cu^+-ATPases ATP7A and ATP7B give rise to Menkes' disease, which produces copper deficiency, and Wilson's disease, which results in copper overload. Both diseases have a variety of phenotypes that typically lead to neurological disorders. When mapped to the CopA structure the residues affected by missense mutations in Menkes' disease affect all aspects of Cu^+-ATPase function (Figure 9.42). Increased activity of human Cu^+-ATPases is associated with Alzheimer's disease.

CONCLUSION

The examples of enzymes in this chapter show a wide variation in how proteins carry out catalysis in the membrane milieu. Each has a fascinating story that will be more completely revealed when higher-resolution structures and/or more structures of mutants or of activated states become available. From the dimerization of OMPLA that enables it to trap its lipid substrate to the thinning of the bilayer at the site of rhomboid, the role of membrane lipids in protein function is clearly important. Thus the challenges that make studying these membrane proteins difficult also provide a wealth of information in the outcomes.

FOR FURTHER READING

Papers with an asterisk (*) present original structures.

Prostaglandin H_2 Synthase

Garavito, R. M., and A. M. Mulichak, The structure of mammalian cyclooxygenases. *Annu Rev Biophys Biomol Struct*. 2003, 32:183–206.

Kulmacz, R. J., *et al*., Comparison of the properties of prostaglandin H synthase-1 and -2. *Prog Lipid Res*. 2003, 42:377–404.

*Loll, P. J., D. Picot, and R. M. Garavito, The structural basis of aspirin activity inferred from the crystal structure of inactivated prostaglandin H2 synthase. *Nat Struct Biol*. 1995, 2:637–643.

*Malkowski, M. G., *et al*., The x-ray structure of prostaglandin endoperoxide H synthase 1 complexed with arachidonic acid. *Science*. 2000, 289:1933–1937.

Rouzer, C. A., and L. J. Marnett, Structural and functional differences between cyclooxygenases: fatty acid oxygenases with a critical role in cell signaling. *Biochem Biophys Res Commun*. 2005, 338:34–44.

OMPLA

Snijder, H. J., and B. W. Kijkstra, Bacterial phospholipase A: structure and function of an integral membrane phospholipase. *Biochim Biophys Acta*. 2000, 1288:91–101.

*Snijder, H. J., I. Ubarretxena-Belandia, M. Blaauw, *et al*., Structural evidence for dimerization-regulated activation of an integral membrane phospholipase. *Nature*. 1999, 401:717–721.

Omptins

Eren, E., M. Murphy, J. Goguen, and B. van den Berg, An active site water network in the Plasminogen Activator Pla from *Yersinia pestis*. *Structure*. 2010, 18:809–818.

Hritonenko, V., and C. Stathopoulos, Omptin proteins: an expanding family of outer membrane proteases in Gram-negative *Enterobacteriaceae*. *Molec Membr Biol*. 2007, 24:395–406.

*Vandeputte-Rutten, L., R. A. Kramer, J. Kroon, N. Dekker, M. R. Egmond, and P. Gros, Crystal structure of the outer membrane protease OmpT from *Escherichia coli* suggests a novel catalytic site. *EMBO J*. 2001, 20:5033–5039.

Rhomboid

Erez, E., D. Fass, and E. Bibi, How intramembrane proteases bury hydrolytic reactions in the membrane. *Nature*. 2009, 459:371–378.

Urban, S., Taking the plunge: integrating structural, enzymatic and computational insights into a unified model for membrane-immersed rhomboid proteolysis. *Biochem J*. 2010, 425:501–512.

Vinothkumar, K. R., The structural basis for catalysis and substrate specificity of a rhomboid protease. *EMBO J.* 2010, 29:3797–3809.

[3]*Wu, Z., N. Yan, L. Feng, *et al.*, Structural analysis of a rhomboid family intramembrane protease reveals a gating mechanism for substrate entry. *Nature Str and Molec Biol.* 2006, 13:1084–1091.

An interactive structure of GlpG is available at www.BiochemJ.org/bj/425/0501/bj4250501add.htm.

Formate Dehydrogenase

*Jormakka, M., S. Tornroth, B. Byrne, and S. Iwata, Molecular basis of proton motive force generation: structure of formate dehydrogenase-N. *Science.* 2002, 295:1863–1868.

Jormakka, M., B. Byrne, and S. Iwata, Formate dehydrogenase: a versatile enzyme in changing environments. *Curr Opin Struct Biol.* 2003, 13:418–423.

Jormakka, M., B. Byrne, and S. Iwata, Proton motive force generation by a redox loop mechanism. *FEBS Lett.* 2003, 545:25–30.

Calcium ATPase

Moller, J. V., C. Olesen, A. M. Winther, and P. Nissen, The sarcoplasmic Ca^{2+}-ATPase: design of a perfect chemi-osmotic pump. *Quart Rev Biophys.* 2010, 43:501–566.

*Oleson, C., M. Picard, A. M. Winther, *et al.*, The structural basis of calcium transport by the calcium pump. *Nature.* 2007, 450:1036–1042.

Palmgren, M. G., and P. Nissen, P-Type ATPases. *Ann Rev Biophys.* 2011, 40:243–266.

Toyoshima, C., How Ca^{2+} ATPase pumps ions across the sarcoplasmic reticulum membrane. *Biochim Biophys Acta.* 2009, 1793:941–946.

[3] Seven other papers were published with structures of GlpG within one year.

*Toyoshima, C., and H. Nomura, Structural changes in the calcium pump accompanying the dissociation of calcium. *Nature.* 2002, 418:605–611.

*Toyoshima, C., M. Nakasako, H. Nomura, and H. Ogawa, Crystal structure of the calcium pump of sarcoplasmic reticulum at 2.6 Å resolution. *Nature.* 2000, 405:647–655.

Animations of SERCA are available at http://octo3.bioxlay.au.dk/gallery/ and www.iam.u-tokyo.ac.jp/StrBiol/resource/movies

Na^+, K^+ ATPase

Bublitz, M., In and out of the cation pumps: P-type ATPase structure revisited. *Curr Opin Struct Biol.* 2010, 20:431–439.

*Morth, J. P., B. P. Pedersen, M. S. Toustrup-Jensen, *et al.*, Crystal structure of the sodium-potassium pump. *Nature.* 2007, 450:1043–1049.

Morth, J. P., B. P. Pedersen, M. J. Buch-Pedersen, *et al.*, A structural overview of the plasma membrane Na^+, K^+ ATPase and H^+-ATPase ion pumps. *Nature Reviews Mol Cell Biol.* 2011, 12:60–70.

Shinoda, T., H. Ogawa, F. Cornelius, and C. Toyoshima, Crystal structure of the sodium-potssium pump at 2.4 Å resolution. *Nature.* 2009, 459:446–450.

Takeuchi, A., N. Reyes, P. Artigas, and D. C. Gadsby, The ion pathway through the opened Na^+,K^+-ATPase pump. *Nature.* 2008, 456:413–416.

H^+-ATPase

*Pedersen, B. P., M. J. Buch-Pedersen, J. P. Morth, *et al.*, Crystal structure of the plasma membrane proton pump. *Nature.* 2007, 450:1111–1114.

Cu^+-ATPase

*Gourdon, P., X. Y. Liu, T. Skjørringe, *et al.*, Crystal structure of a copper-transporting PIB-type ATPase. *Nature.* 2011, 475:59–64.

MEMBRANE RECEPTORS

10

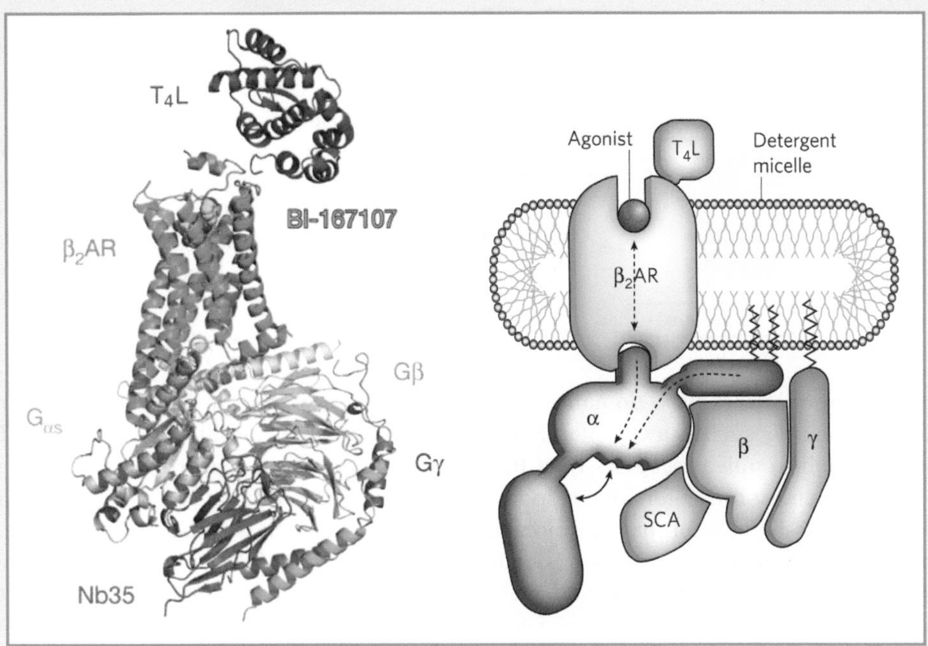

Structure of a signaling complex, the agonist-bound β_2-adrenergic receptor in complex with a G protein. The first snapshot of a GPCR-mediated signal transduction process (left) shows the β_2-adrenergic receptor (green) bound to an agonist (space filling model) and to the heterotrimeric G_s protein, $G\alpha\beta\gamma$ (gold, cyan, and blue, respectively). Extensive biochemical procedures needed to achieve crystals of the complex included modification with T_4 lysozyme (T_4L, magenta) and stabilization with a single-chain antibody, Nb35 (red). The complex was inserted into detergent micelles (cartoon on right). The arrow in the receptor points out the allostery between agonist binding and G protein binding, and the arrows in $G\alpha$ indicate the conformational change upon binding. From Rasmussen, S. G. F., *et al.*, *Nature*. 2011, 477:549–555. Inset from Schwartz, T. W., and T. P. Sakmar, *Nature*. 2011, 477:540–541. © 2011, Nature Publishing Group. Reprinted by permission of Macmillan Publishers Ltd.

Structural biology of membrane receptors has expanded dramatically in recent years. Membrane receptors required for signaling are abundant in the plasma membrane. Most receptors bind signaling molecules (ligands), while some respond to light, touch, odors, even changes in oxygen concentration. A given ligand can elicit many varied responses when it binds to different receptors in different kinds of cells or to the same type of receptor distributed in different tissues. In spite of this complexity, membrane receptors share a basic mode of action. Binding a signaling molecule from the exterior triggers a conformational change in the receptor that either enables the cytosolic domain to bind an effector protein or opens an ion channel. The receptor in the membrane mediates the communication coming from outside the cell and passes it to the cell interior.

Receptors are classified as *ionotropic*, which are ligand-gated ion channels, or *metabotropic*, which trigger signal transduction pathways that release second messengers and other signaling molecules and may link indirectly to ion channels. A metabotropic receptor may have enzymatic activity that initiates the signal, for example

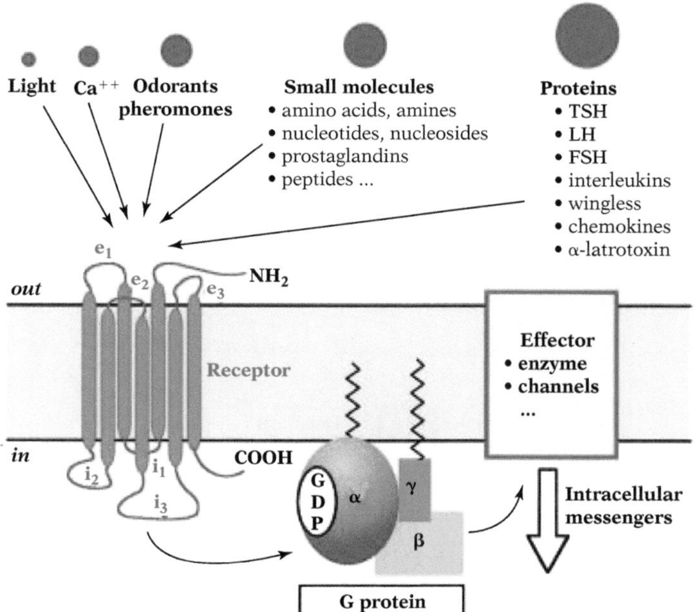

10.1 The generalized function of G protein-coupled receptors. GPCRs respond to a variety of stimulants, including light, odorants, calcium ions, small molecules, and proteins. They trigger activation of a G protein $\alpha\beta\gamma$ complex, which stimulates the release of second messengers or affects channels in the membrane. TSH, thyroid-stimulating hormone; FSH, follicle-stimulating hormone; LH, luteinizing hormone. From Bockaert, J., and J. P. Pin, *EMBO J.* 1999, 18:1723–1729. © 1999. Reprinted by permission of Macmillan Publishers Ltd.

by activation of an intracellular kinase domain, or it may bind an effector molecule, such as a G protein, that relays the signal. G protein-coupled receptors (GPCRs) are a large and important class of receptors that respond to external signals such as light, odorants, hormones, and neurotransmitters, and activate G proteins in signal transduction pathways that alter levels of second messengers and/or control ion channels (Figure 10.1). Their importance is underscored by the award of the 2012 Nobel Prize in Chemistry to Robert Lefkowitz and Brian Kobilka for characterizing these "gateways to the cells." As key elements in sensory physiology and neurobiology, GPCRs are targets of between a quarter and half of drugs currently on the market.

Receptors from both classes are critical to the function of the nervous system, allowing it to pass information between cells and enabling it to respond to hormones. Knowledge from the detailed structures of key examples of each class is rapidly advancing the broader understanding of the roles of these receptors in neurotransmission and membrane signaling.

G PROTEIN-COUPLED RECEPTORS

GPCRs make up the largest family of membrane proteins. Found in eukaryotic organisms from yeast to humans, they carry out vital functions: for example, the receptors for epinephrine, serotonin, and glucagon are all GPCRs. The ~800 GPCRs in humans respond to a large variety of ligands and have a multitude of physiological roles. Many essential roles are still unknown since one-quarter of these are orphans (meaning their ligands are unidentified), most of which are expressed in the central nervous system. With the notable exception of rhodopsin, GPCRs are present at very low abundance. All GPCRs share a common architecture, a bundle of seven TM helices like that of bacteriorhodopsin (bR) (see Chapter 5), and a common mode of action, triggering a change in the activity of the associated G proteins.

Activated GPCRs bind specific G proteins. Most G proteins are heterotrimers consisting of Gα, Gβ, and Gγ subunits and bind membranes via attached lipids on Gα and Gγ (Figure 10.2). They are inactive when binding GDP and become active when it is replaced with GTP, with concurrent dissociation to Gα and G$\beta\gamma$. Activated G proteins turn on kinase activities that lead to phosphorylation cascades of signaling molecules, often producing second messengers such as cyclic AMP, cyclic GMP, 1,2-diacylglycerol, and inositol 1,4,5-triphosphate, or they open ion channels. Activation starts when ligand binding triggers a conformational change in a GPCR, which enables the GPCR to bind the trimeric G protein, allowing the Gα subunit to release its bound GDP and bind GTP. After Gα-GTP and G$\beta\gamma$ dissociate to interact with downstream effector proteins, the GTP is hydrolyzed; the G protein trimer reforms and is ready to interact with another GPCR. GPCRs are regulated by phosphorylation, usually near the C terminus, which allows inhibitors called arrestins to bind and block the binding of G$\alpha\beta\gamma$. Another regulatory mechanism is the removal of GPCRs from the plasma membrane by internalization. Once internalized, GPCR–arrestin complexes can initiate further G protein-independent signaling events.

GPCRs have been thoroughly studied with many biochemical, genetic, and spectroscopic techniques. The seven TM helices are roughly parallel as they cross the membrane and are connected by short loops, with the N terminus outside and the C terminus containing an

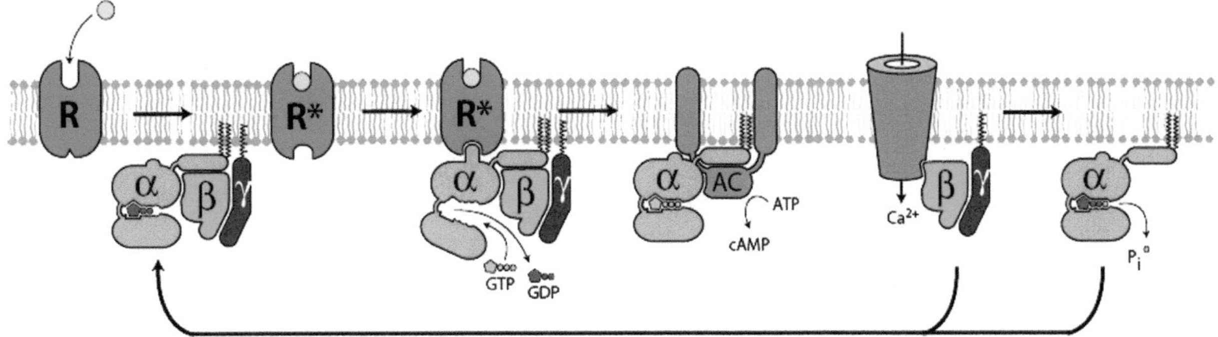

10.2 The G protein cycle in the function of GPCRs. When an extracellular agonist (ligand) binds to the GPCR, a conformational change in the receptor (R) enables it to bind the G protein heterotrimer ($\alpha\beta\gamma$) at its cytoplasmic side. Receptor binding stimulates GDP/GTP exchange in the α subunit of the G protein, which triggers dissociation of the α and $\beta\gamma$ subunits from the receptor. In this example, Gα-GTP stimulates adenylyl cyclase (AC) to synthesize cAMP, while G$\beta\gamma$ opens a Ca^{2+} channel. Hydrolysis of GTP by Gα leads to reformation of the heterotrimer. From Nobel Lecture by Brian K. Kobilka, Dec. 8, 2012 © The Nobel Foundation 2013.

amphipathic helix (H8) inside the cell (Figure 10.3). The compact helical bundles lack segments that can undergo large, dramatic conformational changes; instead their activation involves changes in a few of the TM helices and connecting loops. The seven TM helices of GPCRs have highly conserved amino acids at key positions that form the basis for a generic system for numbering residues. For comparison of different GPCR structures, each residue is given a number for the helix it is in, followed by a number for its position relative to the most conserved residue in the helix, which is assigned an index of 50. For example, a residue in helix TM7 that is seven residues from the most conserved residue in that helix is 7.43; 43 tells its position relative to the most conserved residue in helix 7. (The relative position of a residue number in a particular polypeptide is given in superscript.) GPCRs share three important motifs: E(D)RY at the cytoplasmic end of TM3 where R denotes Arg3.50; CWxP in TM6 where P denotes Pro6.50; and NPxxY$(x)_{5,6}$F at the cytoplasmic end of TM7 where P is Pro7.50 and Phe is in helix 8. Other common features include highly conserved Pro residues in the middle of TM5, TM6, and TM7 helices and a disulfide bond between external Loop 2 (ECL2) and the extracellular end of TM3.

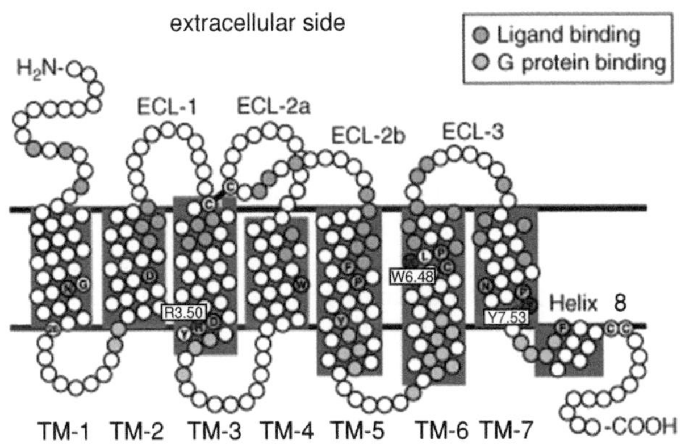

10.3 Generalized topology of GPCRs. A generalized GPCR topology diagram shows the seven TM helices with the eighth amphipathic helix near the C terminus. A highly conserved disulfide bond anchors extracellular Loop 2 (ECL2) to TM3, thus dividing it into ECL2a and ECL2b. Residues of the three common motifs are labeled: DRY of the E(D)RY motif in TM3, CWxP in TM6, and NxxPY$(x)_{5,6}$F in TM7 and H8. Residues in red are molecular switches important in activation. Residues highlighted in purple are typically involved in binding the ligand, while residues highlighted in green are typically involved in binding the α subunit of the G protein. From Nygaard, R., *et al.*, *Trends Pharm Sci*. 2009, 30:249–259. © 2009, Elsevier. Reprinted with permission from Elsevier.

GPCRs are grouped in five families based on sequence and structural similarity: rhodopsin (family A), secretin (family B), glutamate (family C), adhesion, and Frizzled/Taste2. All of the first GPCRs whose structures have been solved at high resolution belong to the rhodopsin family, giving information about the shared structural features of the diverse members of the largest GPCR family. Recent crystal structures of two of these GPCRs in activated states provide exciting insights into the activation mechanisms of all GPCRs.

Rhodopsin, a light-sensitive GPCR

Rhodopsin, the first GPCR with a high-resolution crystal structure, consists of a 40-kDa apoprotein called opsin and a chromophore, 11-*cis*-retinal. Rhodopsin is an abundant photoreceptor in the outer segment of rod cells in the retina of the vertebrate eye.[1] When a photon

[1] Interestingly, to detect colors the cone cells in the retina use three photoreceptors very similar to rhodopsin; slight differences in the opsin structure change the absorption spectrum of the chromophore. As night blindness can be caused by mutations in rhodopsin, color blindness results from mutations in one of the genes for the cone opsins.

BOX 10.1 EFFICIENCY OF LIGHT-INDUCED SIGNAL TRANSDUCTION

Rhodopsin is an example of a protein that is highly enriched in the tissue from which it is extracted. In the mammalian eye, each rod's outer segment contains 1000–2000 stacked discs that originated from the plasma membrane. Greater than 90% of the protein in the disc membranes is rhodopsin. The mouse retina, which has ~5 million discs, contains ~80 000 molecules of rhodopsin per disc. A single photon is sufficient to activate rhodopsin. Each excited rhodopsin can activate >500 molecules of transducin, each of which can activate a molecule of the cGMP phosphodiesterase by binding its inhibitory subunit. Now activated, phosphodiesterase can hydrolyze >4000 cGMP molecules per second. Three cGMPs are required to open one ion channel, so a small change in the concentration of cGMP makes a large change in ion conductance. Thus each photon absorbed by rhodopsin can close 1000 or more channels and change the potential by about 1 mV.

hits rhodopsin it triggers conversion of the bound 11-*cis*-retinal to all-*trans*-retinal (see Figure 10.5), causing conformational changes that stimulate rhodopsin to interact with its G protein, transducin (Gt).

Activated rhodopsin stimulates the exchange of GTP for GDP in transducin, which triggers a series of events leading to perception of the light. First, the Gt heterotrimer dissociates, allowing Gtα–GTP to bind an inhibitory subunit of cGMP phosphodiesterase, thus activating this enzyme to convert cGMP to 5′-GMP. The rapid drop in cGMP levels blocks reentry of Na^+ via cGMP-gated ion channels in the rod cell, which hyperpolarizes the cell (changing the electrical potential from –45 mV to –75 mV) to signal the optic nerve. The typical signaling cascade amplifies the signal at different steps (see Box 10.1). When the GTPase activity of Gtα hydrolyzes the GTP to GDP, the transducin heterotrimer can reform. Adaptation to prolonged exposure to light triggers the phosphorylation of rhodopsin by rhodopsin kinase. Arrestin binds phosphorylated rhodopsin and prevents its interaction with transducin. The slow release of all-*trans*-retinal leads to dissociation of arrestin and prepares rhodopsin for dephosphorylation. After reconstitution with 11-*cis*-retinal the system is ready for a photon to initiate the cycle again.

Ground State Rhodopsin

Since the ground-breaking high-resolution structure of bovine rhodopsin was determined in 2000, other crystal structures of bovine rhodopsin, and recently squid rhodopsin, have been solved, giving very similar results that provide details of its fold in the inactive (ground) state. In addition to now resolving all 348 amino acids of bovine opsin, the structures show the palmitoyl chains on Cys322 and Cys323 that anchor the C terminus to the membrane, the oligosaccharide chains on Asn2 and Asn15 at the N terminus, and the 11-*cis*-retinylidene Schiff base linkage to Lys296$^{7.43}$ (Figure 10.4). As expected, Glu113$^{3.28}$ in TM3 provides a counterion to the protonated Schiff base. The E(D)RY motif consisting of Glu134–Arg135–Tyr136 is at the cytoplasmic end of TM3, and the NPxxY motif is the sequence NPVIY that starts with Asn302 and goes towards the cytoplasmic end of TM7. Cys264–Trp265–Leu266–Pro267 make up the third conserved motif CWxP around the middle of TM6 close to the retinal-binding site.

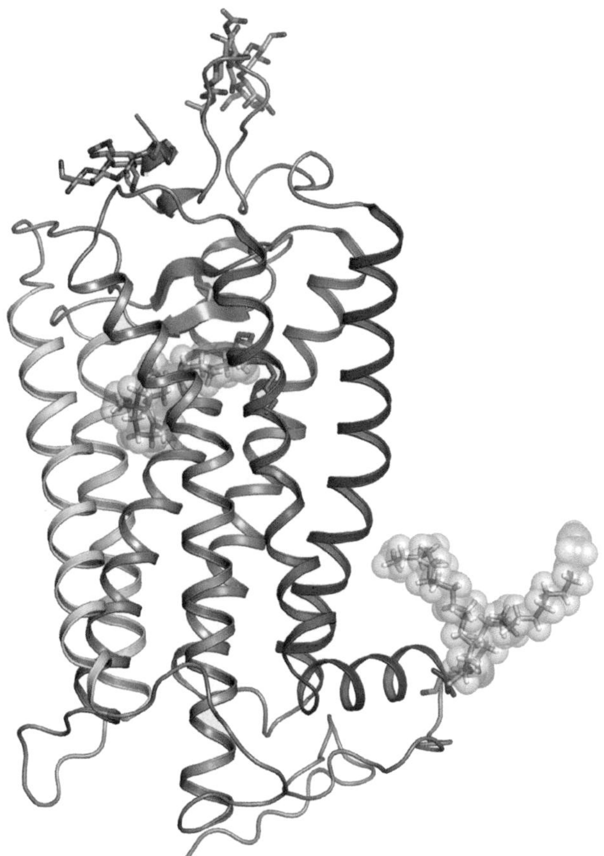

10.4 High-resolution structure of bovine rhodopsin. A lateral view of rhodopsin (ribbon diagram in rainbow coloring) with the extracellular face at the top and the cytoplasmic face at the bottom shows a tight bundle of slightly irregular helices, with a prominent bend in TM6 at the front. The helices are colored according to their primary sequence: TM1, blue; TM2, blue-green, TM3, green; TM4, lime-green; TM5, yellow; TM6, orange; TM7, red; helix 8, purple; with β-strands, gold. The retinal chromophore is colored as a transparent pink surface, acyl chains as transparent white surface, and oligosaccharides as stick models with C atoms gray, N blue, and O red. From Lodowski, D. T., and K. Palczewski *Curr Opin HIV AIDS*, 2009, 4:88–95. © 2009 Lippincott Williams.

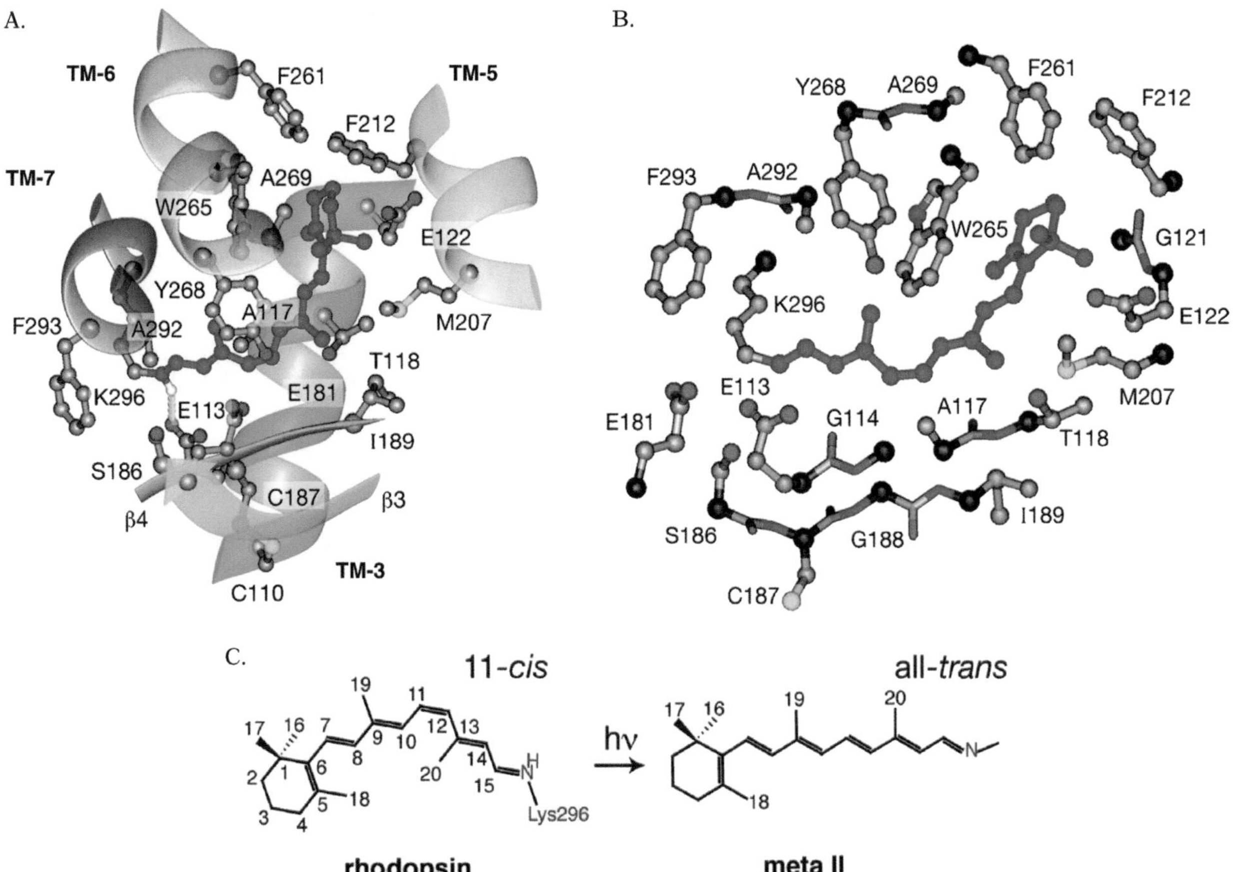

10.5 Retinal-binding pocket in bovine rhodopsin. (A) A side view through portions of TM3, TM5, TM6, and TM7 (colored as in Figure 10.4), with a β-hairpin from ECL2 (blue, at the bottom) reveals the amino acid residues in the vicinity of the 11-*cis*-retinylidene (pink/lavender) bound to the protein. Lys296 on TM7 forms the Schiff base linkage to retinal, and its counterion Glu113 on TM3. (B) A schematic shows the residues within 5 Å of the 11-*cis*-retinylidene group. (C) The isomerization of 11-*cis*-retinylidene to the all-*trans* form is shown with the Lys 296 ε-N in blue. From Palczewski, K., *Annu Rev Biochem*. 2006, 75:743–767. © 2006 by Annual Reviews. Reprinted with permission from the Annual Review of Biochemistry, www.annualreviews.org.

11-*cis*-retinal forms a covalent Schiff base, becoming 11-*cis*-retinylidene, in the extracellular half of the TM bundle, with the entrance to its binding site completely blocked by β-hairpins formed by extracellular Loop 2 (ECL2). The retinal-binding site is a tight pocket with interactions between the β-ionone ring of retinal and Glu122, Phe212, Phe261, and Trp265 (Figure 10.5). Distortions in the helices allow them to pack together closely while enclosing the retinal-binding pocket. Although most of the helices contain highly conserved Pro residues, only Pro267 in TM6 makes a strong kink, while TM2 is bent at a pair of Gly residues (Gly89 and Gly90), TM3 has a kink at Glu113 that forms a salt bridge with the protonated Schiff base, and TM7 is irregular at Lys296 that binds retinal.

The TM helices in rhodopsin are connected by extensive hydrogen bond networks involving polar side chains, some main chain carbonyl groups, and highly ordered water molecules. One hydrogen bond network links the retinal binding pocket to the NPxxY motif where the protein interacts with transducin and may constitute the communication pathway between them (Figure 10.6). Another holds the E(D)RY motif in TM3 close to TM6 with an "ionic lock" salt bridge between Arg135 (TM3) and Glu247 (TM6), strengthened by hydrogen bonds also linking Arg135 with Glu134 (TM3) and Thr251 (TM6). The ionic lock is a feature of other GPCRs as well.

Activated rhodopsin

When a photon of light transforms 11-*cis*-retinal to all-*trans*-retinal, rhodopsin very rapidly converts to short-lived intermediates prior to its active state, called metarhodopsin II (Figure 10.7). Tricks have been used to capture the crystal structures of the short-lived intermediates, but they revealed no structural differences from the ground state structure. Two different approaches have been successful in achieving structures that correspond to metarhodopsin II. When the retinal is removed and a C-terminal peptide fragment of Gtα (residues 340–350, called GαCT) is added, opsin revealed spectral

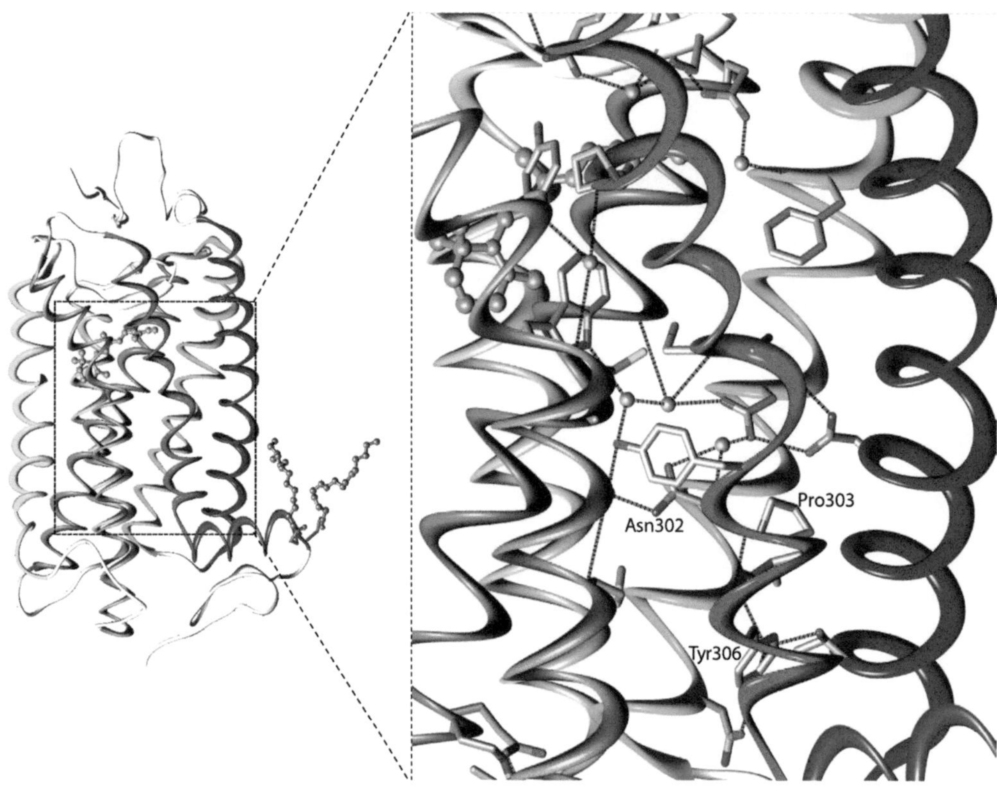

10.6 Hydrogen bonding networks in bovine rhodopsin. One hydrogen bonding network (dotted lines in enlarged figure) stretches from the binding pocket for retinal (pink ball and stick model) through the core of rhodopsin to the NPxxY motif (Asn302, Pro303, and Tyr306) near the Gtα-binding site. The hydrogen bonds utilize polar side chains, such as four Asn residues (at positions 55, 73, 83, and 302) and four Tyr residues (positions 192, 268, 301, and 306), main chain peptide groups (for example, from Gly120, Phe91, and Pro291), as well as six of the ordered water molecules (blue spheres). The TM helices are colored as in Figure 10.4. Kindly provided by Tividar Orban and Krzysztof Palczewski.

Rhodopsin (498 nm)
hυ ↓ 200 fs
Bathorhodopsin (529 nm)
↓ 120 ns
BSI (477 nm)
⇅ 150 ns
Lumirhodopsin (492 nm)
↓ 10 μs
Meta-I (478 nm)
⇅ 1 ms
Meta-II (380 nm) ⟶ Transducin activation
↓
Opsin + all-*trans* retinal

10.7 The photolysis of rhodopsin. Absorption of a photon of light stimulates conversion of rhodopsin into short-lived intermediates with different absorbance maxima (given in parentheses). Following reaction of metarhodopsin II with transducin, the all-*trans*-retinal dissociates from opsin, to be replaced with another 11-*cis*-retinal in the rod cell. From Schertler, G. F. X., *Curr Opin Struct Biol*. 2005, 15:408–415. © 2005, Elsevier. Reprinted with permission from Elsevier.

characteristics of metarhodopsin II and structural changes indicative of activation even in the absence of all-*trans*-retinal. This activated form, called Ops*, is also observed at low pH without GαCT. The second approach involved crystallization of two different rhodopsin mutants with constitutive activity, one affecting the retinal-binding site and the other the G protein-binding site. The E113Q$^{3.28}$ mutant lacks the counterion for the Schiff base to retinal, while the M257Y$^{6.40}$ mutant targets the ionic lock. Together these new structures reveal the features of metarhodopsin II, the activated state of rhodopsin that binds transducin. In particular, the structure of the M257Y mutant (stabilized with an engineered disulfide bond between N2C and D282C and in a complex with GαCT) shows all-*trans*-retinal in its native binding pocket (Figure 10.8).

In the activated state the rhodopsin core composed of TM1–TM4 is unchanged. Striking changes are observed in TM5 and TM6: The cytoplasmic half of TM6 tilts away from the core by about 6 Å, while TM5 is extended by about two turns of the helix and tilts toward TM6 at its cytoplasmic end (Figure 10.9). In addition there are adjustments in the three cytoplasmic loops. These rearrangements open a crevice on the cytoplasmic side

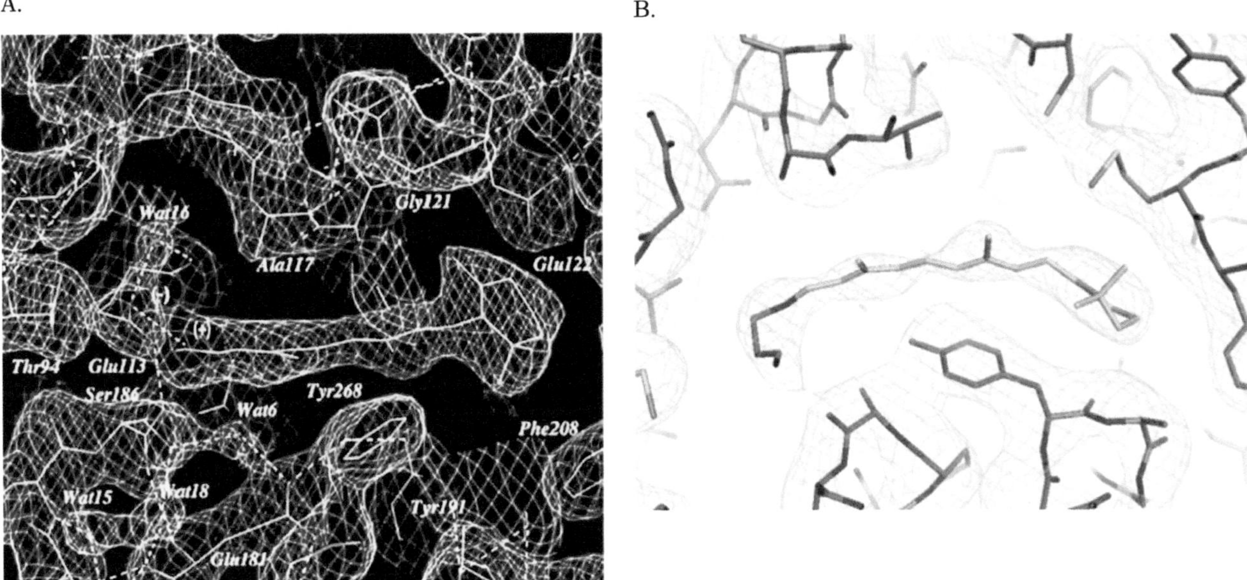

10.8 Binding pockets for retinal in ground state and activated rhodopsin. (A) 11-*cis*-retinal in its binding pocket is defined by the electron density map in bovine rhodopsin. From Li, J., *et al.*, *J Mol Biol*. 2004, 343:1409–1438. (B) Electron density map for the all-*trans*-retinal in the binding pocket of the M257Y constitutive mutant shows a nearly relaxed planar conformation. The same position is seen when opsin crystals are soaked with all-*trans*-retinal. From Deupi, X. *et al.*, *Proc Natl Acad Sci USA*. 2012, 109:119–124, redrawn by authors.

that becomes the binding site for transducin; the structures of Ops* and M257Y with GαCT show an α-helix from GαCT inserted into this crevice (Figure 10.9C). Hydrophobic side chains are exposed from residues of TM5 and TM6 that allow van der Waals interactions with residues of GαCT, while new hydrogen bond networks link polar residues (Cys347 and Lys345) of GαCT to TM3 and TM5 on one side and to TM7 and H8 on the other side.

The large movement of TM6 upon activation breaks the ionic lock between Glu247 (TM6) and Arg135 of the E(D)RY motif (TM3). With TM6 paired with TM5, Glu247 and Thr251 form hydrogen bonds with Lys231, while the side chain of Arg135 (TM3) extends into the central floor of the crevice and interacts with Tyr223 (TM5) (Figure 10.10). The interaction between Glu134 and Arg135 in the ground state is disrupted by protonation of Glu134, which frees Arg135 to interact with the G protein. Tyr306 (TM7) also swings from its position in ground state rhodopsin to form a hydrogen bond with Tyr223 via a water molecule (Figure 10.10B and C). Interestingly, in the M257Y mutant the phenolic side chain at 257 sandwiches between Tyr306 and Tyr223 and forms a hydrogen bond to Arg135, in this way enhancing the stability of the active conformation, which explains the constitutive activity of the mutant (Figure 10.10B). A hydrogen bond network extends from Tyr306 and Tyr223 to Arg135 and then to Cys347 of the GαCT. Thus Arg135, Tyr223 and Tyr306 are molecular switches that allow binding of the G protein. Trp265 of the CWxP motif (TM6) is another important activation switch that in the ground state restricts movement of TM6 through close interactions with 11-*cis*-retinal. Upon retinal isomerization Trp265 is released and facilitates opening of the G protein binding site. Proton transfer from the Lys296–retinal Schiff base to the counterion Glu113 breaks a link between TM7 and TM3, allowing relocation of the NPxxY and E(D)RY motif. From Trp265, as part of the retinal-binding site, to Tyr306 of the NPxxY motif and Arg135 of the E(D)RY motif near the cytoplasm, the molecular switches and hydrogen bond networks thus propagate rearrangements throughout rhodopsin in response to photoactivation.

Rhodopsin as prototype

The elegant structural biology of rhodopsin is supported by a massive amount of biochemical and biophysical data collected over decades. Sophisticated spectroscopic methods, including DEER and solid-state NMR, are giving new insights into conformational flexibility and dynamic interactions during rhodopsin activation. The extensive characterization of rhodopsin makes it a paradigm for other GPCRs. At present all GPCRs with crystal structures are in the rhodopsin family, including the human adenosine A_{2A} receptor, histamine H_1 receptor, dopamine D3 receptor, CXCR4 chemokine receptor,

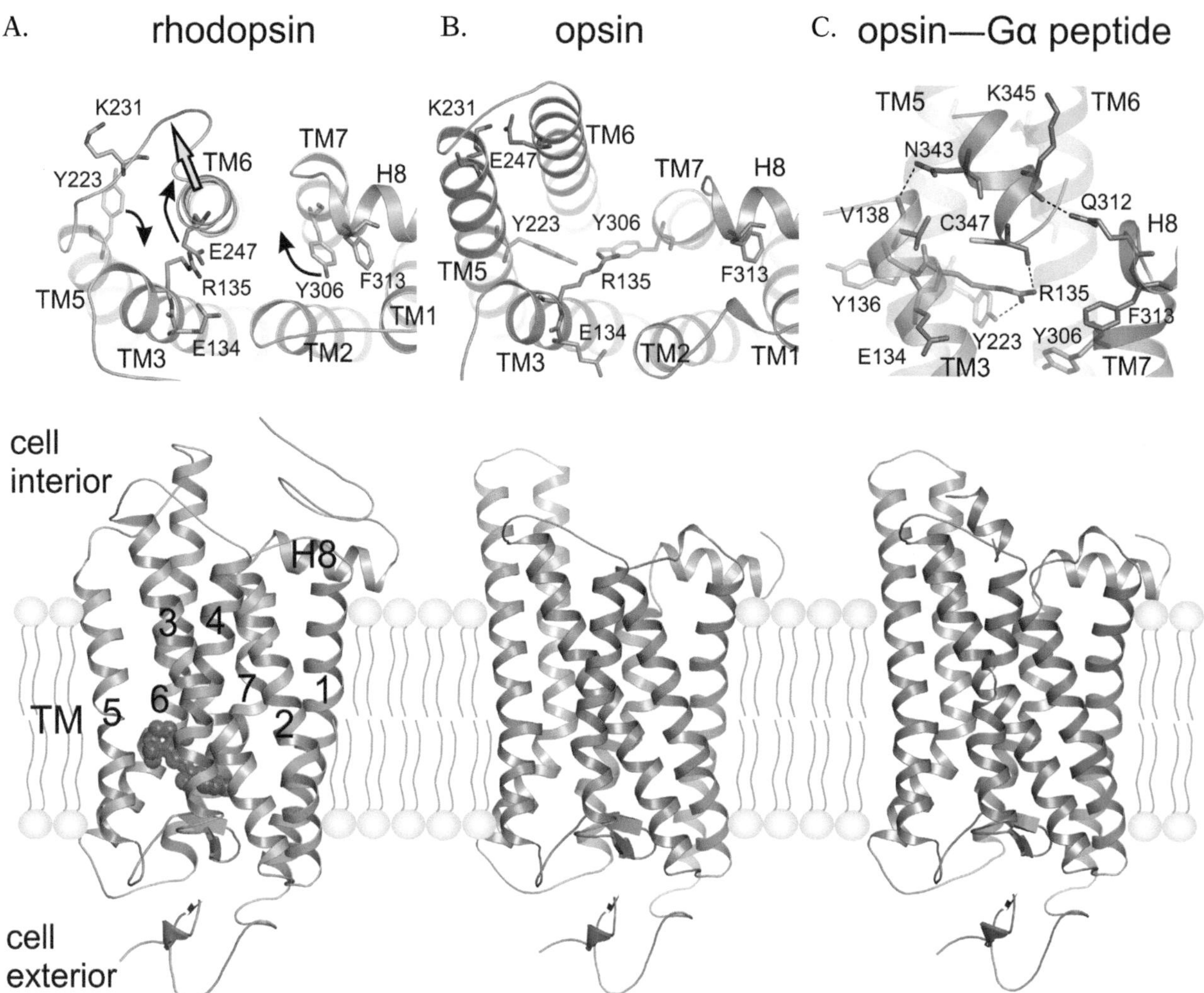

10.9 Crystal structures of ground state rhodopsin and active opsin (Ops*). Ribbon diagrams viewed from the cytoplasm and from the plane of the membrane with the cytoplasmic side at the top allow comparison of ground state rhodopsin (green), opsin (gold), and the complex of opsin (blue) with GαCT (magenta). Residues of the ionic lock between TM3 and TM6 (Glu134, Arg135, and Glu247) and residues that act as molecular switches (Arg135, Tyr223, Try306) plus key residues they interact with are labeled in the cytoplasmic views. (A) Activation of rhodopsin involves a large movement of TM6 (yellow arrow) as it tilts toward TM5, necessitating a break of the ionic lock with side chain movements (black arrows). (B) Rotations of Tyr223, Tyr306, and Arg135 side chains (molecular switches) allow them to form the floor of the cavity between TM5 and TM6 on one side and TM2, TM7, and H8 on the other. (C) A close-up cytoplasmic view reveals the binding of the C-terminal peptide from Gtα (GαCT) within the cavity of Ops* with hydrogen bonds to Arg135 of the E(D)RY motif, as well as Gln312 in the NPxxY(x)$_{5,6}$F motif. From Hoffman, K. P., *et al.*, *TIBS*. 2009, 34:540–552. © 2009, Elsevier. Reprinted with permission from Elsevier.

and the β-adrenergic receptors described next. Their overall architecture is strikingly similar (Figure 10.11A). Even the ordered water molecules are close to the locations seen in rhodopsin, suggesting conserved roles (Figure 10.11B). Functional importance of these water molecules is indicated by their lack of exchange with H_2O^{18} when rhodopsin is subject to radiolytic hydroxyl labeling.

A comparison of bovine rhodopsin, avian β_1-adrenergic receptor (β_1AR), human A_{2A} adenosine receptor (A_{2A}R), and human β_2-adrenergic receptor (β_2AR) shows similarities in the ionic lock region that links the E(D)RY motif in TM3 with E6.30 in TM6 (Figure 10.12A). This stabilizing network is conserved among all GPCRs in the rhodopsin family. Fewer polar interactions are seen in these regions in β_1AR, A_{2A}R, and β_2AR than in rhodopsin (see below), which may explain why they have higher basal levels of activity than rhodopsin.

At the extracellular surface most GPCRs have a ligand-binding site that is open, whereas the retinal-binding site in rhodopsin is a pocket, closed from the outside with a plug formed by two β hairpins (Figure 10.12B; also see Figure 10.4). Furthermore, most GPCRs can bind to several molecules and will respond differently

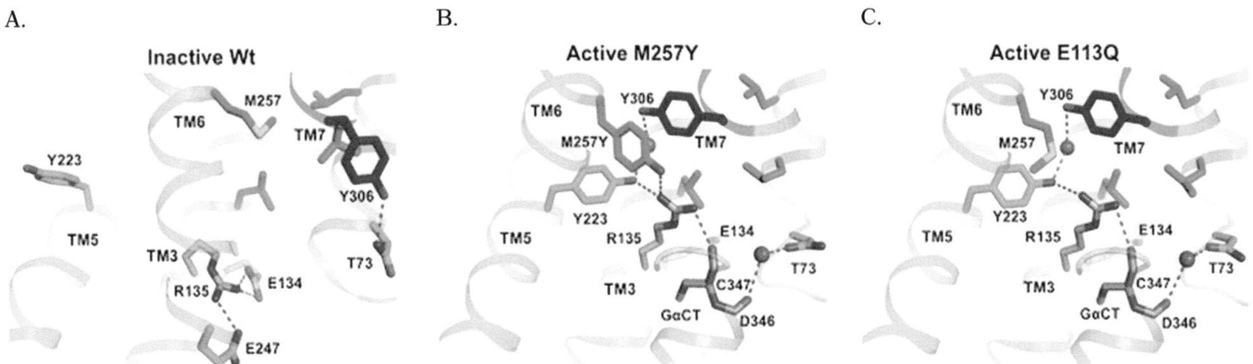

10.10 Ionic lock region in ground state and constitutive mutants of rhodopsin. View of the ionic lock region showing the NPxxY (dark blue) and E(D)RY (salmon) motifs close to the binding site for GαCT (orange). (A) Inactive rhodopsin shows the interactions between Arg135, Glu134, and Glu247 that hold TM6 close to TM3. In addition Tyr306 interacts with Thr73, and Met257 is in a hydrophobic region (green) separating the NPxxY motif from the ionic lock region. (B) The same region in the M257Y mutant shows the side chain of Arg135 extended into the space between TM3 and TM6, forming part of the binding site for GαCT. The phenolic ring at position 257 positions itself between Tyr223 and Tyr306, adding another hydrogen bond to Arg135 in its new position and stabilizing the active conformation in which TM6 is close to TM5. (C) In the E113Q mutant the ionic lock is similarly broken, as Arg135 forms a hydrogen bond to Tyr223. The hydrophobic region is opened, allowing entry of a water molecule (red). From Deupi, X., *et al.*, *Proc Natl Acad Sci USA*. 2012, 109:119–124.

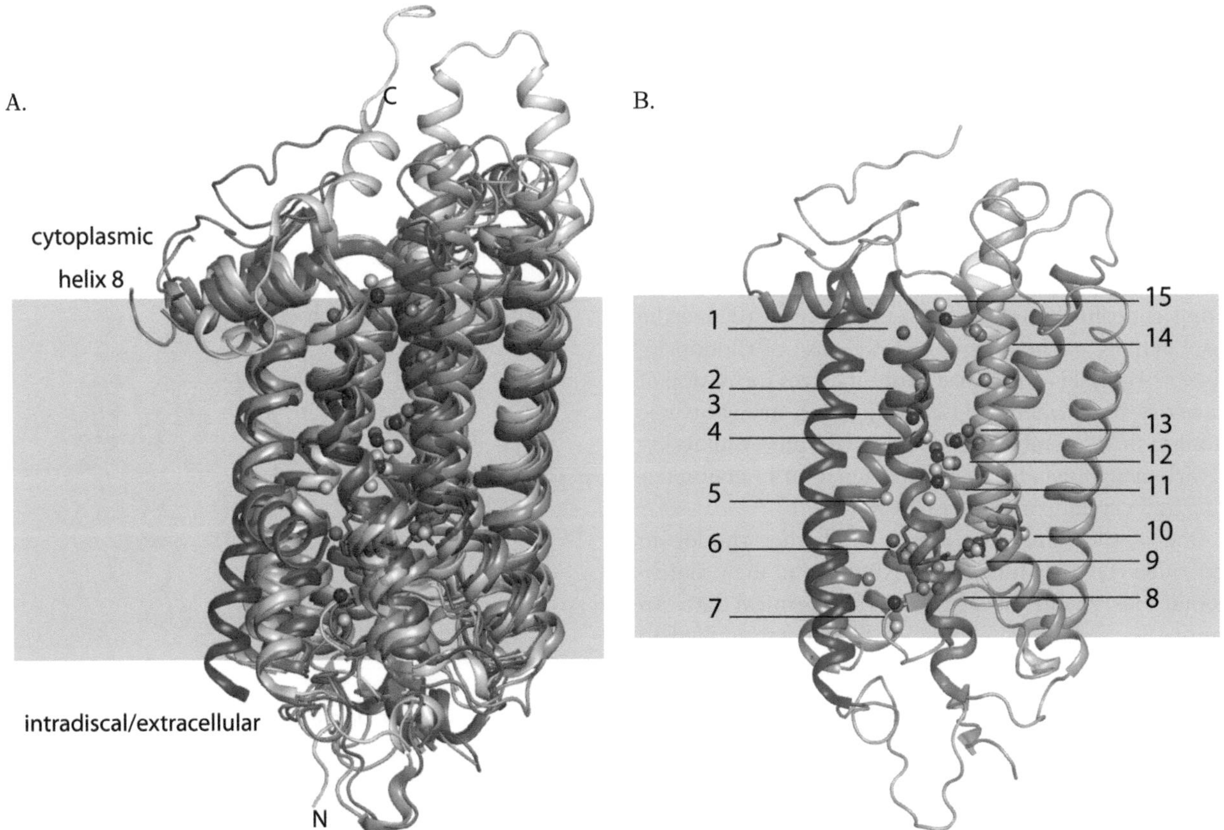

10.11 Comparison of several GPCRs. (A) The similarity of the overall structures of several GPCRs is evident in the superimposed ribbon diagrams of bovine rhodopsin (red), squid rhodopsin (wheat), β_1-adrenergic receptor (β_1AR, light blue), β_2-adrenergic receptor (β_2AR, dark blue), and bovine opsin (gray). (B) The ordered water molecules in the TM regions are co-localized, as seen on the structure of bovine rhodopsin (a ribbon diagram with rainbow helices colored as in Figure 11.4). The observed water molecules are depicted in spheres colored to indicate their source: bovine rhodopsin (red), squid rhodopsin (wheat), bovine opsin (light gray), bovine opsin with Gα peptide (dark gray), β_1AR (light blue), β_2AR (dark blue), and A_{2A} adenosine receptor (orange). Water molecules are localized to 15 regions (numbered), including several in extended hydrogen bond networks. (A) and (B) from Angel, T. E., *et al.*, *Proc Natl Acad Sci USA*. 2009, 106:8555–8560.

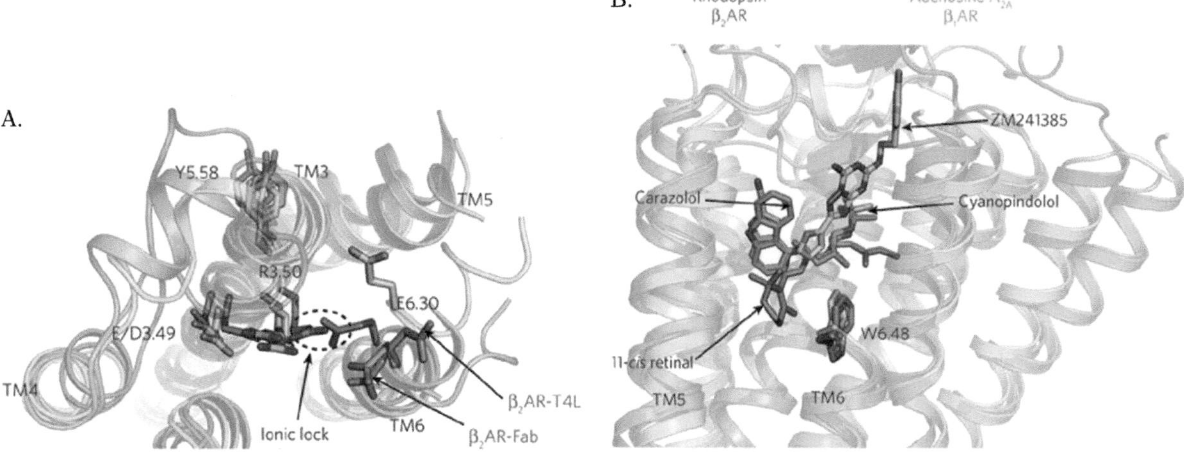

10.12 Comparison of the ionic lock and the ligand-binding sites in four different GPCRs. (A) The residues involved in the ionic lock (R3.50, E/D3.49, and E6.30) occupy very similar positions in bovine rhodopsin (purple), avian β_1AR (orange), human A_{2A} adenosine receptor (green), and β_2AR (blue), although the E6.30 rotamer varies in different crystals of β_2AR. (B) The ligand-binding pockets are shown occupied with ligands: 11-*cis*-retinal (pink) for rhodopsin, cyanopindolol (orange) for β_1AR, ZM241385 (green) for A_{2A} adenosine receptor, and carazolol (blue) for β_2AR. W6.48 (Trp265 in rhodopsin) interacts with all the ligands. (A) and (B) from Rosenbaum, D. M., *et al.*, *Nature*. 2009, 459:356–363. © 2009, Nature Publishing Group. Reprinted by permission of Macmillan Publishers Ltd.

to different ligands. The ligands are classified by their effects: agonists activate the receptor (stabilize the active conformation) and induce its response (mimicking the biological function of a natural molecule such as a hormone); antagonists bind the receptor and block other ligands while eliciting no response; partial agonists produce less than the maximal response even at saturating concentrations; and inverse agonists reduce the basal activity of the receptor. In the case of rhodopsin, all-*trans*-retinal is an agonist, since it drives formation of the active state, whereas 11-*cis*-retinal is a strong inverse agonist, since it stabilizes the inactive state. Availability of various agonists has been instrumental in characterizing and crystallizing other GPCRs.

One of the remaining issues is whether rhodopsin and other GPCRs function as dimers in their native membranes, as indicated by much biochemical data. An impressive variety of techniques have been applied to the question of rhodopsin oligomerization. When rhodopsin is visualized directly in native membranes with EM and AFM, dimers appear in higher-order arrays, which suggests that the dimers seen in some crystal structures may not be artifacts (Figure 10.13). A model was developed that pairs a ground state rhodopsin with a photoactivated rhodopsin in complex with transducin, and a complex consistent with this stoichiometry was purified in the absence of nucleotides. However, several reconstituted systems show activation when the rhodopsin:transducin ratio is 1:1. The question of GPCR dimerization is still open, in view of the structure of a complex of one β_2-adrenergic receptor with one G protein heterotrimer, described below.

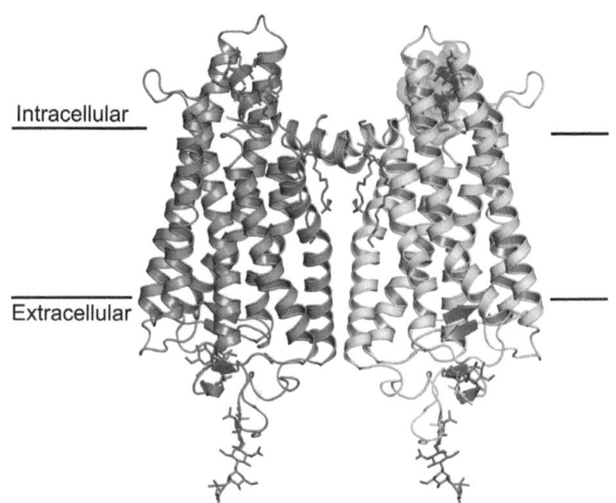

10.13 A crystallographic dimer of activated rhodopsin. A symmetric dimer is seen in the crystal structure when the activated form of rhodopsin, Ops*, is bound by the C-terminal fragment of Gα, GαCT. The Ops* molecules (orange and yellow) and the GαCT peptides (blue) are shown as ribbon diagrams. The GαCT side chains are shown as stick models, and one of the GαCT molecules is shown as a transparent space-filling model. The two antiparallel β-sheets in Ops* are shown as blue arrows, the oligosaccharides at Asn2 and Asn15 and the palmitoyl chains at Cys322 and Cs323 are shown as stick models (purple). From Scheerer, P., *et al.*, *Nature*. 2008, 455:497–502. © 2008, Nature Publishing Group. Reprinted with permission of Macmillan Publishers Ltd.

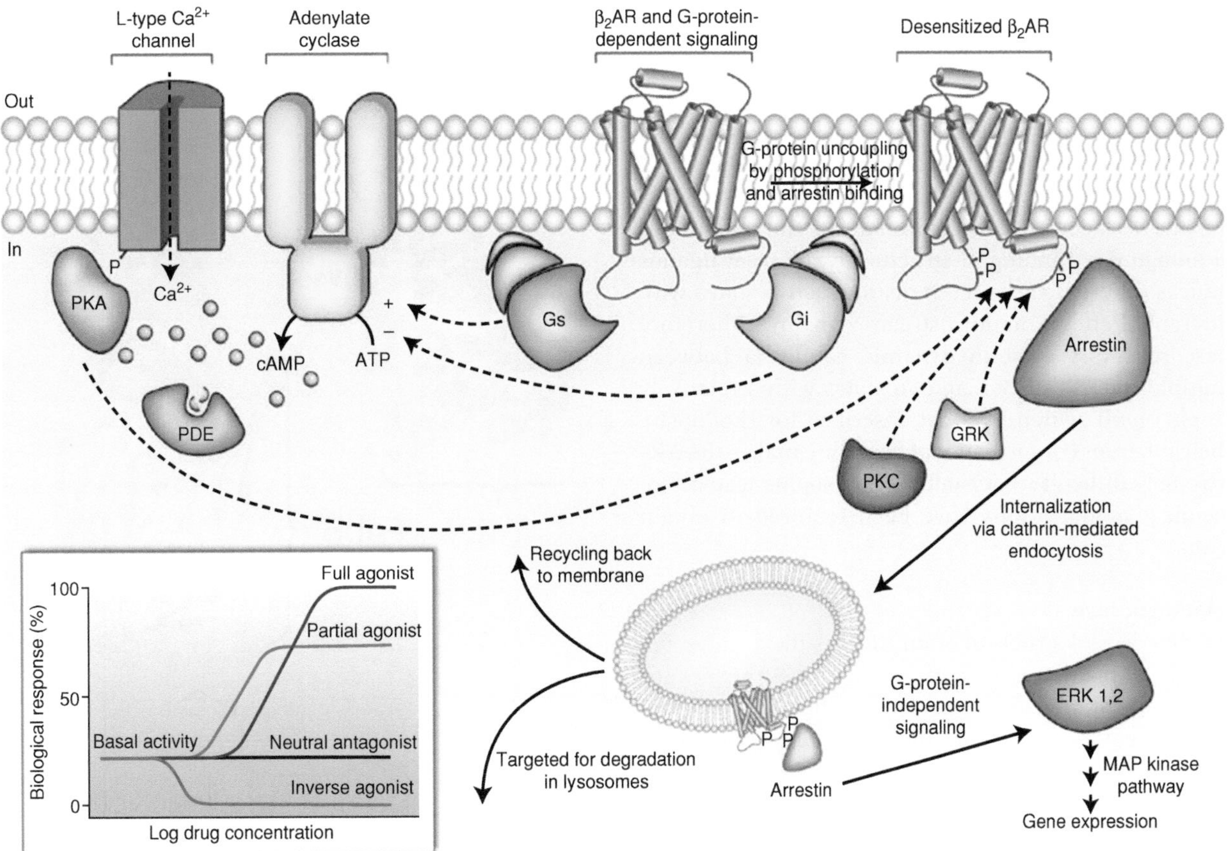

10.14 Role of β_2AR in signal transduction. β_2AR can regulate diverse signaling pathways through its interaction with two G proteins, G_s and G_i, that regulate adenylate cyclase. When G_s stimulates adenylate cyclase, the enzyme makes cAMP, which activates protein kinase A (PKA). PKA regulates β_2AR, as well as other proteins such as the L-type Ca^{2+} channel. Phosphodiesterase (PDE) cleaves cAMP to down-regulate these pathways. A G protein-coupled receptor kinase (GRK), as well as protein kinase C (PKC), can phosphorylate β_2AR, which allows arrestin binding. Arrestin blocks the activation of G proteins and stimulates extracellular signal-regulated kinases (ERK) and the internalization of the receptor in clathrin-coated pits. Inset shows the effects of agonists, partial agonists, and inverse agonists, as described in the text. From Rosenbaum, D. M., *et al.*, *Nature*. 2009, 459:356–363. © 2009, Nature Publishing Group. Reprinted by permission of Macmillan Publishers Ltd.

Adrenergic receptors

Some of the best-characterized metabotropic GPCRs are adrenergic receptors that respond to adrenaline (epinephrine) and noradrenaline to transmit signals from the sympathetic nervous system to the cardiovascular system, triggering the increased heart rate and mobilization of energy known as the fight-or-flight response. Adrenergic receptors (ARs) fall into two groups, α and β, with nine subtypes that differ in their locations and effects. Like all GPCRs, when ARs bind ligands they undergo conformational changes that allow the activated receptor to bind a G protein heterotrimer, exchange its GDP for GTP, and send signals into the cell (Figure 10.14). As described above for rhodopsin, ARs are regulated by phosphorylation and dephosphorylation, by binding arrestin, and by internalization.

AR subtypes also differ in the G proteins they bind when activated. For example, activated α_1-adrenergic receptors bind Gq and increase Ca^{2+} uptake, which stimulates smooth muscle contraction, while activated α_2-adrenergic receptors bind G_i, which inhibits adenylyl cyclase and causes smooth muscle contraction. Activated β_1-, β_2-, and β_3-adrenergic receptors bind G_s, which activates adenylyl cyclase and causes heart muscle contraction, smooth muscle relaxation, and glycogen breakdown. β_2AR can also bind G_i in localized tissues.

With crucial roles in cardiovascular and pulmonary physiology, β-adrenergic receptors have many agonists and antagonists that have medical applications. For example, the asthma drug albuterol (salbutamol) is a β_2AR agonist. Beta blockers long used to treat heart disease and hypertension such as propranolol and metoprolol are βAR antagonists. Unofficially beta blockers are used to reduce anxiety, and they had to be banned from the Olympics. Many other ligands of these receptors are not in use clinically but are useful in the laboratory. Besides adrenaline and noradrenaline, β_2AR responds to other agonists, such as formoterol and isoproteremol, and partial agonists like salbutamol. Carazolol, timolol, and the highly selective ICI-118,551 are inverse agonists, and alprenolol is an example of an antagonist. Binding of structurally different ligands induces different states, even distinct active states with differential effects on downstream signaling. Therefore these receptors exist in dynamic equilibria between multiple states, rather than two states with a relatively simple on/off switch like that described for rhodopsin. Their inherent conformational flexibility makes the ARs extremely difficult to crystallize, and sophisticated biochemical modifications have been required to obtain crystals.

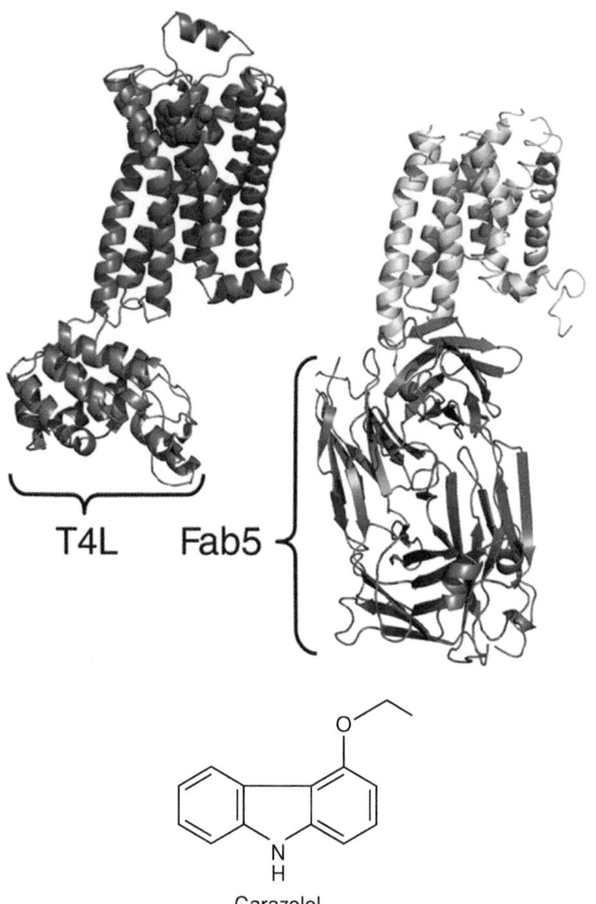

10.15 Crystal structures of β_2AR bound to carazolol. A comparison of the crystal structures of β_2AR bound to Fab5 (yellow) and fused with T4L (blue) shows good agreement between them. The extracellular portions of β_2AR bound to Fab5 are not resolved, and the N terminus is missing from the β_2AR–T4L structure. Carazolol (inset) is modeled to fit the electron density in the β_2AR–T4L fusion protein as the S-(-)-isomer (red spheres). From Rosenbaum, D. M., *et al.*, *Science*. 2007, 318:1266–1273. © 2007, American Association for the Advancement of Science. Reprinted with permission from AAAS.

β_2AR structure

The first crystal structure of an AR was the x-ray structure of the β_2-adrenergic receptor (β_2AR). Strategies for its crystallization had to overcome the flexibility of cytoplasmic regions, especially the C terminus and intracellular loop (ICL3) involved in G protein recognition. One method utilized Fab fragments from a monoclonal antibody specific for ICL3 in the native protein (Fab5), and the other removed ICL3 and replaced it with T4 lysozyme (T4L) (Figure 10.15). In both, β_2AR was truncated about 50 amino acids from the C terminus (corresponding to the location of the C-terminal end of rhodopsin), deglycosylated, and bound to the inverse agonist carazolol. The x-ray structure of β_2AR with Fab5 bound to ICL3 was solved with an anisotropic resolution of 3.4/3.7 Å, quickly followed by a more complete structure of β_2AR–T4L at 2.4 Å resolution.

The β_2AR structure shows the expected seven TM helices with the eighth helix parallel to the cytoplasmic face of the membrane, and can be closely superimposed with the structure of rhodopsin (Figure 10.16). Significant differences are seen in the extracellular region above the ligand-binding pockets, where extracellular Loop 2 (ECL2) is more exposed to solvent compared to how it is buried in rhodopsin and is stabilized by an extra disulfide bond between Cys184$^{4.76}$ and Cys190$^{5.29}$. (The typical disulfide that ties ECL2 to TM3 is Cys191$^{5.30}$–Cys106$^{3.25}$.) Instead of the β-hairpin in rhodopsin, ECL2 has an extra helix that enables it to link the ends of TM3, TM4, and TM5 without completely closing off the binding site. This binding surface contains residues identified by mutagenesis as having roles in agonist binding, such as Asp113, Val114, Phe289, Phe290, and Asn312. Carazolol fits snugly in the site (Figure 10.17), which is consistent with its low K_d (<0.1 nM) and slow off-rate. The tight fit suggests a flexibility that allows induced fit, as NMR data show clear conformational differences in the site when agonists and induced agonists bind.

The cytoplasmic ends of TM3 and TM6 are more open in β_2AR than in rhodopsin, which precludes formation of an ionic lock between Arg131$^{3.50}$ and Glu268$^{6.30}$. This conformational difference may explain the higher level of basal activity of β_2AR-carazolol when compared to the lack of activity of rhodopsin in the dark, as the arrangement is more similar to the positions of TM3 and TM6 in the activated opsin structure (Ops*) described above. While there is much evidence from biochemical and biophysical studies that β_2AR forms dimers in cells, in the x-ray structures with Fab and T4L present the receptor is a monomer. Interestingly, it is also a monomer in the structure of activated β_2AR in complex with a G protein (see below).

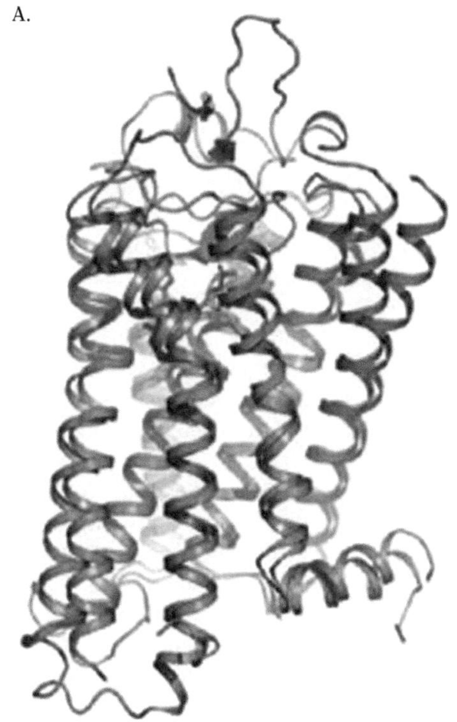

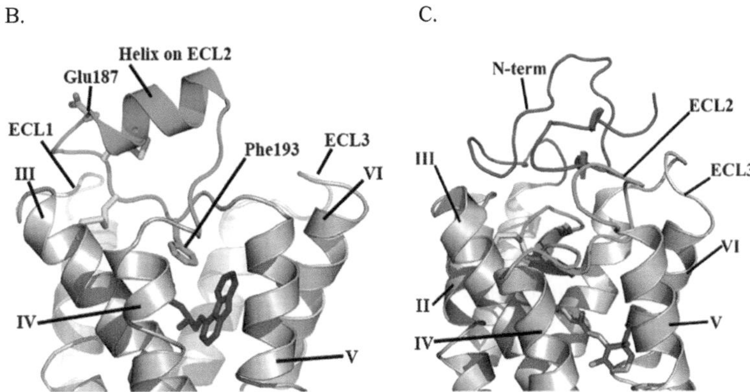

10.16 Comparison of the structures of β_2AR and rhodopsin. (A) Superposition of the ribbon diagrams for β_2AR (blue) and bovine rhodopsin (purple) show close alignment of the TM helices when viewed from the plane of the membrane. The largest differences are in the helix ends and connecting loops, especially at the extracellular end (top). From Rosenbaum, D. M., *et al.*, *Nature*. 2009, 459:356–363. © 2009, Nature Publishing Group. Reprinted by permission of Macmillan Publisher Ltd. Close-up views of the extracellular ends of rhodopsin (B) and β_2AR–T4L (C) show the striking differences in ECL2 (green). In β_2AR–T4L ECL2 is rigidified by the helix and two disulfide bonds (yellow), which keeps it out of the ligand binding site. The space above the carazolol (blue) is quite open, with a single interaction between it and Phe193 on ECL2. In contrast, in rhodopsin ECL2 is lower and forms a β-hairpin that plugs the access to site over retinal (pink). Rhodopsin has the conserved disulfide bond between ECL2 and TM3. The N terminus (purple on rhodopsin) is missing from β_2AR–T4L. From Cherezov, V., *et al.*, *Science*. 2007, 318:1258–1265. © 2007, American Association for the Advancement of Science. Reprinted with permission of AAAS.

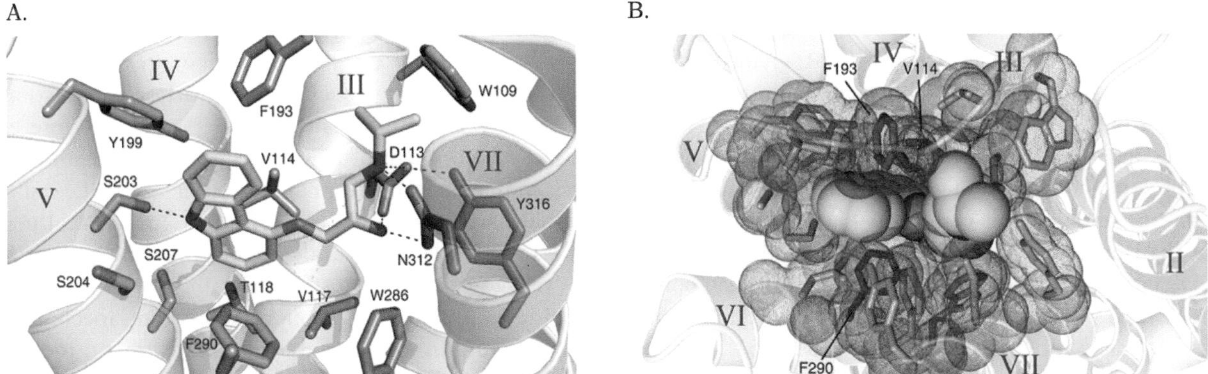

10.17 Ligand binding pocket of β_2AR–TL4 with carazolol bound. (A) Most residues within 4 Å of carazolol (yellow) are shown as sticks, highlighting those making polar interactions (green) (oxygens are red, nitrogens blue). (B) Packing interactions are shown in a view from the extracellular side with carazolol (yellow spheres) nestled snugly among the side chains of nearby residues (sticks within van der Waals dot surfaces). The nonpolar residues Val114$^{3.33}$, Phe193$^{5.32}$, and Phe290$^{6.52}$ (red) make a hydrophobic sandwich around the ligand. (A) and (B) from Rosenbaum, D. M., *et al.*, *Science*. 2007, 318:1266–1273. © 2007, American Association for the Advancement of Science. Reprinted with permission from AAAS.

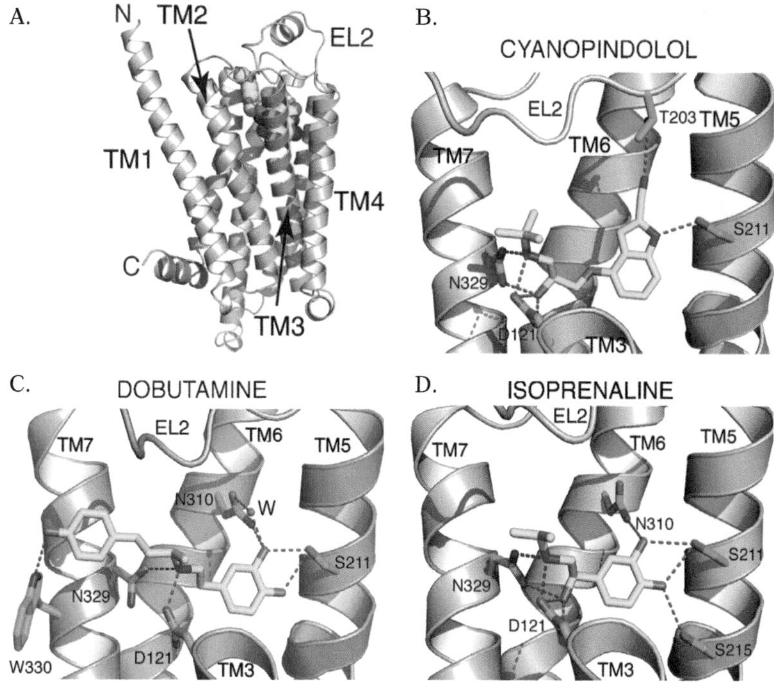

10.18 Structures of the β_1-adrenergic receptor bound to different ligands. (A) A ribbon diagram of the overall structure of β_1AR highlights the location of the binding pocket, bound to the ligand carmoterol (space-filling model). The remaining panels show the binding sites for different ligands (stick models with the same colors) with the side chains they interact with (stick models with C, green; O, red; N, blue) with residues 94–119 and 171–196 removed for clarity. The ligands shown are: (B) the antagonist cyanopindolol; (C) the partial agonist dobutamine; (D) the full agonist isoprenaline, which has an extra hydrogen bond between the ligand and Ser215. From Warne, T., *et al.*, *Nature*. 2011, 469:241–244. © 2011, Nature Publishing Group. Reprinted by permission of Macmillan Publishers Ltd.

β_1AR structure

The human β_1 and β_2 subtypes of the adrenergic receptor are 67% identical in their TM regions and are almost identical in the residues surrounding the ligand-binding pocket, yet they have quite different affinities for a variety of ligands. The question of ligand specificity has been addressed in studies of the β_1-adrenergic receptor (β_1AR). Because the human β_1AR is very unstable in detergent, the x-ray structure of a thermo-stabilized β_1AR from turkey was determined at 2.7 Å resolution. The construct, β_1AR-m23, carries six mutations that do not change its affinity for antagonists while lowering its affinity for agonists, indicating it is more stabilized in the inactive state; however, with agonist bound it is still capable of coupling with G protein. First crystallized with the antagonist cyanopindolol bound, the structure of β_1AR is highly similar to that of β_2AR, giving verification that the β_2AR structure is not significantly perturbed by either Fab5 or T4L.

β_1AR-m23 has now been crystallized bound to many different ligands, including the additional antagonists iodocyanopindolol and carazolol, the agonists carmoterol and isoprenaline, and the partial agonists salbutamol and dobutamine. All of the ligands make hydrogen bonds with Asp121$^{3.32}$, Asn329$^{7.39}$, and Ser211$^{5.42}$. In addition, full agonists make a hydrogen bond with Ser215$^{5.46}$ (Figure 10.18). The only other significant difference is a slight narrowing of the pocket when agonists are bound. Similar results are obtained with bucindolol and carvedilol (not shown), members of a special class of agonists that behave differently with regard to activation of G protein-dependent and G protein-independent pathways. In all cases, the observed conformational changes are small and do not bring the receptor to the fully activated state. However, the structures vary in the cytoplasmic end of TM6, which is either bent or straight (Figure 10.19). In the molecules with TM6 bent, the salt bridge between Arg139$^{3.50}$ and Glu285$^{6.30}$ makes the ionic lock seen in rhodopsin, whereas the lock is "broken" due to a longer distance between TM3 and TM6 in molecules with TM6 in the straight conformation. Most likely the absence of the ionic lock, also seen in the structure of β_2AR-carazolol, facilitates a higher basal activity of the receptor.

With nine new x-ray structures of GPCRs in the year 2012 alone, there are now enough examples to survey their differences. Due to variations in both extracellular loops and TM regions the diversity falls in the outer half when the common seven TM fold is divided along the middle of the bilayer. This is not surprising since this half contains the ligand-binding sites that enable GPCRs to carry out such a wide variety of functions (Figure 10.20). These sites are of paramount interest to the development of new drugs, such as an agonist of the sphingosine-1-phosphate receptor I (S1pI), called fingolimod, recently approved for the clinical treatment of multiple sclerosis. On the other hand, the half of the protein toward the cytosol, where the G proteins bind, is quite conserved and supports a common mechanism of signaling. In a remarkable accomplishment, the first steps of that mechanism have now been elucidated in the structure of an active GPCR bound to its G protein.

Activated β_2AR in complex with G_S

The structure of the ternary complex of agonist-bound β_2AR coupled to a heterotrimeric G protein provides a spectacular first view of a GPCR activated for signaling

TM7
TM2
TM3
TM5
T76
D138
R139
CL2
TM6
Y149
E285
H286
Q237
CL3
H8
A.
B.

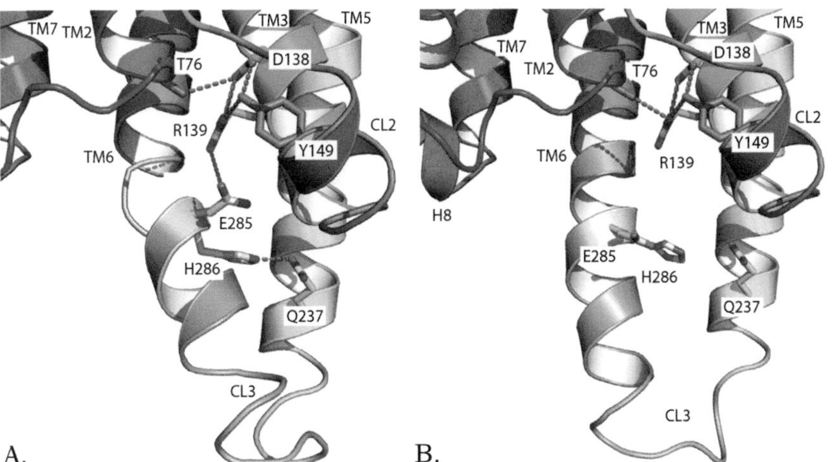

10.19 Effect of ligand binding on TM6 and the ionic lock in β_1AR. (A) The cytoplasmic portion of β_1AR (ribbon representation with rainbow coloring based on the entire molecule, N terminus blue and C terminus red, except the cytoplasmic extension of TM6, gray) contains several residues (sticks) that make interhelical hydrogen bonds. When TM6 is bent near the cytoplasmic face, the ionic lock between Arg139 and Glu285 is intact. (B) When binding of certain ligands (see text) straightens TM6, the ionic lock is broken. From Moukhametzianov, R., *et al.*, *Proc Natl Acad Sci USA*. 2011, 108: 8228–8232.

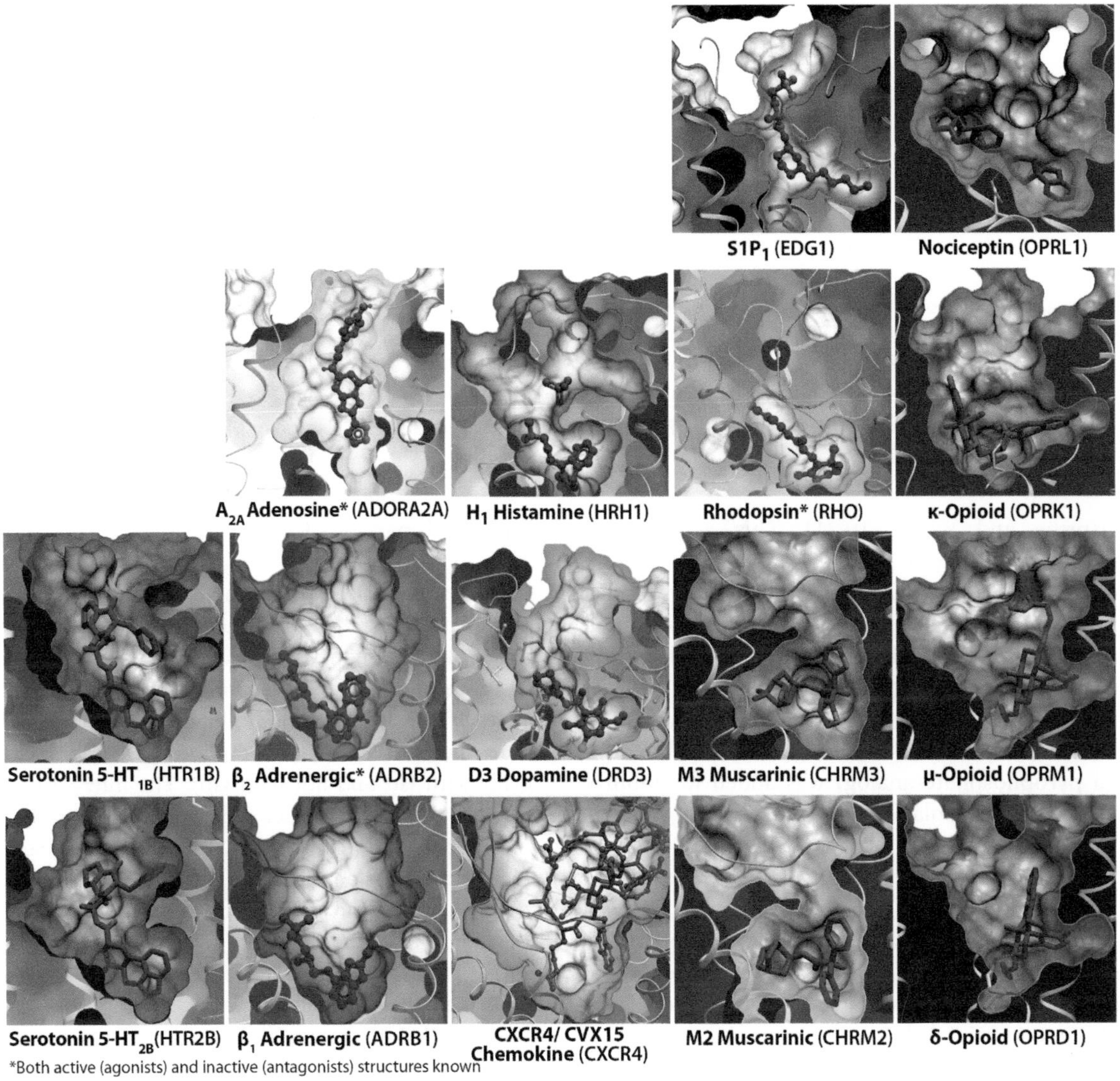

S1P1 (EDG1)
Nociceptin (OPRL1)
A2A Adenosine* (ADORA2A)
H1 Histamine (HRH1)
Rhodopsin* (RHO)
κ-Opioid (OPRK1)
Serotonin 5-HT1B (HTR1B)
β2 Adrenergic* (ADRB2)
D3 Dopamine (DRD3)
M3 Muscarinic (CHRM3)
μ-Opioid (OPRM1)
Serotonin 5-HT2B (HTR2B)
β1 Adrenergic (ADRB1)
CXCR4/ CVX15 Chemokine (CXCR4)
M2 Muscarinic (CHRM2)
δ-Opioid (OPRD1)
*Both active (agonists) and inactive (antagonists) structures known

10.20 Diversity of binding pockets in GPCRs. The binding sites for ligands (molecular surfaces highlighted with various colors) are shown from the same orientation in 16 different GPCRs. Related GPCRs are grouped by colored frames and show more similar binding pockets. For comparison the states shown are the inactive states with antagonists (stick models) bound, even in cases where the active state structures are also known. While most are open to the extracellular solvent, two binding pockets are buried, that for retinal (rhodopsin) and for sphingolipid (sphingosine-1-phosphate receptor I, $S1P_I$). Reprinted with permission from Raymond C. Stevens. A similar figure appears in Katritch, V., *et al.*, *Ann Rev Pharm Tox*. 2013, 53:531–556.

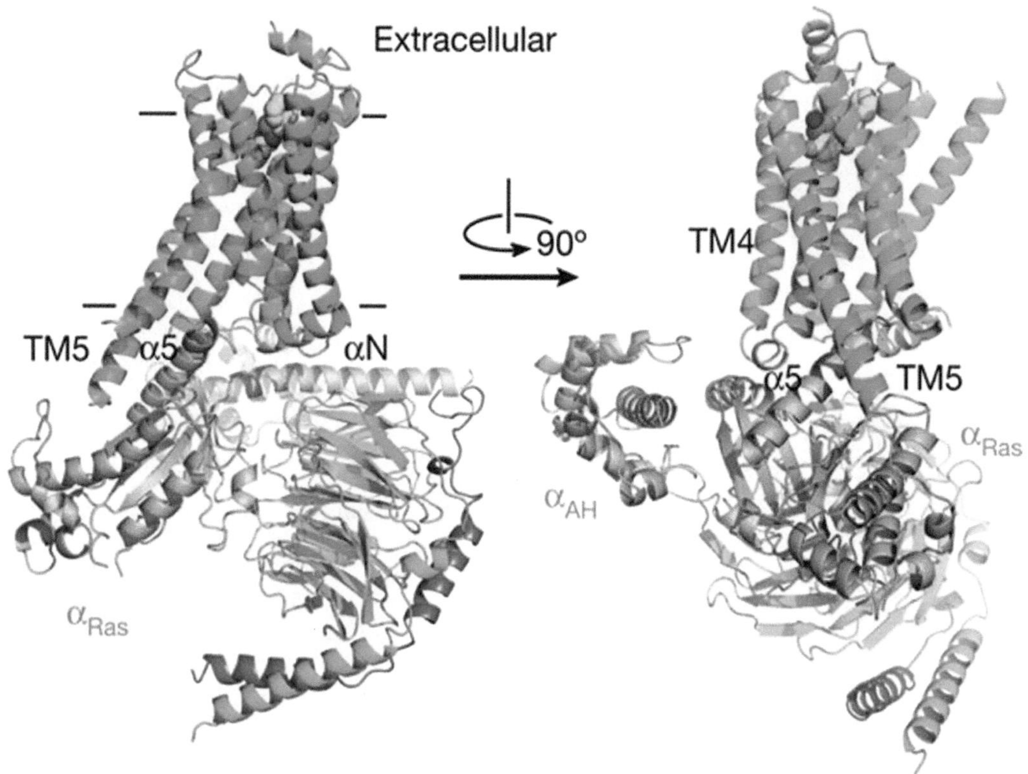

10.21 Overall structure of a ternary complex of β_2AR, carazolol, and G_s. Two views of the crystal structure of the ternary complex omitting crystallization aids (see Frontispiece) show β_2AR (green), the agonist BI-167107 (space-filling model with C, yellow; O, red; N, blue), and the G_s heterotrimer. The two domains of Gαs (gold) are widely spread, with α_{Ras} making extensive contacts with the receptor and α_{AH} not interacting with it. The Gβ (cyan) and Gγ (blue) subunits also do not contact β_2AR. From Rasmussen, S. G. F., *et al.*, *Nature*. 2011, 477:549–555. © 2011, Nature Publishing Group. Reprinted by permission of Macmillan Publishers Ltd.

(Frontispiece and Figure 10.21). This achievement required several tricks to attain a stable complex that crystallized well. β_2AR bound to a high-affinity agonist called BI-167107 was added to GDP-G_s in dodecylmaltoside micelles and later doped with the new detergent maltose neopentyl glycol (MNG-3). Because both GDP and GTP disrupt the interaction between β_2AR and G_s, the GDP was removed with apyrase, and phosphonoformate (a pyrophosphate analog) was added to stabilize G_s. To increase the polar surface for packing contacts in the crystal lattice, the unstructured N terminus was replaced with T4 lysozyme. Finally, a single-chain antibody, or nanobody (Nb35), to the whole complex was obtained from llamas. Examinations of the complexes with single-particle electron microscopy guided these extensive steps toward obtaining crystals of T4L-β_2AR-G_s-Nb35 that gave 3.2 Å diffraction.

The x-ray structure of the ternary complex gives insight into the changes in β_2AR from its inactive state, as well as the changes in activated G_s and their interaction. When active β_2AR in the complex is compared with the structure of carazolol-bound β_2AR, the largest changes occur at the intracellular ends of TM5 and TM6, which move outward (Figure 10.22). In addition, ICL3 becomes disordered while part of ICL2 forms an α-helix. The extracellular portion of the ternary complex is not well defined, and the position of BI-167107 is modeled based on a different crystal structure of β_2AR stabilized in an active state by a G protein mimetic nanobody (Nb80) that has similar positions of the active site residues except Arg131, which interacts with the nanobody (Figure 10.22B).

In the active ternary complex, β_2AR interacts extensively with Gα and not with G$\beta\gamma$. The interaction site is not defined by a consensus sequence for G_s-coupling, perhaps because β_2AR interacts with more than one G protein. At the interface TM5 and TM6 of β_2AR move aside, making a cavity for insertion of the α5 helix of Gα, while Phe139 on ICL2 inserts into a hydrophobic pocket of Gα (Figure 10.23). The interactions between TM2, TM3, and ICL2 that position ICL2 include Asp130 of the E(D)RY motif on TM3, making a structural link between it and the G protein. Of the two domains in the Gα subunit (the Ras domain and a helical domain called AH) the interaction with β_2AR involves groups of the Ras domain: the α5 helix, αN–β1 junction, the top of the β3 strand and the α4 helix. The α5 helix is thrust into the cavity of β_2AR. The activated Gα shows large conformational changes when compared to the structure of GTPγS-bound G_s, with a rotation

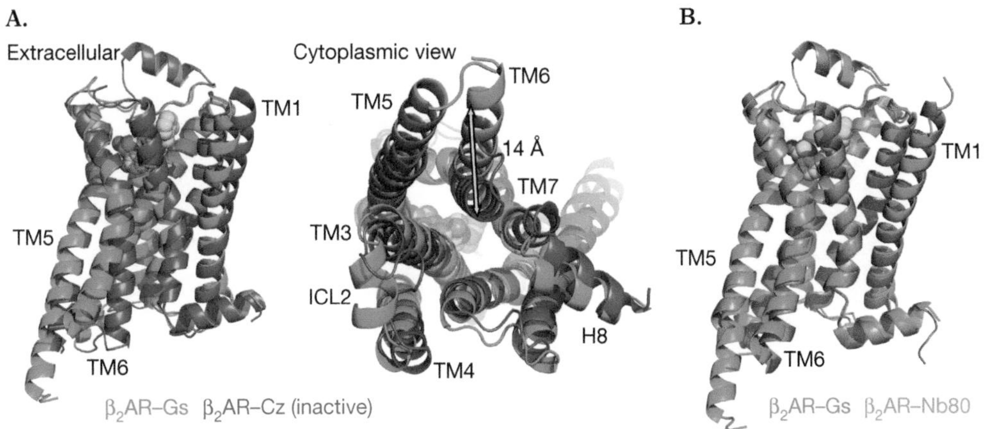

10.22 Comparison of active and inactive β_2AR structures. (A.) The ribbon diagrams for β_2AR from the active β_2AR-Gs structure (green) and the inactive β_2AR-carazolol structure (blue) are superimposed and viewed from the side and from the cytoplasm. With activation TM5 is extended by two helical turns, and TM6 is moved outward by 14 Å measured at Glu268 (yellow arrow in cytoplasmic view), comparable to the changes observed in Ops*. (B.) Superposition of the structure of active β_2AR-Gs structure (green) with the structure of active β_2AR-Nb80 structure (orange) shows that the crystallization aids did not have a significant effect on either the extracellular portion (at T4L) or the cytoplasmic G-binding surface (near Nb35) in the ternary complex. The ligand-binding site from β_2AR-Nb80 was better resolved, which allowed modeling BI-167107 in β_2AR-Gs. From Rasmussen, S.G.F. *et al.*, *Nature.* 2011, 477:549–555. © 2011, Nature Publishing Group. Reprinted by permission of Macmillan Publishers Ltd.

of nearly 130° of the AH domain from the Ras domain, which is consistent with electron resonance (DEER) measurements in light-activated rhodopsin–G_i complexes (Figure 10.24).

The crystal structure of agonist-bound β_2AR G protein complex gives an exciting snapshot of the signaling process between an activated GPCR and its G protein. In spite of the several modifications made to β_2AR, its overall structure is supported by studies of the unmodified receptor using hydrogen–deuterium exchange mass spectrometry. The interaction of Gα with β_2AR in the active ternary complex fits well with the position of the short peptide Gα CT in complex with Ops*. The large conformational change in Gα confirms a prediction made decades ago in the clam-shell model for activation of G proteins. A single snapshot leaves many unanswered questions, such as how different ligands induce different states in β_2AR, what determines selectivity towards G proteins, and how GPCR kinases and arrestins regulate the receptor. Nonetheless the structure of the ternary complex has broad implications for a common understanding of transmembrane signaling by GPCRs.

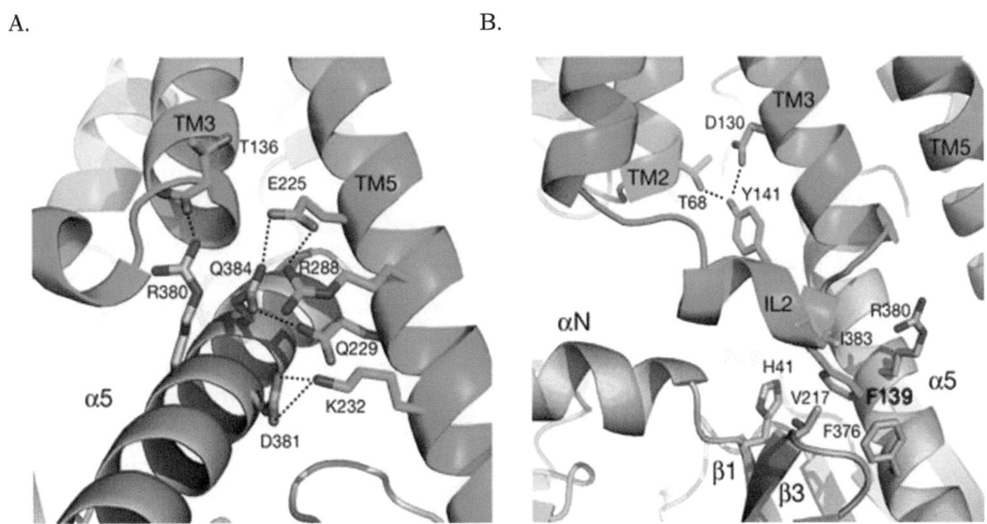

10.23 Interactions between β_2AR and Gα in the ternary complex. (A) A close-up view of the α5 helix of Gαs (gold) docking into the cavity on the cytoplasmic side of the receptor (green) shows specific interactions between residues of the α5 helix and residues in TM5 and TM3. (B) When the α5 helix enters the crevice in β_2AR, Phe139 on ICL2 of the receptor docks into a hydrophobic pocket on the Gα protein. ICL2 is positioned by interactions between Tyr141 and Thr68 (TM2) and Asp130 of the E(D)RY motif (TM3). From Rasmussen, S. G. F., *et al.*, *Nature*. 2011, 477:549–555. © 2011, Nature Publishing Group. Reprinted by permission of Macmillan Publishers Ltd.

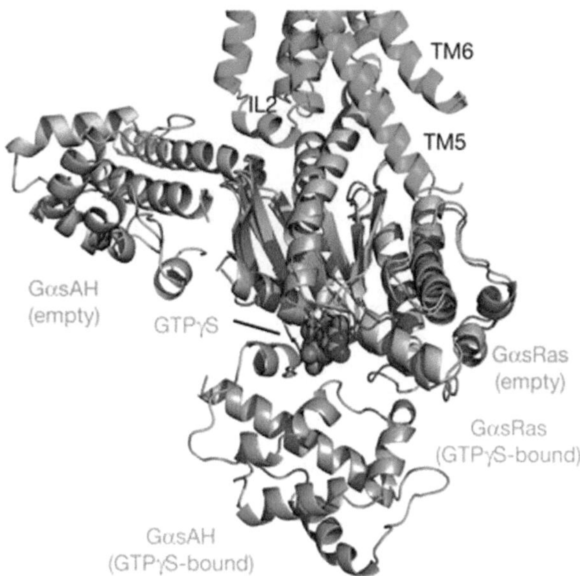

10.24 Conformational changes in activated $G\alpha$ in the ternary complex. A large conformational change takes place in $G\alpha s$ upon activation, as seen in a comparison of $G\alpha s$ in the ternary complex (gold) with GTPγS-bound $G\alpha s$ (gray, with GTPγS shown as spheres). As $\alpha 5$ enters the crevice in β_2AR, the rest of the Ras domain is little changed, while the helical AH domain rotates to the side. From Rasmussen, S. G. F., *et al.*, *Nature*. 2011, 477:549–555. © 2011, Nature Publishing Group. Reprinted by permission of Macmillan Publishers Ltd.

NEUROTRANSMITTER RECEPTORS

The synaptic junction between neurons is especially dense in membrane proteins involved in converting an electrical signal, the action potential, to a chemical signal that can diffuse across the cleft to depolarize or hyperpolarize a neighboring cell. Specific receptors in the postsynaptic membrane bind the released neurotransmitters to transmit their message, while specific transporters in the presynaptic membrane mop up the excess from the synaptic fluid (see "Neurotransmitter Transporters" in Chapter 11). Receptors in the nervous system are either ionotropic, the ligand-gated ion channels, or metabotropic, which link indirectly to ion channels via signal transduction mechanisms that can involve GPCRs (see previous section). The x-ray structures of two very different glutamate receptors provide exciting new insights into the function of ligand-gated ion channels.

Glutamate receptors: GluA2

Glutamate is the predominant neurotransmitter for excitatory synapses in the brain. The family of ionotropic glutamate receptors (iGluRs) is divided into subtypes named for their pharmacological agonists: the GluA subtype

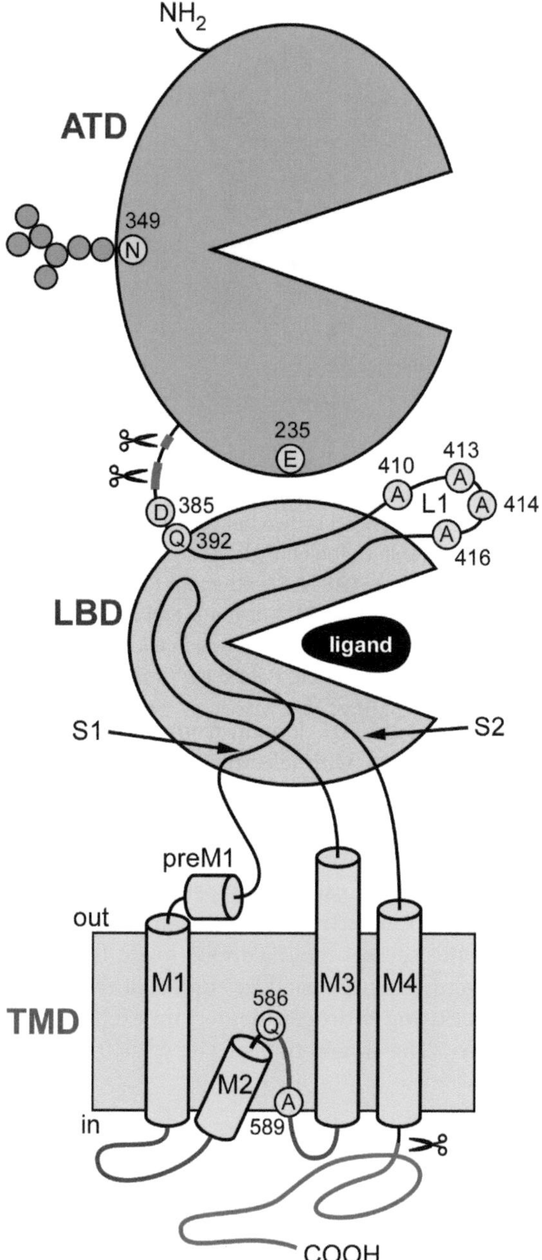

10.25 Illustration representation of a subunit of iGluRs. Amino-terminal domain (ATD), ligand-binding domain (LBD), and transmembrane domain (TMD) are shown in lavender, orange, and light green, respectively. Changes that were made to achieve crystals of the GluA2 construct include deletions in the ATD–LBD linker and of the C terminus (highlighted in pink) and point mutations in the LBD Loop L1 (K410A, E413A, M414A, and E416A) and TMD segment M2 (R586Q and C589A) (highlighted in cyan). Disordered loops M1–M2 and M2–M3 that show no clear density in GluA2 crystal structure are colored blue. Predicted glycosylation sites are highlighted in yellow, three of them mutated (N235E, N385D, and N392Q) and one (N349) showing electron density for *N*-linked carbohydrates (green). From Sobolevsky, A. I., *et al.*, *Nature*. 2009, 462:745–758, supplementary figure 1. © 2009, Nature Publishing Group. Reprinted by permission of Macmillan Publishers Ltd.

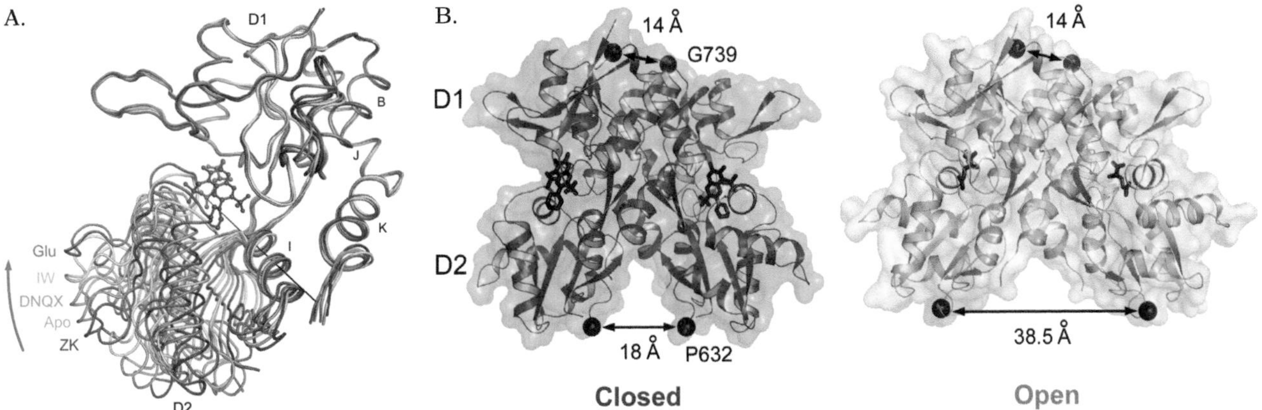

10.26 LBD structures co-crystallized with different ligands. (A) The structure of the isolated LBD in the apo state (cyan) is superimposed with that for complexes with antagonist DNQX20 (6,7-dinitro-2,3-quinoxalinedione, green), partial agonist IW23 (5-I-Willandiine, yellow) and full agonist glutamate (red). The red arrow indicates the clam shell closure as domain 2 (D2) approaches domain 1 (D1) of the isolated LBD. The structure of the LBD in GluA2 bound to ZK (ZK200775, blue) is the most open. (B) Extent of opening is compared for the LBD dimers from GluA2 bound to ZK (purple) and the isolated LBD dimer bound to glutamate (light green). While the distance between Gly739 residues (black spheres at the top) is unchanged, the distance between Pro632 (black spheres at the bottom) is more than doubled, comparing the LBDs bound to glutamate (stick models on light green molecule) with the LBDs bound to ZK (stick models on purple molecule). From Sobolevsky, A. I., *et al*., *Nature*. 2009, 462:745–758, supplementary figures 30 and 31. © 2009, Nature Publishing Group. Reprinted by permission of Macmillan Publishers Ltd.

sensitive to AMPA (α-amino-3-hydroxy-methyl-4-isoxazole propionic acid), GluR, sensitive to kainate, and GluN, sensitive to NMDA (N-methyl-D-aspartate). The iGluRs have different functions but share a common architecture. Each subunit of these tetramers has a large extracellular amino-terminal domain (ATD), an extracellular central ligand-binding domain (LBD), a transmembrane domain (TMD), and a cytoplasmic C-terminal domain that varies in size (Figure 10.25). The modular nature of these domains allows genetic excision of the separate regions for purification of each water-soluble domain.

Numerous crystal structures of LBDs from AMPA iGluRs in combination with different ligands reveal conserved elements of the ligand-binding site and show that agonist efficiency is directly coupled to the extent of domain closure (Figure 10.26). A crystal structure of the ATD from GluA2 shows a dimer that clearly lacks a site for binding amino acids (not shown). With these structures on hand for molecular replacement, the crystal structure of a modified GluA2 from rat in complex with the competitive antagonist ZK200775 (ZK) was solved at 3.6 Å resolution. The structure shows the entire molecule except the portions on the cytoplasmic side of the membrane.

The GluA2 receptor is a dimer of dimers, with its modular domains in layers that broaden out from a narrow base like the letter Y (Figure 10.27). Closed around the bound antagonist, the LBD dimers sit just over the seal of the ion pore of the TMD in this conformation. The structure exhibits two-fold symmetry throughout the ATD, LBD, and TMD, although the subunits are not aligned with the overall axis of symmetry or with each other. Extensive contacts are seen between subunits A–B and C–D in the ATD and between A–D and B–C in the LBD, requiring a subunit crossover between the two domains. The conformation of the connecting polypeptides varies considerably, markedly changing the proximity of ATD and LBD and the space between ATD and LBD pairs (Figure 10.27B).

In contrast, the ion channel domain exhibits four-fold symmetry (Figure 10.28). Each subunit has three TM helices (M1, M3, and M4) and a central pore-like helix (M2), in addition to a disordered pore-lining loop not seen in the x-ray structure. In the tetramer the short pre-M1 helices make a cuff around the ion channel domain that appears to hold the M3 helices together. The very long (52 Å) M3 helices cross near the pre-M1-helix at a highly conserved sequence motif (SYTANLAAF) to occlude the pore. M4 helices make extensive contacts with the adjacent subunit. The three helices M1, M2, and M3 quite closely superimpose with the KcsA potassium channel (see Chapter 12), matching very well its gating machinery and central cavity and lacking its selectivity filter (Figure 10.28C).

Structural details in GluA2 provide a beautiful explanation of how iGluRs function by defining the gating mechanism and allowing predictions of the mechanism of activation and desensitization. Three peptide linkers form the gating machinery that transforms structural changes when ligands bind to LBDs into movements in the TMDs that open and close the pore. These linkers are the strands that mediate the symmetry mismatch

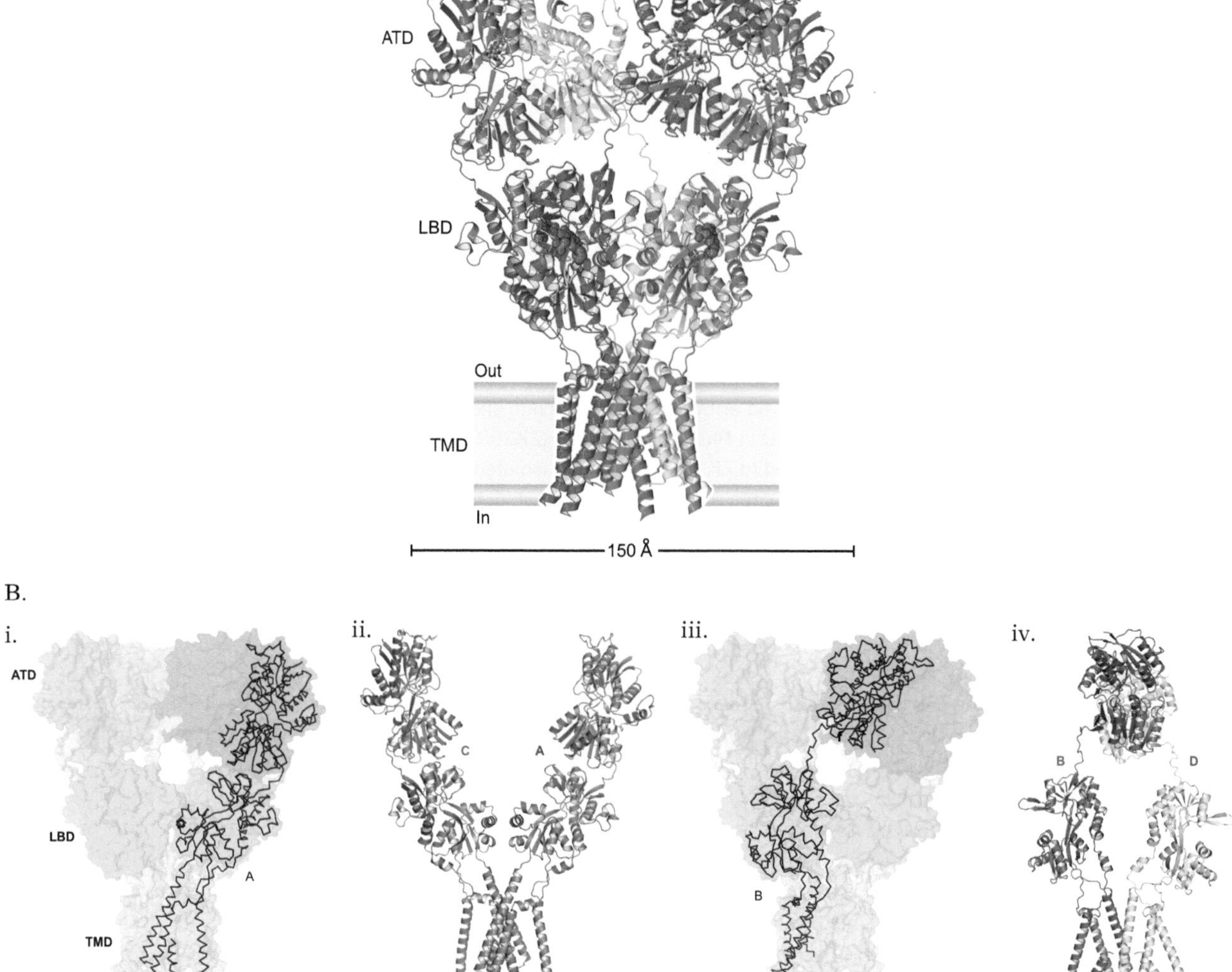

10.27 Overall structure of GluA2. (A) View of the receptor perpendicular to the overall two-fold axis of symmetry, with each subunit a different color. The competitive antagonists (space-filling models) bind to the center of each LBD. The ion channel in the receptor is closed near the extracellular side of the membrane. (B) The crossover of subunits between domains in GluA2 is illustrated on the partially transparent solvent-accessible surface of the entire receptor with A/B ATD dimer (red), C/D ATD dimer (lavender), A/D LBD dimer (blue), and B/C LBD dimer (orange). The trace of subunit A is superimposed on the surface in (i) and of subunit B in (iii). (ii) α-carbon traces of subunits A and C. (iv). α-carbon traces of subunit B and D, which exhibit domain swapping from the ATD to the LBD layers. From Sobolevsky, A. I., *et al.*, *Nature*. 2009, 462:745–758. © 2009, Nature Publishing Group. Reprinted by permission of Macmillan Publishers Ltd.

between the LBDs and TMDs: S1–M1 (Lys506–Gly513), M3–S2 (Val626–Glu634) and S2–M4 (Gly774–Ser788) (Figure 10.29). The conformations of the linker peptides are very similar within the pairs of subunits A–C and B–D, and are very different between the two pairs. For example, in the A–C pair the M3–S2 linker adopts a helical conformation to Met629, whereas in the B–D pair this helix is broken at Val626. As a consequence the Pro632 residues are 50 Å apart in the B–D pair and only 27 Å apart in the A–C pair. For the channel to open, the M3 helices must be pulled apart by conformational changes of these linkers as the LBDs close (Figure 10.30).

To visualize channel opening in response to agonist binding, the structure of the glutamate-bound LBD is superimposed on the antagonist-bound GluA2 LBD (Figure 10.30A). Binding glutamate closes the LBDs, increasing the distance between the bottom of the LBD dimers. This movement is transmitted to the peptide linkers differently in the A–C and B–D subunit pairs. The movements of the M3–S2 linkers are most significant, resulting in 4 Å and 7 Å movements of Pro632 in the

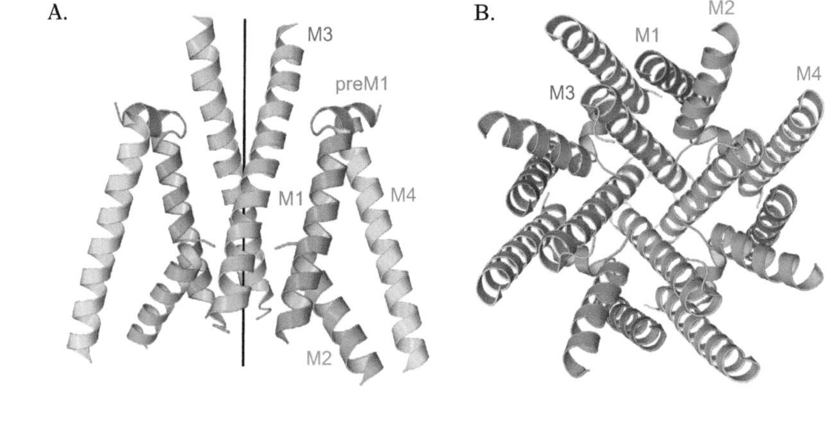

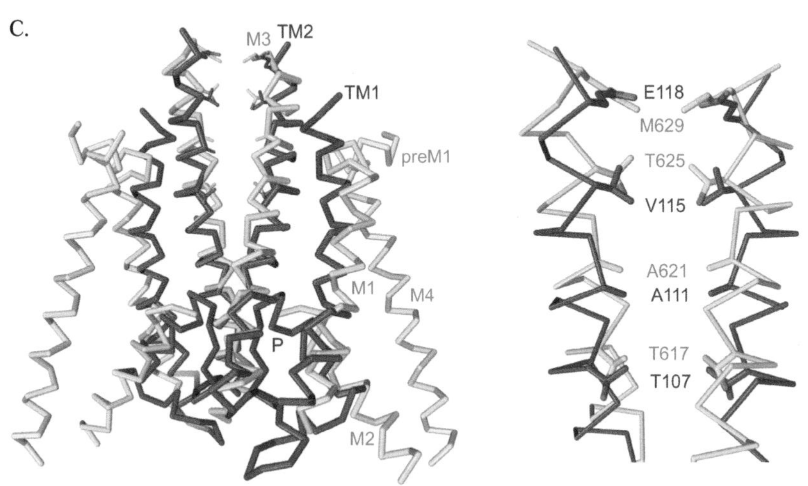

10.28 Transmembrane domain (TMD) of GluA2. Ribbon diagrams show the TMD helices of two subunits with the four helices in different colors, viewed parallel to the membrane (A) and of all four subunits viewed parallel to the four-fold axis (B). The Pre-M1 cuff is visible in orange in (A). (C) Overlay of two subunits of Glu2A (blue) with the inverted KcsA channel (gray) viewed parallel to the membrane. The channels are closed by residues from M3 in Glu2A and the inner helix in KcsA. From Sobolevsky, A. I., *et al.*, *Nature*. 2009, 462:745–758. © 2009, Nature Publishing Group. Reprinted by permission of Macmillan Publishers Ltd.

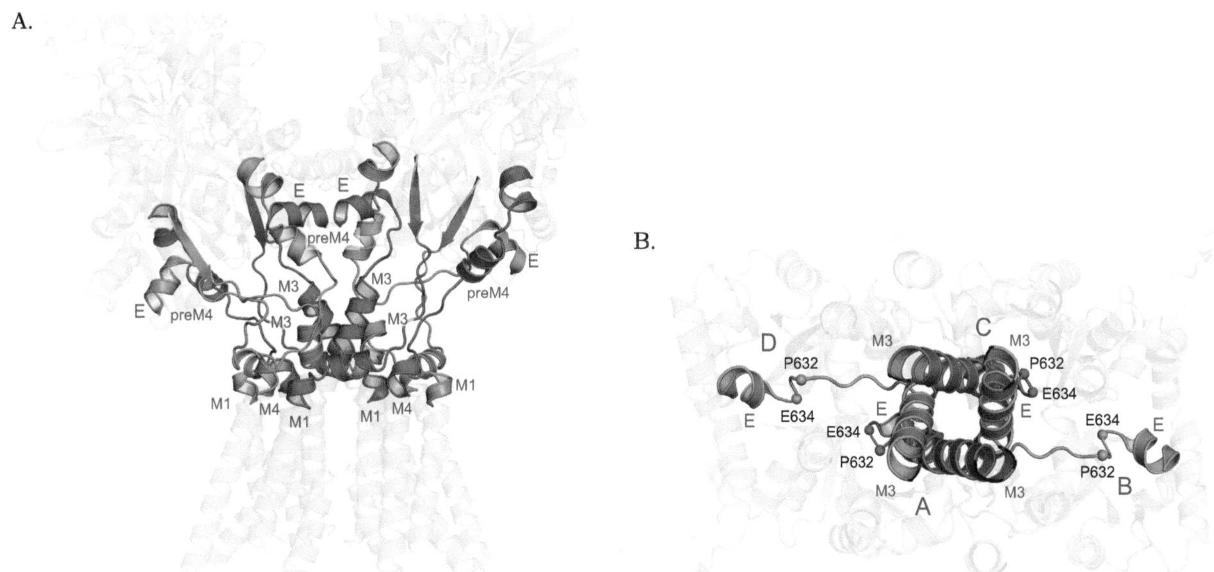

10.29 Gating machinery from symmetry mismatch. The GluA2 structure is highlighted to show the symmetry mismatch between LBDs and TMDs. The peptide linkers S1–M1, M3–S2, and S2–M4 are colored red (subunits A and C) or blue (subunits B or D). (A) All three linkers are seen in the structure viewed perpendicular to the overall two-fold axis of symmetry. (B) An example of the mismatch is clear when the M3–S2 linkers are viewed from the cytoplasm, parallel to the ion channel four-fold axis of symmetry. The unraveled helices in subunits B and D cause large separations between Pro632 residues as compared to the helices in A and C. From Sobolevsky, A. I., *et al.*, *Nature*. 2009, 462:745–758. © 2009, Nature Publishing Group. Reprinted by permission of Macmillan Publishers Ltd.

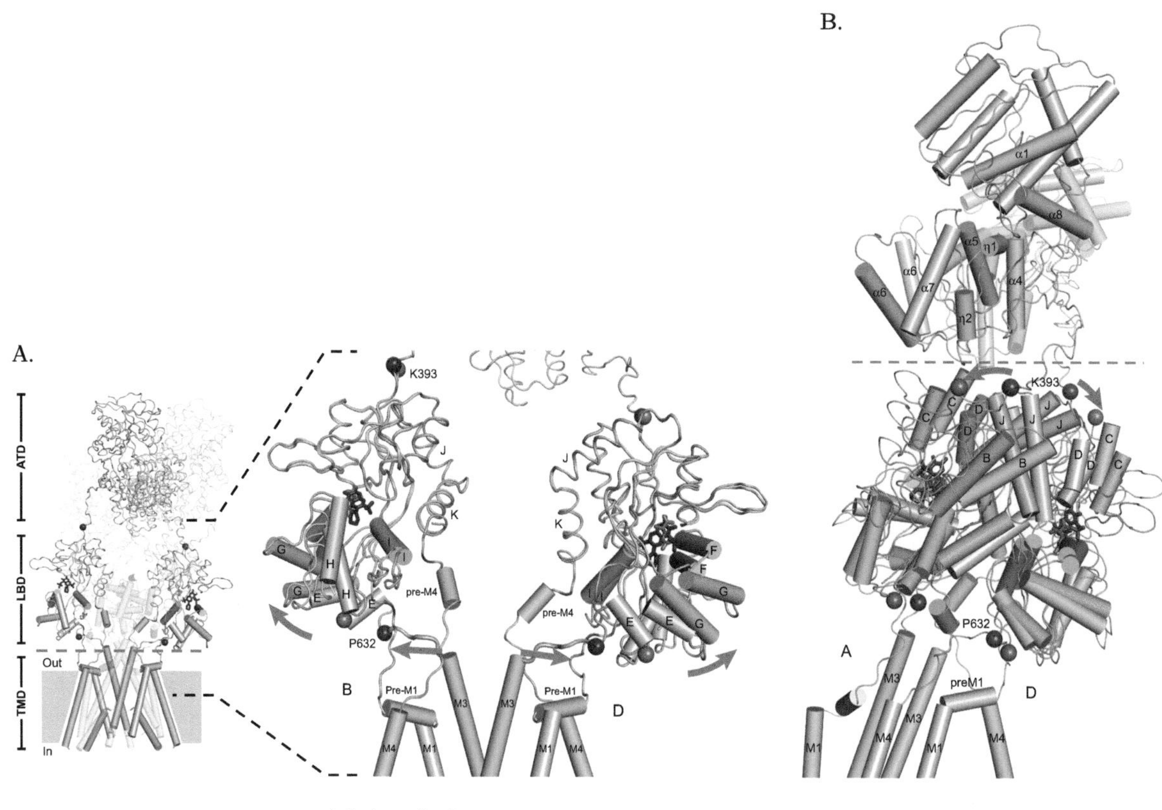

10.30 Projected conformational changes for activation and desensitization. (A) The conformational changes involved in activation are portrayed by superimposing the glutamate-bound LBD on the LBD in the structure of GluA2. On the right, the red dashed line indicates the interface between LBD and TMD and the black dashed lines delineate the portion of the structure viewed in more detail. The close-up view shows subunits B and D, with the glutamate bound LBD (green) superimposed on the GluA2 structure (with ZK bound) (purple). Stick models of ZK200775 and glutamate are shown in purple and green, respectively, and the helices of the ion channel and those that move in the LBDs are shown as cylinders. The red arrows show the movement during activation. Note the position of Pro632 in the two conformations (purple and green spheres) at the interface between LBD and TMD. (B) The conformational changes involved in desensitization are portrayed by superimposing the S729C LBD (orange) on GluA2 (purple). The figure shows subunits A and D, with a red dashed line at the interface between ATD and LBD and stick models of ZK200775 (purple) and glutamate (orange). The red arrows indicate the movement of Lys393 at the ATD and LTD interface upon desensitization. From Sobolevsky, A. I., *et al.*, *Nature*. 2009, 462:745–758. © 2009, Nature Publishing Group. Reprinted by permission of Macmillan Publishers Ltd.

A–C and B–D subunits, respectively, and allowing them to pull apart the M3 helices to open the channel. Thus iGluR activation is stepwise: Closure of the LBDs occurs first and triggers the conformational changes in the peptide linkers that affect the positions of helices in the TMD.

When agonist remains present in the cell, AMPA receptor activation is quickly followed by desensitization, the state in which agonist binds but the channel does not open. Similarly, mutation of Ser729 to Cys desensitizes the receptor so it does not respond to binding glutamate. The crystal structure of the isolated LBD of the S729C mutant shows that glutamate binding does not open the LBD dimer very much. When this structure is superimposed on the LBD of GluA2, the lower lobes of the closed dimers superimpose well enough for the linkers to stay in the closed state conformation (Figure 10.30B). The upper lobes rotate away from each other, indicating that the conformation of the ATDs must also change, and this suggests a mechanism for how modulator ligands that bind ATDs can affect receptor function.

Cys-loop receptors and GluCl

The second receptor for glutamate is an example of a pentameric ligand-gated ion channel very similar to the nicotinic acetylcholine receptor (nAChR) described in Chapter 6. The glutamate-gated chloride channel, GluCl, from *Caenorhabditis elegans* is closely related to the human glycine receptor in the family of Cys-loop receptors and is the first eukaryotic Cys-loop receptor to have a high-resolution x-ray structure. Cys-loop receptors that are channels for cations, like the nAChR as well as the serotonin 5-HT_3 receptor, are excitatory, while Cys-loop receptors that are anion-selective channels, such as those for $GABA_{A/C}$ and glycine, are involved in inhibitory neurotransmission in the brain. Several of the congenital channelopathies are the result of mutations affecting Cys-loop receptors, including hyperekplexia (glycine receptor), juvenile myoclonic epilepsy ($GABA_A$ receptor) and congenital myasthenia (nAChR). Two of the 15 bacterial homologs of the pentameric ligand-gated ion channels have been crystallized: GLIC from *Gloeobacter violaceus* and ELIC from *Erwinia chrysanthemi*. Although they lack the disulfide bridge in the extracellular loop for which the family is named, they have a common fold with both nAChR and GluCl.

The overall structure of Cys-loop receptors consists of a ring of five subunits, each of which has an extracellular domain with a β-sandwich that together make a water-filled vestibule leading to the channel. The transmembrane domain of each subunit has four α-helices, the second of which (M2) lines the channel. The Cys-loop is important in communication across the distance of 50–60 Å between the ion channel and the extracellular neurotransmitter binding sites. To solve the crystal structure of GluCl, the N and C termini were truncated by 41 and six residues, respectively, and an Ala–Gly–Thr tripeptide replaced the M3–M4 loop. Crystals of this construct in complex with Fab fragments, lipids, and an allosteric agonist were solved at 3.3 Å resolution (Figure 10.31). The structure shows the Cys-loop fold and can be superimposed on the bacterial homologs.

The structure of GluCl was solved in complex with ivermectin, a semi-synthetic macrocyclic lactone with broad antiparasite activity that activates GluCl and makes it susceptible to increased activation by glutamate. Ivermectin binds at subunit interfaces near the extracellular side of the membrane, making hydrophobic contacts and hydrogen bonds with the M3 α-helix of one subunit (called +) and the M1 α-helix of a complementary subunit (called –), even contacting the M2 (+) pore-lining helix (Figure 10.32A). Ivermectin also interacts with the M2–M3 loop that is postulated to trigger the large conformational change needed to open the channel. As an allosteric activator, it stabilizes the open-channel state.

In contrast, glutamate binds the receptor in the "classical" neurotransmitter site in the extracellular domain, which could be observed when the GluCl–ivermectin crystal was soaked in glutamate (Figure 10.32B). The glutamate-binding site is like a box with walls formed by two Tyr, Ser, and Thr residues on loops from the (+) subunit and a floor made up of two Arg residues in β-strands of the (–) subunit. In addition, Loop C closes around the ligand. The electropositive contacts for the α- and γ-carboxyl groups of glutamate make the binding pocket selective for small dicarboxylate L-amino acids

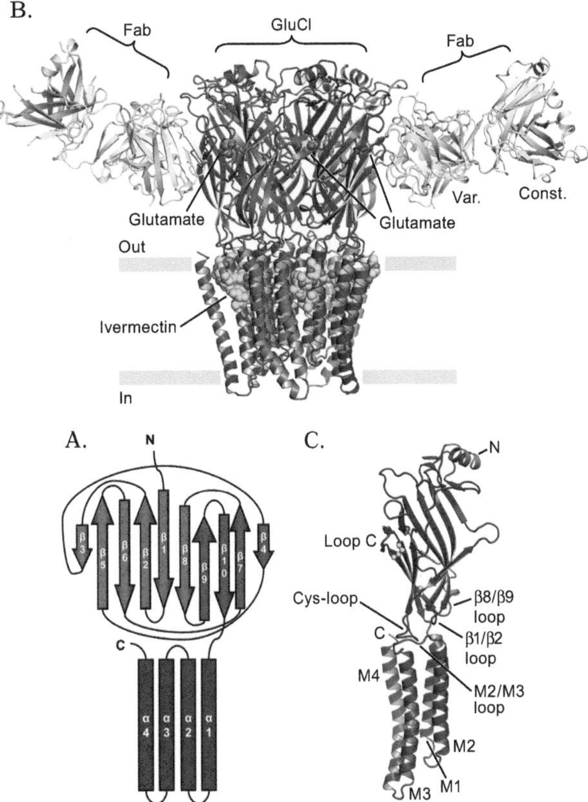

10.31 Architecture of the GluCl–Fab complex. (A) Topology of GluCl and other pentameric ligand-gated ion channels. The strands forming the two sides of the β-sandwich are colored red and green. Connecting loops are named for the strands or helices they link (not shown). (B) The overall structure of GluCl viewed from the plane of the lipid bilayer with only two Fab molecules shown for clarity. Ivermectin and glutamate are represented as spheres with C atoms in yellow, O atoms red, and N atoms blue. (C) A view of a single subunit shows the four TM helices (red), the β-sandwich (blue), the Cys loop and loop C disulfide bonds (yellow spheres), several other important loops, and the N and C termini (labeled). (A) From Hilf, R. J. C., and R. Dutzler, *Curr Opin Struct Biol*. 2009, 19:418–424. © 2009, Elsevier. Reprinted with permission from Elsevier. (B)–(C) From Hibbs, R. E., and E. Gouaux, *Nature*. 2011, 474:54–60. © 2011, Nature Publishing Group. Reprinted by permission of Macmillan Publishers Ltd.

so glutamate analogs bind to the receptor, although with reduced affinity.

Picrotoxin, a tricyclic toxin first isolated from plant seeds and used clinically to counteract barbiturate poisoning, has been shown to block open GluCl channels in voltage clamp experiments in *Xenopus* oocytes. When GluCl crystals were soaked in picrotoxin, new electron density appeared in the ion channel pore close to the cytosolic side of the membrane (Figure 10.32C). Located on the five-fold axis of symmetry, it represents an average of five orientations of picrotoxin bound between Pro and Thr residues from all five subunits.

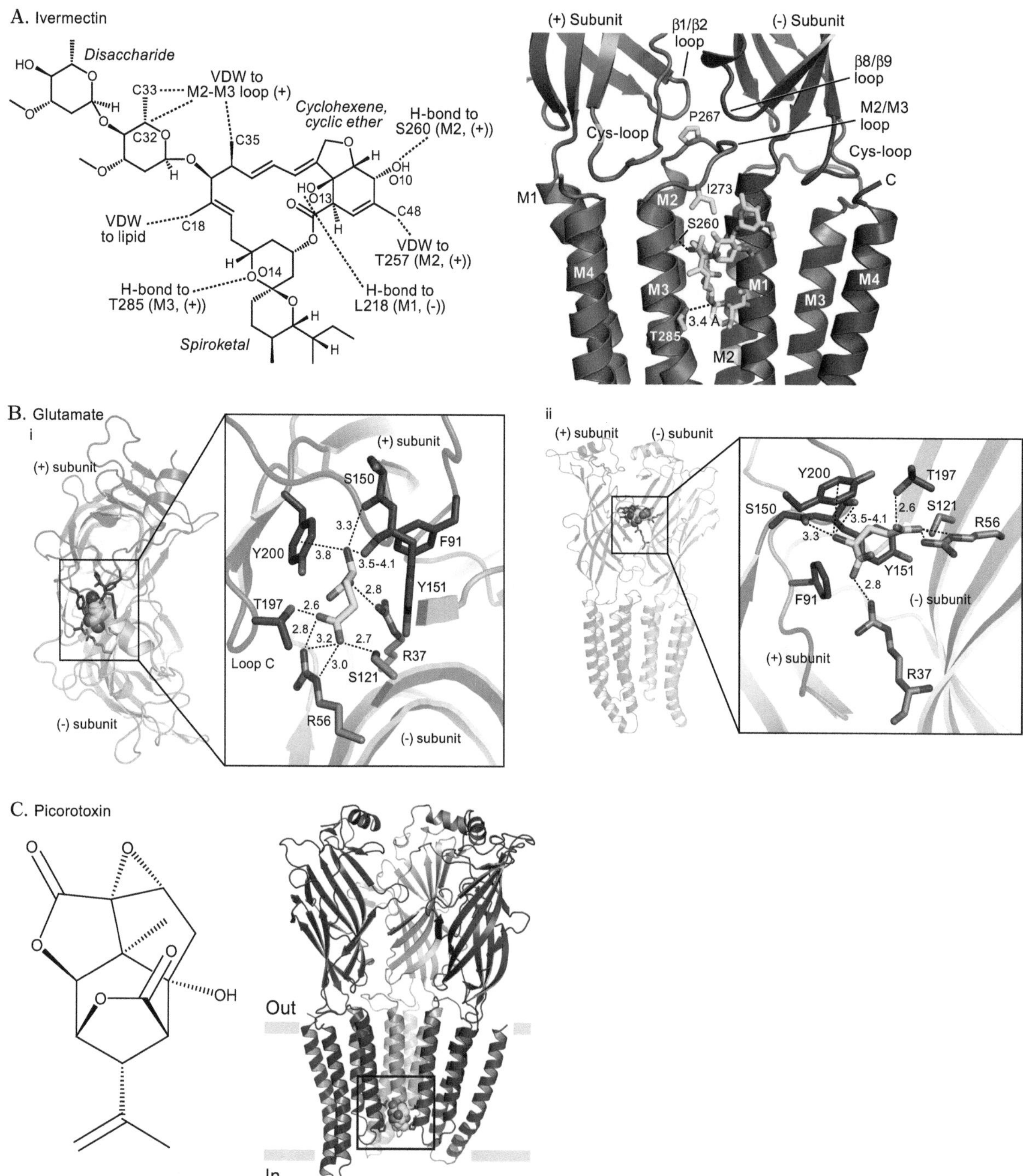

10.32 Ligand-binding sites in GluCl. (A) The allosteric agonist ivermectin binds at the subunit interface at the periphery of the TM domains, viewed from outside the receptor. Two subunits are shown, with complementary subunits (+) and (–) colored green and red, respectively. Ivermectin, represented as a stick model (yellow C atoms and red O atoms), lies between M3 and M1 of the different subunits, with hydrogen bonds (dashed lines) to Ser260 and Thr285, and is adjacent to the M2–M3 loop. (B) The agonist glutamate binds to a site in the extracellular domain, viewed from the extracellular side (i) and looking parallel to the membrane (ii). In (ii), loop C is removed for clarity. Glutamate and important side chains are represented by stick models. Dashed lines indicate hydrogen bonds and one cation–π bond (at Tyr200), with distances given in angstroms. (C) The open-channel blocker picrotoxin binds inside the pore near the cytoplasmic surface among Pro and Thr residues on M2 helices. The view of GluR has the front two subunits removed to reveal the pore. Insets show the chemical structures of ivermectin and picrotoxin. From Hibbs, R. E., and E. Gouaux, *Nature*. 2011, 474:54–60. © 2011, Nature Publishing Group. Reprinted by permission of Macmillan Publishers Ltd.

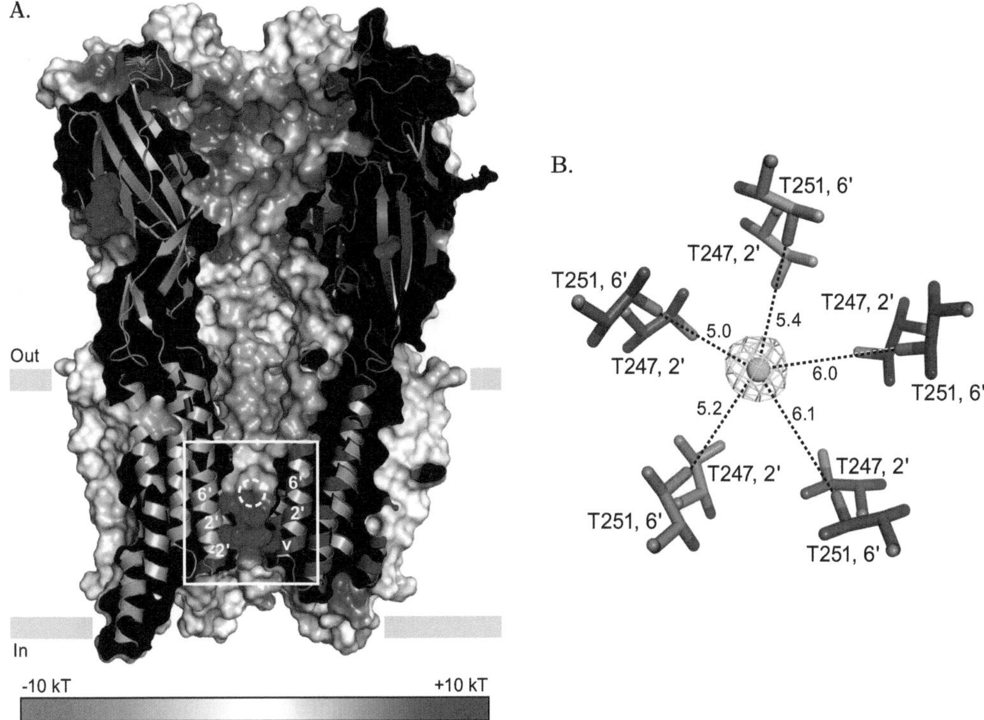

10.33 Selectivity of the ion channel. (A) The front of the receptor is cut away to view the interior surface of the pore, colored to show electrostatic potential and the putative chloride-binding site (dashed circle). (B) Putative chloride site viewed from the extracellular side, with Thr side chains from different subunits in different colors. The chloride is represented by a cyan sphere within a yellow mesh of electron density. The closest distances from each subunit to the Cl^- ion are indicated in angstroms, and they are large enough to allow room for water molecules. From Hibbs, R. E., and E. Gouaux, *Nature*. 2011, 474:54–60. © 2011, Nature Publishing Group. Reprinted by permission of Macmillan Publishers Ltd.

Picrotoxin binding indicates the ion channel is open in the GluCl crystals, which is consistent with a limiting constriction of 4.6 Å (compared to the 1.8 Å diameter of dehydrated Cl^-). None of the pore-lining residues are charged, but the helical dipoles of the M2 helices make the bottom of the pore electropositive, which results in an anion-selective channel (Figure 10.33). Above the picrotoxin-binding site in the pore is a site proposed to bind Cl^- transiently via water-mediated hydrogen bonds to the side chains of Thr residues. In addition, at the base of the pore are five electropositive pockets formed by Pro, Ala, and Ile residues that have been shown to be important in determining pore selectivity. These pockets bind I^- when the GluCl crystals are soaked in iodide, a heavy atom analog of chloride, and may function to increase the local concentration of Cl^- at the cytoplasmic mouth of the pore.

From one structure of activated GluCl with its channel open, the conformational changes that open and close the channel are not at all clear. However, a model can be made based on the structures of the bacterial homologs ELIC (in the basal state) and GLIC (in an active state). The comparison of ELIC and GLIC suggests that channel opening involves a twist in the quaternary structure created by tilting the upper part of each subunit and a rotation of the β-sandwich in each subunit, moving down the $\beta1$–$\beta2$ loop and tilting M2 and M3 away from the central axis (Figure 10.34). More high-resolution structures are needed to investigate the global conformational change and the coupling between agonist binding to sites in the extracellular region and the ion channel in GluCl.

The structure of GluCl sheds light on characteristics of other Cys-loop receptors. Agonist binding is accompanied by Loop C capping in nAChR, as in GluCl. Ivermectin binding is similar in GluCl, GLIC, ELIC, and AChR and involves the M2–M3 loop, which plays a central role in activation of receptors throughout the Cys-loopfamily. Lipophilic modulators of other Cys-loop receptors may behave like ivermectin, binding to exposed sites at the edge of the TM domain. The pore itself can differ little between cation-selective and anion-selective channels, because ion selectivity of the channel is reversed by simply placing negatively charged side chains near the pore constriction point and in the electropositive pockets. Thus common structural elements exist even within the complex diversity due to multiple subunit isoforms and splice variants of many eukaryotic ligand-gated ion channels.

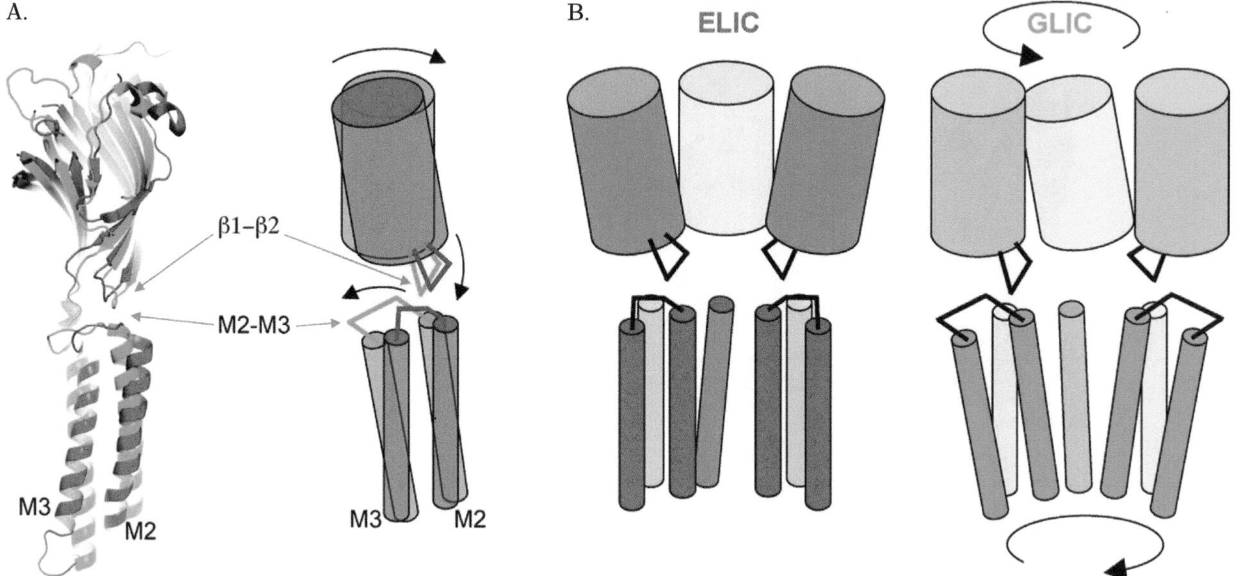

10.34 A model for gating bacterial pentameric ligand-gated ion channels. (A) Superposition of the ribbon diagrams for the subunits of ligand-gated ion channels from *Erwinia chrysanthemi* (ELIC, red) and *Gloeobacter violaceus* (GLIC, green), which are in closed and open states, respectively, reveals a rotation between the extracellular domains that is suggested to initiate pore opening after ligand binds. (B) Gating is proposed to involve a twist-tilt mechanism, in which the twisting of the extracellular domains, which rotates them relative to the rest of the molecule, tilts the two helices of the TM domains to open the pore. From Corringer, P. J., *et al.*, *J Physiol.* 2010: 588:565–572. © 2010, The Authors and the Physiological Society.

CONCLUSION

Structural biology of membrane proteins has provided significant insights into neurochemistry, as shown by the glutamate receptors described here. The crystal structures of ion channels (see Chapter 12) and neurotransmitter transporters (see Chapter 11) with essential roles in nerve impulses help to further a mechanistic understanding of the nervous system. The glutamate receptors are dramatically different from the GPCRs described in this chapter, revealing a diversity in receptor architecture than is not expected from the many highly similar GPCRs. Yet each type of receptor has an elegant design for molecular communication across membranes, which makes the field of membrane receptors one of the most dynamic and exciting today.

FOR FURTHER READING

Papers with an asterisk (*) present original structures.

Rhodopsin

Deupi, X., and J. Standfuss, Structural insights into agonist-induced activation of G-protein-coupled receptors. *Curr Opin Struct Biol.* 2011, 21: 541–551.

Deupi, X., P. Edwards, A. Singhai, *et al.*, Stabilized G protein binding site in the structure of consitutively active metarhodopsin-II. *Proc Natl Acad Sci USA.* 2012, 109:119–124.

Hofmann, K. P., P. Scheerer, P. W. Hildebrand, *et al.*, A G protein-coupled receptor at work: the rhodopsin model. *Trends Biochem Sci.* 2009, 34: 540–552.

Li, J., P. C. Edwards, M. Burghammer, C. Villa, and G. F. Schertler, Structure of bovine rhodopsin in a trigonal crystal form. *J Mol Biol.* 2004, 343:1409–1438.

Palczewski, K., G protein-coupled receptor rhodopsin. *Annu Rev Biochem.* 2006, 75:743–767.

Palczewski, K., Chemistry and biology of vision. *J Biol Chem.* 2012, 287:1612–1619.

*Palczewski, K., T. Kumasaka, T. Hori, *et al.*, Crystal structure of rhodopsin: a G protein-coupled receptor. *Science.* 2000, 289:739–745.

*Park, J. H., P. Scheerer, K. P. Hofmann, H-.W. Choe, and O. P. Ernst, Crystal structure of the ligand-free G-protein-coupled receptor opsin. *Nature.* 2008, 454:183–188.

Scheerer, P., J. H. Park, P. W. Hildebrand, et al., Crystal structure of opsin in its G-protein-interacting conformation. *Nature.* 2008, 455:497–502.

β-adrenergic Receptors

*Cherezov, V., D. M. Rosenbaum, M. A. Hanson, *et al.*, High resolution crystal structure of an engineered human β2-adrenergic G protein-coupled receptor. *Science*. 2007, 138:1258–1265.

Katritch, V., V. Cherezov, and R. C. Stevens, Structure–function of the G protein-coupled receptor superfamily. *Ann Rev Pharm Tox*. 2013, 53:531–556.

*Rasmussen, S. G. F., H.-J. Choi, D. M. Rosenbaum, *et al.*, Crystal structure of the human β2 adrenergic G-protein-coupled receptor. *Nature*. 2007, 450:383–387.

*Rasmussen, S. G. F., B. T. De Vree, Y. Zou, *et al.*, Crystal structure of the β2 adrenergic receptor-Gs protein complex. *Nature*. 2011, 477:549–555.

Rosenbaum, D. M., The structure and function of G-protein-coupled receptors. *Nature*. 2009, 459:356–363.

Rosenbaum, D. M., V. Cherezov, M. A. Hanson, *et al.*, GPCR engineering yields high-resolution structural insights into β2-adrenergic receptor function. *Science*. 2007, 318:1266–1273.

*Warne, T., M. J. Sergano-Vega, J. G. Baker, *et al.*, Structure of a β1-adrenergic G-protein-coupled receptor. *Nature*. 2008, 454:486–491.

Warne, T., R. Moukhametzianov, J. G. Baker, *et al.* The structural basis for agonist and partial agonist action on a β1-adrenergic receptor. *Nature*. 2011, 469:241–244.

Glutamate Receptors

GluA2

Gouaux, E., Structure and function of AMPA receptors. *J Physiol*. 2003, 554:249–253.

Mayer, M., Glutamate receptors at atomic resolution. *Nature*. 2006, 440:456–462.

*Sobolevsky, A. I., M. P. Rosconi, and E. Gouaux, X-ray structure, symmetry and mechanism of an AMPA-subtype glutamate receptor. *Nature*. 2009, 462:745–758.

GluCl

Corringer, P. J., M. Baaden, N. Bocquet, *et al.*, Atomic structure and dynamics of pentameric ligand-gated ion channels: new insight from bacterial homologues. *J Physiol*. 2010, 588:565–572.

*Hibbs, R. E., and E. Gouaux, Principles of activation and permeation in an anion-selective Cys-loop receptor. *Nature*. 2011, 474:54–60.

Hilf, R. J. C., and R. Dutzler, A prokaryotic perspective on pentameric ligand-gated ion channel structure. *Curr Opin Struct Biol*. 2009, 19:418–424.

TRANSPORTERS

11

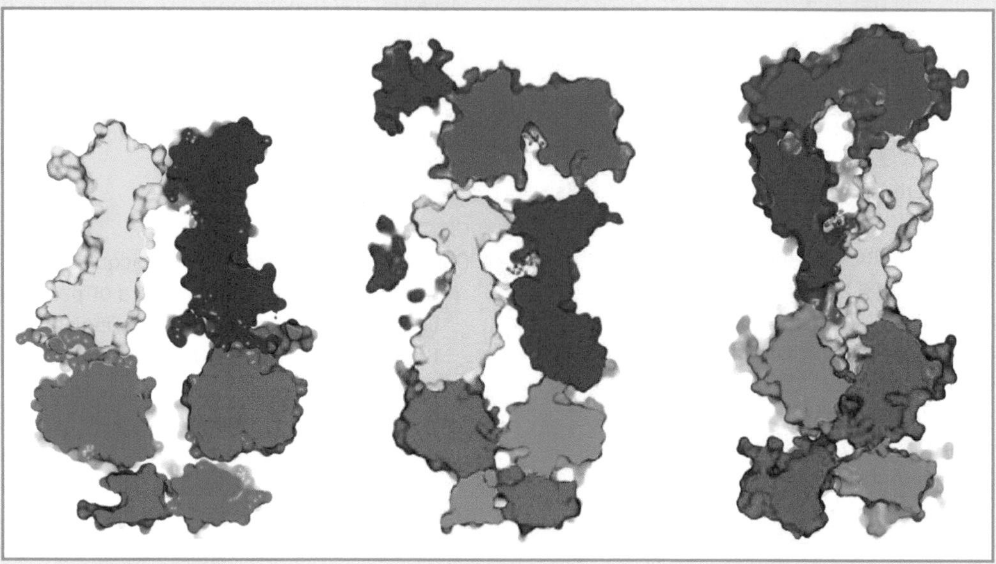

Cross-sectional slabs of the high-resolution structures of the entire maltose transporter in three conformations showing the resting state, a pre-translocation state, and a transition state (see text for details). From Khare, D., *et al.*, *Molecular Cell*. 2009, 33:528–536 and Oldham, M. L., and J. Chen, *Science*. 2011, 332:1202–1205. (Left) © 2009, Elsevier. Reprinted with permission from Elsevier. © 2011, American Association for the Advancement of Science. Reprinted with permission from AAAS.

Transport proteins are required for all cells to take up nutrients and to dispose of toxic substances, as well as to mediate flux of metabolites between intracellular compartments of eukaryotes. Moreover, they are involved in bioenergetics energy conversion, e.g., in mitochondria. Each cell comprises many diverse transporters, some of them highly specific for their respective substrate. For much of the past century transporters have been the object of physiological, genetic, biochemical, biophysical, and bioinformatic investigations that have provided a wealth of data describing how molecules and ions cross biological membranes. In the last two decades they have also become the subject of structural biology, as high-resolution structures have added details from a rich vein of structure–function relationships.

Because of the diversity of mechanisms used to carry out transport (see "Transport Proteins" in Chapter 6), a number of proteins involved in substrate transport across the membrane are described in other chapters (P-type ATPases in Chapter 9, an F-type ATPase in Chapter 13, porins in Chapter 5), and channels are the topic of the following chapter. This chapter presents a variety of well-investigated transporters, from LacY, the long-studied lactose permease encoded in the *lac* operon, and LeuT, the bacterial homolog of neurotransmitter transporters essential to brain function, to ABC transporters and multicomponent transporters that form elaborate drug efflux systems.

SECONDARY TRANSPORTERS

Common principles have emerged from an explosion of new structural information for secondary transporters that couple the "uphill" movement of a substrate with the "downhill" movement of a second substrate, often a proton or a sodium ion. Secondary transporters allow cells to take up or release a wide variety of compounds and fall into many different families based on shared functions and fairly high sequence similarity. However, the structures of numerous secondary transporters from different families show a common fold and imply a common mechanism, in spite of sometimes low or insignificant sequence homology among them. Based on their folds many of these transport proteins are members of superfamilies such as the major facilitator superfamily (MFS) and the LeuT superfamily (also called the Amino Acid-Polyamine-Organocation (APC) family). The architecture of the secondary transporters reveals repeats of two or more sets of TM helices, often oppositely oriented in the membrane, which play essential roles in the transport mechanism (see Chapter 6).

The known structures have crystallized in different transporter states, so together they illustrate the mechanism of alternating access in which the substrate-binding site is in the center of the protein and conformational changes give it access to the inside (Ci, inward-facing) or the outside (Co, outward-facing) compartments (Figure 11.1). The structures support a "rocker switch" mechanism during the alternating access cycle that designates the change between the outward-facing and inward-facing states as the relative movement of two rigid bodies. The conformational change goes through one or more occluded states in which substrate access to the aqueous milieu is closed off, which may be accomplished by individual residues or even flexible helical segments bending along partly unfolded stretches that form gates. In these cases the rigid movements of whole segments – the rocking bundle – are described as "thick gates" while the conformational changes of side chains or flexible helices make "thin gates" (Figure 11.1C). Some of the structures also explain how symporters accomplish the necessary coupling that drives transport. The close proximity of binding sites for substrates and ions enable binding of the coupling ion to affect binding and transport of the substrate. These principles will be demonstrated in an examination of several important secondary transporters, from the first ground-breaking structures to recent triumphs in capturing more than one conformation.

MFS Transporters

Crystal structures were reported in 2003 for both lactose permease and the glucose-3-phosphate transporter of *E. coli*. That same year the structure of the mitochondrial ADP/ATP carrier was solved, giving an example from the mitochondrial carrier family (MCF). Lactose permease (LacY) and the glycerol-3-phosphate transporter (GlpT) of *E. coli* are members of the MFS of secondary active transporters. The largest of all transporter families, this group has 70 members identified in *E. coli* and over 3000 members identified in genomes of organisms from all three kingdoms of life. In spite of low sequence similarity, MFS transporters appear to have a common fold, as predicted by hydrophobicity plots and supported by gene fusion data. For this reason, the high-resolution structures of LacY and GlpT provided information relevant to a vast number of transporters. Structures of EmrD and FucP in different conformations from LacY and GlpT fill out the depiction of MFS transporters.

LacY, a Scrutinized Symporter

The lactose permease, which is the galactoside/H^+ symporter of *E. coli*, simply called the LacY protein, utilizes the electrochemical proton gradient of the inner membrane to drive the 100-fold accumulation of lactose inside the cell. The *lacY* gene, induced when lactose is in the medium, was the first transporter gene to be cloned and

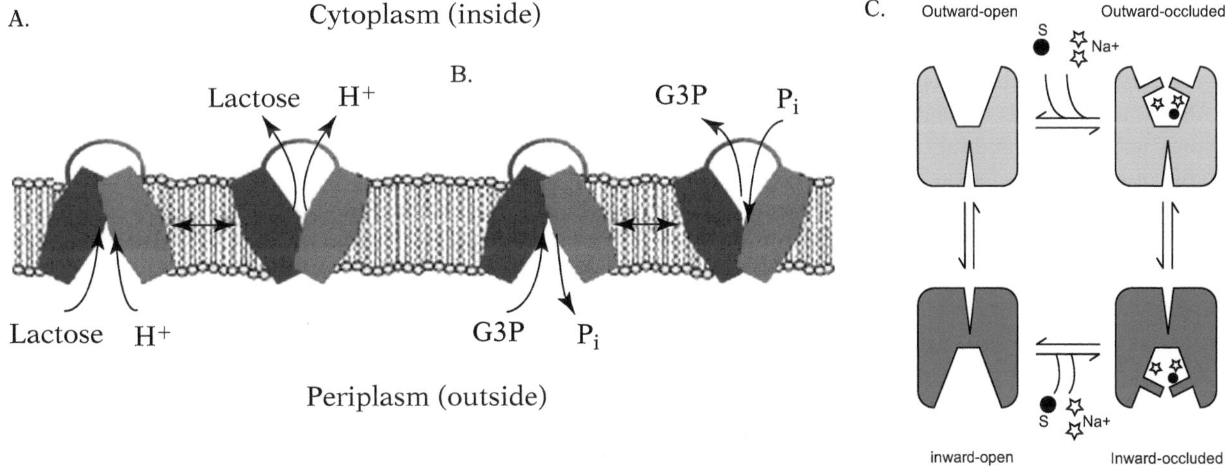

11.1 Alternating access model of transport. The alternating access mechanism allows both symport, illustrated with LacY (A), and antiport, illustrated with GlpT (B). The mechanism alternates between Co, a conformation facing out (to the periplasm) and Ci, a conformation facing in (to the cytoplasm), with a rocker-type switch between them. From Abramson, J., *et al.*, *Curr Opin Struct Biol*. 2004, 14:413–419. © 2004 by Elsevier. Reprinted with permission from Elsevier. (C) Intermediate between the conformations open to the outside and inside are typically occluded states, both outward-facing and inward-facing. In many transporters, thick gates are involved in changing from outward-facing to inward-facing, while thin gates open and close the access, as diagrammed here for LeuT which transports two Na^+ ions (stars) per substrate (filled circle). From Krishnamurthy, H., and E. Gouaux, *Nature*. 2012, 481:469–474.

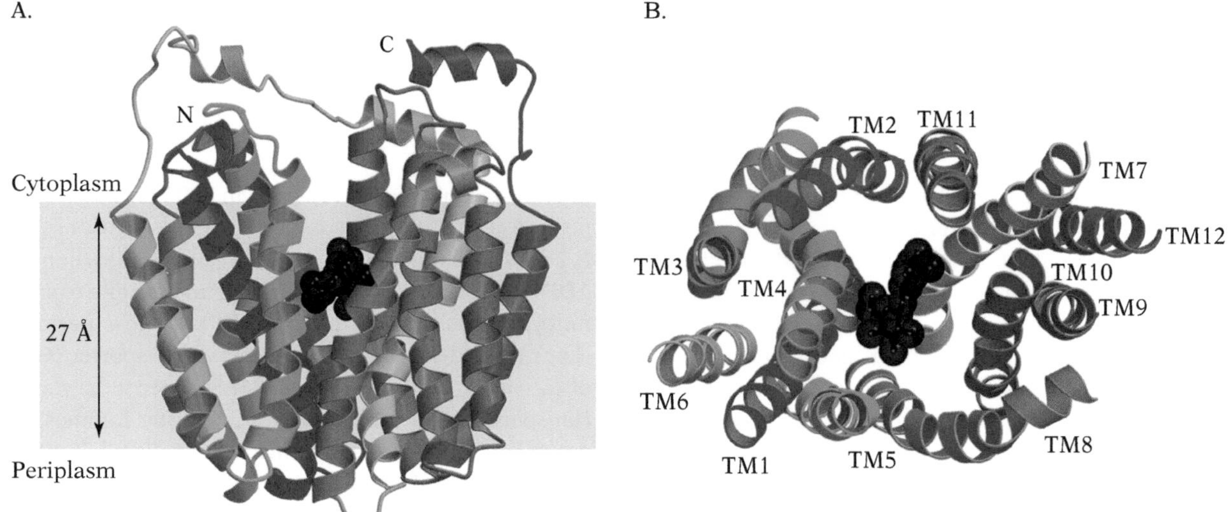

11.2 X-ray structure of LacY. The high-resolution structure of the LacY C154G mutant is shown in ribbon representations viewed from the membrane plane (A) and from the cytoplasm (B), with the 12 helices colored from dark blue at the N terminus to red at the C terminus. The sugar-binding site in the cavity is occupied by the inhibitor p-D-galactopyranosyl 1-thio-p-D-galactopyranoside (TDG), represented with black spheres. In (B), the loops are removed to view the helices normal to the membrane, revealing their curves and kinks. From Abramson, J., *et al.*, *Science*. 2003, 301:610–615. © 2003, American Association for the Advancement of Science. Reprinted with permission from AAAS.

sequenced, ushering in the use of molecular biology techniques to investigate membrane transport. Many of the features of the LacY protein subsequently determined by extensive genetic, biochemical, and biophysical analyses have been confirmed with the x-ray structure, which took ten years to achieve. The first structure was determined at 3.5 Å resolution for a mutant of LacY (C154G) that can bind substrate with high affinity, but not transport it. This phenotype suggested the mutant protein is stabilized in one conformation, which facilitated its crystallization. A crystal structure of wild type LacY protein then provided general confirmation to the overall fold seen in the mutant structure described in detail here.

All 417 amino acids of LacY have now been mutated to study its structure and function, which allows precise identification of critical residues in the crystal structure. As predicted by spectroscopic measurements, LacY is an α-helical bundle almost entirely buried in the membrane, with both N and C termini in the cytoplasm. Its two-fold pseudo-symmetry, which suggests a gene duplication event occurred in its evolution, is typical of many transporters (see Figure 6.20) and produces a heart shape open to the cytoplasm when viewed from the plane of the membrane. The structure consists of two bundles of six α-helices forming N-terminal and C-terminal domains that are linked by a long loop between TM6 and TM7[1] (Figure 11.2). The TM helices are very irregular due to curves and kinks, which may be important in the conformational change involved in the mechanism of transport. The two domains are separated by a large hydrophilic cavity that opens to the cytoplasm and contains a single sugar-binding site. The side chains involved in sugar binding are mainly in the N-terminal helix bundle and the side chains that form a proton-binding site are mainly in the C-terminal bundle.

LacY is highly specific for the galactose moiety of lactose and transports many derivatives of galactose with hydrophobic groups on the anomeric carbon, including the familiar ones used as inducers and dyes. The substrate-binding site in the crystal structure is very well defined by the presence of β-D-galactopyranosyl 1-thio-β-D-galactopyranoside (TDG) (Figure 11.3). Substrate binding involves both hydrophobic and polar interactions. The galactopyranosyl ring sits over the indole ring of Trp151 of TM5 (explaining why substrate binding shifts its fluorescence maximum), and the C6 atom of the galactose ring interacts with Met23 in TM1. Essential polar residues, Arg144 of TM5 and Glu126 of TM4, make hydrogen bonds with the oxygen atoms of the galactose moiety. Another essential residue, Glu269 in TM8, appears to form a salt bridge with Arg144 close to Trp151 and may be a critical link between the two domains.

The x-ray structure represents the protonated form of LacY with the proton on Glu325 in a hydrophobic pocket made by residues from TMs 7, 9, and 10 of the C-terminal domain (Figure 11.4). Proton translocation occurs only when coupled with sugar uptake, and quantitative pH measurements verify that proton binding must be followed by substrate binding. (Interestingly,

[1] Formally TM helices are given Roman numerals to distinguish them from other helices in the protein, although in the literature both Roman and Arabic numerals are used. In the text the helices of LacY are designated TM1, TM2... to be consistent with the rest of the chapter.

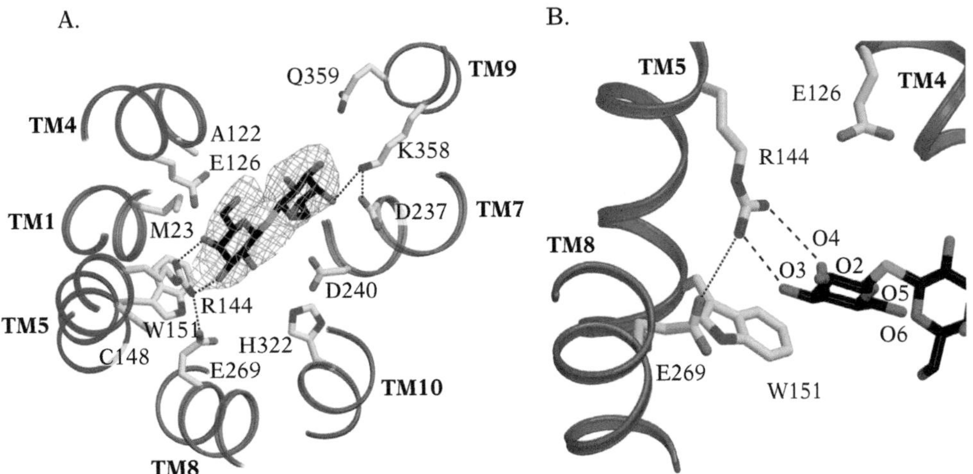

11.3 Substrate-binding site of LacY with TDG present. Interactions between the galactosyl moiety and the N-terminal domain (blue) produce the binding specificity, while the interactions with the C-terminal domain (red) are less specific and give room for more bulky adducts. (A) The residues involved in TDG binding, viewed along the membrane normal from the cytoplasmic side. (B) A close-up view of the N-terminal domain interactions with TDG. From Abramson, J., *et al.*, *Science*. 2003, 301:610–615. © 2003, American Association for the Advancement of Science. Reprinted with permission from AAAS.

LacY catalyzes passive exchange of sugar down its concentration gradient without translocation of protons.) Coupling has been proposed to involve a presumed salt bridge between Arg144 and Glu126 in the absence of substrate, which is disrupted upon binding of substrate. Transfer of a proton from His322 (in TM10) to Glu325 then allows Glu269 to form a salt bridge with Arg144, triggering the proposed conformational change that results in release of both substrate and proton to the cytosol. The location of the salt bridge/hydrogen bond network for proton translocation parallel to the plane of the membrane (instead of crossing it as in bR and other proton pumps) strongly suggests that the proton and sugar are released together via the alternating access conformational rearrangement.

All high-resolution structures of LacY obtained so far are open to the inside of the cell, suggesting the Ci conformation has the lowest energy during crystallization. However, H/D exchange experiments indicate that LacY is very dynamic in the membrane. Numerous biophysical techniques (including FRET and DEER) along with site-directed alkylation and site-directed crosslinking have probed the conformational change involved in alternating access. Insights into the other possible conformations also come from structures of MFS transporters in other conformations.

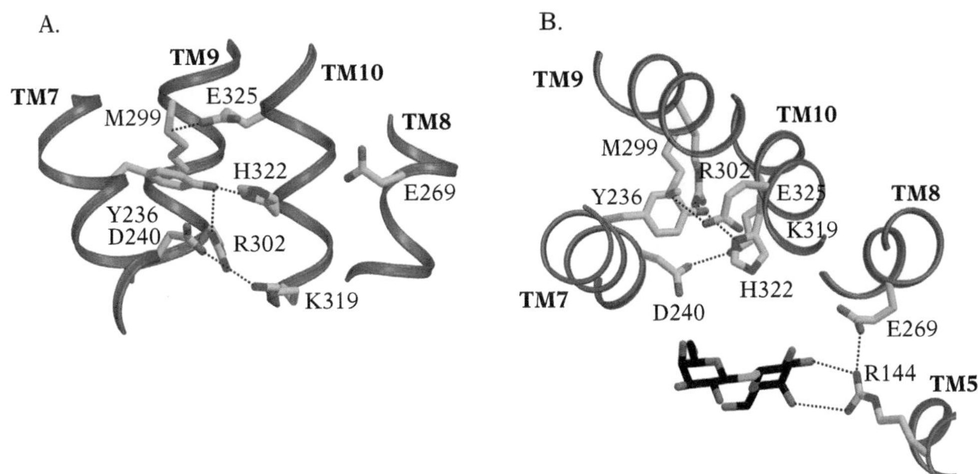

11.4 Residues of LacY involved in proton translocation and coupling. The labeled residues form a network of salt bridges and hydrogen bonds (black broken lines) as described in the text. (A) View from the side. (B) View from the cytoplasm. From Abramson, J., *et al.*, *Science*. 2003, 301:610–615. © 2003, American Association for the Advancement of Science. Reprinted with permission from AAAS.

GlpT, an MFS Antiporter

Simultaneous with publication of the crystal structure of LacY came the high-resolution structure for a second MFS transporter, the antiporter GlpT. The *glpT*gene of *E. coli* is induced by the presence of glycerol-3-phosphate (G3P), which the bacteria use as an energy source and as a precursor for phospholipids. GlpT carries out the exchange of G3P for inorganic phosphate, whose high (millimolar) concentrations inside the cell favor its efflux. Because GlpT recognizes the phosphate moiety of G3P, it will not transport glycerol. It will transport glycerol-2-phosphate, as well as a phosphate-based antibiotic called fosfomycin that is used clinically to treat bacterial urinary tract infections.

Most of the 452 amino acids of GlpT form 12 TM helices, with very short connecting loops except the loop between TM6 and TM7, which connects the N- and C-terminal domains. In the 3.3-Å resolution structure of GlpT these two domains make distinct helical bundles, with a wide opening at the cytoplasmic side (Figure 11.5). At the periplasmic side the two domains are linked by extensive van der Waals interactions, while at the open side they are connected by a long cytoplasmic loop not fully resolved in the structure. Therefore, like the structure of LacY, the GlpT structure has the inward-facing conformation, Ci. The helices have many glycine residues, allowing them to pack closely, along with a number of bulky side chains in the N-terminal side opposite pockets on the C-terminal side. The helices vary in length and tilt angles, with significant curvature in the four at the interface that may facilitate a rocker-switch type of movement to the outward-facing form. The overall folds of LacY and GlpT are nearly superimposable, with all of the helices closely aligned except the TM2/TM7 pair along the hydrophilic cavity, which is likely due to the fact that in the crystals, LacY has a substrate analog bound while GlpT does not (Figure 11.6).

The substrate-binding site in GlpT is identified by the two essential arginine residues, Arg45 in TM1 and Arg269 in TM7, presumed important in binding substrate phosphate groups and located in an otherwise neutral translocation pore (see Figure 11.5C). Within the major facilitator superfamily, GlpT is in the organophosphate:phosphate antiporter family, whose members do share significant sequence homology, with conserved hydrophilic and hydrophobic residues on both sides of the essential arginine residues likely involved in substrate binding. Interestingly, this family

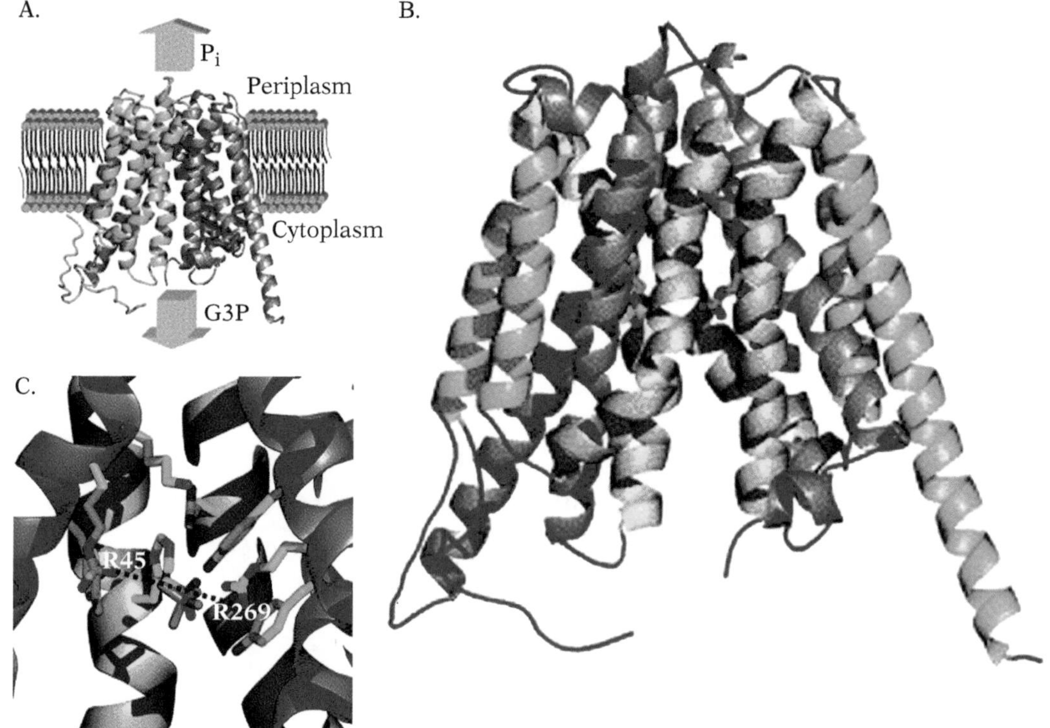

11.5 X-ray structure of GlpT. (A) A ribbon diagram of the two domains (green and purple) shows the trapezoid shape of GlpT and its orientation in the membrane. The directions for translocation of G3P and Pi are shown. (B) The symmetry between the two domains of GlpT is highlighted when its TM segments are colored according to function. Four of the TM helices are not involved in pore formation (TM3, TM6, TM9, and TM12; green), four line the central pore (TM2, TM5, TM8, and TM11; yellow), and the remaining four are central in the structure (purple). The two arginine residues in the transport path are shown. (C) At the substrate-binding site the distance between Arg45 (helix 1) to Arg269 (helix 7) is 9.9 Å. From Lemiuex, M. J., *et al.*, *Curr Opin Struct Biol.* 2004, 14:405–412. © 2004 by Elsevier. Reprinted with permission from Elsevier.

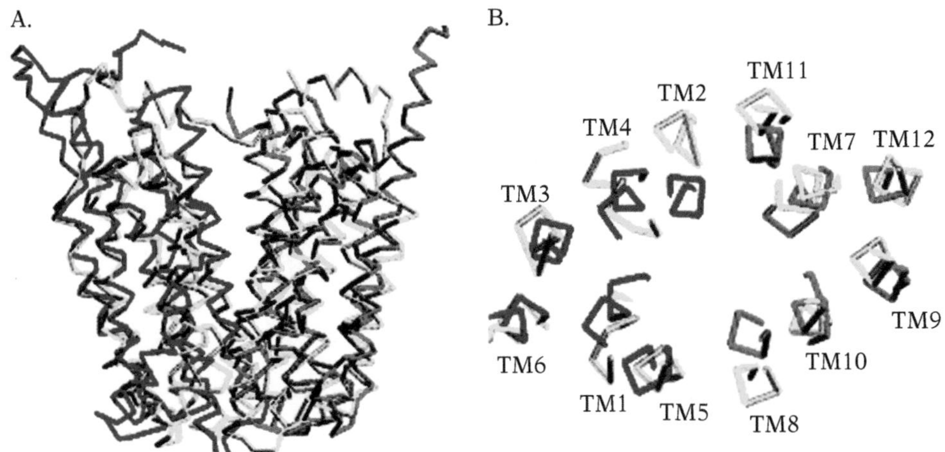

11.6 Superposition of structures of LacY and GlpT. (A) Superposition of the structures of LacY (yellow) and GlpT (blue and red, for the N- and C-terminal domains, respectively) viewed parallel to the membrane. (B) Superposition viewed along the membrane normal from the cytoplasmic side (same color scheme), with helices numbered and loops removed. From Abramson, J., *et al.*, *Curr Opin Struct Biol*. 2004, 14:413–419. © 2004 by Elsevier. Reprinted with permission from Elsevier.

contains human transporters for G3P and for glucose-6-phosphate that are implicated in a variety of conditions, including deafness, hyperglycerolemia, respiratory distress, and seizures.

EmrD, an MFS Exporter in an Occluded Conformation

A transporter that exports a wide variety of amphipathic inhibitors and detergents, EmrD has a role in multidrug resistance (MDR; other examples of MDR proteins are described at the end of this chapter). EmrD operates in the cytoplasmic membrane of *E. coli*, where it can remove deleterious compounds from the inner leaflet and expel them to the periplasm, coupled to the uptake of protons. Since it transports compounds that are unrelated chemically, EmrD does not have a highly specific binding site as seen in LacY. Indeed, the 3.5-Å resolution structure of EmrD shows a quite different cavity even though the rest of its 12 TM helices are organized similarly to LacY and GlpT (Figure 11.7). However, the six N-terminal helices and six C-terminal helices form a compact bundle that lacks an opening to either side of the membrane, thus it represents an occluded conformation of EmrD.

Consistent with the substrate cavities of other MDR proteins, the internal cavity of EmrD is defined by bulky aromatic residues, Ile28, Ile217, Ile253, Tyr52, Tyr56, Trp300, and Phe249 (Figure 11.8). In the x-ray structure two pairs of aromatic groups (Tyr52/Tyr56 and Trp300/Phe249) stack to interact with aromatic portions of drug substrates. In addition, the internal cavity has polar and

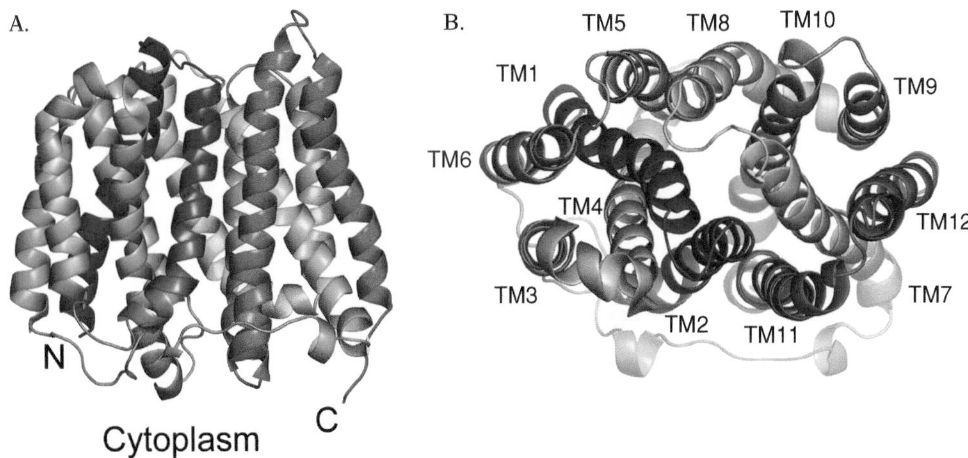

11.7 Structure of EmrD. The N- and C-terminal halves of EmrD, each with six TM helices (in rainbow colors from the N terminal (blue) to the C terminal (red)), form a closed bundle when observed from the plane of the membrane (A) or from the periplasm (B). The symmetry between the two halves is apparent. From Yin, Y., *et al.*, *Science*. 2006, 312:741–744. © 2006, American Association for the Advancement of Science. Reprinted with permission from AAAS.

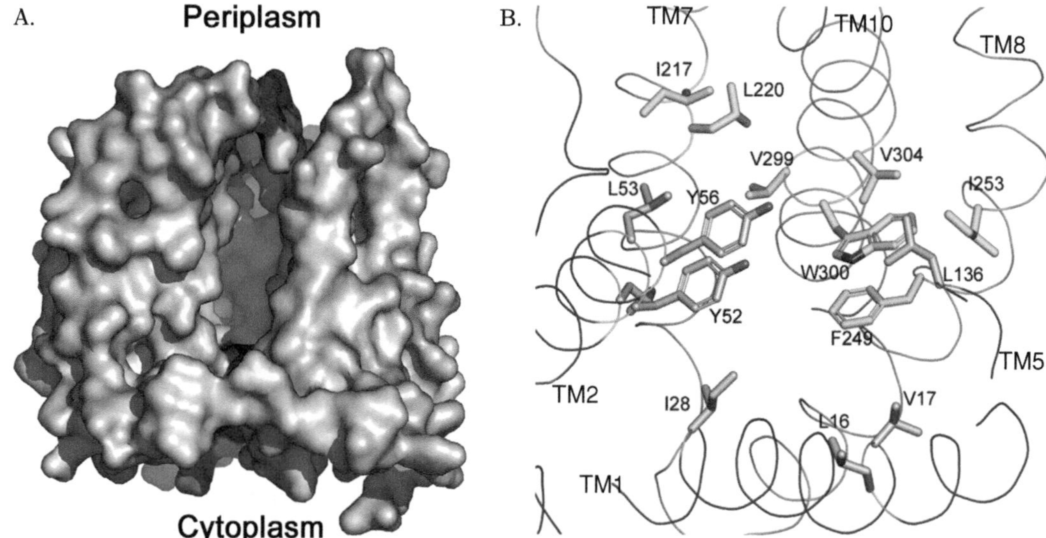

11.8 The inner cavity of EmrD. (A) The internal cavity is shown in a cut-away view of the surface with residues 43–67 omitted. Hydrophobicity is depicted from low (tan) to high (light brown), with charged residues highlighted (positive, blue; negative, red). (B) Hydrophobic residues (sticks) from the N- and C-terminal halves (blue and orange, respectively) line the internal cavity of EmrD, with Tyr52/Tyr56 on the left and Trp300/Phe249 on the right. From Yin, Y., *et al.*, *Science*. 2006, 312:741–744. © 2006, American Association for the Advancement of Science. Reprinted with permission from AAAS.

charged residues on the cytoplasmic side (Gln21, Gln24, Gln46, Gln60, and Arg118) and the periplasmic side (Thr25, Asp33, Glu227) that may play roles in proton translocation. Finally, the cytoplasmic side of the cavity has two long helical regions with a number of charged residues that are important for substrate recognition in homologous MDR proteins of the MFS family.

FucP, an MFS Symporter in Co Conformation

Like other microbes, plants, and animals, *E. coli* takes up L-fucose to use as a constituent of N-linked glycans, as well as an energy source. The *E. coli* fucose/H^+ symporter FucP is a typical MFS transporter with 12 TM helices related by an internal structural repeat and folded into separate N- and C-terminal domains. When crystallized in the absence of substrate to give a structure with 3.1-Å resolution, FucP has a large central cavity at the periplasmic side; thus it is in the Co conformation (Figure 11.9). Although FucP has only limited sequence homology with LacY, their separate N- and C-terminal domains can be superimposed, strongly suggesting they both move as rigid bundles during the Co ↔ Ci conformational change. In FucP TM4 and TM10 are at the center of the outward-open face, whereas in LacY TM1 and TM7 are at the center of the inward-open structure.

The cavity of FucP is amphipathic and also has complementary faces, with a strip of negative electrostatic potential along the side formed by the N-terminal domain and positive charges with a hydrophobic patch on the C-terminal domain side (Figure 11.10). Two essential acidic residues, Asp46 and Glu135, in TM1 and TM4 of the N-terminal domain play roles in transport according to mutant studies. Asp46 is likely to be involved in proton translocation, analogous to Glu325 in LacY, while Glu135 may be involved in binding fucose.

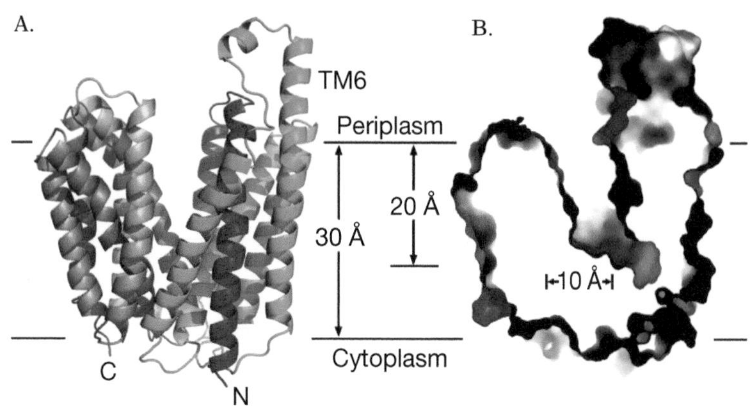

11.9 Structure of FucP. (A) The N- and C-terminal halves of FucP (TM helices in rainbow colors) form a trapezoid with a large cavity open to the periplasm, as seen from the plane of the membrane. (B) A slab of FucP in surface electrostatic potential reveals the depth of the central amphipathic cavity. From Dang, S., *et al.*, *Nature*. 2010, 467:734–737. © 2010, Nature Publishing Group. Reprinted by permission of Macmillan Publishers Ltd.

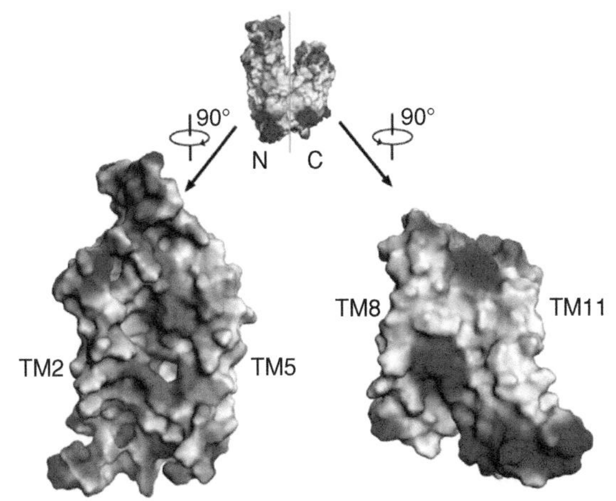

11.10 Complementary faces of the sides of the FucP cavity. When the two halves of FucP (shown on top) are opened to reveal the sides that face the cavity, the surface electrostatic potentials show the dominant negative potential (red) along the N-terminal side and positive potential (blue) along the C-terminal side. From Dang, S., *et al.*, *Nature*. 2010, 467:734–737. © 2010, Nature Publishing Group. Reprinted by permission of Macmillan Publishers Ltd.

Binding of the detergent β-nonyl-glucoside appears to block fucose binding in the crystals.

The high-resolution FucP structure offers a much sought after example of an outward-facing MFS transporter and is consistent with a 6.5-Å resolution cryo-EM structure of the *E. coli* oxalate/formate antiporter OxlT in the Co conformation. Together the structures of these five MFS transporters strongly support the rocking bundle model for alternating access in transport. With additional structures appearing at a rapid pace, they also establish a paradigm for the largest superfamily of transporters.

A Paradigm for MFS Transporters

Prokaryotic MFS proteins consist of 400–600 amino acids in a single polypeptide chain that usually folds into 12 TM helices organized in two domains. Although they need unique binding sites for their diverse substrates, they appear to share a common architecture and mechanism of transport. Overall the sequence homology among MFS members is weak; however, those that transport sugars exhibit conservation at the sugar-binding sites. When bacterial MFS transporters for other sugars are compared with *E. coli* LacY, the sequence homology varies from 35% to 75%. The sequence of a maltose permease from a deep sea bacterium, which has conserved the most essential amino acid residues in LacY (17% identity), was employed in a BLAST search of prokaryotic and eukaryotic genomes (see Chapter 6). Interestingly, the eukaryotic proteins thus identified are predicted to have 12 TM segments and many of the essential residues at their substrate-binding sites, even though in general they are about twice the size of the prokaryotic sugar transporters. The first crystal structure of a eukaryotic MFS transporter, that of the phosphate transporter PiPT with 12 TM helices, shows striking similarities to the bacterial MFS transporters in occluded states.

The structural biology of MFS transporters, in addition to extensive biochemical and biophysical characterizations, has revealed much about their common architecture and transport mechanism, yet questions remain regarding the coupling mechanisms for secondary transport. Deduction of further mechanistic details will require higher resolution in crystal structures, capturing these proteins in more conformations, and dynamic studies employing spectroscopic, biophysical, and computational approaches.

Mitochondrial ADP/ATP Carrier

Like the Gram-negative bacteria from which they evolved, mitochondria are surrounded by two membranes: an outer membrane with pores that allow fairly nonselective passage of molecules and ions up to around 500 Da, and an inner membrane with numerous specific transporters. Prominent in the mitochondrial inner membrane is the ADP/ATP carrier (AAC), whose job is to provide the ADP substrate needed inside the matrix for ATP synthase and to export the ATP it produces. The AAC is an electrogenic antiporter that exchanges one ADP^{3-} for one ATP^{4-} without bound Mg^{2+} ions, with the net export of one negative charge driven by the membrane potential. Indispensable to the generation of ATP by respiration, it is the most abundant of the mitochondrial carriers and makes up to 10% of the protein extracted from the inner mitochondrial membrane.

The mitochondrial AAC was discovered over 40 years ago through investigations of the mechanism of poisoning by atractyloside (ATR), a diterpene heteroglucoside that is produced by a widespread Mediterranean thistle used in herbal medicine (see inset of Figure 11.14). ATR binds to the AAC from the intermembrane side, blocking the entrance of ADP, while another inhibitor, bongkrekic acid (BA; a complex fatty acid derivative), binds from the matrix side to prevent ATP export. Characterization of the binding of these inhibitors and various derivatives of them demonstrated that the ADP/ATP carrier exists in two conformations.

AAC Structure

The mitochondrial carriers are structurally unlike the MFS transporters. The AACs from different species are organized into three homologous repeats of around 100 residues each (Figure 6.20). Each repeat contains a motif common to members of the mitochondrial carrier family (MCF): PxD/ExxK/RxK/R-(20–30 residues)-D/EGxxxxaK/RG, where "x" denotes any amino acid and "a" denotes an aromatic residue. The AACs consist of six TM helices with both N and C termini in the inter-membrane space (IMS) (Figure 11.11). The yeast carrier has been the

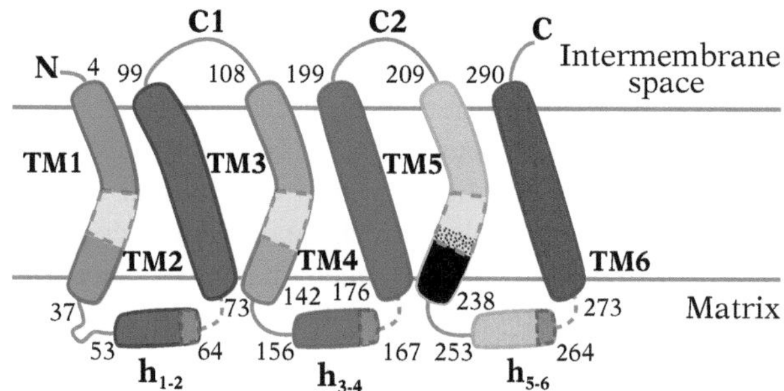

11.11 The topology of the mitochondrial ADP/ATP carriers. Schematic diagram of the secondary structure of the bovine AAC shows the six TM helices with their connecting loops, along with MCF motif residues (gray) and the RRRMMM signature (black). From Nury, H., *et al.*, *Annu Rev Biochem*. 2006, 75:713–741. © 2006 by Annual Reviews. Reprinted with permission from the Annual Review of Biochemistry, www.annualreviews.org.

object of studies employing site-directed mutagenesis to identify critical residues; it shares some of these residues, along with general features, with the bovine AAC that has been crystallized. Purified bovine AAC in detergent was bound to carboxyatractyloside (CATR) to stabilize it in one conformation for crystallization. Two crystal forms have been solved, one at 2.2 Å resolution and a second allowing resolution of three cardiolipin molecules per protein. AAC requires cardiolipin for efficient transport.

The crystal structure shows the 297 amino acids of bovine AAC form a bundle that is closed toward the matrix,with a wide opening toward the IMS (Figure 11.12). The backbone exhibits the three-fold symmetry that is expected from the primary structure. The TM helices are tilted, with TMs 1, 3, and 5 sharply kinked at conserved proline residues. All but TM4 are linked by extensive hydrogen-bond networks. The three

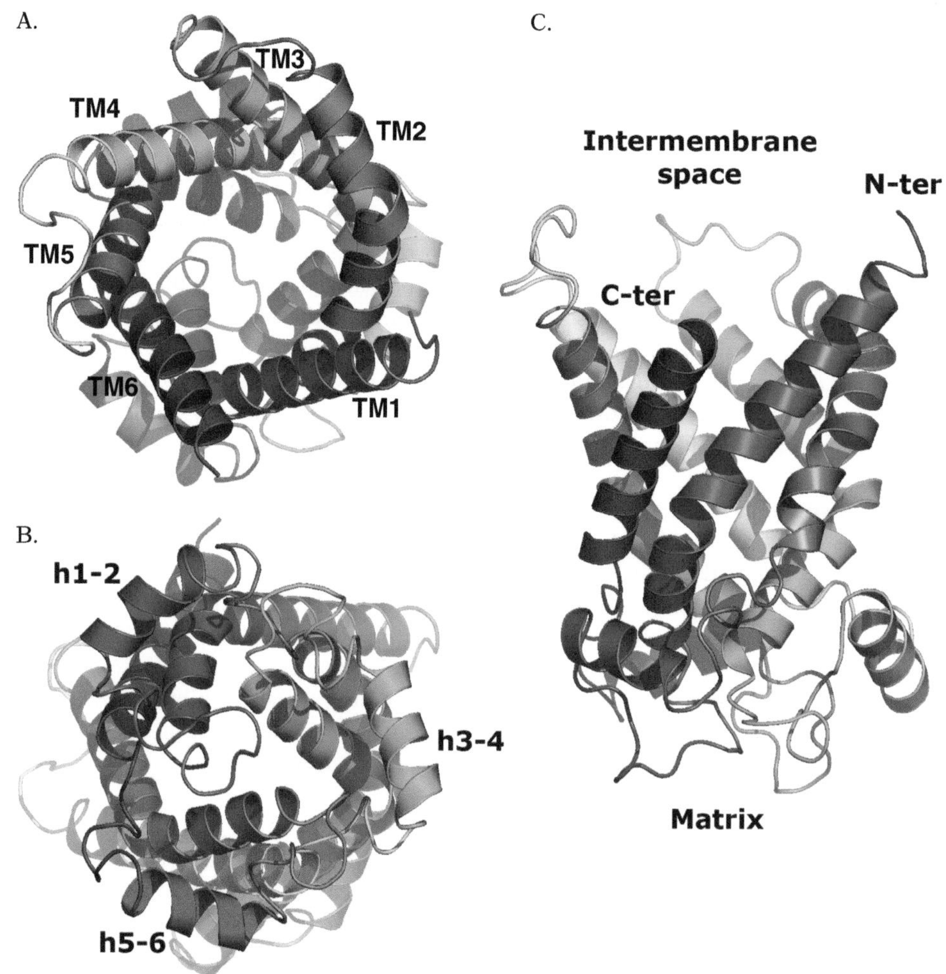

11.12 The overall structure of the bovine ADP/ATP carrier. The ribbon diagrams of AAC viewed from three perspectives are colored to reflect the primary structure (rainbow colors from N terminus, blue, to C terminus, red). The TM helices (TM1–6) and the short amphipathic helices on the end facing the matrix (h_{1-2}, h_{3-4}, and h_{5-6}) are labeled. **(A)** View from the IMS; **(B)** view from the matrix; and **(C)** view from the membrane plane. From Nury, H., *et al.*, *Meth in Molec Biol*. 2010, 654:105–117. © 2010 by Springer-Verlag. With kind permission of Springer Science + Business Media B. V.

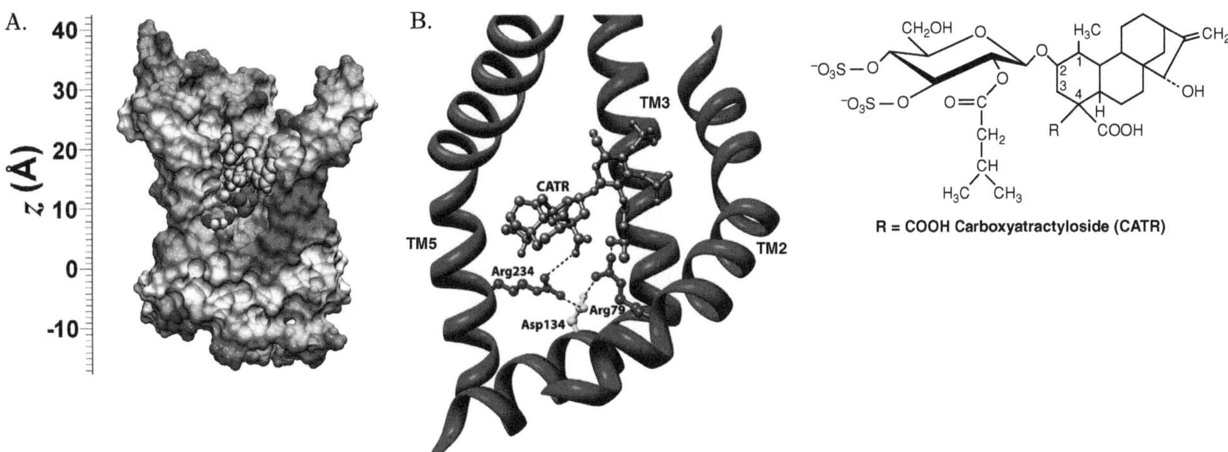

11.13 The AAC cavity and binding site. (A) MD simulation of ADP^{3-} binding to AAC starts with ADP^{3-} (green spheres) bound in the cavity. The surface of the carrier is colored according to the electrostatic potential. From Dehez, F., *et al.*, *JACS*. 2008, 130:12725–12733. © 2008, American Chemical Society. (B) The inhibitor-binding site of AAC. The crystal structure of AAC bound to CATR (gray) reveals salt bridges that stabilize the inhibitor involving Arg79 and Arg234 (blue ball-and-stick), and Asp134 (yellow). From DiMarino, D., *et al.*, *J Struct Biol*. 2010, 172:225–232. The inset shows the structure of CATR. © 2010, Elsevier. Reprinted with permission from Elsevier.

loops on the matrix side contain small amphipathic helices parallel to the membrane surface that close off the deep hydrophilic cavity. The arginine residues of the highly conserved AAC signature sequence RRRMMM contribute to an electrostatic funnel at the bottom of the cavity that MD simulations indicate is important in attracting nucleotides to the binding site (Figure 11.13A).

In the crystal structure CATR fills the cavity, interacting with the protein through many hydrogen bonds and van der Waals contacts and stabilized by salt bridges to Arg79, Arg234, and Asp134 (Figure 11.13B). The funnel-shaped opening of the cavity is thought to orient incoming nucleotide substrates, allowing them to glide along a ladder of tyrosine residues (Tyr194, Tyr190, and Tyr186) on TM4. At the bottom of the cavity is a salt bridge between Arg236 and Glu264 that would be disrupted in an inward-open conformation when the AAC binds ATP^{4-} from the matrix.

The exchange process catalyzed by AAC requires that it bind to and release substrates on opposite sides of the membrane. The large movements of TM helices needed for alternating access most likely involve hinges at the Pro residues that form kinks in the odd-numbered helices. Since these proline residues are part of the highly conserved MCF motif, such hinging may be a general mechanism for mitochondrial carriers.

The crystal structure of AAC indicates that a monomer is sufficient for transport, however many biophysical measurements suggest that mitochondrial carriers function as dimers. A second crystal structure of AAC shows dimers linked by molecules of cardiolipin, which coats the surface and interacts with the MCF motif (Figure 11.14).

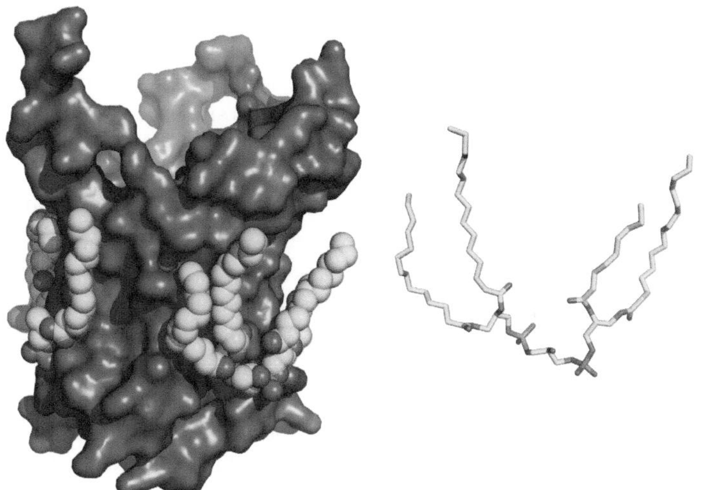

11.14 Cardiolipin-binding sites on AAC. Three molecules of cardiolipin (yellow and red van der Waals spheres) interact closely with AAC (blue surface model), covering much of its surface when viewed from the plane of the membrane. The inset shows the structure of a cardiolipin (yellow and red stick model). Kindly provided by Prof. Eva Pebay-Peyroula.

It is possible that in the mitochondrial membrane AAC forms large complexes with lipids, along with the supercomplexes of the respiratory chain (see Chapter 13).

Neurotransmitter Transporters

Uptake of neurotransmitters at synapses in the brain utilizes other types of secondary transporters. Nerve impulses travel between neurons by release of chemical signals – neurotransmitters such as glutamate, γ-aminobutyric acid (GABA), glycine, acetyl choline, serotonin, dopamine, and norepinephrine. Prior to their release at synapses, neurotransmitters are loaded into synaptic vesicles. Exocytosis of the vesicles in response to an action potential allows a rapid flux of very high concentrations of the neurotransmitters into the synapse, where they can bind to postsynaptic receptors (see Chapter 10). Excess neurotransmitters are taken up by transporters in the presynaptic membrane and the surrounding glial cells in order to ready the synapse for another signal. (Most neurotransmitters are taken up intact, while choline is taken up after acetylcholine is hydrolyzed by acetylcholine esterase.) The transporters that load the vesicles are dedicated specific antiporters that allow protons to move out down the gradient created by a V-type ATPase to drive the uptake of neurotransmitters. The transporters that carry out reuptake at the synapse are symporters that take advantage of ion gradients, typically Na^+, to drive the uptake of neurotransmitters. This action is critical to clear the synaptic junction to prevent continued stimulation, and it also recycles them for reuse.

There are two structurally distinct families of neurotransmitter transporters, one dependent on sodium and chloride (called NSS for neurotransmitter:sodium symporter) and the other dependent on sodium and potassium (called EAATs for excitatory amino-acid transporters). The NSS family includes the transporters for GABA, norepinephrine, dopamine, and serotonin, which have sequence identities as high as 60–70% in humans. Because of their critical roles in normal brain functions, neurotransmitter transporters are important drug targets. Antidepressants like fluoxetine (Prozac), desipramine, and imipramine block reuptake of serotonin and norepinephrine. The main effects of cocaine, which inhibits transporters for norepinephrine, dopamine, and serotonin, are attributed to prolonged signaling by dopamine. After decades of biochemical and physiological studies, understanding of the mechanisms of neurotransmitter transporters surged forward with crystal structures of two prokaryotic orthologs of the brain transporters.[2] GltPh from *Pyrococcus horikoshii* (a thermophilic archaea) is an ortholog of the human glutamate transporter, and LeuT from the thermophile *Aquifex aeolicus* is an ortholog of the serotonin, GABA, and biogenic amine transporters. The two structures differ greatly and thus illustrate two dissimilar types of amino acid transport. Given their importance, each of them merits a detailed description.

[2] Orthologous proteins are coded by genes in different species that originated from the same ancestral gene. They may or may not have identical functions.

Glutamate Transporters and GltPh

Glutamate is the predominant excitatory neurotransmitter in the brain. It is involved in learning and memory and has been implicated in schizophrenia, depression, and stroke. Humans have five subtypes of glutamate transporters, EAAT1–5, which couple the uptake of glutamate to the co-transport of Na^+ and H^+ and the counter transport of K^+, along with an uncoupled Cl^- conductance. GltPh, the first protein in this class of neurotransmitter transporters to have its structure solved, is a sodium/aspartate symporter from *P. horikoshii* that has 36% sequence identity with the human EAAT2.

The topology of GltPh shows eight transmembrane helices with some unusual characteristics (Figure 11.15). TM2 and TM5 are very long, up to 49 residues in length; TM7 spans the membrane in two half-helices; TM8 is amphipathic; and TM4 contains a helix–turn–helix–turn–helix corkscrew structure. In addition, two helical hairpins, called HP1 and HP2, occur between TM6 and TM8 in the primary structure.

The first crystal structure solved at 3.5-Å resolution revealed a novel fold. GltPh is a trimer with a triangular shape when viewed from the extracellular side of the membrane (Figure 11.16). In each subunit the N-terminal helices, TM1–6, mediate the intersubunit contacts, with the C-terminal core bundled within the wedge they form (see also Figure 11.18A). The trimer spans the membrane with a deep, bowl-shaped interior that is lined with hydrophilic residues, giving ready access to substrate from the synaptic fluid.

Each subunit has a substrate-binding site and functions independently. The substrate-binding site consists of portions of the C-terminal core: specifically the tips of HP1 and HP2, the unwound region of TM7 and polar residues of TM8 (Figure 11.17A). In the crystal structure of GltPh with substrate bound, this site is occluded from the surface by an extracellular gate formed by HP2 (see Figure 11.18A). In crystals soaked with the competitive inhibitor D,L-threo-β-benzyloxyaspartate (TBOA), the bulky aromatic group prevents HP2 closure: HP2 has moved 10 Å, allowing access to the substrate-binding site. This open conformation is also observed in averaged density maps of substrate-depleted crystals, strongly supporting the role of HP2 as a gate.

Aspartate binding is sodium dependent, and the structure shows two Na^+ ions are bound in close proximity to the substrate (Figure 11.17B). Na1 is located where TM8 fits into a groove created by unwound portions of TM7

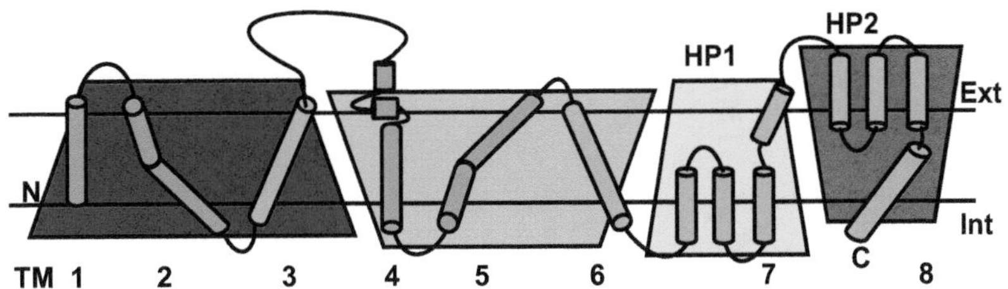

11.15 Topology of GltPh showing the eight TM helices and two helical hairpins that point toward the center of the bilayer from opposite sides. The colored trapezoids show two inverted repeats, with one repeat colored blue and green and the other yellow and red. Modified from Boudker, O. and Verdon, G., *Trends Pharmcol Sci*. 2010, 31:418–426.

and TM1 and is coordinated by carbonyl oxygen atoms of Gly306, Asn310, and Asn401 in TM7 and TM8, a carboxyl group of Asp405 in TM8, and possibly a hydroxyl oxygen of Ser278 in HP1. Na2 is coordinated by four carbonyl oxygen atoms in TM7 (Thr308) and HP2 (Ser349, Ile350, and Thr352) and may be stabilized by the dipole moments of helices TM7a and HP2a. The similarities to the structure of the Na^+-binding site in LeuT (see below) suggest a common characteristic for sodium binding where a break in helix structure opens up the peptide chain to free main chain carbonyl oxygen atoms to coordinate the Na^+, with two carbonyls providing ligands for Na1 and a third carbonyl from an intermediate position providing a ligand for Na2 (Figure 11.17C). This arrangement ties Na^+ binding to conformational changes of the helices likely to occur during transport.

Specificity of GltPh for aspartate is clear when uptake is measured in proteoliposomes: Glutamate is a very

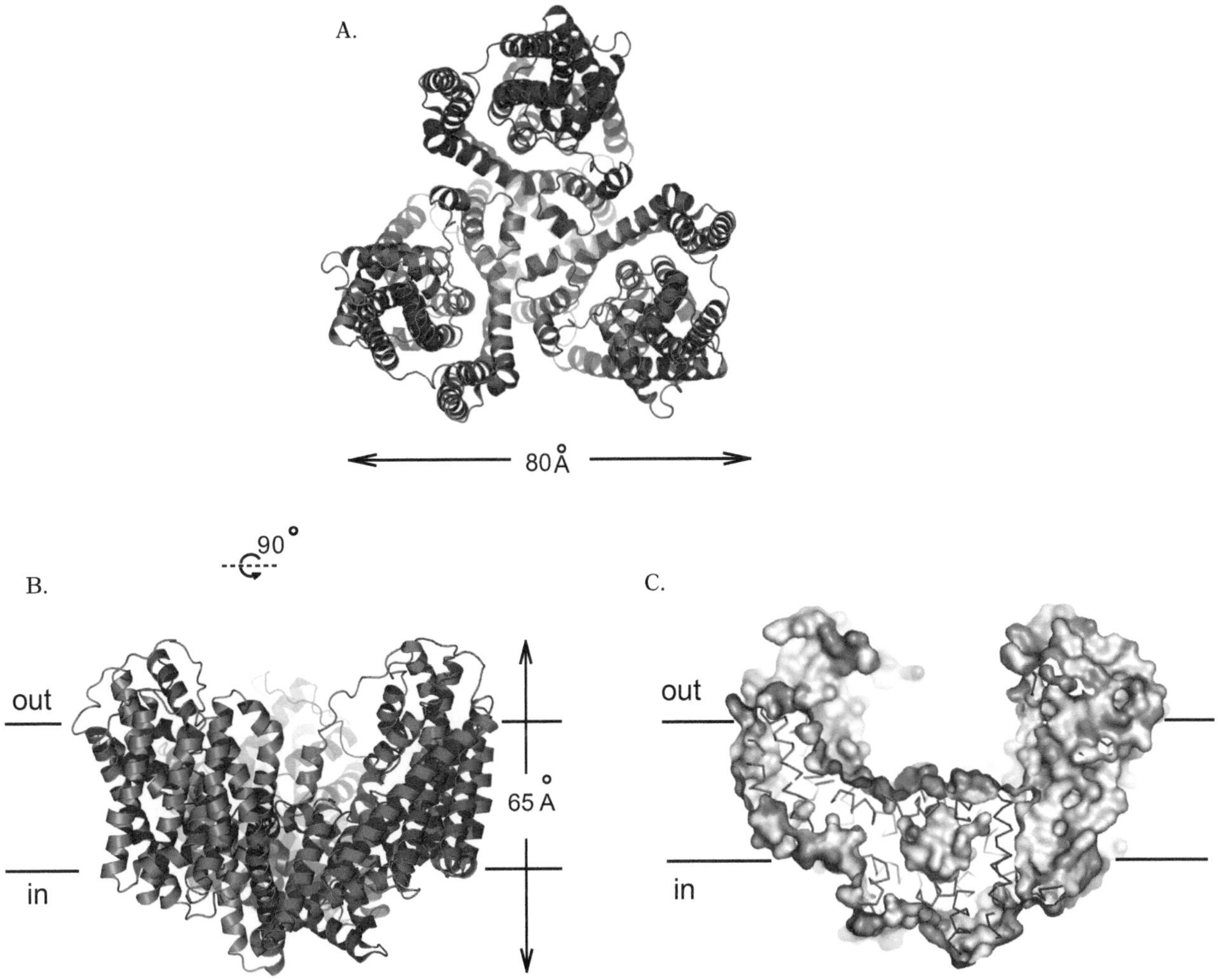

11.16 Representations of the crystal structure of GltPh. (A) Ribbon representation of the trimer viewed from the extracellular side of the membrane. The wedge-shaped subunits are colored red, blue and green. (B) View of the trimer parallel to the membrane, with the same color scheme. The designated boundaries of the lipid bilayer are based on the hydrophobic residues on TM1. (C) Surface representation of the trimer sliced through the center of the basin in the plane of the membrane. Polar residues are colored cyan and apolar residues are colored white. From Yernool, D., *et al.*, *Nature*. 2004, 431:811–818. © 2004, Nature Publishing Group. Reprinted by permission of Macmillan Publishers Ltd.

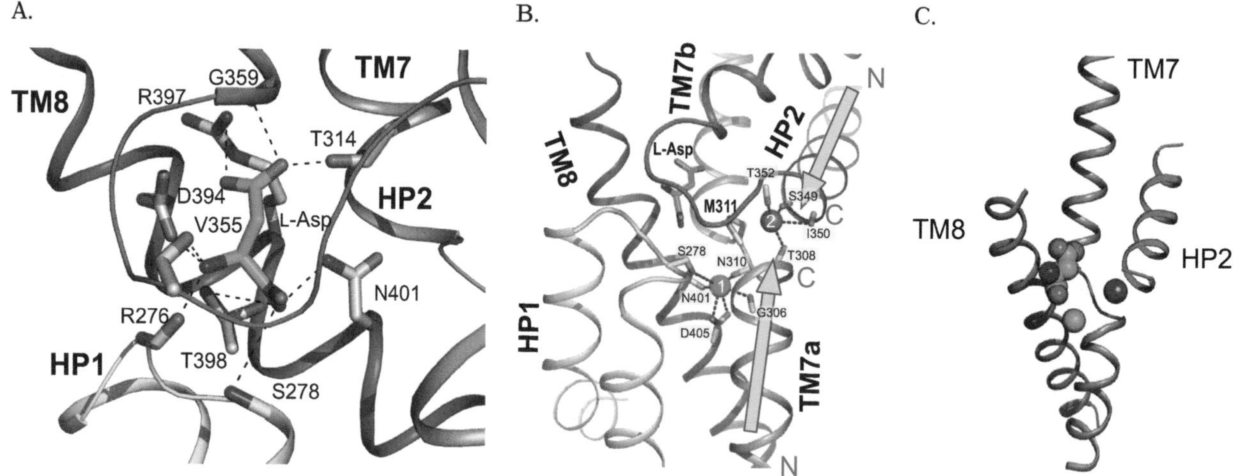

11.17 Substrate- and sodium-binding sites of GltPh. (A) A view of aspartate (stick representation with carbon atoms in green) in the substrate-binding site shows groups from HP1 (yellow), TM7 (orange), HP2 (red), and TM8 (purple) making many polar interactions with aspartate. These include the main chain carbonyls of R276 on HP1 and V355 on HP2, the main chain nitrogen of G359 on HP2, the main chain nitrogen of S278 on HP1, the amide nitrogen of N401, the hydroxyl group of T398, the side chain carboxylate of D394, and the guanidinium group of R397 on TM8, and the hydroxyl of T314 on TM7. (B) The sodium-binding sites are near (but not in contact with) the bound aspartate. The two sodium ions (Na1, green and Na2, purple) are 6.7 Å apart. Na2 is buried below HP2 (red), while Na1 holds together TM7 (orange) and TM8 (magenta). (C) The central role of TM7 in substrate and ion binding is clear in a simplified view of TM7, TM8 and HP2. The colors are the same as in (B), with spheres representing the aspartate, Na1, and Na2. The unwound region of TM7 defines a sodium-binding motif with the main chain carbonyl oxygen atoms at the first and fifth positions as ligands for Na1 and at the third position as a ligand for Na2. From Boudker, O., *et al.*, *Nature*. 2007, 445:387–393.

poor substrate, and Na^+ cannot be replaced effectively by Li^+ or K^+. Symport of two Na^+ ions with each aspartate would produce an electrogenic flux if it were not for the chloride conductance carried out by GltPh. Although permeation by Cl^- is dependent on the presence of glutamate/aspartate and Na^+, it is independent of their rate or direction of transport. The pathway for this chloride channel is not known.

Structures of GltPh are now available in both outward- and inward-facing states, giving insight into the conformational changes involved in the alternating access mechanism of transport. A crystal structure of a GltPh in an inward-facing conformation was achieved by engineering cysteines at positions 55 and 364, predicted to make a disulfide bond that blocks transport activity according to biochemical studies of EAATs, even though they are far apart in the first crystal structure of GltPh. The crystal structure of mercury-crosslinked K55C/A364C in the presence of Na^+ and aspartate was solved at 3.5–3.9 Å resolution using molecular replacement for portions of the protein which can be crosslinked without blocking function. Indeed, comparing the new structure with the first one, the trimerization domain consisting of these helices – TM2, TM4, and TM5 – plus TM1, is essentially unchanged (Figure 11.18). For the conformational change the transport domain, consisting of TM3, TM6, HP1, TM7, HP2, and TM8, makes a large rigid movement relative to the trimerization domain, putting the substrate-binding site at the bottom of a large crevice near the cytosolic surface. The arms formed by TM3 and TM6, which are structurally symmetrical, extend from the trimerization domain to the transport core and enable the hinge movement, aided by flexible loops. In the inward-facing structure the substrate is still occluded by HP1, which is predicted to function as the intracellular gate. Thus two types of molecular movements mediate the transport cycle: the large hinge motions make a thick gate that changes the orientation of the transport domain (illustrated in Figure 11.18), and the gating movements of thin gates – HP2 and HP1 – occlude the substrate-binding site (Figure 11.19).

As a glutamate transporter ortholog, GltPh represents the important family of human secondary transporters called the solute carrier 1 (SLC1) family. Members of this family transport acidic and neutral amino acids and dicarboxylic acids coupled to symport of sodium and/or protons. (The EAATs also carry out antiport of potassium.) Many residues of the transport domain in the C-terminal core of the proteins are highly conserved. Also, the sodium-binding motif is similar in the unrelated transporter LeuT (see below) and thus may be common to sodium-dependent transporters.

Neurotransmitter Sodium Symporters and LeuT

Members of the NSS family transport a diverse group of neurotransmitters that includes the biogenic amines

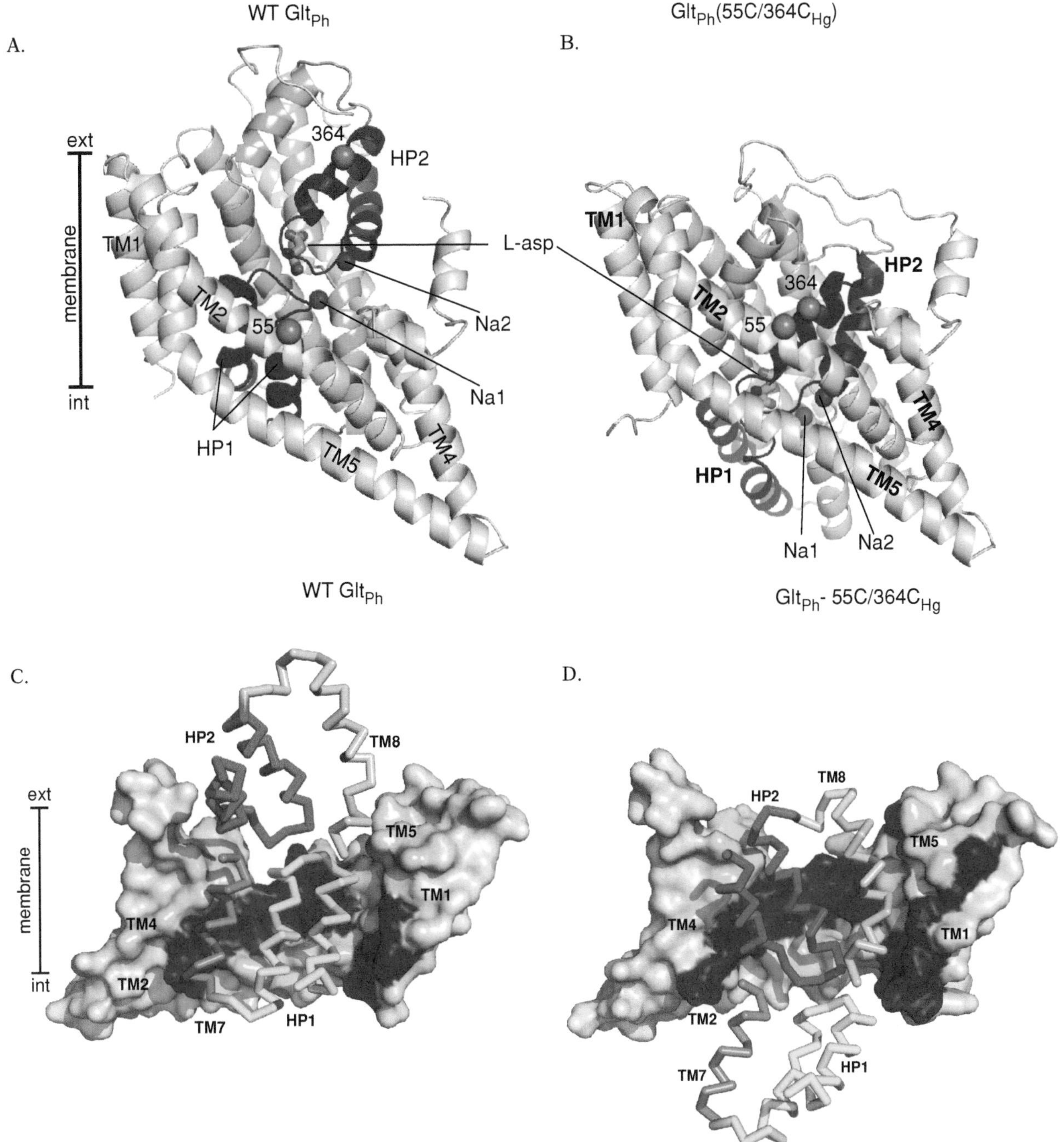

11.18 Monomer and trimer of GltPh in two conformational states. The wild type crystals (represented in (A) and (C)) are in the outward-facing state and the crystals of Glt$_{Ph}$ (55C/364C$_{Hg}$) (represented in (B) and (D)) are in the inward-facing state. (A),(B) Ribbon representations of the monomer with the trimerization domain (TMs 1, 2, 4, and 5) colored wheat, the rest of the monomer light blue and the Na^+ ions as purple spheres. Aspartate (shown as stick representation) is between HP1 and HP2, which are a darker shade of blue. In (B), the large movement of the transport domains has brought together the cysteine residues at 55 and 364 (represented as large red spheres), which were separated by >25 Å in (A). (C),(D) In the trimer the conformational change moves the transport domains through the wedge formed by the trimerization domains via hinge movements, preserving the substrate- and ion-binding sites while moving them 18 Å toward the cytoplasmic surface. To illustrate these movements, the trimerization domains are shown in surface representation, while the core of the transport domains are shown in ribbon representation. The residues involved in domain contacts in the trimerization domain (wheat) are colored blue (TM1 and TM2) and light green (TM4 and TM5). The elements of the transport core – HP1 (yellow), TM7 (orange), HP2 (red), and TM8 (pink) – are essentially superimposable when examined away from the trimerization domain, indicating the transport domain moves as a rigid body with little change in its conformation. From Reyes, N., *et al.*, *Nature*. 2009, 462:880–885. © 2009, Nature Publishing Group. Reprinted by permission of Macmillan Publishers Ltd.

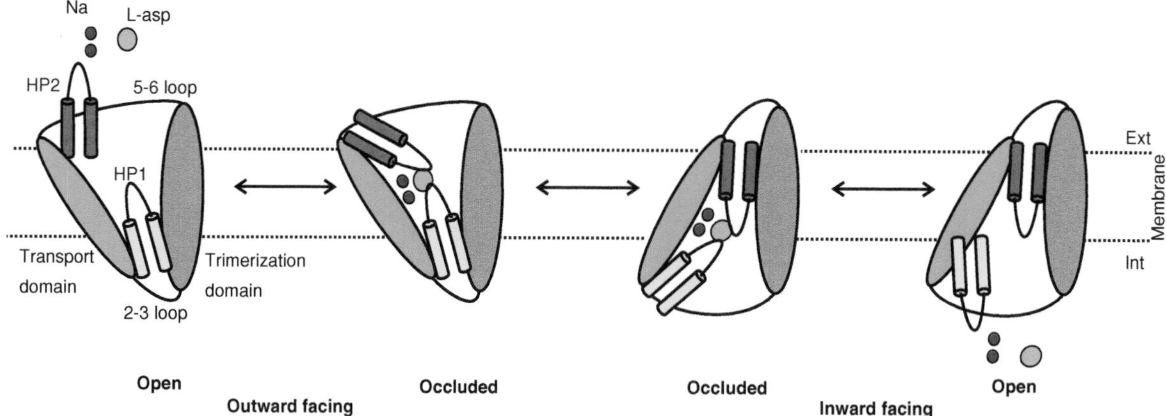

11.19 Schematic transport mechanism showing the reversible steps that take place in one subunit of GltPh. One substrate and two Na^+ ions bind to the open, outward-facing state that is also observed in the presence of the inhibitor TBOA. HP2 (red) closes the external gate to produce the substrate- and sodium-bound occluded outward-facing state observed in the wild type crystals. The movement of the transport domain (blue) relative to the trimerization domain (gray) accomplishes the conformational change to the occluded inward-facing state observed in the 55C/364C$_{Hg}$ mutant. Finally, HP1 (yellow) is predicted to open the internal gate, although this state has not yet been observed. From Reyes, N., *et al.*, *Nature*. 2009, 462:880–885. © 2009, Nature Publishing Group. Reprinted by permission of Macmillan Publishers Ltd.

(serotonin, dopamine, norepinephrine) and amino acids (GABA, glycine), as well as osmolytes (betaine, taurine, and creatine). These secondary transporters use sodium and chloride gradients to drive the uptake of their substrates across the presynaptic membrane. Like the glutamate transporters, these proteins are implicated in neurological problems, including depression, Parkinson's disease, autism, and epilepsy, and they are the targets of anticonvulsants and antidepressants, as well as cocaine and amphetamine. The functional regions of the NSS are highly conserved in bacterial homologs such as LeuT, the nonpolar amino acid transporter of *A. aeolicus*, in spite of an overall sequence identity of only 20–25%. The x-ray structure of LeuT provided the first detailed view of an amino acid transporter as well as an NSS ortholog.

LeuT structure

The first crystal structure of LeuT was solved at very high resolution, 1.65 Å, providing access to exquisite atomic detail. LeuT has 12 TM helices with an internal repeat of TM1–5 and TM6–10 (Figure 11.20). In the LeuT structure the helices are tightly packed, forming an outer scaffold that surrounds a central core, as observed in GltPh. The scaffold is formed by TMs 3–5 and TMs 8–10; the core contains TM1, TM2, TM6, and TM7. Between these domains two inner pairs of helices, TM1/TM6 and TM3/TM8, form the central translocation pathway with substrate- and Na^+-binding sites. The x-ray structure shows one leucine bound, along with two Na^+ ions and a nonessential Cl^- ion.

The central leucine-binding site is formed by the unwound portions of TM1 and TM6 at residues Val23 and Gly24 and from Ser256 to Gly260 (Figure 11.21A). Unwinding the helices exposes main chain α-amino and α-carboxyl groups to form hydrogen bonds with the substrate, and also allows the charged groups of the substrate to associate with helical dipole moments at the ends of the half helices. The hydroxyl group of a strictly conserved tyrosine residue (Tyr108) also forms a hydrogen bond with the substrate carboxyl group. The aliphatic chain of leucine lies in a hydrophobic pocket formed by residues from TM3, TM6, and TM8 that determines the substrate affinity. The close fit is consistent with the high affinity of LeuT for leucine determined by equilibrium binding studies to be 20–50 nM.

The structure shows LeuT in an occluded state, as leucine is bound near the center of the bilayer with no solvent accessibility. The occluded conformation is stabilized by ion pairs between TM1 and TM10 on the extracellular side (Arg30 and Asp404) and between TM1 and TM8 on the intracellular side (Arg5 and Asp369). The substrate is much less shielded by protein groups on the extracellular side than the intracellular side of the membrane, so this structure portrays a closed, outward-facing conformation ("outward-occluded" – see below).

Consistent with the sodium-dependence of leucine uptake by LeuT in liposomes, two dehydrated Na^+ ions are observed near the substrate-binding site (Figure 11.21B). Na1 is coordinated by the α-carboxyl oxygen of the substrate, in addition to carbonyl and hydroxyl groups from residues in TM1, TM6, and TM7 (Ala22, Asn27, Thr254, and Asn286). Na2 is 7 Å away from Na1, coordinated by carbonyl and hydroxyl groups of Gly20, Val23, Ala351, Thr354, and Ser355 between TM8 and the unwound region of TM1. Both coordination sites are too small to accommodate potassium ions; although they can bind lithium ions, transport rates are markedly lower with Li^+. Both sodium ions in the LeuT structure

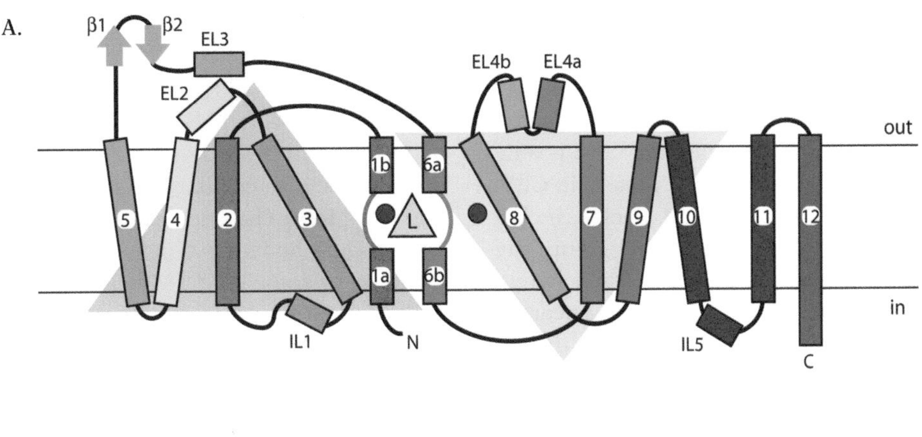

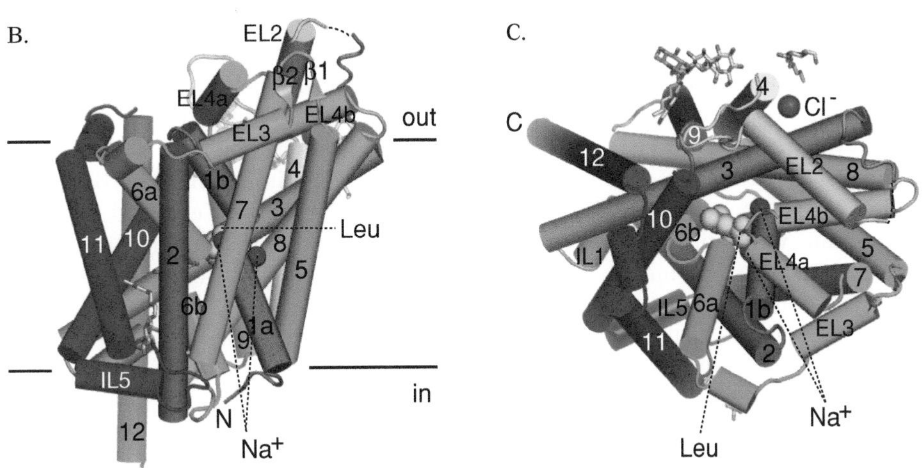

11.20 Structure of LeuT from *A. aeolicus*. (A) Topology of LeuT. LeuT has 12 TM helices, 6 short α-helices in extracellular and intracellular loops (EL and IL), and an external β turn. The inverted repeat units TM1–5 and TM6–10 are highlighted by pink and blue triangles. The positions of substrate and Na^{+} ions are depicted as a yellow triangle and blue circles, respectively. (B)–(C) The overall structure of LeuT is shown from within the plane of the membrane (B) and from the extracellular side (C), with colors and numbers of TM helices and external and internal loops (labeled IL and EL) corresponding to (A). The resolution is sufficient to distinguish not only bound L-leucine (CPK, yellow), but also two sodium ions and a chloride ion (spheres, blue and magenta, respectively), and detergent molecules (β-OG, stick). From Yamashita, A., *et al.*, *Nature* 2005, 437:215–223. © 2005, Nature Publishing Group. Reprinted by permission of Macmillan Publishers Ltd.

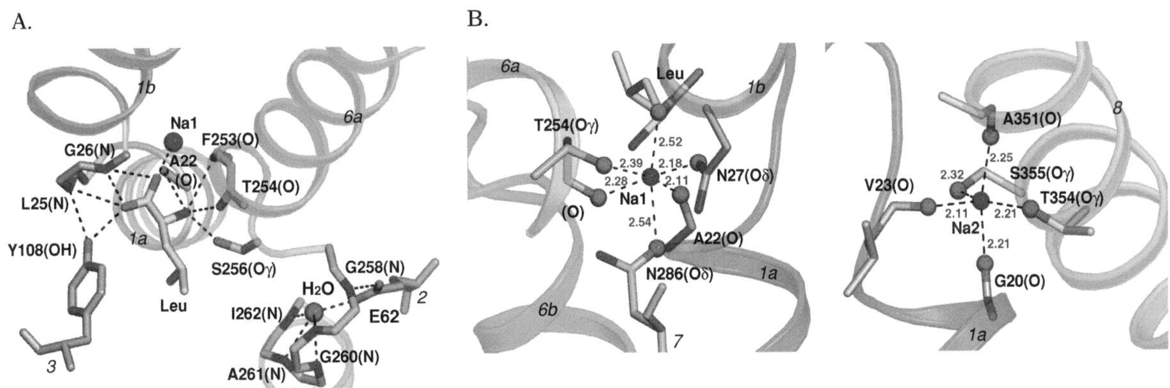

11.21 Substrate- and ion-binding sites of LeuT. (A). Leucine binds to a substrate-binding site where TM1 (pink) and TM6 (green) are unwound. Many hydrogen bonds (dashed lines) are formed between the substrate and the main chain peptide groups, along with the side chain of Tyr108 from TM3 (gold). The α-carboxyl group of leucine interacts with the positively charged end of TM1b helix and the α-amino group is near the negatively charged ends of both TM1a and TM6a. (B) The binding sites for the two Na^{+} ions, Na1 (left) and Na2 (right), are positioned between TM1 (pink), TM6 (green), TM7 (cyan), and TM8 (blue). Coordination of Na1 is octahedral and coordination of Na2 is trigonal bi-pyramidal. The distances in angstroms are given between the sodium ions (blue spheres) and their ligands: the backbone carbonyl oxygen atoms from the unwound portions of TM1 and TM6 (Ala22 and Thr254 for Na1, and Gly20 and Val23 for Na2); side chain carbonyls from Asn27 in TM1 and from Asn286 in TM7 (Na1), and hydroxyl oxygens from Thr254 in TM6 (Na1) and Thr354 and Ser355 in TM8 (Na2). The carboxyl group of the leucine substrate (stick model) is a ligand for Na1 as well. From Yamashita, A., *et al. Nature*. 2005, 437:215–223. © 2005, Nature Publishing Group. Reprinted by permission of Macmillan Publishers Ltd.

are located between the two unwound helices and TM8. Two of three carbonyl oxygen atoms from the unwound portion of TM1 are ligands at Na1 and the third is a ligand at Na2, similar to the characteristic pattern in GltPh (see Figure 11.17C). Their proximity to leucine, including the coordination of Na1 by its carboxyl group, suggests a mechanism for coupling amino acid and sodium ion transport. No Cl^- ions were found to bind near the substrate/sodium binding pocket, consistent with finding that transport is not chloride dependent.

Structures of LeuT in Two Other Conformational States. Based on homology modeling and the architecture of protein structures of similar folds in Ci and Co conformational states, mutations were designed to stabilize those states in LeuT. Introducing the Y108F mutation in a wild type-like variant, K288A ($LeuT^K$), weakens the substrate affinity to stabilize the outward-facing state. Three mutations in $LeuT^K$ are required to stabilize the inward-facing state: weakening the Na2 binding site with T354V and S355A, and perturbing the intracellular gate with Y268A. Conformation-specific antibodies aided crystallization of these LeuT mutants in the absence of substrate (Figure 11.22). The 3.1-Å resolution structure of Y108F $LeuT^K$ is open to the outside due to hinge movements in TMs 1b, 2a, and 6a, which displace EL4 and TM11, compared to the outward-occluded structure. The thin gate in the outward-occluded structure is opened in the Co structure by disruption of the salt bridge between Asp404 and Arg30, as well as rotation of Phe253. With Na^+ ions at both Na1 and Na2, the intracellular thick gate remains closed.

In contrast, the 3.2-Å resolution structure of the triple mutant TSY $LeuT^K$ indicates that large hinge-like movements within the core domain relative to the scaffold domain produce the inward-open conformation (Figure 11.22B). TM1a is tilted by 45° and TM6b is rotated by 17°, while TM1b and TM6a tilt together (slightly >20°) toward the scaffold domain to block the extracellular pathway. This difference is facilitated by bending of the supporting helices, TM2, TM7, and TM5, at Gly or Pro residues. EL4 closes off the extracellular solvent pathway by packing tightly against TM helices from both sides. Interactions between residues of the thick intracellular gate observed in the occluded and Co states are disrupted, allowing it to open, making a cavity lined by TM1a, TM5, and TM8. Separation of TM1a and TM8 obliterates the Na2 site, with only minor alterations at Na1 and substrate-binding sites.

Transport Mechanism of LeuT

The three crystal structures of LeuT show conformational states that support an alternating access mechanism with thick gates and thin gates controlling access (Figure 11.23; see also Figure 11.1C). The thick gates of LeuT are formed by movements of the core bundle (TM1/TM2 and TM6/TM7) relative to the scaffold (TM3/TM4/TM5 and TM8/TM9/TM10). The extracellular thin gate is formed by helical segments TM1b, TM6a, and loop EL4. (An inward-facing occluded structure is lacking, but the intracellular thin gate is predicted to involve TM1a–TM6b.) The thick gates are opened and closed by nearly rigid body movements, with adjustments due to bending of some core TM helices (TM2 and TM7) and independent movements of others (TM1a and TM6b). The hinge-like movements of TM helices form and disrupt the substrate- and sodium-binding sites to facilitate transport when the LeuT structure changes overall conformation.

Binding Sites and Inhibitors. LeuT inhibitors include pharmacologically important compounds, whose mechanisms of action can be illuminated by crystal structures of LeuT in their presence. The tricyclic antidepressants clomipramine, imipramine and desipramine are noncompetitive inhibitors of LeuT, and tryptophan is a competitive inhibitor. Clomipramine binds in the outward-facing vestibule about 11 Å above the substrate. By displacing two waters, it allows direct contact between Arg30 and Asp404, which further stabilizes the closed state (Figure 11.24A). Tryptophan binds to the substrate-binding site with its bulky indole ring wedged between TM3, TM8, and TM10, preventing closure on the extracellular side (Figure 11.24B). Thus noncompetitive inhibitors stabilize LeuT in the occluded state, while competitive inhibitors lock it in an open-out state.

Binding of clomipramine displaces a molecule of detergent (β-octyl glucoside) that occupies the external vestibule in the first LeuT crystal structure. This external vestibule has been proposed to be a second high-affinity substrate-binding site (S2). However, leucine and the readily detected substrate selenomethionine do not bind S2 in crystals of LeuT, even in a crystal structure of LeuT in DMPC/CHAPSO bicelles with no β-octyl glucoside present. Introducing the mutation F253A to weaken leucine binding at the primary (S1) site seemed to produce LeuT that binds leucine only at S2, with competitive inhibition by clomipramine. However, careful equilibrium binding studies show clomipramine to be a noncompetitive inhibitor of wild type LeuT, and the crystal structure of the F253A mutant protein shows that both leucine and selenomethionine still bind only to the primary (S1) site. It is possible that leucine binding to S2 is a transient step on the entrance pathway.

The NSS Family and the LeuT (APC-Fold) Superfamily

The structure of LeuT provides a framework to better understand the mammalian neuronal transporters for serotonin, norepinephrine, and dopamine and for the biogenic amino acids GABA and glycine, along with other members of the NSS family that transport osmolytes (betaine, creatine) and other amino acids (proline,

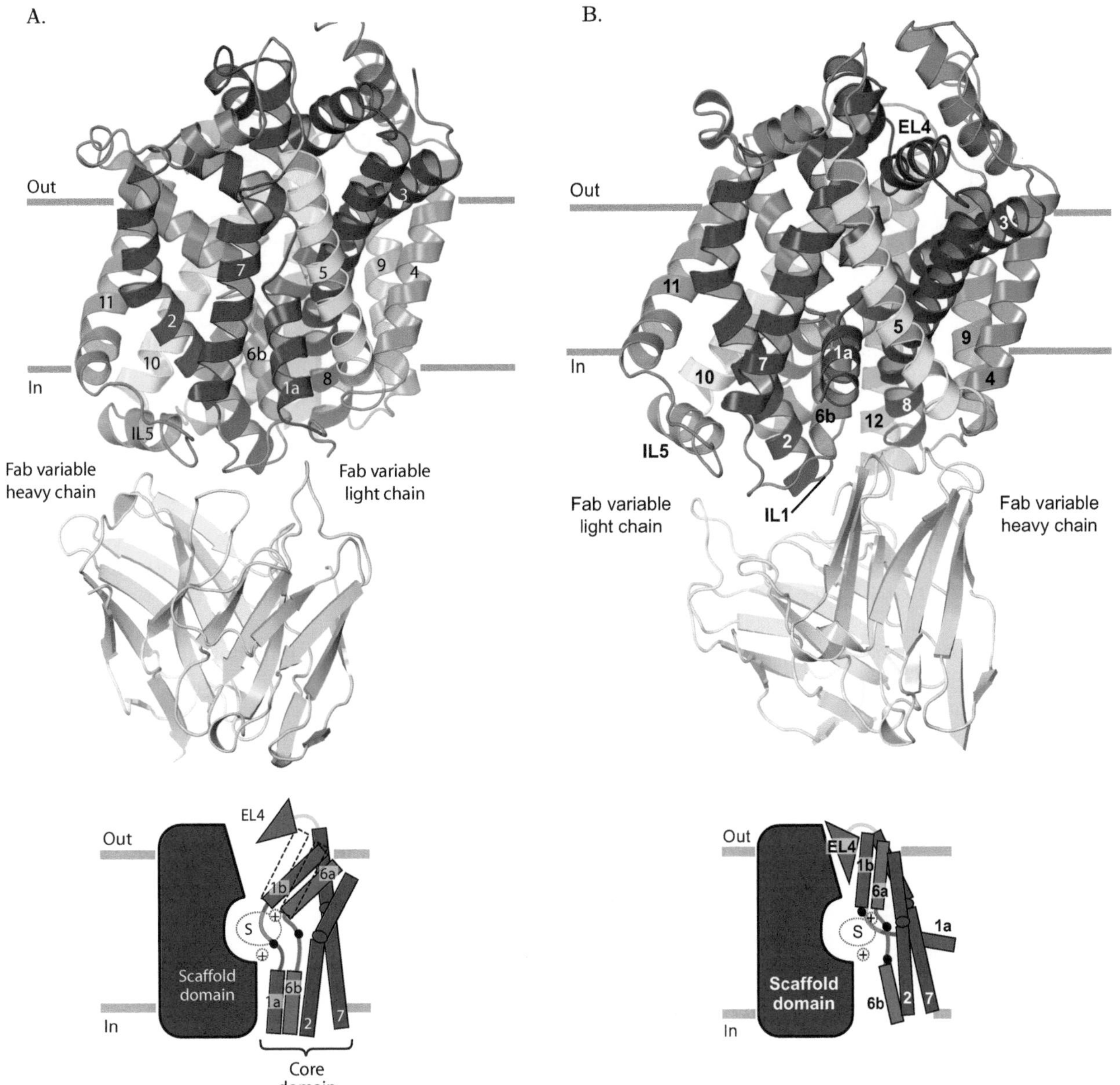

11.22 Crystal structures of LeuT in the open-outward and open-inward conformations. (A) The Y108F LeuT^K mutant crystallized in the Co state when stabilized with the 2B12 Fab. A ribbon representative of the structure seen from the plane of the membrane shows the interaction between LeuT (rainbow coloring from N terminus, red, to C terminus, blue) and the antibody (pink and light green, with only the variable domains of the light and heavy chains shown). A cartoon depicts the scaffold and core domains of the Co structure. EL4 is not in contact with TM helices of the scaffold, due to hinge movements in TM1 and TM6 (red and green, respectively) at their pivot points (solid black circles). The empty substrate-binding site (S, dotted circle) is between Na1 and Na2 (+). (B) The TSY LeuT^K construct crystallized in the Ci state when stabilized with the 6A10 Fab (representation and color scheme as in (A)). A cartoon depicts the core and scaffold domains of the Ci conformation indicative of large hinge-like movements along with bending of support helices. From Krishnamurthy, H., and E. Gouaux, *Nature*. 2012, 481:469–474. © 2012, Nature Publishing Group. Reprinted by permission of Macmillan Publishers Ltd.

taurine). Classified as the solute carrier 6 (SLC6) family, its human members share clusters of high sequence conservation in structurally important regions. Some conserved structural elements include the indispensable Tyr residue in TM3 (Tyr108 in LeuT), an Asp or Gly residue in TM1 that determines whether a monoamine or an amino acid binds (Gly24 in LeuT), and an Asp or Glu residue in extracellular Loop 5/TM10 predicted to be in the external gate.

LeuT differs from the eukaryotic NSS in some important ways. It lacks extended portions of the N and C termini of eukaryotic NSS, which interact with modulator

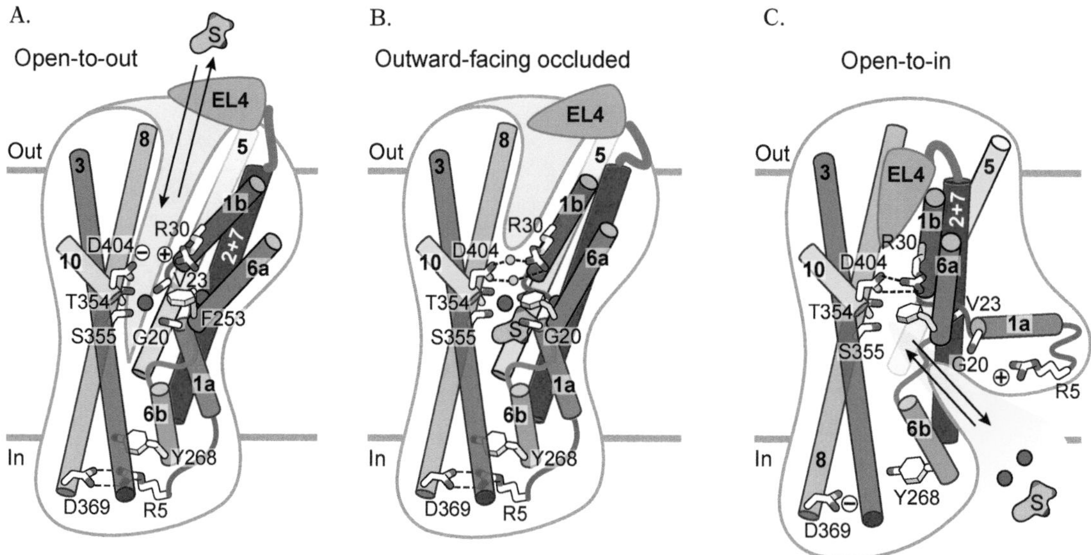

11.23 Transport mechanism of LeuT. A schematic shows structural elements and gating residues that are key to the conformational changes from the Co state (A) to the outward-occluded state (B) and the Ci state (C). From Krishnamurthy, H. and E. Gouaux, *Nature*. 2012, 481:469–474. © 2012, Nature Publishing Group. Reprinted by permission of Macmillan Publishers Ltd.

proteins at the synapse, and it lacks phosphorylation sites on its terminal domains and intracellular loops, important in regulating some of the eukaryotic homologs. The tricyclic antidepressants that are noncompetitive inhibitors of LeuT are competitive inhibitors of the serotonin and norepinephrine transporters. Introduction of 13 mutations in LeuT makes it resemble the human serotonin receptor so that chlomipramine and other drugs such as

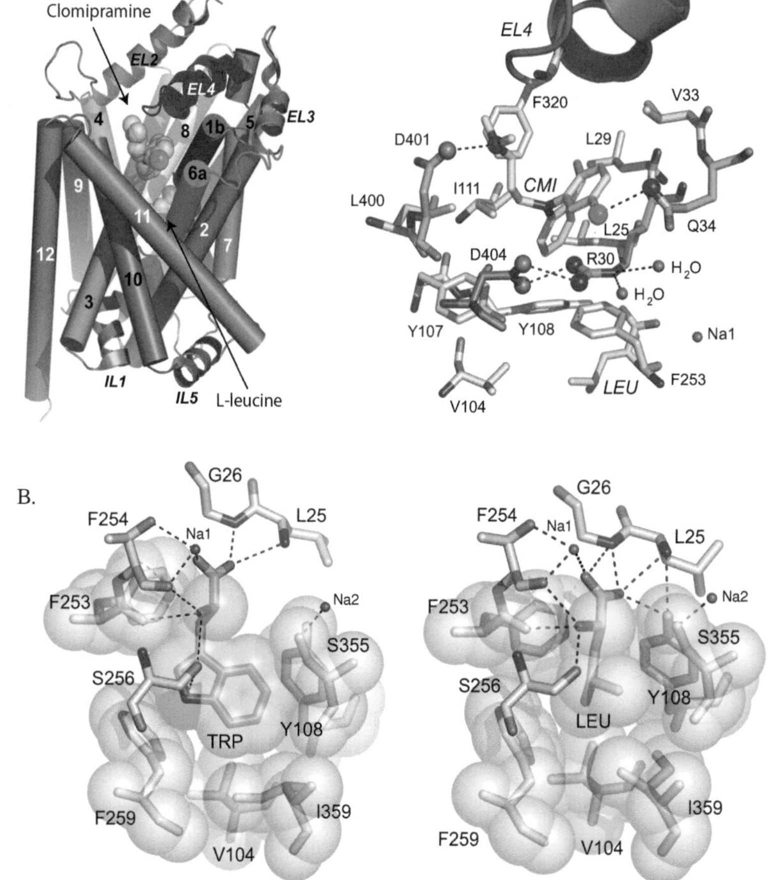

11.24 Inhibitor binding to LeuT. (A) The noncompetitive inhibitor clomipramine binds to the extracellular vestibule of LeuT with no overall change in the occluded structure of the protein. A close-up of the binding site reveals the salt bridges between Asp404 and Arg30 just above Tyr108 that traps the leucine in its binding site. From Singh, S., *et al*., *Nature*. 2007, 448:952–956. (B) The competitive inhibitor tryptophan binds at the substrate-binding site and prevents closure over the site, thus blocking occlusion. This is shown with space-filling models of tryptophan (green) binding on the left (open out), compared to leucine (yellow) binding on the right (occluded). From Singh, S., *et al*., *Science*. 2008, 322:1655–1661. © 2007, Nature Publishing Group. Reprinted by permission of Macmillan Publishers Ltd.

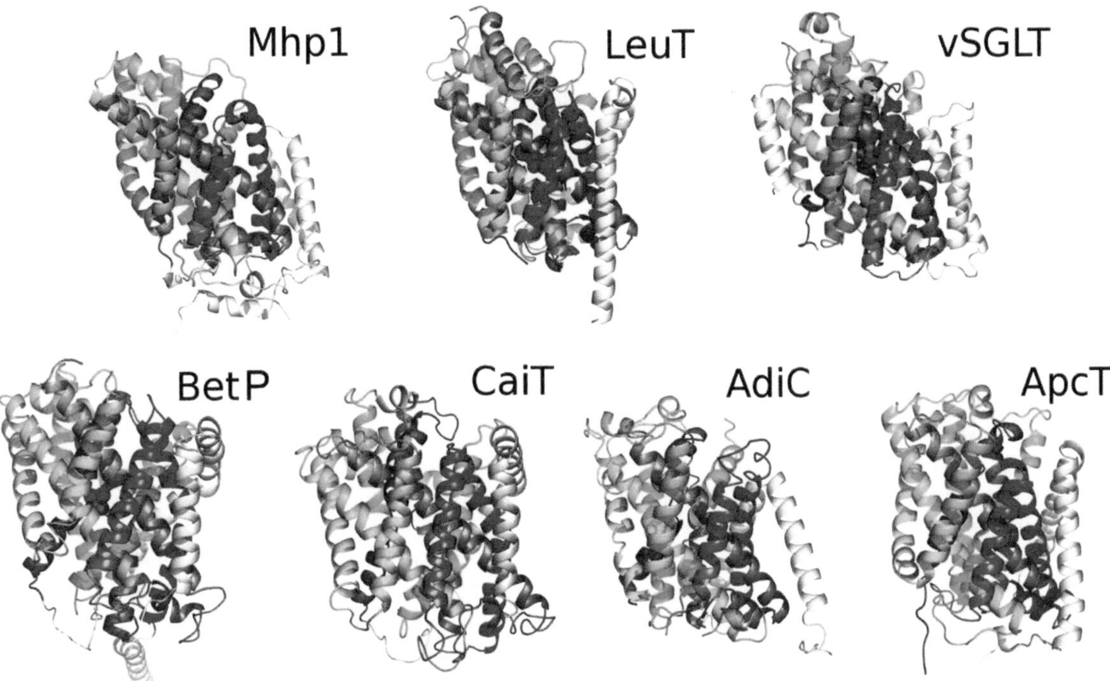

11.25 Seven transporters with the LeuT fold. Ribbon diagrams portraying the structures of Mhp1, LeuT, vSGLT, BetP, CaiT, AdiC, and ApcT are colored to emphasize the similar folds of the internal repeat (rainbow coloring from N terminus, red, to C terminus, blue). The structures vary in the number and positions of TMs that are not part of the ten helix core (gray). From Weyand, S., *J Synchrotron Rad*. 2011, 18:20–23. Figure 1 of Weyand, Shimamura, Beckstein, Sansom, Iwata, Henderson and Cameron (2011). J. Synchrotron Rad. vol 18, pp. 20–23. Reprinted with permission from the International Union of Crystallography.

fluoxetine (Prozac), sentraline (Zoloft), and duloxetine (Cymbalta) now bind to the primary substrate-binding site. Clearly, the detailed structure of LeuT gives important insights into NSS structure and has stimulated many new experiments and homology models.

A significant development is the discovery that structures of other transporters from different families and with dissimilar sequences have the same overall-fold as LeuT. The common architecture of this structural superfamily is based on an inverted repeat motif in ten TM helices that fold into a four-helix bundle at the core with an outer layer scaffold. The two domains are not formed by the N- and C-terminal halves, as in the MFS, but still have corresponding elements in the inverted repeats (see Figure 11.20A). Besides LeuT, the superfamily includes the following transporters with x-ray structures: vSglT, a galactose transporter in the sodium:solute symporter (SCC) family; ApcT and AdiC, amino acid transporters in the amino acid:polyamine:organocation family; Mhp1, the benzyl-hydantoin transporter of the nucleobase cation symporter (NCS1) family; and BetP and CaiT in the betaine-coline-carnitine transporters (BCCT) family (Figure 11.25). The striking implication is that a genetically diverse group of transporters for a wide variety of substrates and organized into several large families share a common fold and mechanism. The structure of one of these transporters, BetP, gives insight into how the activity of membrane proteins can be regulated by environmental conditions.

BetP and Osmoregulated Transport

Microorganisms respond rapidly to many environmental effects, including changes in osmotic pressure. Hyperosmotic conditions stimulate cells to accumulate ions and osmolytes[3] in order to restore normal hydration levels. (Hypo-osmotic conditions induce the mechanosensitive channels described in Chapter 12.) In the soil bacterium *Corynebacterium glutamicum* the activity of the Na^+/betaine symporter BetP is induced by hyperosmotic conditions that shrink the cytosol and thus increase the internal concentration of K^+ ions. The increased cytoplasmic K^+ level activates BetP to take up betaine along with Na^+ ions in a 1:2 ratio. Structural details are emerging that give clues to the mechanism of BetP regulation.

[3] Osmolytes used in biological systems (technically called compatible solutes) are organic solutes that can be accumulated to very high internal concentrations without impairing the cells' vital processes. These highly water soluble compounds are preferentially excluded from the hydration shell of proteins because they interact unfavorably with the peptide backbone. Many have a quaternary ammonium group as in choline, betaine (N,N,N-trimethylglycine), or L-carnitine.

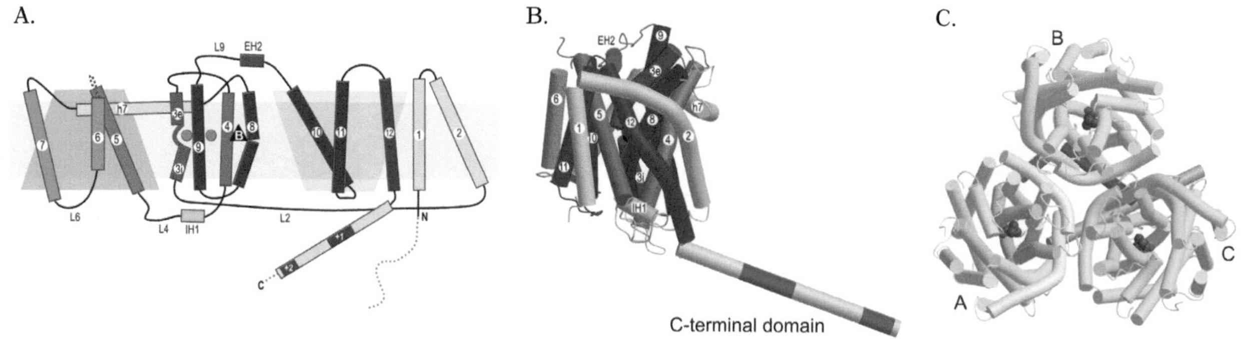

11.26 Structure of BetP. (A) Topology diagram for BetP shaded to show the inverted repeats. Repeat 1 is TM3–TM7 (red/orange, with color intensity decreasing from the N terminus to the C terminus) and repeat 2 is TM8–TM12 (blue with color intensity decreasing in the same way). The remaining helices are TM1, TM2, helix 7, and the C-terminal helix (gray). The substrate-binding site is at the middle of the bilayer, with betaine (black triangle) and two Na^+ ions (green spheres). The N terminus (dotted line) is missing from the crystal structure. (B) A view from the plane of the membrane shows the x-ray structure of the BetP monomer lacking its N-terminal with the periplasmic side at the top and the cytoplasmic side at the bottom. In the ribbon diagram the helices are numbered and colored to reflect the topology shown in (A), with the two basic regions of the C-terminal helix highlighted (blue). (C) The BetP trimer as seen from the periplasm shows three compact cylinders, subunits A (colored as in B but lighter), B (blue), and C (pink), which are linked on the cytoplasmic side (see Figure 11.28). From Ressl, S., *et al.*, *Nature*. 2009, 458:47–52.

The 3.34-Å resolution structure of engineered BetP, which lacks part of the N-terminal domain but is active and osmoregulated, shows a trimer of three nearly cylindrical subunits. Each monomer has 12 TM helices plus a curved amphipathic α-helix (H7) and an extended C-terminal α-helix, both of which contact the neighboring subunits (Figure 11.26). Ten of the 12 TM helices have an internal structural repeat made by a pseudo-two-fold axis in the plane of the membrane between TM3–TM7 and TM8–TM12. In the four-helix bundle TM3 and TM4 nest within TM8 and TM9, while TM5–7 and TM10–12 make the outer scaffold. TM2 and H7 wrap around the exterior near the periplasmic end.

The four-helix bundle is central in the transport pathway. Betaine binds in an aromatic box between TM4 and TM8 near the center of the four-helix bundle (Figure 11.27). Very near the betaine are two putative binding sites for Na^+ ions. In this occluded conformation of BetP betaine clearly lacks a pathway to either the periplasm or the cytoplasm. However, other BetP conformations are seen by fitting the x-ray structure to a cryoEM map of a mutant BetP trimer at 8-Å resolution: Surprisingly, each of the protomers is in a different conformation representing inward-facing, occluded, and outward-facing structures.

The long C-terminal helix is essential for responding to osmotic stress: A BetP mutant that lacks the C-terminal extension has constitutive transport activity and still forms trimers. Not fully resolved in the x-ray structure, the C-terminal helix occupies a different position in each of the subunits (Figure 11.28). Its two basic regions at Arg558–Arg568 and Lys581–Lys587 are likely to form salt bridges with the adjacent subunits within the trimer as well as with Loop 8, which could transmit changes in position of the C-terminal helix to the four-helix bundle. They are also likely to interact with anionic lipids, which are dominant in the membrane of *C. glutamicum* and affect the extent of BetP activation. The x-ray structure reveals sites where lipids could bind to BetP, with the amphipathic helix H7 positioned where it could sense changes in the membrane. Both the interaction of the C-terminal helix with loops 2 and 8 of the adjacent subunit and interactions between its basic regions with charged lipids are likely to affect the conformation of BetP in the mechanism of osmoregulation (Figure 11.29).

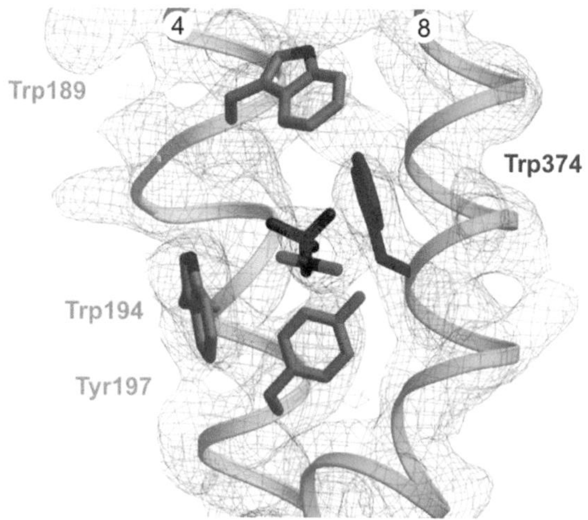

11.27 The binding site for betaine in BetP. Betaine binds in an aromatic box delineated by Trp189, Trp194, Tyr197 (orange sticks) and Trp374 (purple stick). The electron density map contours for TM4 (tan) and TM8 (light blue) are shown, as well as the density for betaine (red, in the center). From Ressl, S., *et al.*, *Nature*. 2009, 458:47–52. © 2009, Nature Publishing Group. Reprinted by permission of Macmillan Publishers Ltd.

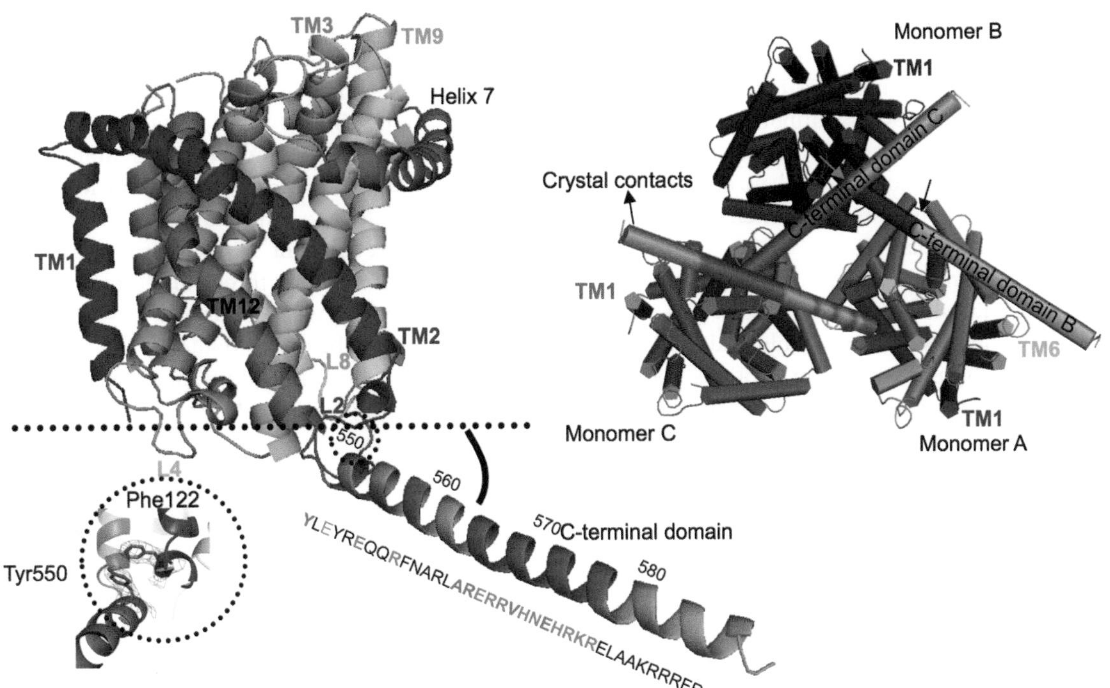

11.28 Interactions of the C-terminal domains in the BetP trimer. (A) In the x-ray structure the C-terminal domain (gray) extends outward from a monomer of BetP (ribbon representation in rainbow colors) when viewed from the plane of the membrane with the cytoplasm at the bottom (dotted line). Basic residues are scattered along the helical extension, along with positions leading to a loss of activity in *E. coli* (red) and *C. glutamicum* (pink). The insert within the dotted circle is a close-up showing the interactions between the C-terminal domain (Tyr550) and Loop 2 (Phe122). (B) The BetP trimer (monomer A, rainbow coloring; monomer B, dark gray; monomer C, brown) is viewed from the cytoplasm. The C-terminal domain of A (colored as in (A)) is best resolved as it mediates the crystal contacts between trimers. The extensions of the C-terminal domains of B and C are based on A as a template. The C-terminal domain of monomer C points toward TM1 of monomer B and probably would interact with the N-terminal domain were it present; the C-terminal domain of monomer B points toward TM6 of monomer A, where it may interact with Loop 4 and Loop 8. From Kramer, R., and Ziegler, C., Regulative interactions of the Osmsensing C-terminal domain in the trimeric glycine betaine transporter BetP from Corynebacterium, Biol. Chem. 2009, 390:685–690, Fig 1 a & b. Reprinted by permission of De Gruyter.

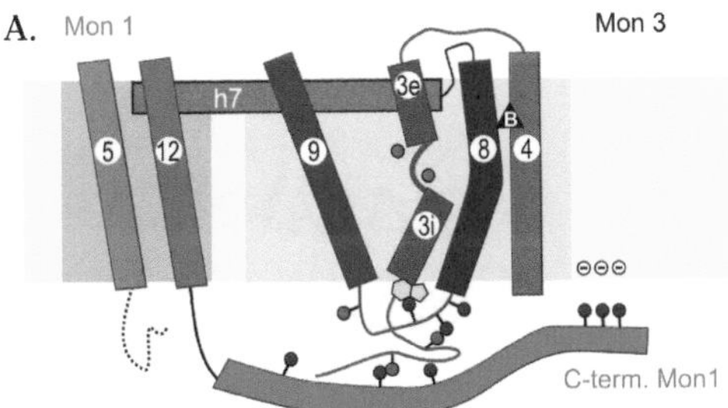

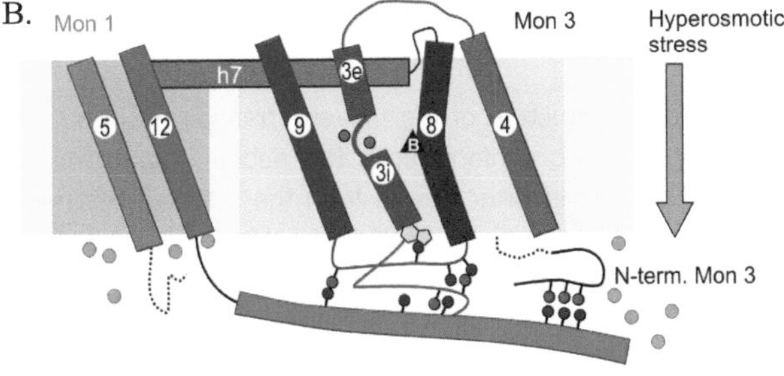

11.29 Model for regulation of BetP. The proposed mechanism for regulation of BetP emphasizes the position of the C-terminal domain. (A) Under low external osmolarity (and low concentrations of K^+ ions in the cytoplasm) BetP is locked in a resting conformation with specific ionic interactions between basic residues on the C-terminal domain of one subunit and charged residues in Loop 2 and Loop 8 of the adjacent subunit, along with specific anionic lipids. (B) With hyperosmotic conditions and elevated levels of K^+ ions (gray spheres) the network of salt bridges alters to trigger a conformational change in the four-helix bundle allowing transport activity. Betaine (black triangle) and Na^+ ions (green spheres) are bound to occluded sites. The role of acidic residues (red spheres) on the N terminus is hypothetical. From Ziegler, C., *et al.*, *Molec Microbiol.* 2010, 78:13–34. © 2010, Blackwell Publishing Ltd.

BetP is a member of the large BCCT family, betaine–choline–carnitine transporters found mainly in the genomes of bacteria and in some halophilic archaea. Crystal structures of the BCCT antiporter CaiT, the carnitine transporter of *E. coli*, show many similarities to BetP. However, CaiT is not involved in osmotic stress response and it lacks the C-terminal helix that plays a crucial role in BetP regulation.

The secondary transporters described above represent huge transporter families, and their similarities suggest that common principles govern the fold and mechanism of transport of very many different proteins across all kingdoms of life. The striking advances in the structural biology of secondary transporters have provided elegant descriptions of the alternating access mechanism of transport as well as insights into the mechanism of coupling substrate and cation permeation. The conformational change accomplished by rigid body movements of two helical bundles also creates alternating access in some primary transporters, such as ABC transporters, where new structures are illuminating how transport is driven by ATP binding and hydrolysis.

ABC TRANSPORTERS AND BEYOND

ATP-binding cassette (ABC) transporters are a well-characterized class of primary transporters that couple the thermodynamically uphill uptake or efflux of a wide variety of substances to the hydrolysis of ATP (see Chapter 6). All ABC transporters have four domains, two TM domains and two nucleotide-binding domains (NBDs) that are synthesized as one to four polypeptides (see Figure 6.8). The NBDs have ATPase activity and share a common fold (see Figure 6.9). A molecular understanding of this important class of transporters is provided by the x-ray structures of the maltose transporter, the vitamin B_{12} transporter, and two transporters involved in drug efflux, the bacterial Sav1866 and a eukaryotic P-glycoprotein. These structures and others show more than one conformation of the transport cycle, supporting an alternating access mechanism and suggesting how transport and ATP hydrolysis are coupled. Unresolved issues remain, such as the stoichiometry of ATP hydrolysis in transport: Are one or both ATPs hydrolyzed to power the conformational change that translocates the substrate?

Maltose Transporter

A continuous stream of genetic, biochemical, and biophysical studies over decades makes the maltose transporter of *E. coli* the best characterized ABC transport system. Its five subunits – MalF, MalG, two copies of MalK, and maltose-binding protein (MBP) – transport maltose and maltodextrins of up to seven glucose residues across the inner membrane after the sugars reach the periplasm via the maltoporin channel (LamB protein, described in Chapter 5). Structural biology of the isolated MalK dimer reveals extensive conformational differences in the NBDs during the catalytic cycle (see Chapter 6), and the x-ray structures of MBP show how its two lobes close to bind maltose with very high affinity. Now three elegant crystal structures exhibit the entire transporter in three different conformations (see Frontispiece), suggesting the basis for conformational coupling between maltose transport and ATP binding and hydrolysis. Additional new structures elucidate the mechanism of hydrolysis of ATP.

The first crystals of the entire maltose transporter contained a mutation in MalK, E159Q, to prevent ATP hydrolysis, which allowed a catalytic conformation with both maltose and ATP bound to be trapped in an outward-facing structure (Figure 11.30). Subsequent structures of the wild type with ADP and phosphate analogs show the mutation did not affect the conformation of the

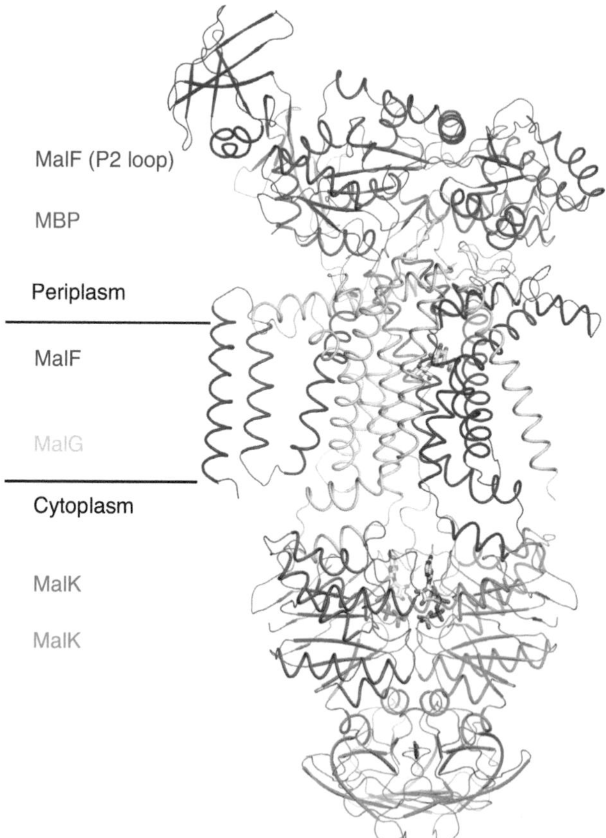

11.30 Crystal structure of the maltose transporter in an outward-facing catalytic conformation. The ribbon diagram of the entire maltose transporter viewed from the plane of the membrane shows MBP (magenta), MalF (blue), MalG (yellow) and the MalK dimer (red and green) with maltose bound in the center cavity and two ADP-VO_4 (all as stick models) at the interface of the MalK dimer. Kindly provided by Michael L. Oldham and Jue Chen.

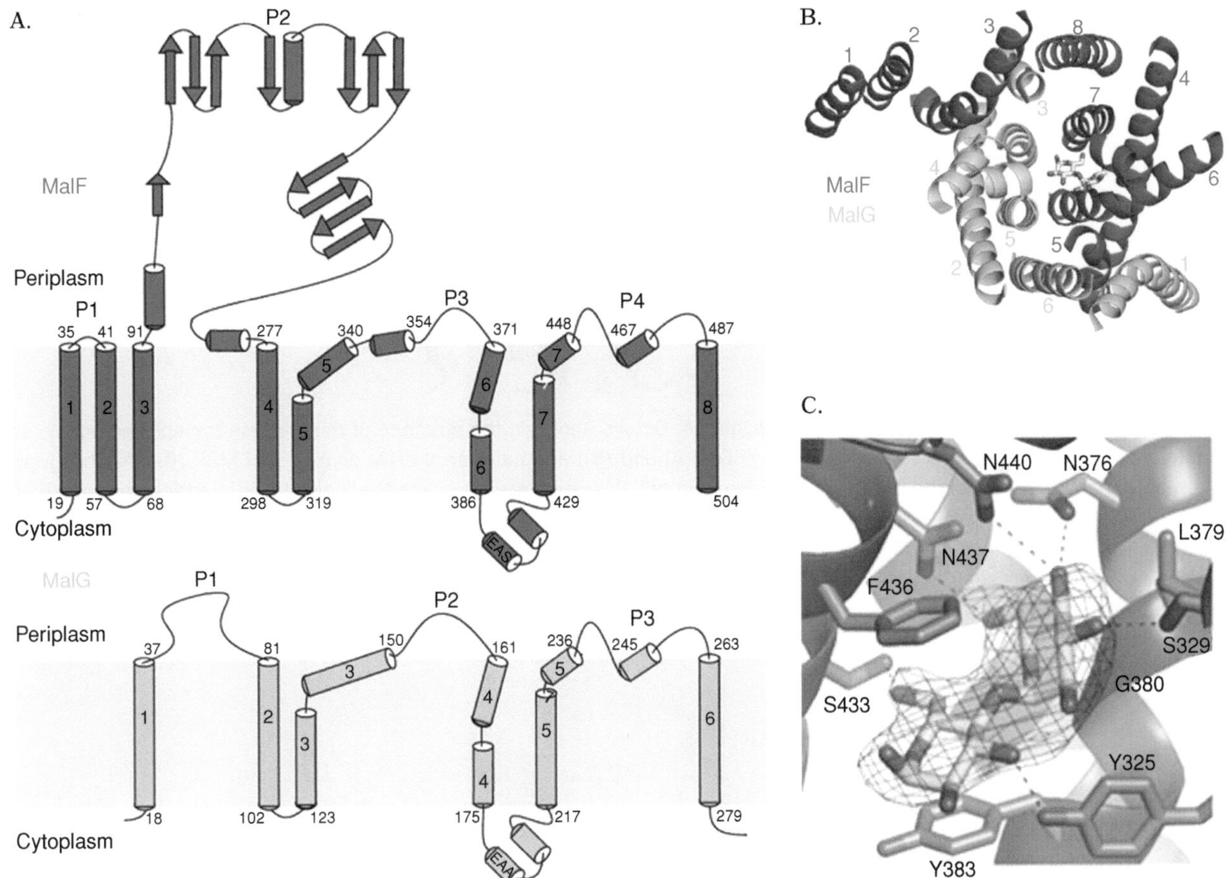

11.31 Structures of MalF and MalG and the maltose-binding site. (A) Topology of the MalF (blue) and MalG (yellow) subunits is very different. (B) A ribbon representation of the 14 TM helices is viewed from the periplasm with loops omitted for clarity. TM3–8 of MalF (blue) and TM1–6 of MalG (yellow) form pseudo-equivalent crescents that face each other with maltose (ball-and-stick model with O atoms red, C atoms gray) in the center. The core is formed by TM4–7 of MalF and TM2–5 of MalG. (C) A close-up view of the maltose-binding site shows the residues making hydrogen bonds and stacking interactions with the substrate (electron density with ball-and-stick model with O atoms red, C atoms gray added). From Oldham, M. L., *et al.*, *Nature*. 2007, 450:515–521. © 2007, Nature Publishing Group. Reprinted by permission of Macmillan Publishers Ltd.

transporter. The 2.8-Å resolution structure provided the first view of the structures of MalF and MalG, which are unequal in length (with eight and six TM helices, respectively) and lack significant sequence homology, even though in most ABC transporters their role is filled by a homodimer. MalF and MalG are assembled symmetrically facing each other with a core composed of TM4–8 of MalF and TM2–5 of MG (Figure 11.31). Peripheral helices including TM1–3 of MalF and TM1 of MalG, cross over the core bundle. Electron density in the large central cavity corresponds to a molecule of maltose, surrounded by residues of MalF (and none of MalG) and making stacking interactions with Phe436 and Tyr383 (Figure 11.31C).

Analysis of the interactions of MalF and MalG with the other subunits shows well-ordered loops packed at the interfaces (Figure 11.32). Between the TM domain and the NBDs, the Q-loop of MalK is well ordered next to the C terminus of MalG. In addition, the two loops of MalF and MalG that contain the conserved EAA motif dock into clefts on the surface of the MalK subunits, each stabilized by both ionic and hydrophobic interactions. On the periplasmic side MBP is docked onto MalF/MalG in an open position and seals the opening of the internal cavity. One periplasmic loop of MalG enters the sugar-binding site of MBP, where it is postulated to "scoop out" the maltose (Figure 11.32C).

To obtain a crystal structure in the inward-facing conformation required deletion of residues 2–35 (ΔTM1) of MalF, which does not interfere with transport or ATPase activity. The crystals of $MalFGK_2$ ΔTM1 in the absence of MBP and nucleotide gave a low resolution (4.5 Å × 6.5 Å), and the structure was determined by placing individual TM helices from the 2.8 Å structure and verifying their positions with 39 Se–Met substitutions (Figure 11.33). Most of the large periplasmic loop (P2) is unresolved. Comparison of this inward facing state with the first structure indicates the needed conformational change

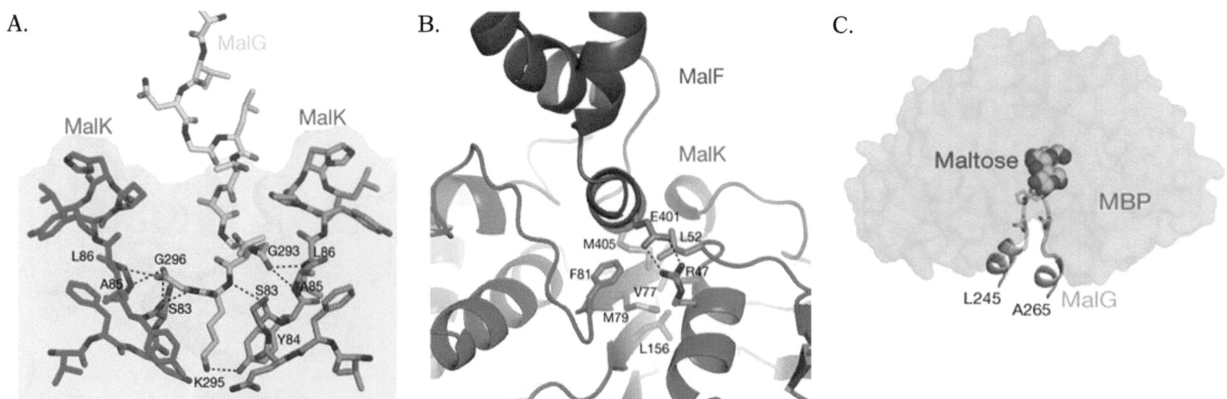

11.32 Interactions of the TM domains with other subunits. Details shown in the structure of the maltose transporter include specific interactions between the TM domain and the NBDs ((A) and (B)) and between the TM domain and MBP (C). (A) The Q-loops of MalK are ordered by the insertion of the MalG C-terminal tail (yellow stick model) into the MalK dimer interface (green and pink transparent surfaces with the Q-loops shown in stick models). Hydrogen bonds and salt bridges (dashed lines) show the interactions of MalG side chains and peptide groups at the interface. (B) The specific interactions between MalF (blue) and MalK (green) include a salt bridge between Glu401 of the EAA motif and Arg47 in an otherwise nonpolar region with Met405 of MalF inserting into hydrophobic pockets on MalK. (C) A loop from MalG (yellow) inserts into the open MBP (surface, magenta), shown with maltose modeled into the binding site. From Oldham, M. L., *et al.*, *Nature* 2007, 450:515–521. © 2007, Nature Publishing Group. Reprinted by permission of Macmillan Publishers Ltd.

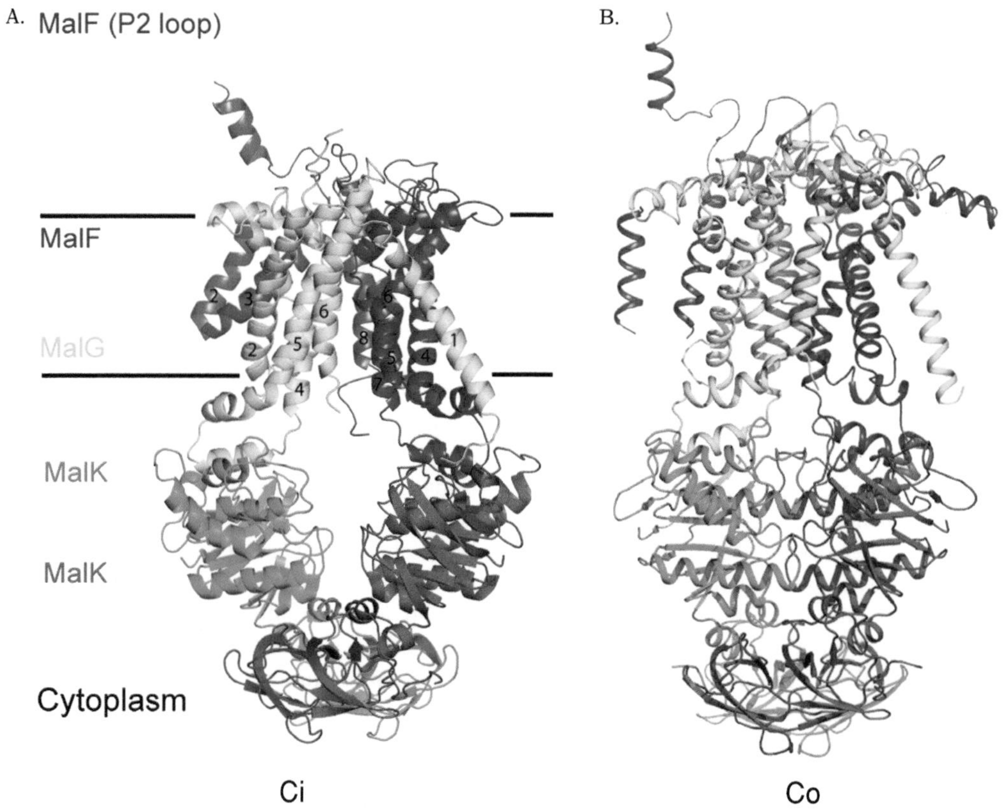

11.33 Crystal structure of the maltose transporter in an inward-facing resting conformation. (A) The ribbon representation of the ΔTM1 maltose transporter (colored as in Figure 11.30) viewed from the plane of the membrane gives the inward-facing conformation (Ci). (B) Rigid-body rotations are needed in both the TM domain (MalF and MalG) and the NBDs of MalK for the transition from Ci to the outward-facing conformation (Co) of the first crystal structure. From Khare, D., *et al.*, *Molecular Cell*. 2009, 33:528–536. © 2009, Elsevier. Reprinted with permission from Elsevier.

(Ci → Co) requires a 22–23° rotation of the core TM helices (TM4–7 of MalF and TM2–5 of MalG). A rocker-switch mechanism between two rigid bodies – TM1–3 of MalF with TM2–6 of MalG, and TM4–8 of MalF with TM1 of MalG – opens the periplasmic end of the central cavity while closing the cytoplasmic end (Figure 11.33B). Access to the maltose-binding site from the periplasm is blocked by a hydrophobic gate composed of four loops each at the bend of a kinked TM helix, in contrast to the Co structure in which a bundle of four TM helices makes a much larger barrier to the cytoplasm.

The inward-facing structure also shows the MalK dimer in its open position (resting state, see Figure 6.9). The conformational change needed to close the NBDs is a 14° rotation of the entire domain. Additional rotations of the helical subdomain within each NBD allow movement relative to the TM domain without loss of contact, like a ball-and-socket joint (Figure 11.34). Thus the rigid-body rotations of the TM domains coincide with closing and opening the NBD interface, a mechanism which couples ATP hydrolysis to transport.

The concerted motions of the TM helical cores and the NBDs have been studied with EPR, using site-directed spin labeling to place labels across the MalK interfaces (with V16C and R129C mutations) to detect closed, open, and semi-open states. In the intact transporter both MBP and ATP are required for closure of the nucleotide-binding interface in the NBD, indicating its

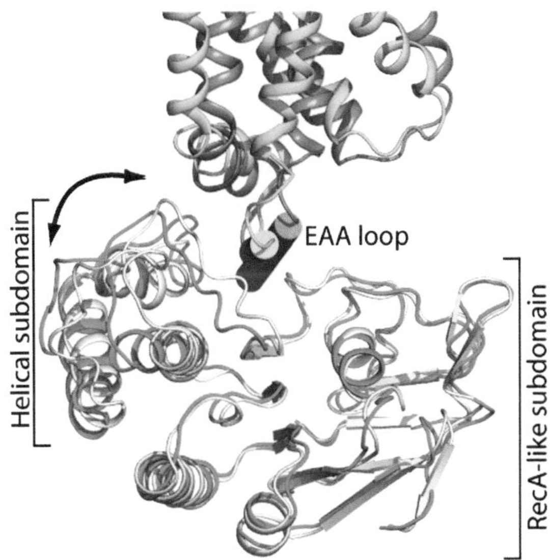

11.34 The conformational change at the interface of the TM domain with the NBDs. Superposition of the inward- and outward-facing structures of the RecA-like and helical subdomains of MalK (Ci, green and Co, gray) and a portion of MalG (Ci, yellow and Co, orange). The arrow shows the rotation of the helical subdomain. The coupling helix (EAA loop) inserts into the cleft in MalK at different angles in the two structures. From Khare, D., *et al.*, *Molecular Cell*. 2009, 33:528–536. © 2009, Elsevier. Reprinted with permission from Elsevier.

closure is coupled to the global conformational changes upon MBP binding that are transmitted through the TM domain. The semi-open conformation occurs after ATP hydrolysis. Spin labels placed on the MalK helical subdomain (Q122C) and RecA-like subdomain (A167C) show only a slight rotation of the helical subdomain upon ATP hydrolysis unless MBP and maltose are present, indicating the TM domains constrain the MalK dimer. The EPR results give strong evidence for conformational coupling based on concerted motions of TM helices and NBDs.

A third, intermediate structure of the maltose transporter is seen in crystals of $MalFGK_2$ with MBP and maltose, lacking nucleotide. In an effort to capture the initial complex formed when MBP carrying maltose docks on the resting-state conformation, a disulfide bridge was engineered (G69C/S337C) that keeps MBP closed; however, the same structure was obtained with crystals of the wild type MBP under the same conditions. This 3.1-Å resolution structure shows MBP in its maltose-bound closed conformation, an occluded maltose-binding pocket in the TM domain, and the MalK dimer in a semi-open state (Figure 11.35). This structure represents a transient state prior to translocation of the maltose; however, maltose is also observed in the internal cavity due to its very high concentrations during crystallization. The maltose-binding site is closed from the periplasm by the gate observed in the inward-facing structure described above, and it is also blocked from the cytoplasm by a shield created by Thr176, Leu221 of MalG and Tyr383 and Leu429 of MalF, while the TM helices remain separated.

The different conformations observed in crystal structures of the maltose transporter give information about three states of the transport cycle (Figure 11.36). The Ci conformation represents a resting state, before MBP binds. The pre-translocation conformation represents the initial step upon MBP/substrate binding, and the Co conformation with MBP and ATP bound is a transition state prior to ATP hydrolysis. The relationship between the subunits changes in a coordinated manner. In the pre-translocation conformation, the orientations of MalK subdomains are like those in the inward-facing state (with different hydrogen bonds between opposing MalK subunits), while the contacts between closed MBP and the TM domains are smaller than those of open MBP in the outward-facing conformation. Interestingly, soaking crystals of the wild type pre-translocation complex in AMP-PNP converts it to the outward-facing conformation, while soaking the crystals of the disulfide-bonded mutant (that cannot open) does not, suggesting that the opening of MBP triggers the conformational changes involved in the Ci → Co transition.

To study the mechanism of ATP hydrolysis by the complete maltose transporter, crystals were formed of $MalFGK_2$ with MBP, maltose, AMP-PNP or ATP plus one of three γ-phosphate mimics – vanadate (VO_4^{3-}), beryllium fluoride (BeF_3^-), or aluminum fluoride (AlF_4^-). The

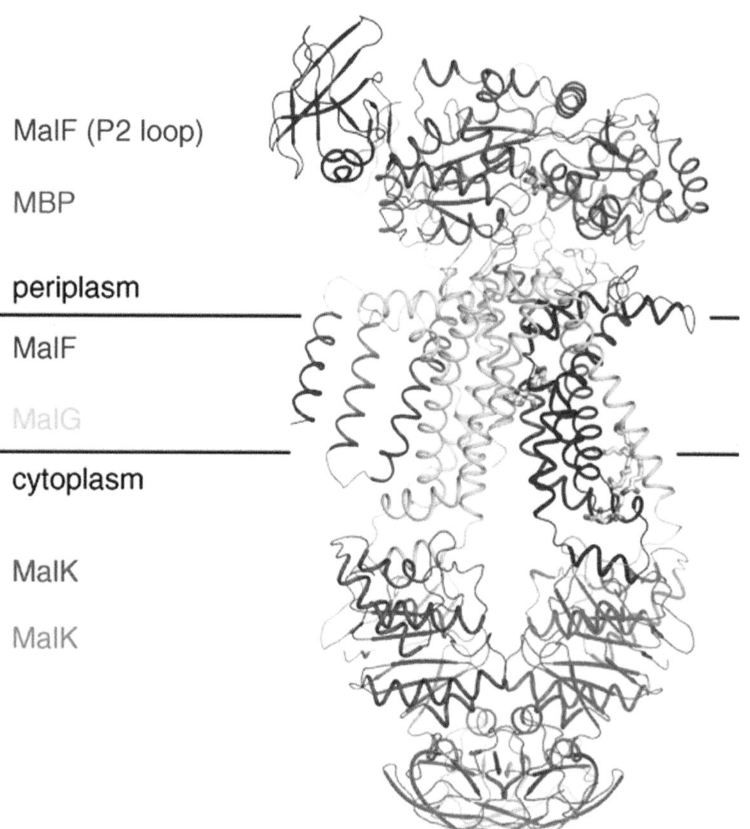

11.35 Crystal structure of the pretranslocation intermediate conformation of the maltose transporter. The ribbon representation (colored as in M1) of the pretranslocation conformation is viewed from the plane of the membrane. Two molecules of maltose (stick model with O atoms red and C atoms gray) can be seen, probably due to the high concentration of maltose during crystallization. From Oldham, M. L., and J. Chen, *Science*. 2011, 332:1202–1205. © 2011, American Association for the Advancement of Science. Reprinted with permission from AAAS.

overall structures of each are essentially identical to the outward-facing conformation described above. The high resolutions obtained (2.2–2.4 Å) show details of the active site. Two of the structures (with AMP-PNP or ADP-BeF_3) show the ground state (tetrahedral) geometry for the leaving phosphate group, while the other two (with ADP-VO_4 or ADP-AlF_4) show trigonal bipyramidal geometry for the transition state during nucleophilic attack by water. The structures support a general base mechanism of ATP hydrolysis.

The architecture and mechanism of the maltose transporter are mirrored in crystal structures of the molybdate transporter $ModB_2C_2A$ in spite of a lack of sequence homology. These and other structures suggest that ABC importers generally share the conformational coupling mechanism, even when the TM domains differ in fold, as seen in the vitamin B_{12} transporter described next.

11.36 Overall comparison of the states of the maltose transporter captured in the three high-resolution structures. Cartoon representation shows protein components MalF, MalG, $MalK_2$, and MBP in different states of the transport cycle. Maltose (solid black ball) is brought by MBP. The ATP-binding sites in the MalK dimer (open or gray circles) are filled upon ATP binding. From Oldham, M. I., and Chen, J. *Science* 2011, 332:1202–1205. © 2011, American Association for the Advancement of Science. Reprinted with permission from AAAS.

The Vitamin B_{12} Uptake System

Vitamin B_{12}, or cyanocobalamin (CN-cbl), is a cofactor produced by some bacteria and archaea and required by a variety of enzymes in most cells. Specific transport systems enable *E. coli* cells to import this large, inflexible molecule via two energized phases: uptake across the inner membrane via the ABC transporter BtuCD-BtuF and uptake across the outer membrane utilizing the specific receptor BtuB coupled with the TonB/ExbB/ExbD system for energy input (Figure 11.37). The structure of the inner membrane transport system BtuCD was the first x-ray structure of an ABC transporter; the structure of BtuB provides insight into TonB-dependent transporters.

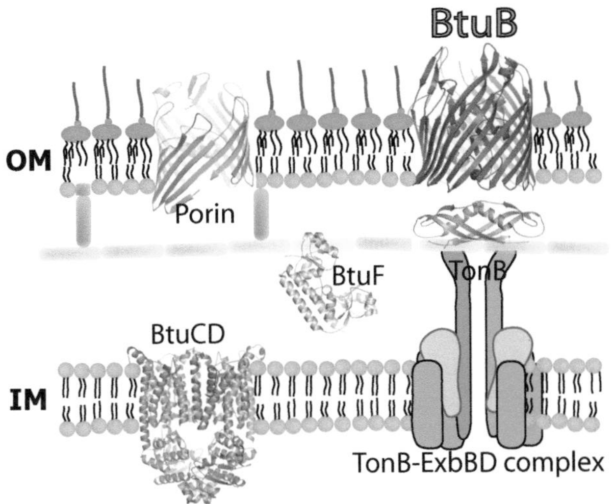

11.37 Components of the transport system for vitamin B_{12} in *E. coli*. Structures for BtuCD, BtuF, and BtuB have been solved, along with the C terminus of TonB, while the structures of ExbB and ExbD are not known. The general porin is included in the outer membrane because it may allow passive diffusion of vitamin B_{12} into the periplasm. From Kadner, R. J., *et al.*, in R. Benz (ed.), *Bacterial and Eukaryotic Porins*, Wiley-VCH, 2004, pp. 237–258. © 2004, John Wiley & Sons, Inc.

BtuCD–BtuF, an ABC Transport System

Transport of vitamin B_{12} across the inner membrane utilizes three components, BtuC, BtuD, and BtuF. BtuF is a periplasmic binding protein that delivers vitamin B_{12} to the BtuCD inner membrane transporter after it enters the periplasm. BtuF avidly binds the cofactor (K_d ~15 nM) between two lobes that each consist of a Rossmann-like fold (a central five-stranded β-sheet surrounded by helices), held by a relatively inflexible helical backbone (Figure 11.38A). BtuF has been reconstituted with BtuCD to give a complex that hydrolyzes ATP and transports B_{12} in proteoliposomes.

A BtuC dimer spans the inner membrane, interacting with BtuD at the cytoplasmic side and with BtuF when it docks on the periplasmic side of the membrane. Each BtuC protein contributes ten TM helices (instead of the six that is typical of ABC transporters). Two BtuD subunits form the NBDs that bind and hydrolyze ATP to drive the transport. The first crystal structure refined at 3.2 Å defines the entire structure of the BtuCD heterotetramer except for 17 C-terminal residues of BtuD and a few residues of the periplasmic loops of BtuC (Figure 11.38B). Both ATP and vitamin B_{12} are absent from the structure, although between the BtuC subunits is a hydrophobic

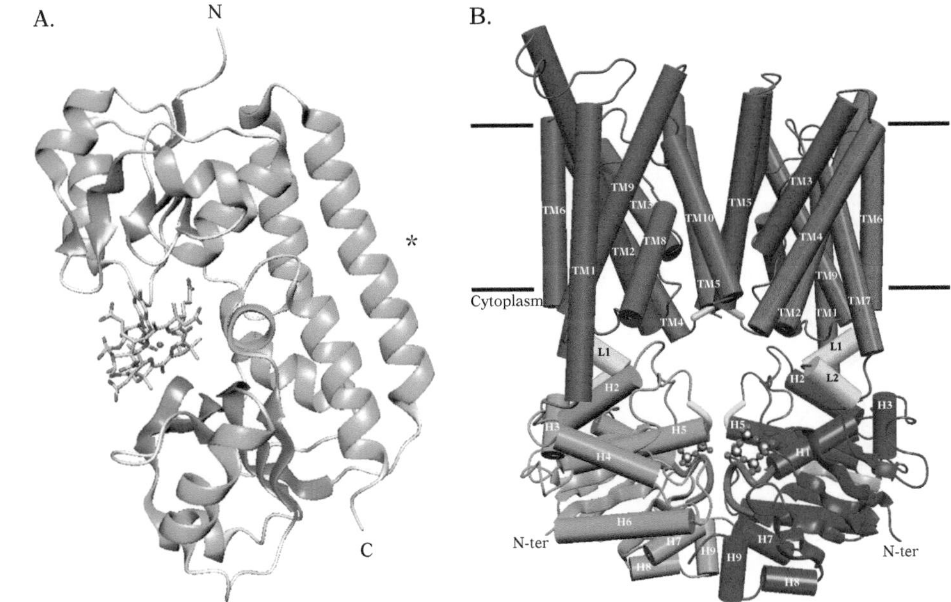

11.38 X-ray structures of the components of the ABC transporter for vitamin B_{12}. (A) BtuF, the periplasmic vitamin B_{12}-binding protein. The ribbon diagram shows β-sheets (blue) at the substrate-binding lobes and α-helices (green) in the lobes and the backbone, with an asterisk denoting the helices that form the backbone. The substrate, vitamin B_{12}, is shown as a ball-and-stick model. From Borths, E. L., *et al.*, *Proc Natl Acad Sci USA*. 2002, 99:16642–16647. © 2002 by National Academy of Sciences, USA. (B) The $BtuC_2D_2$ transporter viewed from the side, with the two BtuC subunits (purple and red) spanning the inner membrane with their L1 and L2 helices (gold) at the cytoplasmic surface. The TM helices are numbered. The two BtuD subunits (green and blue) associate with BtuC at the cytoplasmic side and have cyclotetravanadate molecules (ball-and-stick models) at their ATP-binding sites. From Locher, K. P., *et al.*, *Science*. 2002, 296:1091–1098. © 2002, American Association for the Advancement of Science. Reprinted with permission from AAAS.

cavity large enough to accommodate vitamin B_{12} and open to the periplasm. Two loops at the ends of TM helices 4 and 5 from the BtuC subunits form a gate that closes the channel to the cytoplasm, now called cytoplasmic gate I. Two short helices, L1 and L2, that make an L-shape at the interface with BtuD are likely to mediate interactions with the NBDs. Both the overall fold and the nucleotide-binding sites of BtuD resemble those of other ABC NBDs described in Chapter 6 (see Figure 6.9).

The 2.6-Å resolution structure of BtuCD (lacking cysteines) in complex with BtuF shows an occluded substrate-binding pocket that lacks vitamin B_{12} (Figure 11.39A). The two lobes of BtuF are opened at a hinge in the backbone helix with conserved residues Glu74 and Glu202 of BtuF making salt bridges with Arg56 in BtuC. The contacts between N- and C-terminal lobes of BtuF with different BtuC subunits induce slight asymmetry in the BtuC dimer, especially at the cytoplasmic ends of TM3, TM4, TM5, and helix 5a. The asymmetry is suggestive of an alternating access intermediate with one side aligned with the open-outward conformation of the BtuCD structure and the other side with another ABC transporter that is open-inward. Lacking nucleotides, the NBDs are in an open conformation.

A third crystal structure captures the BtuCD–BtuF complex with an ATP analog (AMP-PNP) bound (Figure 11.39B). The NBDs are held in a closed state by an engineered disulfide bond at N162C in BtuD and inactivated by the E159Q mutation in the Walker B motif (analogous to that in MalK). As expected from other ATP-bound NBD dimer structures, the two AMP–PNP molecules bind at the BtuD interface. As seen in the maltose transporter (above), ATP binding is transmitted to the TMDs through coupling helices that move toward the center, pushing out the gating helices (TM5) and opening cytoplasmic gate I (loops connecting TM4 and TM5). However, the translocation path is closed by cytoplasmic gate II, made by the loops connecting TM2 and TM3 (Figure 11.40). At the periplasmic side, the central cavity is closed by the periplasmic gate (rather than being open to the binding protein). Although no substrate is seen in the crystal structures, in proteoliposomes ^{57}Co-CN-cbl binds to BtuCD with AMP–PNP present. The crystal structures, along with other mutant studies and EPR analysis,

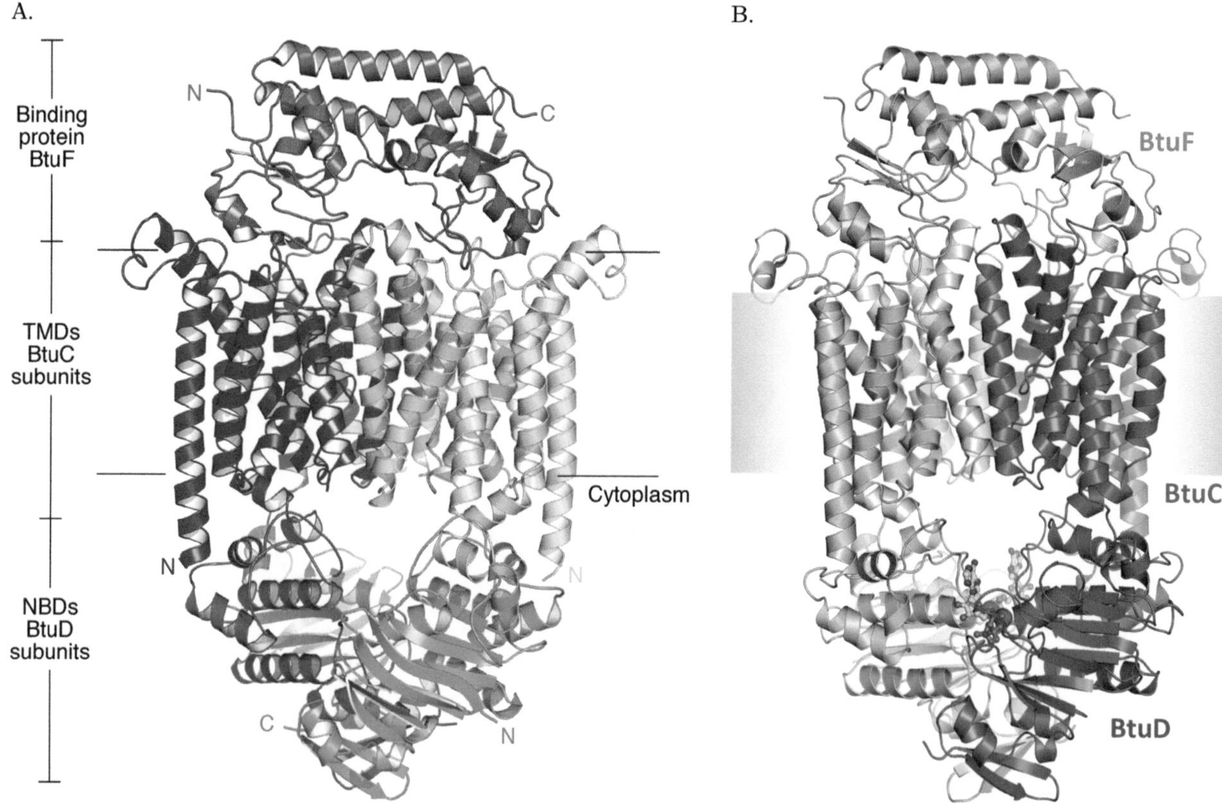

11.39 Two crystal structures of the vitamin B_{12} transporter with the periplasmic binding protein. (A) Ribbon representation shows the BtuCD–BtuF nucleotide-free complex viewed from the plane of the membrane with the periplasm above and the cytoplasm below. The horizontal lines indicate the approximate boundaries of the membrane. From Hvorup, R. N., *et al.*, *Science* 2007, 317:1387–1390. (B) Ribbon representation of BtuCD–BtuF carrying a double mutation in BtuD (E159Q/N162C) shows the state with bound nucleotide AMP–PNP (ball-and-stick) and Mg^{2+} (pink spheres). The five subunits are BtuF (orange), BtuC dimer (marine and light blue), and BtuD dimer (light and dark green). From Korkhov, V. M., *et al.*, *Nature*. 2012, 490:367–372. © 2007, American Association for the Advancement of Science. Reprinted with permission from AAAS.

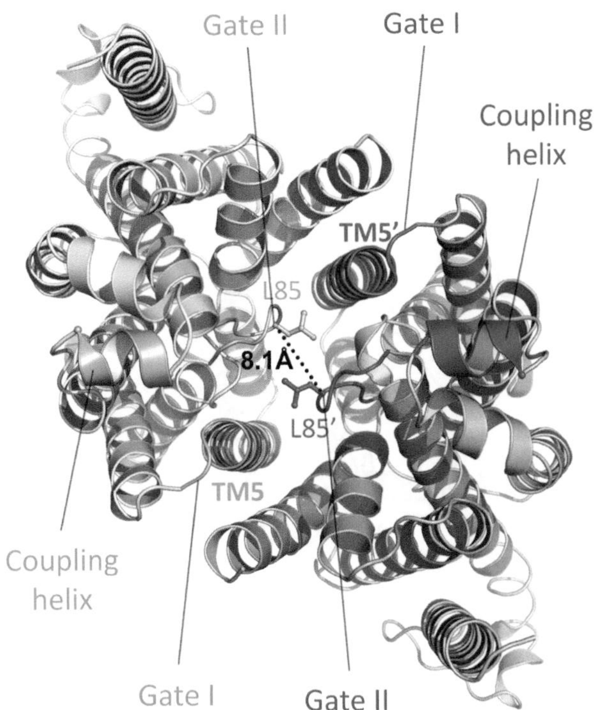

11.40 Cytoplasmic gates and gating helices of BtuC. Portions of the BtuC subunits (light and dark gray ribbons) with coupling helices that interact with BtuD, gate 1, and gate II (light blue for one subunit, marine for the other). The loops forming gate II contain highly conserved Asn83 and Leu85 residues. The closure of gate II brings the Cα atoms of Leu85 (dotted line) closer in the AMP–PNP-bound complex than in the nucleotide-free complex. From Korkhov, V. M., *et al.*, *Nature*. 2012, 490: 367–372. © 2012, Nature Publishing Group. Reprinted by permission of Macmillan Publishers Ltd.

suggest that once BtuF carrying vitamin B_{12} docks at BtuCD, ATP binding triggers release of substrate into the translocation cavity and the conformational change that opens access to the cytoplasm by an alternating access mechanism. However, ATP hydrolysis also occurs in the absence of the B_{12} substrate, giving a stoichiometry of ~100 ATP/B_{12} in proteoliposomes, and ATPase activity is only stimulated two-fold by the presence of BtuF-B_{12}. These results indicate the BtuCD-F system utilizes a mechanism that is distinct from the maltose transporter described above. The details of the protein interactions and conformational changes, along with the mechanism in other ABC transporters, will continue to emerge as other high-resolution structures are obtained.

BtuB, an Outer Membrane Transporter Energized by TonB

Transport of vitamin B_{12} through the outer membrane utilizes the BtuB protein, a β-barrel protein with striking similarities to the FepA and FhuA iron receptors described in Chapter 5. These specific outer membrane transporters carry out active transport only in the presence of the TonB protein (see below), so they are called TonB-dependent transporters (TBDTs). Like other outer membrane proteins, BtuB is also a receptor for colicins and bacteriophage, in this case colicins E and A and phage BF23. Entry of these agents into the cell, which is also TonB-dependent, is poorly understood.

BtuB has a very high affinity for vitamin B_{12}, and its crystal structure has been solved in the presence and absence of bound substrate. The BtuB protein has two domains: a C-terminal domain that makes the β-barrel, and an N-terminal globular domain that folds into the barrel to make a hatch (Figure 11.41). The barrel consists of 22 tilted antiparallel amphipathic β-strands that give it a polar interior and a hydrophobic exterior. The β-strands are connected by short turns on the periplasmic side and longer loops on the outside. When apo- and substrate-bound BtuB structures are compared, some of the external loops appear to fold toward the substrate. Binding vitamin B_{12} by BtuB requires Ca^{2+} ions, which appear to order some of the loops involved in binding substrate. However, the roles of loops are not precisely defined because different loops are disordered in different crystal preparations, including the structure with the highest resolution (1.95 Å), achieved for BtuB crystallized in cubic phase (Figure 11.41B).

The hatch domain, which has a hydrophobic core with a polar exterior to fit inside the barrel, is highly conserved among TBDTs. The 1.95-Å resolution structure gives the most complete view of its core β-sheet connected by loops and encircled by amphipathic α-helices (Figure 11.42A). The loop between the third and fourth strand appears to latch onto the barrel at β-strands 13–15. However, the hatch domain in the internal cavity does not have a close fit to the barrel walls, and the many water molecules in the interface between them mediate half of the hydrogen bonds between the two domains (Figure 11.42B). Two-thirds of the water molecules present are hydrogen bonded to one domain or the other, not both, a characteristic of a transient protein–protein interaction that is suggestive of hatch movement. Thus it is likely that the hatch moves out of the barrel to allow passage of its very large and inflexible substrate.

The tight binding of vitamin B_{12}, along with the absence of a channel in BtuB, indicates a major conformational change is required for transport, energized by its interaction of TonB. In the hatch domain of BtuB are seven residues near the N terminus (Asp6–Thr–Leu–Val–Val–Thr–Ala) that form the "Ton box," defined as the major site of interaction with the TonB protein by genetic and crosslinking studies. In crystal structures of BtuB with no vitamin B_{12} present, the Ton box is inside the barrel near the periplasm. Both x-ray and EPR data indicate that upon binding substrate, the Ton box becomes exposed to the periplasm, where it can interact with TonB.

BtuB and the other TBDTs in the outer membrane are supplied energy from the pmf (proton motive force)

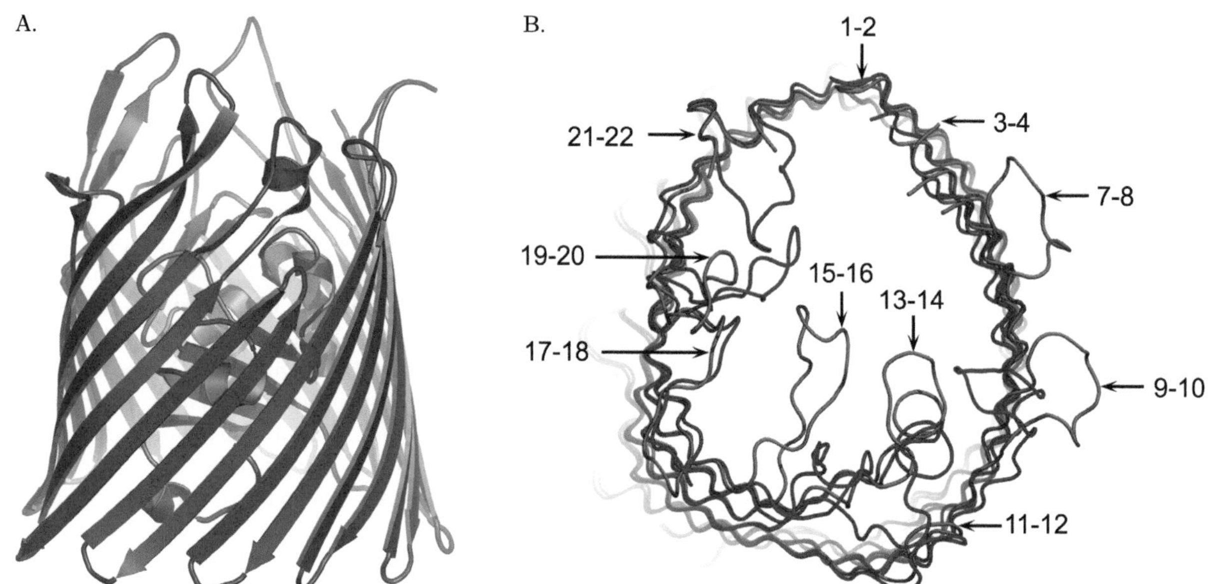

11.41 Crystal structure of BtuB, the outer membrane receptor for vitamin B_{12}. (A) Ribbon diagram of the first high-resolution structure determined for BtuB, in which the 22-stranded β-barrel surrounds the globular hatch domain in the interior, similar to the FepA and FhuA proteins also in the outer membrane (see Figure 5.23). From Chimento, D. P., *et al.*, *Proteins*. 2005, 59:240–251. © 2005, Wiley-Liss, Inc. (B) Differences in the extracellular loops are observed in the higher-resolution structure obtained in lipid cubic phase (red, compared to the first structure in blue). The loops are numbered to show which strands of the β-barrel they connect. The hatch domain is omitted for clarity. From Cherezov, V., *et al.*, *J Mol Biol*. 2006, 364:716–734. © 2006, Elsevier. Reprinted with permission from Elsevier.

across the inner membrane by coupling to TonB and its accessory proteins, ExbB and ExbD, which are anchored in the inner membrane. TonB and ExbD are both predicted to have single TM segments near their N termini, with large periplasmic domains. ExbB is predicted to have three TM segments with a large cytoplasmic loop between the first two. (ExbB and ExbD seem to be similar to MotA/B, which use the pmf to drive the motion

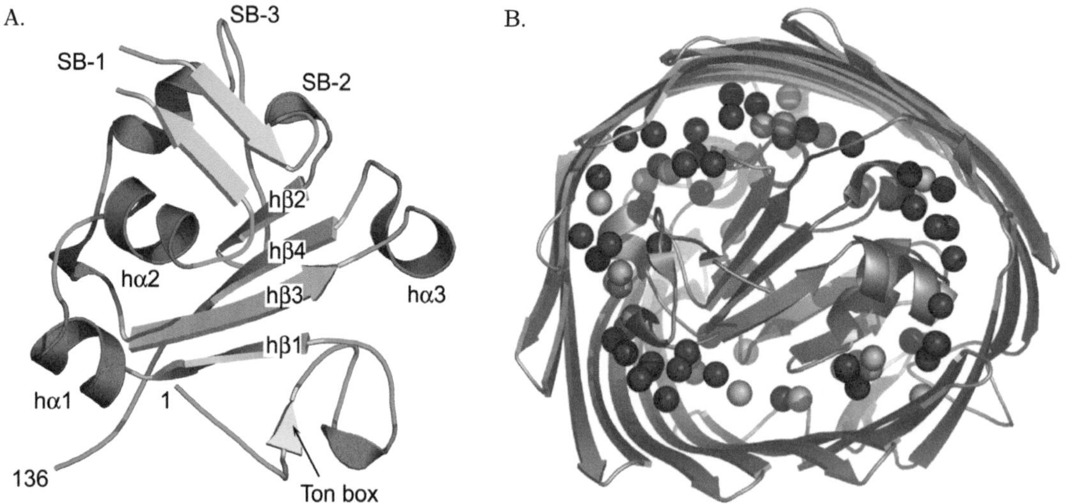

11.42 The hatch domain within BtuB. (A) The hatch domain (residues 1–136) of BtuB is best resolved in the 1.95-Å resolution structure in the absence of vitamin B_{12}. A ribbon diagram shows the hydrophobic β-strands (hβ1–4, chartreuse) in the core, the surrounding amphipathic α-helices (hα1–3, red), and the connecting loops (green). Three loops involved in substrate binding are labeled (SB1–3). The five residues at the N terminus, next to the Ton box (yellow), are resolved; however, residues 57–62 of SB-1are disordered. From Cherezov, V., *et al.*, *J Mol Biol*. 2006, 364:716–734. © 2006, Elsevier. Reprinted with permission from Elsevier. (B) The interface between the barrel and hatch domains of BtuB are filled with many waters. Bridging waters (green) make hydrogen bonds to both domains, while nonbridging waters (blue) make hydrogen bonds to a single domain or to other water molecules. From Chimento, D. P., *et al.*, *Proteins*. 2005, 59:240–251. © 2005, Wiley-Liss, Inc.

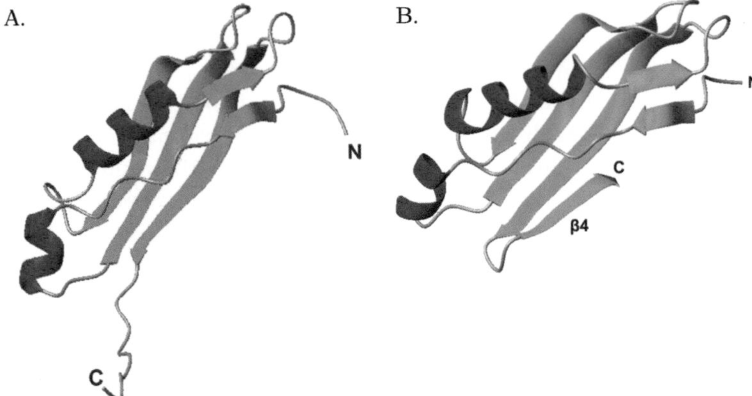

11.43 The C-terminal domain of TonB. Structures of the isolated C-terminal domain of TonB have been determined by x-ray crystallography (A) and NMR (B). While the peptides vary by only one residue (residues 150–239 for x-ray analysis, residues 151–239 for NMR) and give the same overall structure, an additional β-strand ($\beta4$) is evidenced by NMR. From Krewulak, K. C., and H. J. Vogel, *Biochem Cell Biol*. 2011, 89:87–97. © 2011, NRC Research Press.

of bacterial flagella.) In the cell, these proteins form a complex of ~260 kDa with the stoichiometry 1 TonB:2 ExbD:7 ExbB.

TonB has 239 amino acids in three domains: the N-terminal TM domain; a proline-rich periplasmic domain; and a C-terminal domain that contacts outer membrane receptors. The TM domain, an uncleaved signal sequence with a highly conserved sequence along one face of the α-helix, interacts with the TM segment of ExbD. TonB is thus anchored in the inner membrane, but its periplasmic domain enables it to contact the outer membrane. A spacer peptide (residues 66–102) rich in Pro, Lys, and Glu residues and folded into a Proline helix is followed by an extended flexible region (residues 103–149) that together allow it to reach across the periplasmic space, bringing the C-terminal domain to TBDTs in the outer membrane. The isolated C terminus of TonB (residues 150–239) appears as a dimer in the first crystal structure; however, subsequent x-ray and NMR structures show it as a monomer composed of a small β-sheet in front of two α-helices (Figure 11.43).

Models propose the C terminus of an energized state of TonB contacts a TBDT in the outer membrane and alters the conformation of the transporter. Evidence for such a mechanism is provided by a high-resolution structure of BtuB complexed with the C terminus of TonB, which shows the Ton box of BtuB has become a β-strand recruited by the β-sheet in TonB (Figure 11.44). Questions remain about how TonB forms a complex with ExbB and ExbD, how that complex is energized, and how it pulls the hatch domain out of the barrel to open a transport channel for vitamin B_{12}.

These questions apply to numerous other microbial transport systems, since around 100 TBDTs have been described and 4000 have been predicted in the genomes of 350 organisms. Crystal structures of a dozen unique TBDTs with and without substrates show conserved architecture, varying of course in substrate-binding sites. Dynamic approaches will be needed to probe the energy coupling and the role of TonB.

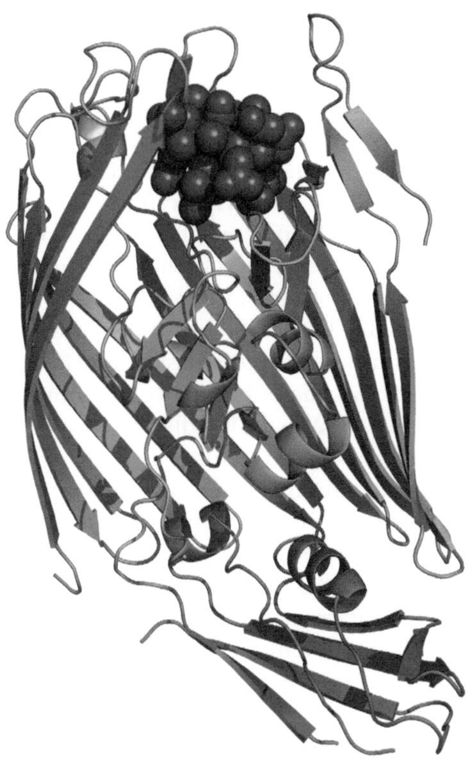

11.44 The ribbon diagram of the complex of BtuB and the C-terminal domain (residues 147–239) of TonB. The front of the BtuB protein, shown with the outer β-barrel (copper) and the inner hatch domain (green), has been removed to show the lumen and the BtuB–TonB interaction. The purified proteins were combined in a molar ratio of 1:5 in the presence of vitamin B_{12} and calcium. In the x-ray structure, the Ton box from BtuB (blue) has been recruited by the β-sheet of the TonB C terminus (magenta). The substrate, vitamin B_{12} (red spheres), is bound in the extracellular region above the hatch domain of BtuB. From Shultis, D. D., *et al*., *Science*. 2006, 312:1396–1399. © 2006, American Association for the Advancement of Science. Reprinted with permission from AAAS.

DRUG EFFLUX SYSTEMS

In contrast to the mysterious TonB/ExbB/ExbD complex that couples the inner and outer membranes for uptake, much more is known about how molecules on their way out of the cell cross the periplasmic space due to the high-resolution structure for the amazing channel-tunnel made by TolC, along with structures of several other proteins involved in efflux. Some of these proteins serve to export lipids from the inner membrane, and some are related to proteins involved in the secretion of proteins such as hemolysin and colicins. But they are best known for their role in drug efflux because the problem of multidrug resistance (MDR) is now limiting treatment options for many cases of cholera, pneumonia, gonorrhea, and tuberculosis. Today, pathogens resistant to almost any antibiotic can arise due to multidrug efflux pumps. Genome analyses predict a role in drug efflux for 6% to 18% of all transporters in bacterial membranes. MDR proteins in humans contribute to the drug resistance of many tumors.

In Gram-negative bacteria, many MDR proteins are organized into tripartite systems, each consisting of an inner membrane transporter and a periplasmic lipoprotein called a membrane fusion protein, in addition to the outer membrane channel. The transporters use either ATP hydrolysis or the pmf as the source of energy, and belong to at least five families in the transporter classification scheme described in Chapter 6. X-ray structures are now available for prokaryotic transporters involved in drug efflux from all of these classes. EmrD, a member of the MFS family (described above), is comparable in size to another 12-TM helix bundle protein, NorM of the MATE (multidrug and toxic compound extrusion) family that was recently shown to have a very different fold. The Sav1866 protein is an ABC transporter related to mammalian MDR proteins, including P-glycoprotein (both described below). The EmrE protein is a member of the SMR (small MDR) family that exports toxic hydrophobic compounds. The secondary transporter AcrB belongs to the RND (resistance nodulation cell division) superfamily. AcrB is part of a well-characterized tripartite drug efflux system that includes the membrane fusion protein, AcrA, and utilizes the remarkable TolC tunnel to export drugs directly to the extracellular space. Representing the diverse mechanisms for drug efflux, these proteins provide important models for the mechanisms that rid cells of unwanted toxic compounds.

Sav1866 and P-glycoprotein, ABC Exporters

Sav1866 is a multidrug transporter from *Staphylococcus aureus* with homology to several human MDR transporters, including CFTR and P-glycoprotein. Sav1866 uses the hydrolysis of ATP to drive efflux of numerous drugs, including the cancer drugs doxorubicin and vinblastine. This ABC transporter is a homodimer, with each subunit contributing one TM domain and one NBD. The 3.0-Å resolution structure of Sav1866 provided the first detailed look at the architecture of an ABC exporter.

Each Sav1866 subunit has an N-terminal TM domain (residues 1–320) and a C-terminal NBD (residues 337–578) with a short linking peptide. The x-ray structure shows an elongated molecule (120 Å long) that extends well into the cytoplasm, with two molecules of ADP bound to the NBDs (Figure 11.45). The NBD structure is similar to those of other well-characterized ABC transporters, with the nucleotide-binding sites at the interface between subunits where the P-loop of one NBD is across from the ABC signature motif of the other (see Figure 6.9). The six TM helices of each subunit show pseudo two-fold symmetry between TM1–3 and TM4–6; however, helices from the two subunits are intertwined at the center of the membrane, where they bend outward with TM1–TM2 from one subunit aligning with TM3–TM6 from the other. Since the two subunits do not form separate lobes of the transporter, they are constrained by their interactions and probably do not act independently.

Sav1866 has a large internal cavity and lacks a well-defined drug-binding site, which is typical of multidrug transporters. In the x-ray structure, the translocation pathway is accessible to the outside and the outward-facing cavity is hydrophilic, with charged residues along its surface. In this conformation extrusion of hydrophobic drugs would occur with simple diffusion from the low-affinity cavity. Biochemical studies suggest that when the cavity is open to the cytoplasm, it has a high affinity for the drugs. Thus the transport mechanism appears to utilize an alternating access mechanism using the binding and hydrolysis of ATP to drive the conformational change.

At the interface between the NBD and the TM domain the long intracellular loops ICL1 and ICL2 each contain a short coupling helix running nearly parallel to the plane of the membrane. Implicated in communication between domains, one of the coupling helices contacts the NBDs of both subunits, while the other one contacts only the opposite NBD. The portions of the NBDs that contact the TM domain are from the Q-loop and a conserved motif called the x-loop near the ABC signature motif of the NBD. The intertwined structure suggests that the coupling between the NBDs and the TM domains is communicated by moderate conformational rearrangements in the NBDs upon ATP binding, rather than the large opening and closing seen in ABC importers.

The high-resolution structure of Sav1866 provides a basis for modeling related proteins of significance to human health. The low-resolution structure of CFTR, the cystic fibrosis transmembrane conductance regulator (see Figure 7.30), obtained by EM could be refined using the Sav1866 structure as a basis. Many MDR proteins, including P-glycoprotein, have also been modeled based on Sav1866.

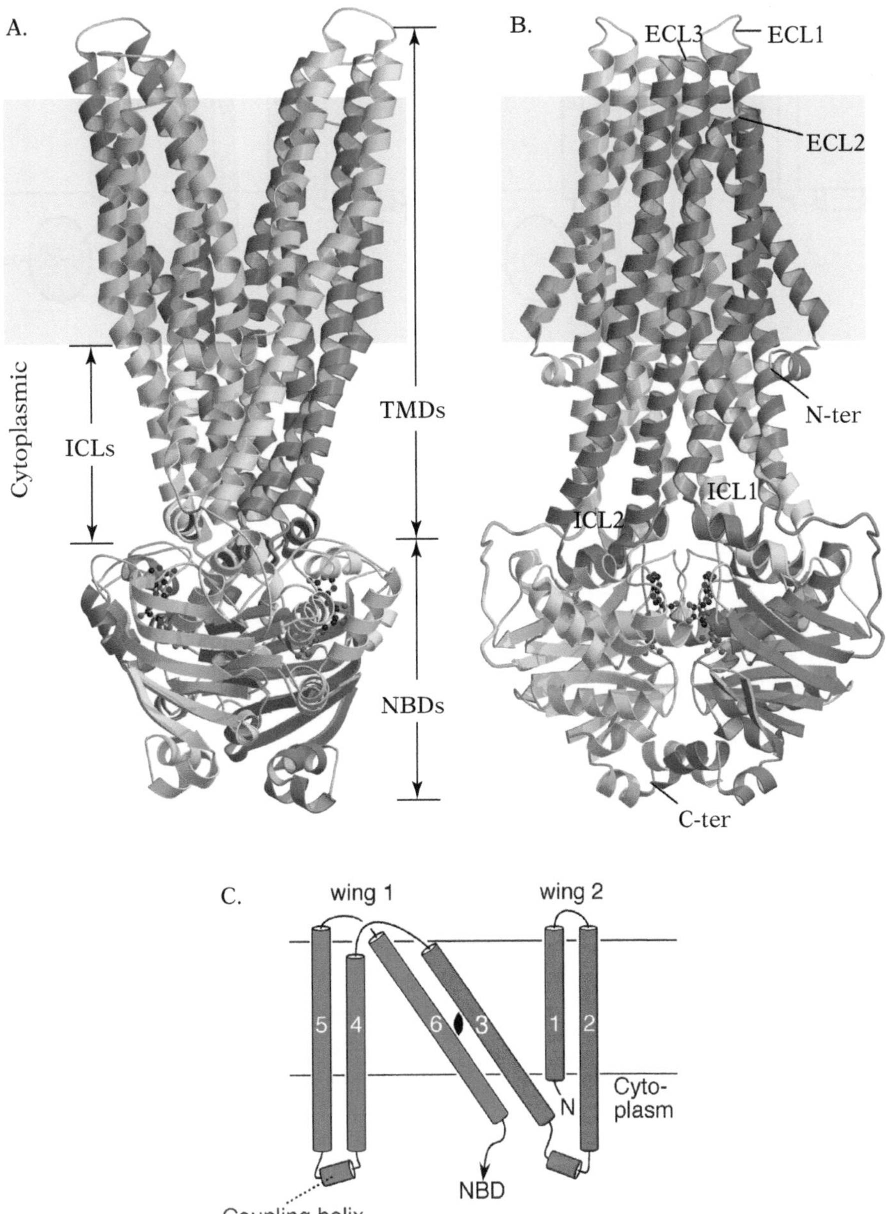

11.45 The structure of Sav1866, an ABC drug efflux transporter in *S. aureus*. The ribbon presentations of the x-ray structure viewed in (A) and (B) are rotated by 90°. The subunits (yellow and green) are intertwined, especially as they cross the membrane (gray shading). Two molecules of bound ADP are visible between the NBDs. The TM segments are connected by short loops on the extracellular side (labeled ECL1, 2, and 3 in (B)) and long loops on the cytoplasmic side (labeled ICL). Based on Dawson, R. J. P., and K. P. Locher, *Nature*. 2006, 443:180–185. © 2005. Adapted by permission of Macmillan Publishers Ltd. (C) The topology of a TM domain of Sav1866 illustrates the pseudosymmetry between TM1–3 and TM4–6. From Locher, K. P., *Phil Trans R Soc B*. 2009, 364:239–245. © 2009, The Royal Society.

P-glycoprotein (P-gp) is an important cause of drug resistance in human cancer cells, as well as a pump for efflux of other drugs and xenobiotics, and has been a focus of many studies. P-gp is a monomer of over 1300 amino acids that folds into the two TM domains and two NBDs typical of ABC transporters. The first x-ray structure of P-gp from mice showed a fold very like Sav1866. A new high-resolution structure of P-gp from *Caenorhabditis elegans* offers a more accurate analysis of the architecture and a good basis for modeling human P-gp, which has 46% sequence identity. The 3.4-Å resolution structure shows P-gp in an inward-facing conformation that is wide open to the cytoplasm, suggesting that drugs have access from both the cytoplasm and the inner leaflet of the membrane (Figure 11.46). Most of the TM helices superpose with those in the mouse structure;

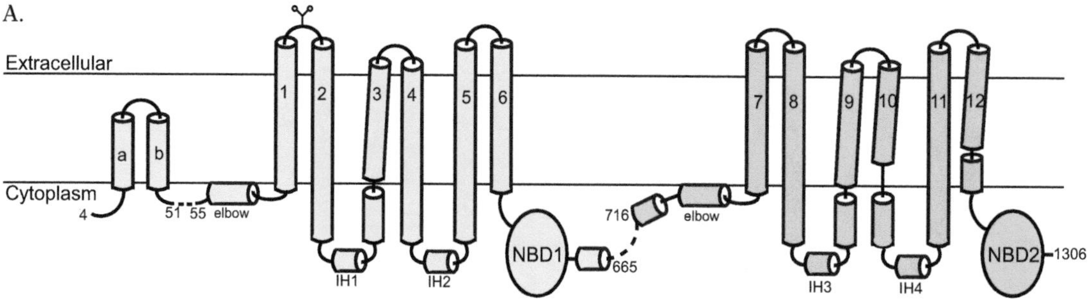

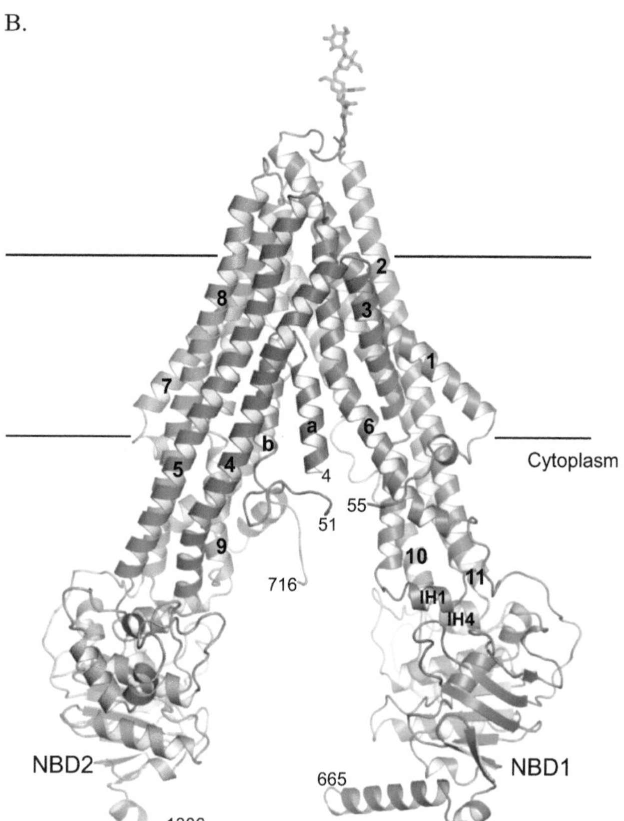

11.46 The structure of P-glycoprotein from *C. elegans*. (A) In P-glycoprotein all four ABC transporter domains are in a single polypeptide chain. The topology diagram for P-glycoprotein shows the two symmetric halves (blue and yellow) each contain a TM domain and an NBD. (B) Ribbon presentation of the x-ray structure (colored as in (A)). From Jin, M. S., *et al*., *Nature*. 2012, 490:566–569. © 2012, Nature Publishing Group. Reprinted by permission of Macmillan Publishers Ltd.

however, TM3, TM4, and TM5 are shifted significantly due to a register shift and are now consistent with results from arginine scanning. The structure is also consistent with residues assigned to the translocation pathway by engineered cysteine crosslinks.

All four of the short helices on the cytoplasmic side of the protein, IH1–4, make important contacts at the domain interfaces in P-gp. The interactions between the TM domains and the NBDs involve insertion of coupling helices IH2 and IH4 into clefts in the NBDs, creating ball-and-socket joints that are stabilized by salt bridges to the other two helices and residues of the NBDs (Figure 11.47; note the similarity to Figure 11.32B). Clearly ABC exporters share many characteristics of ABC importers, in spite of differences such as the lack of periplasmic partners.

EmrE, an Example of Dual Topology

EmrE from *E. coli* is only 12 kDa (110 amino acids), and yet when overexpressed it causes bacteria to be resistant to tetracycline, tetraphenylphosphonium (TPP^+), ethidium bromide, and other antiseptics and intercalating dyes. Found in both Gram-positive and Gram-negative bacteria but not in eukaryotes, it is a proton/drug antiporter that uses the energy of the pmf to drive transport of drugs out of the cytoplasm. Glu14, the only charged residue that is buried in EmrE, stabilizes the positive charge on aromatic cations for export and binds protons for import, thus directly coupling substrate and proton translocation. The hydrophobic environment around Glu14 raises its pK_a to 7.3–8.5, which is crucial to coupled antiport: substitution of Asp in that position, with a

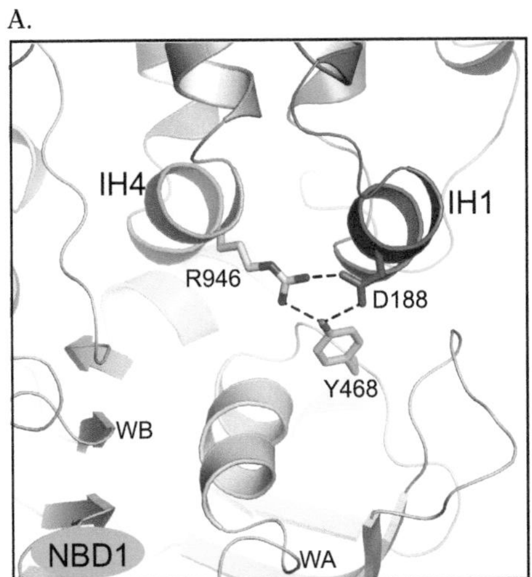

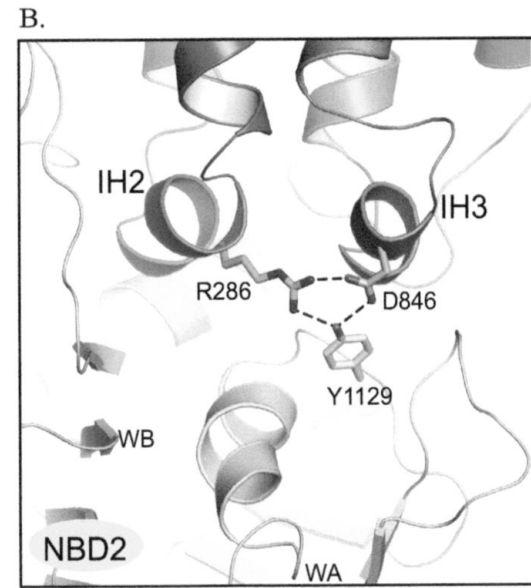

11.47 The interactions between TM domains and NBDs in P-gp. The interfaces between NBD1 (aqua) and the TM domain (A) and between NBD2 (yellow) and the TM domain (B) are stabilized by salt bridges (dashed lines) between Arg residues of the coupling helices IH2 and IH4, Glu residues on the additional helices IH1 and IH3, and Tyr residues of the NBDs. The clefts in the NBDs are near the Walker A motif (WA) and the Walker B motif (WB). From Jin, M. S., *et al.*, *Nature.* 2012, 490:566–569. © 2012, Nature Publishing Group. Reprinted by permission of Macmillan Publishers Ltd.

pK_a of around 6.5, dissipates the proton gradient without pumping drugs.

EmrE forms a dimer of unusual topology, which was for many years the object of controversy. The structure of TPP$^+$-bound EmrE solved by both cryo-EM and x-ray crystallography shows an antiparallel dimer making an eight-helix bundle (Figure 11.48). In each monomer is an axis of pseudo-symmetry that relates TM1–3 of one monomer to TM1–3 of the other. The TPP$^+$ substrate is bound in a central cleft between the Glu14 residues on TM1 of each monomer. When different substrates are bound, both cryo-EM structures and EPR dynamical studies indicate the overall structure is the same, with a kink in TM3 allowing for different helix tilt angles.

The dual topology of EmrE has been studied in mutants and other constructs, with crosslinking, and with various EPR and NMR techniques. The sensitivity of EmrE to its lipid/detergent environment and the presence of higher oligomers at high protein:lipid ratios contributed to inconsistent results. However, the antiparallel topology now has unequivocal support. Solution NMR of TPP$^+$-bound EmrE gives twice the number of peaks (210 for 105 non-proline residues) since each monomer is in a different conformation. Double label NMR experiments called ZZ exchange, which follow shifts in ^{15}N and ^{1}H peaks recorded at different times to observe changes, show equal populations in each of two states and give kinetic parameters for the widespread conformational

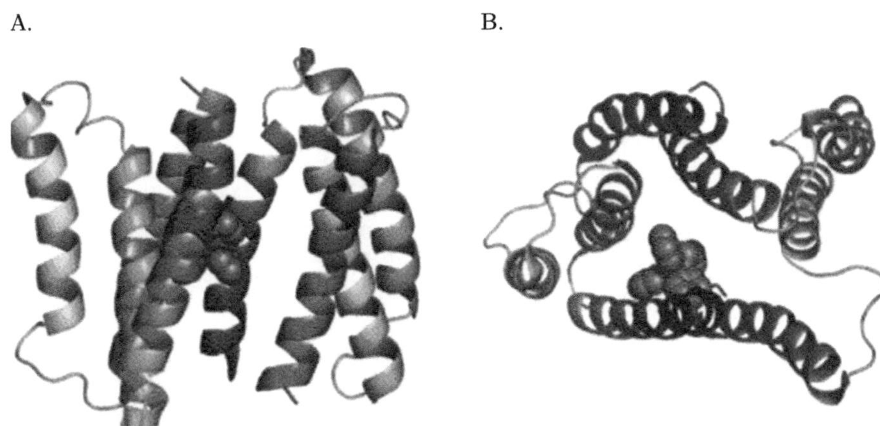

11.48 X-ray structure of the EmrE dimer. Ribbon diagrams of the EmrE structure viewed from the plane of the membrane (A) and perpendicular to the plane of the membrane ((B), top view). Each monomer is shown in rainbow color (N terminus, blue; C terminus, red), with a space-filling model of TPP$^+$ (magenta). From Tate, C. G., *Curr Opin Struct Biol.* 2006, 16:457–464. © 2006 by Elsevier. Reprinted with permission from Elsevier.

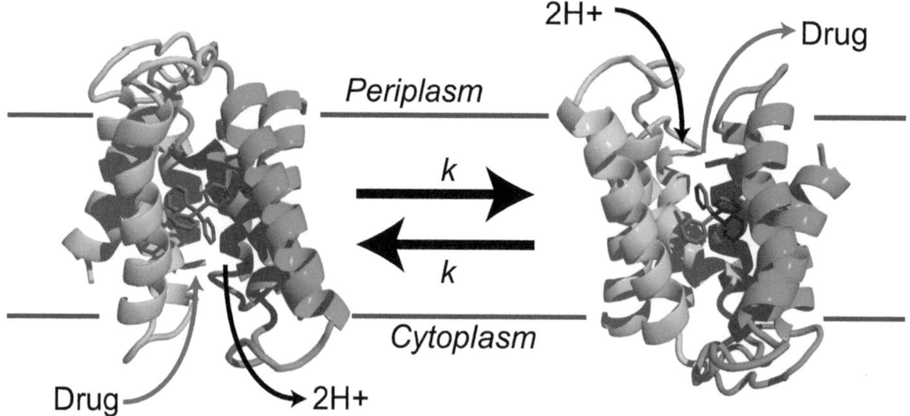

11.49 Conformational exchange between EmrE monomers in alternating access transport. The antiparallel EmrE dimer consists of two monomers (cyan and green) of identical structures in opposite orientations with respect to the membrane. In the alternating access model, each monomer switches to the conformation of the other during transport. When the dimer is open inward (Ci, left) the drug substrate (TPP^+, magenta sticks) is taken up from the cytoplasm; in the Co conformation (right) it is excreted to the periplasm, where protons are taken up. From Morrison, E. A., *et al.*, *Nature*. 2012, 481:45–50. © 2012, Nature Publishing Group. Reprinted by permission of Macmillan Publishers Ltd.

change. Crosslinking that can only occur in antiparallel dimers, using a heterobifunctional crosslinker to link the Cys residue in a S107C mutant with the only Lys residue (K22), occurs with 100% efficiency. Lastly, from single-molecule FRET in which a single position on one face of the monomer is labeled, the efficiency of energy transfer in the dimer gives the observed distance between labeled residues of each monomer (providing donor and acceptor) for the opposite sides of the membrane. Together these data strongly indicate that EmrE forms an antiparallel dimer and support a transport mechanism of alternating access with conformational exchange between monomers (Figure 11.49). The EmrE homodimer functions enough like larger secondary transporters described above to suggest that SMR transporters represent an intermediate point in the evolution of larger transporters with internal dual repeat structure.

Tripartite Drug Efflux via a Membrane Vacuum Cleaner

Some of the drugs that are expelled from the cytoplasm by EmrE, such as ethidium bromide, need the tripartite system AcrAB/TolC to leave the cell. This three-component system spans the cell envelope of *E. coli* to pick up drugs from the inner membrane or periplasm and extrude them directly into the outside medium. A well-characterized MDR system in Gram-negative bacteria, it is composed of AcrB, an inner membrane transporter; AcrA, a periplasmic adaptor protein; and TolC, the channel-tunnel protein (Figure 11.50). An analogous system is made up of the secondary transporter EmrB, the periplasmic lipoprotein EmrA, and TolC. Similarly, the system for exporting hemolysin is made up of HlyB, HlyD, and TolC. In *E. coli* several systems for drug and protein export use TolC, whereas drug efflux systems in other bacteria, such as *Pseudomonas aeruginosa*, have different tunnel components for the different efflux systems (see below). Assay of different transporters *in vitro* demonstrates that substrate specificity of these MDR systems resides in the inner membrane transporters. The AcrAB/TolC system exports a broad variety of mostly amphiphilic compounds that may be positively

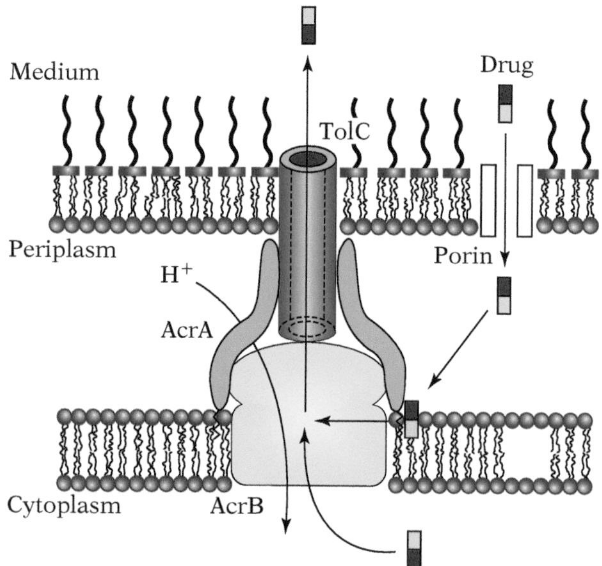

11.50 Schematic representation of the MDR system composed of AcrAB/TolC. Drugs can enter AcrB (yellow) in the membrane for export from either the cytoplasm or the periplasm. AcrA (green) is thought to stabilize the docking of TolC (blue) on AcrB and allow drugs to be transported outside the cell. From Murakami, S., and A. Yamaguchi, *Curr Opin Struct Biol*. 2003, 13:443–452. © 2003 by Elsevier. Reprinted with permission from Elsevier.

or negatively charged, zwitterionic, or neutral, including bile salts, erythromycin, and β-lactams such as ampicillin. Such MDR systems have been called "membrane vacuum cleaners" because they can remove a large number of unwanted substances and prevent their accumulation in either the cytoplasm or the periplasm.

AcrB, a Peristaltic Pump

The inner membrane transporter of a tripartite drug efflux system may belong to one of several different classes of transport proteins. As a member of the RND superfamily, AcrB is a proton/drug antiporter and, like numerous other transporters, its sequence consists of two homologous halves, suggesting they evolved from gene duplication events. The first crystal structure of AcrB revealed a homotrimer in the shape of a jellyfish, with each 110-kDa monomer (1049 residues) providing 12 TM α-helices and also contributing to the large periplasmic headpiece (Figure 11.51). In the bilayer-spanning domain the TM segments are fairly loosely organized around a central cavity of 35 Å diameter, which must be filled with membrane lipids *in vivo* to

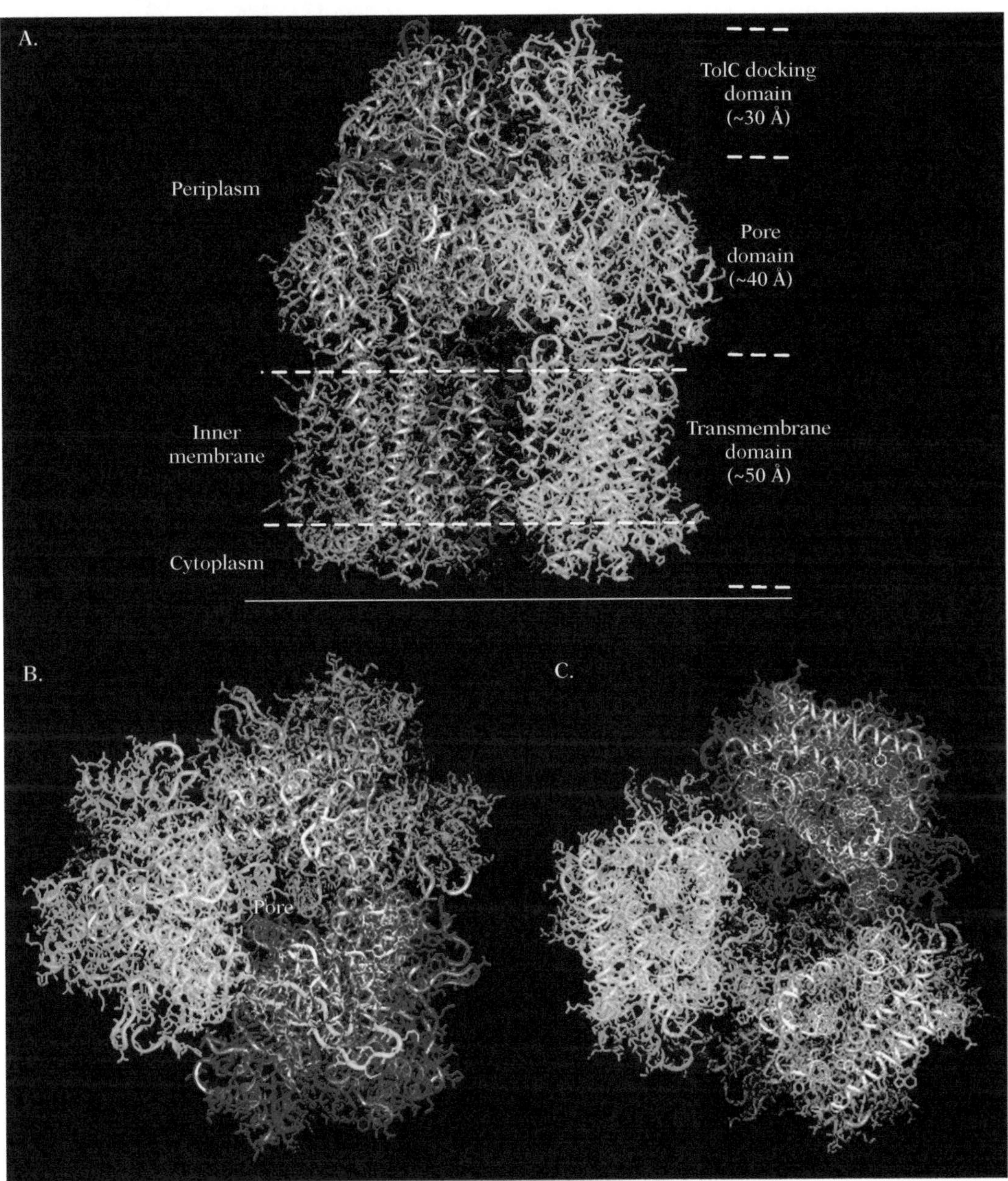

11.51 X-ray structure of the drug efflux pump AcrB. (A) View of the AcrB trimer from the side, oriented with the TM helices on the bottom and the periplasmic headpiece on the top. Two of the three subunits (colored purple, green, and blue) are seen clearly from this angle. (B) The AcrB trimer, viewed from the periplasm, reveals a peptide strand from each subunit that reaches far into the neighboring subunit. (C) The view from the cytoplasm of the AcrB homotrimer shows the large central cavity. From Murakami, S., and A. Yamaguchi, *Curr Opin Struct Biol*. 2003, 13:443–452. © 2003 by Elsevier. Reprinted with permission from Elsevier.

avoid loss of the permeability barrier. In the hydrophobic TM segments of each monomer the central helices, TM4 and TM10, contain the only three charged residues, Asp407, Asp408, and Lys940, which form salt bridges and are assumed to be involved in proton translocation.

The headpiece is divided into two layers: the upper TolC-docking domain and the lower pore domain. The TolC-docking domain opens like a funnel to a diameter of ~30 Å, matching the diameter of the TolC tunnel (see below). Below it, the pore domain has a large central pore lined by three helices, one from each subunit. In the pore domain each subunit has four β-α-β subdomains that together form a large substrate-binding pocket. The central cavity of AcrB is enormous, measuring in total around 5000 Å^3. Between the subunits are three openings to the periplasm, called vestibules. Substrates can therefore access the central cavity either from the periplasm or the bilayer for transport out of the central pore (Figure 11.52).

Alternating Site Mechanism of AcrB

Because the first crystals of AcrB were of the trigonal form, they confined the trimer to exact three-fold symmetry. Solution of the AcrB structure from crystals of different space groups that allowed asymmetry of the monomers revealed three different conformations affecting the access in the pore region. The three conformations are called loose (L, or "access"), tight (T, or "binding"), and open (O, or "extrusion"), and they vary in the binding pocket and pore regions (Figure 11.53). A key difference is the position of the central pore helices, which in the O monomer is inclined nearly 15° toward the T monomer, opening the pore from the O monomer to the exit funnel while contracting the T monomer to form the drug-binding pocket. The lining of the pocket has eight Phe residues, allowing hydrophobic and aromatic interactions with drugs of diverse sizes and structures, as seen in the crystal structures of AcrB with several different drugs.

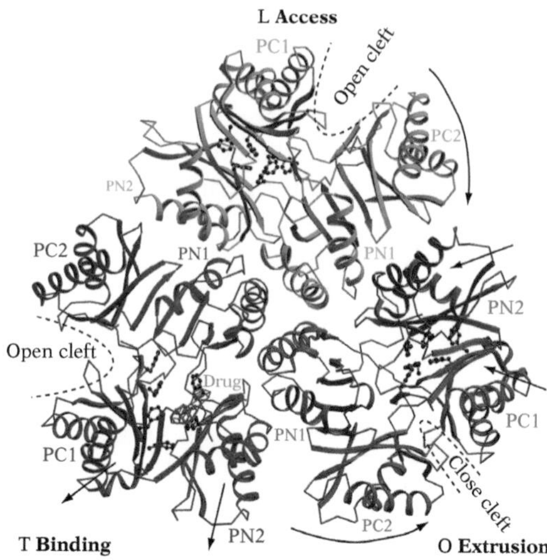

11.53 Illustration of the three conformations of subunits of the AcrB trimer. Each subunit is in a different conformation: the access state (L, green), extrusion state (O, red), and binding state (T, blue) shown in a cut view of the pore domain from the exterior of the trimer. The vestibules are clefts that are open in the L and T conformations and closed in the O conformation, which is open to the exit pore. Phenylalanine residues are shown in ball-and-stick representation, and the binding pocket of the T conformation contains a molecule of the drug minocycline (orange contour labeled "Drug"). The arrows indicate the relative movements of the subdomains, which are labeled PC1, PC2, PN1, and PN2. From Murakami, S., *et al.*, *Nature*. 2006, 443:173–179. © 2006. Reprinted by permission of Macmillan Publishers Ltd.

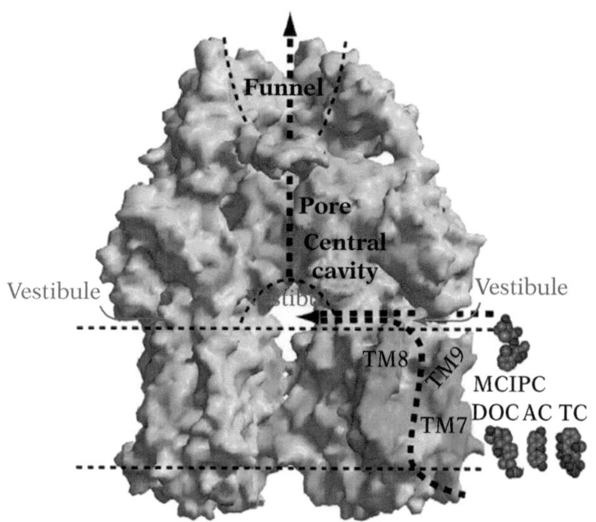

11.52 Solvent-accessible surface of AcrB in a cutaway model. The front subunit has been removed, allowing a view into the central cavity and revealing pore helices (yellow) and TM helices that form grooves (green). Broken lines indicate the putative membrane boundaries and the framework of the funnel and cavity. Broken arrows indicate the postulated translocation pathways for substrates from the membrane, such as deoxycholate (DOC), acriflavine (AC), and tetracycline (TC), as well as substrates from the periplasm/outer leaflet, such as cloxacillin (MCIPC). From Murakami, S., and A. Yamaguchi, *Curr Opin Struct Biol*. 2003, 13:443–452. © 2003 by Elsevier. Reprinted with permission from Elsevier.

The observation of three subunits in three conformations suggests an alternating site, functional-rotation mechanism seen first in the F_1-ATPase (Figure 11.54; see Chapter 13). In the rotation, a monomer first binds substrate in the L conformation, binds it more tightly in the T conformation, and then extrudes it to the exit funnel in the O conformation. The subunits communicate via the central helices. The conformational changes are coupled to proton translocation in the TM domain: Proton uptake is postulated to drive the change from O to L and proton release to accompany the change from T to O. Drug extrusion is accomplished by the diffusion of substrate along a pathway that bulges and occludes as

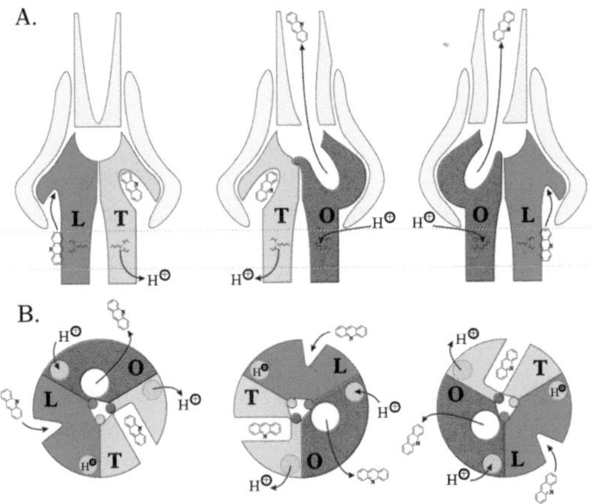

11.54 Schematic representation of the AcrB alternating site functional rotation transport mechanism. The three conformational states are loose (L, access; blue), tight (T, binding; yellow), and open (O, extrusion; red). They are shown as viewed from the side ((A), with dotted lines denoting the membrane) and from the top (B). The side chains presumed to be involved in proton translocation (Asp407, Asp408, and Lys940) are indicated in the TM part of each monomer in (A). An acridine molecule is depicted as substrate, which first binds to the L state, then binds tightly to the aromatic pocket in the T state, and then is extruded toward the funnel in the O state. From Seeger, M. A., *et al.*, *Science*. 2006, 313:1295–1298. © 2006, American Association for the Advancement of Science. Reprinted with permission from AAAS.

it migrates toward the funnel, much like the action of a peristaltic pump.

The funnel at the top of AcrB fits the dimensions of the bottom of the TolC tunnel, and their interaction can be modeled to involve six hairpins at the top of AcrB, docking with six helix–turn–helix structures at the bottom of TolC. Although crosslinks can be inserted between the two molecules, both with chemical crosslinking and with disulfide formation between inserted Cys residues, unmodified TolC does not bind AcrB *in vitro*. In contrast, purified AcrA exhibits micromolar affinity for both AcrB and TolC, which suggests that the docking between AcrB and TolC is reinforced by AcrA. The role of AcrA is likely to be quite dynamic given that the complex formation is transient, allowing TolC to partner with other efflux transporters.

AcrA, a Periplasmic Adaptor Protein

In the tripartite drug efflux systems, adaptor proteins do not contribute to the transport pores but are required to stabilize the interactions of the inner membrane transporter with the channel protein. The adaptor proteins are periplasmic proteins attached to the inner membrane either via lipid acylation of a cysteine residue or with an N-terminal TM segment. However, studies of mutants lacking these regions reveal that membrane attachment is not essential for the drug efflux function of these proteins. AcrA is highly homologous to MexA, from the MexAB/OprM tripartite system of *P. aeruginosa*, the first periplasmic adaptor protein to have a high-resolution structure.

AcrA consists of 397 amino acids, including a cleavable N-terminal signal, residues 1 to 24. Ninety residues at the C terminus that are very sensitive to proteolysis and correspond to a flexible region of MexA were removed from AcrA for crystallization. In addition, to allow incorporation of selenomethionine for phase determination the structure was solved with substitution of four methionine residues. The resulting 2.7-Å resolution structure of AcrA shows residues 53 to 299 in an elongated sickle that consists of three domains: a β-barrel domain, a central lipoyl domain, and a coiled coil α-helical hairpin (Figure 11.55). Nearest the membrane is the β-barrel domain composed of six antiparallel β-strands that include both the N and C termini of the fragment. The β-sandwich in the lipoyl domain contains two half-domains of four β-strands each, typical of binding sites for lipoic acid or biotin. Between the two half-domains in the peptide chain is a long coiled coil with five heptad repeats in each helix. Packing between the helices involves canonical knobs into holes at the first and fourth positions of the heptad repeats. At the base of the helical hairpin is a hinge due to variation in unwinding of the helices, fortuitously observed in the crystal structure of AcrA, in which each of four monomers has a different conformation (Figure 11.55B). The implied flexibility is likely important in the function of AcrA as it links AcrB to TolC.

The x-ray structure of the purified AcrA fragment does not reveal how AcrA interacts with both AcrB and TolC. It is anchored in the inner membrane by lipid modification of Cys25, the first residue of the mature protein. The portion of its N terminus that is missing in the x-ray structure is sufficiently long to bring the β-barrel domain to a position alongside the periplasmic domain of AcrB, putting the coiled coil well into the periplasm, where it can interact with TolC. It has been modeled as an adaptor that overlaps portions of AcrB and TolC, or as a bridging funnel that meets the tip of TolC (see Figure 11.57). While both AcrB and TolC are trimers, the oligomeric state of AcrA is not known; models of the tripartite complex incorporate three to six copies of AcrA.

TolC, the Channel-Tunnel

Drug efflux via the tripartite AcrAB/TolC results in expelling the unwanted compounds into the outside medium because TolC provides a pathway across both the periplasm and the outer membrane. This is evident in the remarkable high-resolution structure of TolC (2.1 Å resolution) that shows a homotrimer with a β-barrel channel domain in the outer membrane connected to a unique α-barrel in the periplasm (Figure 11.56). The α-barrel is 100 Å in length and extends far into the periplasm, allowing TolC to couple with its partners to provide an

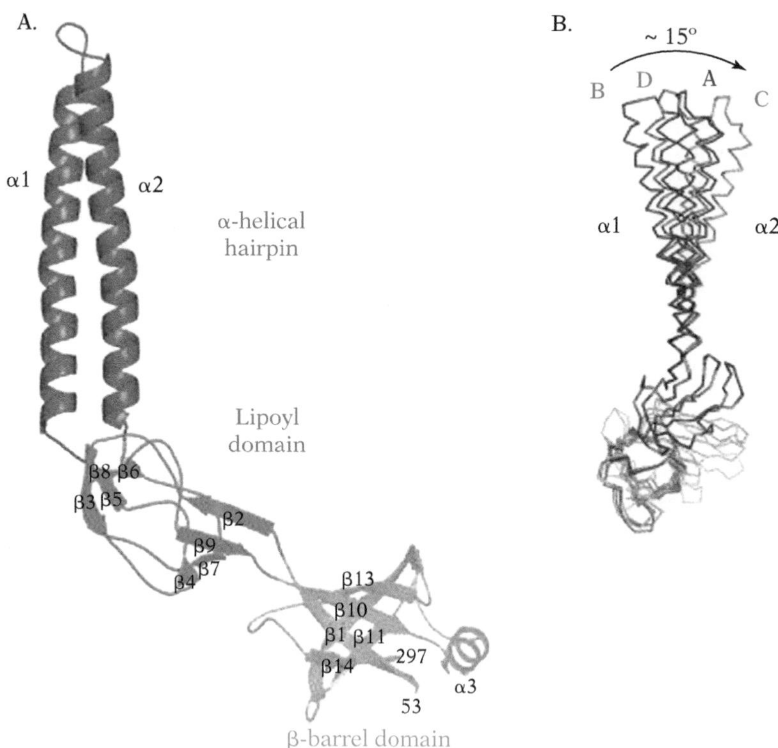

11.55 X-ray structure of a monomer of AcrA. (A) A ribbon representation of a monomer of the stable core fragment of AcrA (residues 45–312) with four methionine substitutions is viewed with the part nearest the inner membrane at the bottom. The AcrA monomer has three domains: helical hairpin (red), lipoyl domain (green), and β-barrel (cyan). From Mikolosko, J., *et al.*, *Structure*. 2006, 14:577–587. © 2006 by Elsevier. Reprinted with permission from Elsevier. (B) Four conformations of AcrA observed in the crystal structure. The crystal form captured each monomer (labeled A, B, C, and D) in a different conformation due to the flexible hinge between the helical hairpin and the lipoyl domain. Superposition of the four conformations reveals a 15° rotation between the most different structures. From Mikolosko, J., *et al.*, *Structure*. 2006, 14:577–587. © 2006 by Elsevier. Reprinted with permission from Elsevier.

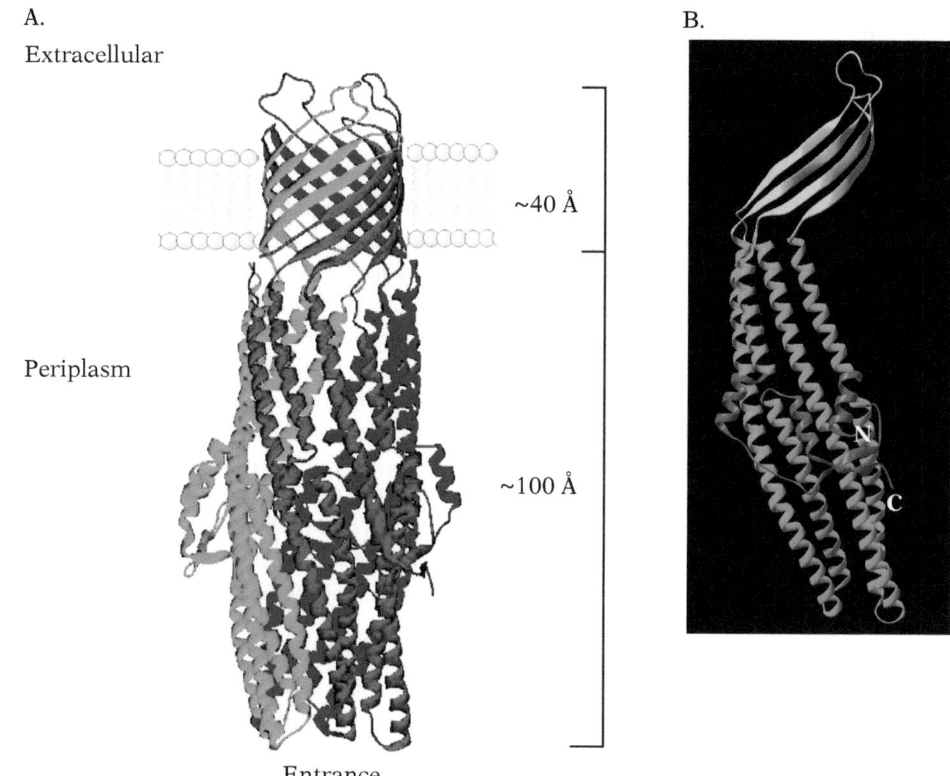

11.56 High-resolution structure of the TolC channel-tunnel. (A) TolC is a homotrimer (each subunit is a different color) that spans the outer membrane with a β-barrel channel and extends into the periplasm with an α-barrel. See text for details. From Koronakis, V., *et al.*, *Annu Rev Biochem*. 2004, 73:467–489. © 2004 by Annual Reviews. Reprinted with permission from the Annual Review of Biochemistry, www.annualreviews.org. (B) Each TolC monomer contributes four strands to the β-barrel (yellow) and uses six helices to span the length of the tunnel (green). It also has an equatorial domain (red) containing both α and β structures. From Koronakis, V., *et al.*, *Nature*. 2000, 405:914–919. © 2000. Reprinted by permission of Macmillan Publishers Ltd.

uninterrupted pathway from the inner membrane to the outside medium. Passage through the TolC channel-tunnel is passive, as the inner membrane transport partners carry out the energized step.

The TolC channel crosses the outer membrane with an antiparallel 12-stranded β-barrel with a right-handed twist and external loops that are sites for colicin and bacteriophage attachment. Aromatic residues delineate the regions located in the lipid bilayer interface (see Chapter 5). TolC differs from other outer membrane β-barrels in significant ways. Rather than having each polypeptide form a barrel, each TolC subunit contributes four β-strands to its single pore. With neither a constriction formed by inward folded loops nor a plug formed by a separate domain, the pore is constitutively open. Finally, the β-barrel is connected directly to the α-barrel, which has a left-handed twist, requiring interdomain linkers with conserved proline residues that make abrupt turns.

The periplasmic tunnel of TolC is also 12-stranded, with α-helices packed uniformly in an antiparallel left-handed superhelical twist. Each subunit contributes two long helices and two pairs of shorter helices stacking end-to-end to span the length of the barrel (Figure 11.56B). Three short helices and two short β-strands thicken the middle of the tunnel at its equatorial domain. Helix–helix interactions are stabilized by knobs-into-holes packing of the side chains. The lower part of the tunnel is less uniform with an inner coiled coil plus a second pair of outer helices. These form a 3.9-Å constriction that closes the tunnel at the cytoplasmic end, presumably in a resting state. Rotation of the inner pair of helices to untwist them is predicted to open the end like opening the aperture of a camera lens, as would be required when TolC interacts with a substrate-carrying inner membrane transporter such as AcrAB (Figure 11.57).

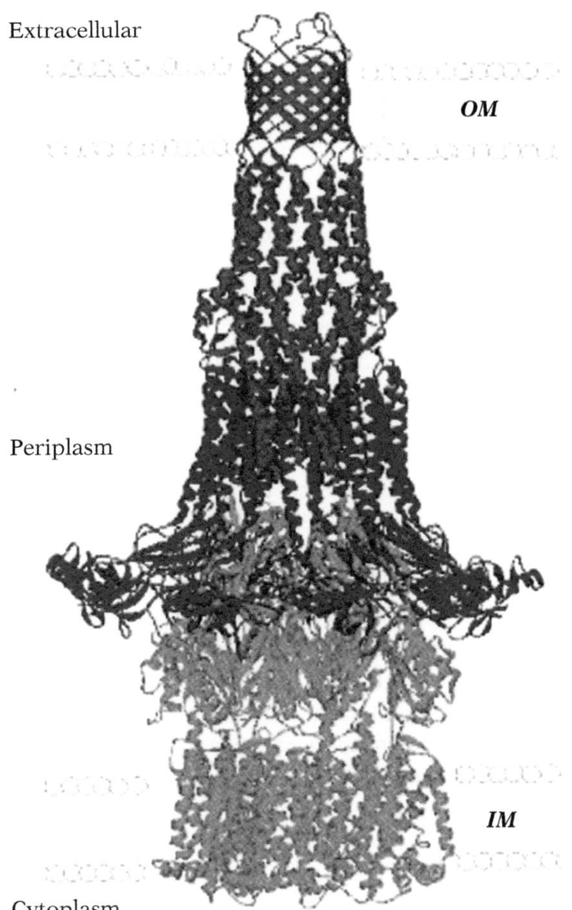

11.57 Model of the tripartite AcrAB/TolC drug efflux system that spans the inner membrane, periplasmic space, and outer membrane. A ribbon model of the AcrAB/TolC system composed of the x-ray structure of TolC modeled to be the open state (red), the x-ray structure of AcrB (green), and a representation of AcrA based on the x-ray structure of its close homolog, MexA (blue) shows overlap between AcrA and TolC. From Eswaran, J., *et al.*, *Curr Opin Struct Biol*. 2004, 14:741–747. © 2004 by Elsevier. Reprinted with permission from Elsevier.

Partners of TolC

TolC joins different *E. coli* inner membrane transporters to form tripartite efflux systems in a dynamic manner. In some bacteria, such as *P. aeruginosa*, specialized homologs of TolC homologs work with different transporters. The structures of two TolC homologs, OprM from *P. aeruginosa* and VceC from *Vibrio cholerae*, have been found to be cylindrical channels very similar to TolC. Homologs of TolC, ubiquitous among Gram-negative bacteria, are used not only for drug export but for efflux of cations and secretion of proteins. TolC is utilized by specialized systems for secretion of hemolysin and colicin V from *E. coli*, leukotoxins from *Pasteurella haemolytica* and *Actinobacillus actinomycetemcomitans*, adenylate cyclase from *Bordetella pertussis*, and several proteases and lipases. Thus the molecular picture provided by the structures of AcrA, AcrB, and TolC furthers the understanding of many important processes related to pathogenesis and drug resistance.

CONCLUSION

Structural biology of transporters has made remarkable progress in recent years. The variety of transporters described in this chapter is astounding: from monomers that vary in size from 110 residues to over 1300 residues, to oligomers that respond to environmental conditions, to assemblies that span two membranes. Whether coupled to ion transport or ATP hydrolysis for active transport, these transporters appear to share the common general mechanism of alternating access. Thus their functions necessitate quite large conformational changes to open to the other side of the membrane. While this mechanism distinguishes them from channels, mutations can stabilize them in an open conformation that allows passive diffusion, giving them channel characteristics.

FOR FURTHER READING

Papers with an asterisk (*) present original structures.

LacY and GlpT

*Abramson, J., I. Smirnova, V. Kasha, G. Verner, H. R. Kaback, and S. Iwata, Structure and mechanism of the lactose permease of *Escherichia coli*. *Science*. 2003, 301:610–615.

Abramson, J., H. R. Kaback, and S. Iwata, Structural comparison of lactose permease and the glycerol-3-phosphate antiporter: members of the major facilitator superfamily. *Curr Opin Struct Biol*. 2004, 14:413–419.

Guan, L., and H. R. Kaback, Lessons from lactose permease. *Annu Rev Biophys Biomol Struct*. 2006, 35:67–91.

*Huang, Y., M. J. Lemieux, J. Song, M. Auer, and D.-N. Wang, Structure and mechanism of the glycerol-3-phosphate transporter from *Escherichia coli*. *Science*. 2003, 301:616–620.

Lemiuex, M. J., Y. Huang, and D.-N. Wang, Glycerol-3-phosphate transporter of *Escherichia coli*: structure, function and regulation. *Res Microbiol*. 2004, 155:623–629.

Lemiuex, M. J., Y. Huang, and D.-N. Wang, The structural basis of substrate translocation by the *Escherichia coli* glycerol-3-phosphate transporter: a member of the major facilitator superfamily. *Curr Opin Struct Biol*. 2004, 14:405–412.

EmrD and FucP

*Dang, S., L. Sun, Y. Huang, *et al.*, Structure of a fucose transporter in an outward-open conformation. *Nature*. 2010, 467:734–737.

*Yin, Y., X. He, P. Szewczyk, T. Nguyen, and G. Chang, Structure of the multidrug transporter EmrD from *Escherichia coli*. *Science*. 2006, 312:741–744.

Mitochondrial ADP/ATP Carrier

Nury, H., I. Blesneac, S. Ravaud, and E. Pebay-Peyroula, Structural approaches of the Mitochondrial Carrier Family. *Meth in Molec Biol*. 2010, 654:105–117.

Nury, H., C. Dahout-Gonzalez, V. Trézéguet, G. J. Lauquin, G. Brandolin, and E. Pebay-Peyroula, Relations between structure and function of the mitochondrial ADP/ATP carrier. *Annu Rev Biochem*. 2006, 75:713–741.

*Pebay-Peyroula, E., C. Dahout-Gonzalez, R. Kahn, V. Trézéguet, G. J. Lauquin, and G. Brandolin, Structure of the mitochondrial ADP/ATP carrier in complex with carboxyatractyloside. *Nature*. 2003, 426:39–44.

Neurotransmitter Transporters: Glutamate Transporter

Boudker, O., R. Ryan, D. Yernool, *et al.*, Coupling substrate and ion binding to extracellular gate of a sodium-dependent aspartate transporter. *Nature*. 2007, 445:387–393.

Reyes, N., C. Ginter, and O. Boudker, Transport mechanism of a bacterial homologue of glutamate transporters. *Nature*. 2009, 462:880–885.

*Yernool, D., O. Boudker, Y. Jin, and E. Gouaux, Structure of a glutamate transporter homologue from *Pyrococcus horikoshii*. *Nature*. 2004, 431:811–818.

LeuT

Claxton, D. P., M. Quick, L. Shi, *et al.*, Ion/substrate-dependent conformational dynamics of a bacterial homolog of neurotransmitter:sodium symporters. *Nature Struct Molec Biol*. 2010, 17:822–829.

Krishnamurthy, H., C. L. Piscitelli, and E. Gouaux, Unlocking the molecular secrets of sodium-coupled transporters. *Nature*. 2009, 459: 347–355.

Singh, S. K., LeuT: a prokaryotic stepping stone on the way to a eukaryotic neurotransmitter transporter structure. *Channels*. 2008, 2:380–389.

*Yamashita, A., S. K. Singh, T. Kawate, Y. Jin, and E. Gouaux, Crystal structure of a bacterial homologue of Na^+/Cl^- dependent neurotransmitter transporters. *Nature*. 2005, 437:215–223.

Zhao, Y., D. S. Terry, L. Shi, *et al.*, Substrate-modulated gating dynamics in a Na^+-coupled neurotransmitter transporter homologue. *Nature*. 2011, 474:109–113.

A video illustrating the LeuT conformational changes is available at http://www.nature.com./nature/journal/v481/n7382/extref/nature10737-s2.mov

BetP

Krämer, R., and C. Ziegler, Regulative interactions of the osmosensing C-terminal domain in the trimeric glycine betaine transporter BetP from *Corynebacterium glutamicum*. *Biol Chem*. 2009, 390:685–691.

*Ressl, S., T. van Scheltinga, C. Vonrhein, V. Ott, and C. Ziegler, Molecular basis of transport and regulation in the Na^+/betaine symporter BetP. *Nature*. 2009, 458:47–52.

Ziegler, C., E. Bremer, and R. Krämer, The BCCT family of carriers: from physiology to crystal structure. *Molec Microbiol*. 2010, 78:13–34.

Maltose Transporter

*Oldham, M. L., D. Khare, F. A. Quiocho, A. L. Davidson, and J. Chen, Crystal structure of a catalytic intermediate of the maltose transporter. *Nature*. 2007, 450:515–521.

*Khare, D., M. L. Oldham, C. Orelle, A. L. Davidson, and J. Chen, Alternating access in maltose transporter

mediated by rigid-body rotations. *Molecular Cell*. 2009, 33:528–536.

*Oldham, M. L., and J. Chen, Crystal structure of the maltose transporter in a pretranslocation intermediate step. *Science*. 2011, 332:1202–1205.

Orelle, C., T. Ayvaz, R. M. Everly, C. S. Klug, and A. L. Davidson, Both maltose-binding protein and ATP are required for nucleotide-binding domain closure in the intact maltose ABC transporter. *Proc Natl Acad Sci USA*. 2008, 105:12837–12842.

Chen, J., Molecular mechanism of the *Escherichia coli* maltose transporter. *Curr op Str Biol*. 2013, 23:492–498.

Oldham, M. L. and Chen, J., Structural basis for substrate specificity in the *Escherichia coli* maltose transport system. *Proc Nat Acad Sci US*. 2013, 110:18132–18137.

Vitamin B_{12} ABC Transporter

Borths, E. L., K. P. Locher, A. T. Lee, and D. C. Rees, The structure of *Escherichia coli* BtuF and binding to its cognate ATP binding cassette transporter. *Proc Natl Acad Sci USA*. 2002, 99:16642–16647.

*Hvorup, R. N., B. A. Goetz, M. Niederer, K. Hollenstein, E. Perozo, and K. P. Locher, Asymmetry in the structure of the ABC transporter-binding protein complex BtuCD–BtuF. *Science*. 2007:1387–1390.

*Korkhov, V. M., S. A. Mireku, and K. P. Locher, Structure of AMP-PNP-bound vitamin B_{12} transporter BtuCD-F. *Nature*. 2012, 490:367–372.

*Locher, K. P., A. T. Lee, and D. C. Rees, The *E. coli* BtuCD structure: a framework for ABC transporter architecture and mechanism. *Science*. 2002, 296:1091–1098.

Locher, K. P., Structure and mechanism of ABC transporters. *Curr Opin Struct Biol*. 2004, 14:426–431.

BtuB and TonB-dependent transporters

Chang, C., A. Mooser, A. Plückthun, and A. Wlodawer, Crystal structure of the dimeric C-terminal domain of TonB reveals a novel fold. *J Biol Chem*. 2001, 276:27535–27540.

*Chimento, D. P., Structure-induced transmembrane signaling in the cobalamin transporter BtuB. *Nature Struct Biol*. 2003, 10:394–401.

Chimento, D. P., R. J. Kadner, and M. C. Wiener, Comparative structural analysis of TonB-dependent outer membrane transporters: implications for the transport cycle. *Proteins*. 2005, 59:240–251.

Noinaj, N., M. Guillier, T. J. Barnard, and S. K. Buchanan, TonB-dependent transporters: regulation, structure and function. *Ann Rev Microbiol*. 2010, 64: 43–60.

Drug Efflux

*Dawson, R. J. P., and K. P. Locher, Structure of a bacterial multidrug ABC transporter. *Nature*. 2006, 443:180–185.

*Jin, M. S., M. L. Oldham, Q. Zhang, and J. Chen, Crystal structure of the multidrug transporter P-glycoprotein from *Caenorhabditis elegans*. *Nature*. 2012, 490:566–569.

*Koronakis, V., A. Sharff, E. Koronakis, B. Luisi, and C. Hughes, Crystal structure of the bacterial membrane protein TolC central to multidrug efflux and protein export. *Nature*. 2000, 405:914–919.

Koronakis, V., J. Eswaran, and C. Hughes, Structure and function of TolC. *Annu Rev Biochem*. 2004, 73: 467–489.

*Ma, C., and G. Chang, Structure of the multidrug resistance efflux transporter EmrE from *Escherichia coli*. *Proc Natl Acad Sci USA*. 2004, 101: 2852–2857.

*Mikolosko, J., K. Bobyk, H. I. Zgurskaya, and P. Ghosh, Conformational flexibility in the multidrug efflux system protein AcrA. *Structure*. 2006, 14: 577–587.

Morrison, E. A., G. T. De Koster, S. Dutta, *et al.*, Antiparallel EmrE exports drugs by exchanging between asymmetric structures. *Nature*. 2012, 481:45–50.

Murakami, S., and A. Yamaguchi, Multidrug-exporting secondary transporters. *Curr Opin Struct Biol*. 2003, 13:443–452.

*Murakami, S., R. Nakashima, E. Yamashita, and A. Yamaguchi, Crystal structure of bacterial multidrug efflux transporter AcrB. *Nature*. 2002, 419: 587–593.

Murakami, S., R. Nakashima, E. Yamashita, T. Matsumoto, and A. Yamaguchi, Crystal structures of a multidrug transporter reveal a functionally rotating mechanism. *Nature*. 2006, 443:173–179.

Seeger, M. A., A. Schiefner, T. Eicher, F. Verrey, K. Diederichs, and K. M. Pos, Structural asymmetry of AcrB trimer suggests a peristaltic pump mechanism. *Science*. 2006, 313:1295–1298.

Yu, E. W., G. McDermott, H. I. Zgurskaya, H. Nikaido, and D. E. Koshland, Structural basis of multiple drug-binding capacity of the AcrB multidrug efflux pump. *Science*. 2003, 300:976–980.

Zgurskaya, H. I., and H. Nikaido, Multidrug resistance mechanisms: drug efflux across two membranes. *Mol Microbiol*. 2000, 37:219–225.

CHANNELS

12

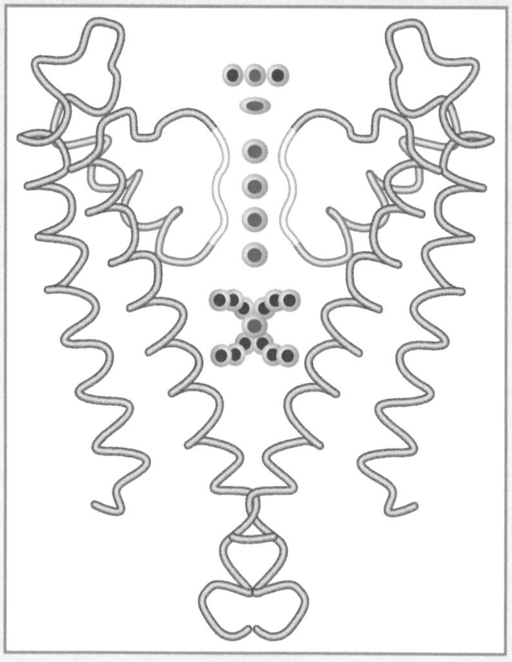

The high-resolution structure of the KcsA channel provided the first detailed view of the ion conduction pore common to all potassium channels, here exposed in a cutaway view of two of its four subunits to reveal the positions of the pore-forming helix (red) and the selectivity filter (gold). From MacKinnon, R., *FEBS Lett.* 2003, 555:62–65. © 2003 by the Federation of European Biochemical Societies. Reprinted with permission from Elsevier.

In contrast to the flexibility of many transporters described in the last chapter, channels do not require large conformational changes to allow their substrates to cross the membrane, since they contain conduction pores. The passage of ions and molecules through the pores is fast, but not uncomplicated. Speed and selectivity depend on the nature of the pores, including special features like vestibules (wide entryways on both ends) that attract ions and selectivity filters that exclude other ions from the pores. Channels vary in their degree of discrimination. Some ion channels are permeable to cations and not anions, or vice versa. Highly selective ion channels usually dehydrate the ions inside the pore, replacing the water molecules with side chains or dipoles within the selectivity filter. Further, many channels have gating mechanisms that enable them to open in response to outside signals communicated via ligand binding, electric potentials, pH, even temperature and pressure, and adaptive mechanisms that close them for periods of time (Figure 12.1).

Channels are essential for many biological processes – in fact, every physical and mental activity involves diverse ion channels. Of the 340 genes identified so far that code for ion channels in humans, mutations in >60 of them are known to cause diseases, called channelopathies. Many pharmaceutical drugs, including local anesthetics, antianxiety agents, and sedatives, target ion channels.

Given their importance it is not surprising that channels have been studied for many decades, and a large body of electrophysiology, biochemistry, biophysics, and

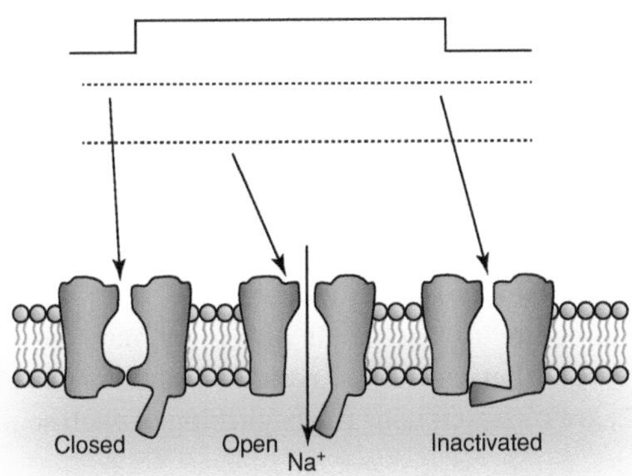

12.1 Schematic illustration of an ion channel. Opening and closing of ion channels, as detected in single-channel recordings (top), occurs randomly but may be increased by ligand binding or transmembrane voltage. For example, the sodium channel involved in the action potential of nerve and muscle cells opens in response to depolarization and then enters an inactive state. From Ashcroft, F. M., *Nature*. 2006, 440:440–447. © 2006, Nature Publishing Group. Reprinted by permission of Macmillan Publishers Ltd.

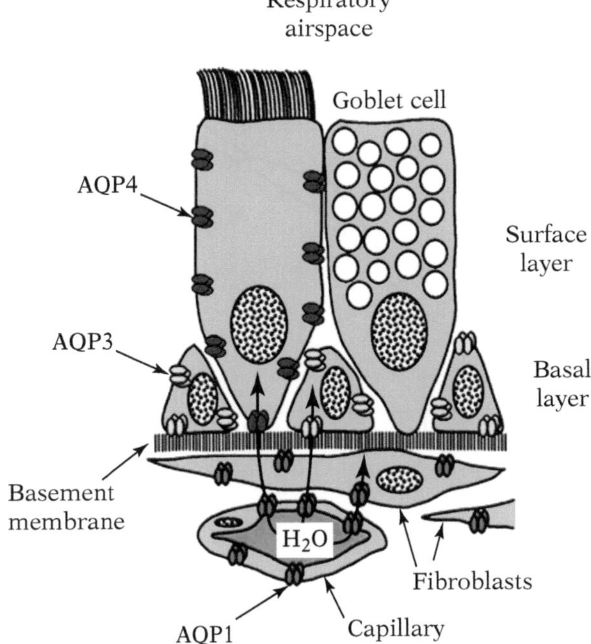

12.2 Localization of different human aquaporins. An example of aquaporin distributions is the epithelial tissues of the lungs, where AQP4 is expressed in surface columnar cells, AQP3 in basal cells, and AQP1 in underlying fibroblasts and capillaries, while no AQPs are expressed in goblet cells. From Agre, P., *Proc Am Thorac Soc*. 2006, 3:5–13. © 2006 by American Thoracic Society. Reprinted with permission from Proceedings of the American Thoracic Society.

genetics has contributed to understanding their properties. Some have structures very well characterized by electron microscopy, such as the cation channel of the nicotinic acetyl choline receptor (see Chapter 6). Other receptors that are ion channels have been described in Chapter 10. This chapter will explore channel characteristics with high-resolution structures of channels of diverse properties, starting with the aquaporins and the KcsA potassium channel structure (see Frontispiece) that were the basis for awarding the 2003 Nobel Prize in Chemistry to Peter Agre and Roderick MacKinnon.

The channels described in this chapter vary in architecture and form oligomers from dimers to heptamers. Often the pore is in the center of the oligomer, consisting of TM segments from each subunit, but in some oligomeric channels each subunit has a pore.The channels are gated by various forces: for example, there are potassium channels controlled by pH, voltage, Ca^{2+} ions, certain lipids, even temperature. The aquaporins transport water in response to an osmotic pressure gradient, and the mechanosensitive channels open in response to tension in the membrane. By definition channels carry out passive transport, so it was very surprising to learn that a closely related family of "chloride channels" includes some transporters.

AQUAPORINS AND GLYCEROAQUAPORINS

Although predicted in the 1950s, water channels in biological membranes were not discovered until 1992, when a protein purified from the erythrocyte membrane was reported to greatly increase the water flux in response to a gradient of osmotic pressure. Since the discovery of this first aquaporin (AQP), over 350 different AQPs have been identified in all forms of life. Mammals have 13 isoforms, designated AQP0 to AQP12, that fall into two classes, AQPs and glyceroaquaporins. The latter group allows entry of glycerol along with urea, DL-glyceraldehyde, and glycine, in addition to water. Four of the human AQPs (AQP3, AQP7, AQP9, and AQP10) are glyceroaquaporins.

The physiological roles of the different human AQPs vary widely because they differ in cellular locations as well as in modes of regulation. For example, in secretory glands, AQP5 is specifically expressed in the apical membrane where water passes into secretions such as tears, saliva, and sweat, while in respiratory epithelia of the lungs different cell types express different AQPs (Figure 12.2). The role of AQPs in the lungs was explored in the interesting case of a few asymptomatic AQP1-null individuals, whose pulmonary capillaries swelled normally in response to saline but their surrounding tissue did not accumulate fluid (Figure 12.3). In addition to the basic need for all cells to keep water in balance, the AQPs are involved in several illnesses, including abnormalities of kidney function, loss of vision, onset of brain edema, and starvation.

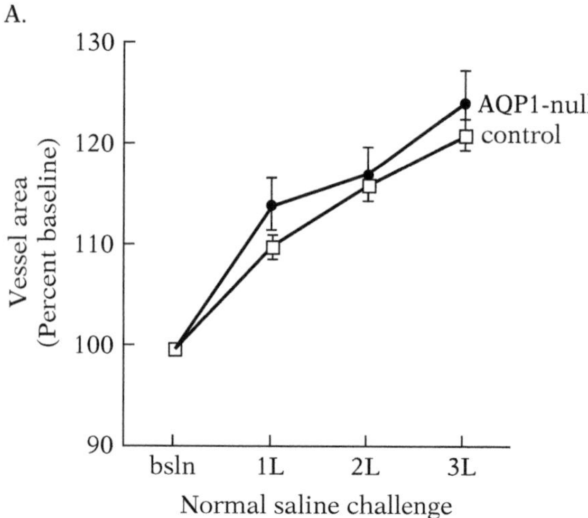

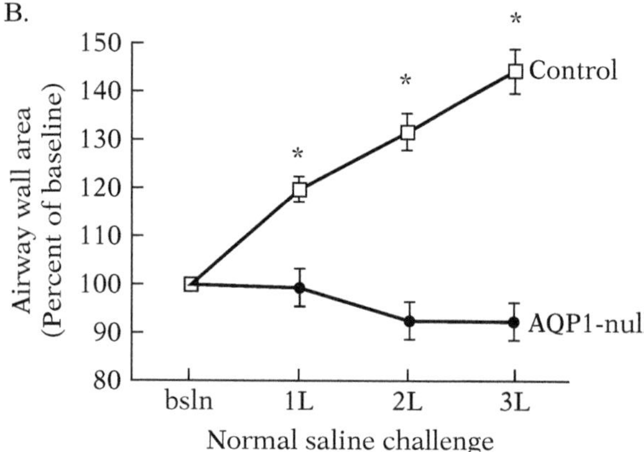

12.3 Decreased pulmonary vascular permeability in AQP1-null humans. Computed tomography scans of the lung before and after intravenous infusion of saline revealed differences in water permeability of lung tissue. AQP1-null individuals and normal controls received infusions of up to 3 L of physiologic saline, and images of their bronchioles and adjacent venules were recorded and quantified. In both groups, the vessel wall of the bronchiole became thickened due to the accumulation of fluid (A); however, the surrounding area did not accumulate fluid in the APQ1-null individuals (B) since they lack the AQP to secrete water. King L. S., *et al.*, *Proc Natl Acad Sci USA*. 2002, 99:1059–1063. © 2002 National Academy of Sciences, USA.

Water channels are typically bidirectional, allowing influx and efflux of water molecules in response to changing osmotic conditions. They are extremely fast: Water flows through a single AQP1 molecule at a rate of three billion molecules per second. And they are very selective, allowing the passage of water molecules without protons or other ions. Glyceroaquaporins may conduct water relatively slowly and even demonstrate stereoselectivity, as indicated by the ten-fold higher rates of transport of ribitol than of D-arabitol, both of which are simple five-carbon reduced sugars.

The high sequence homology among the AQPs suggests that they use a common architecture to accomplish selective water transport. The similarity between their N- and C-terminal segments was detected by superimposing the two halves of GlpF after a rotation at the quasi-two-fold axis (see Figure 12.6). This early recognition of inverted topology repeats found in many secondary transporters (see Chapters 6 and 11) indicates a structure that arose from a gene duplication event in evolution. The highly conserved sequence motif NPA is repeated near the center of each half of the primary structure, and several other conserved residues, such as Glu14 and Glu152, are repeated near the beginning of each segment.

Structure of Aquaporins

Early images of AQP1 in lipid bilayers obtained by both AFM and EM gave evidence for pores. The structure of AQP1 from human red blood cells was determined by EM at a resolution of 3.8 Å in the same year that the crystal structure of GlpF, the glyceroaquaporin from *E. coli*, was reported at 2.2-Å resolution. These structures were quickly followed by that of the bovine AQP1 at 2.2 Å and AqpZ, the *E. coli* aquaporin, at 2.5-Å resolution. Other AQP structures, including the sheep and bovine AQP0 and the rat and human AQP4, have now been determined. The structures all share a unique AQP fold consisting of six TM α-helices with two additional half-helices that each span half the bilayer. The TM helices cross in a right-handed twist at a 30° tilt to form an hourglass-shaped pore, with the two half-helices meeting in the center of the membrane (Figure 12.4). Both N and C termini are cytoplasmic, with extracellular loops connecting the N- and C-terminal segments of

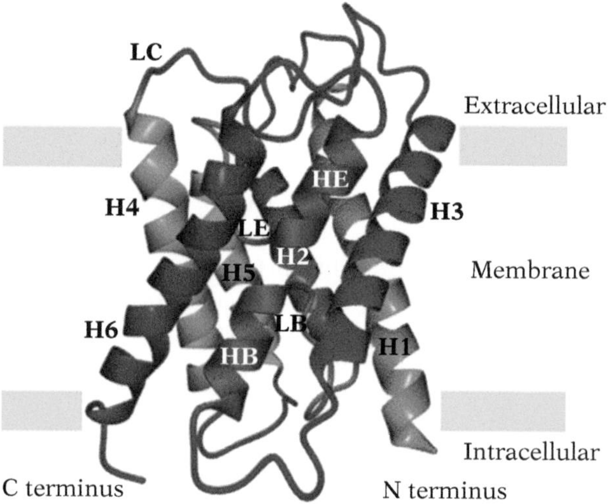

12.4 X-ray structure of AQP1. Each TM helix (colored differently) tilts about 30° off the bilayer normal. Two half-helices (HB and HE) form one of the seven TM segments. The helices are twisted into a right-handed bundle. From Fujiyoshi, Y., *et al.*, *Curr Opin Struct Biol*. 2002, 12:509–515. © 2002 by Elsevier. Reprinted with permission from Elsevier.

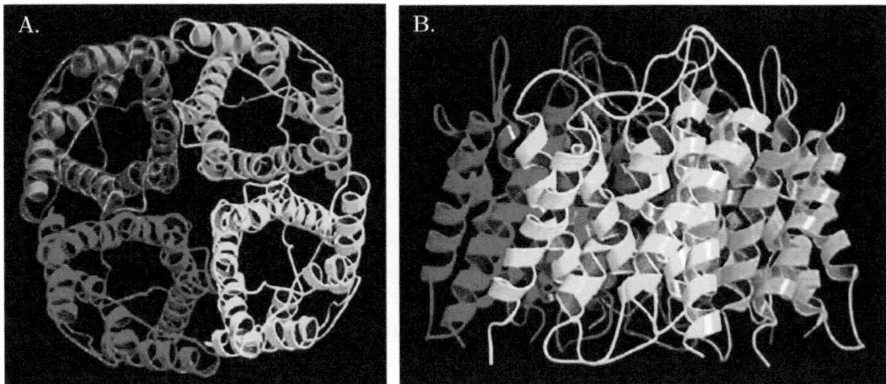

12.5 Tetramers of AQP1. Each monomer contains a pore, and hydrophobic interactions between monomers stabilize the tetrameric assembly, which is viewed from the top (A) and the side (B), with each monomer colored differently. From DeGroot, B. L., and H. Grubmuller, *Curr Opin Struct Biol*. 2005, 15:176–183. © 2005 by Elsevier. Reprinted with permission from Elsevier.

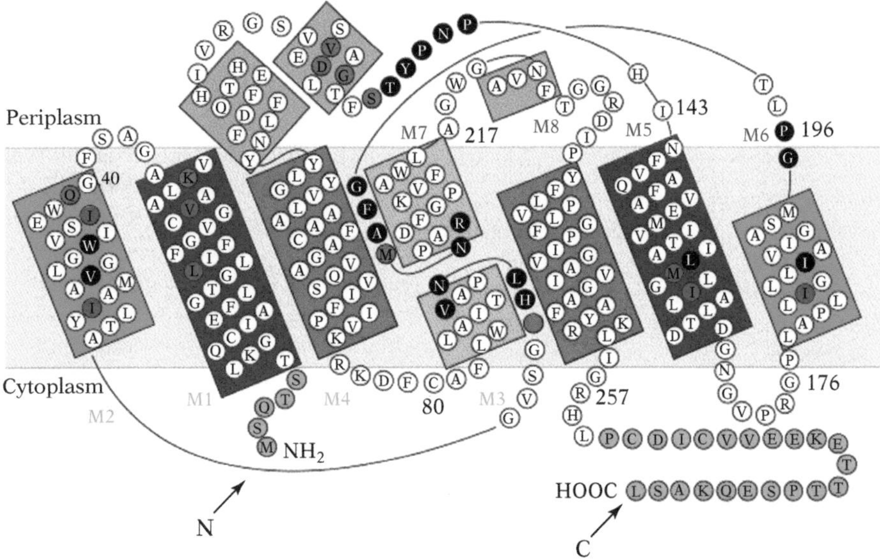

12.6 Topology of GlpF. From the N terminus in the cytoplasm, the peptide chain makes three and a half helices in the membrane before reversing the pattern, ending with the C terminus in the cytoplasm. Two short helices contribute to the periplasmic portion of GlpF. The two segments of GlpF are colored to show their symmetry, with highlighted residues to show which interact with the glycerol molecules in the channel (black), contribute carbonyls to the central channel (red), and contribute hydrocarbon to the channel (purple). From Stroud, R. M., *et al.*, *Curr Opin Struct Biol*. 2003, 12:424–431. © 2003 by Elsevier. Reprinted with permission from Elsevier.

varying lengths. The NPA signature sequence for the AQPs is repeated in each of the half-helices and makes the interface between them. Purified AQPs assemble into tetramers both in the crystal lattice (Figure 12.5) and when reconstituted into lipid bilayers, but clearly each monomer has a pore. Since there is no evidence for cooperation between subunits, the functional unit of AQP is a monomer. Insights into channel selectivity have come from details of the structure of GlpF, the first high-resolution AQP structure, and more recently AQP4, the aquaporin in the human brain.

Glyceroaquaporins: GlpF

The GlpF protein provides a channel for passive diffusion of glycerol into *E. coli*. In the cytosol glycerol is immediately converted to G3P, ensuring that the inside concentration of glycerol is low, which drives its uptake. The topology diagram for GlpF shows the symmetry between the N- and C-terminal segments (residues 6–108 and 144–254), with similar TM segments on each side of the two half-helices (M3 and M7) (Figure 12.6). In three dimensions the two segments form two inverted halves of the channel and are linked by a protruding periplasmic region (Figure 12.7). The two half-helices form an important junction in the center of the membrane, held by van der Waals interactions between the proline residues of the NPA signature motifs. The NPA motifs cup each other between the proline and alanine side chains of the opposite helix, with their orientation stabilized by other very highly conserved residues. In addition,

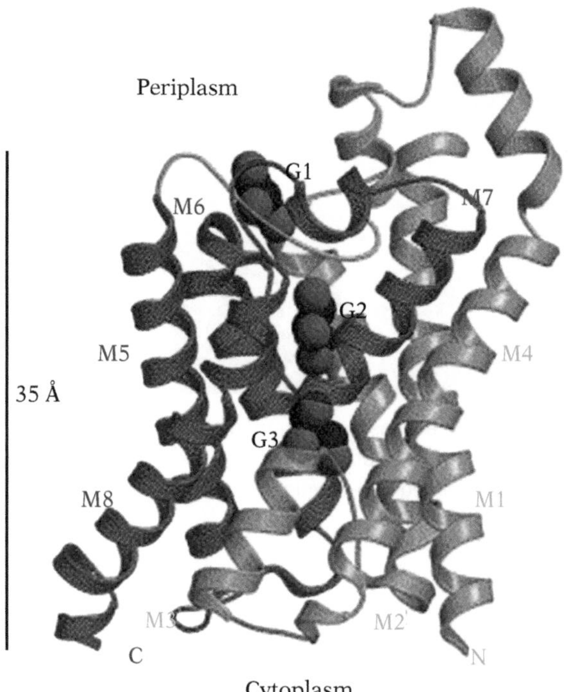

12.7 X-ray structure of GlpF. The ribbon diagram of the GlpF structure viewed from the membrane plane shows the tilted helices make a bundle, each half of which is derived from the N- or C-terminal segment of the molecule (gold and blue, respectively). Three glycerol molecules (red space-filling models) are observed transiting the channel. From Stroud, R. M., *et al.*, *Curr Opin Struct Biol*. 2003, 12:424–431. © 2003 by Elsevier. Reprinted with permission from Elsevier.

conserved glycine residues allow close contact between the α-helices where they cross near the center of the bilayer, stabilized by CH–O hydrogen bonds.

The channel in GlpF starts on the outer surface with a wide vestibule and then constricts to form a selective channel that is 28 Å long. The crystal structure shows three glycerol molecules in transit (see Figure 12.7). The narrowest point in the channel defines the selectivity filter, with a very close fit for glycerol that involves hydrophobic interactions at the corner between Trp48 and Phe200 and hydrogen bonds between the hydroxyl groups on C1 and C2 of the glycerol and Arg206, Gly19, and Phe200.

The channel is lined by nonpolar residues, except along one side. The polar side has a striking row of carbonyl oxygens along the channel, with four on the periplasmic side and four on the cytoplasmic side separated by a gap of ~3 Å. From the cytoplasm, these are the carbonyl groups from Gly64, Ala65, His66, and Leu67, and from the periplasm they are from Phe200, Ala201, Met202, and Asn203 (Figure 12.8). In each row the four residues are in an extended conformation and have alternating right-handed and left-handed helical backbone configurations, making a ladder that is maintained by hydrogen bonds to the conserved buried glutamates, Glu14 and Glu152. These carbonyls provide hydrogen-bonding acceptors for a line of water molecules in the channel, which orient in opposite directions outward from the center. The only hydrogen bond donors in the water channel are Asn68 and Asn203 from the NPA signatures

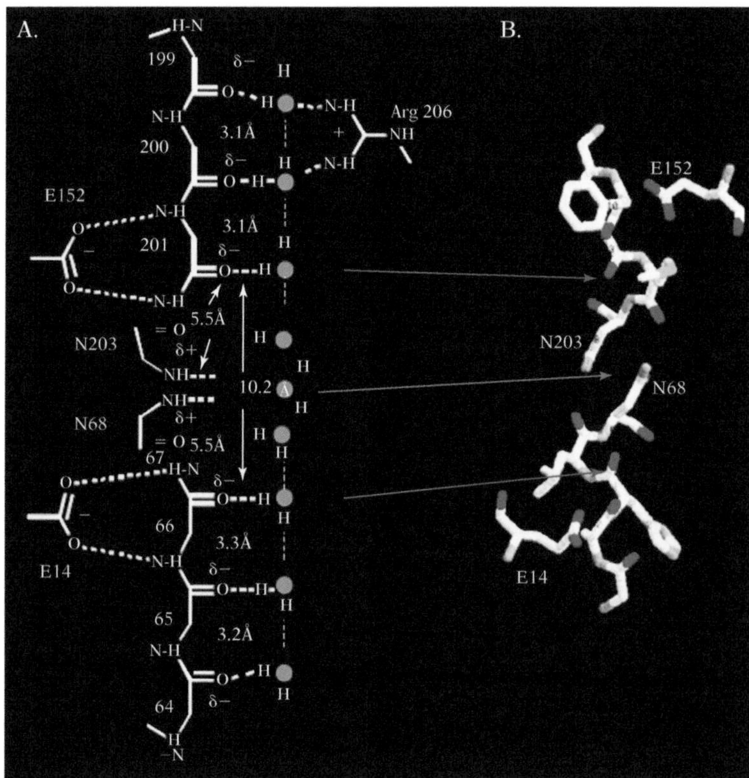

12.8 Orientation of water molecules in the channel of GlpF. (A) The carbonyl groups within the channel are arranged in close proximity to conduct a line of water molecules. (B) A close-up view of the hydrogen-bond acceptors and hydrogen-bond donors in the center of the channel shows the region where the orientation of the water molecule flips. From Stroud, R. M., *et al.*, *Curr Opin Struct Biol*. 2003, 12:424–431. © 2003 by Elsevier. Reprinted with permission from Elsevier.

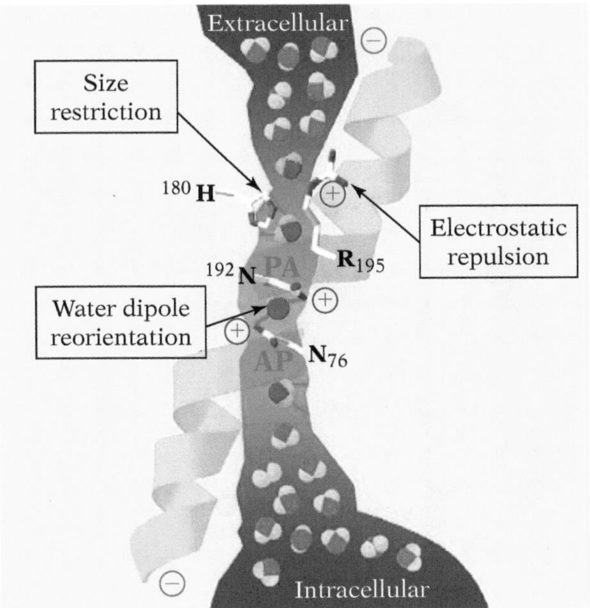

12.9 Reorientation of water molecules in the GlpF channel. Selectivity of the AQP channel is shown schematically. The internal pore (blue) of the AQP allows passage of water molecules and not protons and other ions by reorienting the water molecules during passage. Other important factors are size restriction and electrostatic repulsion. From Agre, P., and D. Kozono, *FEBS Lett.* 2003, 555:72–78. © 2003 by Elsevier. Reprinted with permission from Elsevier.

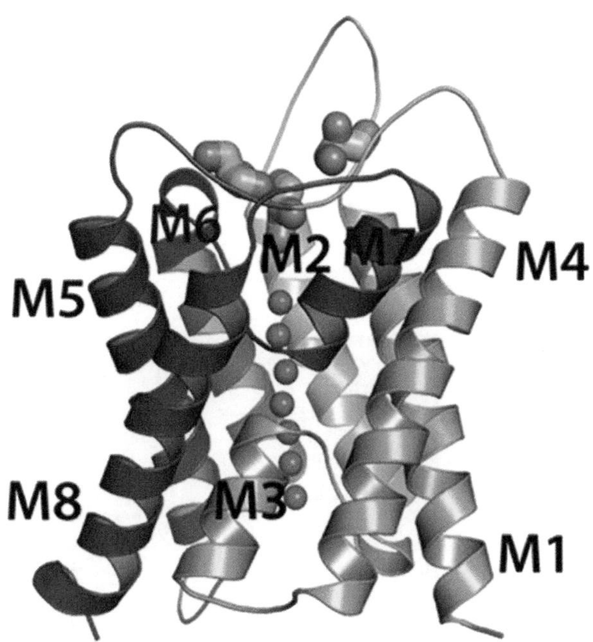

12.10 High-resolution structure of human AQP4. A ribbon diagram of a monomer of trypsinized AQP4 is colored to show the N-terminal (gold) and C-terminal (brown) halves of the peptide chain. A chain of water molecules (red spheres) extends through the permeability channel, while three glycerol molecules (red and green space-filling models) are confined to the outer vestibule. From Ho, J. D., *et al.*, *Proc Natl Acad Sci USA.* 2009, 106:7437–7442.

at the center. As each water molecule reaches the center, it forms hydrogen bonds to the two asparagine residues and then flips its dipole orientation to continue out the channel (Figure 12.9). In this way each water molecule is hydrogen bonded in the channel, while avoiding a continuous network of water molecules of the same orientation, which explains how AQPs conduct water but do not leak protons.

Normally protons move along a line of hydrogen-bonded waters in a concerted manner, such that a proton attaches to one end of a water molecule and another leaves from the other end, as long as the dipoles of all the waters "face" the same way. Called the Grotthuss mechanism, this proton transfer can leak protons across a membrane without generating any charge in the channel. The change in polarization at the center of the line of water molecules in the AQP channel prevents proton conductance by the Grotthuss mechanism, as observed in molecular dynamics simulations. This is of fundamental importance since cells need to be able to transport a lot of water without losing their membrane potential.

Human Aquaporins: AQP4

While the high-resolution structure of GlpF reveals fundamental properties of all aquaporins, the differences between channels of AQPs and glyceroaquaporins can be discerned from a high-resolution structure of human AQP4. Crystals of AQP4 giving 1.8-Å resolution were obtained after trypsin treatment that did not significantly alter its channel activity in proteoliposomes. The structure of human AQP4 shows water molecules throughout the channel, with glycerol excluded from the pore even though it was crystallized in 5% glycerol (Figure 12.10). Passage of glycerol is prevented by constriction of the channel diameter to ~1.5 Å at His201 in the selectivity filter (Figure 12.11). Like GlpF and other aquaporins, the channel is amphipathic, with hydrophobic residues along one side and carbonyls from eight residues providing hydrogen bonds for water molecules. At the top of the channel Arg216 is hydrogen-bonded to the sterically excluded glycerol, as well as other residues and two water molecules (Figure 12.12; see also Figure 12.11). It is proposed that the slower passage of water through GlpF is due to the lower degree of hydrogen bonding of the corresponding residue, Arg206, which gives the guanidinium a stronger positive charge that allows it to hold transiting water molecules more tightly.

Human AQP4 is localized in brain cells in contact with the blood–brain barrier and plays a crucial role in homeostasis of cerebral water. Since brain damage often results from the increased pressure caused by edema, AQP4 is a logical target for therapeutic intervention for

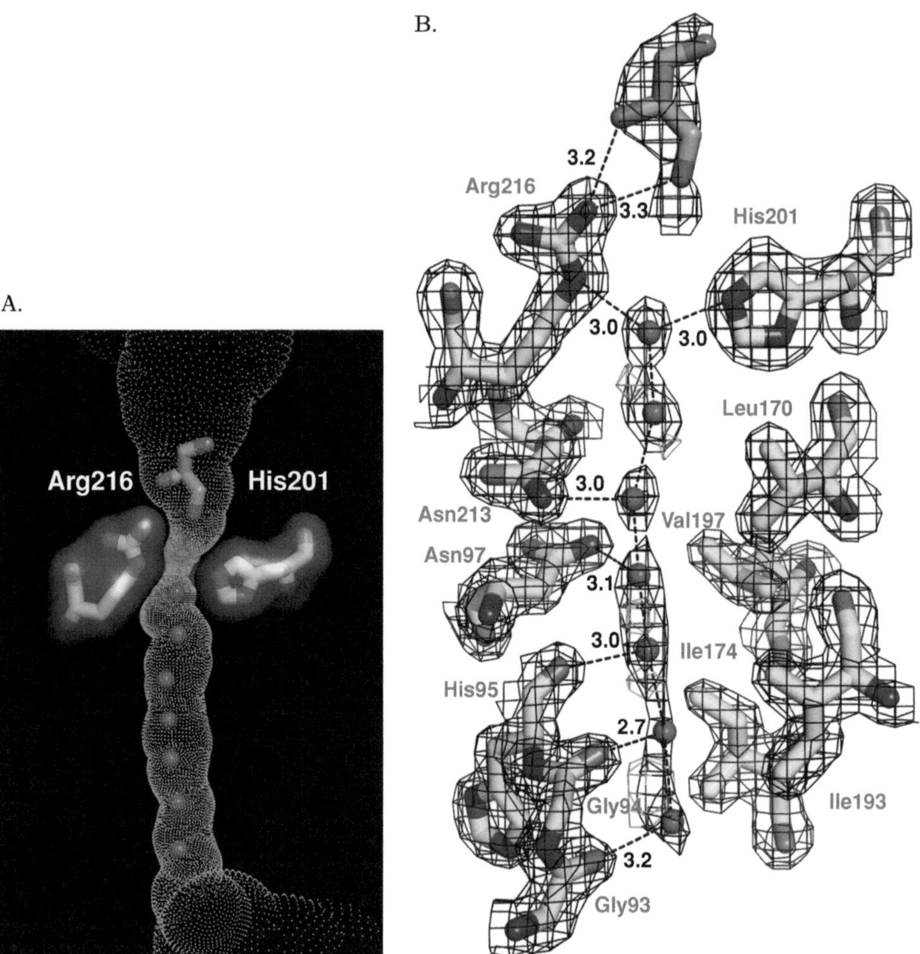

12.11 Details of the channel in human AQP4. (A) The trace of the pore inner surface (cyan) reveals the constriction at Arg216 and His201 (stick models with surfaces in purple). A glycerol molecule (stick model) is unable to pass the constriction. (B) Electron density map of the walls of the pore details the residues involved (stick models with black mesh for 2Fo-F_1 density), with hydrogen bonds (dashed lines) to water molecules (red spheres) and glycerol (stick model). The closeness of most water molecules in the channel is indicated (green mesh for Fo-F_1 density), with gaps at Asn213 and Asn97, which each bind a water molecule. Both from Ho, J. D., *et al.*, *Proc Natl Acad Sci USA*. 2009, 106:7437–7442.

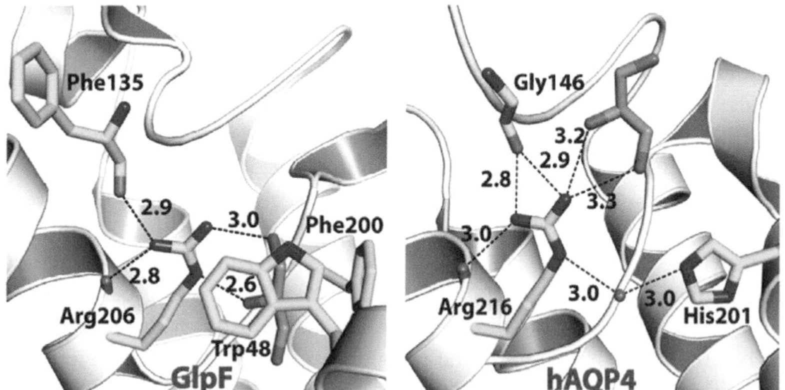

12.12 Efficiency of water conductance. Comparison of the hydrogen bond network of the selectivity filter arginine in GlpF (left) and human AQP4 (right) illustrates the increased number of hydrogen bonds between Arg216 and other residues (stick models) and water molecules (red spheres) in AQP4. With fewer hydrogen bonds to Arg206 in the structure of GlpF, the guanidium residue is able to more tightly bond to water in the channel. From Ho, J. D., *et al.*, *Proc Natl Acad Sci USA*. 2009, 106:7437–7442.

victims of stroke, epilepsy, brain tumors, and traumatic head injury. Another AQP of vital importance to human health is the sole aquaporin of the malarial parasite, *Plasmodium falciparum*. PfAQP takes up both glycerol and water, and both are seen in the channel in its crystal structure. Glycerol uptake is crucial for the synthesis of lipids during the rapid reproduction of *P. falciparum* in the host's bloodstream.

The channels of all aquaporins exclude ions and charged solutes because they are too small for the passage of hydrated ions. Since much of the surface of the channel walls is hydrophobic, residues in the channel do not have groups to replace the hydrogen bonds, making it too costly to strip waters from hydrated ions. This contrasts with other ion channels. The KcsA potassium channel (also a tetramer) is completely different in overall architecture, and yet its pore also features lines of carbonyl groups from residues with alternating right- and left-handed α-helical configurations.

POTASSIUM CHANNELS

Like the AQPs, potassium channels are both selective and fast: Their selectivity for K^+ over Na^+ is usually over 1000-fold, and their conduction rates of ~10^8 ions/sec are close to diffusion limited. They are passive channels that typically allow K^+ ions to flow out of a cell in response to the electrochemical gradient created by the Na^+, K^+ ATPase, which pumps three Na^+ out and two K^+ into the cell (see Chapter 9). One group of voltage-sensitive potassium channels, called inward-rectifying channels, opens in response to a drop in the membrane potential to allow K^+ to flow into the cell. The diversity of potassium channels surpasses that of any other channel family. In excitable cells, such as neurons, they set the resting potential and regulate the action potential. In the cardiovascular system they regulate the heartbeat. In epithelial cells they help balance the passage of salts and water. Many cells have several types of potassium channels, some that respond to changes in the membrane potential or in pH and others that are triggered by binding ions, such as calcium ions; ligands, such as neurotransmitters, cyclic nucleotides, and lipids; or even other proteins, such as certain G proteins.

While potassium channels differ in gating mechanisms, they share a common architecture with an ion conductance pathway down the central interface between four identical subunits. Their signature sequence forms the selectivity filter of the channel (see Frontispiece). For the channel to be gated, allowing the pore to open and close in response to different signals, sensors must transmit information to the pore domain. Voltage-sensitive channels have a voltage sensor consisting of four additional TM segments. Thus the membrane domains of potassium channel subunits vary from two to seven TM helices, plus a pore helix that does not cross the membrane (Figure 12.13). Larger K^+ channels have an additional cytoplasmic domain with one or two regulators of K^+ conductance (RCK) gating domains typically attached to the intracellular C terminus, although RCK domains are also expressed as independent proteins. Thus the basic channel design appears to have evolved by addition of modular gating domains conferring sensitivity to voltage and/or ligands to generate the diverse potassium channels.

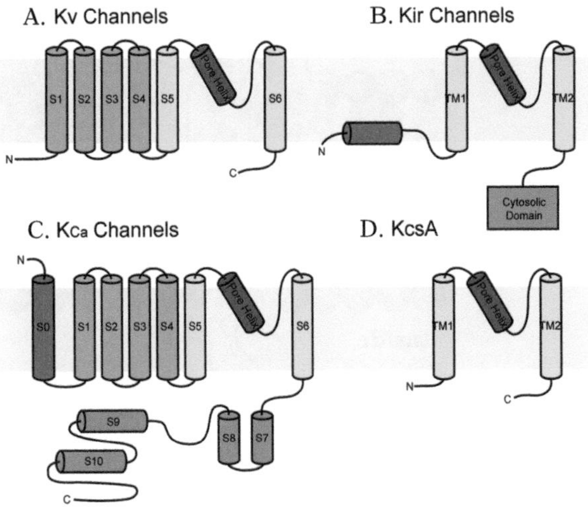

12.13 Topology diagrams for different K^+ channels. Representative channel types Kv, Kir, Kca, and KcsA vary in both the number of TM helices and the presence of domains outside the membrane. All have two pore TM helices (cyan) and a helix within the P loop (pink) that carries the signature sequence. When they have additional TM helices (tan), the numbering convention is S1, S2, etc. Helices S3 and S4 make the voltage sensor in Kv channels (A). Cytosolic domains typically involved in regulation may contain helices (green). The cytosolic domain from KcsA removed for crystallization is not shown. From McCoy, J. G., and C. M. Nimigean, *Biochim Biophys Acta*. 2012, 1818:272–285.

Over 40 years after experiments measuring the flux of $^{42}K^+$ across the membrane from squid axon suggested the single-file movement of K^+ ions occurs through a putative membrane channel, the first high-resolution structure of a potassium channel, KcsA from *Streptomyces lividans*, verified that model. The KcsA channel represents the pore domain of all potassium channels, as first indicated by its ability to bind the inhibitor charybdotoxin and its comparable selectivity among cations ($K^+ > Rb^+ > NH_4^+ >> Na^+ > Li^+$). KcsA is proton-dependent, active at pH below 5. The next K^+ channels to have structures solved represent other kinds of gating: MthK from *Methanobacterium thermoautotrophicus* is calcium-activated, and KvAP from the thermophilic archaea *Aeropyrum pernix* is voltage-gated. These crystal structures captured the channels in either open or closed conformations, allowing comparisons and structures of mutant channels to address mechanisms of channel opening, gating, and inactivation.

Tremendous progress from numerous crystal structures of prokaryotic potassium channels paved the way to approach the more complex potassium channels in higher organisms. Voltage gating in KvAP shares many features with eukaryotic Kv channels such as the family of Shaker channels (named for the characteristic leg motions of the *Drosophila* mutant). The Shaker Kv1.2 K^+ channel from rat brain has been crystallized in complex with an oxido-reductase β subunit that binds to an

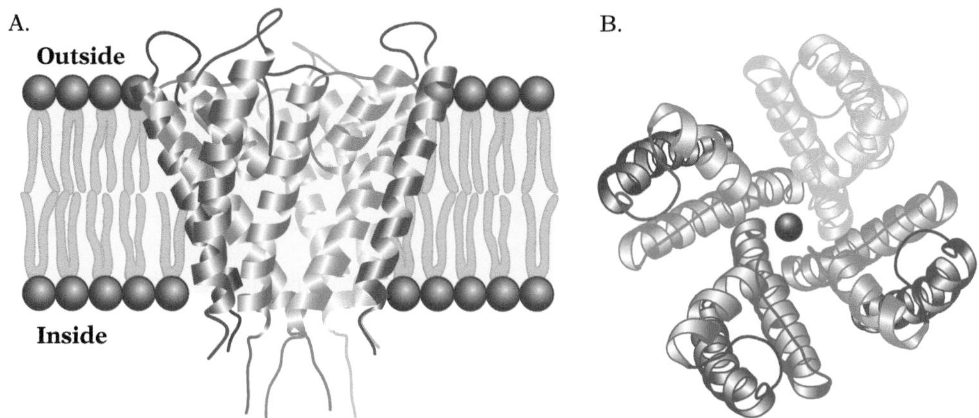

12.14 X-ray structure of KcsA channel. Each subunit of the tetramer contributes two TM helices in addition to a pore half-helix, making a cone shape when viewed from the side (A). The view from the top (B) shows a K^+ ion (green sphere) in the channel at the subunit interfaces. Redrawn from Nelson, D. L., and M. M. Cox, *Lehninger Principles of Biochemistry*, 4th ed., W. H. Freeman, 2005, p. 410. © 2005 by W. H. Freeman and Company. Used with permission.

unusual N-terminal T1 domain. The structure of a well-studied "paddle chimera" constructed from rat Kv1.2 with a voltage sensor from rat Kv2.1 reveals interactions with stabilizing lipids. The structure of a bacterial inward-rectifying K^+ channel, Kirbac, was followed by structures of the eukaryotic Kir2.2 from chicken and the G protein-sensitive Kir3.2 (or GIRK2) channel from mouse. One of the newest structures is a lipid-gated human potassium channel called TRAAK, described below. Even with the added complexities in the eukaryotic proteins, the preservation of the pore design in all the potassium channels makes KcsA a structural as well as historical paradigm.

KcsA Structure and Selectivity

The first crystallization of KcsA required removal of its flexible C-terminal domain by cleavage with chymotrypsin, resulting in an x-ray structure of 3.2-Å resolution that allowed detection of added Rb^+ or Cs^+ ions in the pore. When binding monoclonal Fab fragments to KcsA improved the resolution to 2.0 Å, the K^+ ions could be detected. The overall shape of the KcsA protein is a cone within a cone, constricted at the cytoplasmic side and opening at the extracellular side (Figure 12.14). The channel is lined by four pairs of helices: Each subunit contributes an outer helix and an inner helix that span the bilayer. In addition, a shorter tilted half-helix from each subunit fills in the top of the cone to make the pore. Each pore helix is oriented with its C-terminal end inside, putting the negative end of its dipole toward the center. A narrow pore 12 Å long defines the selectivity filter; it starts on the extracellular side and ends at a large water-filled cavity in the center (Figure 12.15 and Frontispiece). A fully hydrated K^+ ion is observed inside the cavity. Within the pore there are four binding sites for dehydrated K^+ ions. Both the water-filled cavity and the orientation of the pore helices help overcome the electrostatic repulsion for the positively charged ions entering the channel. Indeed, in a similar arrangement in the Cl^- channel, the dipole of the pore helices is reversed, which places their N termini inside to contribute positive charge that attracts the anion (see below).

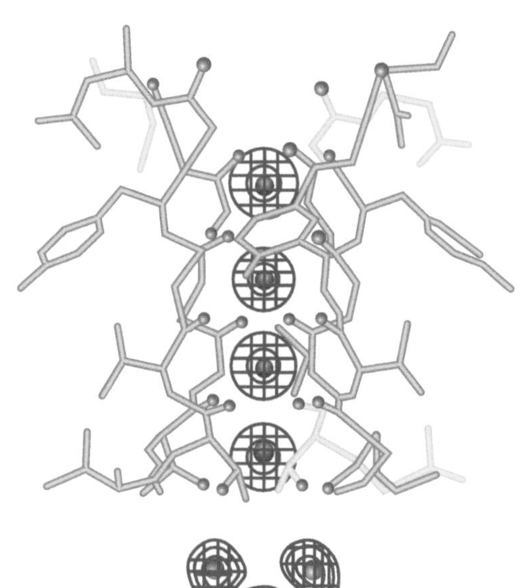

12.15 The selectivity filter and potassium-binding sites of KcsA. A close-up view of the selectivity filter in the KcsA ion channel with the extracellular surface at the top shows the electron density (blue mesh) observed for dehydrated K^+ ions at positions 1 to 4, in addition to a hydrated K^+ ion in the central cavity. The chapter Frontispiece shows the location of the selectivity filter in the ion channel. From MacKinnon, R., *FEBS Lett.* 2003, 555:62–65. © 2003 by the Federation of European Biochemical Societies. Reprinted with permission from Elsevier.

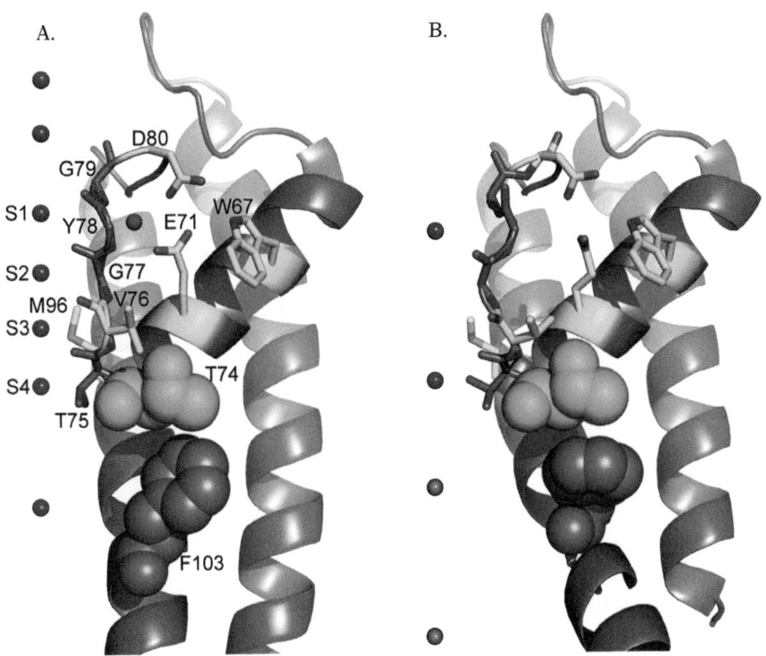

12.16 KscA and KcsA mutant in an open form. (A) The residues behind the selectivity filter are important for the conformation of the filter, with four K^+ sites well defined in wild type KcsA. (B) The filter collapses to have only S1 and S4 sites in a partially open KcsA mutant lacking N and C termini, with only one charged residue in the proton sensor (see text). A hydrogen-bonding network involving D80/E71/W67/V76/M96 (yellow sticks) stabilizes the selectivity filter. Upon disrupting the bundle crossing to open the activation gate (seen in the change of angle of inner helix in (B)), the movement of the inner helix causes F103 (orange space-filled) to rotate to avoid collision with T74 (cyan space-filled), at the bottom of the selectivity filter. Both (A) and (B) show K^+ (purple spheres) and H_2O (red spheres) in the filter. From McCoy, J. G. and C. M. Nimigean, *BBA*. 2012, 1818:272–285. © 2012, Elsevier. Reprinted with permission from Elsevier.

In the selectivity filter of KcsA, each signature sequence of TVGYG provides four evenly spaced layers of carbonyl oxygen atoms and a single layer of threonine hydroxyl oxygen atoms to create the four K^+-binding sites (Figure 12.16A). Both glycine residues and the threonine have dihedral angles that are allowed for a left-handed α-helix. Alternating between right-handed and left-handed helical configurations enables the backbone carbonyl groups to line up. This puts at each binding site eight oxygen atoms at the vertices of a cube (or twisted cube) that corresponds to the arrangement of the water molecules surrounding the hydrated K^+ observed in the central cavity. Therefore, the role of the selectivity filter is to mimic the waters of hydration for a queue of K^+ ions, which pays the energetic cost of their dehydration. While the lines of oxygen atoms provide good binding sites, binding is not so strong that it inhibits the fast rate of flux through the channel. Binding is weakened by electrostatic repulsion from neighboring cations in the queue, as well as by a conformational change that takes place upon potassium binding. At very low (nonphysiological) potassium concentrations, with only one K^+ bound per tetramer, the selectivity filter collapses inward; binding a second K^+ causes the channel to straighten out. This conformational change takes up some of the binding energy, making the binding weaker.

The stoichiometry of binding measured experimentally with Tl^+ (thallium ion, an excellent replacement for K^+ in laboratory work) shows the channel binds two ions. Computational experiments indicate that it has two states of equal energy with S1/S3 occupied or S2/S4 occupied, so the flux of K^+ through the channel involves a concerted movement in response to the approach of another K^+ ion, either from the internal cavity or from outside the cell.

Recent discovery of a channel with the same basic architecture but a less selective channel demonstrates that the binding sites of the selectivity filter are essential for the discrimination of potassium channels. The nonselective channel, called NaK, allows Na^+, Cs^+, and even divalent cations including Ca^{2+} to pass, and yet its structure may be closely superposed on KcsA (Figure 12.17). The main difference is a wide vestibule in place of binding sites S1 and S2 of the NaK selectivity filter; S3 and S4 are identical to the sites in KcsA. When crystallized in the presence of Na^+, the NaK structure shows Na^+ accompanied by water molecules in the filter (where K^+ would be dehydrated in KcsA), with one Na^+ at S4 bound lower in the site than K^+ binds.

Gating and Activation

Activated by low pH, KcsA shows a sharp drop in conductance (or $^{86}Rb^+$ uptake) above pH 5–6 *in vitro*. The crystal structure of the KcsA pore domain shows a closed channel, sealed below the central cavity by a bundle of the inner helices. This closure is referred to as the gating ring or the activation gate. Activation (channel opening) would require a hinged motion in the inner helices that disrupts the bundle crossing at the gating ring seen in the x-ray structure. The collapsed conformation of KcsA in low K^+ is also inactive, which indicates the selectivity filter itself is a second gate (see Figure 12.16). Inactivation can occur in KcsA due to either closure of the inner helix bundle (seen in the crystal structure) or collapse of the selectivity filter.

How does KcsA respond to pH? The proton sensor is predicted to be a cluster of charged residues (REQERRGH, starting from Arg117) that surround

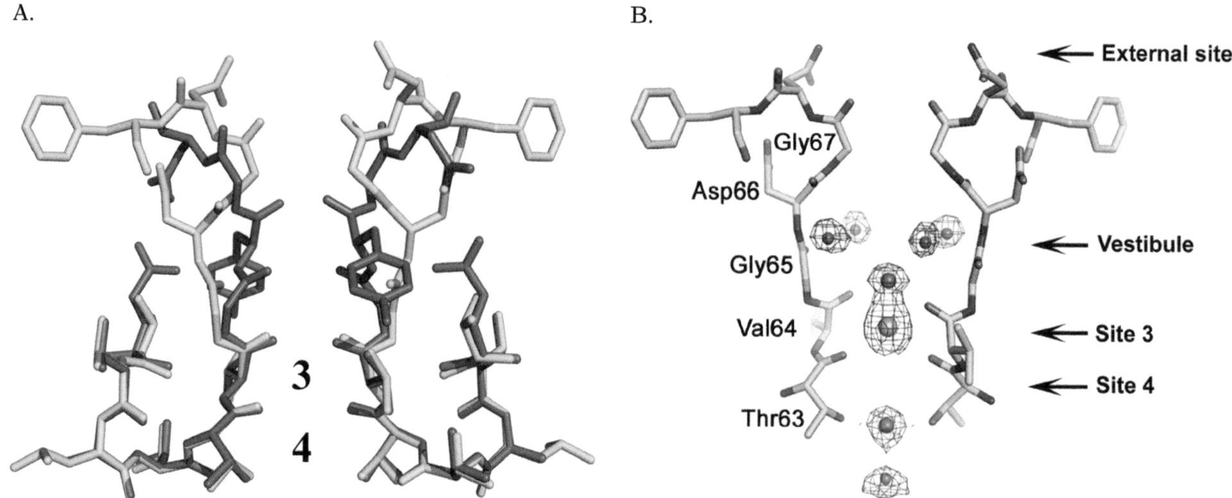

12.17 NaK channel selectivity filter. (A) Superposition of the NaK selectivity filter (yellow) in presence of K^+ with the selectivity filter of KcsA (green) shows near identity of S3 and S4, while NaK lacks S1 and S2. (B) Residues in the selectivity filter of NaK with electron density (blue mesh) for Na^+ ions (green spheres), water molecules in the vestibule and in the central cavity (red spheres) and a contaminating ion at site 3 (orange sphere). In (A) and (B) the front and back subunits of NaK have been removed for clarity. From Alam, A., and Y. Jiang , *Nature Struct Molec Biol*. 2009, 16:35–41. © 2009, Nature Publishing Group. Reprinted by permission of Macmillan Publishers Ltd.

the inner helix bundle. Changes in pH would shift their attractive and repulsive forces and thus stabilize/destabilize the inner bundle. Logically, removal of the charges by mutation could abolish the gating. However, gate opening needs electrostatic repulsion, so the sensor was engineered to contain one negative charge at E118 (QEQQQGQ) along with mutating the partner to E118, H25Q. The resulting mutant KcsA showed conductance at pH 8, but only a slight opening of the bundle at the activation gate. Larger openings were seen in a series of heterogeneous crystals containing the single charge at E118 and truncating the N and C termini. As the distance across the activation gate increases, the diameter of the selectivity filter constricts. This coupling is attributed to rotation of Phe103, which interacts with the neighboring subunit just below the selectivity filter (see Figure 12.16).

Spectroscopic techniques including NMR, EPR, and surface plasmon resonance indicate that conformational changes upon channel opening are likely to involve both the N terminus and the proteolytically removed C-terminal domain of KcsA. Indeed, recent high-resolution structures of full-length KcsA show the C-terminal domain interacts directly with the inner helix bundle gate, contributing to gating (Figure 12.18). A second gate, followed by a bulge region, is seen in comparison of the open and closed states of full-length KcsA, and affects the degree of opening measured with current traces.

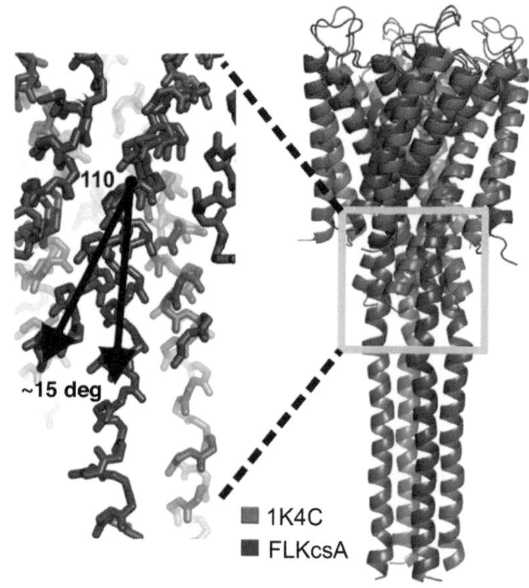

12.18 The x-ray structure of full-length KcsA. The truncated KcsA structure (red ribbons) is superposed on the structure of full-length KcsA in a closed state (blue ribbons). The C-terminal four-helix bundle projects ~70 Å into the cytoplasm. The inset shows the effect on the inner helix bundle gate, with a 15° tilt that tightens its diameter at a lower position in the truncated structure. From Uysal, S., *et al.*, *Proc Natl Acad Sci USA*. 2009, 106:6644–6649.

A different kind of gating is observed in MthK, a Ca^{2+}-activated K^+-channel from *Methanobacterium thermoautotrophicum*, even though the pore domain is very similar to that of KcsA. When crystallized in the presence of Ca^{2+}, MthK reveals an open pathway from the cytoplasm up to the selectivity filter, in contrast to the closed KcsA channel (Figure 12.19). The TM2 helices of the MthK pore domain are hinged at a conserved glycine residue, Gly83, about halfway across the membrane. In addition, a conserved glycine or alanine is found at a position five amino

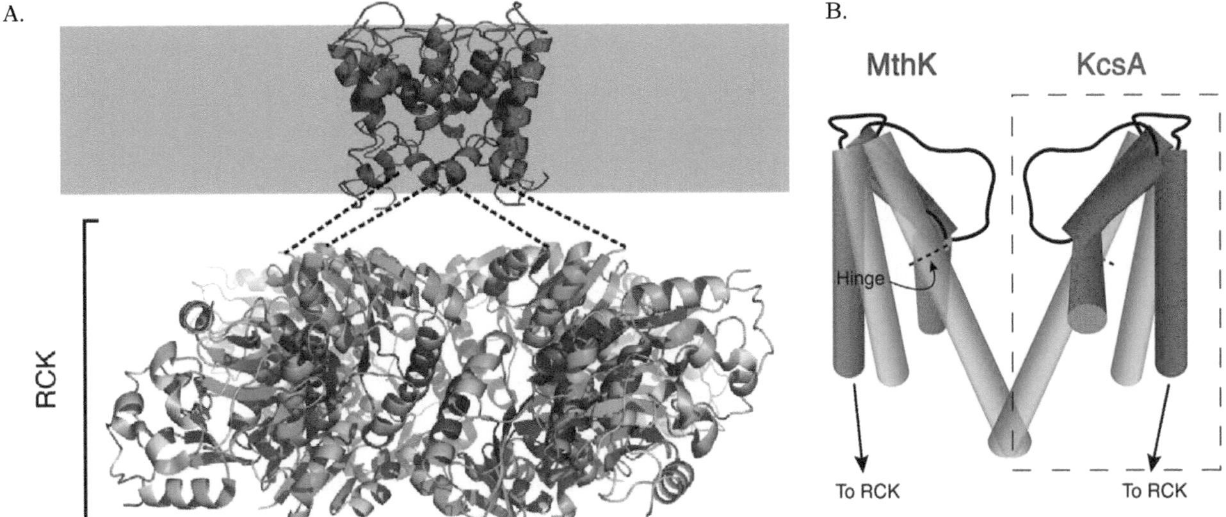

12.19 The x-ray structure of MthK in the presence of Ca^{2+}. (A) Ribbon diagram of MthK viewed from the plane of the membrane. The tetrameric pore (red) is open with a large vestibule. The RCK domains (blue, green, and gold) form a gating ring in the cytoplasm. The connections between them (residues 99–115) are not resolved in the structure. (B) The TM domains of the open pore in MthK (red cylinders) and the closed pore in KcsA (blue cylinders) are compared to show the movement needed to open a channel. Gly38 in TM2 of MthK (corresponding to the position of the dashed line in KcsA) forms a hinge that allows the C-terminal half of TM2 to swing up and away from the pore. From Chakrapani, S., and E. Perozo, *Nature Struct Molec Biol*. 2007, 14:180–182. © 2007, Nature Publishing Group. Reprinted by permission of Macmillan Publishers Ltd.

acids toward the C terminus, below the hinge where a larger side chain would block the open pore. Conservation at these two positions in a wide range of K^+ channels suggests that the role of the hinge in pore opening and closing is conserved in the different K^+ channels.

Calcium sensitivity is conferred to MthK by the RCK domains that form a large gating ring on the cytoplasmic side of the membrane. Interestingly, while the pore is a tetramer, the gating ring is an octomer, made up of four copies of the RCK domain at the C terminus of each monomer in the pore in addition to four more copies of the RCK domain lacking the TM domain. The isolated RCK domains have been crystallized in the presence and absence of Ca^{2+} and have an $\alpha\beta$ fold similar to dehydrogenase enzymes. In the cleft between two RCK domains two Ca^{2+} ions bind to carboxylate groups of Glu210, Glu212, and Asp184. The mechanism of gating is proposed to involve movement of a flexible hinge near the Ca^{2+}-binding site, which pushes two adjacent rigid domains to rotate, opening or closing the pore.

Voltage gating

Conformational changes in response to voltage differences across the cell membrane require a voltage sensor that transfers electric charge (the gating charge) across the membrane. In voltage-dependent potassium channels the gating domains are integral to the membrane and surround the pore. A subunit of a typical voltage-gated K^+ channel has six TM segments, numbered S1 to S6 from the N terminus. Lining the pore are S5 and S6, which correspond to the outer and inner helices of KcsA and differ very little from them. The other four segments, S1 to S4, form the voltage sensor, with the residues carrying the gating charge located on S4 (Figure 12.20). Measurements of the transient current associated with voltage sensor movement indicate the gating charge of the tetramer is 12–14 electron charges, making the channel quite sensitive to voltage.

The structure of KvAP, the first crystal structure of a voltage-dependent K^+ channel, shows the expected tetramer of the pore helices (S5 and S6), outside of which are the voltage sensor domains (S1 to S4, Figure 12.21). The voltage paddle (S3 and S4), a helix–turn–helix structure ending at a flexible loop in the middle of S3, is in contact with the bilayer, allowing it to be influenced by lipids in the native membrane environment. The pore helix S6, which corresponds to TM2 in KcsA and MthK, has the conserved glycine residue thought to enable bending at a hinge during gating. A critical contact between the voltage sensor and the bottom of the pore, the S4–S5 linker, couples the two domains. The model for gating suggests that the upward movement of the S3–S4 paddle pulls on the S4–S5 linker helix to trigger opening of the channel formed by S5 and S6.

Numerous studies employing electrophysiology, scanning mutagenesis, fluorescence, and other techniques indicate arginine residues R1 to R4 on S4 in the voltage sensor are gating charges that respond to the membrane

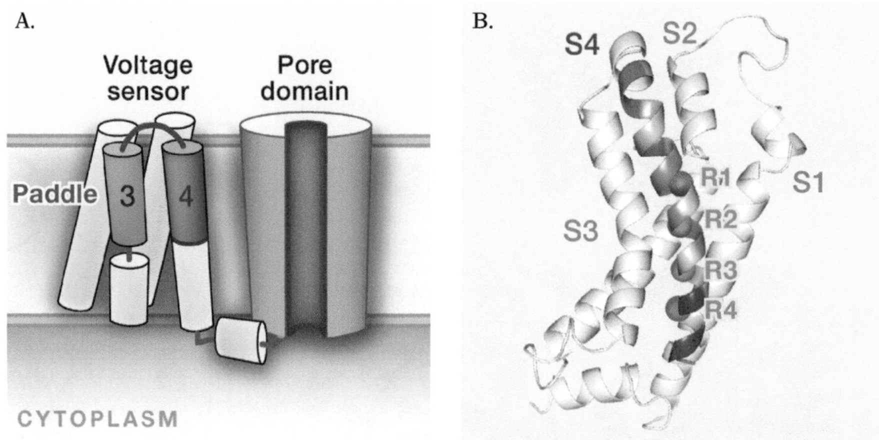

12.20 Common structural organization of voltage-dependent channels. (A) A general feature of voltage-dependent channels is the voltage sensor domain adjacent to the central pore domain. It contains a voltage-sensing paddle at the helix–turn–helix motif (green-blue loop) between TM segments traditionally labeled S3 and S4. (B) In a typical voltage sensor domain (this one from Kv1.2) the S4 helix contains four gating charges (R1–R4, orange spheres) in a predominantly hydrophobic region, with hydrophilic residues (red) at both ends. From Hulse, R. E., *et al.*, *Cell*. 2010, 142:515–516. © 2010 Elsevier. Reprinted with permission from Elsevier.

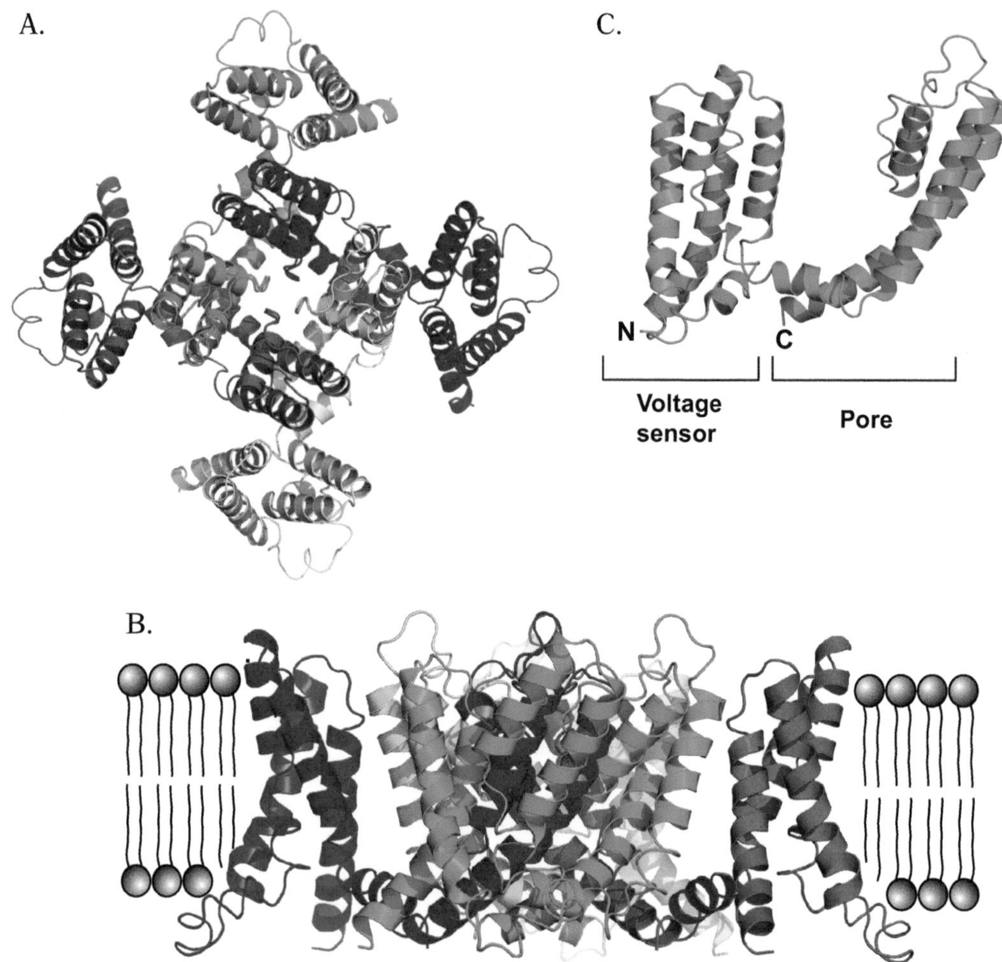

12.21 X-ray structure of voltage-gated KvAP. The KvAP tetramer is viewed from the extracellular side of the membrane (A) and the plane of the membrane (B), with each subunit a different color. (C) A single subunit viewed from the plane of the membrane is oriented with the intracellular solution on the bottom. The voltage-sensor paddle domain (left) is made up of S1 to S4 and the pore (right) is made up of S5 and S6. From Devaraneni, P. K., *et al.*, *Biochemistry* 2011, 50:10442–10450. © 2011 by American Chemical Society.

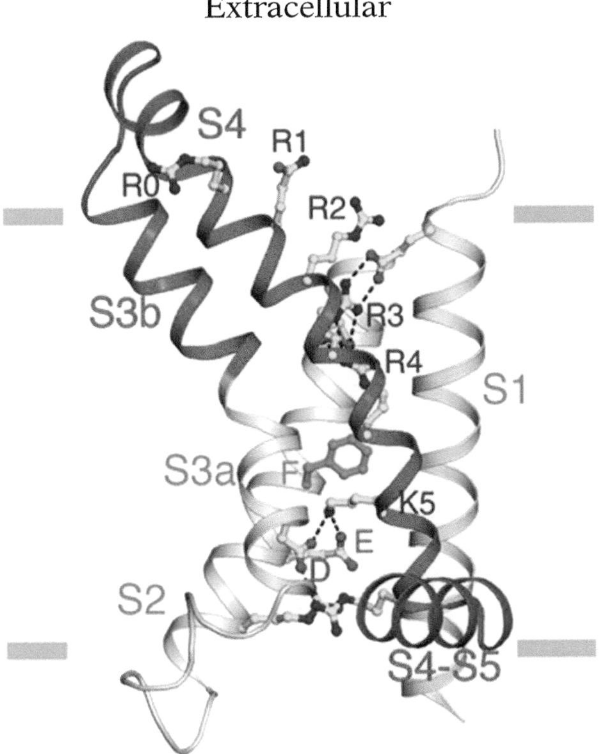

12.22 Voltage sensor gating charge transfer. A ribbon diagram of the Kv paddle chimera shows the position of the S4–S5 linker helix (red) near the S1 and S2 pore helices in open conformation. Along the voltage paddle on S4 positive residues R0, R1, R2, R3, and R4 (stick side chains) are in contact with the aqueous compartment (including the crevice between the paddle and the rest of the molecule), while K5 is sequestered by the side chain of phenylalanine. Both R3 and K5 make ionizing hydrogen bonds (black dashed lines) to acidic residues. From Tao, X., *et al*., *Science*. 2010, 328:67–73. © 2010, American Association for the Advancement of Science. Reprinted with permission from AAAS.

potential (see Figure 12.22). The extent of movement of the gating paddle has been debated: For example, FRET experiments indicate it is small, while avidin binding to biotinylated KvAP indicates some gating charges become accessible from the opposite side of the membrane upon channel opening. The simplest interpretation is that paddle movement pushes two or more arginine residues across the hydrophobic barrier in response to voltage. Other possible mechanisms for responding to voltage include moving the gating charges past countercharges in other parts of the sensor domain and small conformational changes that allow water cavities to change the local dielectric. It is likely the mechanism for sensing and responding to voltage is shared by other voltage-sensitive channels (and enzymes) since the voltage sensor domains are conserved. In a voltage-sensitive phosphatase that has been crystallized in both positions, the critical charge movement past a hydrophobic constriction requires a move of ~5 Å.

An informative set of mutants affecting the voltage sensors of the Shaker-type Kv channels clarifies the role of the gating residues, including an occluded fifth site for a basic residue (a lysine) that is shielded by a highly conserved phenylalanine residue (Figure 12.22). The crystal structure of a Kv chimera called Kv1.2-Kv2.1 with an open pore shows the charges on R1–R4 on S4 are paired with countercharges on S1 and/or S3. Movement of the paddle would switch the residue pairs, leaving the Arg residue on each end exposed to water in the "up" or "down" position. Dissection of voltage sensor movements in the mutants reveals that the pore only opens when the positive charge reaches the occluded fifth site.

Gating in Human Potassium Channels

The first complete structure of a human K^+ channel is that of TRAAK, a member of a two-pore family called K2P. TRAAK (TWIK-related arachidonic acid-stimulated K^+ channel) carries "leak" currents that set the resting potential in cells of the nervous system. While "leak" currents may not seem important, TRAAK channels are essential and mutations in TRAAK cause inherited diseases including certain familial migraines. They are also targets of antidepressents, inhalational anesthetics, neuroprotective agents, and respiratory stimulants.

The K2P family consists of dimers with each subunit contributing two pores to give the channel four pore domains, Pore 1 and Pore 2 from each subunit. The 3.8-Å resolution structure of the TRAAK dimer reveals four inner, outer, and pore helices and selectivity filters very similar to KcsA, giving it close to the four-fold symmetry of other K^+ channels (Figure 12.23). However, there are differences between Pore 1 and Pore 2. The selectivity filter of each Pore 2 is unusually long and forms a bridge to the inner helix, and the inner helix of Pore domain 1 has a midbilayer kink stabilized by a conserved Pro residue that is absent in Pore 2 domains (Figure 12.23D). TRAAK has a unique feature not seen in any other known channel structures, called the helical cap. Extending ~35 Å above the membrane, the helical cap protects the mouth of the channel, blocking K^+ access from above so its pathway is bifurcated (Figure 12.24). The helical cap explains why K2P channels are not sensitive to known K^+ channel blockers and toxins.

TRAAK channels are highly sensitive to the chemical composition of the lipid bilayer, as well as pressure and even temperature. As the name implies, TRAAK is activated by arachidonic acid, in addition to other polyunsaturated fatty acids and lysophosphatidic acid. The crystal structure shows spaces for lipids to bind at lateral

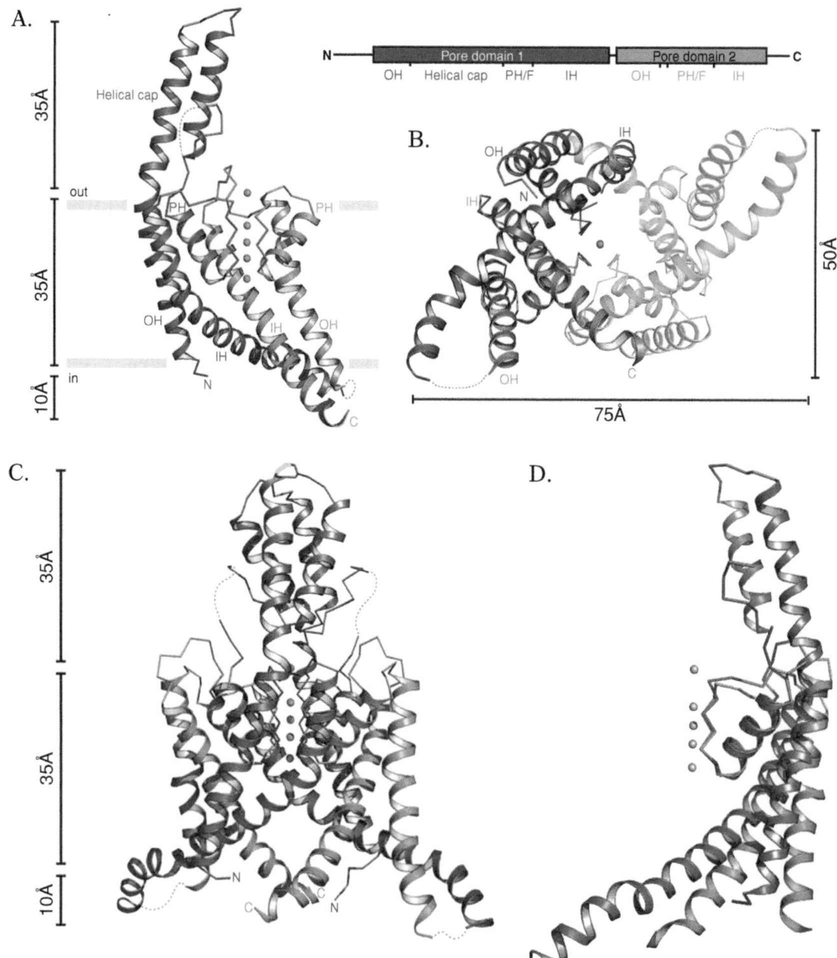

12.23 The high-resolution structure of human TRAAK, a member of the K2P family. (A) Left: a ribbon representation of a single TRAAK protomer viewed from the membrane plane with the extracellular solution above shows the Pore Domain 1 (blue) and Pore Domain 2 (orange), with potassium ions (green spheres). Right: Illustration of the organization of the TRAAK protomer is labeled to show the outer helix (OH), helical cap, pore helix (PH), selectivity filter (F), and inner helix (IH). (B) View of the TRAAK dimer from the cytoplasmic solution with the second protomer half transparent shows the nearly four-fold symmetry with a K^+ ion (green sphere) in the channel. (C) View of TRAAK dimer from the membrane plane. The apex of the helical cap is bridged by a disulfide bond (yellow). The cytoplasmic extension (residue 180–187) of protomer B is modeled based on the same region in protomer A. Unresolved loops are indicated by gray dashed lines. (D) Differences between the two pore domains in a subunit are shown by superposition of Pore Domain 1 (blue) with Pore Domain 2 (orange). Because of the differences between the inner helices and in the angles made by the outer helices and the inner helices, the molecule does not have four-fold symmetry. From Brohawn S. G., *et al.*, *Science*. 2012, 335:436–441. © 2012, American Association for the Advancement of Science. Reprinted with permission from AAAS.

openings in gaps between the two subunits (see Figure 12.24). Patch clamp studies indicate that TRAAK is also sensitive to pressure, activated by positive (and not negative) pressure. The likely sensor of the bilayer state is an amphipathic helix that occurs after the kink in Pore domain 1 inner helix. The amphipathic helix approaches the cytoplasmic membrane surface with five basic residues pointing toward the membrane–cytoplasm interface. Thus it could interact with both acidic lipid head groups and hydrophobic tails to sense the properties of the membrane. Such mechanosensitivity is also seen in bacterial ion channels described at the end of this chapter.

Potassium channels belong to a huge superfamily of tetrameric cation channels found in bacteria, archaea, and eukaryotes. Voltage-gated channels for Na^+ and for Ca^{2+} share many of the features of voltage-gated K^+ channels, although they are much larger proteins (~2000 amino acids). They consist of four domains, each of which is homologous to the K^+ channel TM domain, so they are considered pseudo-tetramers and are expected to have folds similar to those of the K^+ channel tetramer. Like the potassium channel, these passive channels open to allow the ions to flow down gradients created by active ion pumps, the Na^+, K^+ ATPase and the Ca^{2+}-ATPase.

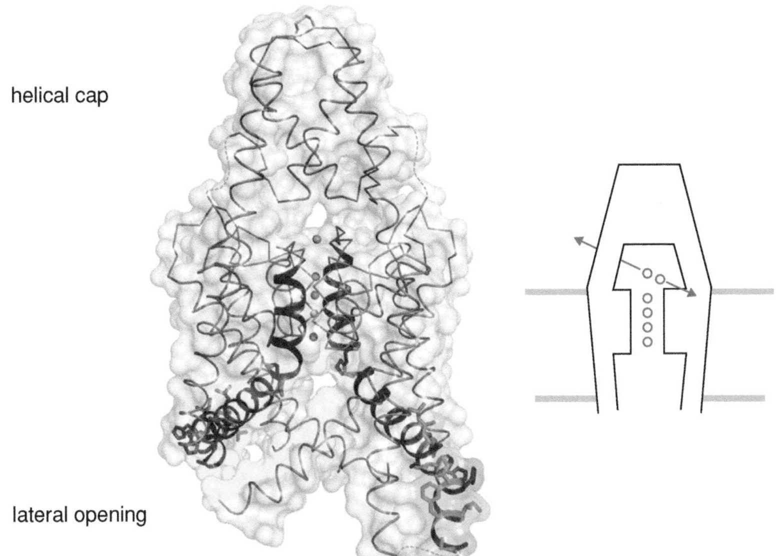

12.24 Features of the TRAAK structure. TRAAK is shown as a wire with inner helices from Pore Domain 1 as ribbons, with a semi-transparent surface representation. Pore Domain 1 (blue) and Pore Domain 2 (orange) with K^+ as green spheres. The helical cap at the top of TRAAK blocks accessibility to K^+, causing a bifurcated pathway as diagrammed in the inset, and also prevents toxins from binding. The hydrophobic surface (green) and basic residues (blue) of the amphipathic segment are exposed near the cytoplasm. One of the lateral openings allowing lipid accessibility (yellow) is on the central cavity. From Brohawn S. G., *et al.*, *Science*. 2012, 335:436–441.

CHLORIDE CHANNELS AND THE CLC FAMILY

For decades chloride channels received less scrutiny than channels for cations. The housekeeping roles of Cl^- in the secretion of mucous, sweat, and other fluids seem to merit less attention than exciting functions such as carrying messages in brain cells or signals in muscle cells. In fact, along with other physiological roles, chloride controls the resting potential in muscle and some neurons, and it is important enough to warrant numerous protein types involved in its uptake or release, including the CFTR protein described in Chapter 6.

The CLC family of "*Cl* Channels" is a ubiquitous family found in plasma membranes and organelles of eukaryotes, with nine isoforms in humans. CLC homologues in prokaryotes are required for resistance of the bacteria to strongly acidic conditions. Unlike most other chloride channels, CLC proteins exhibit preference for Cl^- over both Br^- and I^-, and they are not related to any other channels in either their electrophysiology or their structure. Family members share a common architecture as homodimers containing two pores, and scattered in their sequences are highly conserved residues involved in coordinating Cl^- ions. They respond to different signals, including voltage and H^+, Ca^{2+}, and Cl^- ion concentrations. The well characterized ClC-0 from *Torpedo* electric rays exhibits fast and slow gating of chloride conductance in single channel recording (Figure 12.25). Slow gating closes both pores within the dimer, whereas fast gating allows each pore to open and close independently. Structural and mutational analyses have explored the basis of gating and its voltage sensitivity. Although the eukaryotic proteins are very difficult to express and purify, a crystal structure of a eukaryotic

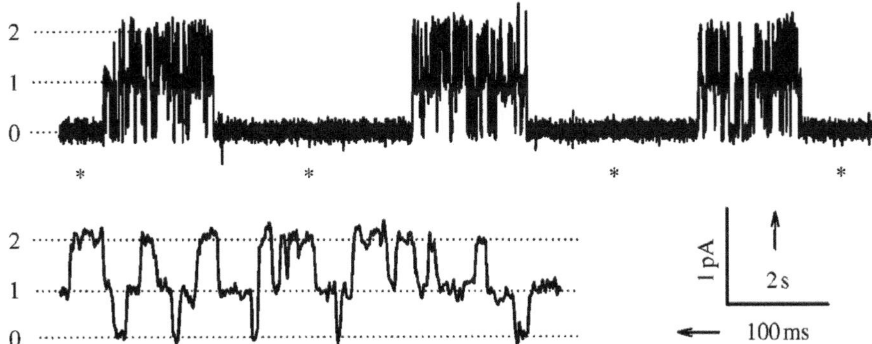

12.25 Fast and slow gating of CLCs. Features of ClC-0 gating are shown in a representative single-channel recording made in 120 mM Cl^- at pH_{int} of 7.5, pH_{ext} of 8.5 and –70 mV. Conductance level 0, both pores are closed; conductance level 1, one pore is closed and one pore is open; conductance level 2, both pores are open. Slow-gate closures (asterisks) indicate periods on time scales of ~sec when both pores are closed. Bursts of fast-gating involve independent opening and closing of the two pores, as seen more clearly in the lower trace with expanded time scale. From Lisal, J., and M. Maduke, *Phil Trans R Soc B*. 2009, 364:181–187. © 2009, The Royal Society.

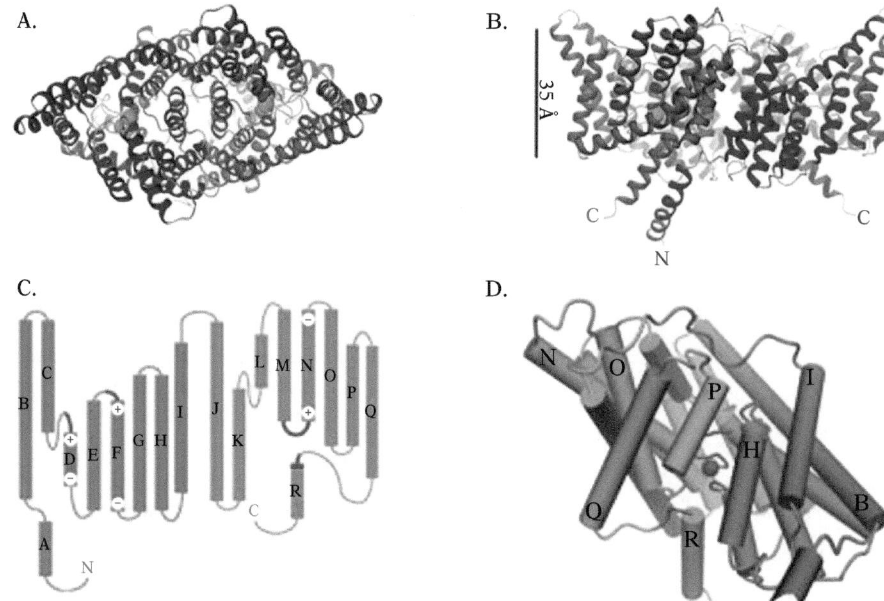

12.26 Topology and structure of ClC-ec1. The x-ray structure of the CLC from *E. coli* was determined at 3.5-Å resolution and that from *Salmonella typhimurium* determined at 3.0-Å resolution. (A later structure of ClC-ec1 in complex with a Fab fragment determined at 2.5-Å resolution is closely similar to both structures.) (A) The two subunits of ClC-ec1 (red and blue) are viewed from the extracellular side, where their triangular shape is evident. Two chloride ions are bound (green spheres). (B) The view from within the membrane is shown with the extracellular side on top and the intracellular side on the bottom. A vertical line denotes the approximate thickness of the membrane. Helix B is tilted by about 45° (gray shading). (C) Topology of the protein with α-helices (cylinders) labeled A to R. The two domains within the subunit are shown (green and cyan) with stretches of residues encoding the selectivity filter (red), which match the alignment of these stretches in different ClC channels. The partial charges due to the helix dipoles are indicated by + and – where they affect the Cl^- binding region. (D) A view of a ClC-ec1 subunit from the dimer interface shows the pseudo-symmetry between the two domains (colored as in (C)). The two bound Cl^- ions are shown (red spheres). From Dutzler, R., *et al.*, *Nature*. 2002, 415:287–294. © 2012, Nature Publishing Group. Reprinted by permission of Macmillan Publishers Ltd.

CLC has now been obtained from a thermophilic alga. The breakthrough in structural biology of the CLC family came with the crystal structure of the *E. coli* CLC. Sophisticated biochemical characterization of this CLC and others presents a surprising insight into the evolution of channels and transporters.

ClC-ec1

The first prokaryotic CLC discovered is ClC-ec1, the 50-kDa CLC in *E. coli* that is activated by low pH. The structure of ClC-ec1 determined at 3.5-Å resolution and then at 2.5-Å resolution in complex with a Fab fragment shows two triangular subunits when viewed from the outside (Figure 12.26). Each subunit consists of 18 TM helices of varying lengths (labeled A to R) with an inverted repeat that produces an antiparallel architecture around a pseudo-two-fold symmetry axis, as seen in aquaporins and many transporters. The two domains wrap around the center to bring together residues from disparate regions, including the conserved motifs GSGIP (106–110) in helix D, G(K/R)EGP (146–150) before the beginning of helix F, GXFXP (335–359) before helix N, and Tyr445 in helix R. This convergence brings together the positive ends of helices D, F, and N, making the pore attractive for anions. The presence of these motifs in other CLCs along with a wealth of functional data strongly indicates that the structure is conserved. Thus ClC-ec1 is an excellent structural model for other CLC channels, even though it turns out to be a transporter, not a channel (see below).

It is not clear why ClC-ec1 is a homodimer with independent pores in each subunit and extensive contact between the subunits. In elegant studies to probe this question, ClC-ec1 was designed to form stable monomers by insertion of two tryptophan residues on the subunit interface near the level of the lipid headgroups (I201W/I422W). The double mutant gives less than half the Cl^- efflux of the wild type dimer and crystallizes as a monomer with almost no structural alterations. ClC-ec1 engineered to have numerous cysteine–cysteine or cysteine–lysine crosslinks across the dimer interface functions normally, ruling out involvement of large quaternary rearrangements in function. Thus the properties of ClC-ec1 do not clarify how other CLC dimers close both pores together in slow gating.

The ion pores in the center of each subunit are 39 Å from each other in the homodimer. Each pore is shaped like an hourglass with a selectivity filter in a narrow

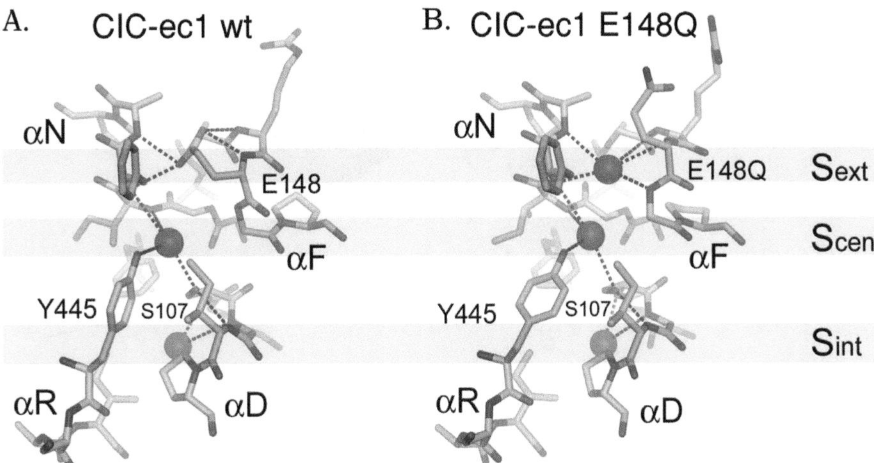

12.27 Nature of the ClC-ec1 pore. The residues where Cl^- ions (red spheres) bind are shown for the wild type (A) and mutant E148Q (B). Dashed lines represent coordination of the Cl^- ions and hydrogen bonds. In both structures the Cl^- at the center site (Scen) is coordinated by amide nitrogen atoms from Ile356 and Phe357 (not labeled) and side chains of Ser107 and Tyr445. In (A) the side chain of Glu148 is fully hydrogen bonded with the peptide, closing the pore or access to Scen. In contrast, in (B) the uncharged glutamine side chain in the E148Q mutant is rotated toward the exterior, allowing a third Cl^- ion to bind at Sext. This change opens the exterior gate and is presumed to mimic the protonated form of the wild type. From Dutzler, R., *FEBS Letts.* 2004, 564:229–233. © 2004, Elsevier. Reprinted with permission from Elsevier.

region 15 Å long in the middle of the membrane bilayer. Two Cl^- ions are bound in the selectivity filter at sites at the center (Scen) and interior (Sint) of the membrane (Figure 12.27). The dehydrated Cl^- ion at Scen is coordinated by amide nitrogen atoms from Ile356 and Phe357 and by the side chains of Ser107 and Tyr445, and lacks bonds to strongly basic Lys or Arg side chains, which would bind it too tightly. The Cl^- ion at Sint interacts with free backbone amide groups on the loop preceding helix D and is still solvated where it is exposed to the aqueous vestibule. Above Scen the pore is blocked by the side chain of Glu148. A third Cl^- ion near the exterior of the membrane site (Sext) is observed in the E148Q mutant, when neutralization of the glutamate side chain allows it to be displaced for ion binding. Thus Glu148 is considered the gating glutamate at the external end of the pore, which opens when it is protonated. This mechanism is a general feature of CLCs, since mutation of this residue in ClC-0, ClC-1, and ClC-2 abolishes voltage-dependent gating of the pores, and is accepted as the basis of fast gating.

Careful electrophysiology measurements revealed a second ion makes an asymmetric contribution to the current. Identification of this ion as the proton led to the surprising conclusion that ClC-ec1 carries out H^+–Cl^- countertransport with a stoichiometric ratio of 2 Cl^-/H^+ and thus can use the proton gradient to pump chloride ions (and vice versa). Glu148 is required to couple protons to Cl^- permeation: When mutant proteins carrying E148A or E148Q are reconstituted in liposomes Cl^- flux is maximum and independent of pH, and H^+ flux is abolished. A second glutamate residue, Glu203, is also required for H^+ transport and appears to function as an internal gate for protons. Since Glu203 is not required for Cl^- flux, the pathways of H^+ and Cl^- diverge (Figure 12.28). The role of ClC-ec1 as an antiporter is supported by the lack of a clear channel through the protein in the x-ray structures. Scen can be described as the active site rather than the selectivity filter, and the conformational change of Glu148 in response to proton binding viewed as the access to that site, rather than the gate of the channel. These characteristics are in marked

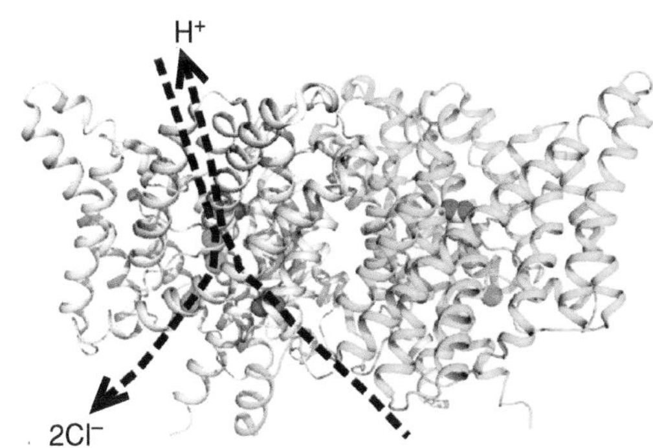

12.28 Diverging paths for H^+ and Cl^- ions in ClC-ec1. The ribbon diagram for the mutant E148Q homodimers (subunits colored gray and yellow) shows three bound ions (green spheres) below the position of Glu148 (space-filled) along the pathway for Cl^- (dashed arrow on one subunit). The H^+ pathway diverges to go through Glu107 (space-filled) near the interior. From Miller, C., and W. Nguitragool, *Phil Trans R Soc B.* 2009, 364:175–180. © 2009, The Royal Society.

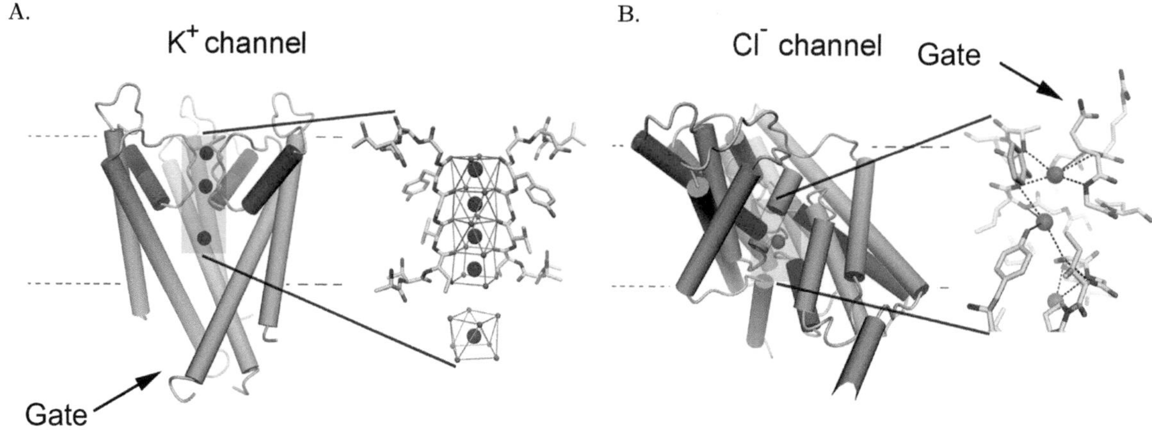

12.29 Comparison of ClC and KcsA channels. (A) In the KcsA K^+ channel, shown with the front subunit removed, the pore helices (red and blue) define the pore with the selectivity filter near the external surface of the bilayer. The enlargement shows the K^+ ions (blue spheres) in the open pore. (B) In contrast, the Cl^- ions (red spheres) bind to the ClC-ec1 E148Q subunit in the center of the subunit and in the middle of the membrane bilayer. Even with the gate held open (due to the mutation), the protein lacks a pore all the way through the membrane. From Dutzler, R., *FEBS Letts*. 2004, 564:229–233. © 2004, Elsevier. Reprinted with permission from Elsevier.

contrast to the K^+ channel (Figure 12.29). Although further details of the ion paths through the protein are unknown, it is clear that ClC-ec1 is a secondary transporter, not a channel.

CLC Channels as "Broken Transporters"

The example of ClC-ec1 clouds the usually stark delineation between channels and transporters. In fact, it takes only two mutations in ClC-ec1 to convert it from a transporter to a channel. Since replacement of Tyr445 at Scen with Ala or Ser opens the Cl^- pathway to the intracellular side, the loss of both steric gates in a E148A/Y445S double mutant gives >20-fold increase in the Cl^- transport rate. Is the double mutant a fast transporter or a slow channel? Because it has a continuous path across the membrane and carries out passive equilibration of Cl^- down its concentration gradient, with no coupling to H^+, the double mutant fits the definition of a channel.

In a remarkable outcome of evolution the CLC family is split between transporters and channels. In eukaryotes the CLC channels are localized to the plasma membrane and the CLC exchangers to intracellular membranes. Thus ClC-0, ClC-1, ClC-2, ClC-Ka, and ClCKb are channels, while ClC-4 and ClC-5 are antiporters. Among the 400 sequences for prokaryotic CLCs, about half lack a residue equivalent to Glu203 of the proton exit pathway and are therefore likely to be channels. Structural analysis may resolve the question, since an open pore through the protein is a clear indication of a channel that does not allow movements of different ions to be coupled. However, subtle changes might be sufficient to convert a transporter to a channel, albeit a slow one as the double mutant of ClC-ec1 exemplifies. Because all the CLCs share the same basic architecture, they illustrate that mechanistic differences can arise from very similar structures. The CLC family brings a new ambiguity to the classical distinction between channels and transporters.

Eukaryotic CLCs differ from ClC-ec1 in having a large cytoplasmic domain at the C terminus that is required for proper trafficking and is also involved in gating. Crystal structures of the separate C-terminal domains of ClC-0 and ClC-5 were followed by a 3.5-Å resolution structure of the CLC from a thermophilic red alga *Cyanidioschyzon merolae*, CmCLC (Figure 12.30). The TM domain of CmCLC is very close to the structure of ClC-ec1 and its C-terminal domain shares the fold of the cytoplasmic domains of the other eukaryotic CLCs. The extensive interface between the TM domain and the cytoplasmic domain is highly complementary, as much so as an antibody–antigen interface. Residues at this interface are targets in some human genetic diseases (Figure 12.30B). The pathologies associated with CLC defects vary from hypertension and kidney disease to bone problems.

The structure of CmCLC suggests an intermediate state in the exchange of two Cl^- for H^+ because it reveals a third position for the gating glutamate (Glu210 in CmCLC, which corresponds to Glu148 in ClC-ec1). Electron density due to Br^- and indicative of Cl^- binding is present at Sext and Sint, while the gating glutamate occupies Scen (Figure 12.31). Swinging in and out of the membrane, the gating glutamate can pick up protons from either the exterior or interior during the transport cycle. The role of this glutamate defines the fine line between these antiporters and pH-gated channels of the CLC family. Both for their physiology and for their

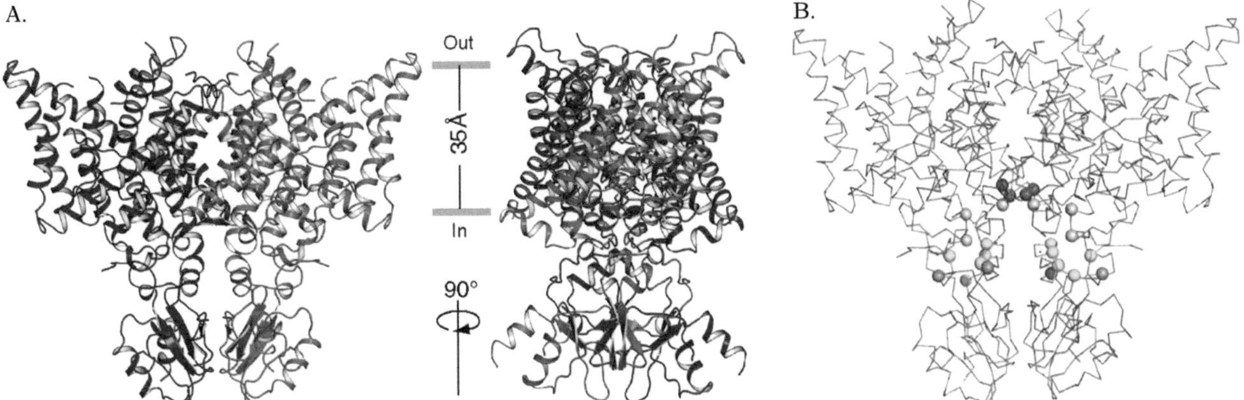

12.30 Structure of a eukaryotic CLC. (A) Ribbon diagrams of CmCLC dimer (subunits colored red and blue) are viewed from the side of the membrane with the extracellular solution above. To achieve the crystals, 86 N-terminal and 7 C-terminal amino acids were deleted. (B) The large interface between the TM domains and the cytoplasmic domains is the site of numerous mutations associated with genetic diseases. The mutations in human ClC-1, ClC-0, and ClC-7 are mapped on the structure of CmCLC based on sequence alignment. They include ClC-1 and ClC-0 mutations in the cytoplasmic domain that affect gating (green spheres), Thomsen's disease mutations in ClC-1 (red spheres), and osteopetrosis mutations in ClC-7 cytoplasmic domain and loops (yellow and pink spheres, respectively). From Feng, L., *et al.*, *Science*. 2010, 330:635–641. © 2010, American Association for the Advancement of Science. Reprinted with permission from AAAS.

unique position straddling the divide between channels and transporters, the members of the CLC family will be objects of much future research.

MECHANOSENSITIVE ION CHANNELS

Mechanosensitive (MS) channels respond to mechanical stresses applied to the membrane or to membrane-attached elements of the cytoskeleton, enabling organisms to respond to touch, sound, pressure, and gravity. They fall into two broad classes, depending on whether or not cytoskeletal elements are involved (Figure 12.32). MS transduction in vertebrate auditory hair cells provides examples of fast and sensitive responses involving the cytoskeleton. The prokaryotic MS channels that respond to decreased osmotic pressure exemplify the class that does not involve the cytoskeleton, instead sensing membrane pressure directly. Crystal structures of two of the prokaryotic MS channels illustrate the best understood models for how membrane tension can drive conformational change and lead to channel opening.

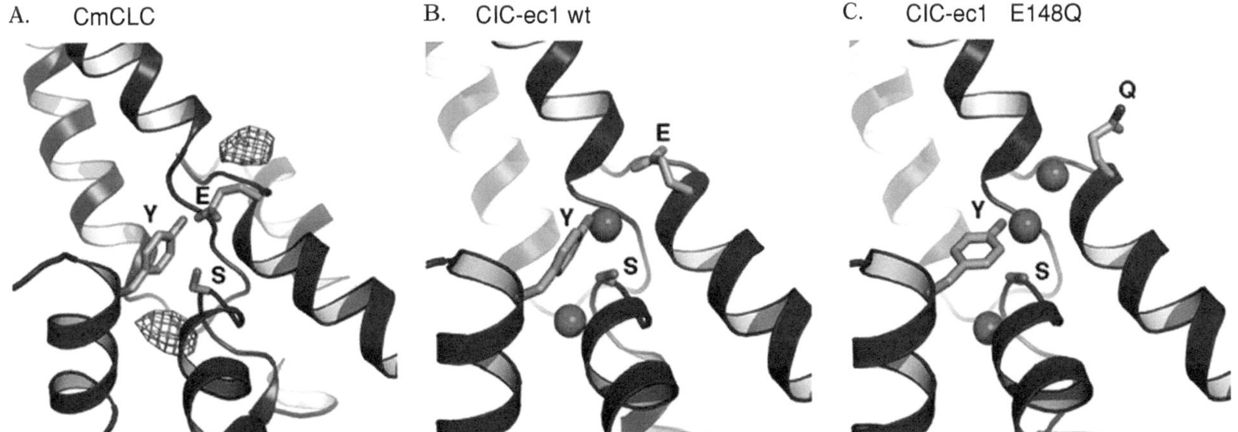

12.31 Three positions of the gating glutamate. For a view of the ion transport path of CmCLC, wild type ClC-ec1, and the E148Q mutant of ClC-ec1 (from left to right), the foreground helices were removed. Active site residues (stick models), Cl^- ions (pink spheres) and electron density for bromine (red mesh) show the locations of Sext, Scen, and Sint (see text for a description of the transport sites). The side chain of the gating glutamate swings from Scen in the membrane interior in CmCLC (A) to Sext in wild type ClC-ec1 (B) to the exterior aqueous compartment E148 mutant of ClC-ec1 (C). From Feng, L., *et al.*, *Science*. 2010, 330:635–641. © 2010, American Association for the Advancement of Science. Reprinted with permission from AAAS.

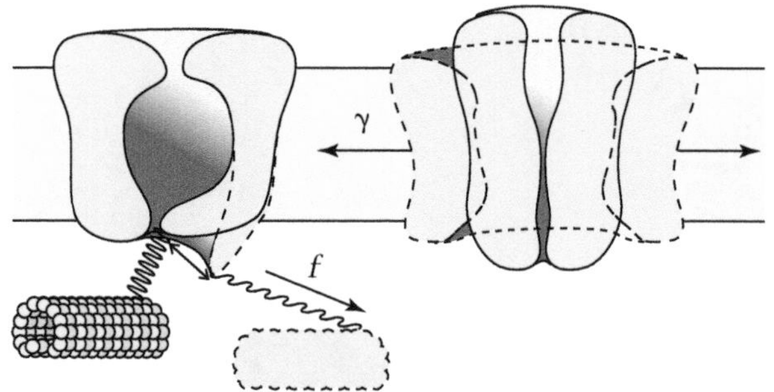

12.32 Two classes of MS channels. In many vertebrate cells, the channel has a swinging gate that is unlocked by tension from the cytoskeleton (left). In contrast, some channels respond directly to pressures in the bilayer and can be envisioned as an expandable barrel (right). Redrawn from Sukharev, S., and A. Anishkin, *Trends Neurosci*. 2004, 27:345–351. © 2004 by Elsevier. Reprinted with permission from Elsevier.

Bacteria maintain a slight outward turgor pressure as they respond to varying osmotic conditions. When osmotic pressure increases, the cells accumulate solutes such as betaine, proline, and potassium ions to balance the pressure and minimize water efflux. When osmotic pressure drops, they avoid rupture by opening MS channels for solute efflux. MS channels were discovered in 1987 in patch clamp experiments on giant bacterial spheroplasts (see Chapter 3). *E. coli* has three types of nonspecific MS channels, named for their different single-channel properties: MscL (large), MscS (small), and MscM (mini), along with a small potassium-dependent channel MscK (Figure 12.33). The smallest channels, MscM, open at low tensions, are not essential, and are poorly characterized. *E. coli* has six species of the small MS channels, including the best characterized MscS and the much larger MscK, which contains a subdomain very similar to MscS. The large nonselective channel MscL opens at high tension, near that which would rupture the cell. Double mutants that lack both MscL and MscS do not survive osmotic

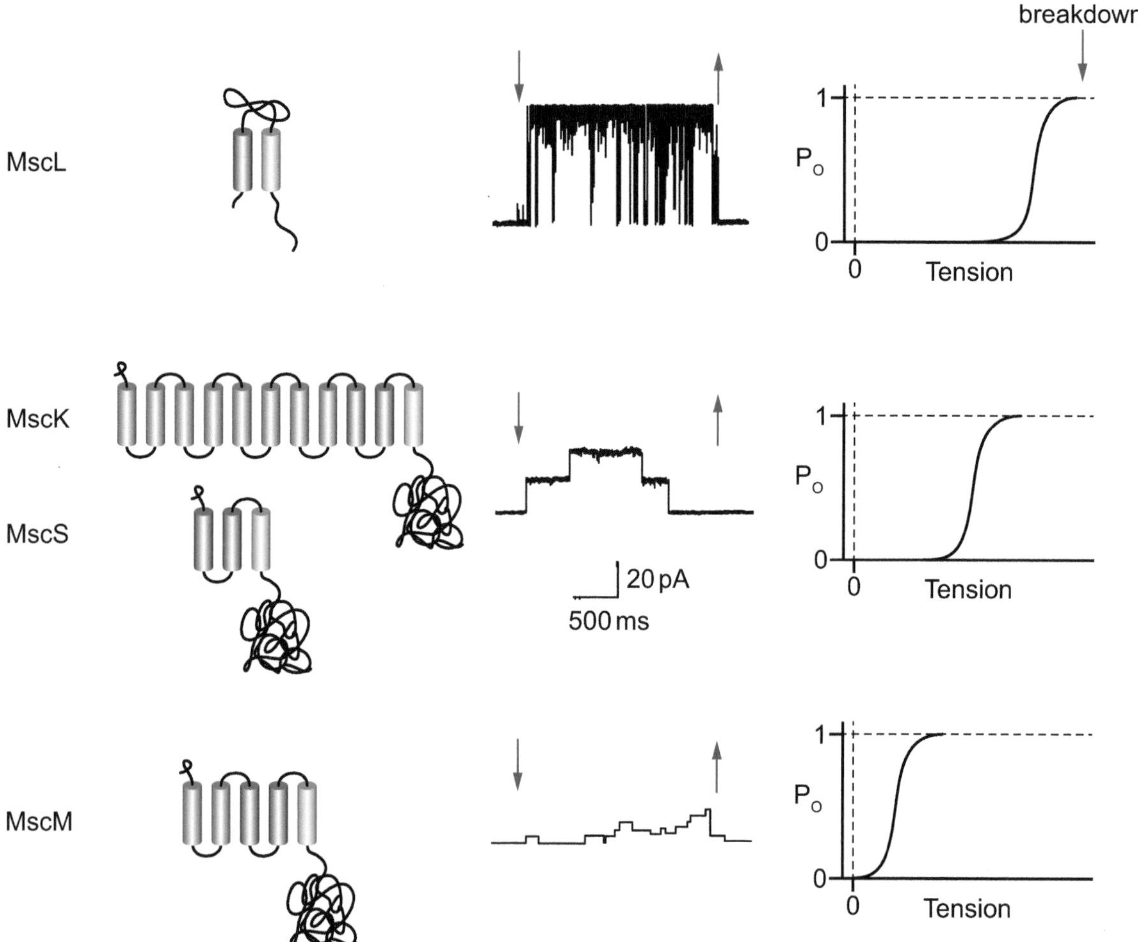

12.33 The four types of MS channels detected in *E. coli* fall into three groups based on their conductance and gating thresholds. The topology of each channel listed on the left is followed by their single-channel conductance characteristics (center) and their response to pressure (right). Based on a figure from Perozo, E., and D. C. Rees, *Curr Opin Struct Biol*. 2003, 13:432–442. © 2003 by Elsevier. Reprinted with permission from Elsevier.

BOX 12.1 MECHANOSENSITIVITY

While other membrane proteins may be sensitive to pressures in the membrane (see discussions of hydrophobic mismatch and curvature in Chapters 2, 4, and 7), MS proteins uniquely respond to physiological pressure changes by significant conformational changes that result in opening channels. The larger the channel opening in response to tension, the more mechanosensitive the channel. Tension (measured in milliNewtons per meter) is created in patch clamp systems by sucking on the pipette holding the patch: 10 mN m^{-1} tension is generated by a pressure of 0.05 atm in a patch with a diameter of 1 μm. The critical gating tension (the tension at which a channel opens) depends on the thickness and curvature of the membrane *in vitro* and is also affected by the presence of the cell wall *in vivo*. The size of the channel is proportional to its conductance (measured in nanoSiemens): For current of 1 nS, the current flow through the channel at an applied voltage of 0.1 V would be 100 pA or ~6 × 10^8 ions/sec.

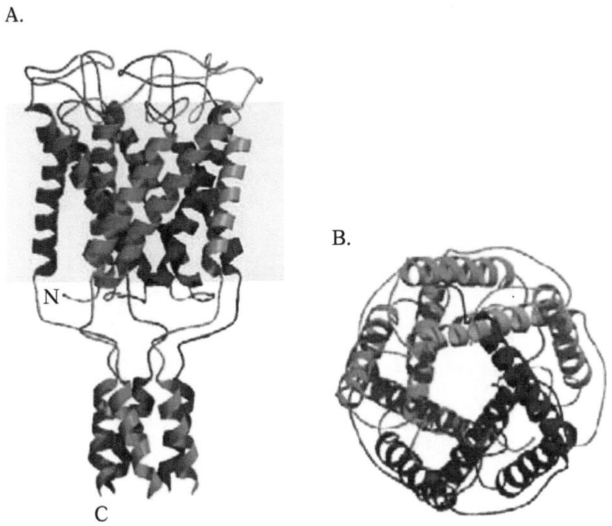

A.

B.

N

C

12.34 X-ray structure of MscL from *Mycobacterium tuberculosis*. (A) The side view shows the separate cytoplasmic and TM domains, with shading to indicate the membrane. (B) Each monomer contributes two TM helices to the TM domain, as seen in the view of the TM domain from the top. In the homopentamer the pore is lined with five TM1 helices. From Perozo, E., and D. C. Rees, *Curr Opin Struct Biol*. 2003, 13:432–442. © 2003 by Elsevier. Reprinted with permission from Elsevier.

downshock. Furthermore, introducing MscL from *E. coli* into marine bacteria gives them increased resistance to drops in osmotic pressure. MscS is distributed widely in eukaryotic fungi and plants: For example, *Arabidopsis thuriana* has ten MscS-related proteins.

MscL

MscL, the large MS channel of the *E. coli* inner membrane, forms a nonselective channel that is activated by quite high levels of membrane tension (~10 mN m^{-1}). The MscL pore is very large, with a diameter of 30–40 Å measured in sieving experiments, and gives a conductance of 2.5–3 nS *in vitro*. Opening such a large pore in response to severe osmotic downshock releases ions, nutrients, and even small peptides to relieve the osmotic stress and avoid complete rupture of the bacterial cell.

Crystallization attempts with a dozen MscL homologs from different bacteria produced the first crystals of the MscL from *Mycobacterium tuberculosis* in dodecylmaltoside, and its structure was solved at 3.5 Å resolution. This protein, called Tb-MscL, has 37% sequence identity with the *E. coli* MscL, Eco-MscL, and its structure is consistent with crosslinking and EM data on Eco-MscL. Tb-MscL is a homopentamer consisting of two domains, a TM domain and a cytoplasmic domain (Figure 12.34). Each monomer has two TM helices that tilt about 30° from the bilayer normal and a cytoplasmic helix that tilts about 15°, making two bundles with five-fold symmetry. The first TM helix, TM1, lines the pore and is quite buried, as it is in contact with two TM1 helices from other subunits, TM2 from the same subunit, and another TM2 from another subunit. The TM2 helices from neighboring subunits are separated by ~20 Å. The periplasmic loops between TM1 and TM2 form a flap that makes the extracellular surface, and part of each loop folds into the pore. Overall, the TM domain of the pentamer has a funnel shape with the typical location of aromatic residues at the polar/nonpolar interfaces.

The pore helix (TM1) is one of the most highly conserved regions in the sequences of MscL proteins. It contains GxxxG motifs for close helix packing in the sequence AxGXxxGAAxG at residues 20 to 30 (where x is a residue at the pore constriction). Several mutations in TM1 alter the channel gating (see below). The pore of Tb-MscL is lined with a series of hydrophilic residues, with two hydrophobic residues near the cytoplasmic end, Ile14 and Val 21, creating a constriction. Because the radius of the pore is only 2 Å at its narrowest, the Tb-MscL structure is considered to portray a closed channel (Figure 12.35). To investigate channel opening, cysteine scanning mutagenesis of MscL was used to place nitroxide labels along both TM1 and TM2, allowing comparison of EPR spectra in reconstituted membranes that trapped MscL in an open and a closed intermediate state (see below). The results, along with computational simulations, suggest the helices tilt out in an open state to give a pore as large as 25 Å in diameter.

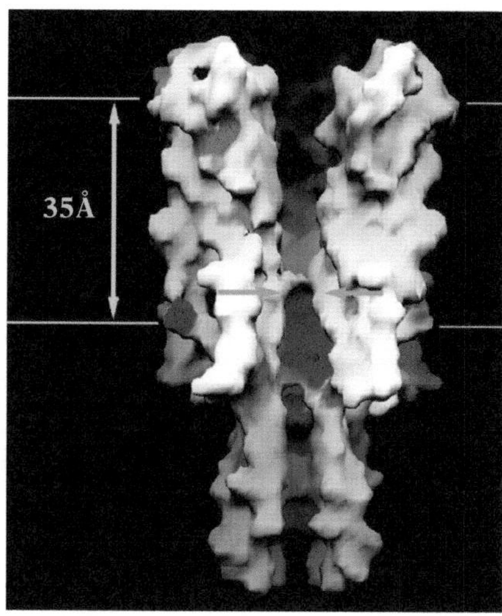

12.35 Pore structure in Tb-MscL. In the MscL homopentamer, the pore lining has many polar and charged residues from TM1. A cutaway side view of the molecular surface of the pore shows areas with charged residues (blue, basic; red, acidic). The green arrows show where the channel is closed. From Chang, G., *et al.*, *Science*. 1998, 282:2220–2226. © 1998. Reprinted with permission from AAAS.

Genetic experiments indicate the cluster of charged groups at the cytoplasmic end of the MscL pentamer is important to channel function. The cytoplasmic domain is a compact left-handed five-helix bundle that might act to filter the solutes permeating an open channel upon osmotic downshock. The C termini beyond the bundle are disordered in the crystal and are of varying length in the different MscL homologs.

MscS

The family of small MS channels found in some yeasts and plants, as well as bacteria and archaea, is highly diverse: It includes proteins that vary from 286 residues (in *E. coli*) to over 1000 residues in length. *In vitro* the *E. coli* MscS shows a slight preference for anions; its conductance of ~1 nS predicts a pore size of 14–16 Å in diameter. The conductance of MscS is still several orders of magnitude larger than that of voltage-gated K^+ channels (see above). Mutations that introduce charges close to the pore do not alter its ionic selectivity, so the preference for anions can be attributed to rings of positive charge within the vestibule. MscS is gated by voltage as well as pressure: It requires less tension to open as the membrane is depolarized.

The first x-ray structure of *E. coli* MscS at 3.9-Å resolution reveals an N-terminal TM domain and a C-terminal cytoplasmic domain, but is otherwise very distinct from the structure of MscL. MscS is a homoheptamer, in which each subunit corkscrews around the pore axis (Figure 12.36). Each subunit contains three TM helices and a much more extensive cytoplasmic region. In the membrane-spanning region, TM1 and TM2 pack closely, while TM3 has a kink at residue G113. TM3a (S95 to L111) from each subunit lines the pore, while TM3b (G113 to M126) angles out to connect to the cytoplasmic domain.

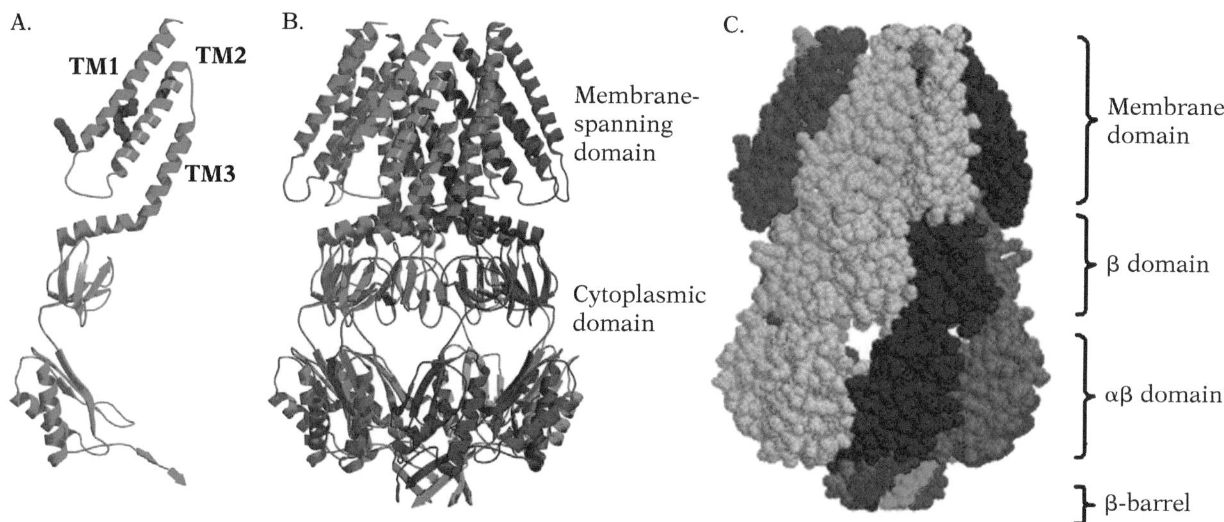

12.36 X-ray structure of MscS from *E. coli*. (A) The fold of a single MscS monomer shows the three TM helices (labeled) and the mixed α/β structures of the cytoplasmic domain. The N terminus is at the top and the C terminus at the bottom of the peptide, as drawn. Two arginine residues are shown in the TM region as space-filling models. (B) The MscS heptamer viewed from the side with each subunit in a different color. (C) Space-filling representation of the MscS heptamer. With each subunit a different color, this view emphasizes how they corkscrew around the central axis. In addition, the cytoplasmic domain is divided into a β domain near the membrane, an α/β domain, and a β-barrel at the end. (A) and (B) from Bass, R. B., *et al.*, *FEBS Lett.* 2003, 555:111–115. © 2003 by the Federation of European Biochemical Societies. Reprinted with permission from Elsevier. (C) From Edwards, M. D., *et al.*, *Curr Opin Microbiol.* 2004, 7:163–167. © 2004 by Elsevier. Reprinted with permission from Elsevier.

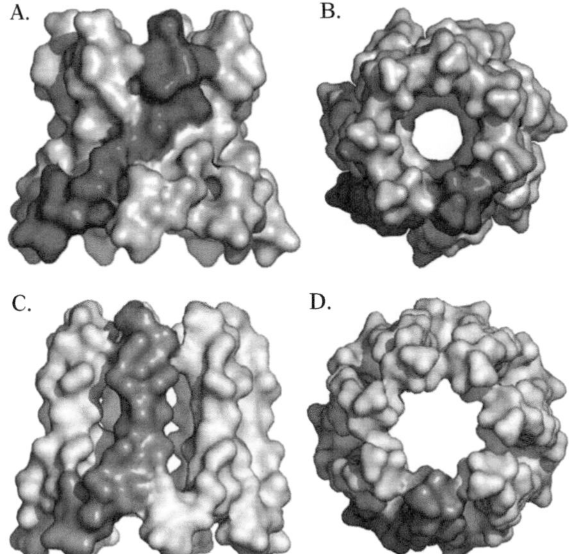

12.37 Open and closed MscS pores. To visualize the pores formed by TM3a (residues S95 to L111), residues T93 to N117 are shown in space-filling models from the side and top. (A) and (B), wild type (pink) with a single subunit highlighted (dark pink), and residue A106 marked (green). (C) and (D), A106V mutant (yellow), with a single subunit highlighted (orange) and residue V106 marked (green). The helices are tightly packed in the closed wild type structure (A), whereas in the mutant the spaces between them are large enough for interactions with lipids (C). The opening of the pore is 4.8 Å in the wild type channel (B) and 13 Å in the A106V mutant (D). From Wang, W., *et al.*, *Science*. 2008, 321:1179–1183. © 2008, American Association for the Advancement of Science. Reprinted with permission from AAAS.

TM3b is an amphipathic helix that likely resides in the lipid headgroup region where two Arg residues (R128 and R131) may interact with phosphate headgroups. The cytoplasmic domain has a large interior chamber formed by a β subdomain and an $\alpha\beta$ subdomain and contiguous to the TM3a pore. This large vestibule (40 Å in diameter) has seven side openings of 10 × 8 Å that provide access to the central cavity from the cytoplasm. At the bottom of the cavity is a seven-stranded β-barrel that is filled by side chains and therefore is not part of the channel.

In the TM domain of MscS the seven TM3a helices are tightly packed (Figure 12.37A). TM3 has a conserved sequence starting at Ala97 in *E. coli* MscS and consisting of AxxGAAGXAxGXAxyG (where x is a hydrophobic residue, X is a hydrophobic residue at the pore constriction, and y is a hydrophilic residue). The interhelical A–G pairs that result allow close packing without locking the helices in place, as demonstrated in mutant studies. MscS proteins with Ala substituted for Gly in this region gate more easily, while the reverse (Gly substituted for Ala) inhibits its gating. MscS with a double mutation such as A106G/G108A has normal gating and conductance, indicating knobs and holes can be interchanged with little effect.

Although the structure of MscS has a visible pore, which suggests that it is the open conformation (Figure 12.37B), MD simulations indicate it is a nonconducting state. The pore is lined with hydrophobic residues and impeded at the narrowest part near the cytoplasmic surface by two rings of Leu105 and Leu109. The hydrophobicity of the pore prevents wetting, creating a "vapor lock" that blocks the passage of ions. Final evidence that this structure represents the closed state comes from a new crystal structure of MscS in an open state.

By making a knob that didn't fit in the hole with the mutation A106V, MscS could be stabilized and crystallized with the pore open. In the 3.45-Å structure of A106V the cytoplasmic domain along with TM3b shows little difference from the wild type structure, while TM1 and TM2 are rotated and TM3a is tilted and slightly rotated. The TM3a helices are not tightly packed, and removal of the two leucine side chains from the center of the pore creates a pore diameter of 13 Å (Figure 12.37C, D). However, the conductivity of open MscS is larger than that predicted from the size of the A106V pore.

MS Channel Gating

Portions of the MS channels involved in opening and closing the pores have been identified by the characterization of mutants that affect channel gating. Loss-of-function (LOF) mutants either require a higher pressure threshold to open the channel or cannot open it at all, in which case the bacteria do not survive osmotic downshock. Gain-of-function (GOF) mutants have a lower threshold for opening or staying open, generally resulting in slow growth rates. When mapped onto the crystal structures of MscL and MscS, the residues affected by GOF mutations cluster in the permeation pathway and on the outer helices (Figure 12.38). LOF mutations affect loops and regions of the protein that interact with lipid headgroups, which suggests that sensor regions are

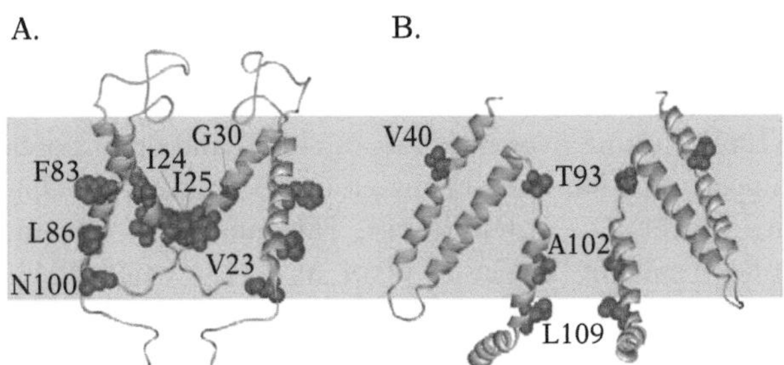

12.38 Structural determinants in gain-of-function mutants in MscL (A) and MscS (B). Mutants were produced by random mutagenesis and screening of growth rates, potassium ion retention, or overall survival. Affected residues are thought to play a critical role in permeation and/or gating. From Perozo, E., and D. C. Rees, *Curr Opin Struct Biol*. 2003, 13:432–442. © 2003 by Elsevier. Reprinted with permission from Elsevier.

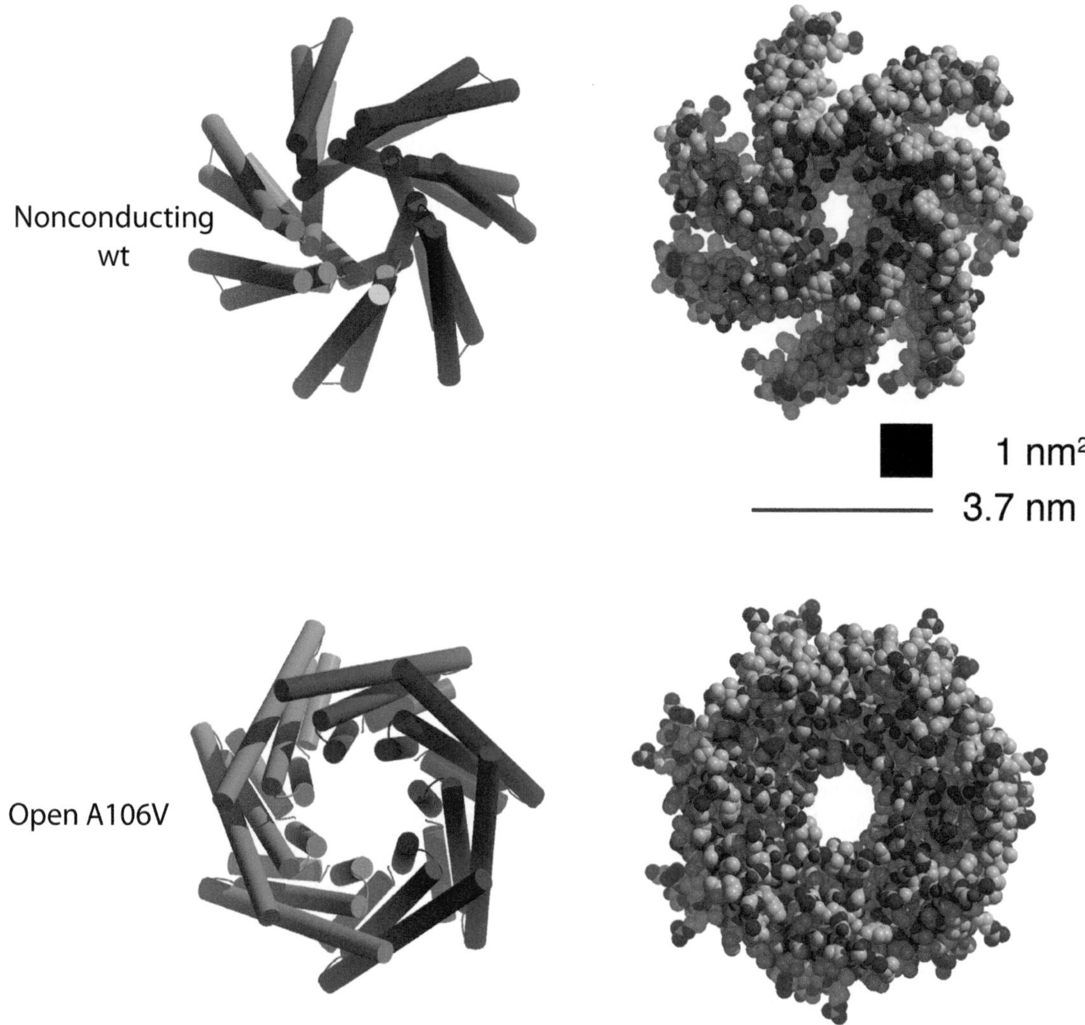

12.39 Gating in MscS. The nonconducting and open states of the TM regions from the x-ray structures of MscS wild type and the A106V mutant are shown using cylinders for the TM helices (left, each subunit a different color) and space-filling representations (right). As the outer helices (TM1 and TM2) tilt inwards, the pore helices (TM3a) are pulled out to enlarge the pore opening. The bar shows length scale and the black square area scale for the space-filling models. From Haswell, E. S., *et al.*, *Structure*. 2011, 19:1356–1369. © 2011, Elsevier. Reprinted with permission from Elsevier.

located at the interfacial regions of the membrane, where the lateral pressure is higher than it is in the center.

The crystal structures of MscS wild type and the A106V mutant give snapshots of closed and open pores, suggesting what must change for channel opening (Figure 12.39). The outer helices, TM1 and TM2, tilt and rotate, causing TM3a to pivot at G113, while TM3b remains anchored to the lipid bilayer (Figure 12.40A). Both MscS and MscL have a central set of helices flanked by sets of outer helices that are less tightly packed. The outer helices are likely to be the sensors when the channels respond to tension in the bilayer. Analyses of the lateral pressure profile of the lipid bilayer indicate that when the membrane is stretched, the pulling force concentrates at the interfacial regions. Increased tension at the inner and outer rims of the MS channel barrel could tilt TM helices to allow the channel to expand like the opening of the iris of a camera.

To investigate such changes in a lipid environment the direct effect of lateral pressure on MS proteins has been tested with the spin-labeled MscL channels described above reconstituted into vesicles of varying lipid compositions for patch clamp experiments. The results indicate that varying the acyl chain length from 10 to 20 carbon atoms is not sufficient to open the MscL channel, although the threshold for activation is lower in vesicles made with short-chain phospholipids. However, introduction of lysophosphatidylcholine into the external leaflet to modify the lateral pressure triggers channel opening. Using the same approach, spin-labeled MscS in lysoPC has given a model of the open structure based on exposure of the nitroxide labels to oxygen, Ni^{2+}, and

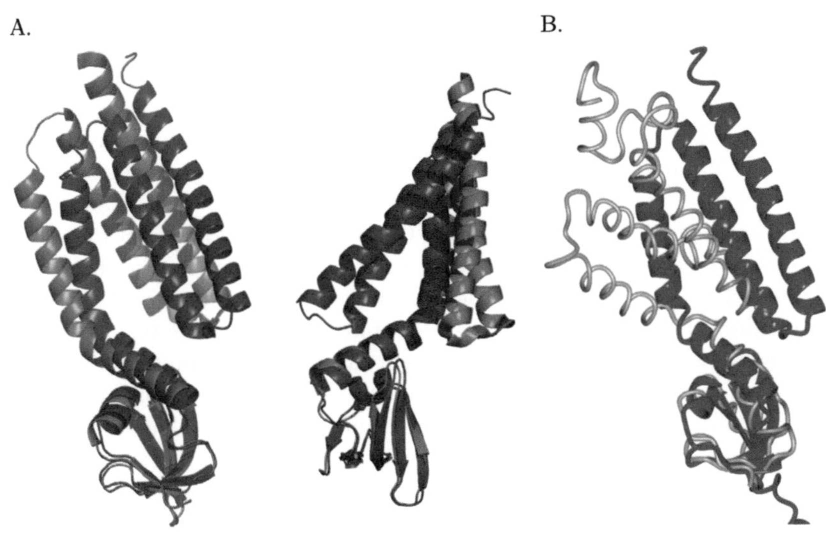

12.40 Movement of the TM helices of a monomer of MscS. (A) Superposition of the TM helices aligned from the β-sheets shows the difference between the closed wild type (magenta) and open A106V (blue) x-ray structures of a monomer of MscS seen from two angles. TM1 and TM2 tilt from approximately normal to the membrane to almost 45°, pulling on TM3a while TM3b shows little change. (B) Comparison of the EPR-derived open structure that includes the N terminus (pale blue) with the A106V crystal structure (blue) shows significant differences in the helix tilts due to further rotation of helix TM3a in the EPR results. From Naismith, J. H., and I. R. Booth, *Ann Rev Biophys*. 2012, 41:157–177. © 2012 by Annual Reviews. Reprinted with permission from the Annual Review of Biophysics, www.annualreviews.org.

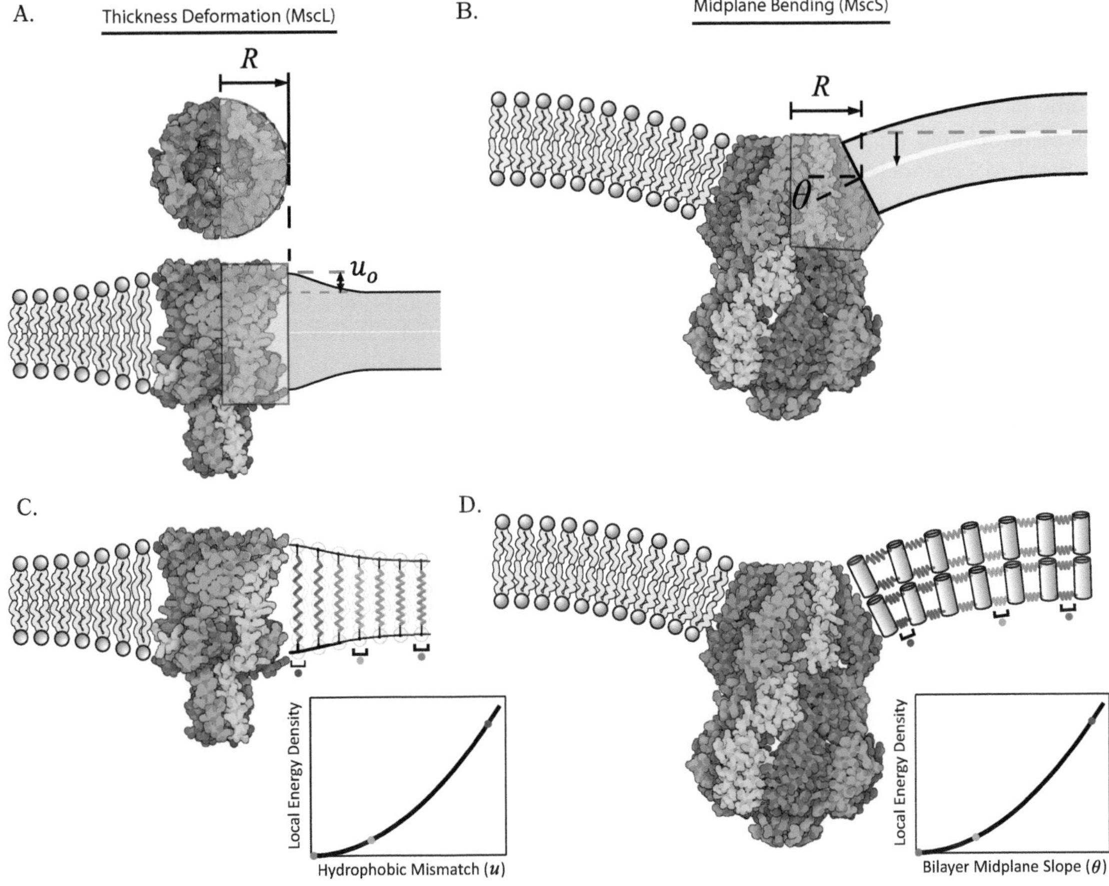

12.41 Summary of effect of mismatch and bending of lipid bilayer on MSCs. (A) The structure of MscL is idealized as a rigid cylinder with a hydrophobic mismatch at the protein–lipid interface. R is the effective radius of the channel and μ_0 is the hydrophobic mismatch between the protein and bilayer thicknesses. (B) The structure of MscS is idealized as a wedge with a slope that glues continuously onto the surrounding lipids. θ is the midplane bending angle at the protein–lipid interface. Distortion by MscL (in (C)) and MscS (in (D)) of the surrounding lipid bilayer is illustrated, portraying the lipids as springs that are color-coded to indicate their strain. The graph in (C) shows the deformation energy as a function of hydrophobic mismatch at three points, and the graph in (D) shows the deformation energy as a function of bilayer slope at three points. From Phillips, R., *et al*., *Nature*. 2009, 459:379–385. © 2009, Nature Publishing Group. Reprinted by permission of Macmillan Publishers Ltd.

Ni^{2+} linked to lipids to measure how much is embedded in lipid, in contact with water and located at the lipid–water interface, respectively. The EPR model of MscS in the presence of lysoPC is similar to the A106V mutant structure (Figure 12.40B). TM2 is closer to TM3 than in the crystal structure, while TM3 is still kinked at Gly113. TM3a undergoes a one-quarter rotation and translation about its axis to open the channel.

One advantage of the EPR studies is that the entire molecule is present, including the N terminus missing in the x-ray structures that is known to interact with lipids. It is clear that bilayer thickness and bending can affect MS channels, although the effects on MscS and MscL are different (Figure 12.41). Computational work has furthered the model of MS proteins using pairs of stiff helices connected by spring-like loops that expand or contract their pores in response to changes in membrane tension.

The helix tilt mechanism of MS channels is only one mechanism for gating of channels formed by α-helices. The nicotinic acetylcholine receptor, a ligand-gated ion channel, opens by rotation of the channel-lining helices to displace bulky residues from the pore (see Chapter 6). And some potassium channels open by bending the inner helices at specific hinge points. Given the wide diversity among MS channels, it is possible that they utilize different mechanisms to open and close in response to membrane tension.

GAP JUNCTION CHANNELS

Gap junctions play roles in signaling, cell adhesion, and cell motility. As the only direct mechanism for cell-to-cell communication they are crucial to development and differentiation and to processes in the nervous, reproductive, and immune systems. A gap junction is an array containing tens to hundreds, even thousands of channels formed by the docking of hemichannels from each cell (Figure 12.42). The hemichannels consist of connexins in vertebrates (innexins in invertebrates) of several types. Humans have 21 connexin (Cx) isoforms with 40% sequence identity, a shared general structure, and distinct physiological roles. Each hemichannel, or connexon, has six Cx subunits that may be of the same or different types, as most cells express more than one. Further complexity results when they dock with another hemichannel, which may be of the same type or a different type. Different connexins respond to different regulatory factors, including pH, Ca^{2+} ion concentration, phosphorylation, and transmembrane voltage, as well as transjunctional membrane potential (the potential difference between the two cells). Connexins also interact with a diverse array of other proteins such as adherins, cytoskeletal proteins, and caveolin, and are associated with lipid rafts.

Early structural studies of these large, three-dimensional structures revealed hexagonal arrays that were named connexons. Higher-resolution EM studies of a purified Cx43 mutant (truncated from its 43-kDa size) showed four TM helices per subunit, 24 per connexon. The first x-ray structure of a 26-kDa connexin is a groundbreaking achievement. The 3.5-Å resolution structure of Cx26 shows the entire protein except for a short cytoplasmic loop (residues 110–124) and the C terminus (residues 218–226). The Cx26 subunit has the expected four TM segments, making a typical four-helix bundle, plus two extracellular loops connected by disulfide bonds and a short N-terminal helix (Figure 12.43A). TM1 is the major pore-lining helix and ends in a 3_{10} helix, TM2 is kinked at Pro87, and TM3 and TM4 protrude into the cytoplasm. The extracellular loops are quite structured,

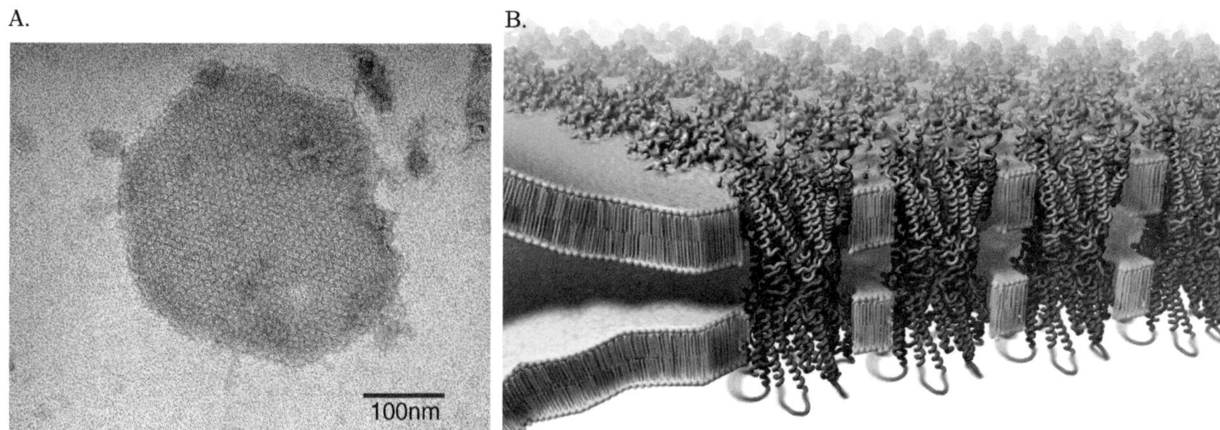

12.42 Gap junctions contain clusters of channels. (A) A reconstituted gap junction prepared with purified Cx26 connexin reveals hexagonal arrays of ~90 Å in the electron micrograph. (B) Illustration of a gap junction shows Cα traces of hemichannels from two adjoining membranes separated by a space of 40 Å. From Nakagawa, S., *et al.*, *Curr Opin Struct Biol.* 2010, 20:423–430. © 2010, Elsevier. Reprinted with permission from Elsevier.

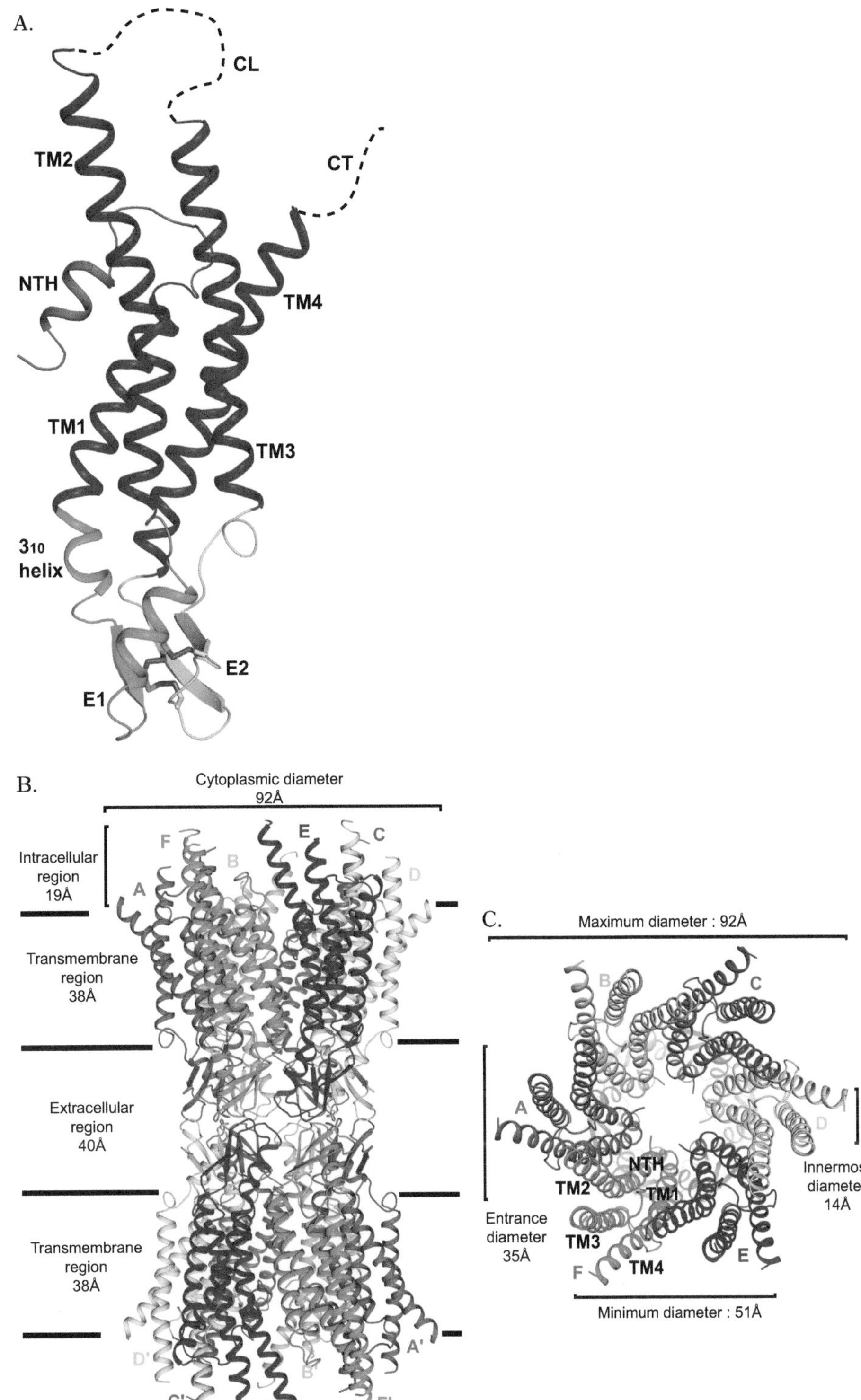

12.43 High-resolution structure of Cx26 connexon. (A) The ribbon representation of a subunit includes TM1–TM4 (blue), the N-terminal helix (red), external loop E1 that connects TM1 and TM2 (green), external loop E2 that connects TM3 and TM4 (yellow) and the disulfide bonds (gray). The dashed lines represent the unresolved cytoplasmic loop (CL) and C terminus (CT). (B) The structure of a Cx26 gap junction channel consists of two hemichannels. In the ribbon representation the corresponding subunits in the two hemichannels are shown in the same color and viewed from the plane of the membrane. (C) The view looking down the Cx26 channel (TM helices only) from the extracellular side shows the channel dimensions at each end, along with the pore diameters. From Maeda, S., *et al.*, *Nature*. 2009, 458:597–602. © 2009, Nature Publishing Group. Reprinted by permission of Macmillan Publishers Ltd.

containing a short α-helix and a β-sheet, and are stabilized by three disulfide bonds between them. All six cysteine residues are essential for function.

Two connexons make a channel of 155 Å in length that spans two membranes separated by 40 Å, where the extracellular loops from two hemichannels interdigitate to overlap by 6 Å (Figure 12.43B). Important hydrogen bonds between highly conserved residues on the two connexons stabilize the tight contacts between hemichannels. Each connexon is larger at the cytoplasmic side than at the extracellular side (Figure 12.43C). In the x-ray structure the pore is constricted to 14 Å at its narrowest; however, spectroscopic studies indicate it is flexible, and gap junction channels generally allow ions and molecules up to 1 kDa to pass. The channel entrance to the pore at the cytoplasmic side contains 11 positively charged residues, making it favor negatively charged solutes, yet the path toward the extracellular side is negatively charged. At the cytoplasmic side is a funnel formed by the short N-terminal helices, stabilized by hydrogen bonds between Asp2 and Thr5 on neighboring subunits and hydrophobic interactions between Trp3 and Met34 with the neighbor on other side.

Each connexin type may have several gating mechanisms. While Cx26 is unusual in that it is not phosphorylated, probably because of its short C terminus, it is gated by voltage. Cx26 channels are closed by inside negative potential. Asp2 in the funnel is involved in gating since the closely related Cx32 has an Asn in that position and has the opposite gating (closed by inside positive). Since protonation of Asp2 would break its hydrogen bond to Thr5, it is likely to disrupt the funnel sufficiently to allow the N-terminal helices to form a plug. Detailed understanding of gating mechanisms will greatly contribute to clarifying the physiological actions of connexins. A huge medical literature describes the role of gap junctions in bone, kidney, retina, brain, heart, and epithelial tissues, along with the connexin mutations linked to deafness, dental disease, cataracts, neuropathy, skin disorders, and heart problems. Cx26 is one of the main connexins involved in deafness, and over 50 mutations in the Cx26 gene lead to inherited hearing impairment. Some of the deafness mutations can now be pinpointed on the structure, such as Met34Thr in the funnel and others at the hemichannel interfaces.

CONCLUSION

From aquaporins to connexins, the structures of channels give elegant examples of the way new details gleaned from structure can help explain function. The variation in size and architecture is striking and perhaps unanticipated in view of the simple concept of a passive pore across the membrane. Also unexpected is the merging of the classifications of transporter and channels seen in the chloride family. The sensitivity to their environment exhibited by some potassium channels, such as TRAAK, and the MS channels that respond to lateral pressure in the membrane is most likely a general characteristic of many membrane proteins, even if to a lesser degree. Finally, every type of channel discussed in this chapter has unique and essential physiological roles, making it clear that structural biology of channels has clinical importance, as well as being crucial to fundamental membrane processes.

FOR FURTHER READING

Papers with an asterisk (*) present original structures.

Aquaporins

Agre, P., Aquaporin water channels (Nobel lecture). *Angew Chem Int Ed*. 2004, 43:4278–4290.

DeGroot, B. L., and H. Grubmuller, The dynamics and energetics of water permeation and proton exclusion in aquaporins. *Curr Opin Struct Biol*. 2005, 15:176–183.

*Fu, D. X., A. M. Libson, L. J. W. Miercke, *et al.*, Structure of a glycerol-conducting channel and the basis for its selectivity. *Science*. 2000, 407:599–605.

Fujiyoshi, Y., K. Mitsuoka, B. L. de Groot, *et al.*, Structure and function of water channels. *Curr Opin Struct Biol*. 2002, 12:509–515.

*Ho, J. D., R. Yeh, A. Sanstrom, *et al.*, Crystal structure of human aquaporin 4 at 1.8 Å and its mechanism of conductance. *Proc Natl Acad Sci USA*. 2009, 106:7437–7442.

*Murata, K., K. Mitsuoka, T. Hirai, *et al.*, Structural determinants of water permeation through aquaporin-1. *Nature*. 2000, 407:599–605.

Stroud, R. M., *et al.*, Glycerol facilitator GlpF and the associated aquaporin family of channels. *Curr Opin Struct Biol*. 2003, 12:424–431.

*Sui, H., B. G. Han, J. K. Lee, *et al.*, Structural basis of water-specific transport through the AQP1 water channel. *Nature*. 2001, 414:872–878.

Potassium Channels

*Alam, A., and Jiang, Y., High-resolution structure of the open NaK channel. *Nature Struct Molec Biol*. 2009, 16:30–34.

*Brohawn S. G., J. del Mármol, and R. MacKinnon, Crystal structure of the human K2P TRAAK, a lipid- and mechano-sensitive K^+ ion channel. *Science*. 2012, 335:436–441.

*Doyle, D. A., J. M. Cabrai, R. A. Pfuetzner, *et al.*, The structure of the potassium channel: molecular basis of K^+ conduction and selectivity. *Science*. 1998, 280:69–77.

Gouaux, E., and R. MacKinnon, Principles of selective ion transport in channels and pumps. *Science*. 2005, 310:1461–1465.

*Jiang, Y., A. Lee, J. Chen, M. Cadene, B. T. Chait, and R. MacKinnon, Crystal structure and mechanism of a calcium-gated potassium channel. *Nature*. 2002, 417:523–526.

*Jiang, Y., A. Lee, J. Chen, *et al.*, X-ray structure of a voltage-dependent K^+ channel. *Nature*. 2003, 423:33–41.

*Long, S. B., E. Campbell, R. Mackinnon, *et al.*, Crystal structure of a mammalian voltage-dependent Shaker family K^+ channel. *Science*. 2005, 309:897–903.

*Long, S. B., X. Tao, E. B. Campbell, and R. MacKinnon, Atomic structure of a voltage-dependent K^+ channel in a lipid membrane-like environment. *Nature*. 2007, 450:376–382.

MacKinnon, R., Potassium channels and the atomic basis of selective ion conduction (Nobel lecture). *Angew Chem Int Ed*. 2004, 43:4265–4277.

Tombola, F., M. M. Pathak, and E. Y. Isacoff, How does voltage open an ion channel?. *Annu Rev Cell Dev Biol*. 2006, 22:23–52.

*Zhou, Y, J. H. Morais-Cabral, A. Kaufman, and R. MacKinnon, Chemistry of ion coordination and hydration revealed by a K^+ channel–Fab complex at 2.0 Å resolution. *Nature*. 2001, 414:43–48.

CLC Family of "Chloride Channels"

Accardi, A., and C. Miller, Secondary active transport mediated by a prokaryotic homologue of ClC Cl^- channels. *Nature*. 2004, 427:803–807.

Accardi, A., and A. Picollo, CLC channels and transporters: proteins with borderline personalities. *Biochim Biophys Acta*. 2010, 1798:1457–1464.

*Dutzler, R., E. Campbell, M. Cadene, *et al.*, X-ray structure of a ClC chloride channel at 3.0 Å reveals the molecular basis of anion selectivity. *Nature*. 2002, 415:287–294.

Dutzler, R., E. B. Campbell, and R. MacKinnon, Gating the selectivity filter in ClC chloride channels. *Science*. 2003, 300:108–112.

*Feng, L., E. B. Campbell, Y. Hsiung, *et al.*, Structure of a eukaryotic CLC transporter defines an intermediate state in the transport cycle. *Science*. 2010, 330:635–641.

Lisal, J., and M. Maduke, Proton-coupled gating in chloride channels. *Phil Trans R Soc B*. 2009, 364:181–187.

Matulef, K., and M. Maduke, The CLC "chloride channel" family: revelations from prokaryotes. *Molec Membr Biol*. 2007, 24:342–350.

Miller, C., ClC chloride channels viewed through a transporter lens. *Nature*. 2006, 440:484–489.

Mechanosensitive Channels

*Bass, R. B., P. Strop, M. Barclay, and D. C. Rees, Crystal structure of *Escherichia coli* MscS, a voltage-modulated and mechanosensitive channel. *Science*. 2002, 298:1582–1587.

*Chang, G., R. H. Spencer, A. T. Lee, *et al.*, Structure of the MscL homolog from *Mycobacterium tuberculosis*: a gated mechanosensitive ion channel. *Science*. 1998, 282:2220–2226.

Haswell, E. S., R. Phillips, and D. C. Rees, Mechanosensitive channels: what can they do and how do they do it?. *Structure*. 2011, 19:1356–1369.

Naismith, J. H., and I. R. Booth, Bacterial mechanosensitive channels – MscC: evolution's solution to creating sensitivity in function. *Ann Rev Biophys*. 2012, 41:157–177.

Peerozo, E., and D. C. Rees, Structure and mechanism in prokaryotic mechanosensitive channels. *Curr Opin Struct Biol*. 2003, 13:432–442.

Sukharev, S., and A. Anishkin, Mechanosensitive channels: what can we learn from "simple" model systems? *Trends Neurosci*. 2004, 27:345–351.

*Wang, W., S. S. Black, M. D. Edwards, *et al.*, The structure of an open form of the *E. coli* mechanosensitive channel at 3.45 Å resolution. *Science*. 2008, 321:1179–1183.

Gap junctions

Maeda, S., and T. Tsukihara, Structure of the gap junction channel and its implications for biological functions. *Cell Molec Life Sciences*. 2011, 68:1115–1129.

*Maeda, S., S. Nakagawa, M. Suga, *et al.*, Structure of the connexin 26 gap junction channel at 3.5 Å resolution. *Nature*. 2009, 458:597–602.

Nakagawa, S., S. Maeda, and T. Tsukihara, Structural and functional studies of gap junction channels. *Curr Op in Struct Biol*. 2010, 20:423–430.

Wei, C. J., X. Xu, and C. W. Lo, Connexins and cell signaling in development and disease. *Ann Rev Cell Dev Biol*. 2004, 20:811–838.

ELECTRON TRANSPORT AND ENERGY TRANSDUCTION

13

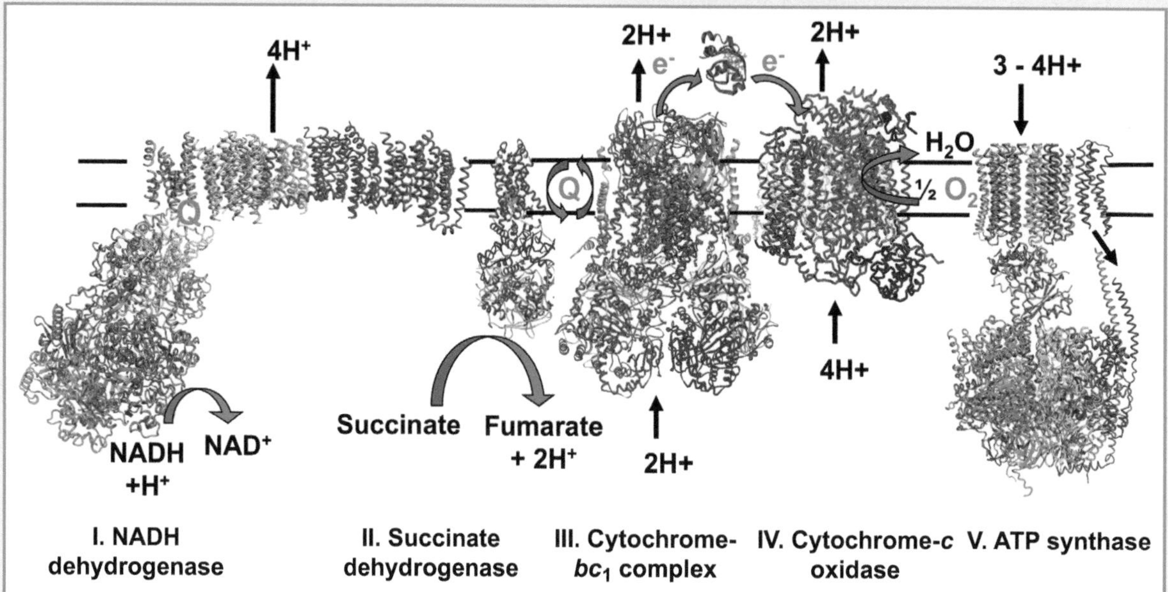

The components of the respiratory chain in mitochondria. High-resolution structures give a detailed picture of the electron transport chain complexes I to IV and the ATP synthase or complex V (left to right). Electrons enter the chain on NADH or succinate and transfer through the membrane as reduced quinone (Q) to complex III. The peripheral protein cytochrome *c* carries them to complex IV, which reduces O_2 to H_2O. Complexes I, III, and IV couple electron transfer to the export of protons, creating a proton gradient that drives the synthesis of ATP. The complexes are truly molecular machines: They range up to nearly one million Daltons in mass and utilize different mechanisms to harvest the energy of the reducing power. Kindly provided by Ferguson-Miller, S., *et al.*, based on *Annu Rev Biochem*. 2006, 75:165–168. © 2006 by Annual Reviews. Reprinted with permission from the Annual Review of Biochemistry, www.annualreviews.org.

Energy transduction typically occurs in large, supramolecular organizations of protein and lipid components – nanomachines in the membrane. Supercomplexes of respiratory enzymes, called respirasomes, have been observed by native polyacrylamide gel electrophoresis and cryo-EM and likely exist to boost the efficiency of oxidative pathways while minimizing formation of reactive oxygen species. The components of the respiratory chain include dimers that each contain many subunits, such as cytochrome bc_1 oxidase, and very large enzymes composed of many subunits, such as ATP synthase.

The large size is not essential for energy transduction. Chapter 9 describes an anaerobic pathway for coupling electron transport to proton translocation, the nitrate respiratory pathway that consists of nitrate reductase and formate dehydrogenase (FDH). The high-resolution structure of FDH, a dimer of heterotrimers, elucidates an electron path involving a chain of four Fe-S centers and two heme prosthetic groups terminating in the reduction of menaquinone. Although smaller than the complexes described in this chapter (the FDH heterotrimer is 255 kDa), these enzymes accomplish the same goal: generating a proton gradient that drives the ATP synthase.

The respiratory chains in mitochondrial membranes of eukaryotes and prokaryotic cell membranes comprise a sequence of large complexes of enzymes and redox cofactors that eject protons as they transfer electrons to carriers of higher reduction potentials. The proton gradient thus created powers the synthesis of ATP along with secondary transport processes, as proposed in the chemiosmotic theory by Peter Mitchell, the recipient of the 1978 Nobel Prize in Chemistry. A huge body of literature describes the composition and function of the respiratory complexes. Structural biology is now illuminating details of their architecture that reveal different mechanisms for coupling electron transfer with proton translocation.

COMPLEXES OF THE RESPIRATORY CHAIN

The respiratory complexes utilize the electrons extracted from various reduced metabolites and transport them via coupled redox carriers to terminal acceptors, usually O_2 in aerobic organisms. The familiar electron transport chain consists of complex I, which transfers electrons from NADH to ubiquinone; complex II, which transfers electrons from the $FADH_2$ of succinate dehydrogenase to ubiquinone; complex III, which transfers electrons from ubiquinol to cytochrome *c*; and complex IV, which transfers electrons from cytochrome *c* to O_2, picking up protons to reduce oxygen to water (see Frontispiece). Additional enzymes feed electrons from other sources such as acyl-CoA into the chain, reducing ubiquinone to ubiquinol, so they also provide the substrate for complex III. Three of the complexes (I, III, and IV) pump protons across the membrane to create a TM proton electrochemical potential gradient, sometimes called the proton motive force (pmf). If the system is "uncoupled" by the action of compounds that allow protons to leak back across the membrane, electron transfer proceeds without creating or maintaining a proton gradient and the energy is dissipated as heat.

When physically separated by membrane solubilization and ion exchange chromatography in detergents, the respiratory complexes can retain their functions *in vitro*. This property has allowed their extensive characterization by kinetic, spectroscopic, and electrochemical techniques over decades of research. As structures of respiratory complexes isolated from both eukaryotic and prokaryotic organisms became available, the field moved toward a detailed understanding of the principles of redox-driven proton transfer. This section focuses on describing that transfer process in light of the high-resolution structures of complexes I, III, and IV.

COMPLEX I

Electrons from NADH enter the respiratory chain at complex I, which pumps four protons per electron pair to contribute ~40% of the proton flux that drives the ATP synthase. The eukaryotic complex I is huge, with 44 subunits and a total mass of 980 kDa. The prokaryotic complex I has 14 "core" subunits that are conserved from bacteria to humans, with a total mass of ~550 kDa. The shape of complex I, first revealed by EM, is bipartite: One lobe of its L-shape corresponds to the membrane domain and the other is a hydrophilic domain that protrudes into the mitochondrial matrix or the bacterial cytoplasm. In addition to this architectural similarity, the eukaryotic and prokaryotic complex I have equivalent redox components and a high degree of sequence conservation in the core subunits, thus they are presumed to share a mechanism for coupling proton pumping to the flow of electrons. The question of whether this coupling is direct, with the redox pathway driving the proton translocation, or indirect, with conformational changes that mediate the two processes, has finally been answered with details from a high-resolution structure of the entire prokaryotic complex.

This tremendous achievement was attained after several advances in the structural biology of complex I. First came the 3.3-Å resolution x-ray structure of the hydrophilic domain of the complex from *Thermus thermophilus*, followed by low-resolution x-ray structures of the entire complex from *T. thermophilus* and yeast mitochondria. The structure of the *E. coli* membrane domain was solved at 3.9-Å resolution and then 3.0-Å resolution, and finally the structure of the entire 536-kDa *T. thermophilus* complex was determined at 3.3-Å resolution. (Refer to Table 13.1 to compare the composition in the two organisms.)

In complex I a covalently bound flavin cofactor, FMN, accepts two electrons as a hydride from NADH, then passes single electrons to Fe-S centers, which pass them along to a quinone. The details of the eight subunits of the hydrophilic domain of *T. thermophilus* reveal the electron transfer pathway from NADH to FMN, then through seven conserved Fe-S clusters to the quinone-binding site, or Q-site (Figure 13.1). The pathway for electrons is NADH–FMN–N3–N1b–N4–N5–N6a–N6b–N2–quinone (menaquinone in *T. thermophilus*), with a change in reduction potential from –320 mV (NADH) to –80 mV (menaquinone). In addition, N1a can temporarily accept an electron from FMN and pass it back to N3 (via FMN) to minimize the lifetime of the flavin free radical, thus acting as an antioxidant. The ninth Fe-S center, N7, which occurs in some bacteria, appears to be off the pathway. The structure shows a solvent-exposed binding site for FMN next to a cavity large enough to bind NADH and at the top of a chain of Fe–S clusters (Figure 13.2). The last Fe–S center in the chain, N2, is a high-potential cluster that donates electrons to the quinone. Crystal structures of reduced complex I show small but significant changes that occur upon binding NADH, which are propagated to helices at the interface with the membrane domain, in particular helices H1 and H2 of Nqo6 and a four-helix bundle of Nqo4.

The high-resolution crystal structure of the *E. coli* membrane domain of complex I, achieved with SeMet-derived protein, allowed assignment of 1952 residues

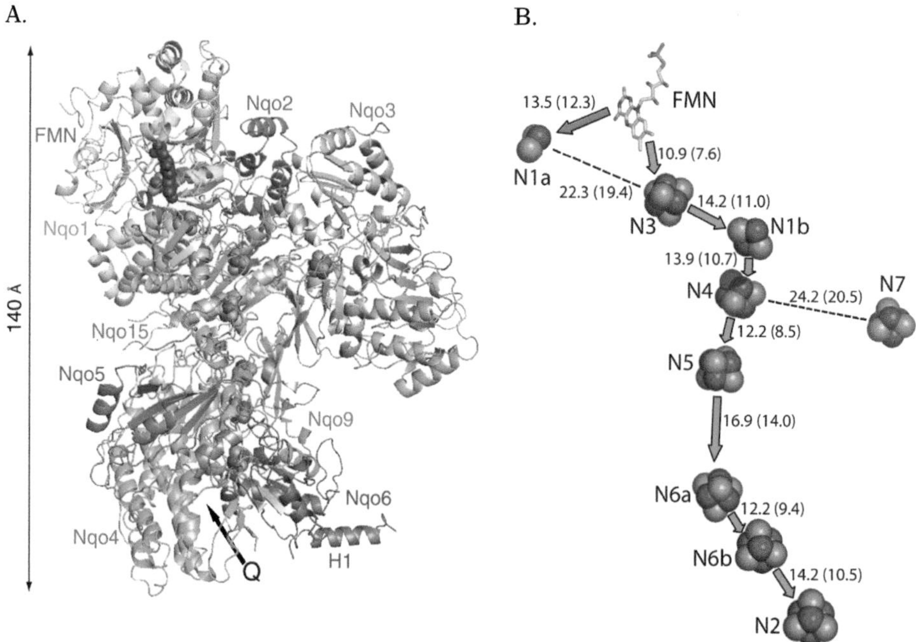

13.1 Structure of the hydrophilic domain of complex I from *T. thermophilus*. (A) The view from the side extending up from the membrane shows the eight subunits in different colors, along with FMN (magenta spheres) and Fe–S clusters (Fe atoms, red spheres; S atoms, yellow spheres). The arrow indicates a possible quinone-binding site (Q). (B) The redox centers are tilted slightly from the view in (A). The main pathway of electron transfer (blue arrows), and the diversion to N1a (green arrows) show the short distances between centers (calculated center-to-center, with edge-to-edge distances shown in parantheses). From Sazanov, L. A., and P. Hinchliffe, *Science*. 2006, 311:1430–1436. © 2006, American Association for the Advancement of Science. Reprinted with permission from AAAS.

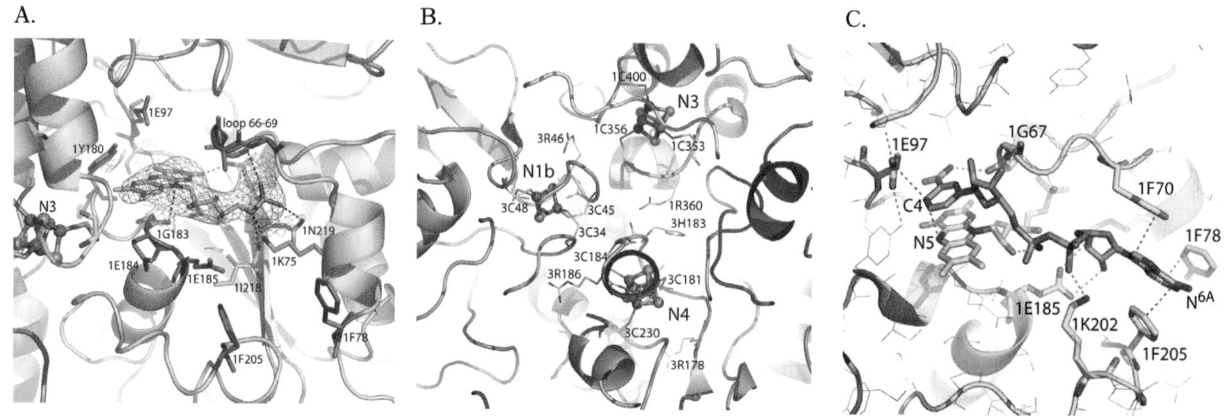

13.2 Cofactor binding sites in the hydrophilic domain of complex I from *T. thermophilus*. Binding sites of different cofactors are shown in close-up views of subunits colored as in Figure 13.1, with residues in their environments named using prefixes to indicate the subunit number. (A) FMN (stick, with electron density contours) binds at the bottom of a cavity (viewed from the solvent-exposed side) to residues (stick representation with carbon in yellow) in Nqo1. Residues likely to be involved in NADH binding (stick representation with carbon in magenta) are also highlighted. Hydrogen bonds are indicated by dotted lines. Cluster N3 is visible to the left. (B) Binding sites of Fe–S clustered named N3, N1b, and N4 are shown on subunits Nqo1 and Nqo3 (yellow and pink, respectively). N3 is a [4Fe–4S] cluster coordinated by Cys354, Cys356, Cys359, and Cys400 from loops between helices of Nqo1. N1b is a [2Fe–2S] cluster coordinated by Cys34, Cys45, Cys48, and Cys83 in a ferredoxin-like fold, while N4 is a [4Fe–4S] cluster coordinated by Cys181, Cys184, Cys187, and Cys230 in the largest subunit, Nqo3. Both (A) and (B) are from Sazanov, L. A., and P. Hinchliffe, *Science*. 2006, 311:1430–1436. © 2006, American Association for the Advancement of Science. Reprinted with permission from AAAS. (C) A view of the NADH-binding site from the solvent-exposed side reveals FMN and residues binding NADH (sticks with carbon in yellow). NADH (sticks with carbon in salmon) is bound as seen in reduced Complex I crystals. Hydrogen bonds (green dotted lines) and stacking interactions (gray dotted lines) exist all along the length of the nucleotide. The path of hydride transfer from the nicotinamide ring to the flavin ring is indicated (red dotted line). From Berrisford, J. M., and L. A. Sazanov, *J Biol Chem*, 2009, 284:29773–29783. © 2009, by the American Society for Biochemistry and Molecular Biology.

TABLE 13.1 NOMENCLATURE OF SUBUNITS IN RESPIRATORY COMPLEX I IN *T. THERMOPHILUS** AND *E. COLI*

Nqo1	Nqo2	Nqo3	Nqo4	Nqo5	Nqo6	**Nqo7**	**Nqo8**	Nqo9	**Nqo10**	**Nqo11**	**Nqo12**	**Nqo13**	**Nqo14**
NuoF	NuoE	NuoG	NuoD/C#	NuoC/D#	NuoB	**NuoA**	**NuoH**	NuoI	**NuoJ**	**NuoK**	**NuoL**	**NuoM**	**NuoN**

Membrane subunits are in bold.
* *T. thermophilus* also has a unique subunit, Nqo15.
# Portions of NuoC and NuoD are fused to make the corresponding *T. thermophilus* subunits.

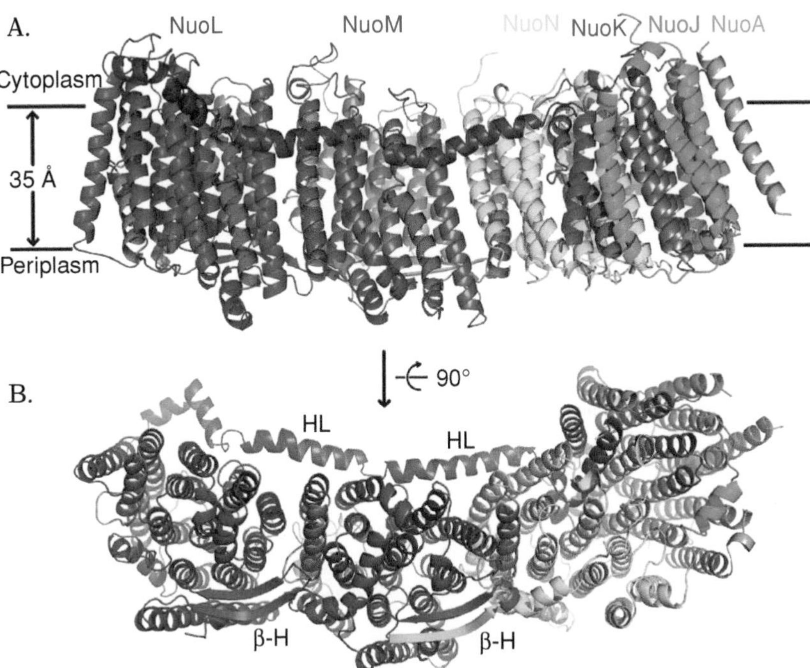

13.3 The high-resolution structure of the membrane domain of *E. coli* complex I. (A) The view from the membrane plane shows 55 TM helices, from six membrane subunits (colored as labeled), highlighting the discontinuous helices TM7 (red) and TM12 (orange). The position of the lipid bilayer (arrow at left) is estimated from the positions of surface-exposed Tyr and Trp residues, as well as the hydrophobic residues in between. (B) The view from the periplasm in the same subunit colors reveals the connecting elements. The lateral helix HL (purple) interacts with residues on NuoL, NuoM, and NuoN along the top. The connecting *β*-H element (purple and blue) consists of *β*-hairpins and connecting helices and runs along the bottom. Helices involved in conformational changes include TM7, TM12, and TM8 (red, orange, and yellow, respectively) in NuoL, M, and N (see text). One helix of NuoJ (green) contacts NuoN (yellow). From Efremov, R. G., and L. A. Sazanov, *Nature*. 2011, 476:414–420. © Nature Publishing Group. Reprinted by permission of Macmillan Publishers Ltd.

(out of 2038) of the six subunits NuoL, M, N, A, J, and K (see Table 13.1). It lacks NuoH, which diffuses readily from the complex. The 222-kDa complex has 55 TM helices and a length of ~160 Å along the plane of the membrane (Figure 13.3). As expected, it does not contain any redox cofactors. It is stabilized by helix HL, a C-terminal extension of NuoL that runs almost the entire length of the domain, and on the opposite side by two *β*-hairpins between amphipathic helices, called a *β*-hairpin-helix (*β*H) element (Figure 13.3B). HL is positioned to be a mechanical link, with a proposed role as the piston driving the conformational change as discussed below.

Helix HL connects the three largest subunits, NuoL, M, and N, each with 14 TM helices and sequence homology with the Na^+/H^+ antiporter (Figure 13.4). Each has an inverted symmetry of TM4–8 and TM9–13 (see Figure 13.4B), and they form two half-channels with face-to-back contact. The two broken helices TM7 and TM12, which are not close in the structure, are each interrupted in the middle of the bilayer by a loop of 5–7 residues. At the tip of each loop is a conserved Pro residue. TM8 is also partly unwound in the middle with a kink (lacking Pro). The discontinuity of the helices adds flexibility and charge that are likely involved in proton transport. There are several essential charged residues in this central region, including lysine–glutamate pairs whose side chains point into putative proton channels (see below). In all three subunits TM7 interacts with

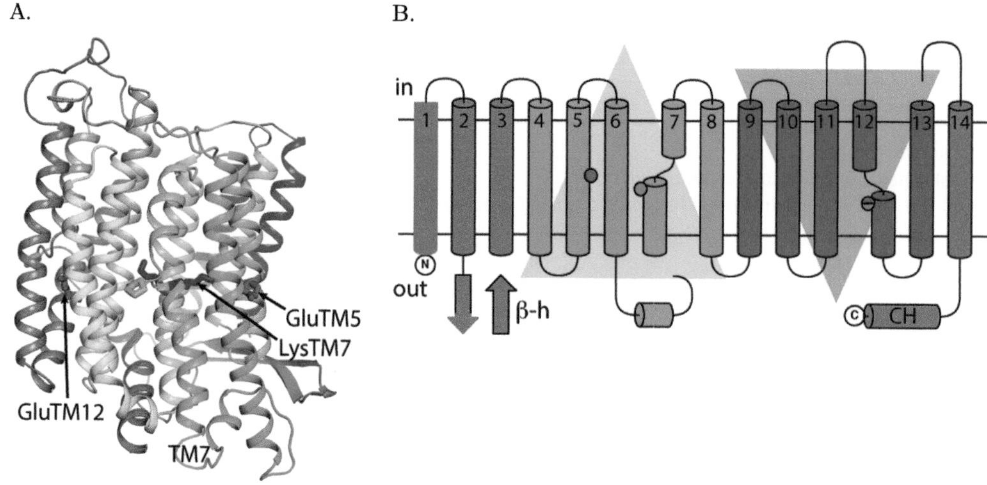

13.4 Structure of antiporter-like subunits, NuoL, M, and N. (A) The fold of the antiporters-like subunits is illustrated in a side view of NuoM (blue to red from N to C terminus), with the cytoplasmic side up. Essential charged residues (sticks) include Glu in TM5 and Lys in TM7 in half-channel 1, and Glu in TM12 in half-channel 2, as well as the conserved Pro residues in intra-helical loops. (B) The relation of the two symmetry-related domains (green and magenta), which form half-channels, is indicated on the topology diagram for the antiporter-like subunits. The location of crucial charged residues in the primary structure are indicated: Glu in TM5 (red sphere), Lys in TM7 (blue sphere), and Lys/Glu in TM12 (red/blue sphere). From Efremov, R. G., and L. A. Sazanov, *Nature*. 2011, 476:414–420. © 2011, Nature Publishing Group. Reprinted by permission of Macmillan Publishers Ltd.

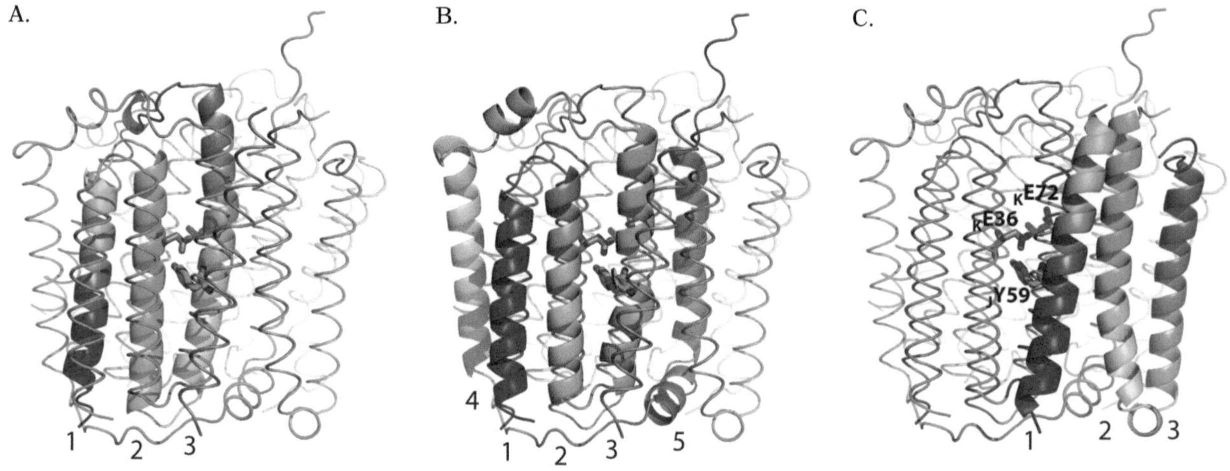

13.5 Structure of NuoK, J, and A. Subunits NuoK, J, and A are integrated into an 11-helical bundle, traced here in views in the membrane plane with the cytoplasmic side up and conserved essential residues (sticks) highlighted. (A) NuoK (ribbon representation, blue to red from N to C terminus) has three TM helices in a linear array, with Glu36 and Glu72 at the center of the bilayer. (B) NuoJ (same representation) has three linearly arranged N-terminal helices that border NuoK and two more TM helices on the opposite side of the domain. Tyr59 is also near the center of the bilayer. (C) NuoA (same representation) has three TM helices that interact with NuoK, J, and N and probably with NuoH, which is missing from the structure. From Efremov, R. G., and L. A. Sazanov, *Nature*. 2011, 476:414–420. © 2011, Nature Publishing Group. Reprinted by permission of Macmillan Publishers Ltd.

helix HL, which is unwound near TM7 of NuoM, allowing interactions via backbone and side chains, while hydrogen bonds are formed between HL and TM7 of NuoL and N.

Stabilization by helix HL and the βH element is important because the interactions between NuoL and M and NuoM and N are not extensive and are mostly hydrophobic. Helix HL also anchors them tightly to NuoJ and K, which are part of the 11-helix bundle at the end of the membrane domain (see Figure 13.5). NuoK has two conserved Glu residues, and NuoJ has a conserved Tyr near a π-bulge in TM3, all three implicated in proton transport (see below).

The 3.3-Å resolution structure of the entire complex I from *T. thermophilus* shed light on the interface between the hydrophilic domain (nine subunits) and the membrane domain (seven subunits). The membrane domain consists of Nqo12, 13, 14, 10, 11, 7, and 8, with a total of 64 TM helices visible in the structure (Figure 13.6). Sequence conservation is high for the membrane core of all subunits and poor for helix HL, except at its critical contacts with other subunits. The three antiporter

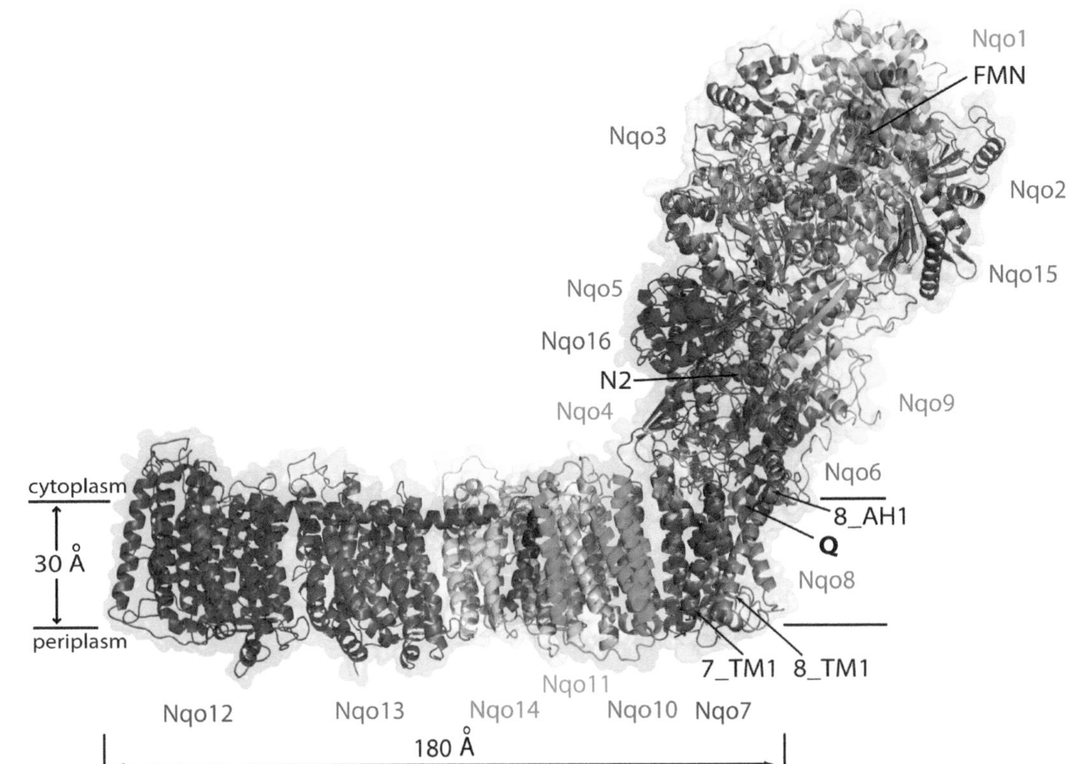

13.6 The high-resolution structure of the entire complex I from *T. thermophilus*. A view from the membrane plane reveals how the hydrophilic domain connects to the membrane domain of complex I. At the interface, TM1 of Nqo8 (orange) interacts with TM1 of Nqo7 (blue) in the membrane and amphipathic helix 1 of Nqo8 interacts with Nqo4, 6, and 9 (dark green, red, and cyan) in the hydrophilic domain. The antiporter-like subunits (Nqo12, magenta; Nqo13, blue; and Nqo14, yellow) are very similar to the corresponding subunits in *E. coli*, and are also spanned by the lateral helix HL. The quinone-binding site (Q) is indicated. From Baradaran, R., *et al.*, *Nature*. 2013, 494:443–448. © 2013, Nature Publishing Group. Reprinted by permission of Macmillan Publishers Ltd.

subunits, Nqo12, 13, and 14, have similar structures, key residues, and links to helix HL already seen in *E. coli* complex I. Previously unseen in crystal structures, Nqo8 has nine TM helices and also interacts closely with TM1 from Nqo7 (NuoA). At the interface between domains, Nqo8 forms many salt bridges and hydrogen bonds to the hydrophilic subunits Nqo4, 6, and 9. Interactions with the membrane domain involve extensive contact with Nqo7, including a cytoplasmic loop that wraps around Nqo8 and contacts made by a well-conserved amphipathic helix in Loop 1 of Nqo8.

The quinone-binding site lies at the interface of Nqo8 with Nqo4 and Nqo6, and contains charged and polar residues on Loop 3, another well-conserved loop of Nqo8. Soaking quinone analogs into the crystals reveal them binding about 15 Å from the membrane surface, putting the headgroup ~12 Å from the Fe–S cluster N2 that donates electrons to the quinone and the tail in a 30-Å long enclosed chamber. The unexpected distance between this site and the membrane bilayer requires a significant movement of the quinone out of the membrane. Very similar features are seen in the 6.3-Å resolution structure of 40 subunits of the massive mitochondrial complex I from the yeast *Yarrowia lipolytica* (chosen because *S. cerevisiae* lacks complex I). The long distances rule out direct coupling from the redox chain to proton transport sites across the membrane and suggest the lateral helix HL could be a critical transmission element in conformational coupling.

Conformational Coupling Mechanism

To deduce the mechanism for coupling of electron transport with proton pumping from the structure of complex I requires identification of the proton channels. Proton pathways are visible in each of the three antiporter-like subunits, although none fully traverse the membrane (Figure 13.7). Rather, each subunit has two half-closed symmetry-related channels. The channel formed by a cavity surrounding the Lys in TM7 is closed from the periplasm by large hydrophobic residues, while the channel near Lys/Glu in TM12 is closed to the cytoplasm and accessible to the periplasm via short polar networks. The two half-channels are connected by conserved charged and polar residues in the middle of the membrane. All water molecules observed in the structure are located in either the channels or the connection. A fourth proton pathway, called the E-channel

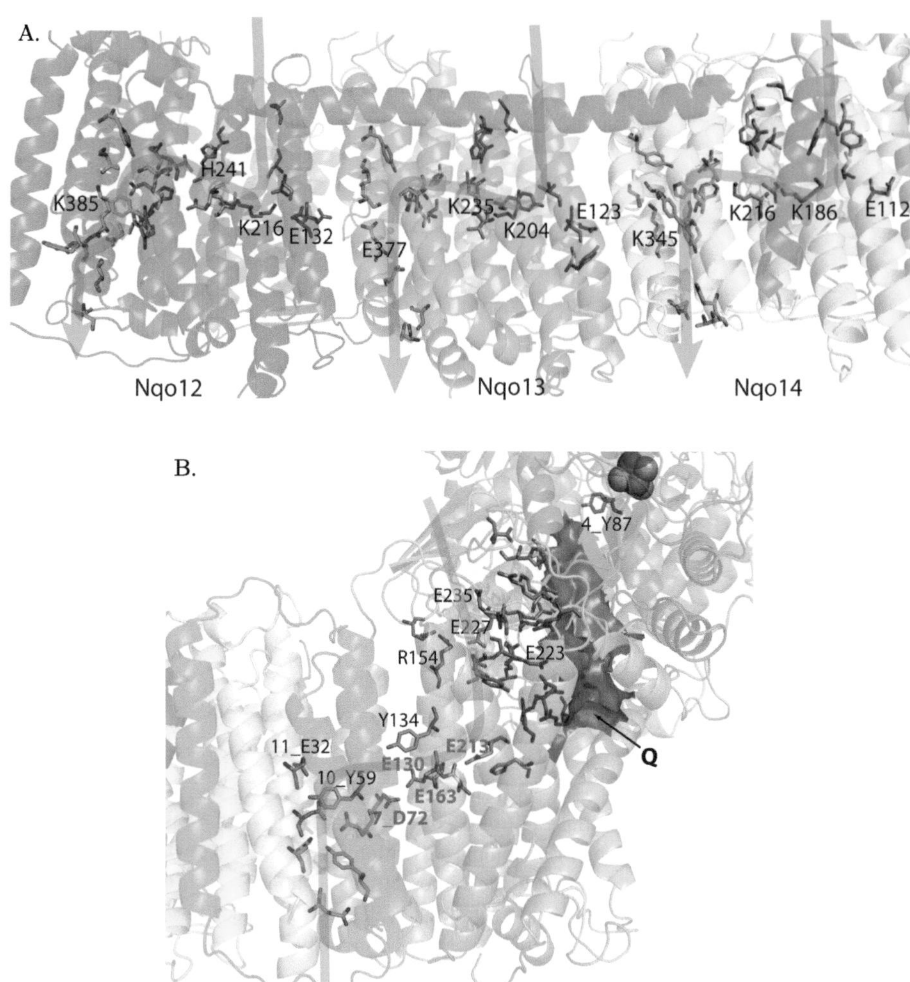

13.7 Proton translocation channels in the membrane domain of complex I. **(A)** Putative proton channels (blue arrows) in the antiporter-like subunits (equivalent to NqoL, M, N in *E. coli*) are illustrated in portions of the subunits (colored as in Figure 13.6) viewed from the plane of the membrane with the cytoplasm on top. The two half-channels of each proton path are clear. Essential charged residues (sticks, carbon in dark blue for the first half channel, green for the second half channel, and orange for connecting residues) are implicated in proton transport. A few water molecules (red spheres) can be seen along the proton paths. From Efremov, R. G., and Sazanov, L. A., *Nature*, 2011, 476:414–420. **(B)** The fourth proton channel also consists of two half-channels formed by three of the membrane subunits, Nqo 8, 10 and 11 (Nqo H, J, K) . A network of charged residues connects this site to the quinone-binding site (Q). From Baradaran, R., *et al.*, *Nature* 2013, 494:443–448. © 2013, Nature Publishing Group. Reprinted by permission of Macmillan Publishers Ltd.

for the abundance of Glu residues, is formed by a half-channel from Nqo8 and another half-channel formed by Nqo10 and Nqo11 (Figure 13.7B). A network of polar residues connects the E-channel to both the Q site and the cytoplasm. This network is further continued all the way to the tip of the membrane domain as a remarkable central hydrophilic axis of charged and polar residues. It is likely to be flexible, as most key residues sit on the breaks in TM helices.

How can conformational changes in the hydrophilic domain that occur upon reduction be transmitted to the membrane domain of complex I to open the proton channels? The structure suggests a beautiful mechanism. First, given the closed cavity in which it binds, quinone upon reduction can only pick up protons from nearby charged residues, affecting key electrostatic interactions that trigger conformational changes. Conformational changes are initiated at the interface to Nuo8, 7, 10, and 11 (NuoH, A, J, and K), and may then be relayed to Nuo12, 13, and 14 (NuoL, M, and N) either by helix HL or by the central hydrophilic axis. Helix HL would coordinate with the βH element to control the conformational changes in the half-channels: specifically, HL would drive TM7 in the first half and βH would drive TM12 in the second half. Helix HL (or the central axis, or both) could thus be the piston (coupling rod) that

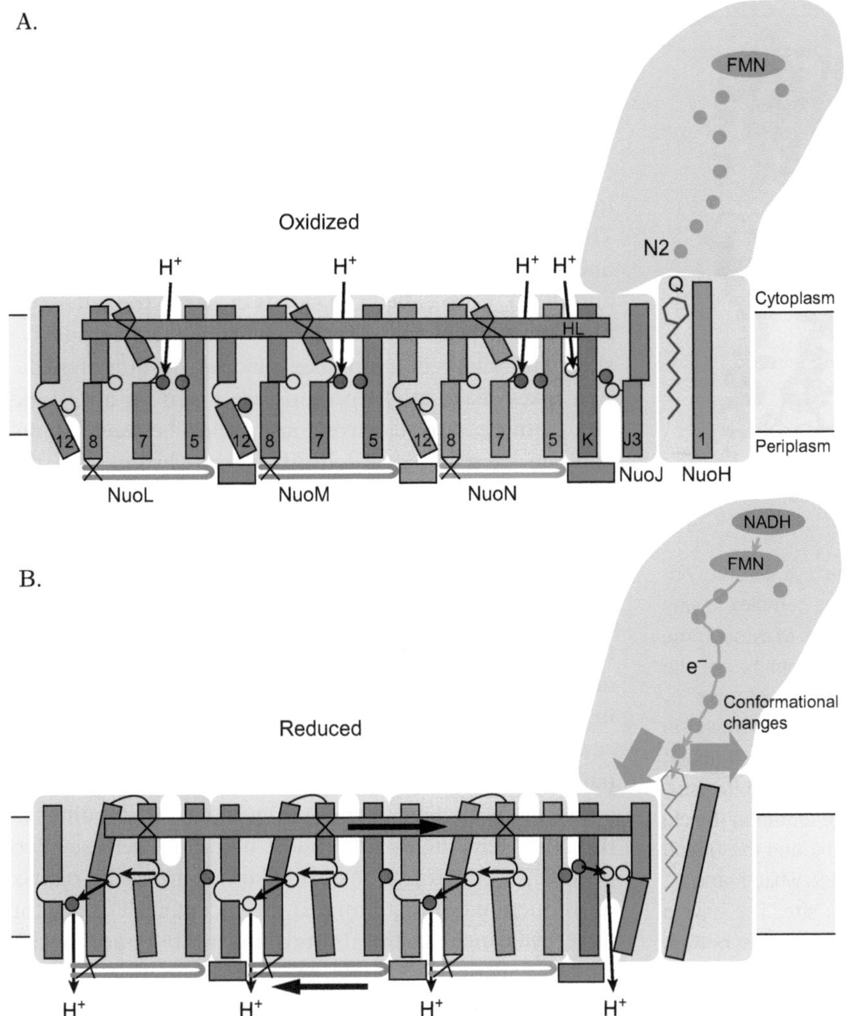

13.8 Proposed mechanism for conformational coupling in complex I. A cartoon of the conformational coupling between electron transfer and proton translocation shows the oxidized state (A) and reduced state (B). TM helices with important roles are numbered in (A), with helix HL along the top (cytoplasmic side) and the βH element along the bottom (periplasmic side). The cartoon shows the following crucial charged residues: Glu in TM5, Lys in TM7, and Lys/His in TM8 in NuoL, M, and N, and Glu72 and Glu36 in NuoK with Tyr59 in NuoJ (protonated Glu, red circles; protonated Lys or His, blue circles; deprotonated residues empty circles). In the first half-channel Lys in TM7 is assumed protonated in the oxidized state. Upon reduction it donates its proton to the connecting Lys/His in TM8, then to Lys/Glu in TM12 at the second half-channel. Lys/Glu in TM12 ejects its proton into the periplasm when the complex returns to the oxidized state. The fourth proton is translocated at the interface of NuoN, K, and J. From Efremov, R. G., and L. A. Sazanov, *Nature*. 2011, 476:414–420. © 2011, Nature Publishing Group. Reprinted by permission of Macmillan Publishers Ltd.

drives the concerted movement coupled to the flow of electrons through the hydrophilic domain (Figure 13.8). The essential charged Lys and Glu residues in the antiporter-like subunits would cycle through protonation/deprotonation as follows: In the oxidized state the Lys in TM7 is protonated, stabilized by the nearby Glu in TM5, and the Lys/Glu in TM12 is deprotonated. Upon reduction conformational changes move the Glu in TM5 away from the Lys in TM7; then the Lys in TM7 would donate its proton to residues between the two half-channels and eventually to Lys/Glu in TM12. When the complex returns to an oxidized state, Glu in TM5 moves back, allowing the Lys in TM7 to pick up a proton from the cytoplasm, while Lys/Glu in TM12 ejects its proton into the periplasm. With each cycle the three subunits translocate three protons from the cytoplasm to the periplasm. The conformational changes occurring in the redox cycle would also change the environment of TM3 in Nqo10 (NuoJ), allowing protonation and deprotonation of Glu36 to pass the fourth proton. In this way the connecting elements allow the redox energy from the hydrophilic domain to power the movement of six symmetrical structural domains plus one asymmetrical domain in this veritable "steam engine of the cell."

CYTOCHROME BC_1

Complex III is cytochrome bc_1, or ubiquinol-cytochrome *c* reductase, which catalyzes electron transfer from ubiquinol to cytochrome *c* coupled to the translocation of two protons across the membrane per electron transferred through the complex. It is a dimeric multimer, in which each half contains up to 11 protein subunits. Three of the subunits form the catalytic core: They contain the redox centers and make up the functional unit (Figure 13.9). This core contains both *b*- and *c*-type cytochrome subunits: Cytochrome *b* has two *b*-type hemes, called b_H and b_L, and cytochrome c_1 has a *c*-type heme. (*a*-, *b*-, and *c*-type hemes differ in their substituents off the tetrapyrrole ring and their linkages to the proteins.) In addition to the two cytochromes, the catalytic core has an iron–sulfur protein subunit of the Rieske type, a [2Fe–2S] cluster in which one Fe is coordinated by two histidine

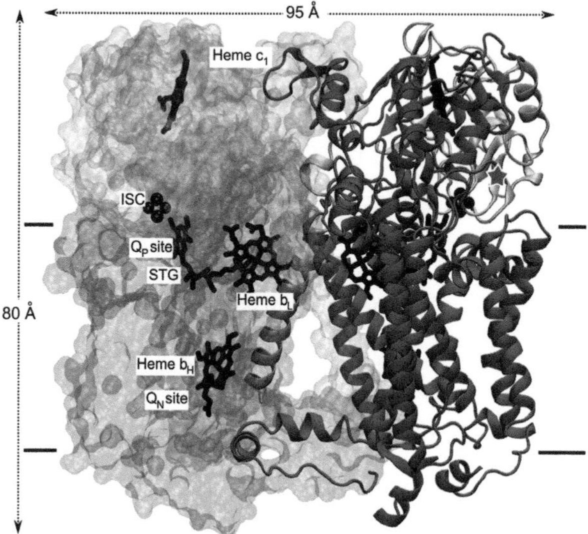

13.9 The 2.7-Å resolution structure of the cytochrome bc_1 complex from *P. denitrificans*. The *P. denitrificans* complex III was crystallized after deletion of the acidic N terminus of cytochrome c_1 that is specific to this organism. Its three subunits are the conserved catalytic core: cytochrome *b* (red), cytochrome c_1 (blue), and the Rieske Fe–S protein (ISC, yellow) portrayed in a transparent surface representation (left protomer) and ribbon diagram (right protomer). Cofactors and the bound inhibitor stigmatellin (STG) are shown in stick-and-ball representation (black) and labeled on the left protomer only. Heme b_H and the bound ubiquinone are at the Q_n site (also called Q_i site), while heme b_L, bound ubiquinol and STG define the Q_p (or Q_o) site. The view is from the bilayer plane with the cytoplasmic side at the bottom with black horizontal lines to indicate the position of the membrane. The N terminus of cytochrome c_1 is marked (blue asterisk). From Kleinschroth, T., *et al.*, *BBA*. 2011, 1807:1606–1615. © 2011, Elsevier. Reprinted with permission from Elsevier.

residues and the other by two cysteine residues. The redox sites in these cytochrome *b* subunits are buried in the membrane interior, and the 2Fe–2S cluster impinges on the hydrophobic domain because the quinone substrates are very hydrophobic, with isoprenoid side chains 30–50 carbons in length. In addition, redox reactions of quinones involve a semiquinone intermediate that needs to be shielded from the aqueous environment to avoid formation of damaging reactive oxygen species.

The Q Cycle

In the overall reaction carried out by cytochrome bc_1, two electrons from ubiquinol (ubihydroquinone) reduce two molecules of cytochrome *c* with uptake of two protons from inside (mitochondrial matrix or bacterial cytosol) and release of four protons to the outside. Unlike other proteins that pump protons, such as complexes I and IV (and bacteriorhodopsin, see Chapter 5), the cytochrome bc_1 complex does not provide a direct proton path across the membrane. Instead, it uses a special mechanism called the Q cycle. The Q cycle requires two spatially separated quinone-binding sites in the cytochrome *b* subunit that lead to two electron transfer paths: a high-potential chain consisting of the [2Fe–2S] cluster and cytochrome c_1 and terminating at cytochrome *c*; and a low-potential chain containing the two *b*-type hemes. The Q_n (also called Q_i) site is closer to the inner surface and, with heme b_H, forms the N center (for negative side), and the Q_p (also called Q_o) side is closer to the outer surface and with heme b_L forms the P center (for positive side). Quinone reduction takes place at Q_n, quinol oxidation takes place at Q_p, protons are taken up at Q_n and exit from Q_p, and electrons are cycled between quinol and the hemes before delivery to cytochrome *c* at the outer surface (Figure 13.10). In brief, the first ubiquinol binds to the P center and is oxidized, with its two electrons taking divergent paths. One electron is transferred to the Rieske Fe/S center, and from there to cytochrome c_1 and then to cytochrome *c*. The other is transferred to the b_L heme, and from it to the b_H heme and then to a ubiquinone that binds to the Q_n site at the N center, making ubisemiquinone. A second ubiquinol binds to the P center, and again the two electrons follow these different paths, resulting in the reduction of a second cytochrome *c* and the reduction of the semiquinone to ubiquinol at the N center. Each oxidation of ubiquinol releases two protons, and two protons are taken up from the matrix for reduction of ubiquinone at the N center. Evidence for the Q cycle includes the ability of the inhibitor antimycin, which binds to the Q_n site, to inhibit oxidation of heme b_H and stop all electron transfer to cytochrome *c*.

High-Resolution Structures

The first x-ray structures of cytochrome bc_1 were obtained for bovine, chicken, and yeast complexes, followed by prokaryotic complexes from *Rhodobacter capsulatus* and *Rb. sphaeroides*, and most recently from *Paracoccus denitrificans*. The prokaryotic dimers comprise two three-subunit cores, sometimes with an additional protein, while the eukaryotic complexes contain additional proteins that form a large domain extending into the matrix. The additional seven or eight proteins likely play roles in regulation, assembly, and stability. The fold of the catalytic core is well conserved in the larger eukaryotic complexes, as expected from the high (~40–50%) sequence identity, and shows clearly the separation of the two quinone-binding sites of the Q cycle.

In the 2.7 Å *P. denitrificans* structure a bound stigmatellin inhibitor defines the Q_p site on the side facing the periplasm (equivalent to the intermembrane space in mitochondria). The two cytochrome *b* subunits form a hydrophobic core, each having two four-helix bundles with the two *b*-type hemes in the first bundle (see

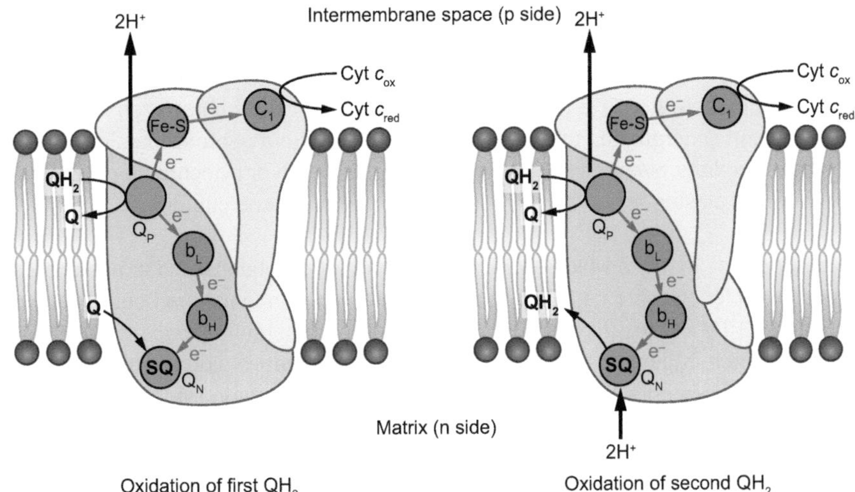

13.10 The mechanism of the Q cycle as originally proposed by Mitchell. The flow of electrons (red arrows) and protons (black arrows) in the two stages of the Q cycle is illustrated in a schematic representing cytochrome bc_1 viewed from the plane of the membrane. The b-type hemes (b_L and b_H, pink) are near the quinone-binding sites, and the c-type heme (c_1) is near the exterior where cytochrome c docks. Quinones (Q and QH_2) bind to two sites, Q_p and Q_n, located at the P side (intermembrane space, top) and N side (matrix, bottom). In the first stage Q is reduced to the semiquinone radical, $\cdot Q^-$ (SQ) at the Q_n site and QH_2 is oxidized to Q at the Q_p site. In the second stage a second QH_2 molecule is oxidized at the Q_p site and the $\cdot Q^-$ at the Q_n site is reduced, essentially replenishing one of the QH_2 molecules oxidized at the Q_p site. Redrawn from Mazat, J.-P., and Ransac, S., *Medecine/Sciences* 2010, 26:1079–1086. Reprinted with permission.

Figure 13.9). The Rieske Fe–S protein crosses the dimer, with its TM helix anchored to one monomer and its head domain interacting with the other. A hinge between them, which allows the [2Fe–2S] cluster to move between orientations toward the cytochrome *b* and cytochrome c_1, plays an essential role in electron transfer.

The x-ray structures of mitochondrial cytochrome bc_1 complexes all portray a symmetric, pear-shaped dimer that protrudes from the membrane in both directions, ~75 Å into the matrix on the inside and ~35 Å into the intramembrane space on the outside (Figure 13.11). The position of the lipid bilayer is clearly delineated by the bound

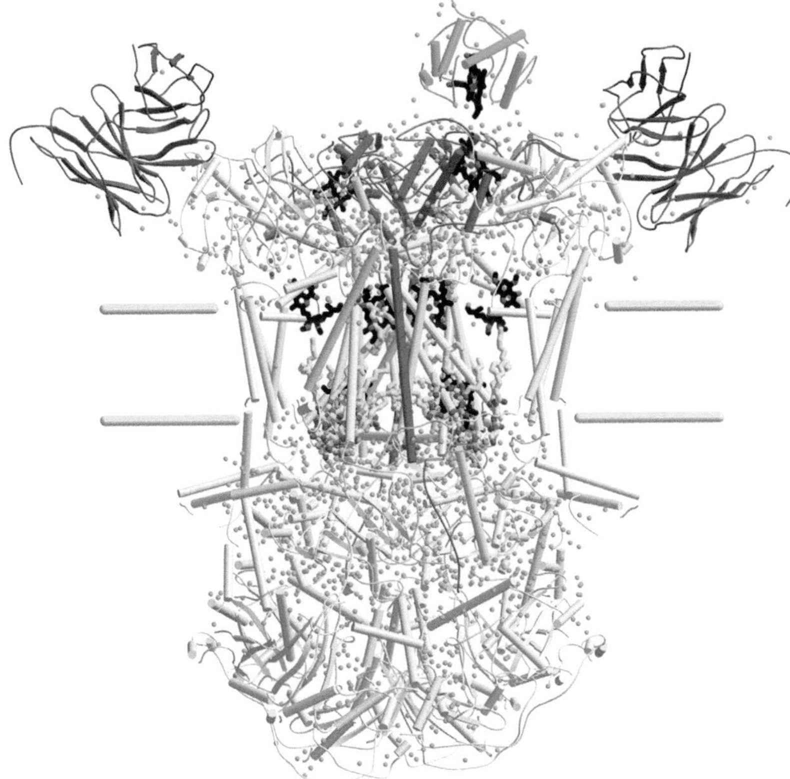

13.11 The 1.9-Å resolution structure of the reduced yeast cytochrome bc_1 complex bound to cytochrome *c*. After reduction with ascorbic acid and crystallization with Fv fragments (dark gray), the structure shows a monovalent complex of cytochrome bc_1 (light gray) and cytochrome *c* (green), with cytochrome *c* bound to one protomer only. Cytochrome *c* interacts most with cytochrome c_1 (pink). Heme groups and stigmatellin (carbon atoms, black; oxygen atoms, red), and lipid and detergent molecules (yellow, with oxygen atoms, red) are shown in ball-and-stick form. Many water molecules (cyan) are resolved. The position of the membrane is indicated by the yellow lines, with the intermembrane space at the top and the matrix at the bottom. From Solmaz, S. R. N., and C. Hunte, *JBC*. 2008, 283:17542–17549. © 2008, by the American Society for Biochemistry and Molecular Biology.

BOX 13.1 A MODIFIED Q CYCLE IN CYTOCHROME b_6f

Another example of the Q cycle is found in cytochrome b_6f, the central electron transport and proton translocating complex, which connects the photosystem 1 and photosystem 2 reaction center complexes in oxygenic photosynthetic membranes and generates up to two-thirds of the energy from photosynthesis stored in the membrane. The b_6f complex oxidizes plastoquinol and reduces plastocyanin or cytochrome c_6. Like ubiquinol, plastoquinol can be oxidized to plastoquinone by two one-electron transfers and can diffuse laterally through the membrane bilayer.

Cytochrome b_6f is a symmetric dimer in which each protomer has eight subunits and a mass of 225 kDa. Crystal structures have been obtained for cytochrome b_6f in its native state and with inhibitors bound. Each monomer has seven bound cofactors and 13 TM helices (Figure 13.1.1). Three of the prosthetic groups in the conserved core of the complex, two *b*-type hemes, b_p and b_n on the electropositive and negative sides of the complex and a high-potential [2Fe–2S] cluster, are arranged similarly to the corresponding cofactors in cytochrome bc_1 complex, with two sites for quinone binding near the positions of heme b_p and heme b_n. A difference is that b_n in b_6f is connected to a unique heme c_n, which serves as the quinone-binding site.

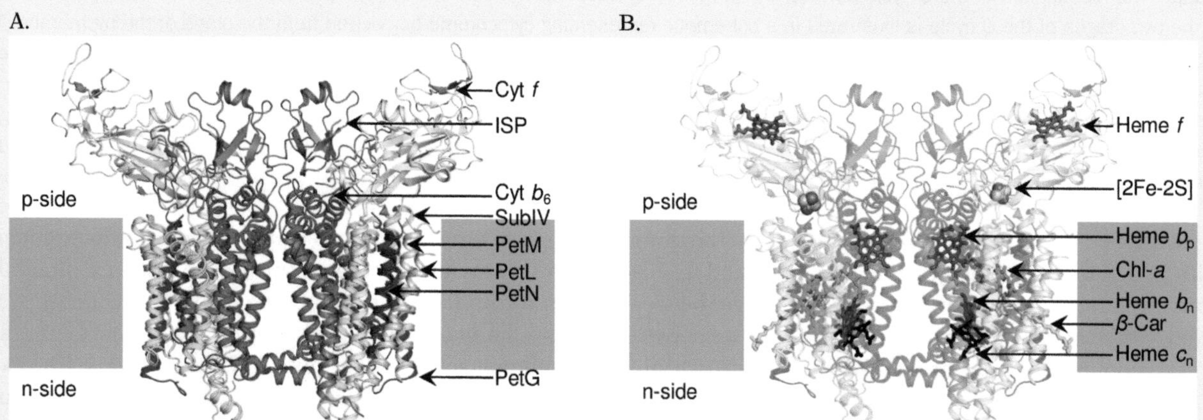

13.1.1 Crystal structure of the cytochrome b_6f dimer from cyanobacteria. (A) The ribbon diagram of the dimer shows the polypeptide composition of the complex: cytochrome b_6 (red), subunit IV (yellow), cytochrome *f* (cyan), the Fe–S Rieske protein (green), and the small subunits PetG, L, M, and N (blue, wheat, light gray, and dark gray, respectively). A lipophilic cavity is seen between the two monomers. The electrochemically negative (n) and positive (p) sides of the lipid bilayer (green) are labeled. (B) The cytochrome b_6f structure is shown with the electron transfer prosthetic groups indicated. Each TM portion of cytochrome b_6 includes three hemes, heme b_p (blue) at the Qo site, heme b_n (blue) and heme c_n (black) at the Qi site. The soluble domain on the p-side has a water channel near the [2Fe–2S] center (light and dark brown spheres) along with heme *f* (red, on cytochrome f). From Hasan, S. S., *et al.*, *Proc Natl Acad Sci USA*. 2013, 110:4297–4302.

On the p-side, cytochrome *f* (*f*, feuille, leaf, Fr.) has the same function as cytochrome c_1 in the bc_1 complex, but their structures are almost completely different. The following charge events define a modifed Q (quinone) cycle that allow two H^+ to be translocated to the p-side aqueous phase for every electron transferred through the complex.

1 **p-side** oxidation of plastoquinol on the p-side, transferring one electron to the Fe/S center and one to heme b_p, releasing two H^+ to the aqueous phase;
2 transfer of the electron from the Fe–S cluster to cytochrome *f*, to plastocyanin, and then to the PSI reaction center;
3 transfer of one electron across the membrane from heme b_p to heme b_n;
4 transfer of an electron from heme b_n to its attached heme c_n, that forms a plastoquinone binding site different from the n-side ubiquinone binding site in the bc_1 complex;
5 **n-side** bound plastoquinone is reduced by a two-electron transfer from hemes b_n/c_n, or by consecutive transfer of electrons through hemes b_n/c_n, picking up two H^+ from the n-side aqueous phase;
6 alternatively, hemes b_n/c_n may be reduced by a "cyclic" electron transfer pathway that carries electrons from the reducing side of photosystem I.

phospholipids, with additional lipids bound to the complex interior (see also Figure 8.23). Delipidation of the complex inactivates it, and reconstitution requires cardiolipin.

More than half the mass of the complex is in the matrix portion, including the misnamed subunits Core 1 and Core 2 that function in mitochondrial peptide processing. Each half of the dimer has eight TM helices from cytochrome *b* and one each from cytochrome c_1 and the Fe–S protein, along with two or three additional TM helices from small subunits that surround the functional unit (Figure 13.12). The hinge in the Rieske Fe/S protein that allows the [2Fe–2S] center to be shuttled between cytochromes *b* and c_1 enables a large conformational change rotating the extrinsic domain 60° and moving the [2Fe–2S] cluster 16 Å. Mutations that limit the hinge movement lower the activity of the complex.

A 1.9-Å resolution structure of reduced complex III with bound cytochrome *c* provides insight into how their interaction can be very specific and yet transient enough for the 100 sec^{-1} turnover rate. At the surface where cytochrome *c* docks is a core of hydrophobic residues surrounded by a semicircle of polar residues (Figure 13.13). The polar residues are too far from oppositely charged residues on cytochrome *c* to make electrostatic bonds; instead, their Coulombic interactions provide orientation to "steer" the docking. The interface has no hydrogen bonds, but plenty of water molecules (1.5 times the normal amount at protein–protein interfaces).

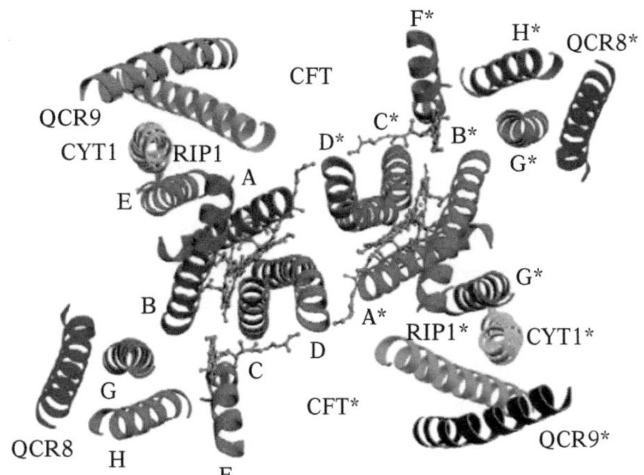

13.12 TM helices of cytochrome bc_1 homodimer. Segments of one subunit are designated with asterisks. The TM helices from cytochrome *b* (red) are labeled A to H, those from cytochrome c_1 (yellow) are labeled CYT1, and those from the Rieske Fe/S protein (green) are labeled RIP1. The additional TM helices from small subunits QCR8 and 9 (blue and gray) are also shown. The dimer interface has two large hydrophobic clefts, labeled CFT. The hemes and the headgroups of the quinone (behind helix A) and stigmatellin (between B and C) are shown in ball-and-stick models (gray). From Hunte, C., *et al.*, *Structure*. 2000, 8:669–684. © 2000 by Elsevier. Reprinted with permission from Elsevier.

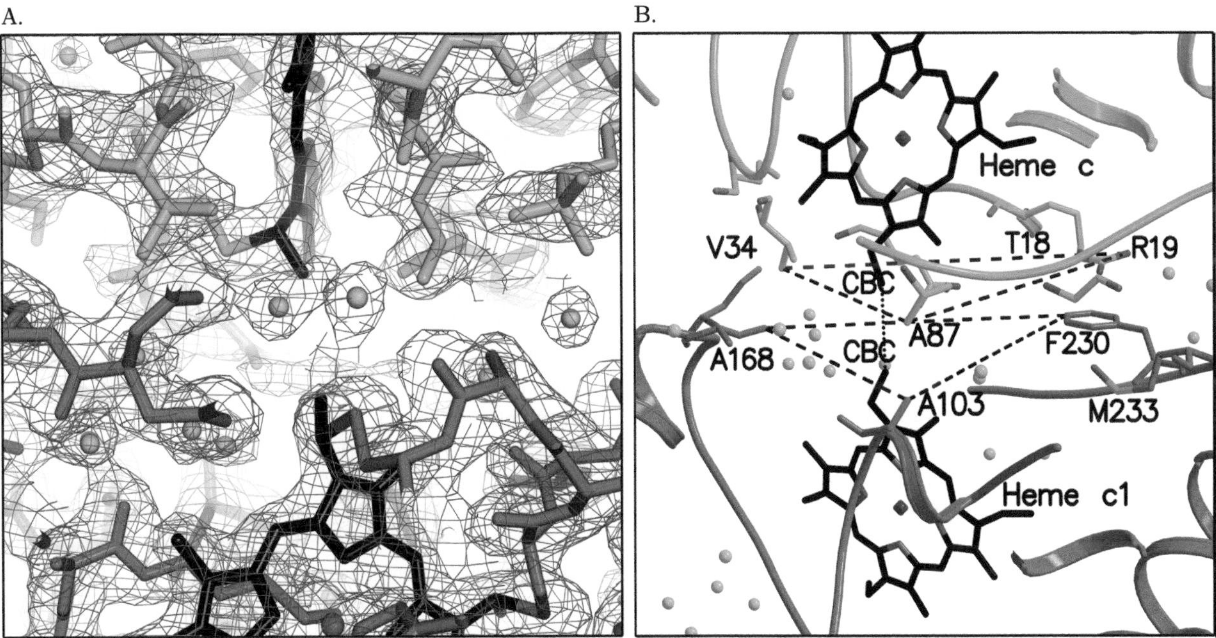

13.13 The interface of cytochrome *c* and cytochrome c_1. The high resolution of the reduced complex with cytochrome *c* bound allows close examination of the interaction between cytochrome *c* (green) and cytochrome c_1 (pink). (A) The electron density map (blue mesh) of the interface shows the short distance (9 Å) between hemes (black) that allows very fast electron transfer. (B) The hydrophobic core interface (dashed lines), which largely excludes water molecules (cyan) is defined by four residue pairs (labeled). The contact between the heme CBC atoms is indicated by the dotted line. From Solmaz, S. R. N., and C. Hunte, *JBC*. 2008, 283:17542–17549. © 2008, by the American Society for Biochemistry & Molecular Biology.

Comparison of the protomers with and without cytochrome c bound shows binding causes almost no conformational change in cytochrome c_1, and small changes in the head domain of the Fe/S Rieske protein, as well as the acidic subunit called QCR6p that is known to aid in binding cytochrome c.

The crystallization of the yeast cytochrome bc_1 complex in the presence of substrate and inhibitors has provided insight into the mechanisms of electron transfer and proton conduction. In the crystal structures one molecule of ubiquinone is bound to Q_n at the N center. At the Q_p site, the inhibitor stigmatellin binds in the same position that ubiquinone is expected to bind, while the inhibitor 5-*n*-heptyl-6-hydroxy-4,7-dioxobenzothiazole (HHDBT) is a hydroxyquinone anion that resembles an intermediate step of ubiquinol oxidation (Figure 13.14). Critical residues at the P center include His181 on the Rieske Fe/S protein and Glu272 of cytochrome *b*. In the first step of the Q cycle ubiquinol is hydrogen bonded to both of these residues to form an electron donor complex, which allows essentially simultaneous electron transfer to the Rieske cluster and the b_L heme. Oxidation of ubiquinol to ubiquinone breaks this complex, allowing the ubiquinone to diffuse out to the medium or possibly to the N center of the opposite protomer. In addition, formation of the complex increases the pK_a on the imidazole nitrogen of His181, allowing it to take up a proton from ubiquinol; when the complex dissipates and the Rieske Fe–S center moves away, the pK_a is lowered and the proton is released. The second proton from the ubiquinol is transferred to Glu272 after electron transfer to the b_L heme. Rotation of protonated Glu272 allows it to hydrogen bond with a water molecule that is hydrogen bonded to a heme propionate side chain. From there the proton is released to the surface via a hydrogen-bonded water chain associated with four charged residues of cytochrome *b*, Arg79, Asn256, Glu66, and Arg70.

While protons are released from the P center of cytochrome bc_1, they are taken up at the N center. The x-ray structure of the yeast complex reveals two distinct pathways for proton uptake, called the E/R pathway and the cardiolipin/K pathway (Figure 13.15). The two pathways are located at either end of the substrate-binding site, indicating a quinone can be reduced on both ends of the molecule without changing positions. The E/R path runs from Glu52 of one of the minor subunits and is gated by Arg218 of cytochrome *b*. A molecule of cardiolipin is at the entrance of the other path, which is gated by Lys228 of cytochrome *b*. Upon reduction of ubiquinone, a proton is abstracted from each of the gating residues (Arg218 and Lys228) and is replenished with a proton from the matrix via hydrogen-bonded networks to those residues. The structures have provided an elegant explanation for the coupling of proton movements to electron transfer in the Q cycle.

A.

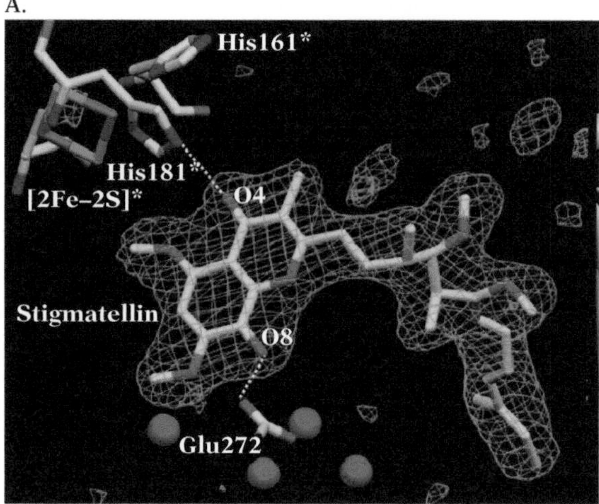

B.

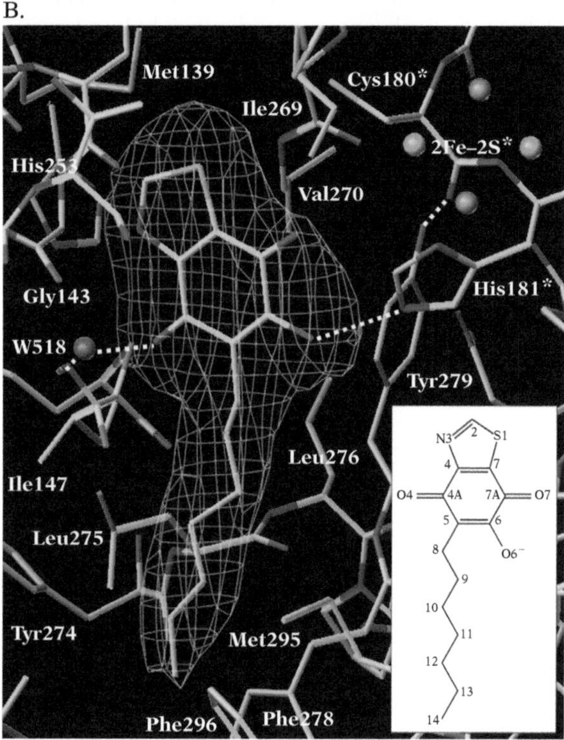

13.14 Binding of different inhibitors to the Q_p site of yeast cytochrome bc_1. **(A)** Electron density fitted to the structure of stigmatellin at the Q_p binding site between Glu272 and the [2Fe–2S] center of the Rieske protein coordinated by His181 and His161. Hydrogen bonds are shown to the carbonyl oxygen (O4) and hydroxyl group (O8) of stigmatellin. From Hunte, C., *et al.*, *Structure*. 2000, 8:669–684. © 2000 by Elsevier. Reprinted with permission from Elsevier. **(B)** Electron density fitted to the structure of the inhibitor 5-*n*-heptyl-6-hydroxy-4,7-dioxobenzothiazole (HHDBT), whose structure is given in the inset. Residues that stabilize the binding are labeled, and hydrogen bonds to the carbonyl oxygen (O4) and the deprotonated hydroxyl oxygen (O6) are shown. From Palsdottir, H., *et al.*, *J Biol Chem*. 2003, 278:31303–31311. © 2003, by the American Society for Biochemistry & Molecular Biology.

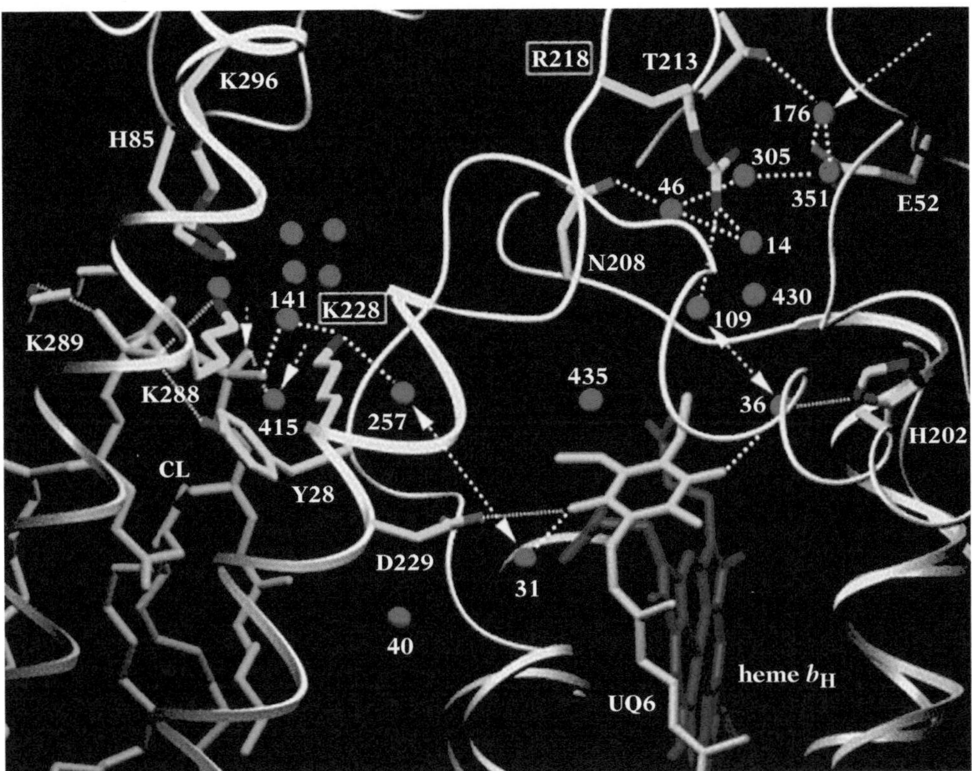

13.15 The suggested pathways for proton uptake at the N center of the yeast cytochrome bc_1 complex. Two distinct pathways connect the solvent at the matrix side with the quinone-binding pocket at the N center. The E/R pathway (right side of the figure) has an entrance at Glu52 of the Qcr7 subunit and is gated by Arg218 (yellow side chain with blue nitrogen atoms). The cardiolipin/K pathway has a cardiolipin (CL, green) positioned at its entrance and is gated by Lys228 (yellow side chain with blue nitrogen atom). Water molecules (red spheres) are associated with charged residues, and hydrogen bond interactions (dotted and dashed lines) are indicated. The arrows indicate the access sites from the bulk solvent, and double-headed arrows indicate proton transfer between the residues and the ubiquinone (UQ6, cyan). From Hunte, C. *et al.*, *FEBS Lett.* 2003, 545:39–46. © 2003 by the Federation of European Biochemical Societies. Reprinted with permission from Elsevier.

CYTOCHROME *C* OXIDASE

Complex IV of the respiratory chain, cytochrome *c* oxidase, carries out the reduction of O_2 to H_2O using four electrons coming from four molecules of reduced cytochrome *c* while pumping a total of eight protons, four "substrate" protons that combine with the oxygen atoms and four "pumped" or "vectorial" protons that contribute to the electrochemical gradient. Like complex III, complex IV is a dimeric multimer with each half containing 3–4 subunits in prokaryotes and 8–13 subunits in eukaryotes. All the redox centers are contained in subunits I and II: two *a*-type hemes called *a* and a_3, and two Cu-containing centers, Cu_A, which has two copper ions, and Cu_B. In addition, the complex has a Mg^{2+} ion that is not involved in redox and a site for binding Ca^{2+} or Na^+. The four redox active metals are organized in two sites: the binuclear copper site, with Cu_A and heme *a*, and the O_2 reduction site with heme a_3 and Cu_B. Spectroscopic studies revealed the path of electron transfer is from cytochrome *c* to Cu_A to heme *a*, then to heme a_3 and Cu_B, and finally to O_2 (Figure 13.16). Elegant laser flash spectroscopy, along with structural biology, has revealed how the reduction of O_2 to water occurs by a four-electron transfer reaction without accumulating reactive oxygen species as intermediates. On the other hand, the mechanism that couples O_2 reduction to proton pumping is still not clear.

High-Resolution Structures

Crystal structures have been obtained for cytochrome *c* oxidase from *Rhodobacter sphaeroides*, *Paracoccus denitrificans*, and bovine mitochondria. Both bacterial complexes have four subunits and the bovine complex has 13 subunits, of which subunits I, II, and III are very similar to the corresponding subunits in the prokaryotes. Subunits I and II contain the binuclear Cu site and the O_2 reduction site. The role of subunit III appears to contribute to stability and act as a proton antenna, as it has many charged residues on the surface and the rate of proton uptake is reduced by 50% when subunit III is lacking.

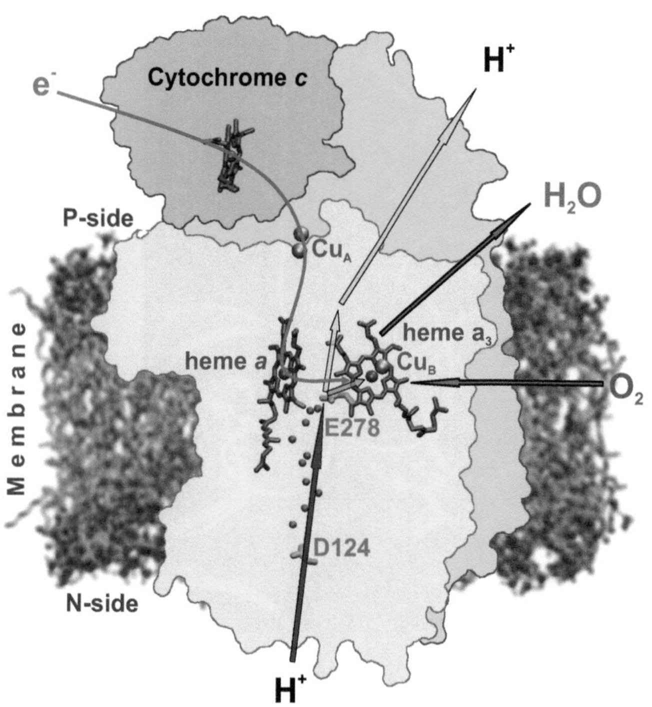

13.16 Schematic diagram showing the paths of proton and electron transfer in cytochrome c oxidase. Subunits I (yellow) and II (green) are depicted in the lipid bilayer along with docked cytochrome c (blue). The chemical reaction of O_2 reduction to water is coupled with the translocation of four protons. Electrons (thin red arrow) flow from reduced cytochrome c docked on the P-side to the Cu_A center, and from there to heme *a* and then to the heme a_3–Cu_B center, where they reduce oxygen (heavy red arrow). Protons from the N-side (cytoplasm or matrix) are either shuttled to the heme a_3–Cu_B site (gray arrow) and consumed in the production of water, or are translocated across the membrane (blue arrow). The water molecules (red spheres) depicted here are crystallographically observed in the D pathway involving Asp124 and Glu278, as described in the text. From Belevich, I., *et al.*, *Nature*. 2006, 440:829–832. © 2006, Nature Publishing Group. Reprinted by permission of Macmillan Publishers Ltd.

Complex IV has an ellipsoid shape that protrudes 32 Å beyond the lipid bilayer into the P phase outside the membrane (Figure 13.17). The proteins cross the membrane with 21–28 TM helices: 12 from subunit I, two from subunit II, and seven from subunit III. The seven additional subunits in the bovine complex are type II membrane proteins (each with a single TM helix and the N terminus inside) that together surround subunits

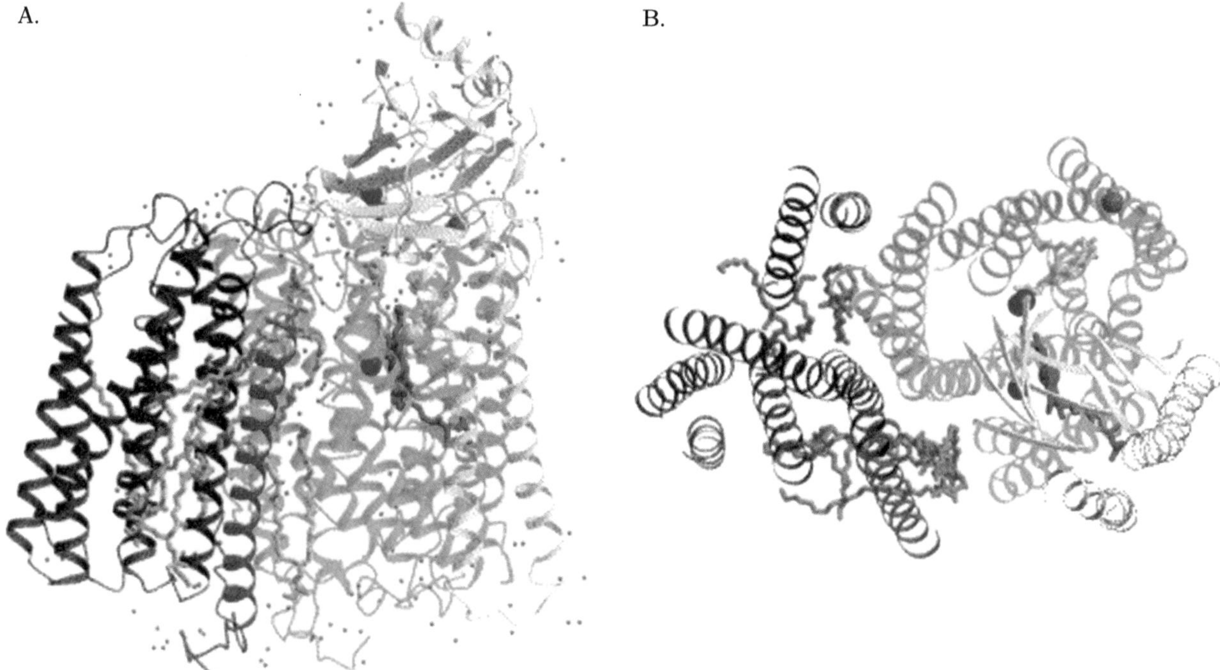

13.17 X-ray structure of cytochrome c oxidase. The crystal structure of cytochrome c oxidase from *Rb. sphaeroides* was determined at 2.3-Å resolution. The subunits are closely packed: subunit I (green), subunit II (light gray), subunit III (dark gray), and subunit IV (magenta). The redox centers include heme a (light blue), heme a_3 (red), and Cu_A and Cu_B (dark blue), and have Mg (light green), calcium (pink), lipids (orange), and water molecules (red spheres). **(A)** Ribbon diagram shows the structure viewed from the plane of the membrane. **(B)** The TM helices in cytochrome c oxidase are viewed from the P-side. From Svensson-Ek, M., *et al.*, *J Mol Biol*. 2002, 321;329–339. © 2002 by Elsevier. Reprinted with permission from Elsevier.

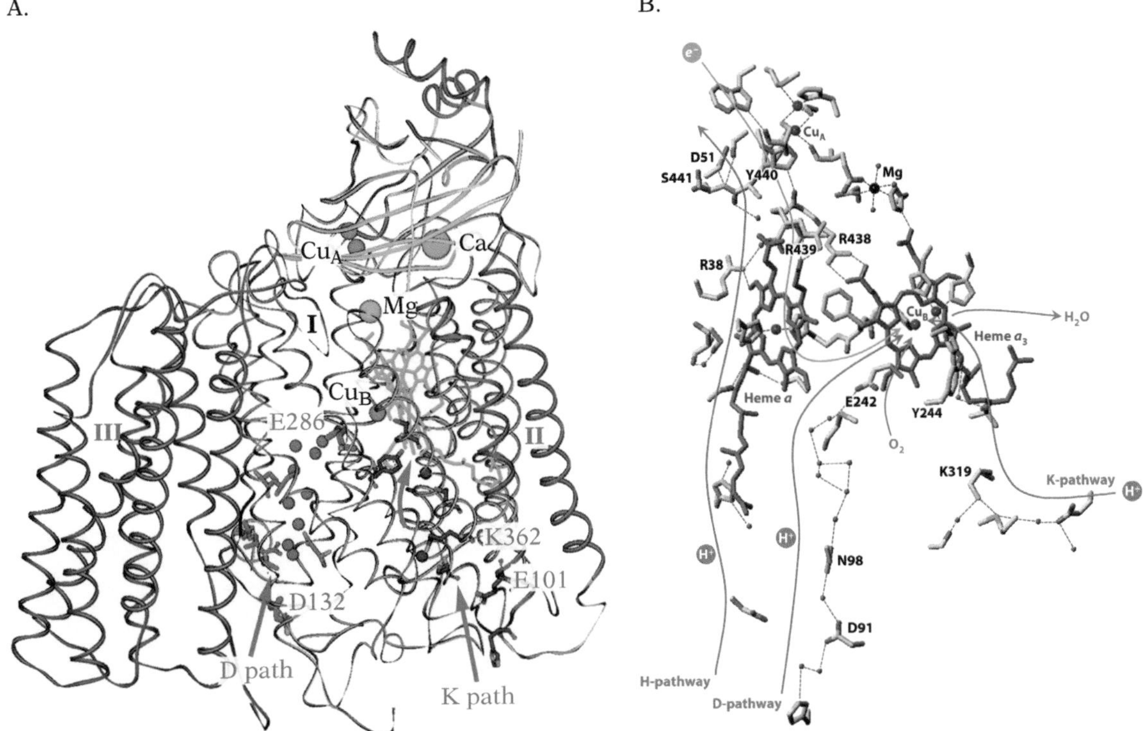

13.18 Proton uptake pathways in cytochrome c oxidase. (A) The ribbon diagram of *Rb. sphaeroides* cytochrome c oxidase shows subunits I, II, and III with heme a and a_3 (green) as stick structures, with Ca and Mg metals (green spheres) and Cu metals (orange spheres). The two paths for water molecules are the D path (red) and the K path (blue), with water molecules colored accordingly. From Hosler, J. P., *et al.*, *Annu Rev Biochem*. 2006, 75:165–187. © 2006 by Annual Reviews. Reprinted with permission from the Annual Review of Biochemistry, www.annualreviews.org. (B) A possible third pathway shown in a close-up view of parts of subunits I and II of bovine cytochrome c oxidase, the H pathway terminates at the critical but unconserved residue Asp51. The electron pathway (blue arrow) is included. Observed water molecules (red spheres) are indicated and critical residues are labeled. From Yoshikawa, S., *et al.*, *Ann Rev Biophys*. 2011, 40:205–223. © 2011 by Annual Reviews. Reprinted with permission from the Annual Review of Biophysics, www.annualreviews.org.

I, II, and III. In the bacterial complex the single TM helix of subunit IV is set off from the rest and almost completely surrounded by lipids. Specific phospholipid molecules are resolved in all the structures, indicating there are conserved lipid-binding sites.

The two hemes are buried in the membrane interior with the heme plane perpendicular to the membrane plane, coordinated by side chains from residues of subunit I, and sufficiently close to allow electron tunneling. The Cu_A center has two cysteine residues as shared ligands to the two copper ions (analogous to [2Fe–2S]) and binds to a globular domain of subunit II on the outside surface. This globular domain meets the surface of subunit I, contributing to the area where cytochrome *c* likely binds, as there are ten acidic residues that can interact with the ring of lysine residues on cytochrome *c* (see Figure 4.2). The O_2 reduction site is connected to the N-phase (the bacterial cytoplasm or mitochondrial matrix) by two conserved hydrogen-bond networks called the K- and D-pathways described below. Mutations of critical amino acids in the pathways block both proton pumping and O_2 reduction. Bovine cytochrome *c* oxidase has a third possible pathway, the H-pathway, located near heme *a* and postulated to be the path for pumped protons (Figure 13.18).

Oxygen Reduction

The reaction can be viewed as starting with a metal reduction phase, when each cytochrome *c* molecule docks on the surface of cytochrome *c* oxidase and transfers its electron to Cu_A. The electrons are then transferred to heme *a* and then to heme a_3/Cu_B, where the reduction of O_2 takes place. The characteristics of the O_2 reduction site, heme a_3 and Cu_B, have been determined in crystals of both fully oxidized and fully reduced enzyme. These structures, along with results from time-resolved spectroscopic studies, indicate that O_2 binds first to the Fe^{2+} of heme a_3 and quickly picks up four electrons from the Fe^{2+} and the $Cu_B{}^{1+}$ and likely Tyr288, creating a tyrosyl radical to form the oxoferro state – $Fe^{4+} = O^{2-}$ and $Cu_B{}^{2+}$ – OH^-. This four-electron reaction occurs sufficiently rapidly that neither the peroxy nor hydroperoxy intermediates are detected spectroscopically (Figure 13.19). The

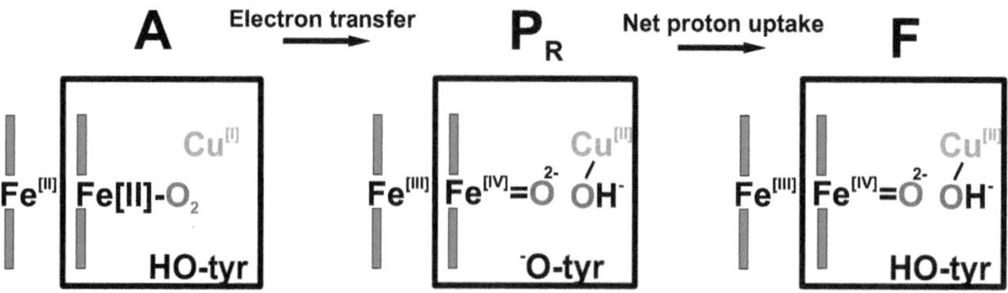

13.19 Diagram of early steps in electron transfer and proton uptake. Starting from fully reduced cytochrome *c* oxidase, the early intermediates upon addition of O_2 may be separated by different kinetics and spectroscopy. In this diagram elements of the O_2 reduction site (heme a_3, Cu_B and Tyr) are enclosed in a square, with heme *a* outside the site and parallel to heme a_3. The intermediate on the left, called A, is the oxygen adduct formed when the binuclear site first binds O_2. Within microseconds, electron transfer from heme *a* allows the O=O bond to split, creating the unstable ferryl/cupric intermediate called P_R. Uptake of the first proton by the Tyr is slower (msec) and produces intermediate F. From Belevich, I., *et al.*, *Nature*. 2006, 440:829–832. © 2006, Nature Publishing Group. Reprinted by permission of Macmillan Publishers Ltd.

next two steps are slower, limited by the time it takes the protons to reach the binuclear center. A proton reacts with the OH^- on Cu_B to release the first water; then two protons react with the O^{2-} on Fe^{4+} of heme a_3 to release the second water. The electrons to re-reduce Tyr288 and heme a_3^{4+} to a_3^{3+} are supplied by additional molecules of cytochrome *c* operating through Cu_A and heme *a*.

Proton Pathways

Cytochrome *c* oxidase from eukaryotes and most prokaryotes have two or three pathways to take up protons from the matrix or cytoplasm to the bimetallic center, revealed in the x-ray structures as series of hydrogen-bonded water molecules linked to residues that are essential for pumping protons. In the *Rb. sphaeroides* complex the D-pathway goes from Asp132 on the surface to Glu286 between heme *a* and heme a_3, a distance of ~26 Å. The K-pathway begins at Glu101 on subunit II and passes in sequence Ser299, Lys362, Thr359, a farnesyl side group of heme a_3, and Tyr288 of subunit I (see Figure 13.18). Either the D- or K-pathway is used to conduct "substrate" protons to the heme a_3/Cu_B site. Since genetic alteration of the D-pathway eliminates proton pumping in *Rb. sphaeroides*, the "pumped" protons use only the D-pathway in the bacterial complexes. The bovine complex may also pump protons via the H-pathway, a hydrogen-bond network connected to the P phase and a water channel to the N phase. The critical Asp51 of the H-pathway is missing in cytochrome *c* oxidase from other organisms. The lack of conservation of this proton pathway may reflect the fact that proton pumping, which is accomplished in various ways in numerous proteins, is relatively simple compared to the problem of reducing O_2 without releasing active oxygen species.

Research to define the exit pathway for "pumped" protons from Glu286 to the outside is ongoing. MD simulations show a single-file column of water molecules

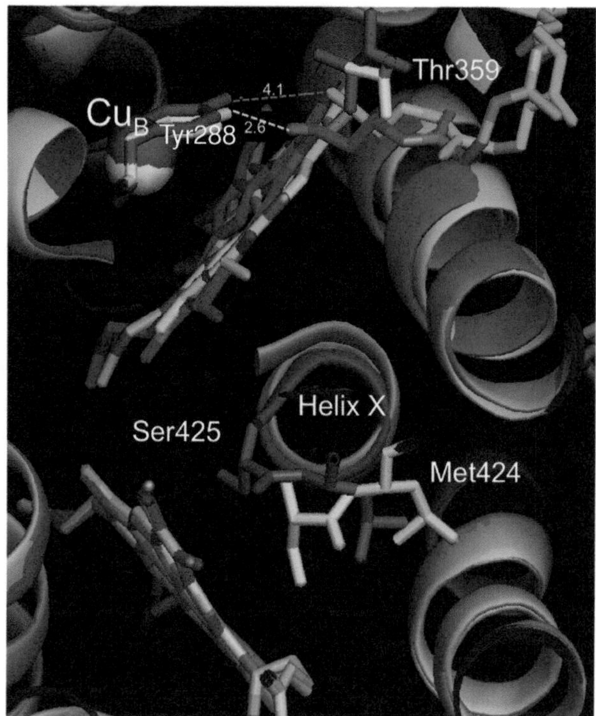

13.20 Conformational change near heme *a*. Higher-resolution (2.0 Å) crystals of the two core subunits I and II of cytochrome *c* oxidase from *Rb. sphaeroides* (missing subunits III and IV) show differences between the reduced and oxidized states at the O_2 reduction site. In this overlay of the cross sections from inside the membrane, significant movement is observed for Cu_B, Tyr288, Met424, and Ser425 (red, reduced; yellow, oxidized). Heme *a* is in the lower left and heme a_3 is in the top center (cyan and green, reduced; blue, oxidized) just above the central helix X (blue, reduced; turquoise, oxidized). The change in distances between the hydroxyl of Tyr288 and the heme a_3 farnesyl-OH from 4.1 Å (reduced, red dotted line) to 2.6 Å (oxidized, yellow dotted line) is sufficient for opening and closing a gate to the heme. A similar conformational change is observed in crystals of complete *R. sphaeroides* (all four subunits) as well as bovine cytochrome *c* oxidase. From Ferguson-Miller, S., *et al.*, *Biochim Brophys Acta*. 2012, 1817:489–494. © 2012, Elsevier. Reprinted with permission from Elsevier.

can form through a hydrophobic cavity to the Mg^{2+} ion. Somewhere along the path for proton pumping must be a mechanism that couples it to O_2 reduction, but it is not accompanied by a large conformational change as seen in complex I. Electron transfer from heme *a* to the oxygen reduction site initiates the proton pump mechanism. Recent high-resolution structures of both the bovine and bacterial complexes reveal conformational changes in the vicinity of heme *a* and heme a_3 upon reduction (Figure 13.20). An attractive hypothesis is that small conformational changes may open a gate at the K-pathway for substrate protons while closing the D-pathway for pumped protons.

A number of other questions about complex IV are being explored, such as what controls the directionality of proton pumping? What is the role of bound lipids? What mechanisms regulate the efficiency of cytochrome *c* oxidase? Yet tremendous progress has been made in understanding the mechanism of this vital enzyme complex that produces water from O_2 while coupling electron transfer with proton pumping.

The mechanism of coupling is completely different in each of the respiratory complexes described: the piston-driven large conformational changes in complex I, the Q cycle in complex III, and what appears to be a small gating movement in complex IV. The pmf they generate is the driving force for the synthesis of ATP in aerobic bacteria and mitochondria. In photosynthetic organisms the proton gradient is produced by light-driven proton pumps, while an Na^+ ion gradient serves the same purpose in anaerobic bacteria. In all cases, the energy in the ion gradients is harvested by another exquisite molecular machine, the ATP synthase, which couples ATP synthesis to the flow of protons (or Na^+ ions) down their electrochemical gradient.

F_1F_0-ATP SYNTHASE

Complex V of the respiratory chain is the ubiquitous F_1F_0-ATP synthase that executes the last step of the multistep process for the generation and utilization of the proton motive force (pmf), an electrochemical gradient across the membrane. It carries out the vast majority of ATP synthesis in all cells, replenishing the ATP that was hydrolyzed to ADP and inorganic phosphate during the energy-consuming reactions of metabolism. Thus it is not surprising that the ATP synthase has been the subject of thousands of biochemical and biophysical studies and more recently structural studies, which have elucidated details of its sophisticated mechanism.

Familiar for decades for its knob-on-a-stalk appearance, the F_1F_0-ATP synthase is named for its two major structural domains, F_1 and F_0 (Figure 13.21). This fascinating complex is also called the F_1F_0-ATPase because it can hydrolyze ATP to ADP and phosphate while reversing the flow of protons across the membrane. Found in the plasma membrane of bacteria, the mitochondrial

A.

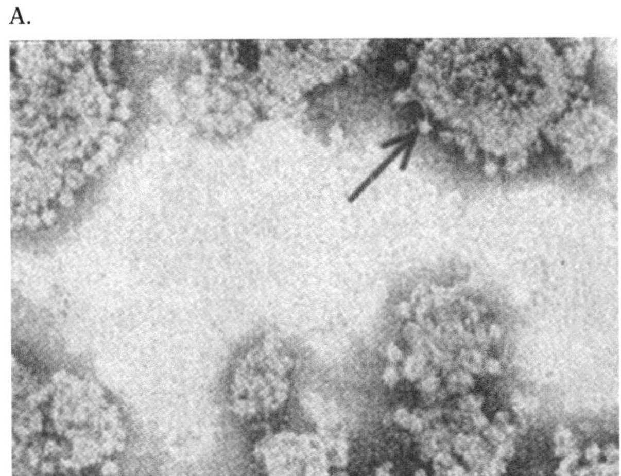

B.

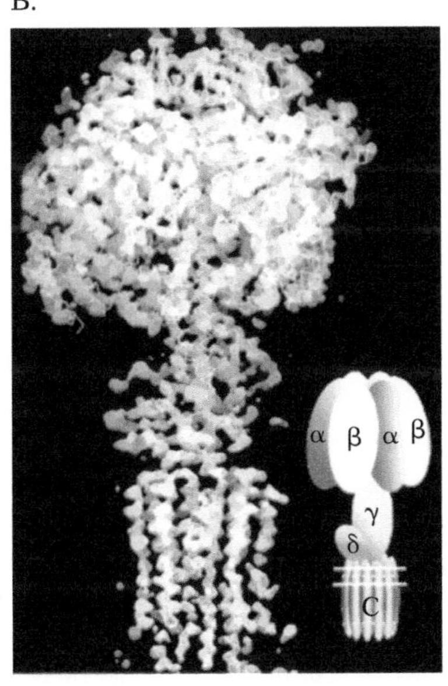

13.21 (A) ATP synthase molecules observed on inside-out vesicles of bovine heart mitochondria. The knob-on-stalk appearance of the ATP synthase is seen in negative staining EM. From Walker, J. E., *Angew Chem Int Ed.* 1998, **37**:2308–2319. © 1998 WILEY-VCH Verlag GmbH, Weinheim, Fed. Rep. of Germany. (B) Low-resolution x-ray structure of the F_1F_0-ATPase from yeast mitochondria. The electron density map of the F_1-c_{10} complex from *Saccharomyces cerevisiae* at 3.9-Å resolution shows the architecture of the complex from the side, with insets identifying the subunits. From Stock, D., *et al.*, *Curr Opin Struct Biol.* 2000, 10:672–679. © 2000 by Elsevier. Reprinted with permission from Elsevier.

inner membrane in eukaryotes, and the chloroplast thylakoid membrane in plants, the ATP synthase is the major producer of ATP in cells using either oxidative phosphorylation or photosynthesis to generate a pmf. The importance of ATP generation is clear to humans, who each use around 40 kg of ATP each day of a sedentary life (an athlete uses much more!). With around 100 mmol of adenosyl nucleotides in the body, this rate of consumption means that each molecule of ADP must be phosphorylated about 1000 times per day to provide the needed ATP. The reverse reaction by the F_1F_0-ATPase is also important under conditions when the utilization of ATP is needed to replenish the proton gradient. How the flow of protons across the membrane is coupled to the catalytic reaction is a question that is fundamental to life. Biochemical data on the catalytic mechanism of the F_1F_0-ATPase gained beautiful support from the first high-resolution structures of its components. This was recognized by the award of the 1997 Nobel Prize in Chemistry to Paul Boyer and John E. Walker for "the elucidation of the enzymatic mechanism underlying the synthesis of ATP." Continued experimentation and structural biology have contributed to a nearly complete understanding of this marvelous mechanism.

Unlike the calcium pump described in Chapter 9, which as a single polypeptide couples ion flux to the hydrolysis of ATP, the ATP synthase is a molecular machine with many different polypeptide components. With 8 (prokaryotic) to 18 (mammalian) different subunits, the complex has a molecular mass of 550–650 kDa. The overall shape of the F_1F_0 complex is shown beautifully in low-resolution images of the yeast mitochondrial F_1 with part of F_0 first published in 1999 (Figure 13.21B).

The soluble F_1 domain is the catalytic domain that carries out the synthesis and hydrolysis of ATP. Forming the knobs in Figure 13.21, F_1 is extrinsic to the membrane, so it can be removed by mild treatments and function as a soluble ATPase. *E. coli* has the simplest F_1 domain, with the composition $\alpha_3\beta_3\gamma\delta\varepsilon$. The F_0 membrane domain is typically composed of $ab_2c_{(10-15)}$, although some complexes have two different *b* subunits instead of a homodimer, and the number of *c* subunits varies in different organisms. When the F_1 domains are stripped off, F_0 by itself

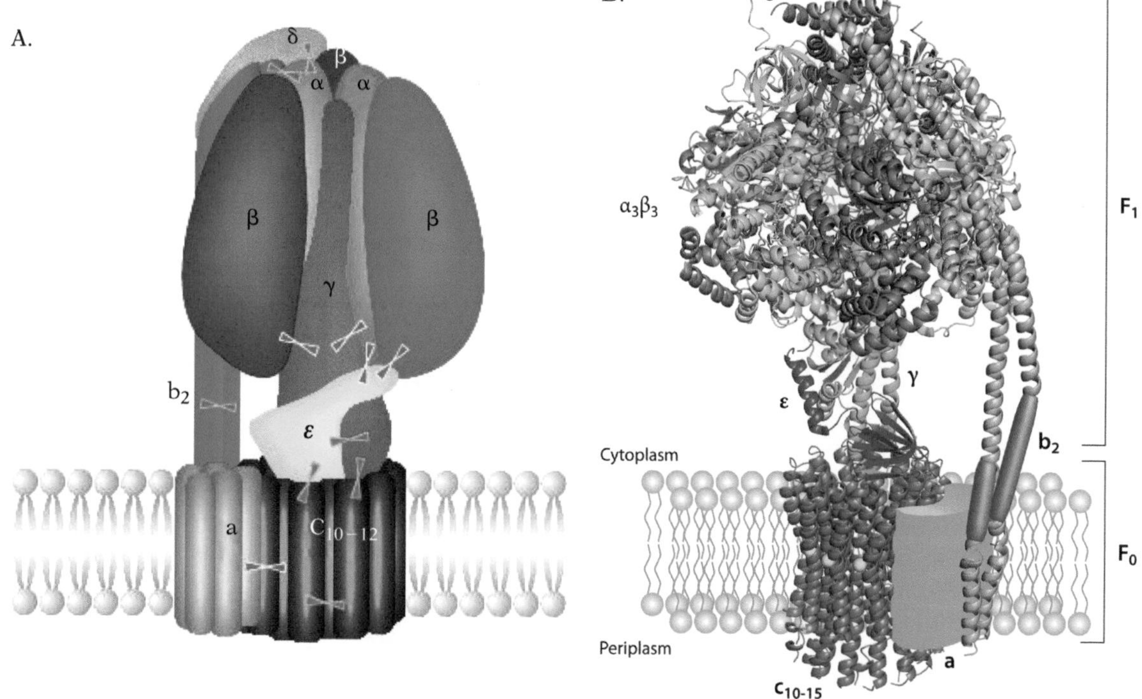

13.22 (A) The structural organization of the F_1F_0-ATP synthase. The arrangement of the subunits making up F_0 and F_1 in *E. coli* is diagrammed with one α subunit removed to reveal the γ subunit down the center. The three pairs of $\alpha\beta$ subunits of F_1 (magenta/pink, dark blue/light blue, and green; the third α subunit is removed) surround the γ (red) of the stalk, with ε (yellow) at its base and δ (orange) along the back. The c subunit ring (black) of F_0 is linked to the central stalk ($\gamma\varepsilon$) as well as to the peripheral stalk composed of b_2 (green) and δ. The five predicted TM helices of the *a* subunit are represented (gold). Note: the bows show some of the crosslinks that either have little or no effect on (green) or inhibit (red) the activity, which demonstrated that *c*, γ, and ε rotate together. From Capaldi, R. A., and R. Aggeler, *Trends Biochem Sci.* 2002, 27:154–160. © 2002 by Elsevier. Reprinted with permission from Elsevier. (B) Model of the high-resolution structure of the F_1F_0-ATP synthase. The model is a composite of structures of individual subcomplexes, including the c ring of *I. tartaricus* (blue), the F1 $\alpha_3\beta_3$ of *E. coli* (light and dark green), the δ subunit of *E. coli* (red), and the peripheral stalk from bovine mitochondria (b_2 in light orange). From von Ballmoos, C., *et al.*, *Ann Rev Biochem.* 2009, 78:649–672. © 2009 by Annual Reviews. Reprinted with permission from the Annual Review of Biochemistry, www.annualreviews.org.

TABLE 13.2 EQUIVALENT SUBUNITS IN ATP SYNTHASES IN BACTERIA, CHLOROPLASTS, AND BOVINE MITOCHONDRIA

Type	Bacteria	Chloroplasts	Mitochondria[a]
F_1	α	α	α
	β	β	β
	γ	γ	γ
	δ	δ	OSCP[b]
	ε	ε	δ
	–	–	ε
F_0	a	a (or x)	a (or ATPase 6)
	b	b and b′ (or I and II)	b
	c	c (or III)	c

[a] Mitochondrial ATP synthase has additional subunits that have no equivalents in bacteria or chloroplasts.
[b] Oligomycin sensitivity conferring protein.
Source: Walker, J. E., ATP synthesis by rotary catalysis (Nobel lecture). *Angew Chem Int Ed*. 1998, 37:2308–2319.

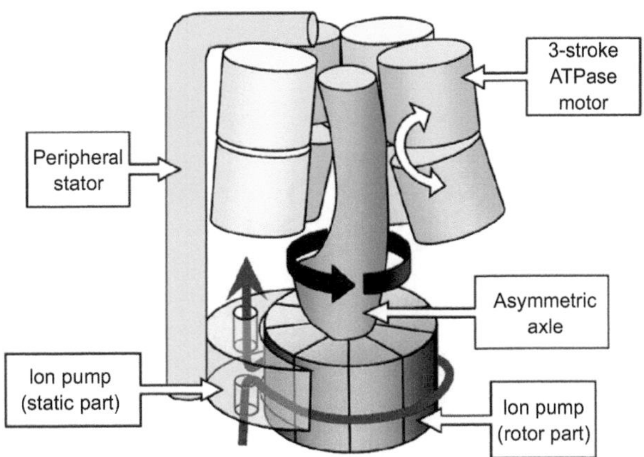

13.23 A generic model for rotary ATPases. The soluble domain (upper portion) houses a three-stroke motor: the ATP hydrolyzing/synthesizing enzyme with three catalytic sites (one has been removed for clarity). An asymmetric axle connects it to the integral membrane domain (lower portion) housing an ion pump that has a rotating ring of subunits connected to a static subunit (stator). The stator allows the relative movement of one pump to drive the rotation of the other. ATP synthesis is driven by the flow of protons (red arrow) down their electrochemical gradient via two half-channels as the rotation of the axle (wide gray arrow) affects the conformations of the catalytic subunits. From Muench, S. P., *et al.*, *Quart Rev Biophys*. 2011, 44: 311–356. © Cambridge University Press 2011.

forms a passive proton pore. The F_0 domain is sensitive to the inhibitor oligomycin, for which it was originally named. In the complex, these two structural domains are connected by two stalks: a central stalk made up of γ and ε subunits ($\gamma\delta\varepsilon$ in mitochondria) and a peripheral stalk made up of the δ and *b* subunits (Figure 13.22). The additional subunits in mammalian F_1F_0-ATPases are mostly in the stalk regions (see Table 13.2). The oligomycin sensitivity conferring protein (OSCP subunit) in the mitochondrial peripheral stalk is equivalent to the bacterial δ subunit. The δ subunit in mitochondria has the role of the ε subunit in bacteria, and the small mitochondrial ε subunit (only 50 amino acids) at the foot of the central stalk has no counterpart in bacteria or chloroplasts. While the division of subunits into F_1 and F_0 is based on their location in or at the membrane, crosslinking studies indicate a different grouping (see Figure 13.22A). As described below, the ATP synthase is a molecular machine composed of two rotary motors whose rotor consists of γ, ε, and c_{10-15}, while the stator and catalytic center are made up of $\alpha_3\beta_3$, δ, *a*, and b_2 (*E. coli* composition). These two motors are typical of rotary ATPases (Figure 13.23).

Subunit Structure and Function

F_1 Domain

The catalytic F_1 domain has alternating α and β subunits forming a spherical hexamer, like sections of an orange around a central cavity, first seen at high resolution in the ATP synthase of bovine mitochondria (Figure 13.24). The α and β subunits have ~20% sequence identity and very similar folds. Both can bind nucleotides, but the catalytic sites are located on the β subunits at the α/β interfaces. The x-ray structures show the three active sites in three different conformations, depending on whether they bind ATP (the closed TP conformation = T, tight-binding), or ADP (the partly open DP conformation = L, loose-binding), or are empty (the E conformation = O, open). The γ subunit has very long N- and C-terminal helices, which form an antiparallel α-helical coiled coil that traverses the central cavity of the α/β hexamer and continues as the central stalk to the F_0 domain. The two long helical domains of γ are connected by a globular domain that protrudes at the base of the stalk, visible in the higher-resolution structure of the complete bovine mitochondrial F_1 ATPase (Figure 13.24C). This globular domain has an α/β structure that wraps around the other two helices and contacts δ and ε subunits. Rotation of γ, which unwinds the lower part of the coiled coil, is driven in one direction by hydrolysis of ATP and is driven in the opposite direction by the pmf. The position of the γ subunit is asymmetric with respect to the $\alpha\beta$ hexamer, and rotation of γ causes conformational changes in β subunits (Figure 13.25).

At the base of the γ subunit where it contacts F_0 is the ε subunit (δ in mitochondria), which has two domains, an N-terminal β-sandwich and a helix–turn–helix C terminus. The N-terminal domain of ε makes contact with F_0 and is essential for coupling. In different crystal structures the ε subunit exhibits different conformations, a closed conformation and an open partially unfolded conformation, suggesting considerable flexibility that may be important in regulation (see below). Data from

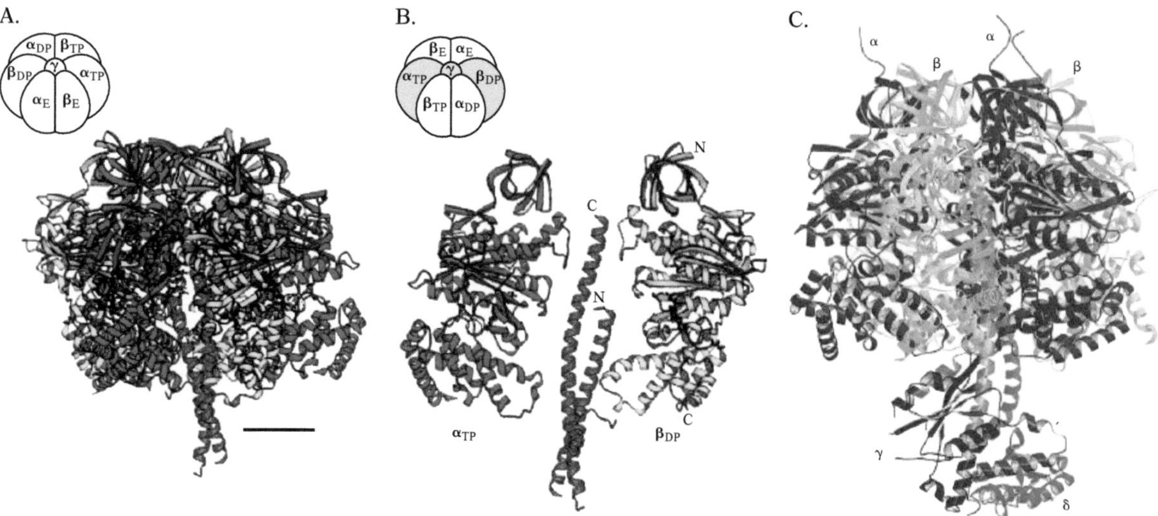

13.24 The first high-resolution structures of the F_1-ATPase from bovine mitochondria. (A) and (B) Ribbon diagrams show the α (red), β (yellow), and γ (blue) subunits of the F_1-ATPase as identified in the schematic drawing accompanying each figure. (A) The entire F_1 particle shows the coiled coil of the γ subunit in the central cavity between the α and β subunits, with bound nucleotides (black ball-and-stick representation). The $\alpha\beta$ pairs are labeled E for empty, TP for binding ATP, and DP for binding ADP, as described in the text. (B) A cutaway view showing only three subunits: α_{TP}, γ, and β_{DP} (shaded in the diagram above the structure). From Walker, J. E., *Angew Chem Int Ed*. 1998, 37:2308–2319. © 1998 WILEY-VCH Verlag GmbH, Weinheim, Fed. Rep. of Germany. (C) The 2.4-Å resolution structure of the complete F_1-ATPase from bovine mitochondria shows the α and β subunits (red and yellow, respectively) alternating around the γ subunit (blue) with the δ (green) and ε (magenta) at the base of the central stalk. From Stock, D., *et al*., *Curr Opin Struct Biol*. 2000, 10:672–679. © 2000 by Elsevier. Reprinted with permission from Elsevier.

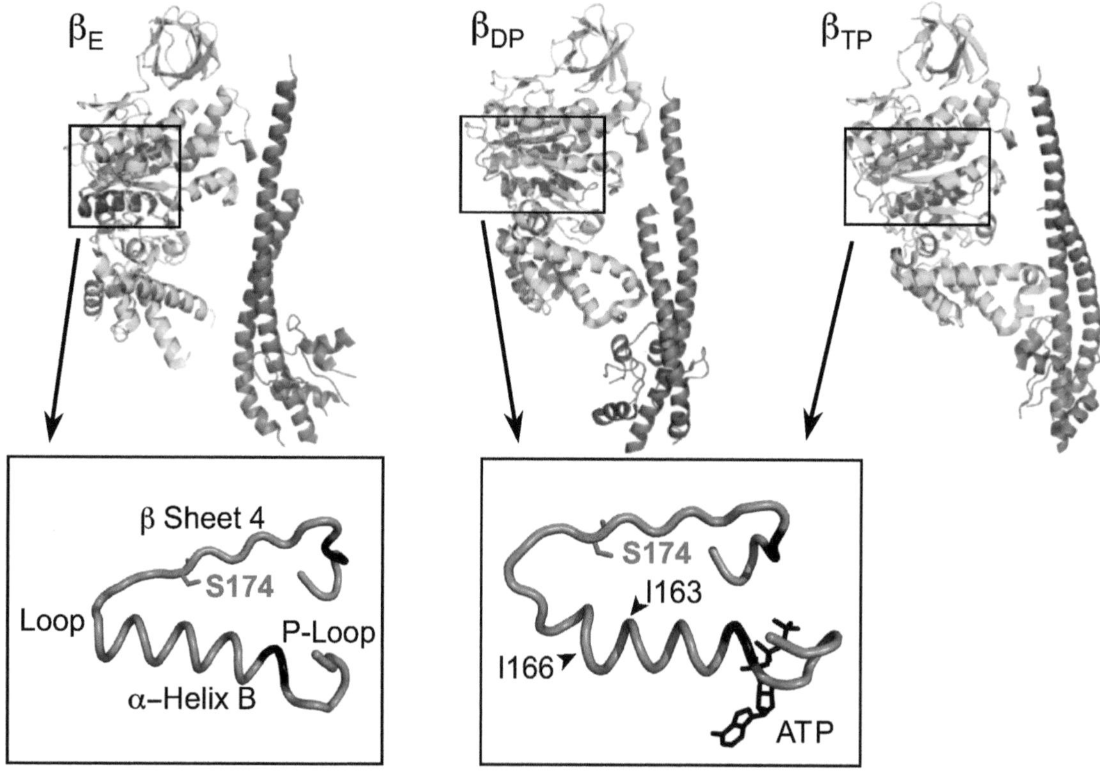

13.25 Different conformations observed in β and γ subunits. The catalytic β subunits (yellow) with different nucleotide occupancies have different conformations and different relations to the γ subunit (green). The expanded figures (orange) emphasize the dramatic changes in the conformation of the β hinge domain (orange with black regions for the location of the catalytic residues Lys155, Thr156, Glu181, and Arg182 (*E. coli* numbering for β subunit)). From Nakanishi-Matsui, M., *et al*., *Biochim Biophys Acta*. 2010, 1797:1343–1352. © 2010, Elsevier. Reprinted with permission for Elsevier.

crosslinking studies indicate that ε rotates with γ and c, because both ATP synthesis and ATP hydrolysis coupled to proton pumping occur when all three are covalently linked (see Figure 13.22A).

The δ subunit is required to bind F_1 to F_0, hence its historic designation in mitochondria as the oligomycin sensitivity conferring protein. The structure of the δ subunit isolated from *E. coli* has been determined by NMR (see Figure 13.22B). With the b_2 dimer from F_0, δ forms the peripheral stalk that holds the $\alpha_3\beta_3$ knob in place while γ rotates. For this reason $b_2\delta$ has been called a stator, the stationary part of a machine in which a rotor (the central stalk and ring of c subunits) revolves (see Figure 13.23).

F_0 Domain

The F_0 structural domain consists of a ring of c subunits connected via the a subunit to the b dimer that forms the peripheral stalk. The number of c subunits varies, with 12 in *E. coli*, 14 in chloroplasts, and 10 in bovine mitochondria. Each small (8000 Da) c subunit forms a coiled coil of two hydrophobic TM helices. The x-ray structure of the c ring from the ATP synthase of *Ilyobactor tartaricus*, which uses an Na^+ gradient, shows an hourglass shape, with Na^+ ions bound at the center (Figure 13.26). The N-terminal helices of the c subunits form a tightly packed inner ring and are connected to the C-terminal helices by a loop of highly conserved Arg, Gln, and Pro residues. Both helices are bent in the middle, causing the hourglass shape, where the Na^+ ions bind to Glu65 (corresponding to Asp61 in *E.coli*). The ring of c subunits is contacted by ε and γ at the central stalk from F_1, and half the polar loops contact the mitochondrial δ subunit (see Figure 13.24C).

The very hydrophobic a subunit, predicted to form five TM helices, connects the ring of c subunits to the b subunit of the stator and contains entry and exit sites for the proton channel. A critical residue in the a subunit is Arg210, which is essential for ATP-driven proton translocation but not for passive proton transport when F_1 has been removed. The interaction of aArg210 with cAsp61 indicates there is a direct connection between the a and c subunits where the protons access the rotating c ring. The a subunit has other basic and acidic residues essential for proton translocation, including His245, Glu196, and Glu219.

The b subunit is usually a homodimer, although some bacterial species have a b–b' heterodimer instead. It is very elongated and anchored in the membrane by its hydrophobic N-terminal region. The rest of the protein is hydrophilic and highly charged, except for a short stretch of hydrophobic amino acids near the C terminus, which is required for assembly of the ATP synthase. Portions of the mitochondrial b subunit are resolved in structures of the stator subcomplex (see Figure 13.22B). The two b subunits may form a coiled coil and can be linked by disulfide bonds when cysteine residues are engineered at many different positions. The length of the *E. coli b* subunit may be shortened or lengthened without loss of activity.

Regulation of the F_1F_0-ATPase

The F_1F_0-ATPase is a highly efficient motor. Attempts to determine its thermodynamic efficiency (the ratio of free energy gained by pumping protons to the free energy expended in the phosphorylation of ADP to make ATP)

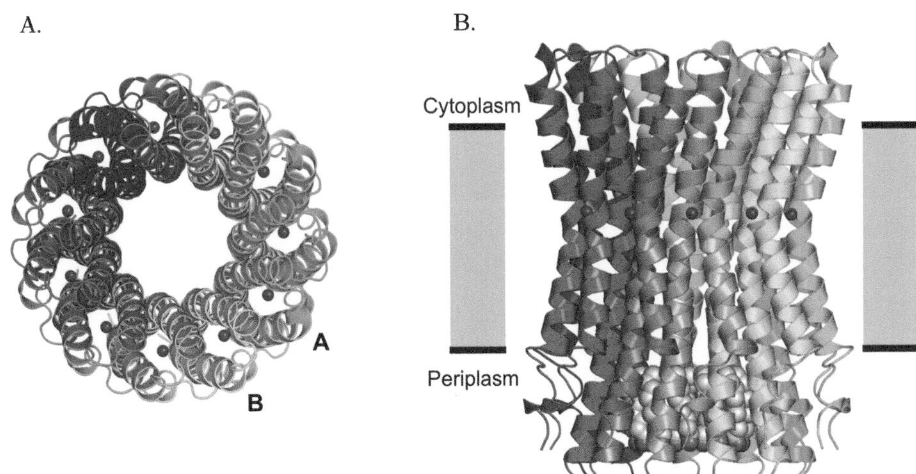

13.26 Structure of the c ring of the *I. tartaricus* ATP synthase. Subunits are shown in different colors and viewed perpendicular to the membrane from the cytoplasm (A) and from the plane of the membrane (B). Na^+ ions (blue spheres) bind at the center of the c ring and some detergent molecules (gray spheres) bind where the c ring protrudes from the membrane (gray shaded bar). From Meier, T., *et al.*, *Science*. 2005, 308:659–662. © 2005, American Association for the Advancement of Science. Reprinted with permission from AAAS.

from the measured torque of the F_1 domain suggest an efficiency near 100%. What, then, is the purpose of regulation? Cells benefit by alteration of the efficiency of F_1F_0-ATPase because if ATP concentrations in respiring cells are high, decreasing the net rate of ATP synthesis is beneficial, whereas if ADP concentrations are high, the cell needs rapid ATP synthesis. Crosslinking studies suggest that regulation of the activity of F_1F_0-ATPase could (partly) depend on the flexibility of the ε subunit described above. The rate of crosslinking between ε and β subunits is affected by ATP/ADP concentrations: In the presence of ATP this rate increases, whereas in the presence of ADP it decreases. Thus the position of the ε subunit varies depending on the needed net rate of ATP synthesis. Recent studies show that ε is able to sample alternate conformations on a time scale of seconds.

Do the conformations themselves suggest a possible mechanism? The conformation of the ε subunit seen in the *E. coli* structure allows its C-terminal helices to wind around the γ subunit and extend toward the $\alpha_3\beta_3$ hexamer, with the central axis of the β-sandwich roughly parallel to the N- and C-terminal helices of the γ subunit. A different conformation observed in the mitochondrial structure has the central axis of the β-barrel shifted to be roughly perpendicular to the N- and C-terminal helices of the γ subunit, with the C-terminal helices flattened against the side of the β-sandwich and away from F_1. These two states have been proposed to work as a rachet to affect the catalytic efficiency of F_1, which is supported by crosslinking studies showing different results when nucleotides are bound to F_1. Crosslinking studies also show that differential effects on hydrolysis and synthesis activities are possible. For example, formation of a crosslink between the C terminus of subunit ε and the γ subunit inhibited ATP hydrolysis by 75% and ATP synthesis by 25%, indicating that in some conformations, ATP synthesis is allowed when ATP hydrolysis is not. Finally, crosslinking results indicate that subunit ε can span the region of the central stalk and interact simultaneously with a β subunit of F_1 and subunit c of F_0.

Catalytic Mechanism of a Rotary Motor

Kinetic studies of the F_1F_0-ATPase established a number of characteristics of its mechanism: (1) Isotope exchange data are consistent with the idea that the enzyme does not release ATP at one active site until substrate is available to bind at another active site. (2) The exchange of ^{18}O showed that all three catalytic sites on the three β subunits are equally capable of carrying out the reaction. (3) Because release of products is rate limiting, the P_i and ADP that remain bound can reversibly resynthesize ATP, resulting in the incorporation of ^{18}O in different positions when the reaction is carried out in $H_2{}^{18}O$. (4) The catalytic sites on each of the β subunits in F_1 differ in affinity for ATP (when measured in excess ATP so ATP binds to all three sites): Kd for the first is <1 nM, for the second is ~1 μM, and for the third is 30 μM. Clearly, substrate binding shows negative cooperativity. However, when more than one site is occupied, the rate of ATP hydrolysis goes up 10^4- to 10^5-fold, so catalysis shows positive cooperativity. Whether this cooperativity requires nucleotide binding to two or three sites is still controversial.

Rotational Catalysis

The kinetic results are consistent with the Boyer binding-change mechanism that states that each of the three binding sites in F_1 cycles between three different conformations, closed (β_{TP} or T), partly open (β_{DP} or L), or open (and empty, β_E or O; Figure 13.27). The site in the open conformation is ready to bind substrate. When it binds nucleotide it closes, triggering conformational changes in the other two so that the closed one becomes partly open and the partly open one becomes fully open. Each site rotates through the three states in a cycle for both forward and reverse reactions. The reaction for ATP synthesis involves (1) binding of ADP and P_i to the partly open site; (2) conformational change at that site that converts it to the closed site, where catalysis occurs (while changing the other two sites to the open and partly

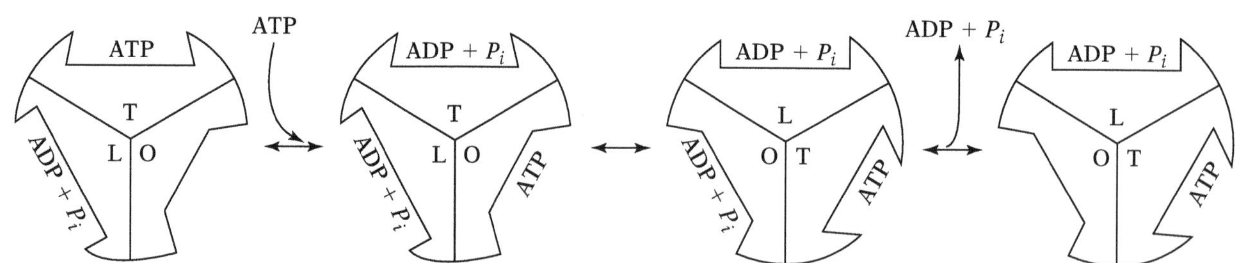

13.27 Sequential stages in the binding-change mechanism. Each catalytic site of the ATP synthase can have one of three conformations, designated O for open (empty) state, T for tight (ATP-bound) state, and L for loose (ADP + P_i occupied) state. The figure shows one iteration through the states, which is coupled to a 120° rotation of the $\gamma\varepsilon$ central stalk. This is also called an alternating sites hypothesis as it involves cycling of each of the three catalytic sites through three states. From Capaldi, R. A., and R. Aggeler, *Trends Biochem Sci.* 2002, 27:154–160. © 2002 by Elsevier. Reprinted with permission from Elsevier.

A.

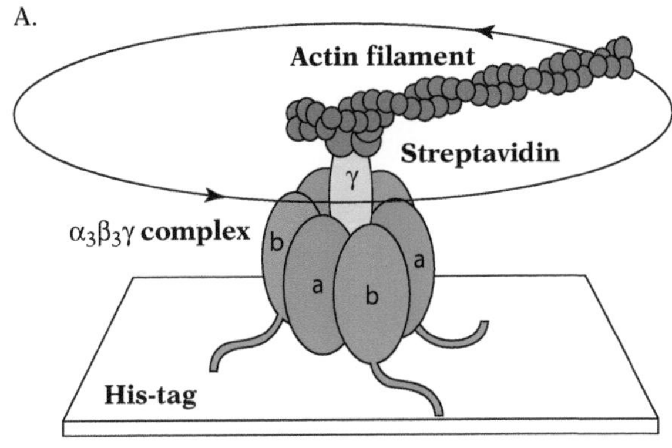

B.

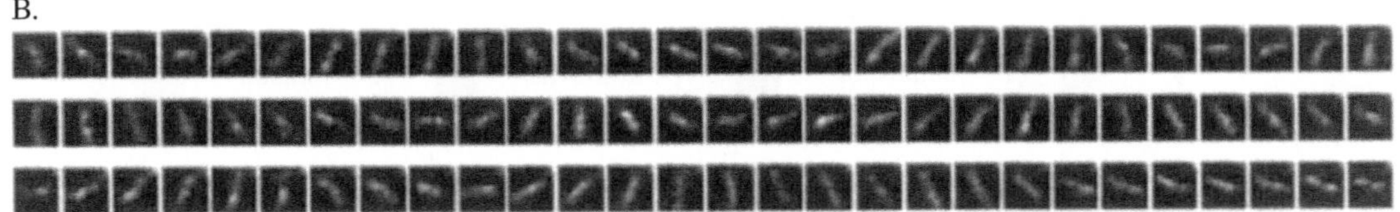

13.28 Demonstration of rotation coupled to ATP hydrolysis by video fluorescence microscopy. (A) Design of the experiment involved engineering ten histidine residues onto the N terminus of each β subunit of F_1 to bind the F_1-ATPase to a nickel plate and using biotin-streptavidin binding to attach to the γ subunit an actin filament carrying a fluorescent label. (B) The fluorescence pattern observed with the rotation axis in the middle of the filament when ATP is provided shows a continuous rotation of the actin in 120° increments, with a time interval between images of 33 msec. Scale bar is 5 μm. Redrawn from Noji, H., *et al.*, *Nature*. 1997, 386:299–302. © 1997. Reprinted by permission of Macmillan Publishers Ltd.

open sites); and (3) a second conformational change to convert that site to the open site that allows dissociation of the ATP formed. In each cycle the rotating γ subunit interacts with one of the β subunits to drive its conformational change from closed (β_{TP}) to open (β_E), which in turn triggers the conformational changes in the other two subunits (see Figure 13.25).

Exciting support for the rotary mechanism came from video fluorescence microscopy using the attachment of fluorescent-labeled actin filaments to the γ subunits. With time, the fluorescent label rotated 360° in three steps of 120° (Figure 13.28). The rotation was quite slow, so more recent experiments used smaller gold particles to tag the γ subunit, and a faster rate was observed. Careful analysis of these single-molecule experiments detects two substeps during ATP hydrolysis, one of 80° attributed to ATP binding and one of 40° for catalysis and product release (Figure 13.29). Substeps are of interest because coupling can be broken down into several mechanochemical steps: In the direction of hydrolysis these are ATP binding, transition state formation, bond cleavage, P_i release, and ADP release.

In the full F_1F_0-ATP synthase, these conformational changes in F_1 are coupled to proton translocation through F_0. For ATP synthesis, the rotation is generated by the passage of protons, driven by the pmf, through F_0 to cross the membrane; in the reverse direction, energy released by hydrolysis of ATP drives the rotation in the opposite cycle and reverses the flow of protons. A mechanism has been proposed for how proton translocation drives rotation (Figure 13.30). Protons enter a hydrophilic channel between TM segments of both *a* and *c* subunits. When each proton enters, it binds to residue Asp61 of a *c* subunit and breaks the interaction between Asp61 of the *c* subunit and Arg210 of the *a* subunit, which releases the *c* subunit and allows it to move. Rotation of the *c* ring is Brownian; however, when Asp61 faces the lipid bilayer it is protonated and when it faces the stator it is deprotonated. Assuming two access channels connect this site to either side of the membrane but are not connected to each other, an entering proton will bind to Asp61 and relieve the electrostatic constraint so the ring can move. Movement of the ring brings the proton attached to another *c* subunit to the exit channel so it can leave (see Figure 13.30). This ring movement generates a torque on γ where it bulges toward the bottom of the β subunits. Rotation causes the bulge to move to the next β subunit, which stimulates conformational changes that determine nucleotide occupancy. The peripheral stalk ($ab_2\delta$) holds the $\alpha_3\beta_3$ knob to prevent its spinning with the movement of the rotor. Some excellent animations of this mechanism are available (see Further Reading).

Since F_0 rotates in smaller increments than the cycling of the $\alpha_3\beta_3$ domain, coupling requires that several movements of the *c* ring accompany one $\alpha\beta$ mechanistic

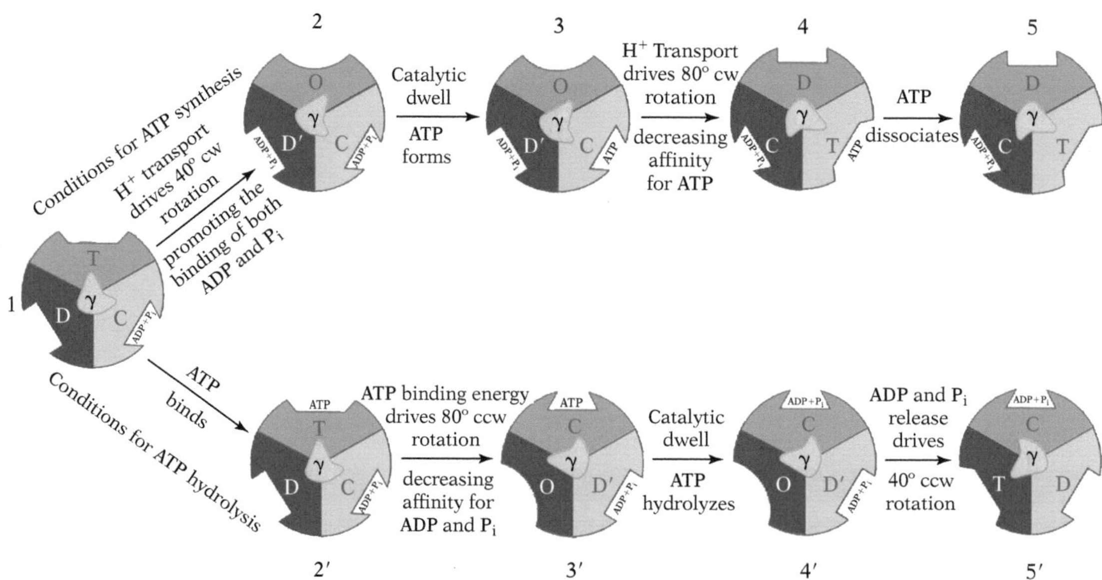

13.29 Proposed substeps in the mechanisms of ATP synthesis and hydrolysis by F_1F_0-ATPase. The three catalytic sites of F_1 (colored green, cyan, and light green) alternate between different conformations, which are labeled C, high-affinity catalytic site (previously called tight-binding site); D, ADP-binding site and D′ for binding both ADP and P_i; T, ATP-binding site; and O, low-affinity site. Steps for ATP synthesis are accompanied by clockwise (cw) rotation (top), while steps for ATP hydrolysis drive counterclockwise (ccw) rotation (bottom). Note that this model indicates binding to only two sites is sufficient, as supported by studies using beef heart mitochondrial F_1; however, studies using prokaryotic F_1 suggest binding to all three sites is required for cooperativity. From Milgram, Y. M., and R. L. Cross, *Proc Natl Acad Sci USA*. 2005, 102:13831–13836. © 2005 by National Academy of Sciences, U.S.A.

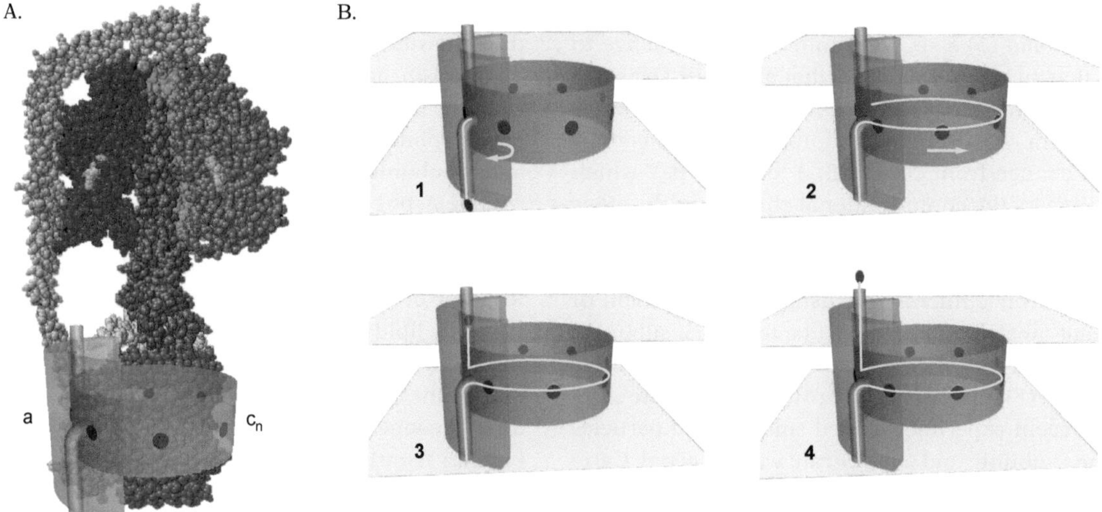

13.30 Model for the coupling of proton translocation in the rotary motor. (A) A space-filling model of *E. coli* F_0F_1 with one α subunit and two β subunits removed to expose the γ subunit (red) in the center of the $\alpha_3\beta_3$ knob (α, blue, β, green). The bulge of the γ subunit is not visible as it points away. Two nucleotides (light gray) are seen bound to the two α subunits. At the base of the central stalk is ε (yellow) and the c_{10} ring (magenta). The peripheral stalk (*b* subunits) along with δ and *a* are shown in the same color (dark gray) to show the shape of the stator. A schematic representation of the assembly of *a* with the c ring is overlaid on the structure. (B) Four still images from an animation showing torque generation by Brownian rotary motion and directed ion flow. In the images the c ring (barrel with spots showing protonated sites in blue and unprotonated sites in red) faces subunit *a* (green, on the left). The proton access and exit channels are offset laterally. The path of the proton is indicated by a yellow line. In step 1, the c ring rotates back and forth in the plane of the membrane (yellow arrow) as the charged site avoids contact with the hydrophobic membrane milieu. In step 2, protonation has occurred, allowing rotation by one step to take place (yellow arrow). Steps 3 and 4 show the proton in the exit channel of subunit *a* and then a return to the initial state by further rotation of the c ring. From Junge, W., *et al.*, *Nature*. 2009, 459:364–370. © 2009, Nature Publishing Group. Reprinted by permission of Macmillan Publishers Ltd.

rotation. Since the number of *c* subunits in the ring varies in different organisms, proton flux through the ring that drives its rotation is not necessarily related by an integer to the 120° rotation of F_1 at each stage. (In other words, the number of protons per ATP is nonintegral. For example, a ring with ten *c* subunits would give a ratio of 3.3 protons/ATP.) This "symmetry mismatch" could be important to avoid a deeper energy minimum that would result if the stages of both were closely matched.

A tremendous amount of evidence supports this mechanism, in addition to the kinetic data, crosslinking results, and numerous structures of increasing resolution and detail. Subunit structures and interactions have also been probed using epitope mapping with monoclonal antibodies, NMR, and FRET. Computational studies of the ATP synthase have simulated the effect of the γ subunit on the β subunits and calculated the free energy for hydrolysis at the three sites to model how the energy for rotation might be stored in conformations of the subunits. The interplay between computational biology and structural biology will continue to examine the determinants of torque generation and the thermodynamic efficiency. Questions of assembly and the role of lipids are being studied. Finally, higher-resolution structures of the entire complex in different conformational states will lead to a complete understanding of the coupling of rotation to catalysis in this elegant and essential molecular machine.

CONCLUSIONS

Determining high-resolution structures of all the major components of the mitochondrial and bacterial respiratory chain is a remarkable achievement. The structures reveal starkly different ways to couple proton pumping with electron transfer and catalysis – from the large movements of the piston in complex I and the rotary motors of ATP synthase to the separate spatial arrangements in the Q cycle in complex III (and cytochrome b_6f) and subtle conformational changes in complex IV. Together these proteins form the powerhouse of cells, working in vast supercomplexes in the membrane to create and utilize proton gradients as proposed in the Chemiosmotic Theory. The structures also provide a myriad of details that lead to further studies of inhibitors, lipid effects, genetic alterations, and human pathologies.

FOR FURTHER READING

Papers with an asterisk (*) present original structures.

Respiratory Complexes

Hosler, J. P., S. Ferguson-Miller, and D. A. Mills, Energy transduction: proton transfer through the respiratory complexes. *Annu Rev Biochem*. 2006, 75:165–187.

Complex I

*Baradaran, R., J. M. Berrisford, G. S. Minhas, and L. A. Sazanov, Crystal structure of the entire respiratory complex I. *Nature*. 2013, 494:443–448.

Berrisford, J. M., and L. A. Sazanov, Structural basis for the mechanism of respiratory Complex 1. *J Biol Chem*. 2009, 284:29773–29783.

*Efremov, R. G., R. Baradaran, and L. A. Sazanov, The architecture of respiratory complex I. *Nature*. 2010, 465:441–445.

*Efremov, R. G., and L. A. Sazanov, Structure of the membrane domain of respiratory complex I. *Nature*. 2011, 476:414–420.

*Hunte, C., V. Zickermann, and U. Brandt, Functional modules and structural basis of conformational coupling in mitochondrial Complex I. *Science*. 2010, 329:448–451.

*Sazanov, L. A., and P. Hinchliffe, Structure of the hydrophilic domain of respiratory Complex I from *Thermus thermophilus*. *Science*. 2006, 311:1430–1436.

Cytochrome bc_1

*Hunte, C., J. Koepke, C. Lange, and T. Rossmanith, Structure at 2.3 Å resolution of the cytochrome bc_1 complex from the yeast *Saccharomyces cerevisiae* with an antibody Fv fragment. *Structure*. 2000, 8:669–684.

Hunte, C., H. Palsdottir, and B. L. Trumpower, Protonmotive pathways and mechanisms in the cytochrome bc_1 complex. *FEBS Lett*. 2003, 545:39–46.

*Kleinschroth, T., M. Castellani, C. H. Trinh, *et al*., X-ray structure of the dimeric cytochrome *bc*1 complex from the soil bacterium *Paracoccus denitrificans* at 2.7 Å resolution. *Biochim Biophys Acta*. 2011, 1807:1606–1615.

Solmaz, S. R. N., and C. Hunte, Structure of complex III with bound cytochrome c in reduced state and definition of a minimal core interface for electron transfer. *J Biol Chem*. 2008, 283:17542–17549.

*Xia, D, C. A. Yu, H. Kim, *et al*., Crystal structure of the cytochrome *bc*1 complex from bovine heart mitochondria. *Science*. 1997, 277:60–66.

Cytochrome b_6f

Baniulis, D., E. Yamashita, H. Zhang, S. S. Hasan, and W. A. Cramer, Structure-function of the cytochrome *b6f* complex. *Photochem Photobiol*. 2008, 84:1349–1358.

Cramer, W. A., S. S. Hasan, and E. Yamashita, The Q cycle of cytochrome *bc* complexes: a structural perspective. *Biochim. Biophys Acta*. 2011, 1807:788–802.

*Kurisu, G., H. Zhang, J. L. Smith, and W. A. Cramer, Structure of the cytochrome b_6f complex of oxygenic photosynthesis. *Science*. 2003, 302:1009–1014.

*Stroebel, D., Y. Choquet, J. L. Popot, and D. Picot, An atypical haem in the cytochrome b_6f complex. *Nature*. 2003, 426:413–418.

Yamashita, E., H. Zhang, and W. A. Cramer, Structure of the cytochrome b_6f complex: quinone analogue inhibitors as ligands of heme c_n. *J Mol Biol*. 2007, 370:39–52.

Cytochrome *c* oxidase

Ferguson-Miller, S., C. Hiser, and J. Liu, Gating and regulation of the cytochrome c oxidase proton pump. *Biochim Biophys Acta*. 2012, 1817:489–494.

*Iwata, S., C. Ostermeier, B. Ludwig, and H. Michel, Structure at 2.8 Å resolution of cytochrome *c* oxidase from *Paracoccus denitrificans*. *Nature*. 1995, 376:660–669.

*Svensson-Ek, M., J. Abramson, G. Larsson, S. Törnroth, P. Brzezinski, and S. Iwata, The x-ray crystal structures of wild type and EQ (I286) mutant cytochrome *c* oxidases from *Rhodobacter sphaeroides*. *J Mol Biol*. 2002, 321: 329–339.

*Tsukihara, T., H. Aovama, E. Yamashita, *et al*., The whole structure of the 13-subunit oxidized cytochrome c oxidase at 2.8 Å. *Science*. 1996, 272:1136–1144.

Yoshikawa, S., K. Muramoto, and K. Shinzowa-Itoa, Proton-pumping mechanism of cytochrome *c* oxidase. *Ann Rev Biophys*. 2011, 40:205–223.

F_1F_0-ATP synthase

Boyer, P. D., A research journey with ATP synthase. *J Biol Chem*. 2002, 277:39045–39061.

Junge, W., H. Sielaff, and S. Engelbrecht, Torque generation and elastic power transmission in the rotary F_0F_1-ATPase. *Nature*. 2009, 459:364–370.

*Kabaleeswaran, V., N. Puri, J. E. Walker, A. G. Leslie, and D. M. Mueller, Novel features of the rotary catalytic mechanism revealed in the structure of yeast F1 ATPase. *EMBO J*. 2006, 25:5433–5442.

Karplus, M., and Y. Q. Gao, Biomolecular motors: the F1-ATPase paradigm. *Curr Opin Struct Biol*. 2004, 14:250–259.

Kinosita, K., Jr., K. Adachi, and H. Itoh, Rotation of F1-ATPase: how an ATP-driven molecular machine may work. *Annu Rev Biophys Biomol Struct*. 2004, 33:245–268.

*Meier, T., P. Polzer, K. Diederichs, W. Welte, and P. Dimroth, Structure of the rotor ring of F-type Na^+-ATPase from *Ilyobacter tartaricus*. *Science*. 2005, 308:659–662.

Nakanishi-Matsui, M., The mechanism of rotating proton pumping ATPases. *Biochim Biophys Acta*. 2010, 1797:1343–1352.

Noji, H., R. Yasuda, M. Yoshida, and K. Kinosita, Jr., Direct observation of the rotation of F1-ATPase. *Nature*. 1997, 386:299–302.

*Stock, D., A. G. W. Leslie, and J. E. Walker, Molecular architecture of the rotary motor in ATP synthase. *Science*. 1999, 286:1700–1705.

von Bullmoos, C., A. Wiedenmann, P. Dimroth, Essentials for ATP synthesis by F_1F_0 ATP synthases. *Ann Rev Biochem*. 2009, 78:649–672.

Narrated animation of ATP synthase showing coupling between proton flow and ATP synthesis: www.dnatube.com/video/104/ATP-synthase-structure-and-mechanism

IN PURSUIT OF COMPLEXITY

14

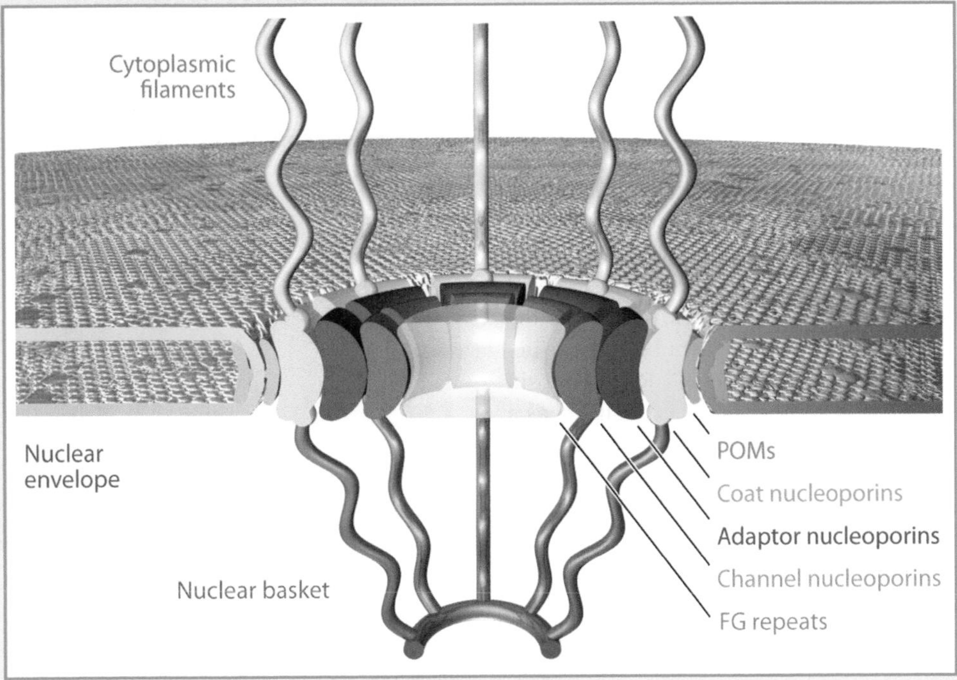

The nuclear pore complex. The nuclear pore, which is the only portal into the nucleus from the cytoplasm, spans both the inner nuclear membrane and the outer nuclear membrane. It is composed of about 30 well-conserved protein species, multiple copies of which assemble into an immense complex as large as 125 Mda in some eukaryotes. This model of it is based on cryo-EM tomography of the nuclear pore complex from *Dictyostelium discoideum*. The three layers of the pore are made up of coat nucleoporins (yellow), adaptor nucleoporins (red), and channel nucleoporins (orange), surrounded by the main integral membrane proteins (called POMs, green). Above and below the pore are the cytoplasmic filaments (cyan) and the nuclear basket (purple), and the center is plugged with a barrier (transparent) consisting of unfolded phenylalanine-glycine (FG) repeats from nucleoporins. From Hoelz, A., *et al.*, *Ann Rev Biochem*. 2011, 80:613–643. © 2011 by Annual Reviews. Reprinted with permission from the Annual Review of Biochemistry, www.annualreviews.org.

The two recent Nobel Prizes for membrane protein structures (2003 and 2012) shine the spotlight on impressive accomplishments in membrane research. The rapid progress in membrane structural biology builds on a foundation of biochemistry, biophysics, genetics, cell biology, and computational biology amassed over decades. When crystallographers initially turned their attention to membrane proteins, it took years to solve some of the first crystal structures and provide the long-awaited high-resolution images. These first glimpses of their beautiful structural details changed the level of comprehension of membrane proteins, providing new insights into their mechanisms of action. At present over 400 unique membrane proteins can be viewed at high resolution. They include examples from every functional class of membrane protein, and they provide models for many related proteins in their families. Because of the many challenges in getting these structures, often efforts were made to simplify the systems being studied.

Now membrane research is tackling even more difficult problems, which means studying more complex proteins – eukaryotic proteins tend to be larger than their prokaryotic homologs; more complex assemblies – many tasks in the cell use multiprotein complexes (see below for examples); and more complex actions – snapshots of more conformations and dynamic tools to evaluate and extend them. In every case, the complexity of the membrane itself cannot be overlooked, thus the interactions of proteins with lipids add additional questions to investigate.

COMPLEX FORMATION

The oligomeric state of many membrane proteins, which is important for their function, is not always clear in crystal structures. Protein associations may require lipids that were lost in detergent solubilization for crystal preparation. On the other hand, nonnative protein contacts can result in crystallization artifacts. However, crystal structures can define quaternary associations in many cases where an active site or transport channel is formed by more than one subunit. For example, the x-ray structures of some channel-forming proteins such as the potassium channels, the mechanosensitive channels, and the TolC channel-tunnel, show how different subunits contribute to one central channel. Less certainty may arise for channels and transporters that have pores in each subunit. The mitochondrial ATP/ADP carrier, for example, is reported to form dimers but appears as monomers in most crystal structures. For others, such as the trimeric porins and the tetrameric aquaporins, there is agreement between structures and functional data. Among the GPCRs, rhodopsin has been characterized as a dimer but is crystallized as a monomer. On the other hand, the cytokine receptor is a dimer in every crystal type, an observation that will stimulate further study.

Some hetero-oligomers form stable complexes that do not dissociate in detergent and thus can be isolated as complexes from the biological source. Examples include the respiratory complexes described in Chapter 13 and the photosynthetic systems. Photosystem I with 36 proteins and 381 cofactors is the largest membrane protein complex to have a crystal structure to date. Furthermore, these stable complexes appear to cluster into supercomplexes in specialized areas of their native membranes. While the intricacies of describing such complexes at the atomic level present many challenges, it can be even harder to investigate large complexes held in the membrane by weak interactions that may be disrupted during the process of solubilization. Typically the most important components of these are targeted for purification and characterization in the absence of the rest of their partners, which may include other integral membrane proteins, peripheral proteins, cofactors, and specific lipids. In order to study large protein complexes, attempts can be made to overexpress all of the components and copurify them. The alternative is reconstitution of the complex from its separate purified components, with uncertainties regarding whether all are present and whether they have reassembled correctly.

Given the many crucial roles of multiprotein complexes in biological systems, they are objects of intense research in spite of these challenges. Membrane complexes vary with respect to their composition. Some large assemblies are arrays containing many copies of one or a few species of molecules – for example, the light-harvesting complexes described in Chapter 5 and the gap junctions described in Chapter 12. The receptors used for chemotaxis of bacteria are another example. Dense patches containing many copies of the chemoreceptors are distributed along the cell surface and can be detected by cryo-EM (Figure 14.1). The chemoreceptor proteins form trimers of homodimers with binding sites for attractants and repellants from outside the cell and partners to send signals into the cytoplasm, ultimately to affect flagellar rotation (Figure 14.2). By increasing the area covered, each of these assemblies enhances their sensitivity and efficiency.

In contrast, other complexes are composed of many different components. A prominent example is the huge number of constituents of the nuclear pore complex (Frontispiece and Figure 14.3). A total of ~500–1000 proteins assemble to make a multi-layered pore with

A.

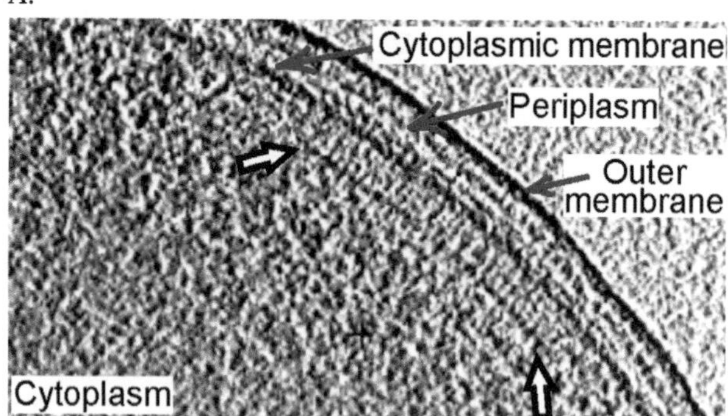

B.

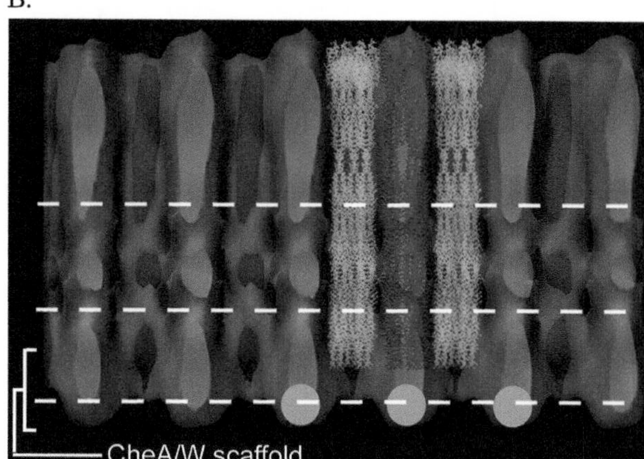

14.1 Chemoreceptors in the bacterial cytoplasmic membrane. (A) Chemoreceptors embedded in the cell membrane in patches (between white arrows) are visualized on whole cells of *E. coli* using cryo-EM tomography. The compartments of the cell are indicated. From Hazelbauer, G., *et al.*, *Trends Biochem Sci.* 2008, **33**:9–20. © 2008, Elsevier. Reprinted with permission from Elsevier. (B) At lower resolution the chemoreceptor molecules forming a patch in the membrane of *C. crescentus* are visualized. The structure of the chemoreceptors is docked onto the center molecules (cyan) and the lower densities represent the associated CheA and CheE proteins (yellow spheres). From Hazelbauer, G., and W. C. Lai, *Curr Opin Microbiol.* 2010, 13:124–132. © 2010, Elsevier. Reprinted with permission from Elsevier.

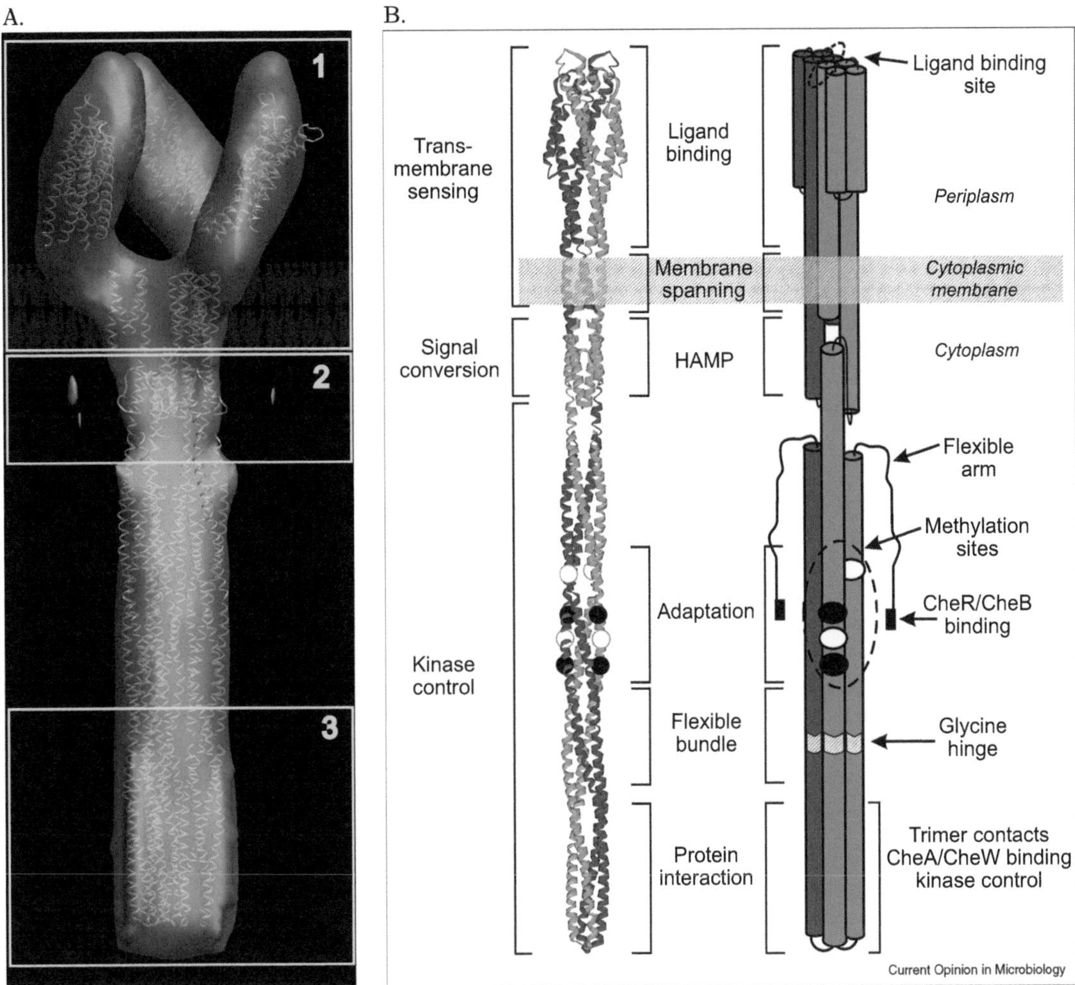

14.2 Structure of chemoreceptors. (A) The structure of three receptor dimers is docked onto a conformation of the trimer detected by cryo-EM tomography. (B) The modular segments of a chemoreceptor are illustrated with a ribbon diagram (left) and a schematic drawing of the three-dimensional organization based on the shape of the membrane-embedded receptor observed by EM. The ribbon diagram includes the x-ray structure of a cytoplasmic fragment and an NMR structure of the HAMP domain that connects it to the membrane. From Hazelbauer, G., and W. C. Lai, *Curr Opin Microbiol*. 2010, 13:124–132. © 2010, Elsevier. Reprinted with permission from Elsevier.

eight-fold symmetry plugged with unfolded peptides and decorated with cytoplasmic filaments and a nuclear basket. Isolation of segments of the nuclear pore complex, such as the Y-shaped Nup84 subcomplex purified from yeast, allows x-ray structures of their components to be fit to EM maps of their shape (Figure 14.4). This divide-and-conquer strategy used to characterize the nuclear pore is a paradigm for structure determination of macromolecular assemblies. Tackling a large multiprotein complex starts with visualization of the whole by EM and with the study of its isolated parts. Once backbones of the components are defined by x-ray crystallography they can be docked onto the EM envelope. Further structural details come with crystallography at higher resolution and spectroscopic analysis of individual components.

The lifetime of membrane complexes adds another aspect to their study. Although stable complexes like the respiratory supercomplexes and the nuclear pore complexes undergo continuous remodeling, their long lifetimes certainly aid their characterization. Other complexes form in the membrane in response to a transient condition, which limits when and how they might be isolated. Examples include the various assemblies needed for cell fusion and fission events involved in cell division, membrane trafficking, mitochondrial remodeling, neurotransmission, virus entry and exit, and more. In most cases (with the exception of mitochondria), membrane fusion occurs in eukaryotes by a conserved mechanism utilizing a family of proteins called SNAREs (soluble N-ethylmaleimide sensitive factor attachment protein receptor). SNAREs are long helical proteins that assemble into a bundle to bring two membranes (or two regions of a membrane) together for fusion (Figure 14.5). Two of the SNARE proteins, Syntaxin1A and Synaptobrevin2,

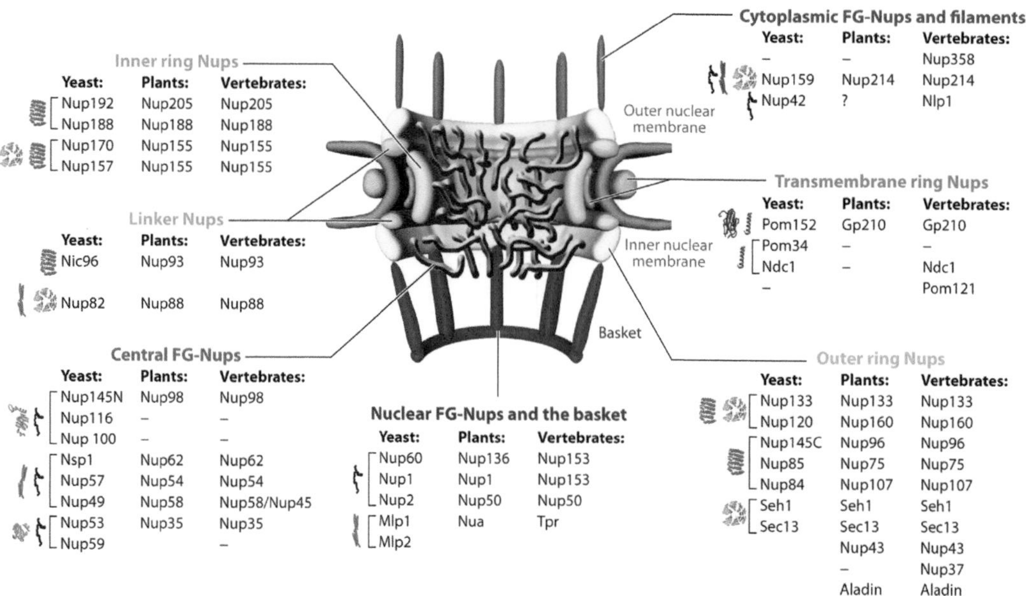

14.3 The protein components of the nuclear pore complex. The approximately 30 species of proteins conserved in the nuclear pore complex in yeasts, plants, and vertebrates are grouped according to their location in the pore and their structural folds. The symmetrical core consists of outer ring nucleoporins (Nups), linker Nups, inner ring Nups, the TM ring Nups, and the central FG Nups. The cytoplasmic FG-Nups and filaments, and the nuclear FG-Nups and basket form the asymmetric parts of the pore. From Grossman, E., *et al.*, *Ann Rev Biophys*. 2012, 41:557–584. © 2012 by Annual Reviews. Reprinted with permission from the Annual Review of Biophysics, www.annualreviews.org.

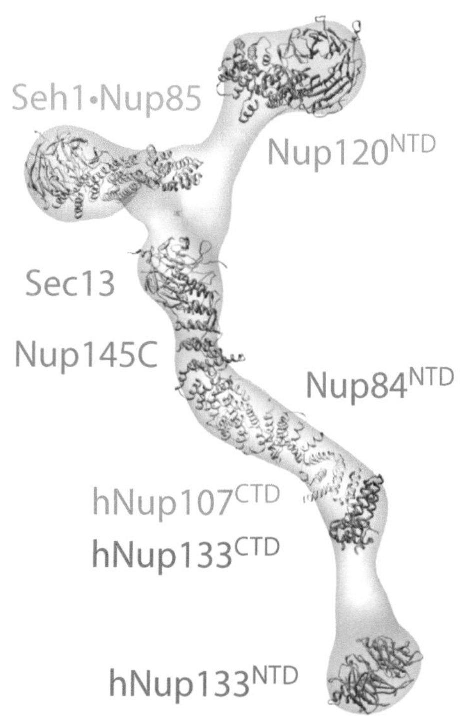

14.4 Architecture of the nuclear pore complex. The structure of the Nup84 complex shows an example of docking crystal structures into the EM envelope. Interestingly, the hinge region at the C-terminal domain (CTD) of Nup133 (red) is extended in the EM image of Nup84 in a different conformation. From Hoelz, A., *et al.*, *Ann Rev Biochem*. 2011, **80**:613–643. © 2011 by Annual Reviews. Reprinted with permission from the Annual Review of Biochemistry, www.annualreviews.org.

have helical TM regions that span the membrane. In addition to assembling into the bundle that draws the two membranes together, SNAREs associate with regulatory proteins and ATP-hydrolyzing enzymes to control when and where fusion takes place, as well as a set of proteins needed to dissemble the complexes afterwards. Combining genetic, biochemical, and biophysical approaches has led to the enumeration of the proteins and lipids involved in fusion in systems ranging from the neuronal synapse to the budding yeast.

Another transient complex is formed by numerous components of the immune system for the translocation of antigens during the immune response. Cytotoxic T-cells recognize the major histocompatibility complex class I (MHC1) in complex with peptide antigens produced by the proteasome and exported to the ER lumen by the transporter associated with antigen processing (TAP). TAP is a heterodimeric ABC transporter (see Chapters 6 and 11) transiently associated with other components of the complex by multivalent, low affinity interactions. In the MHC1 peptide loading complex, TAP is linked to quality control machinery, including several ER chaperones and peptidases (Figure 14.6). The N terminus of each TAP subunit binds tapasin, a type I membrane glycoprotein linked to an oxidoreductase, which binds the MHC1 molecule associated with calreticulin. Coexpression of the two TAP subunits has been optimized to purify TAP from a number of vertebrates. It is possible that the peptide loading complex is even larger, since its composition depends on the detergent used to solubilize it.

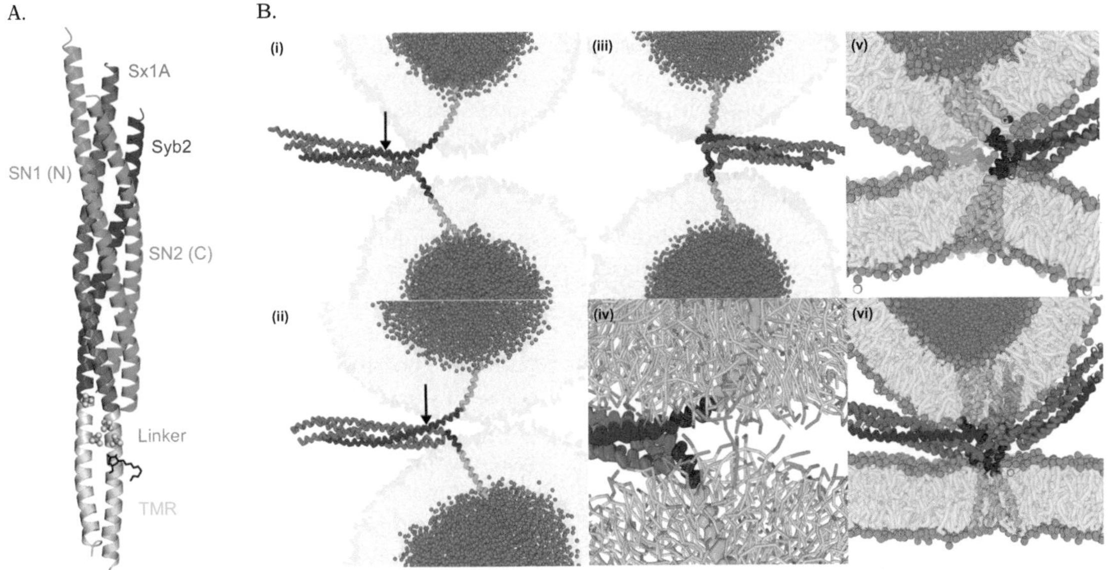

14.5 Role of SNARE proteins in membrane fusion. (A) Ribbon diagram from the x-ray structure of a neuronal SNARE complex. The four-helix bundle is composed of syntaxin1A (red), synaptobrevin2 (blue), each of which end in a TM region (TMR, gold) and two copies of SNAP-25 (green). Sulfate ions (spheres) and two glycylglycylglycine molecules (black sticks) associate at the linker regions (white). From Stein, A., *et al.*, *Nature*. 2009, 460:525–528. © 2009, Nature Publishing Group. Reprinted by permission of Macmillan Publishers Ltd. (B) Simulation of fusion events based on x-ray structure of the neuronal SNARE complex (colored as in (A)). (i) and (ii) Before fusion, zipping of SNARE proteins (at arrows) brings two membrane vesicles into close proximity. (iii) Flexible linkers (black) allow the SNARE complex to position at the fusion site. (iv) Onset of stalk (cyan) formation is followed by (v) SNARE-induced membrane curvature and fusion at the hydrophobic site (cyan). In (vi) an unstructured segment of synaptobrevin (blue) interfered with fusion. From Risselada, H. J., and H. Grubmüller, *Curr Opin Struct Biol*. 2012, 22:187–196. © 2012, Elsevier. Reprinted with permission from Elsevier.

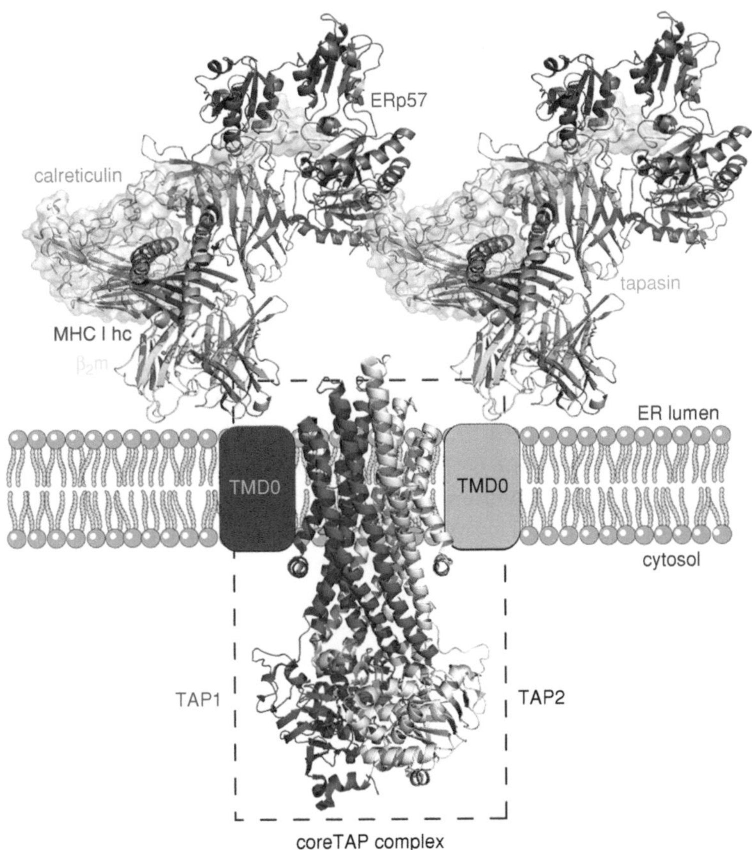

14.6 A model for the TAP transporter in the MHC I peptide loading complex. A model for the structure of TAP (TAP subunit 1, blue; TAP subunit 2, gray) based on the crystal structure of Sav1866 is shown from the plane of the membrane with the cytoplasm below and the ER lumen above. Molecular docking simulations are the basis for the portrayed interactions of TAP with other components of the peptide loading complex: the MHC I heavy chain (green) and β_2-microglobulin (yellow), the oxidoreductase ERp57 (magenta), tapasin (salmon), and calreticulin (light gray). The N-terminal subdomains (TMD0) of TAP are denoted with boxes. Structural elements involved in ATP hydrolysis and/or interdomain communication in TAP are highlighted: the Q-loop (green), the X-loop (cyan), coupling helix 1 (orange), and coupling helix 2 (magenta). Two bound molecules of AMP–PNP are shown as balls and sticks. From Hinz, A. and R. Tampè, *Biochemistry*. 2012, 51:4981–4989.

The challenge of designing ways to trap transient complexes is an important new frontier for membrane structural biology. The success achieved with the β_2 adrenergic receptor illustrates that it is possible: The ternary complex formed when a signaling molecule binds a GPCR and activates it to bind its G protein, which has a short lifetime in the cell, could be stabilized and crystallized (Chapter 10). In their reaction cycles GPCRs go on to associate with other molecules, arrestins, and kinases, providing more protein–protein interactions to study. Transient or not, composed of many protein species or a few, membrane protein complexes are the focus of many structural biology labs.

CONFORMATIONAL CHANGES AND DYNAMICS

Numerous membrane proteins have now been crystallized in more than one conformation, offering snapshots of their mechanisms as well as insight into how membrane proteins accommodate conformational changes. From the first structures of SERCA obtained with different ATP analogs and inhibitors (Figure 14.7), the number of conformations has grown to portray almost every step of its complicated reaction cycle (Chapter 9). High-resolution structures of transporters in the LeuT superfamily show three stages of the alternating access mechanism, outward-facing, occluded, and inward-facing (Chapter 11). Three different stages of the rotating site model for catalysis are seen in the $\alpha_3\beta_3$ domain of the F_1F_0-ATPase (Chapter 13), as well as the drug exporter AcrB (Chapter 11). ABC transporters, including the maltose transporter and the vitamin B_{12} transporter, have been crystallized in different states of the transport cycle (Chapter 11). Closing in on the goal of capturing the active state of proteins are high-resolution structures of the different states of the bacteriorhodopsin photocycle (Chapter 5), the structure of opsin* (the active state rhodopsin), and the complex of the agonist-bound β_2A receptor with its G protein (both in Chapter 10). Similarly, a structure of the SecYEG translocon shows how it opens to allow peptides to move laterally into the bilayer (Chapter 7).

Detailed comparisons of a protein's structure in different conformations clarify how the protein fold facilitates structural movements. Such movements include gates, which may result from rotation of helices or even from one or a few charged residues that form different salt bridges in the different conformations; hinges, typically at bends or unfolded regions of TM α-helices; rocker-switch mechanisms, with bundles of TM helices opening and closing; and even ball-and-socket joints, in which loops from one subunit fit into holes on another, allowing them to move without coming apart. The architecture of the protein folds also reveals designs for stabilization of the overall structure during large movements, such as several TM helices that form a scaffold around the core that undergoes conformational change.

Crystallography can make only a limited approach to understanding conformational change. While a few proteins have been crystallized under a range of conditions to provide different snapshots, other membrane proteins have been crystallized only in the presence of a substrate analog or inhibitor to push the equilibrium to favor one conformation, as in the case of the transporters LacY and mitochondrial AAC. Often the most flexible portion of the protein must be removed to allow crystallization, eliminating the possibility of detecting its different conformations. In some cases interactions in the crystal lattice fix exterior loops or larger domains into nonnative positions that may prevent adopting alternate conformations.

As a method that analyzes proteins in solution and allows dynamic assessments over time, NMR addresses these limitations of x-ray crystallography. The size limitations of solution NMR are being pushed higher, allowing structures of polytopic membrane proteins to be determined by NMR (Chapter 4). Hence the number of membrane proteins to have structure determinations by both NMR and x-ray crystallography is growing. The most flexible regions of the protein evident in NMR spectra correspond to the portions with the highest B value in the x-ray structure (if they show up at all). EPR and other spectroscopic methods can follow structural shifts, and FRET and DEER can precisely measure distances in different conformations. EM studies and even atomic force microscopy can also detect different conformations. MD simulations can be used to characterize and interpret the movements taking place in the proteins and stimulate further experiments.

Combining all these technologies in a multidisciplinary approach will push the frontiers of membrane research even faster in the future. Much of the early success in obtaining structures of membrane proteins focused on prokaryotic sources for advantages of expression and stability. Turning to the human membrane proteome, predicted to have over 7300 proteins, a number of high throughput laboratories increased the number of high-resolution structures of human membrane proteins from three to four per year, to 11 in 2012. Given the medical importance of many membrane proteins highlighted throughout this book, the high-resolution structures of more human membrane proteins provide important targets for future drug design. With new bioinformatic techniques to identify targets of interest and sophisticated genetic methods to modify and express them, there will be no shortage of membrane proteins to study in the future. Good biochemistry will continue to be essential in order to isolate and characterize them. Today some of the best understood membrane proteins – LacY, GPCRs, the maltose transporter – are those that were very thoroughly studied in the biochemistry lab before they became the subject of structural biology and other biophysics methods.

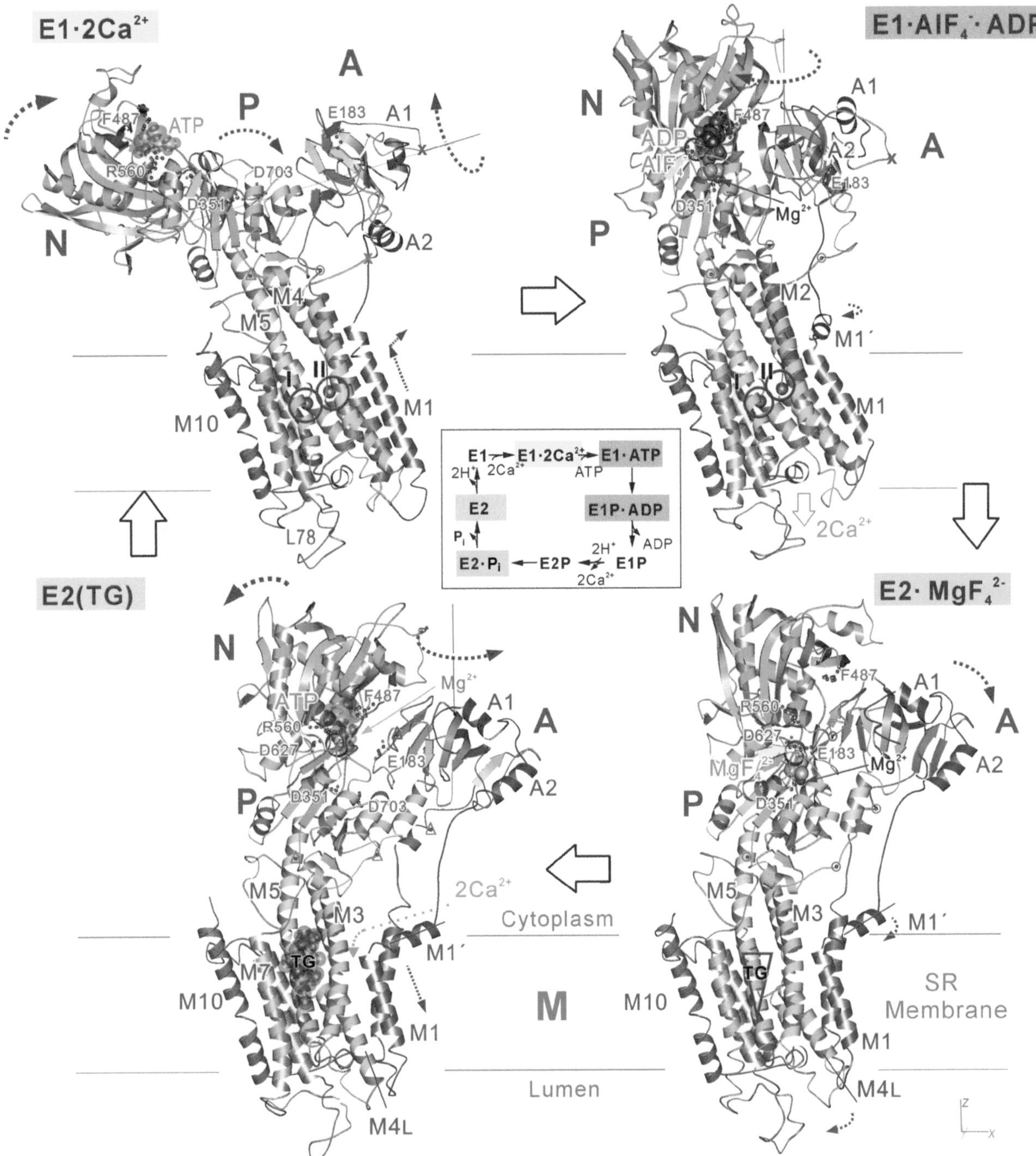

14.7 Early view of structural transitions during the reaction cycle of SERCA1. Four conformational changes during the Ca^{2+}-ATPase transport cycle are shown with structures from major stages of the E1–E2 cycle (shown in center). Motions of the head group domains are indicated by dashed arrows. Color changes gradually from the N terminus (blue) to the C terminus (red). Key residues, including R560, F487, E183, D351, and D703, are shown in ball-and-stick representation. When present, ATP, ADP, AIF4, and TG are shown in space-filling molecules and Ca^{2+} is circled. From Inesi, G., *et al*., *Biochemistry*. 2006, **45**:13769–13788. © 2005 by Elsevier. Adapted with permission from Elsevier. © 2006 by American Chemical Society. Adapted with permission from American Chemical Society.

FOR FURTHER READING

Progress and Directions

Fisher, L. S., A helical processing pipeline for EM structure determination of membrane proteins. *Methods*. 2011, 55:350–362.

White, S. H., Biophysical dissection of membrane proteins. *Nature*. 2009, 459:344–346.

Nuclear Pore Complex

Fernandez-Martinez, J., and M. P. Rout, A jumbo problem: mapping the structure and functions of the nuclear pore complex. *Curr Opin Cell Biol*. 2012, 24: 92–99.

Grossman, E., O. Medalia, and M. Zwerger, Functional architecture of the nuclear pore complex. *Ann Rev Biophys*. 2012, 41:557–584.

Hoelz, A., E. W. Debler, and G. Blobel, The structure of the nuclear pore complex. *Ann Rev Biochem*. 2011, 80:613–643.

Bacterial Chemotaxis

Hazelbauer, G., and W. C. Lai, Bacterial chemoreceptors: providing enhanced features to two-component signaling. *Curr Opin Microbiol*. 2010, 13:124–132.

Sourjik, V., and N. S. Wingreen, Responding to chemical gradients: bacterial chemotaxis. *Curr Opin Cell Biol*. 2012, 24:262–268.

Membrane Fusion

Moeller, A., C. Zhao, M. G. Fried, E. M. Wilson-Kubalek, B. Carragher, and S. W. Whiteheart, Nucleotide-dependent conformational changes in the N-ethylmaleimide sensitive factor (NSF) and their potential role in SNARE complex dissembly. *J Struct Biol*. 2012, 177:335–343.

Stein, A., G. Weber, M. C. Wahl, and R. Jahn, Helical extension of the neuronal snare complex into the membrane. *Nature*. 2009, 460:525–528.

Wickner, W., Membrane fusion: five lipids, four SNAREs, three chaperones, two nucleotides and a Rab, all dancing in a ring on yeast vacuoles. *Ann Rev Cell Dev Biol*. 2010, 26:115–136.

Wickner, W., and R. Shekman, Membrane fusion. *Nat Struc Mol Biol*. 2008, 15:658–664.

MHC-1 Peptide Loading Complex

Hinz, A., and R. Tampè, ABC transporters and immunity: mechanism of self defense. *Biochemistry*. 2012, 51:4981–4989.

APPENDIX I

ABBREVIATIONS

αHL a-hemolysin
ΔGt_r free energy change for the transfer of a substance from one solvent to another
AAC ADP/ATP carrier
ABC transporters a large class of transporters named for their ATP-binding cassettes
AChR acetylcholine receptor
AFM atomic force microscopy
ANK repeated domain of ankyrins
AQP aquaporin
AQP4 aquaporin 4, a human aquaporin
ATR atractyloside, an inhibitor of the mitochondrial ATP/ADP carrier
β_1AR β_1 adrenergic receptor
β_2AR β_2 adrenergic receptor
BA bongkrekic acid, an inhibitor of the mitochondrial ATP/ADP carrier
BC1 bacteriochlorophyll
BetP Na^+/betaine symporter
BO bacterio-opsin, which lacks the retinal cofactor
BPh bacteriopheophytin
bR bacteriorhodopsin
CATR carboxyatractyloside, an inhibitor of the mitochondrial ATP/ADP carrier
CFTR cystic fibrosis transmembrane conductance regulator, a chloride channel
CHAPS 3-[3-(cholamidopropyl) dimethyl-ammonio]-l-propanesulfonate
CHAPSO 3-([3-cholamidopropyl]dimethylammonio)-2-hydroxy-1-propanesulfonate
CL cardiolipin
ClC-ec1 chloride channel/transporter from *Escherichia coli*
CMC critical micellar concentration
cmCLC chloride channel from *Cyanidioschyzon merolae*
CMT critical micellar temperature
CN-cbl cyanocobalamin, or vitamin B_{12}
CT cytidyl transferase
CTAB cetyltrimethylammonium bromide
cyt b_6f cytochrome b_6f
DAG diacylglycerol
DAGK diacylglycerol kinase
DDM dodecyl β-D-maltoside
DEPC dieleidoyl phosphatidylcholine
DHA docosahexaenoic acid
DHPC dihexanoyl phosphatidylcholine
DLPC dilauroyl phosphatidylcholine
DLPE dilauroyl phosphatidylethanolamine
DMB 2,3-dimethyl-benzimidazole
DMPC dimyristoyl phosphatidylcholine
DMPE dimyristoyl phosphatidylethanolamine
DMPG dimyristoyl phosphatidylglycerol
DMPS dimyristoyl phosphatidylserine
DOPC dioleoyl phosphatidylcholine
DOPE dioleoyl phosphatidylethanolamine
DPPC dipalmitoyl phosphatidylcholine
DPPE dipalmitoyl phosphatidylethanolamine
DRMs detergent-resistant membranes
DSC differential scanning calorimetry
DSPC distearoyl phosphatidylcholine
DT diphtheria toxin
EDTA ethylenediamine tetraacetic acid
EF edema factor, a component of anthrax toxin
EGF epidermal growth factor
EM electron microscopy
EmrD multidrug transporter
EPR electron paramagnetic resonance
ER endoplasmic reticulum
FDH formate dehydrogenase
FdlL fatty acid transporter
FRAP fluorescence recovery after photobleaching
FRET fluorescence resonance energy transfer
FTIR Fourier transform infrared spectroscopy
FucP fucose transporter
G3P glycerol-3-phosphate
GdmCl guanidinium chloride
GlpG rhomboid protease
GltPh glutamate transporter from *Pyrococcus horokoshii*
GluA2 glutamate receptor at the synapse
GluCl Cys-loop glutamate receptor at the synapse
GOF gain of function
GPCR G protein-coupled receptor
GPI glycosylphosphatidylinositol
GUVs giant unilamellar vesicles
H_I hexagonal phase with nonpolar centers and polar groups and water outside
H_{II} inverted hexagonal phase (with polar centers and nonpolar exteriors)
HasR-HasA heme receptor – hemophore
HDL high-density lipoprotein

HHDBT	5-*n*-heptyl-6-hydroxy-4,7-diozobenzothiazole, an inhibitor of the cytochrome bc_1 complex
HiPIP	high-potential iron–sulfur protein
HMM	hidden Markov model
HPr	histidine-containing phosphocarrier protein
ICL	intracellular loop
IMS	intermembrane space of mitochondria
K_a	binding constant
K_d	dissociation constant
K_r	lipid association constant describing selective retention of lipids by proteins)
L_α	lamellar liquid crystalline, also called L_d (or *ld*, liquid-disordered)
$L_{\beta'}$	lamellar gel in which the chains are tilted
L_β	lamellar gel, also called s_o (ordered solid)
L_c	lamellar crystalline
LC	liquid condensed, a condensed phase in a lipid monolayer
LDAO	lauryldimethylamine oxide, or dodecyldimethylamineoxide
LE	liquid expanded, an expanded phase in a lipid monolayer
Lep	leader peptidase
LeuT	amino acid transporter
LF	lethal factor, a component of anthrax toxin
LH1, LH2	light-harvesting complexes
L_o	liquid-ordered lipid state that occurs when PLs, sterols, and/or sphingomylins are present
LOF	loss of function
LUVs	large unilamellar vesicles
MBD	membrane-binding domain
MC	Monte Carlo
MCF	mitochondrial carrier family
MD	molecular dynamics
MDOs	membrane-derived oligosaccharides
MDR	multidrug resistance
MFS	major facilitator superfamily
MLVs	multilamellar vesicles
MPoPS	1-myristyl 2-palmitoleoyl phosphatidylserine
MQ	menaquinone
MS	mechanosensitive
MSPs	membrane scaffold proteins
N	Newton, a unit of force
NBD	nucleotide-binding domain
NBD-	cholesterol labeled with the cholesterol: nitrobenzoxadiazolyl group
NBD-DLPE	DLPE (a PL) labeled with the nitrobenzoxadiazolyl group
NMR	nuclear magnetic resonance
NOE	nuclear Overhauser effect (in NMR)
NSAIDs	nonsteroidal anti-inflammatory drugs
octyl-POE	octyl-polyoxyethylene
OG	octyl β-D-glucoside (sometimes put as βOG)
OMPLA	outer membrane phospholipase A
OmpT	outer membrane protease
OSCP	oligomycin sensitivity conferring protein, a subunit of ATP synthase
PA	phosphatidic acid
PAA	protective antigen, a component of anthrax toxin
PC	phosphatidylcholine
PCC	protein-conducting channel, also called the translocon
PCR	polymerase chain reaction
PDB	Protein Data Bank
PE	phosphatidylethanolamine
PEP PTS	phosphoenolpyruvate-dependent phosphotransferase system
PG	phosphatidylglycerol
PGH_2	prostaglandin H_2
PGHS	prostaglandin H_2 synthase (also called cyclooxygenase, COX)
Pgp	P-glycoprotein
PH	pleckstrin homology domain that binds phosphoinositol
PI	phosphatidylinositol
pK_a	pH at which an acidic or basic functional group is 50% protonated
PKC	protein kinase C
PL	phospholipid, glycerophospholipid
PLA_2	phospholipase A_2
POPC	1-palmitoyl-2-oleoyl phosphatidylcholine
POPG	1-palmitoyl-2-oleoyl phosphatidylglycerol
POX	peroxidase
PrP	prion protein
PS	phosphatidylserine
R	the radius of curvature of the lipid–water interface
RC	reaction center (photosynthetic)
RCK	regulators of conductance of K^+
RND	Resistance Nodulation cell Division, a superfamily of drug efflux proteins
R_o	the intrinsic value of R for each lipid species
S	shape parameter for lipids, lipid volume / (cross-sectional area of polar headgroup × lipid length)
SDS	sodium dodecylsulfate, also called sodium laurylsulfate
SDSL	site-directed spin labeling
SDS-PAGE	SDS polyacrylamide electrophoresis
SecDF	membrane-integrated chaperone with role in protein translocation through SecYEG
SERCA	sarcoplasmic reticulum Ca^{2+}-ATPase
SLUVs	short-chain/long-chain unilamellar vesicles
SM	sphingomyelin
SMS	an analog of somatostatin that is an amphiphilic peptide with a positive charge

sn stereochemical numbering
SOPC 1-stearol-2-oleoyl phosphatidylcholine
SOPE 1-stearol-2-oleoyl phosphatidylethanolamine
SRP signal recognition particle
STGA 3-HSO_3-Galpβ1-6Manpα1-2Glcpα1-archaeol
SUVs small unilamellar vesicles
TBDTs TonB-dependent transporters
TC transport classification
TDG β-D-galactopyranosyl 1-thio-β-D-galactopyranoside
TEMPO a small lipid-soluble spin probe whose nitroxide group has an unpaired electron
TIM translocatase across the inner mitochondrial membrane
T_{LH} transition temperature for lipid transitions from lamellar to hexagonal phases, typically L_α–H_{II}
T_m melting temperature for fatty acids and pure lipids; also used as transition temperature for lipid transitions from L_β to L_α
TM transmembrane
TNBS trinitrobenzenesulfonic acid
TOM translocase across the outer mitochondrial membrane
TRAAK human K^+ channel (TWIK-related arachidonic acid-stimulated K^+ channel)
TRAM translocating chain-associated membrane protein
TROSY transverse relaxation optimized spectroscopy (in NMR)
Tsx nucleoside transporter and receptor for phage T-six
VDAC voltage-dependent anion channel
Wza translocon for capsular polysaccharide

APPENDIX II

SINGLE-LETTER CODES FOR AMINO ACIDS

A	alanine
C	cysteine
D	aspartate
E	glutamate
F	phenylalanine
G	glycine
H	histidine
I	isoleucine
K	lysine
L	leucine
M	methionine
N	asparagine
P	proline
Q	glutamine
R	arginine
S	serine
T	threonine
V	valine
W	tryptophan
Y	tyrosine

INDEX